Astrobiology: Exploring Life on Earth
and Beyond

Habitability of the Universe before Earth

Astrobiology: Exploring Life on Earth
and Beyond

Habitability of the Universe before Earth

Series Editors

Pabulo Rampelotto
Federal University of Rio Grande do Sul, Porto Alegre, Brazil

Richard Gordon
Gulf Specimen Marine Laboratory, Panacea, FL, United States
Wayne State University, Detroit, MI, United States

Joseph Seckbach
The Hebrew University of Jerusalem, Jerusalem, Israel

Volume Editors

Richard Gordon
Gulf Specimen Marine Laboratory, Panacea, FL, United States
Wayne State University, Detroit, MI, United States

Alexei A. Sharov
National Institute on Aging, NIH, Baltimore, MD, United States

ACADEMIC PRESS
An imprint of Elsevier

Academic Press is an imprint of Elsevier
50 Hampshire Street, 5th Floor, Cambridge, MA 02139, United States
525 B Street, Suite 1800, San Diego, CA 92101-4495, United States
The Boulevard, Langford Lane, Kidlington, Oxford OX5 1GB, United Kingdom
125 London Wall, London EC2Y 5AS, United Kingdom

ISBN: 978-0-12-811940-2
ISSN: 2468-6352

For information on all Academic Press publications
visit our website at https://www.elsevier.com/books-and-journals

Working together
to grow libraries in
developing countries

www.elsevier.com • www.bookaid.org

Publisher: Candice Janco
Acquisition Editor: Marisa LaFleur
Editorial Project Manager: Hilary Carr
Production Project Manager: Vijayaraj Purushothaman
Cover Designer: Matthew Limbert

Typeset by SPi Global, India

Dedications

Each of us arrived at this book via very different paths and split the work on it, RG mostly recruiting and cajoling authors and AS mostly handling the reviews. So we have different things to say in dedicating this book. Both of us would like to thank our anonymous reviewers, who helped to improve the quality of all chapter contents, and the personnel of Elsevier, especially editors Hilary Carr and Marisa LaFleur, and production manager Vijayaraj Purushothaman, who worked hard to keep the deadlines and bring this book into your hands. We would like to thank our series editor, Pabulo Henrique Rampelotto, for handling the reviews of our mutual chapter (Sharov and Gordon, 2017).

Richard Gordon (in part condensed and updated from: Gordon (2015)).

I would like to dedicate this book to John Musgrave (Fig. 1) and Edward Anders (Fig. 2). I met John Brent Musgrave probably in my last (3rd) year of high school at the University of Chicago Laboratory School. He was then an undergraduate at the University of Chicago and a member of the Astronomy Club (still in business: Baker et al., 2017). The then president Tom Lisco made me recite some constellations (a difficult task for me, being allergic to memorizations) to join the club. I learned them once and promptly forgot most, though Big and Little Dippers, Cassiopeia, and Orion still stick. By the time I entered UC in 1959 as a 16-year-old student, skipping the senior year in high school (John had done the same, I learned from his sister Jackie after his sudden death in 2015), John was the President, I became the Treasurer, and there were no other members and no membership fees. It was typical of John's wry humor, which I went along with, that he gave me this null responsibility for all our money. I enjoyed the irony.

John lived in the crawl space between the roof and the ceiling, just high enough to sit up on the mattress he set up in there, above our club room atop the Ryerson Physics Building from which we could look east across the campus, accessed by a narrow helical staircase. On the roof was the observatory containing a 6-in. refracting telescope. Daytimes, I would occasionally sketch projected images of the sun's spots. At night, I sometimes brought a date up there. I tried simultaneously photographing the same meteors with my younger brother Dan at our home in southwest Chicago and me at

FIG. 1

John Brent Musgrave.

Habitability of the Universe Before Earth, editors: Richard Gordon & Alexei Sharov, Volume 1 in the series:
Astrobiology: Exploring Life on Earth and Beyond, series editors: Pabulo Henrique Rampelotto,
Joseph Seckbach & Richard Gordon. ISSN 2468-6352. https://doi.org/10.1016/B978-0-12-811940-2.09985-8

FIG. 2

Edward Anders.

UC on the Ryerson roof. I had built a strobing device consisting of a flat fan-like blade that went in front of the camera, so that the images of meteors would be a sequence of dashes, allowing calculation of their velocity and triangulation of their height. But the film cracked in the cold of night when advanced in the camera, and I didn't think about how to overcome that problem, perhaps foreshadowing my career as a theoretician. Dan later became an excellent amateur astronomer and inventor (Gordon, 1989) with his own observatory, a prize winning astronomy photographer, and Educator at McDonald Observatory at the University of Texas (Cheri, 2016; Gordon, 2014; Rickey and Gordon, 2004).

Astronomy overnights with John meant we played with an ancient brass calculating machine we had inherited, listened to classical music on WFMT, and browsed through negatives of galaxies photographed by previous members. That space became my second campus home, my first being my own lab over the central lecture theater in the Kent Chemistry Building, where I kept a cot and cooked canned spaghetti in a beaker. I prepared specimens for students to analyze for the quantitative chemistry course, after taking that course my first summer before I entered UC. Ed Anders taught it, so I met both men that same year. I later worked with Ed on organic matter in meteors, my first "publication" being a footnote in a *Science* article on an Orgueil meteorite hoax perpetuated in 1864, but unfortunately for the hoaxer, not discovered until 1964 (Anders et al., 1964).

John majored in history of science, and via him, I gained an appreciation for that history. I recall sitting in on a course on modeling in science, undoubtedly because of John. We shared a love for books, both amassing eclectic collections of thousands. John questioned everything, especially having to do with authority, and I owe much of my professional and daily skepticism to discussions with him. He told me about Vulcan, the planet deduced from its perturbations of Mercury's orbit, and that was sometimes observed, between Mercury and the sun (Baum and Sheehan, 1997). It was later explained away by relativistic effects on Mercury's orbit. We discussed phenomena such as people seeing lights on the Moon. John told me that when he was a kid, in broad daylight he used his telescope to watch something in the sky that had parts twirling around and going in and out, unlike any aircraft with which he was familiar. He later wrote a short book on UFO sightings in Canada (Musgrave, 1979). Again, I learned from him open-mindedness about things, raising questions, but not jumping to conclusions based on scant evidence. He was not a believer in UFOs, just open to their possibility. After a visit 7 years ago that Natalie and I paid to the UFO museum in Roswell, New Mexico, I made sure they got a copy of his book into their extensive library.

John came to my home, where my parents Jack and Diana Gordon got to know and like him. This probably made it easier on them when at 17 I moved out and shared an apartment with John and one other fellow in Hyde Park, north of the University of Chicago. John's pacifism wore off on me. Draft Board and John: "Do you believe in the use of force? Well, I do use a can opener."—Always a wry point of view. It helped shape my views during the Vietnam War.

His father, a West Point graduate living in the San Diego area, had done a lot of reading on witch-craft, which influenced John (Musgrave and Houran, 2000, 2003). After some work at a planetarium (Musgrave, 1978), John drifted far from science, he and his Colombian wife Consuelo having fostered many First Nations kids, and he got involved in the history of local First Nations affairs in British Columbia (Musgrave, 2003).

John was one of those fine intellects who could never have made it in academia. I squeaked through, despite the attitudes I learned from John and concurred with. In retrospect, though I didn't think of him that way, he was a fine, exemplary big brother for an aspiring scientist.

Edward Anders is still very much with us. I dedicated a paper to him (Gordon and Hoover, 2007) and discussed his careful and fair role in the analysis of the "organized elements" in the Orgueil meteorite (Gordon and McNichol, 2012). His impact is mentioned in a recent dramatization (Mansfield, 2016). I learned quantitative chemistry from him with his penchant for precise measurements and quantitative conclusions, with total honesty in one's interactions with others and with nature, and respect for other scientists with whom one might be in disagreement. He was quite a role model. His own works on me-teorites are numerous and publicly recorded in the standard science databases, so I'll just list his reviews here (Anders, 1962, 1964, 1991; Anders and Zinner, 1993). What our readers are less likely to be aware of is his devotion to victims of the Holocaust in Latvia, including most of his family (Anders, 2010a,b; Anders and Dubrovskis, 2001, 2003, 2008; Esse, 2010; Goldman and Anders, 1997).

Alexei Sharov

I dedicate my contributions in this book to Alexander Levich (1945–2016) (Levich, 1995, 1996; Fig. 3), Julius Schreider (1927–98) (Shreider and Sharov, 1982; Wikipedia, 2017a; Fig. 4), and Sergei Viktorovich Meyen (1935–87) (Fig. 5) who helped me to develop a life-long interest in theoretical and evolutionary biology. Alexander Levich started his career in the field of theoretical and mathemat-ical physics, but then he switched to biology in a hope of developing the foundations of theoretical biology using formal principles, similar to those in physics. He was also deeply interested in the principles of human relations that promote productivity and innovation, and his favorite author was Dale Carnegie who wrote the book *How to Win Friends and Influence People* (Carnegie, 1936). Combining these two interests, in 1974, when Levich became employed at the Biological Faculty of Moscow State University, he founded a "Working group for constructive studies in theoretical biology," or shortly "Group β." Everyday group activities (e.g., questionnaires, brainstorming, and sem-inars) were designed following advice from Dale Carnegie. It was a very inspiring time for me that opened a door into a different world of theoretical thought with mathematics and philosophy as guiding threads. Within a couple of years, we studied most of the "Principia Mathematica" by the Bourbaki group

FIG. 3

Alexandr P. Levich.

From Kull, K., 2016. Alexandr Levich (1945–2016) and the Tartu-Moscow Biosemiotic Nexus. Sign Systems Stud. 44(1–2), 255–266.

FIG. 4

Yuly Anatolievich Shreider.

From Wikipedia, 2017. Шрейдер, Юлий Анатольевич [Shreider, Yuly Anatolievich]. https://ru.wikipedia.org/wiki/Шрейдер,
_Юлий_Анатольевич.

(Wikipedia, 2017b), *From Being to Becoming* by Ilya Prigogine (Prigogine, 1980), and the *Towards a Theoretical Biology* series edited by Conrad Waddington (Waddington, 1968, 1969, 1970, 1972).

Activities in theoretical biology brought us in contact with two other Russian scientists deeply interested in theoretical principles of biology: Julius Schreider—a mathematician and semiotician, and Sergei Meyen—a paleobotanist and evolutionary biologist (Meyen, 1987; Sharov, 1995). They both were followers of Aleksandr Lyubishchev (Shishkin, 2006; Svetlova, 1982), who is known for his

FIG. 5

Sergei Viktorovich Meyen.
From Geological Memorials of Perm District. Hermann, A.B., 2009. Meyen, Sergey Viktorovich (1935–1987), paleobotanist.
http://www.mi-perm.ru/pk/pale-2.htm.

critical analysis of neo-Darwinism and development of the formal theory of classification. They worked together developing structural approaches to systematics that included the analysis of arche-types in taxonomy and language and argued that a structural approach is justified in the systematics of organisms even if it occasionally contradicts lineage reconstructions. Schreider realized that there is a deep parallelism between systematics and semiotics, a theory of signs. Description of a new taxon is similar to the development of a new notion in language, because both are based on similarity relation-ships among objects. This idea turned his interests toward the mathematical theory of relations and models. Sergei Meyen's contribution to the theory was a notion that similarity should be extended to homologous parts that belong to multiple levels of hierarchy. Taxonomical decisions (i.e., whether an organism belongs to a certain taxon) are based on comparisons of a large number of characteristics with different degrees of similarity. The overall level of similarity can then be evaluated based on the majority of characteristics, as in public voting systems where the majority determines the outcome. Hence, Schreider developed an elegant theory of majority structures based on set theory formalism.

REFERENCES

Anders, E., 1962. Meteorite ages. Rev. Mod. Phys. 34 (2), 287–325.

Anders, E., 1964. Origin, age, and composition of meteorites. Space Sci. Rev. 3 (5-6), 583–714.

Anders, E., 1991. Organic matter in meteorites and comets: possible origins. Space Sci. Rev. 56 (1-2), 157–166.

Anders, E., 2010a. Amidst Latvians During the Holocaust. Occupation Museum Association of Latvia, Latvia.

Anders, E., 2010b. Liepāja, Encyclopaedia of the Holocaust. http://www.liepajajews.org/LGhetto.pdf.

Anders, E., Dubrovskis, J., 2001. Jews in Liepāja, Latvia, 1941-45. Anders Press, Burlingame, CA.

Anders, E., Dubrovskis, J., 2003. Who died in the Holocaust? Recovering names from official records. Holocaust Genocide Stud. 17 (1), 114–138.

Anders, E., Dubrovskis, J., 2008. Jews in Liepāja/Latvia, 1941-45: A database of victims and survivors. http://www.liepajajews.org/db.htm.

Anders, E., DuFresne, E.R., Hayatsu, R., DuFresne, A., Cavaillé, A., Fitch, F.W., 1964. Contaminated meteorite. Science 146 (3648), 1157–1161.

Anders, E., Zinner, E., 1993. Interstellar grains in primitive meteorites: Diamond, silicon carbide, and graphite. Meteoritics 28 (4), 490–514.

Baker, C., Volpert, C., Misener, W., Cambias, E., Cohen, M., Wintersmith, M., Corso, A.J., Toole, J., 2017. Ryerson Astronomical Society. http://astro.uchicago.edu/RAS/.

Baum, R., Sheehan, W., 1997. In Search of Planet Vulcan: The Ghost in Newton's Clockwork Universe. Plenum Trade, New York.

Carnegie, D., 1936. How to Win Friends and Influence People. Simon and Schuster, New York, NY.

Cheri, 2016. Fort Davis, Texas: The Mind-Bending McDonald Observatory. https://randomcurrents.com/fort-davis-texas-the-mind-bending-mcdonald-observatory/.

Esse, L., 2010. Latvian Cultural Evening at Stanford. http://library.stanford.edu/news/2014/05/latvian-cultural-evening-stanford.

Goldman, R., Anders, E., 1997. Interview with Edward Anders, February 28, 1997, RG-50.030*0451 [https://collections.ushmm.org/oh_findingaids/RG-50.030.0451_trs_en.pdf]. United States Holocaust Memorial Museum, Washington, DC.

Gordon, D., 1989. For Starry-Eyed Photographers. Popular Photography(December), 200.

Gordon, D., 2014. Out of Atmosphere. http://outofatmosphere.com/index-en.html.

Gordon, R., 2015. Intersections with John Brent Musgrave. http://embryogenesisexplained.org/2015/10/.

Gordon, R., Hoover, R.B., 2007. Could there have been a single origin of life in a Big Bang universe? Proc. SPIE 6694, https://www.spiedigitallibrary.org/conference-proceedings-of-spie/6694/669404/Could-there-have-been-a-single-origin-of-life-in/10.1117/12.737041.short.

Gordon, R., McNichol, J., 2012. Recurrent dreams of life in meteorites. In: Seckbach, J. (Ed.), Genesis—In the Beginning: Precursors of Life, Chemical Models and Early Biological Evolution. Springer, Dordrecht, pp. 549–590.

Levich, A.P. (Ed.), 1995. On the Way to Understanding the Time Phenomenon: The Constructions of Time in Natural Sciences. Part 1. Interdisciplinary Time Studies. World Scientific, Singapore.

Levich, A.P., 1996. On the Way to Understanding the Time Phenomenon: The Constructions of Time in Natural Science. Part 2. The "Active" Properties of Time. World Scientific, Singapore.

Mansfield, A., 2016. The Meteorite and the Hidden Hoax. http://www.bbc.co.uk/programmes/b06wg805.

Meyen, S.V., 1987. Fundamentals of Palaeobotany. Chapman and Hall, London.

Musgrave, J.B., 1978. Alberta's Roving Planetarium. Sky Telescope 56 (6), 487–489.

Musgrave, J.B., 1979. UFO Occupants & Critters: The Patterns in Canada http://www.openisbn.com/preview/0787312665/. Global Communications, New York.

Musgrave, J.B., 2003. Smallpox as a weapon of genocide in the Okanagan and Similkameen? Rep. Okanagan Historical Soc. 67, 41–43.

Musgrave, J.B., Houran, J., 2000. Flight and abduction in witchcraft and UFO lore. Psychol. Rep. 86 (2), 669–688.

Musgrave, J.B., Houran, J., 2003. The Witches' Sabbat in legend and literature. Lore Language 17, 157.

Prigogine, I., 1980. From Being to Becoming, Time and Complexity in the Physical Sciences. W.H. Freeman and Co., San Francisco.

Rickey, D.W., Gordon, R., 2004. Unitron 60 mm Refractor. http://www.cloudynights.com/item.php?item_id=243.

Sharov, A.A., 1995. Analysis of Meyen's typological concept of time. In: Levich, A.P. (Ed.), On the Way to Understanding the Time Phenomenon: The Constructions of Time in Natural Sciences. Part 1. Interdisciplinary Time Studies. World Scientific, Singapore.

Sharov, A.A., Gordon, R., 2017. Life before Earth. In: Gordon, R., Sharov, A.A. (Eds.), Habitability of the Universe Before Earth. In: Rampelott, P.H., Seckbach, J., Gordon, R. (Eds.), Astrobiology: Exploring Life on Earth and Beyond. Elsevier B.V., Amsterdam, p. 265–296.

Shishkin, M.A., 2006. Development and lessons of evolutionism. Russ. J. Dev. Biol. 37 (3), 146–162.

Shreider, J.A., Sharov, A.A., 1982. Systems and Models. Radio i sviaz, Moskva (in Russian).

Svetlova, P.G. (Ed.), 1982. Aleksandr Aleksandrovich Lyubishchev 1890-1972. Nauka, Leningrad (in Russian).

Waddington, C.H. (Ed.), 1968. Towards a Theoretical Biology. Edinburgh University Press, Edinburgh.

Waddington, C.H., 1969. Towards a theoretical biology. 2. Sketches. In: An International Union of Biological Sciences Symposium.

Waddington, C.H. (Ed.), 1970. Towards a Theoretical Biology, 3. Drafts, An International Union of Biological Sciences Symposium on General Biology.

Waddington, C.H. (Ed.), 1972. Towards a Theoretical Biology 4. Essays. Edinburgh University Press, Edinburgh.

Wikipedia, 2017a. Julius Anatolyevich Schrader. https://en.wikipedia.org/wiki/Julius_Anatolyevich_Schrader.

Wikipedia, 2017b. Nicolas Bourbaki. https://en.wikipedia.org/wiki/Nicolas_Bourbaki.

FURTHER READING

Hermann, A.B., 2009. Meyen, Sergey Viktorovich (1935–1987), paleobotanist. http://www.mi-perm.ru/pk/pale-2.htm.

Kull, K., 2016. Alexandr Levich (1945–2016) and the Tartu-Moscow Biosemiotic Nexus. Sign Systems Stud. 44 (1-2), 255–266.

L'Heureux, J., 2017. LASR *Remembered*. http://lasr.happyones.com/alumni-list.htm.

Wikipedia, 2017. Шрейдер, Юлий Анатольевич [Shreider, Yuly Anatolievich]. https://ru.wikipedia.org/wiki/Шрейдер,_Юлий_Анатольевич.

Contents

PART 2   PREDICTING HABITABILITY

PART 3 LIFE IN THE COSMIC SCALE

Life Before Earth 265
Alexei A. Sharov, Richard Gordon

Earth Before Life 297
Caren Marzban, Raju Viswanathan, Ulvi Yurtsever

The Drake Equation as a Function of Spectral Type and Time 307
Jacob Haqq-Misra, Ravi K. Kopparapu

Are We the First: Was There Life Before Our Solar System? 321
Pauli E. Laine, Sohan Jheeta

Life Before its Origin on Earth: Implications of a Late Emergence of Terrestrial Life 343
Julian Chela-Flores

PART 4   SYSTEM PROPERTIES OF LIFE

Symbiosis: Why Was the Transition from Microbial Prokaryotes to Eukaryotic Organisms a Cosmic Gigayear Event?

George Mikhailovsky, Richard Gordon

Coenzyme World Model of the Origin of Life

Alexei A. Sharov

Contributors

Ximena C. Abrevaya
Instituto de Astronomía y Física del Espacio (UBA–CONICET), Buenos Aires, Argentina;
Washburn University, Topeka, KS, United States

Dorian Aur
University of Victoria, Victoria, BC, Canada

Fernando J. Ballesteros
Astronomical Observatory, University of Valencia, Paterna, Spain

Peter L. Biermann
Max Planck Institute for Radio Astronomy, Bonn; Inst. for Nucl. Phys., Karlsruhe Institute for Technology, Karlsruhe; Department of Physics and Astronomy, University of Bonn, Germany; Dept. of Physics & Astronomy, University of Alabama, Tuscaloosa, AL, United States

Julian Chela-Flores
The Abdus Salam ICTP, Trieste, Italy; IDEA, Caracas, Venezuela

Aditya Chopra
The Australian National University, Canberra, ACT, Australia

Nikolai D. Denkov
Sofia University, Sofia, Bulgaria

Chaitanya Giri
Earth-Life Science Institute, Tokyo Institute of Technology, Tokyo, Japan; Geophysical Laboratory, Carnegie Institution of Washington, Washington, DC, United States

Richard Gordon
Gulf Specimen Marine Laboratories, Panacea, FL; Wayne State University, Detroit, MI, United States

Michael G. Gowanlock
Massachusetts Institute of Technology, Haystack Observatory, Westford, MA; Northern Arizona University, Flagstaff, AZ, United States

Martin M. Hanczyc
Università degli Studi di Trento, Povo, Italy

Jacob Haqq-Misra
Blue Marble Space Institute of Science; NASA Astrobiology Institute's Virtual Planetary Laboratory, Seattle, WA, United States

Sohan Jheeta
Network of Researchers on Horizontal Gene Transfer and the Last Universal Common Ancestor, Leeds, United Kingdom

Ravi K. Kopparapu
Blue Marble Space Institute of Science; NASA Astrobiology Institute's Virtual Planetary Laboratory, Seattle, WA; NASA Goddard Space Flight Center, Greenbelt; University of Maryland, College Park, MD, United States

Pauli E. Laine
University of Jyväskylä, Jyväskylä, Finland

Charles H. Lineweaver
The Australian National University, Canberra, ACT, Australia

Bartolo Luque
E.T.S.I. Aeronautics and Space, Politechnical University of Madrid, Madrid, Spain

Caren Marzban
University of Washington, Seattle, WA, United States

Paul A. Mason
New Mexico State University, Las Cruces, NM, United States

George Mikhailovsky
Global Mind Share, Norfolk, VA, United States

Jill A. Mikucki
University of Tennessee, Knoxville, TN, United States

Ian S. Morrison
Swinburne University of Technology, Hawthorn, VIC, Australia

Jeffrey M. Robinson
Howard University, Washington, DC; National Institutes of Health, Bethesda, MD, United States

Alexei A. Sharov
National Institute on Aging (NIA/NIH), Baltimore, MD, United States

Stoyan K. Smoukov
Queen Mary University of London, London; University of Cambridge, Cambridge, United Kingdom; University of Sofia, Sofia, Bulgaria

Brian C. Thomas
Instituto de Astronomía y Física del Espacio (UBA–CONICET), Buenos Aires, Argentina; Washburn University, Topeka, KS, United States

Mary A. Tiffany
Bainbridge Island, WA, United States

Karla de Souza Torres
CEFET-MG, Curvelo, Brazil

Jack A. Tuszynski
University of Alberta, Edmonton, AB, Canada

Raju Viswanathan
Technome, Clayton, MO, United States

Branislav Vukotić
Astronomical Observatory, Belgrade, Serbia

Othon C. Winter
FEG-UNESP, Guaratinguetá, Brazil

Ulvi Yurtsever
MathSense Analytics, Altadena, CA, United States

Gerard A.J.M. Jagers op Akkerhuis
Wageningen Environmental Research (Alterra), Wageningen, The Netherlands

Preface: Life as a Cosmic Phenomenon by Alexei A. Sharov & Richard Gordon

The assumption that life originated on Earth has dominated science since the emergence of the theory of evolution. It became the cornerstone of all major scenarios of the origin of life by Charles Darwin (Darwin, 1871), Alexander Oparin (Oparin, 1936), Stanley Miller (Miller, 1953), John Haldane (Haldane, 1929), Sidney Fox (Fox, 1964), and others (Damer and Deamer, 2015; Gallori et al., 2006; Rauchfuss and Mitchell, 2008; Sojo et al., 2016). Presumed habitats of primordial life on Earth include Darwin's "warm little pond" (Damer, 2016; Follmann and Brownson, 2009; Klotz, 2012; Spaargaren, 1985), clays (Cairns-Smith, 1982), and hydrothermal vents (Dodd et al., 2017; Sojo et al., 2016). Thus, challenging the dogma of the origin of life on Earth has been mostly considered nonscientific. Indeed, hypothetical scenarios of living organisms being transferred between planets (McNichol and Gordon, 2012) were often naïve such as the assumption of Arrhenius (Arrhenius, 1908) that living cells can be propagated in space by radiation pressure. However, recent data on extremophile organisms indicates that some microbes can survive and even reproduce in very harsh environments (Canganella and Wiegel, 2011; Moissl-Eichinger et al., 2016), indicating that microbial life may exist on many planets and asteroids in space. Moreover, planets can become contaminated with microbial life from chaotically moving medium-sized space bodies (Belbruno et al., 2012). Thus, it makes sense to reconsider the old premise that life originated on Earth and introduce a new vision of life as a cosmic phenomenon, which is presented in this book.

The seed for the book was planted when Alexei Sharov (AS) gave an online talk in the Embryo Physics Course (Gordon, 2013a) on his extrapolation of a complexity measure for life back in time, suggesting that life may have arisen before the Earth was formed. It melded with Richard Gordon's (RG) view of the unquestioned and unjustified presumption that life began on Earth as possibly just the last anthropocentrism. Together, we posted a draft article (Sharov and Gordon, 2013), which got some popular press (Physics arXiv Blog, 2013) and led to a rebuttal by statistician Caren Marzban and colleagues (Marzban et al., 2014).[1] It helped improve our article, which is now this book's eleventh chapter (Sharov and Gordon, 2017). Extrapolation is always risky in science, but it nevertheless can provide incentives for further research (as in pointing a way to solve breast cancer, for instance (Sivaramakrishna and Gordon, 1997; Vinh-Hung and Gordon, 2005)). We may trip while thus entering the great unknown, but perhaps we'll at least fall in the right direction, if we're lucky.

The Carnegie Institute in Washington, DC, United States, sponsored a symposium on the origin of life at which RG presented a poster inviting people to participate in this book (Gordon, 2015), which gave us (AS and RG) a chance to meet in person and to meet and recruit many of the people who decided to write chapters for this book exploring the vast block of time that preceded Earth's appearance.

The Big Bang is estimated to have occurred 13.8 billion years ago (Ade et al., 2016), and the Earth is dated at 4.55 billion years (Braterman, 2013). At present, evidence for life on Earth goes back to at most

---

[1]It is reproduced here as Chapter 12 (Marzban et al., 2017).

Habitability of the Universe Before Earth, editors: Richard Gordon & Alexei Sharov, Volume 1 in the series:
Astrobiology: Exploring Life on Earth and Beyond, series editors: Pabulo Henrique Rampelotto,
Joseph Seckbach & Richard Gordon. ISSN 2468-6352. https://doi.org/10.1016/B978-0-12-811940-2.09987-1

4.28 (Dodd et al., 2017) or 4.3 (Perez-Jimenez et al., 2011) billion years (reviewed in Mikhailovsky and Gordon (2017)). That leaves a period of about 9.5 billion years, during which life may have originated on, evolved, and been transported between numerous exoplanets. Our authors consider the times and places before Earth that might have provided suitable environments for life. As the universe changed considerably during this vast epoch, there is much time and space to which we can apply our imaginations and observations in estimating where, when, and in what circumstances life might have arisen.

Here we briefly survey and comment on our chapters to whet the appetite of the reader.

PART I. PHYSICAL AND CHEMICAL CONSTRAINTS

1. Fernando J. Ballesteros Roselló & Bartolo Luque. **Gravity and life** (Ballesteros Roselló and Luque, 2017).

Fernando Ballesteros and Bartolo Luque nicely review effects of gravity on organisms and our current understanding of the constraints that gravity places on their sizes and shapes. They have discovered by analyzing data on the known exoplanets that those of Earth size and super-Earths are on a plateau in a plot of surface gravity versus mass that extends from one Earth mass M_E to 100 M_E. Thus, we can anticipate that much terrestrial intelligent life, whenever it might have evolved on planets with some rocky surface, might be about our size.

2. Ximena C. Abrevaya & Brian C. Thomas. **Radiation as a constraint for life in the universe** (Abrevaya and Thomas, 2017).

Ximena Abrevaya and Brian Thomas review the many forms of radiation from outer space that can impact on the origin and evolution of life and survey the ways living organisms protect themselves or even utilize such radiation, which may be what drives mutation, and thus evolution. From these authors, we learn that the slower particles from supernovae can spread their effects out over thousands of years. Radiation may be important in the origin of life too. For example, the halophilic Archaea use salt both within and without their membranes and polyploidy to minimize radiation damage. We will see later that halophilic Archaea may be the closest extant organisms to protocells (Gordon et al., 2017), the latter which could have been subject to and modified by radiation (Griffith et al., 2012; Snytnikov, 2010).

3. Karla de Souza Torres & Othon Cabo Winter. **The when and where of water in the history of the universe** (de Souza Torres and Winter, 2017).

Karla de Souza Torres and Othon Winter give a fascinating survey of the universe from the viewpoint of one molecular compound: water. We learn that it is just about everywhere one looks for it, usually in substantial quantities. But most of it is frozen or in a gaseous phase. Except for Mars, where evidence of seasonally flowing brines is present, we have yet to detect liquid water anywhere off Earth, though its presence has for instance been deduced on Europa and Ganymede (Hand and Carlson, 2015; Hussmann et al., 2016), Enceladus (Waite Jr. et al., 2009), Ceres (Nathues et al., 2017), and some comets (Sheldon, 2015). Detection of liquid water is a challenge for exoplanet exploration, which perhaps will be met (Visser and van de Bult, 2015).

4. Aditya Chopra & Charles H. Lineweaver. **The cosmic evolution of biochemistry** (Chopra and Lineweaver, 2017).

Aditya Chopra and Charles Lineweaver review nucleosynthesis through the first three "generations" of stars and conclude that "...the earliest elemental bottleneck for the emergence of life was not the low abundance of the life elements (C, O, N) but rather the low abundance of the elements that make rocky planets (Fe, Si, Mg) for life to live on." Thus, planets that could support life have been around for the past 11 billion years. They speculate that the earliest forms of life elsewhere would be similar to the earliest forms on Earth, suggesting that von Baer's laws of embryology (Wikipedia, 2017c) can be generalized to astrobiology. They raise the interesting question of whether the abiotic feedback system on a planet can interrupt the persistence of life (cf. Chopra and Lineweaver, 2016). Venus may be an example.

5. Paul A. Mason & Peter L. Biermann. **Astrophysical and cosmological constraints on life** (Mason and Biermann, 2017).

Paul Mason and Peter Biermann give us a detailed survey of the mechanisms by which radiation and cosmic rays are generated and distributed in and between galaxies, including effects of the expansion of the universe. They estimate that parts of the universe became habitable for at least microorganisms about 8 billion years ago, or perhaps a bit earlier, and position complex terrestrial life about 2 billion years later, both before the Earth formed. Overall, the whole universe at every level seems to be playing Whack-a-mole (Wikipedia, 2017d) with life. Just as we now are beginning to be worried about a mere asteroid collision with Earth in nonfiction books for both children and adults (Belton, 2004; Cefrey, 2002; Cox and Chestek, 1998; Nelson, 2008; Solem, 1992) and about a dozen fictional movies (Lambie, 2011), we may soon wake up to the prospects of planet sterilization events. There may be some parallels between the massive amount of death that evolution on Earth is based on (Cairns, 2014) and the deaths of life on whole planets in the course of evolution of the universe, especially if panspermia occurs.

6. Chaitanya Giri. **Primitive carbon: Before Earth and much before any life on it** (Giri, 2017).

Chaitanya Giri plunges us back to "shortly" after the Big Bang to find traces of carbon in the early universe, including the possibility of all-carbon planets. Combined with water (de Souza Torres and Winter, 2017), at least the ingredients of life may have been present early on. Giri suggests we need and could collect a lot more astrochemical data from the galaxies at the edge of the observable universe.

PART II. PREDICTING HABITABILITY

7. Michael G. Gowanlock & Ian S. Morrison. **The habitability of our evolving galaxy** (Gowanlock and Morrison, 2017).

Michael Gowanlock and Ian Morrison treat us to an overview of models of the Galactic Habitable Zone, for our and other galaxies. While there are contradictions between the models, which are based on alternative sets of assumptions, there is a consensus that opportunities for life at the surface of rocky planets have been widespread throughout the universe and for at least the past 10 billion years. The major limitation seems to be supernovae explosions. All of their estimates are underestimates, because they presume that supernovae would have little effect on aquatic life. While there may have been some life on early Earth microcontinents (Mikhailovsky and Gordon, 2017), most higher organisms remained in the seas until 0.5 billion years ago, so in this sense life on Earth, and perhaps elsewhere, may have been protected from supernovae for 90% of its existence. If this applies to other exoplanets, then their numbers potentially bearing life may be even greater. In terms of life of our intelligence, or better, they estimate that others could have had a head start on us of 2–5 billion years.

8. Branislav Vukotić. **N-body simulations and galactic habitability** (Vukotić, 2017).

Branislav Vukotić wishes he could simulate all $N = 100$ billion "particles" (mostly stars) in our Milky Way. Perhaps, someday Moore's Law (Wikipedia, 2017b) will catch up to his ambition. For now only 1 million particles are practical, each representing roughly 100,000 stars. This speaks to the vastness of even our small corner of the Universe. It also means that clever approximations had to be designed to make up for our presently inadequate computers. Vukotić finds that habitability, defined by not too many in the way of nearby supernovae explosions, is widespread in our galaxy and its satellite galaxies. Curiously, the Earth is a latecomer in the appearance of Earth-like planets, perhaps by up to 4 billion years. His review of his and others' simulations suggests that satellite galaxies could affect their hosts in terms of habitability, and that dynamics trumps metallicity in determining habitability. He focuses on spiral galaxies like ours, pointing out that habitability in elliptical galaxies, whose dynamics is much more chaotic, remains to be investigated. Vukotić emphasizes that habitability for an individual solar system may vary substantially with time, which therefore figures in how fast or episodic the drive towards increasing complexity of life may be.

9. Jeffrey M. Robinson & Jill A. Mikucki. **Occupied and empty regions of the space of extremophile parameters** (Robinson and Mikucki, 2017).

The notion of occupied and empty portions of parameter space was first introduced by David Raup in a study of the paleontological record of the shells of invertebrates with spiral growth, such as snails (Raup, 1966). Jeffrey Robinson and Jill Mikucki take us on a tour of the depths of oceans and surface rocks of Earth to survey the range of conditions under which life thrives or at least survives. The space of possibilities is multidimensional, starting with the parameters of temperature, salinity, pressure, pH, etc. Only portions of this multidimensional space are occupied by known Earth organisms, but Robinson and Mikucki show that those portions are well-represented in planets and moons of our own solar system. This extends habitability through a much greater range of distances from the sun than a simple model based on stability of a surface ocean of liquid water would predict. It also is a warning that it could be so easy for us to contaminate these places via our exploration of them.

10. Dorian Aur & Jack A. Tuszynski. **The emergence of structured, living, and conscious matter in the evolution of the universe: A theory of structural evolution and interaction of matter** (Aur and Tuszynski, 2017).

Once again we encounter N-body systems, not in whole galaxies and clusters of them (Vukotić, 2017), but this time at the level of solar systems, to brains, down to single protein molecules. Such a sweeping view is important, because the origin of life may be synonymous with the origin of perception (Gordon and Gordon, 2016; Martin and Gordon, 2001). Dorian Aur and Jack Tuszynski review the evidence of perception by single cells, which may extend to the one-cell stage of embryos (Tuszynski and Gordon, 2012). They discuss intelligence at many levels, even in plants. But intelligence includes the question of purpose. The concept of purpose may be dissected into many levels by considering organisms in the light of cybernetics (Gordon and Stone, 2016). Differentiation waves, which have electrical components, may be the conveyers of information from cell to cell in developing organisms (Gordon and Stone, 2016), much as neuronal electrical "spikes" may convey structural information from cell to cell, as Aur and Tuszynski postulate. It may be a matter of timescale, i.e., of speed of the waves, whose wide spectrum spans four orders of magnitude (Jaffe, 2002; Jaffe and Créton, 1998), with the possibility of a common underlying mechanism. Given the possibility that life originated at planetary boundaries between geospheres, of high redox (electron transfer) potential (Smith and Morowitz, 2016), the wide scope from the origin of life to our level of intelligence (and perhaps beyond) may be tied together in this chapter. It offers an alternative to the view that the universe is a huge digital simulation (Wikipedia, 2017a) and perhaps will lead to a new way to understand the origin of life and the circumstances under which it is possible.

PART III. LIFE IN THE COSMIC SCALE

11. Alexei A. Sharov & Richard Gordon. **Life before Earth** (Sharov and Gordon, 2017).

What's in common between computers and life? According to Alexei Sharov and Richard Gordon, they both show an exponential increase of their complexity over history following Moore's Law (Wikipedia, 2017b). Extrapolation of this trend for organisms back to earlier times suggests that life originated approximately 9.7 billion years ago, long before the formation of Earth. Functional complexity of organisms is assessed by the size of the nonredundant functional fraction of the genome, which doubled in size every 340 million years. Several positive feedback mechanisms including gene cooperation and duplication with subsequent specialization may have caused the exponential growth. The reconstructed Last Universal Common Ancestor (LUCA) of all organisms on Earth may represent a community of microbial organisms that reached early Earth on rogue planets, asteroids, or comets from distant cosmic sources. This scenario has deep implications for astrobiology and experimental approaches to emulating the origin of life. For example, the environments in which life originated and evolved initially may have been quite different from those envisaged on Earth, and the Drake equation for predicting the number of civilizations in the universe is likely wrong, as intelligent life may have just begun appearing in the universe. Sharov and Gordon also discuss the accelerated evolution of advanced organisms that employ a range of additional information-processing systems, such as epigenetic memory, continuing differentiation during development, nervous system, brain, language, books, and computers.

12. Caren Marzban, Raju Viswanathan & Ulvi Yurtsever. **Earth before life** [Reprint of Marzban et al. (2014)] (Marzban et al., 2017).

Caren Marzban with coauthors refined the statistical analysis of data used by Sharov and Gordon to infer the age of life from the increase of genome complexity over time. The analysis includes interval estimates (e.g., confidence or prediction intervals) and application of measurement error models (Buonaccorsi, 2010), which yield wide prediction intervals that include the origin of life on Earth. The authors conclude that the appearance of life after the formation of the Earth is consistent with the data set under examination.

13. Jacob Haqq-Misra & Ravi Kopparpu. **The time-dependent Drake equation** (Haqq-Misra and Kopparpu, 2017).

Jacob Haqq-Misra and Ravi Kopparpu extend the time scale of our endeavor out to 100 billion years, by tying the timing of advanced civilization to the lifetimes of the star types their planets orbit in their dissection of the Drake equation. They implicitly assume that the rate of evolution, in terms of complexity, depends on the type of the host star, i.e., that the slope of the semilog plot of organism complexity versus time (Sharov and Gordon, 2017) varies with stellar type. While Haqq-Misra and Kopparpu cannot definitively conclude whether we are more recent arrivals in our galaxy or among the first, they favor the latter.

14. Pauli E. Laine & Sohan Jheeta. **Are we the first? 10 billion years of evolution before Earth** (Laine and Jheeta, 2017).

Pauli Laine and Sohan Jheeta review the major theories for the origin of life, in the context of the apparently short time between the formation of the Earth and evidence for cellular life. Assuming that this represents an origin of life on Earth, they emphasize the resulting quick transition to cellular life (Gordon, 2008; Lazcano and Miller, 1994, 1996) and conclude from this and the number of exoplanets and exomoons that life should be widely distributed in the universe.

15. Julian Chela-Flores. **Life before its origin on Earth: Implications of a late emergence of terrestrial life** (Chela-Flores, 2017).

Julian Chela-Flores asks what happens to our cultural and religious perspectives if we give up the idea of life being solely on Earth, looking at our tree of life as but one such tree in a cosmic forest of life. He allows, like an aspen forest, that some of these trees may be clones, rooted to one another via panspermia. He elaborates on the idea that life might have started as early as 15 million years after the Big Bang (Loeb, 2014), showing plausible means by which such life could have survived until today. Chela-Flores takes seriously the prospect that we may soon have to face a conclusion about life on Earth "as a latecomer in cosmic evolution." Certainly, we are presently in a state of mind of great anticipation.

PART IV. SYSTEM PROPERTIES OF LIFE

16. George E. Mikhailovsky & Richard Gordon. **Symbiosis: Why was the transition from microbial prokaryotes to eukaryotic organisms a cosmic gigayear event?** (Mikhailovsky and Gordon, 2017).

George Mikhailovsky and Richard Gordon puzzle over one long step in the evolution of life on Earth, trying to find answers to the question of why it took so long for the hypothesized LUCA cell (last universal common ancestor) to give rise to an Archaea cell and a bacterial cell that fused to produce LECA (last eukaryotic common ancestor). They review the many astrophysical and Earthly events that occurred during these 2.5 billion years or so and conclude that none of them explain the delay. This led them to propose a tentative biological model for the delay, involving probabilities of novelty generation, horizontal gene transfer, and cell-cell fusion, which is presently being explored by computer simulation.

17. Alexei A. Sharov. **Coenzyme world model of the origin of life** [Reprint of: Sharov (2016)] (Sharov, 2017).

Alexei Sharov considers that traditional scenarios of the origin of life such as RNA-world or autocatalytic sets of peptides are not satisfactory because the proposed molecular systems cannot be supplied with sufficient quantities of monomers (nucleotides or amino acids) to support polymerization reactions. Also, these systems easily dissipate via diffusion. Thus, Sharov develops an alternative coenzyme-world scenario of the origin of life where coenzyme-like molecules (CLMs) colonized surfaces of oil (hydrocarbon) droplets in water. CLMs modify surface properties of droplets creating conditions that are favorable for self-reproduction of CLMs and spreading of CLMs to other oil droplets. Such niche-dependent self-reproduction is a necessary condition for cooperation between different kinds of CLMs because they have to coexist in the same oil droplet and either succeed or perish together. This model resembles lipid-world models (GARD, graded autocatalysis replication domain: Gross et al. (2014); Segré et al. (1998)) that assume growth and reproduction of hydrophobic molecular assemblies in water. Both models support primitive forms of heredity without nucleic acids. However, the coenzyme-world model has several advantages. First, it does not require abundant lipid-like resources; instead CLMs use oil as a source of carbon to producing other types of organic molecules such as glycerol and lipids. Second, oil droplets can be transformed into membrane-bound cells by engulfing water. And third, polymerization of CLMs may have resulted in the emergence of RNA-like replicons. In summary, life originated from simple but already functional molecules, and its gradual evolution towards higher complexity was driven by natural selection.

18. Richard Gordon, Martin M. Hanczyc, Nikolai D. Denkov, Mary Ann Tiffany & Stoyan K. Smoukov. **Emergence of polygonal shapes in oil droplets and living cells: The potential role of tensegrity** (Gordon et al., 2017).

This chapter stems from the excitement of one of us (Gordon, 2016) in the 2015 discovery by Nikolai Denkov and Stoyan Smoukov and their colleagues that oil droplets are not always round, but when slowly cooled in water in the presence of a surfactant become flat polygons. Two possibilities immediately came to mind: that shaped droplets might have been the original protocells, and that they might also explain the shapes of polygonal diatoms. In the context of this book, diatoms are youngsters, being at most 200 million years old. But extant halophilic Archaea and Bacteria, perhaps close to LUCA, are sometimes flat polygons. Thus, the challenge ahead will be to fill in the steps by which shaped droplets transitioned to living cells.

19. Gerard A.J.M. Jagers op Akkerhuis. **Why on theoretical grounds it may be likely that "life" exists throughout the universe** (Jagers op Akkerhuis, 2017).

Gerard Jagers op Akkerhuis tackles the problem of "What is life?," which has had many answers in over 100 publications bearing that title. Samples: (Coustenis and Encrenaz, 2013; De Pablos, 2006; Deamer, 2010; Dürr et al., 2001; Gaskell, 1928; Haldane, 1947; Kauffman, 2015; Kerbe, 2016; Margulis and Sagan, 1995; Morange, 2012; Morowitz, 1990; Murphy and O'Neill, 1995; Schrödinger, 1945). He suggests that without a consensus definition, agreement on detection of life elsewhere in the universe may be problematic. Certainly, this very question has plagued those who claimed to have already discovered extraterrestrial life. Jagers op Akkerhuis provides an answer in Operator Theory, of which continuing differentiation (Sharov and Gordon, 2017) may be an example.

SUMMARY AND EXTRAPOLATIONS

We have added a Glossary that has been worked on by most of our authors, and thus represents a joint effort.

We are now in an exponential phase of discovery of new exoplanets, learning of their immense variety. We are tooling up with methods for detection of their atmospheres (Mawet et al., 2017; Wang et al., 2017), including lightning (Hodosán et al., 2016), oceans (Visser and van de Bult, 2015), continents (Gómez-Leal et al., 2016), etc. This is an exciting explosion of knowledge of our universe, with many people anticipating the first discovery of confirmed extraterrestrial life, be it by detection of biomarkers or direct communication with intelligent beings.

It is now quite clear that life alters the habitability of its own planet(s), including creating many of the very minerals that abound in its presence (Hazen, 2012; Hazen et al., 2014). We could ask if life itself sometimes gets into runaway positive feedbacks with its planet that wipes out or severely limits life's further development. Perhaps, it is indeed time to develop Gaia-like models (Alicea and Gordon,

2014) for these interactions on cosmic time scales. If life can alter a planet, as is presumed in the Great Oxygenation Event (GOE), a kind of prehuman terraforming: (Ardelean et al., 2009; Banerjee and Sharma, 2005; Friedmann and Ocampo-Friedmann, 1995; Thomas et al., 2008), it may not always be to the benefit of future complex life. At the other end, once life of at least our level of intelligence has developed or arrived on a planet, various doomsday scenarios can be envisaged, from the original term in Drake's equation for self-destruction, as via nuclear war, to environmental catastrophism (Ward, 2009).

Much of the discourse on the origin of life nowadays revolves around extremophiles. Labeling organisms as extremophiles is of course anthropocentric. Many of them thrive or at least survive under conditions that would kill us (Robinson and Mikucki, 2017). But then, with the aid of our technologies, we are able to enter the environments of many extremophiles. This broadens the range of the parameters under which we can thrive. A measure of intelligence might be this extension of the range of habitability.

It may be necessary to distinguish planets on which life could originate from those that are habitable. The conditions needed for the origin of life might be quite different from those necessary for its subsequent evolution. Put another way, the obstacles to increasing complexity may require transfer of life from a planet suitable for "originability" to one suitable for evolvability (Gordon, 2013b; Kirschner, 2013; Kirschner and Gerhart, 1998; Pigliucci, 2008). Such transfers could possibly be one of the bottlenecks that makes the increase in complexity initially so slow (Sharov and Gordon, 2013, 2017). Life may have to terraform a planet to advance to subsequent steps of complexity, and conditions, including astrophysical ones (Mason and Biermann, 2017), may not always be suitable for this to happen. Thus, the concept of habitability may have to be refined in terms of what kinds of organisms a planet can support, and whether it can support increases in complexity of some of its lineages. The extrapolation that started this book (Sharov and Gordon, 2013, 2017) may thus imply a requirement for planet hopping, i.e., panspermia, not just because Earth was formed relatively recently in the history of our universe, but because different kinds of planets may be needed for some of the steps in the origin and evolution of life. This viewpoint may also suggest that steps after our generation, in which we extend the "artificial habitability" of presently hostile planets (Pearson III, 2014a,b,c), might lead to some surprises.

One of the quandaries raised by our backwards extrapolation of complexity (Sharov and Gordon, 2017) is indeed whether this is evidence against an origin of life on Earth? There are three alternatives we can see:

1. Our measure of complexity is inappropriate.
2. Our measure is appropriate, and Earth life indeed started elsewhere.
3. Our measure is appropriate, but there was a tremendous initial acceleration of the rate of increase in complexity.

The problem is that the time interval for an origin of life on Earth keeps getting squeezed between the probably sterilizing event of formation of our Moon and the ever-older evidence for prokaryotic life on Earth (Mikhailovsky and Gordon, 2017). The transition from the abiotic state to LUCA (Last Universal Common Ancestor) may have been sudden (Carter, 2008; Flambaum, 2003; Gordon, 2008; Laine and Jheeta, 2017; Lazcano and Miller, 1994, 1996; Lineweaver and Davis, 2002), a biological analogy to the inflationary period after the Big Bang (Guth et al., 2014; Ijjas et al., 2014; Linde, 2008), or it may have taken a long time. In the former case, our extrapolation (Fig. 1 in Sharov and Gordon, 2017) which kicked off this book would take a deep dive from its semilog linear fit. We need some testable ideas on how to distinguish these two cases (Davies, 2003).

Assuming that our extrapolation is correct, that the slope of the semilog straight line in Fig. 1 of Sharov and Gordon (2017) is universal, and that the earliest origin for life could have been when the universe first cooled to a temperature permitting liquid water (Chela-Flores, 2017; Loeb, 2014), organisms of our intelligence could have evolved as early as 4 billion years earlier than us, but no earlier. On this basis, we would infer that there was no intelligent life anywhere before Earth formed, obviating the directed panspermia hypothesis (Crick and Orgel, 1973). This would also narrow our search for extraterrestrial intelligence (SETI) (Haqq-Misra and Kopparpu, 2017; Jagers op Akkerhuis, 2017; Turnbull and Tarter, 2003a,b; Vukotić, 2017) to a sphere of radius 4 billion lightyears, or $z \sim 0.5$ (Fig. 2 in Chopra and Lineweaver, 2017). If we further suppose that panspermia transfers are only important up to the beginning of prokaryotic life, since later transfers might have trouble competing with established microorganisms, then we could further confine SETI to those Earth-like exoplanets that are at least about 4 billion years old (cf. Safonova et al., 2016).

It would be nice to have a "fossil" record of the whole history of the universe in terms of the life it might have contained. Such a record may be available for our own solar system, in terms of dwarf planets or smaller objects that have stayed away from the sun (Delbo et al., 2017; Sharov and Gordon, 2017; Volk and Malhotra, 2017) and are small enough to lack any heating of their own by impacts, atomic decay, or tidal effects. Some of the 10^5 rogue planets in the Milky Way (Strigari et al., 2012) would predate our solar system. These could have preserved any panspermia that fell on them and recorded their encounters with supernovae and larger events via mineral alterations and depositions (Bishop and Egli, 2011; Fry et al., 2016; Ludwig et al., 2016), including intergalactic gas flows (Anglés-Alcázar et al., 2017).

One theme that comes through loud and clear is that wherever life is feasible, it may indeed be present. This optimism is based on the exponentially increasing discoveries of exoplanets in our galaxy, many in their star's habitable zone, however defined. New notions that metabolism is universal and produced by a series of perhaps inevitable equilibrium and nonequilibrium phase transitions (Smith and Morowitz, 2016) add to this optimism. The contrary view that Earth is perhaps unique or a rare case (Conway Morris, 2003; Waltham, 2014; Ward and Brownlee, 2003) is in the minority. Our in between view is that originability, evolvability, and habitability may not always coincide on each planet. Until we have positive evidence of life elsewhere, and if so, what kind, the issue remains unsettled. What this book offers is the reasoned suggestion that much of the universe is habitable now, and perhaps has been for billions of years before the Earth materialized in our nascent solar system, though our authors do not fully agree on the latter point. Enjoy the ride and stay tuned.

REFERENCES

Abrevaya, X.C., Thomas, B.C., 2017. Radiation as a constraint for life in the universe. In: Gordon, R., Sharov, A.A. (Eds.), Habitability of the Universe Before Earth. In: Rampelotto, P.H., Seckbach, J., Gordon, R. (Eds.), Astrobiology: Exploring Life on Earth and Beyond. Elsevier B.V., Amsterdam, pp. 27–46 (Chapter 2).

Ade, P.A.R., Aghanim, N., Arnaud, M., Ashdown, M., Aumont, J., Baccigalupi, C., Banday, A.J., Barreiro, R.B., Bartlett, J.G., Bartolo, N., Battaner, E., Battye, R., Benabed, K., Benoit, A., Benoit-Levy, A., Bernard, J.P., Bersanelli, M., Bielewicz, P., Bock, J.J., Bonaldi, A., Bonavera, L., Bond, J.R., Borrill, J., Bouchet, F.R., Boulanger, F., Bucher, M., Burigana, C., Butler, R.C., Calabrese, E., Cardoso, J.F., Catalano, A.,

Challinor, A., Chamballu, A., Chary, R.R., Chiang, H.C., Chluba, J., Christensen, P.R., Church, S., Clements, D.L., Colombi, S., Colombo, L.P.L., Combet, C., Coulais, A., Crill, B.P., Curto, A., Cuttaia, F., Danese, L., Davies, R.D., Davis, R.J., de Bernardis, P., de Rosa, A., de Zotti, G., Delabrouille, J., Desert, F.X., Di Valentino, E., Dickinson, C., Diego, J.M., Dolag, K., Dole, H., Donzelli, S., Dore, O., Douspis, M., Ducout, A., Dunkley, J., Dupac, X., Efstathiou, G., Elsner, F., Ensslin, T.A., Eriksen, H.K., Farhang, M., Fergusson, J., Finelli, F., Forni, O., Frailis, M., Fraisse, A.A., Franceschi, E., Frejsel, A., Galeotta, S., Galli, S., Ganga, K., Gauthier, C., Gerbino, M., Ghosh, T., Giard, M., Giraud-Heraud, Y., Giusarma, E., Gjerlow, E., Gonzalez-Nuevo, J., Gorski, K.M., Gratton, S., Gregorio, A., Gruppuso, A., Gudmundsson, J.E., Hamann, J., Hansen, F.K., Hanson, D., Harrison, D.L., Helou, G., Henrot-Versille, S., Hernandez-Monteagudo, C., Herranz, D., Hildebrandt, S.R., Hivon, E., Hobson, M., Holmes, W.A., Hornstrup, A., Hovest, W., Huang, Z., Huffenberger, K.M., Hurier, G., Jaffe, A.H., Jaffe, T.R., Jones, W.C., Juvela, M., Keihanen, E., Keskitalo, R., Kisner, T.S., Kneissl, R., Knoche, J., Knox, L., Kunz, M., Kurki-Suonio, H., Lagache, G., Lahteenmaki, A., Lamarre, J.M., Lasenby, A., Lattanzi, M., Lawrence, C.R., Leahy, J.P., Leonardi, R., Lesgourgues, J., Levrier, F., Lewis, A., Liguori, M., Lilje, P.B., Linden-Vornle, M., Lopez-Caniego, M., Lubin, P.M., Macias-Perez, J.F., Maggio, G., Maino, D., Mandolesi, N., Mangilli, A., Marchini, A., Maris, M., Martin, P.G., Martinelli, M., Martinez-Gonzalez, E., Masi, S., Matarrese, S., McGehee, P., Meinhold, P.R., Melchiorri, A., Melin, J.B., Mendes, L., Mennella, A., Migliaccio, M., Millea, M., Mitra, S., Miville-Deschenes, M.A., Moneti, A., Montier, L., Morgante, G., Mortlock, D., Moss, A., Munshi, D., Murphy, J.A., Naselsky, P., Nati, F., Natoli, P., Netterfield, C.B., Norgaard-Nielsen, H.U., Noviello, F., Novikov, D., Novikov, I., Oxborrow, C.A., Paci, F., Pagano, L., Pajot, F., Paladini, R., Paoletti, D., Partridge, B., Pasian, F., Patanchon, G., Pearson, T.J., Perdereau, O., Perotto, L., Perrotta, F., Pettorino, V., Piacentini, F., Piat, M., Pierpaoli, E., Pietrobon, D., Plaszczynski, S., Pointecouteau, E., Polenta, G., Popa, L., Pratt, G.W., Prezeau, G., Prunet, S., Puget, J.L., Rachen, J.P., Reach, W.T., Rebolo, R., Reinecke, M., Remazeilles, M., Renault, C., Renzi, A., Ristorcelli, I., Rocha, G., Rosset, C., Rossetti, M., Roudier, G., d'Orfeui, B.R., Rowan-Robinson, M., Rubino-Martin, J.A., Rusholme, B., Said, N., Salvatelli, V., Salvati, L., Sandri, M., Santos, D., Savelainen, M., Savini, G., Scott, D., Seiffert, M.D., Serra, P., Shellard, E.P.S., Spencer, L.D., Spinelli, M., Stolyarov, V., Stompor, R., Sudiwala, R., Sunyaev, R., Sutton, D., Suur-Uski, A.S., Sygnet, J.F., Tauber, J.A., Terenzi, L., Toffolatti, L., Tomasi, M., Tristram, M., Trombetti, T., Tucci, M., Tuovinen, J., Turler, M., Umana, G., Valenziano, L., Valiviita, J., Van Tent, F., Vielva, P., Villa, F., Wade, L.A., Wandelt, B.D., Wehus, I.K., White, M., White, S.D.M., Wilkinson, A., Yvon, D., Zacchei, A., Zonca, A., Planck, C., 2016. Planck 2015 results XIII. Cosmological parameters. Astron. Astrophys. 594, A13.

Alicea, B., Gordon, R., 2014. Toy models for macroevolutionary patterns and trends. BioSystems Special Issue: Patterns of Evolution. 25–37.

Anglés-Alcázar, D., Faucher-Giguère, C.-A., Kereš, D., Hopkins, P.F., Quataert, E., Murray, N., 2017. The cosmic baryon cycle and galaxy mass assembly in the FIRE simulations. Mon. Not. R. Astron. Soc. 470 (4), 4698–4719.

Ardelean, I.I., Moisescu, C., Popoviciu, D.R., 2009. Magnetotactic bacteria and their potential for terraformation. In: Seckbach, J., Walsh, M. (Eds.), From Fossils to Astrobiology. Springer Science + Business Media B.V, Dordrecht, pp. 335–350.

Arrhenius, S.A., 1908. Worlds in the Making; The Evolution of the Universe. Harper, New York.

Aur, D., Tuszynski, J.A., 2017. The emergence of structured, living and conscious matter in the evolution of the universe: a theory of structural evolution and interaction of matter. In: Gordon, R., Sharov, A.A. (Eds.), Habitability of the Universe Before Earth. In: Rampelotto, P.H., Seckbach, J., Gordon, R. (Eds.), Astrobiology: Exploring Life on Earth and Beyond. Elsevier B.V., Amsterdam, pp. 231–262 (Chapter 10).

Ballesteros Roselló, F.J., Luque, B., 2017. Gravity and life. In: Gordon, R., Sharov, A.A. (Eds.), Habitability of the Universe Before Earth. In: Rampelotto, P.H., Seckbach, J., Gordon, R. (Eds.), Astrobiology: Exploring Life on Earth and Beyond. Elsevier B.V., Amsterdam, pp. 3–26 (Chapter 1).

Banerjee, M., Sharma, B.D., 2005. Mars terraformation: role of Antarctic cyanobacterial cryptoendoliths. Natl. Acad. Sci. Lett. India 28 (5–6), 155–160.

Belbruno, E., Moro-Martín, A., Malhotra, R., Savransky, D., 2012. Chaotic exchange of solid material between planetary systems: implications for lithopanspermia. Astrobiology 12 (8), 754–774.

Belton, M.J.S., 2004. Mitigation of Hazardous Comets and Asteroids. Cambridge University Press, Cambridge.

Bishop, S., Egli, R., 2011. Discovery prospects for a supernova signature of biogenic origin. Icarus 212 (2), 960–962.

Braterman, P.S., 2013. How science figured out the age of Earth. https://www.scientificamerican.com/article/how-science-figured-out-the-age-of-the-earth/.

Buonaccorsi, J.P., 2010. Measurement Error: Models, Methods, and Applications. Chapman & Hall/CRC Interdisciplinary Statistics, Taylor & Francis Group, Boca Raton, FL.

Cairns, W., 2014. Death Rules: How Death Shapes Life on Earth, and What it Means For Us. BookBaby, Fremantle, Western Australia.

Cairns-Smith, A.G., 1982. Genetic Takeover and the Mineral Origins of Life. Cambridge University Press, Cambridge.

Canganella, F., Wiegel, J., 2011. Extremophiles: from abyssal to terrestrial ecosystems and possibly beyond. Naturwissenschaften 98 (4), 253–279.

Carter, B., 2008. Five- or six-step scenario for evolution? Int. J. Astrobiol. 7 (2), 177–182.

Cefrey, H., 2002. What if an Asteroid Hit Earth? Children's Press, New York, NY, USA.

Chela-Flores, J., 2017. Life before its origin on Earth: implications of a late emergence of terrestrial life. In: Gordon, R., Sharov, A.A. (Eds.), Habitability of the Universe Before Earth. In: Rampelotto, P.H., Seckbach, J., Gordon, R. (Eds.), Astrobiology: Exploring Life on Earth and Beyond. Elsevier B.V., Amsterdam, pp. 343–351 (Chapter 15).

Chopra, A., Lineweaver, C.H., 2016. The case for a Gaian bottleneck: the biology of habitability. Astrobiology 16 (1), 7–22.

Chopra, A., Lineweaver, C.H., 2017. The cosmic evolution of biochemistry. In: Gordon, R., Sharov, A.A. (Eds.), Habitability of the Universe Before Earth. In: Rampelotto, P.H., Seckbach, J., Gordon, R. (Eds.), Astrobiology: Exploring Life on Earth and Beyond. Elsevier B.V., Amsterdam, pp. 75–87 (Chapter 4).

Conway Morris, S., 2003. Life's Solution, Inevitable Humans in a Lonely Universe. Cambridge University Press, Cambridge.

Coustenis, A., Encrenaz, T., 2013. What is life and where can it exist? In: Life Beyond Earth: The Search for Habitable Worlds in the Universe, pp. 19–83, Cambridge University Press, Cambridge.

Cox, D.W., Chestek, J.H., 1998. Doomsday Asteroid: Can We Survive? Prometheus Books, Amherst, NY, USA.

Crick, F.H.C., Orgel, L.E., 1973. Directed panspermia. Icarus 19 (3), 341–346.

Damer, B., 2016. A field trip to the Archaean in search of Darwin's warm little pond. Life 6, 21.

Damer, B., Deamer, D., 2015. Coupled phases and combinatorial selection in fluctuating hydrothermal pools: a scenario to guide experimental approaches to the origin of cellular life. Life 5 (1), 872–887.

Darwin, C.R., 1871. To J. D. Hooker, 1 February. http://www.darwinproject.ac.uk/DCP-LETT-7471.

Davies, P.C.W., 2003. Does life's rapid appearance imply a Martian origin? Astrobiology 3 (4), 673–679.

De Pablos, J.L.S., 2006. Schrodinger's dilemma: 'What is life'? (Erwin Schrodinger). Pensamiento 62 (234), 505–520.

de Souza Torres, K., Winter, O.C., 2017. The when and where of water in the history of the universe. In: Gordon, R., Sharov, A.A. (Eds.), Habitability of the Universe Before Earth. In: Rampelotto, P.H., Seckbach, J., Gordon, R. (Eds.), Astrobiology: Exploring Life on Earth and Beyond. Elsevier B.V., Amsterdam, pp. 47–73 (Chapter 3).

Deamer, D., 2010. Special collection of essays: what is life? Introduction. Astrobiology 10 (10), 1001–1002.

Delbo, M., Walsh, K., Bolin, B., Avdellidou, C., Morbidelli, A., 2007. Identification of a primordial asteroid family constrains the original planetesimal population. Science 357 (6355), 1026–1029.

Dodd, M.S., Papineau, D., Grenne, T., Slack, J.F., Rittner, M., Pirajno, F., O'Neil, J., Little, C.T.S., 2017. Evidence for early life in Earth's oldest hydrothermal vent precipitates. Nature 543 (7643), 60–64.

Dürr, H.-P., Popp, F.-A., Schommers, W., 2001. What is Life? Scientific Approaches and Philosophical Positions. World Scientific, Singapore.

Flambaum, V.V., 2003. Comment on "Does the rapid appearance of life on earth suggest that life is common in the universe?" Astrobiology 3 (2), 237–239.

Follmann, H., Brownson, C., 2009. Darwin's warm little pond revisited: from molecules to the origin of life. Naturwissenschaften 96 (11), 1265–1292.

Fox, S.W., 1964. Thermal polymerization of amino-acids and production of formed microparticles on lava. Nature 201 (4917), 336–337.

Friedmann, E.I., Ocampo-Friedmann, R., 1995. A primitive cyanobacterium as pioneer microorganism for terra-forming Mars. Adv. Space Res. 15 (3), 243–246.

Fry, B.J., Fields, B.D., Ellis, J.R., 2016. Radioactive iron rain: transporting ^{60}Fe in supernova dust to the ocean floor. Astrophys. J. 827 (1), 48.

Gallori, E., Biondi, E., Branciamore, S., 2006. Looking for the primordial genetic honeycomb. Orig. Life Evol. Biosph. 36 (5–6), 493–499.

Gaskell, A., 1928. What is Life? C.C. Thomas.

Giri, C., 2017. Primitive carbon: before Earth and much before any life on it. In: Gordon, R., Sharov, A.A. (Eds.), Habitability of the Universe Before Earth. In: Rampelotto, P.H., Seckbach, J., Gordon, R. (Eds.), Astrobiology: Exploring Life on Earth and Beyond. Elsevier B.V., Amsterdam, pp. 127–145 (Chapter 6).

Gómez-Leal, I., Codron, F., Selsis, F., 2016. Thermal light curves of Earth-like planets: 1. Varying surface and rotation on planets in a terrestrial orbit. Icarus 269, 98–110.

Gordon, R., 2008. Hoyle's tornado origin of artificial life, a computer programming challenge. In: Seckbach, J., Gordon, R. (Eds.), Divine Action and Natural Selection: Science, Faith and Evolution. World Scientific, Singapore, pp. 354–367.

Gordon, R., 2013a. Conception and development of the Second Life® Embryo Physics Course [invited]. Syst. Biol. Reprod. Med. 59, 131–139.

Gordon, R., 2013b. The differentiation tree as a source of novelty and evolvability, comment on: "Beyond Darwin: evolvability and the generation of novelty" by Marc Kirschner. BMC Biol. 11, #110. https://bmcbiol. biomedcentral.com/articles/10.1186/1741-7007-11-110/comments.

Gordon, R., 2015. Book proposal: habitability of the universe before Earth (poster). In: Walker, S. (Ed.), Re-conceptualizing the Origin of Life, November 9–13, 2015. Carnegie Institution for Science, Washington, DC.

Gordon, R., 2016. Shaped droplets, diatoms and the origin of life. http://embryogenesisexplained.org/2016/01/11/shaped-droplets-diatoms-and-the-origin-of-life/.

Gordon, N.K., Gordon, R., 2016. Embryogenesis Explained. World Scientific Publishing, Singapore.

Gordon, R., Stone, R., 2016. Cybernetic embryo. In: Gordon, R., Seckbach, J. (Eds.), Biocommunication: Sign-Mediated Interactions Between Cells and Organisms. World Scientific Publishing, London, pp. 111–164.

Gordon, R., Hanczyc, M.M., Denkov, N.D., Tiffany, M.A., Smoukov, S.K., 2017. Emergence of polygonal shapes in oil droplets and living cells: the potential role of tensegrity in the origin of life. In: Gordon, R., Sharov, A.A. (Eds.), Habitability of the Universe Before Earth. In: Rampelotto, P.H., Seckbach, J., Gordon, R. (Eds.), Astrobiology: Exploring Life on Earth and Beyond, vol. 1. Elsevier B.V., Amsterdam, pp. 427–490 (Chapter 18).

Gowanlock, M.G., Morrison, I.S., 2017. The habitability of our evolving galaxy, In: Gordon, R., Sharov, A.A. (Eds.), Habitability of the Universe Before Earth. In: Rampelotto, P.H., Seckbach, J., Gordon, R. (Eds.), Astrobiology: Exploring Life on Earth and Beyond. Elsevier B.V., Amsterdam, pp. 149–171 (Chapter 7).

Griffith, E.C., Tuck, A.F., Vaida, V., 2012. Ocean-atmosphere interactions in the emergence of complexity in simple chemical systems. Acc. Chem. Res. 45 (12), 2106–2113.

Gross, R., Fouxon, I., Lancet, D., Markovitch, O., 2014. Quasispecies in population of compositional assemblies. BMC Evol. Biol. 14, 265.

Guth, A.H., Kaiser, D.I., Nomura, Y., 2014. Inflationary paradigm after Planck 2013. Phys. Lett. B 733, 112–119.

Haldane, J., 1929. The origin of life. Ration. Annu. 3, 148–153.

Haldane, J.B.S., 1947. What is Life? Boni and Gaer, New York, NY, USA.

Hand, K.P., Carlson, R.W., 2015. Europa's surface color suggests an ocean rich with sodium chloride. Geophys. Res. Lett. 42 (9), 3174–3178.

Haqq-Misra, J., Kopparpu, R., 2017. The Drake equation as a function of spectral type and time. In: Gordon, R., Sharov, A.A. (Eds.), Habitability of the Universe Before Earth. In: Rampelotto, P.H., Seckbach, J., Gordon, R. (Eds.), Astrobiology: Exploring Life on Earth and Beyond. Elsevier B.V., Amsterdam, pp. 307–319 (Chapter 13).

Hazen, R.M., 2012. The Story of Earth: The First 4.5 Billion Years, From Stardust to Living Planet, Kindle ed. Penguin Publishing Group, New York.

Hazen, R.M., Liu, X.-M., Downs, R.T., Golden, J., Pires, A.J., Grew, E.S., Hystad, G., Estrada, C., Sverjensky, D.A., 2014. Mineral evolution: episodic metallogenesis, the supercontinent cycle, and the coevolving geosphere and biosphere. [Special Publication 18], In: Kelley, K.D., Golden, H.C. (Eds.), Building Exploration Capability for the 21st Century. Society of Economic Geologists, Littleton, Colorado, USA, pp. 1–15.

Hodosán, G., Helling, C., Asensio-Torres, R., Vorgul, I., Rimmer, P.B., 2016. Lightning climatology of exoplanets and brown dwarfs guided by Solar system data. Mon. Notices R. Astron. Soc. 461 (4), 3927–3947.

Hussmann, H., Shoji, D., Steinbrügge, G., Stark, A., Sohl, F., 2016. Constraints on dissipation in the deep interiors of Ganymede and Europa from tidal phase-lags. Celestial Mech. Dyn. Astr. 126 (1–3), 131–144.

Ijjas, A., Steinhardt, P.J., Loeb, A., 2014. Inflationary schism. Phys. Lett. B 736, 142–146.

Jaffe, L., 2002. On the conservation of fast calcium wave speeds. Cell Calcium 32 (4), 217–229.

Jaffe, L.F., Créton, R., 1998. On the conservation of calcium wave speeds. Cell Calcium 24 (1), 1–8.

Jagers op Akkerhuis, G.A.J.M., 2017. Why on theoretical grounds it is likely that "life" exists throughout the universe. In: Gordon, R., Sharov, A.A. (Eds.), Habitability of the Universe Before Earth. In: Rampelotto, P.H., Seckbach, J., Gordon, R. (Eds.), Astrobiology: Exploring Life on Earth and Beyond. Elsevier B.V., Amsterdam, pp. 491–505 (Chapter 19).

Kauffman, S., 2015. What Is Life? Isr. J. Chem. 55 (8), 875–879.

Kerbe, W., 2016. What is life-in everyday understanding? A focus group study on lay perspectives on the term *life*. Artif. Life 22 (1), 119–133.

Kirschner, M., 2013. Beyond Darwin: evolvability and the generation of novelty. BMC Biol. 11 (1), 110.

Kirschner, M., Gerhart, J., 1998. Evolvability. Proc. Natl. Acad. Sci. U. S. A. 95 (15), 8420–8427.

Klotz, I., 2012. Did life start in a pond, not oceans? https://www.livescience.com/31204-life-start-pond-oceans.html.

Laine, P.E., Jheeta, S., 2017. Are we the first: was there life before our solar system? In: Gordon, R., Sharov, A.A. (Eds.), Habitability of the Universe Before Earth. In: Rampelotto, P.H., Seckbach, J., Gordon, R. (Eds.), Astrobiology: Exploring Life on Earth and Beyond. Elsevier B.V., Amsterdam, pp. 321–341 (Chapter 14).

Lambie, R., 2011. 10 deadly comets, asteroids and meteorites in the movies. http://www.denofgeek.com/us/movies/17596/10-deadly-comets-asteroids-and-meteorites-in-the-movies.

Lazcano, A., Miller, S.L., 1994. How long did it take for life to begin and evolve to cyanobacteria? J. Mol. Evol. 39 (6), 546–554.

Lazcano, A., Miller, S.L., 1996. The origin and early evolution of life: prebiotic chemistry, the pre-RNA world, and time. Cell 85 (6), 793–798.

Linde, A., 2008. Inflationary cosmology. Lect. Notes Phys. 738, 1–54.

Lineweaver, C.H., Davis, T.M., 2002. Does the rapid appearance of life on Earth suggest that life is common in the universe? Astrobiology 2 (3), 293–304.

Loeb, A., 2014. The habitable epoch of the early Universe. Int. J. Astrobiol. 13 (4), 337–339.

Ludwig, P., Bishop, S., Egli, R., Chernenko, V., Deneva, B., Faestermann, T., Famulok, N., Fimiani, L., Gomez-Guzman, J.M., Hain, K., Korschinek, G., Hanzlik, M., Merchel, S., Rugel, G., 2016. Time-resolved 2-million-year-old supernova activity discovered in Earth's microfossil record. Proc. Natl. Acad. Sci. U. S. A. 113 (33), 9232–9237.

Margulis, L., Sagan, D., 1995. What Is Life? Simon & Schuster, New York.

Martin, C.C., Gordon, R., 2001. The evolution of perception. Cybern. Syst. 32 (3–4), 393–409.

Marzban, C., Viswanathan, R., Yurtsever, U., 2014. Earth before life. Biol. Direct 9, 1.

Marzban, C., Viswanathan, R., Yurtsever, U., 2017. Earth before life [Reprint of: Marzban, C., Viswanathan, R., Yurtsever, U., 2014. Earth before life. Biol. Direct 9, 1.]. In: Gordon, R., Sharov, A.A. (Eds.), Habitability of the Universe Before Earth. In: Rampelotto, P.H., Seckbach, J., Gordon, R. (Eds.), Astrobiology: Exploring Life on Earth and Beyond. Elsevier B.V., Amsterdam, pp. 297–305 (Chapter 12).

Mason, P.A., Biermann, P.L., 2017. Astrophysical and cosmological constraints on life. In: Gordon, R., Sharov, A.A. (Eds.), Habitability of the Universe Before Earth. In: Rampelotto, P.H., Seckbach, J., Gordon, R. (Eds.), Astrobiology: Exploring Life on Earth and Beyond. Elsevier B.V., Amsterdam, pp. 89–126 (Chapter 5).

Mawet, D., Ruane, G., Xuan, W., Echeverri, D., Klimovich, N., Randolph, M., Fucik, J., Wallace, J.K., Wang, J., Vasisht, G., Dekany, R., Mennesson, B., Choquet, E., Delorme, J.R., Serabyn, E., 2017. Observing exoplanets with high dispersion coronagraphy. II. Demonstration of an active single-mode fiber injection unit. Astrophysical Journal 838 (2), #92.

McNichol, J., Gordon, R., 2012. Are we from outer space? A critical review of the panspermia hypothesis. In: Seckbach, J. (Ed.), Genesis—In the Beginning: Precursors of Life, Chemical Models and Early Biological Evolution. Springer, Dordrecht, pp. 591–620.

Mikhailovsky, G.E., Gordon, R., 2017. Symbiosis: why was the transition from microbial prokaryotes to eukaryotic organisms a cosmic gigayear event? In: Gordon, R., Sharov, A.A. (Eds.), Habitability of the Universe Before Earth. In: Rampelotto, P.H., Seckbach, J., Gordon, R. (Eds.), Astrobiology: Exploring Life on Earth and Beyond. Elsevier B.V., Amsterdam, pp. 355–405 (Chapter 16).

Miller, S.L., 1953. A production of amino acids under possible primitive earth conditions. Science 117 (3046), 528–529.

Moissl-Eichinger, C., Cockell, C.S., Rettberg, P., 2016. Venturing into new realms? Microorganisms in space. FEMS Microbiol. Rev. 40 (5), 722–737.

Morange, M., 2012. The recent evolution of the question "What is life?" Hist. Philos. Life Sci. 34 (3), 425–436.

Morowitz, H.J., 1990. What is life? Now is the time to find out. Hosp. Pract. 25 (6), 137–140.

Murphy, M.P., O'Neill, L.A.J., 1995. What is Life? The Next Fifty Years, Speculations on the Future of Biology. Cambridge University Press, Cambridge.

Nathues, A., Platz, T., Thangjam, G., Hoffmann, M., Mengel, K., Cloutis, E.A., Le Corre, L., Reddy, V., Kallisch, J., Crown, D.A., 2017. Evolution of Occator crater on (1) Ceres. Astron. J. 153 (3), 112.

Nelson, J., 2008. Collision Course: Asteroids and Earth. Rosen Publishing Group, New York, NY, USA.

Oparin, A.I., 1936. The Origin of Life, 1953 reprint ed. Dover Publications, Mineola, NY.

Pearson III, P.E., 2014a. Addendum to extending the artificial habitability zone to Pluto? Version: 1. doi:https://doi.org/10.6084/m9.figshare.1119113.

Pearson III, P.E., 2014b. Artificial habitable zones of stars shown through a hypothetical planet. https://figshare.com/articles/Artificial_Habitable_Zones_of_Stars_Shown_Through_a_Hypothetical_Planet_v1_1/1019947.

Pearson III, P.E., 2014c. Extending the artificial habitability zone to Pluto? https://ndownloader.figshare.com/files/1864296.

Perez-Jimenez, R., Inglés-Prieto, A., Zhao, Z.M., Sanchez-Romero, I., Alegre-Cebollada, J., Kosuri, P., Garcia-Manyes, S., Kappock, T.J., Tanokura, M., Holmgren, A., Sanchez-Ruiz, J.M., Gaucher, E.A., Fernandez, J.M., 2011. Single-molecule paleoenzymology probes the chemistry of resurrected enzymes. Nat. Struct. Mol. Biol. 18 (5), 592–596.

Physics arXiv Blog, 2013. Best of 2013: Moore's Law and the Origin of Life. As life has evolved, its complexity has increased exponentially, just like Moore's law. Now geneticists have extrapolated this trend backwards and found that by this measure, life is older than the Earth itself. http://www.technologyreview.com/view/522866/best-of-2013-moores-law-and-the-origin-of-life/.

Pigliucci, M., 2008. Opinion: is evolvability evolvable? Nat. Rev. Genet. 9 (1), 75–82.

Rauchfuss, H., Mitchell, T.N., 2008. Chemical Evolution and the Origin of Life. Springer, Berlin.

Raup, D.M., 1966. Geometric analysis of shell coiling: general problems. J. Paleontol. 40 (5), 1178–1190.

Robinson, J.M., Mikucki, J.A., 2017. Occupied and empty regions of the space of extremophile parameters. In: Gordon, R., Sharov, A.A. (Eds.), Habitability of the Universe Before Earth. In: Rampelotto, P.H., Seckbach, J., Gordon, R. (Eds.), Astrobiology: Exploring Life on Earth and Beyond. Elsevier B.V., Amsterdam, pp. 199–230 (Chapter 9).

Safonova, M., Murthy, J., Shchekinov, Y.A., 2016. Age aspects of habitability. Int. J. Astrobiol. 15 (2), 93–105.

Schrödinger, E., 1945. What is Life? The Physical Aspect of the Living Cell. Cambridge University Press, Cambridge.

Segré, D., Lancet, D., Kedem, O., Pilpel, Y., 1998. Graded autocatalysis replication domain (GARD): kinetic analysis of self-replication in mutually catalytic sets. Orig. Life Evol. Biosph. 28 (4–6), 501–514.

Sharov, A.A., 2016. Coenzyme world model of the origin of life. BioSystems 144, 8–17.

Sharov, A.A., 2017. Coenzyme world model of the origin of life [Reprint of: Sharov, A.A., 2016. Coenzyme world model of the origin of life. BioSystems 144, 8–17.]. In: Gordon, R., Sharov, A.A. (Eds.), Habitability of the Universe Before Earth. In: Rampelotto, P.H., Seckbach, J., Gordon, R. (Eds.), Astrobiology: Exploring Life on Earth and Beyond. Elsevier B.V., Amsterdam, pp. 407–426 (Chapter 17).

Sharov, A.A., Gordon, R., 2013. Life before Earth. http://arxiv.org/abs/1304.3381.

Sharov, A.A., Gordon, R., 2017. Life before Earth. In: Gordon, R., Sharov, A.A. (Eds.), Habitability of the Universe Before Earth. In: Rampelotto, P.H., Seckbach, J., Gordon, R. (Eds.), Astrobiology: Exploring Life on Earth and Beyond. Elsevier B.V., Amsterdam, pp. 265–296 (Chapter 11).

Sheldon, R.B., 2015. Wet comet model: Rosetta redux. Proc. SPIE 9606, #96061a.

Sivaramakrishna, R., Gordon, R., 1997. Detection of breast cancer at a smaller size can reduce the likelihood of metastatic spread: a quantitative analysis. Acad. Radiol. 4 (1), 8–12.

Smith, E., Morowitz, H.J., 2016. The Origin and Nature of Life on Earth: The Emergence of the Fourth Geosphere. Cambridge University Press, Cambridge.

Snytnikov, V.N., 2010. Astrocatalysis-abiogenic synthesis and chemical evolution at pregeological stages of the Earth's formation. Paleontol. J. 44 (7), 761–777.

Sojo, V., Herschy, B., Whicher, A., Camprubi, E., Lane, N., 2016. The origin of life in alkaline hydrothermal vents. Astrobiology 16 (2), 181–197.

Solem, J.C., 1992. Interception of Comets and Asteroids on Collision Course With Earth. United States Department of Energy, Washington, DC.

Spaargaren, D.H., 1985. Origin of life: oceanic genesis, panspermia or Darwin's warm little pond. Experientia 41 (6), 719–727.

Strigari, L.E., Barnabè, M., Marshall, P.J., Blandford, R.D., 2012. Nomads of the Galaxy. Mon. Not. R. Astron. Soc. 423 (2), 1856–1865.

Thomas, D.J., Eubanks, L.M., Rector, C., Warrington, J., Todd, P., 2008. Effects of atmospheric pressure on the survival of photosynthetic microorganisms during simulations of ecopoesis. Int. J. Astrobiol. 7 (3-4), 243–249.

Turnbull, M.C., Tarter, J.C., 2003a. Target selection for SETI. I. A catalog of nearby habitable stellar systems. Astrophys. J. Suppl. Ser. 145 (1), 181–198.

Turnbull, M.C., Tarter, J.C., 2003b. Target selection for SETI. II. Tycho-2 dwarfs, old open clusters, and the nearest 100 stars. Astrophys. J. Suppl. Ser. 149 (2), 423–436.

Tuszynski, J.A., Gordon, R., 2012. A mean field Ising model for cortical rotation in amphibian one-cell stage embryos. BioSystems 109 (3, Special Issue on Biological Morphogenesis), 381–389.

Vinh-Hung, V., Gordon, R., 2005. Quantitative target sizes for breast tumor detection prior to metastasis: a prerequisite to rational design of 4D scanners for breast screening. Technol. Cancer Res. Treat. 4 (1), 11–21.

Visser, P.M., van de Bult, F.J., 2015. Fourier spectra from exoplanets with polar caps and ocean glint. Astron. Astrophys. 579, A21.

Volk, K., Malhotra, R., 2017. The curiously warped mean plane of the Kuiper Belt. https://arxiv.org/abs/1704.02444.

Vukotić, B., 2017. N-body simulations and galactic habitability. In: Gordon, R., Sharov, A.A. (Eds.), Habitability of the Universe Before Earth. In: Rampelotto, P.H., Seckbach, J., Gordon, R. (Eds.), Astrobiology: Exploring Life on Earth and Beyond. Elsevier B.V., Amsterdam, pp. 173–197 (Chapter 8).

Waite Jr., J.H., Lewis, W.S., Magee, B.A., Lunine, J.I., McKinnon, W.B., Glein, C.R., Mousis, O., Young, D.T., Brockwell, T., Westlake, J., 2009. Liquid water on Enceladus from observations of ammonia and 40Ar in the plume. Nature 460 (7254), 487.

Waltham, D., 2014. Lucky Planet: Why Earth is Exceptional—and What That Means for Life in the Universe. Icon Books Limited, London, UK.

Wang, J., Mawet, D., Ruane, G., Hu, R., Benneke, B., 2017. Observing exoplanets with high dispersion coronagraphy. I. The scientific potential of current and next-generation large ground and space telescopes. Astron. J. 153 (4), #183.

Ward, P., 2009. The Medea Hypothesis: Is Life on Earth Ultimately Self-Destructive? Princeton University Press.

Ward, P., Brownlee, D., 2003. Rare Earth: Why Complex Life Is Uncommon in the Universe. Copernicus, Springer-Verlag, New York.

Wikipedia, 2017a. Digital physics. https://en.wikipedia.org/wiki/Digital_physics.

Wikipedia, 2017b. Moore's law. https://en.wikipedia.org/wiki/Moore%27s_law.

Wikipedia, 2017c. von Baer's laws (embryology). https://en.wikipedia.org/wiki/Von_Baer%27s_laws_(embryology).

Wikipedia, 2017d. Whac-A-Mole. https://en.wikipedia.org/wiki/Whac-A-Mole.

PHYSICAL AND CHEMICAL CONSTRAINTS

GRAVITY AND LIFE

Fernando J. Ballesteros*,[1], Bartolo Luque[†]

*Astronomical Observatory, University of Valencia, Paterna, Spain
[†]E.T.S.I. Aeronautics and Space, Politechnical University of Madrid, Madrid, Spain

CHAPTER OUTLINE

1 INTRODUCTION

As physicists, the first relationship between gravity and biology that comes to mind is the apocryphal story of the discovery of universal gravitation by Newton when an apple fell on his head. The apple could fall from the tree because the apple tree was on the planet Earth. Had this tree been native of a world with a much greater gravity, surely the apple would have grown on the ground and would not have fallen. Although this we do not believe would have been an impediment for Newton to discover gravitation.

---

[1]Dedicated to the memory of our daughter Helena, who passed away during the writing of this chapter.

Habitability of the Universe Before Earth, editors: Richard Gordon & Alexei Sharov, Volume 1 in the series:
Astrobiology: Exploring Life on Earth and Beyond, series editors: Pabulo Henrique Rampelotto,
Joseph Seckbach & Richard Gordon. ISSN 2468-6352. https://doi.org/10.1016/B978-0-12-811940-2.00001-0

Gravity is one of the most relevant factors for the appearance of life in the universe, and for its existence. Organisms have grown up fighting this force and have evolved and adapted their form not only to survive its presence, but also to use it for their own benefit. The larger sizes that have existed, in the records, are possibly giving us clues about which are the maximum size limits that our planet's gravity can allow. Lower surface gravities may possibly allow larger sizes, as free-falling experiments seem to indicate. On the other hand, the very appearance of environments where life is created, that is, planets with seas and atmosphere heated by a star, is also intimately linked to gravitation. The gravitational collapse is the motor for the formation of stellar systems and its court of planets. And these exist because they are inside those islands of gravitation we call galaxies. If there were no gravity, or even if its intensity were different from what it is, there might not be beings in the universe who could read these lines.

2 GRAVITY AS SOURCE OF COMPLEXITY

Gravity, the electromagnetic force, and the weak and strong nuclear interactions are the four fundamental forces of nature (Feynman, 1967). In spite of the preponderant role of gravity as architect of the greatest structures of the universe, its future, and its final destiny, it has a discrete intensity: it is 37, 25, and 39 orders of magnitude weaker than the electromagnetic force, the weak nuclear interaction, and the strong nuclear interaction, respectively. Nowadays, the intensity of the fundamental forces, the masses of the different elementary particles, or the values for the constants of nature are not specified by any theory. Quantum field theory (Weinberg, 2005) is not able to determine these magnitudes and, on paper, it is possible to give them different values, thus obtaining other coherent physical laws. However, not all those possible alternatives would be pleasant for us. For example, if the electromagnetic force were much stronger, the electrons would not prowl around the nucleus but would fall inside, and chemistry as we understand it would have no existence; being it only a little bit weaker, the hydrogens bridges of the water molecule would not exist and it would not be a liquid, but a gas. A weaker strong nuclear force would make many of the atoms radioactive and only the lightest ones would remain stable. If the gravity force were more intense, the stars would be much smaller and their nuclear fuel would be wasted in a time too short to allow the emergence of life on planets; on the contrary, if it were weaker, the expansion of the universe would not have been counteracted locally, the galaxies would have been disintegrated, and there would be only few lonely stars scattered in the cosmos. It seems that the slightest modification of the universal constants would radically change the reality of our universe (Rees, 2001).

In the 1930s, Paul Dirac pointed out a curious relationship: the ratio of the electromagnetic field intensity to the gravitational field intensity was the same as the ratio of the observable universe size to the classical electron radius; an enormous number of the order of 10^{37}. As the size of the universe grows over time, while the other values are constant, the cosmologist Robert Dicke thought that gravity (and the electromagnetic field) was indicating a very particular moment in the age of the universe: the moment in which the universe can allow the existence of intelligent observers. Dicke set up the first statement of the so-called weak anthropic principle (Mosterín, 2005): life can only arise and exist during a certain period in the age of the universe, just when the equality discovered by Dirac is fulfilled. In some way, it is reasonable: no complex beings could exist at such an early age as the inflationary age. The universe must first unfold all its complexity to reach the necessary conditions for the emergence of life. In 1973, another cosmologist, Brandon Carter, enunciated a tougher version of this principle (Carter, 1974), the so-called strong anthropic principle: the laws of the universe must be such as to allow the existence of observers. According to Carter, it is not only a problem of observational selection; it is necessary to keep the laws and constants of nature within very limited ranges to allow the appearance of life in the universe. Consciousness is possible in a very special type of universe: ours. It seems that the existence of consciousness has to force the real universe to be this and not another. The strong anthropic principle, with quasimystical scents, remains a hotly debated point in cosmology. For some, any other choice for the values of the fundamental constants would result in an impossible and contradictory universe. So our universe would be the only coherent possibility, a statement that closely resembles Leibnitz's famous aphorism: "Our world is the best of all possible worlds."

On the other hand, the discoveries in complex systems (Waldrop, 1992) undermine the credibility of the strong version of the anthropic principle. Today, we know that we can design very strange and exotic systems, with simple laws that have no commonality with the physics of our universe and in which, however, complex unforeseeable structures emerge. Consider, for example, a classic model of artificial life (Adami, 1998) - the famous "game of life" designed by the British mathematician John Conway in 1970, where insultingly simple rules yield complex structures capable of moving through the board, reproducing, storing and managing information… Analogously, our knowledge of physics is not yet ready to deduce from the properties and interactions of elementary particles, the existence of galaxies, volcanoes, Darwinian evolution, or social behavior. For the moment, it seems out of our reach to determine whether a universe with other physical laws would preclude the emergence of life or determine what kind of unsuspected structures could occur in different universes.

Speculations and possible universes aside, we know that from its beginning our universe has increased its complexity. However, one of the fundamental laws of physics, the second law of thermodynamics, states that in any isolated system, entropy (disorder) always grows. Isolated systems become increasingly disordered until they reach their maximum entropy. If a system increases its complexity (thus decreasing its entropy) it cannot be isolated, it must have an external energy input. But our universe is an isolated system, it is the isolated system *par excellence*. And the universe in its beginnings was a hot soup of radiation in thermal equilibrium, uniform, simple, and without the smallest structure. Its primitive entropy was maximal. In such a case, and if entropy in an isolated system cannot diminish, how were more and more complex structures in the universe arising? How is it that entropy did not impede the evolution of the universe? Part of the answer lies in the expansion. If the universe had not suddenly grown in size during the inflationary epoch, there would have been no cosmological evolution. However, as it expanded it became cooler. The amount of initial entropy, which was maximal for the primitive universe, continued growing in accordance with the second law of thermodynamics. But as the universe grew, the maximum amount of entropy that it could admit became larger and larger, at a rate much faster than the actual growth of entropy. As a result, the universe was in a state of entropy very low, relative to its size, which allowed its subsequent development and evolution. The universe was wound up during the Big Bang (Penrose, 2005).

After the recombination epoch, the homogeneous and uniform universe began to become increasingly tenuous and areas with matter density higher and lower than the average appeared. The matter present in the less dense areas migrated by gravitational attraction towards the zones with greater density, causing the density differences to intensify and the process accelerated. For the first time in the universe, the rich became richer at the expense of the poor getting poorer, starting an aggressive free market gravitational behavior. The effect of gravitation is not usually considered in classical thermodynamic calculations, but is capable of dramatically altering the picture. We are unconsciously accustomed to the concept of entropy in terms of gas mechanics. But between the ideal gases and the original plasma that filled the universe, there is a very important difference: the molecules of ideal gases do not attract each other. The second law of thermodynamics is statistical, it states that a system tends to evolve towards the most probable state. Thus, if we have smoke in a room whose molecules move erratically, it is extremely improbable for these molecules to cluster in a corner of the room. Statistically, it is more likely to be scattered throughout the chamber. That is the most probable state, with greater entropy, from the standpoint of classical, nongravitational thermodynamics. However, when gravity is involved, the opposite occurs precisely because matter is attracted to matter. If we have two masses at rest in space, they will begin to attract each other until they are together; and that is the most likely state. Matter subjected to and generating gravitational fields is continually attracted; it cannot move erratically like the molecules of a gas, but it tends to cling. Thanks to the gravity force, the most probable state, the clumps of matter, has greater entropy. Without this seemingly "antientropic" quality of gravitation, the universe would still be a container full of gas as in its origins. We can therefore be grateful to gravitation, ultimately the greatest motor of the universe along with its own expansion, for it is the root cause of all the complex structures it contains, including stars and planets.

3 THE PLANET-BUILDER FORCE

Today, we think that planets are an indispensable link in the chain of cosmic events that make life possible. They are places large and stable enough for the chemical elements and molecules, formed after stellar alchemy, to interact in high concentrations and produce complex chemical reactions. That is why knowing how common planets are is directly related to the possibility of life in the universe.

Planetary formation is linked to star formation. This is convenient for our search for life, as the light and heat provided by a star is an excellent source of energy for its nourishment. And the engine that moves this formation of the planets and stars is gravitation. As we said, gravitation is the main architect of all structures in the universe, including galaxies: it is the only force capable of counteracting locally the expansion of the universe and increasing its complexity at large scales. Today, we know that in the early stages of the universe, when it was only about 500 million years old and it was smaller, younger, and denser, there was an era of wild star formation before the appearance of galaxies. But given that the universe was chemically much simpler than now, those planets that might have formed around those first stars were not rocky worlds but sterile gas giants, balls of hydrogen and helium that would lack more complex elements, making life impossible. Since then, star formation outside galaxies is almost impossible. With a much less dense and more scattered universe, gas in intergalactic space does not have enough density, and today galaxies are the only habitat where stars can be formed, maintaining with their gravitational attraction regions with sufficient gas density so that new suns may be born. Without these oases of gas and dust, the expansion of the universe, diluting the material in an increasingly faint void, would have prevented the appearance of stars with rocky planets and life.

Our theoretical models tell us that a planetary system begins when gravitation causes the contraction of a molecular cloud, a region of neutral hydrogen and dust at $-260°C$, and a density of about 1000 particles/cm^3, greater than that of interstellar medium. This gas is relatively stable: if we contract a part on itself, it will increase its temperature and internal pressure, and as a consequence, it will expand again, returning to its original state. But when the amount of contracted gas exceeds a certain threshold, the gravitational field of this contracted gas bulk may become intense enough to defeat the pressure and the contraction continues (Luque et al., 2009). This threshold can be estimated by equating the gravitational potential energy GM^2/R with twice[1] the thermal energy Mc_s^2, where c_s is the velocity of sound in the cloud, G is the universal gravitational constant, $\lambda = 2R$ is the typical diameter of the cloud, and M is the mass of the molecular cloud. Combining these expressions with $M = \rho\ (4\pi/3)\lambda^3$, where ρ represents the density of the cloud, we obtain:

$$\lambda_J \sim \left(c_s^2/G\rho\right)^{0.5} \tag{1}$$

This threshold distance is called the Jeans length. The corresponding threshold mass, or Jeans mass, the matter enclosed in a sphere with radius equal to half Jeans length, amounts to about ten thousand solar masses. If the mass within that volume is greater than the Jeans mass, a critical threshold is exceeded and the gas begins to collapse, something that can occur for example by the influence of a nearby supernova explosion. As it contracts, the cloud will break internally into small pieces (Vazquez-Semadeni et al., 2016). These pieces will continue to contract and will break again in turn. Eventually, the initial cloud will have decomposed into innumerable shrinking fragments. From each of these fragments, one or several stars (depending on its mass) will be born later. The range of masses for these clouds presents a notable dispersion, but frequently they used to be between one and two solar masses.[2]

Let us consider the case of a single star. During the collapse process, each of these fragments of radius r soon differentiates into a nucleus and an envelope. As the fragment always begins its contraction with some angular momentum, L, the more the nucleus contracts, the faster the rotation is, due to the conservation of momentum. As the nucleus shrinks, the centrifugal force per unit mass L^2/R^3 increases more quickly than the gravitational force GM/R^2, causing that part of the central zone of the interstellar matter cloud to become a flattened disk. The final size of this disk is given by the radial distance in which both forces are balanced:

$$R \sim L^2/GM.$$

For the values that are usually estimated for a nebula of a solar mass, it gives approximately $R = 50$ astronomical units (au). This disk is especially populated with solid particles that tend to have an orbital velocity higher than the one of gas. Their collective drag produces turbulences that favor the growth of solid bodies by collisions and cohesion phenomena. A process that will end with the formation of several planets.

Although this process is qualitatively easy to describe, it is very difficult to model in detail. In addition to gravitation which is the main actor, computer simulations have to take into account the effects of the aforementioned turbulences, magnetic fields, and stellar wind among other coupled factors. However, the results of current population synthesis models (see, for example, Mordasini et al., 2015) agree to predict a large abundance of rocky planets smaller or equal to Earth. This is good news, as rocky planets are

---

[1]The virial theorem states that, to keep the equilibrium, the gravitational potential energy must equal twice the internal thermal energy.

[2]Massive stars as O and B class stars use to be in multiple stellar systems, but these are the less frequent kind of stars. On the other hand, M class stars are the most frequent type of star (around 85% of the stars in our Galaxy) and almost 75% of them are single stars. Therefore most stellar systems formed in the Galaxy are single and not binary, as has been often asserted. Indeed, in the current epoch two-thirds of all main-sequence stellar systems in the Galactic disk are composed of single stars (Lada, 2006).

places especially suitable for harboring large quantities of liquids. We believe that liquids, because of their solvent properties, are ideal for the complex chemical reactions that are characteristic of life. Solid substrates barely offer mobility to molecules, and chemical reactions in them are slow. On the other hand, gases have an active dynamics, but their molecular components are more separated, which also has the consequence of slowing these reactions. And while a biology based on a gaseous or purely solid environment cannot be ruled out (see, for example, Carl Sagan and Edwin Salpeter's proposal for a biosphere in a Jovian world, Sagan and Salpeter, 1976; Sagan, 1980), liquids offer the best of both worlds: mobility to facilitate chemical reactions, and closeness for them to occur frequently (and refrigeration to absorb heat generated by exothermic reactions). Therefore, the search for unintelligent life outside our planet focuses on rocky planets, with solid surfaces where large masses of liquid can be deposited, and having an atmospheric density that generates an elevated pressure to keep liquids stable and to avoid their total evaporation.

Until a few years ago, observations of exoplanets seemed to point to an abundance of rocky planets, also called telluric planets (Ida and Lin, 2004b). For ease of detection, the first planets that were discovered were gaseous giants, with masses larger than those of Jupiter. As statistics increased and more data were collected, it was seen that the more massive the planets were, the less frequent they were, and that this distribution fitted very well into a power law (Ida and Lin, 2004a). The inference was obvious: therefore, the less massive the planet is, the more frequent it would have to be. This is not surprising given that we see similar power laws in the sizes of asteroids in the Solar System, which are more abundant the smaller they are, or in stars, where the mass function obtained from observations shows that the smallest stars, called red dwarfs, are the most abundant. Thus, although a few years ago there were almost no detections of telluric planets, the distribution found for planets more massive than Jupiter, extrapolated to low masses, led to the prediction of a large number of rocky planets. All in accordance with simulations.

But the Kepler space mission (Basri et al., 2005) changed everything: this power law that fits so well for giant planets is not fulfilled for masses smaller than the Jupiter mass. Today, we have discovered about 3500 exoplanets, most of them thanks to this mission. These discoveries have been published in several open-access databases, the Exoplanet Orbit, the Extrasolar Planets Encyclopaedia, the NASA Exoplanet Archive, or the Open Exoplanet Catalog, among others (Wright et al., 2011; EO, 2016; EPE, 2016; NEA, 2016; OEC, 2016). Such databases compile several parameters of extrasolar planets reported in the peer-reviewed literature, with an easy-to-use interface to filter and organize data. For example, the NASA Exoplanet Archive has mass measurements, or estimates of them from their size, for almost 1200 exoplanets. The frequency distribution of the mass of exosolar planets is shown in Fig. 1:

The distribution of abundances is bimodal, with a peak around planets with one Jupiter mass and another peak around 10 Earth masses. Possibly, part of the decrease towards low masses is due to selection effects, because planets smaller than the Earth are difficult to see, and the peak on the left will probably be higher and wider. However, the incompleteness of the data cannot be used to justify the gap between the two peaks. There seems to be no doubt that the most abundant planets are gaseous giants like Jupiter on the one hand, and super-Earths and Neptunes on the other. In fact, our Solar System is representative of this trend. Recent simulations (Malhotra, 2015) have attempted to explain this phenomenon and conclude that the most frequent exoplanets have masses around the mass of Jupiter. Behind this phenomenon, there seems to be the competition for attracting the material from the disk, which may cause a correlation in the mass distribution between adjacent pairs of planets. In fact, this

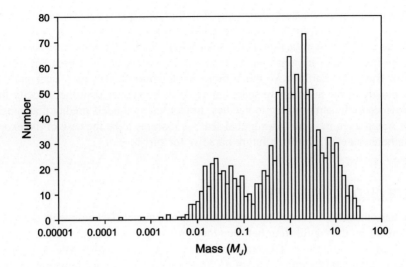

FIG. 1

Histogram of masses for exoplanets discovered up to November 2016, plotted by the authors using mass data from the NASA Exoplanet Archive.

phenomenon is found in our Solar System in the cases of Venus and Earth, Neptune and Uranus, and to a certain extent, Jupiter and Saturn. These correlations between consecutive orbits and the need for long-term stability tend to favor large masses. Thus, planets with masses similar to Earth or smaller may not be as abundant as was thought a few years ago. In contrast, the super-Earths (possibly the greatest planetary novelty since we studied other planetary systems, as none exists in our Solar System) are relatively abundant. However, it cannot be ruled out that the previous distribution is trimodal and that there exists a third peak at lower masses, undetectable with the current technology. After all, there are millions of asteroids in the Solar System.

Anyway, astrobiologically the abundance of super-Earths is good news. Plate tectonics plays an essential role in the maintenance of the biosphere on our planet, by allowing the renewal of atmospheric CO_2 and the existence of a carbon cycle, essential for life. Of all the rocky bodies of the Solar System, only our planet possesses this curious surface dynamics, which works thanks to the high internal heat of our planet, the most massive and dense of the rocky planets of our planetary system. The more massive a planet is, the hotter its interior is, because it has enclosed more gravitational energy and more radioactive materials, so we expect active tectonics in these massive super-Earths. This active volcanism is a good source of energy and materials for the hypothetical biospheres of these planets.

The extra mass of super-Earths also provides stability to the rotation axis. In the case of Earth, the Moon has played an important role in keeping our planet's rotation axis stable. The axis of Mars has undergone enormous oscillations throughout its history, because the gravitational pull of the giant planets Jupiter and Saturn induces in its axis a chaotic dynamics. Without the effect of the Moon, our planet would probably have suffered a fate similar to that of Mars, which would have had devastating consequences for life. But the extra mass that super-Earths have makes the presence of a giant stabilizing satellite unnecessary, since the more massive is a planet, the more stable its rotation.

Finally, its greater gravity also makes it easier to maintain a dense permanent atmosphere, which allows the existence of superficial liquids. Lighter worlds such as Mars slowly suffer the loss of their atmosphere due to the constant action of the solar wind. Also in the long run, Titan, the satellite of Saturn, is expected to lose its atmosphere, made of gases that are still escaping from its nucleus (Glein, 2015). But super-Earths have much more stable atmospheres. Its greater gravitation causes the escape velocity at the atmosphere-space interface to be greater than the velocity that the star's radiation provides to the atmospheric molecules. In fact, many models predict that super-Earths will have seas of water, even in some cases global seas, as it seems to be the case of the exoplanet Gliese 1214 b (Charbonneau et al., 2009), a future paradise for surf lovers.

4 THE GOLDILOCKS GRAVITY

Gravity also has other effects related to the habitability of a planet. To begin with, the temperature of a star is determined by the equilibrium between the force of gravity and the pressure of the gas compressed and heated by this force. The numerical resolution of the hydrostatic equilibrium equations shows that in general the larger the mass of a star, the higher its surface temperature and the shorter its duration, basically because the higher pressure accelerates the nuclear reactions and the hydrogen in the nucleus is depleted sooner. This puts an upper bound on how much mass a star should have if we want a planet around it to develop life. The duration t in years of a star in the main sequence is given by:

$$t = 10^{10}M^{-2.5} \tag{2}$$

where M is the mass of the star measured in solar masses. In the case of Earth, life appeared when the Sun was about 1000 million years old. If we take this estimate as a representative average, we should discard those stars with a mass greater than 2.5 solar masses, since the time of their rapid extinction would be too short for the appearance and consolidation of life.

But low mass stars can also be an inconvenience to the emergence of life. The habitability zone (Huang, 1959) is usually defined as the region around a star where temperatures are between 0°C and 100°C, taking into consideration only solar radiation. It is therefore the area where one might expect to have liquid water. It is easy to estimate the boundaries of this region. For this, we ignore the possible effect of the planet albedo or the presence of a greenhouse effect, and we will approximate the emission of the star to that of a black body. Using Stephan-Boltzmann's law, the emission of the star will be $L = \sigma T_s^4 4\pi R_s^2$, where R_s is the radius of the star, σ is the Stephan-Boltzmann constant, $4\pi R_s^2$ will then be the surface, and T_s its surface temperature. Suppose a planet is at a distance D from the star. The energy arriving per unit area decreases with the distance squared according to $P = L/4\pi D^2$. For the entire planet, the total energy received will be $E_{in} = \pi R_p^2 P$, where R_p is the radius of the planet and πR_p^2 its section. This arriving energy will be in equilibrium with the thermal energy that radiates the planet, $E_{out} = \sigma T_p^4 4\pi R_p^2$ where T_p is the temperature of the planet. Equating E_{in} and E_{out} and clearing, we obtain:

$$D = \frac{1}{2} R_s \left(T_s / T_p \right)^2 \tag{3}$$

Knowing the temperature and radius of the star, and using the values of 0°C and 100°C (that is, 273 and 373 K) for T_p, we obtain the distances to the star D_0 and D_{100} that limit the habitability zone. It is easy to see from Eq. (3) that for less massive stars, and therefore less hot, the habitability zone will be closer to them. In the case of red dwarfs, which is the most abundant type of star, the sizes are a few tenths the size of the Sun, with temperatures of 2500–4000 K. This gives a habitability zone of approximately 0.02–0.05 au. But a planet so close to a red dwarf star would undergo a gravitational anchor due to tidal forces (Leconte et al., 2015). This type of tidal anchoring is common: we find it on our Moon, on the small satellites of Mars, and also on the main satellites of Jupiter, Saturn, Uranus, and Neptune, which always show the same face to the planet they orbit. A planet gravitationally anchored to its star will always show the same face to the star, having extremely high temperatures on the illuminated face and extremely cold ones on the night face, which seems hardly compatible with the existence of stable seas and life as we know it.

Thus, the abundant red dwarf stars do not seem to be a good place to host life. But do not throw in the towel so soon. As we have seen, the giant planets are the most abundant type of planet. And in our Solar System, they all have a large cohort of telluric satellites. Since most of these satellites are gravitationally anchored to the planet, even if the hypothetical planet were located in a habitability zone very close to the star, its satellites would not always show the same face to the star, but to the planet. The different regions of the satellite would therefore have day and night as the satellite rotates around the planet (although of course there may be spin-orbit resonances in simple fractions), which in principle does not compromise its possible habitability as the previous case. However, their main problem could be that they will not be able to retain a long-term atmosphere in general, given that they use to have low masses. In fact, in our Solar System the majority of satellites of the giant planets show dim or nonexistent atmospheres.

Still other sources of pressure can exist, in addition to the atmospheric pressure, which can help to maintain stable masses of liquid. Europa, the smallest of Jupiter's four moons discovered by Galileo Galilei, has been a source of speculation over its possible habitability since it was spotted by the Voyager space probes. Since the discovery of the immense masses of water ice that completely cover its surface, it was postulated that the friction between internal layers of the planet, caused by the gravitational forces of the tides induced by Jupiter, could warm the interior of this satellite to the extent of allowing liquid water under the ice cover. On the other hand, the pressure produced by the weight of the ice sheets could provide enough pressure to keep this water stable in the absence of an atmosphere.

The observations made in the first half of the 1990s by the Galileo space probe led us to conclude that perhaps this is really happening. On the one hand, the high resolution images that this probe took from the surface of Europa showed some ice formations with the appearance of icebergs that at a given moment floated on a frozen sea. It seems that, at some point, the satellite dynamics would have broken the ice crust, and for a short time the fractured ice blocks would have floated freely on liquid water, moving and rotating, until again the underlying water was frozen again (the surface temperature of Europa is around 200°C below zero), restoring the ice crust and sealing the fractures.

On the other hand, the Galileo space probe discovered that Europa has a weak magnetic field, which changes its direction as the moon orbits around Jupiter, aligning with the much more powerful magnetic field of the latter. But the surprising mobility of Europa's magnetic field can only be explained by the presence of some electrically conductive liquid as the cause of the plasticity of this magnetic field. The best candidate is salt water. Therefore, both facts together with other indirect evidences point out that, under the icy crust of Europa, there are large accumulations of liquid salt water. In fact, the characteristics of this magnetic field point to an underground ocean of almost global extent. Perhaps, little sunlight can cross the ice to illuminate the Europan sea, but the gravitation of Jupiter can take the place of the Sun as a source of energy for a hypothetical biosphere. Warming due to Jovian tidal forces is likely to trigger active underwater volcanism, which will generate hydrothermal chimneys in the bottom of this Europan ocean, supplying energy and essential chemical compounds. Europa is not just an isolated case; other satellites in the Solar System also show strong signs of harboring underground seas, like Ganymede, another moon of Jupiter, or Enceladus, a satellite of Saturn.

Furthermore, we can think of other habitability zones, of other liquids that can play the role of water at lower temperatures, such as ammonia, methane, hydrogen sulfide, or hydrogen cyanide among other possible candidates, molecules more or less abundant in the universe. The same calculations as above for the case of water may be performed, but using instead the temperature ranges in which these compounds are liquids. For example, in the case of methane, we consider the temperature range between its melting and boiling points, from 90.58 to 111.65 K (HSDB, 2016). For a typical red dwarf star, with a size 0.2 times that of the Sun and a temperature of 3000 K, we obtain that its "habitability" zone would reach up to 0.5 au. A planet located in this area would already be far from the gravitational anchorage zone and could have stable methane seas, and perhaps a biosphere with a really interesting biochemistry.

If we make the calculation for the Solar System, we find the region where we can expect liquid methane to be between 6.7 and 10.2 au, and just inside this region we find Titan, the largest satellite of Saturn. With a size comparable to Mars and a surface temperature of 180°C below zero, it has a mysterious atmosphere much denser than Earth's one, composed of nitrogen, argon, and methane; a very reductive atmosphere, which possibly generates an interesting and rich organic chemistry. Part of the mystery of Titan's atmosphere lies in the presence of this atmospheric methane. Certainly, methane abounds in the universe and undoubtedly there was methane in the primordial nebula that gave rise to our Solar System. Where is the mystery, then? The mystery is that there is still methane, as it is a molecule that solar ultraviolet radiation breaks up easily. These broken methane molecules recombine to produce more complex hydrocarbons, therefore atmospheric methane should gradually disappear. Since this has not happened, another source should be replenishing it. There has been some speculation with hypothetical organisms that, like some terrestrial bacteria, would release methane into the atmosphere. But it could also have a geological origin.

To solve the enigma of Titan's methane, a European probe called Huygens landed there on January 14, 2005. The Huygens traveled to Saturn carried by the Cassini spacecraft. When Cassini reached Titan, it dropped the Huygens probe towards the moon's surface. The Huygens probe was taking atmospheric data and images throughout the descent. The images it took showed a world with landscapes strikingly similar to those of the Earth, with mountains, valleys and river channels produced by some liquid substance. The spacecraft's sensors also found abundant atmospheric methane and superficial liquid methane: after the impact, the probe detected an increase in methane levels of 40%, revealing the presence of liquid methane mixed with the surface material. The probe had landed on methane mud.

We now know that Titan has a complex methane cycle, analogous to the Earth's water cycle. On Titan, it rains methane. And although the presence of life cannot be ruled out, combined data from the Cassini spacecraft and the Huygens lander show that the origin of methane is not biological, but is replenished by geological processes. A type of cryovolcanism releases methane from the interior, a methane that would have been trapped when this satellite was formed. On the other hand, the images obtained from space by the radar of the Cassini spacecraft show a fascinating world that we have not yet finished mapping. It has revealed a complex network of lakes and even seas, in which we have been even able to see waves. The largest of these seas, the Kraken Sea, is almost as large as half of the Mediterranean sea.

5 GRAVITATIONAL BIOLOGY

Gravitational biology is the study of the effects of gravity on living organisms. We began the introduction of this chapter with the apocryphal story of Newton's apple. Although first example may seem simple and crude, the consequences of the fall of an organism from a great height are critical. The terminal velocity of a falling body is proportional to the square of the ratio mass/cross section area. Living beings have more or less the same density, thus among them this ratio is approximately proportional to the ratio volume/area $\sim$ length. So bigger bodies reach higher limit speeds and that can be fatal for an organism. The biologist John B.S. Haldane wrote it this way in 1928: "You can drop a mouse down a thousand-yard mine shaft; and, on arriving at the bottom, it gets a slight shock and walks away, provided that the ground is fairly soft. A rat is killed, a man is broken, a horse splashes." (Haldane, 1926).

Probably, the gravity of our planet has been the only environmental feature that has remained constant during the evolution of life. Its effect can be seen in the design of a large part of the organisms, which have a symmetry axis in the direction of the vector of the gravitational force exerted by our planet. In particular, most plants have axial symmetry. But the effect of gravity is not only recorded in the structure, but also in the function. In the case of plants, we observe that they display gravitropism: the roots grow in the sense of the gravitational attraction, whereas the branches and stems grow in the opposite direction. Until recently, we did not know that terrestrial organisms have developed a great diversity of systems of graviperception: in some unicellular organisms their membranes respond to the gravity vector (Bräucker et al., 2002); multicellular organisms contain "heavy bodies," like calcium crystals, that are essential for equilibrium, as statoliths in the case of plant roots and most invertebrates, or otoliths in the case of vertebrates; Arthropods possess also body extensions to perceive gravity (Häder et al., 2005), etc. One of the reasons for this ubiquity and diversity is that gravity acts as a key environmental factor for orientation, balance, and postural control. For example, bees are able to report to their counterparts the location of food through figure-eight dance, where the angle between the dance axis and the direction of gravity is roughly equal to the angle between the food and the sun from the hive position.

The diversity of sizes and masses of organisms that inhabit our planet is extraordinary, ranging from more than one hundred thousand kg for the blue whale to only 10^{-13} g for *Mycoplasma*, a bacteria that lacks a cell wall, spanning 21 orders of magnitude (Schmidt-Nielsen, 1975). As the density of all animals is close to 1 g/cm^3, it is practically indifferent to speak about volume, size, or weight. In Fig. 2, we can compare some records that give some clues about the relationship between gravity and size (McMahon and Bonner, 1983).

Let us look at the difference between flying and nonflying birds. In the first case, the weight and intrinsic problems of flight seem to limit their sizes compared to nonflying ones. The albatross (1) in Fig. 2, currently the largest flying bird, only reaches a maximum wingspan of about 3.4 m (genus *Diomedea*). One must go to the past to find a larger flying organism: the *Pheranodon* (7), which was one of the largest flying reptiles. But the record, with a wingspan of about 12 m, is held by the disappeared pterodactyloid pterosaur *Quetzalcoatlus*, named after the Aztec deity. When birds do not fly, they can reach higher sizes, as in the case of *Aepyornis* (12), which was the largest ratite ever; Known as the elephant bird of Madagascar, *Aepyornis* reached up to 3 m in height, weighted over 500 kg and their eggs reached up to 1 m in circumference. Compare its size with an actual horse (18) or ostrich (24).

Still at ground level, the largest terrestrial mammal in history was the *Baluchitherium* (3), a giant animal related to rhinoceroses about 5 m tall, 8 m long, and weighting about 20 t. Humans,

FIG. 2

The largest organisms in the biosphere drawn to the same scale for comparison. See main text for details.

From McMahon, T.A., Bonner, J.T., 1983. On Size and Life. Scientific American Books-W. H. Freeman & Co, New York.

who are smaller by several orders of magnitude, never cohabitated with *Baluchitherium*. The largest terrestrial mammal today, the African elephant, reaches up to 12,000 kg. To find larger terrestrial animals, one has to go farther back to the past. The longest dinosaur of which a complete skeleton has been found was the *Diplodocus*. Thanks to its enormous neck and tail, it reached about 30 m long. But, probably, it was surpassed by the *Argentinosaurus*, which could have weighed about 100 t, the *Seismosaurus*, about 50 m long, and the *Sauroposeidon*, 18 m tall. In comparison, our current tallest terrestrial animal, the giraffe (4), which can reach about 5 m in height and 900 kg in weight, seems small. The *Tyrannosaurus rex* (6), the tyrant lizard, was the largest terrestrial predator (or scavenger, the debate among paleontologists is still alive). Compare its 13 m long and 8000 kg with the measures of a polar bear, the current largest terrestrial predator, which at most reaches 2.5 m long and 500 kg.

But the largest animals are present in the oceans. The Atlantic tarpon (21) is a large fish, but the largest fish is the whale shark (17) that reaches up to 12 m in length, which in turn is surpassed by the giant squid (23) that can reach more than 15 m. But the record, reaching 33 m long and 180 t, goes to the blue whale (2). Given its body mass, the blue whale is the largest animal that has ever existed. The reason that the largest animals in the history have been or are marine is due to the fact that in water the body undergoes a buoyant force that compensates the gravity force. One of the reasons why whales die when they are stranded is because their own weight crushes their lungs to suffocation. A mass that supposes death by compression on land, in the water is sustained without difficulty. Thus, the effective gravity the organism undergoes in water is given by $g' = g(1 - 1/d)$ where d is the density of the organism (taking water density as 1). For many cases, $d \sim 1$, then the effective gravity gives $g' \sim 0$; in principle, there doesn't seem to be size limits inside water (this is true as a first-order correction, but note that buoyant force acts over the body surface, meanwhile weight acts over the volume: water does not make you completely immune to gravity).

According to these examples, it seems evident that gravity acts as a limiting factor for the size of organisms. It is easy to understand why. The ratio between the lengths of any two homologous segments of similar shapes has a constant ratio known as the scale factor f. In Fig. 3, the scale factor would be: $f = a'/a = b'/b$. In the days of Galileo Galilei (1564–1642), it was known that large machines, scaled from smaller machines that were very effective, tended to fail. The issue was known as the problem of scale. Galileo found the cause, the so-called square-cube law, which he put forward as follows: "When an object is scaled up without a change of shape, its surface increases by the square of some characteristic length (for example its height) while the volume increases by the cube of that quantity" (Galilei, 1638). For example, a cube with and edge of n meters has a surface of $6n^2$ square meters and a volume of n^3 cubic meters. So its surface grows as the square of the length of the edge, while its volume grows like the cube of that length.

What can we deduce from this law? There are physical properties of a body that depend on its surface and others on its volume. According to the square-cube law, volume-dependent characteristics will grow faster than surface-dependent ones. Consequently, a variation of size may involve noticeable differences. For example, the weight of any body is proportional to its volume. If we duplicate the size

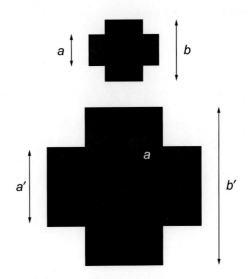

FIG. 3

Two isometric figures. The pairs of homologous segments a-a' and b-b' satisfy the relation: $f = a'/a = b'/b$, where f is called the scaling factor.

of an apple, its weight will increase eight times. The resistance of a column, for example the peduncle that holds the apple, is proportional to its cross section, which is an area. So by doubling the size, the resistance increases only four times. That is why apples hang from trees and watermelons grow on the ground.

The size of living beings has always triggered our imagination: let us remember Rabelais' Gargantua, Voltaire's Micromegas, or Swift's Gulliver, giants from our universal literature. In science fiction in particular, giants can be considered a subgenre (let your imagination fly with Fig. 4). Are these beings viable with the gravity of our planet? Let us consider a couple of famous examples, colonel Glenn Manning (from "The Amazing Colossal Man," 1957) and Nancy (from "Attack of the 50 Foot Woman," 1993). Both grew, by requirements of the script, from a normal size of 1.75 m to about 15 m (astonishing the spectators with the elasticity of their underwear). In this case, the scale factor is approximately $f = 15/1.75 = 8.57$. If we assume an original weight of 80 kg, the weight of these giants would be: $80 \times (8.57)^3$ or about 50 t, more than twice the *Baluchitherium* weight, the largest terrestrial mammal ever. On the other hand, the surface of the soles of their foot would only increase $(8.57)^2$ times,

FIG. 4

Film writers introduced giant versions of all kinds of bugs: tarantulas, octopuses, ants, crabs, lobsters, mantis, vultures . . . and even a leech! In Occident, probably the most famous giant beast is King Kong (1933), the giant gorilla of the hidden Skull Island. However, Godzilla (1954), with its 120 m of height, bears the size record of these giant creatures. His Japanese name is Gojira, a mixture of words *gorira* (ゴリ) "gorilla" and *kujira* (鯨, くじら) "whale."

so the soles would have to endure a pressure 8.75 times greater. As the size increases, these gigantic humans should withstand more and more weight per unit surface, a resistance that increases with the cross section of their bones in a quadratic way and therefore much slower than the weight increases (which grows cubically). To some extent, evolution has bypassed this problem by thickening the leg bone of large mammals. For example, in the case of *Baluchitherium*, metacarpal bone fossils of about 14 cm thick have been found. The compressive stress such a bone can support, in this case the weight acting on the cross section, is about 1800 kg/cm^2. A simple calculation shows that the *Baluchitherium* bones could support about 280 t, more than ten times their estimated real weight. This safety factor of ~10 is approximately what is found in humans, and it is a necessary margin as the maximum stress that the bones support does not occur in static position, but in the accelerations and decelerations of the moving organism. Using the same safety factor for the previous pair of cinematographic giants, a simple calculation leads to conclude that their slenderness is impossible.

And what about giant insects or crustaceans, recurrent movie stars in science fiction films? Gravity imposes limits more severe on insect exoskeletons than on mammalian endoskeletons. Especially, if we consider growth without distortions of form. The stronger limiting effect of gravity in their case is also evident when compared with organisms in the marine environment: the Japanese giant crab (number 19 in Fig. 2), with legs of one meter and a half, is currently the largest crustacean and the missing marine arthropod *Euriptéridus* (20) had enormous sizes compared to the largest terrestrial insects.

Which is the viable size limit for an animal on planet Earth, then? Biology does not yet provide a satisfactory upper bound. As for unicellular organisms, their average size seems the result of a balance between gravity, diffusion, and cytoskeleton. Surprisingly, under experimental conditions where gravity is varied, there is a gravity-size relationship that appears again and again: By increasing the gravitational force, it appears that single-celled organisms tend to reduce their size (Feric and Brangwynne, 2013) and, conversely, under microgravity conditions, tend to increase it (Kim et al., 2014). Something that mimics what we have seen in multicellular beings.

6 A SCALABLE LIFE?

The size of an organism critically affects both its structure and its functions in a complicated way. In the case of our cinematographic giants, size would not only affect the thickness of bones, but would generate severe problems in respiration, blood circulation, kidneys filtering, etc. The gravitational force has influenced throughout the evolution in the musculoskeletal systems, but also in the systems of fluids distribution. The largest tree in the world is the giant redwood, reaching heights taller than 100 m (numbered as 25 in Fig. 2; its first 32 m are superimposed on a 30 m larch). These gigantic organisms have to make their vital fluids circulate to their upper branches. Similar problems are found in extremely long animals: the distance between the heart and the brain of a giraffe is about 2.8 m and reached up to 8 m in the case of some herbivorous dinosaurs, requiring enormous pressures to make blood reach the brain.

Among all vertebrates, probably the best studied example in this sense are snakes. In Fig. 2, number 8 represents the largest extinct serpent. Today, the longest snake is the reticulated python, which can measure up to 10 m. Snakes stand out for the variability of adaptations to gravity that their cardiovascular system have (Lillywhite, 1987). Depending on their habitat, the location of the heart in relation to the head and tail varies ostensibly. In the case of aquatic snakes, the heart is far away from the head, while in the case of terrestrial snakes they are very close. This is especially noticeable in arboreal snakes, which require climbing against the gravitational vector. For example, a corn snake (*Pantherophis guttatus*) can climb vertically the trunk of a tree in search of eggs; A tree boa (*Corallus hortulanus*) can hang from a tree head down while chasing a prey. Both animals move in a way that would be unthinkable without a cardiovascular system adapted to maintain proper blood circulation when the body goes to vertical position. Tree snakes have better capabilities to regulate blood pressure in vertical position than nonclimbing species. How is the blood flow maintained in the tail and how do the snake avoids excessive head pressure when it is tipped down? By adjusting its cardiovascular system to work in the opposite direction: the snake's heart rate decreases and the smooth muscle enveloping the blood vessels in the vicinity of the head relaxes, thus compensating for the pressure increase (Lillywhite, 1987).

While we cannot experiment by changing the size of animals, we can do it with artificial objects, as in engineering or architecture. There is, for example, a limit in the use of bricks for construction determined by the compression effort. So in the construction of skyscrapers, the material must be changed to steel. But one has to consider also design changes to consider not only compression, but also the tension of the support structures. We find something similar in nature: new designs that allow increases of size. For example, typical unicellular cilia movement is restricted to small sizes, so larger organisms are forced to develop very different propulsion mechanisms. It has also been suggested that nutrition is a limitation for size (Blanckenhorn, 2000). By the square-cube law, smaller animals lose more heat per unit weight than larger animals. Thus, animals like mice or shrews need to eat continuously, while larger animals like humans or elephants can be kept without food for a long time.

As we saw, in isometric bodies (with the same shape), we can relate a characteristic length L, surface S, and volume V simply through the square-cube law: $S \propto L^2$, $V \propto L^3$ and $S \propto V^{2/3}$. However, organisms in general do not follow this simple isometry, but rather they follow allometries, that is, nonisometric scaling. Many morphological and physiological variables scale with weight according to allometric power laws: $y \propto x^\alpha$. For example, animal locomotion systems in water, earth, and air seem to have their own characteristic allometric functions (Alexander, 2004; Biewener, 1983). When we represent the amount of energy used to move a unit of weight a unit of distance, we observe a potential decrease with body weight (Schmidt-Nielsen, 1975).

Let us consider an example widely studied and still discussed: the relation between metabolic rate and weight of an organism. The basal metabolic rate B (measured in kJ/h) is the minimum energy expended daily by an animal in thermoneutral conditions to keep its metabolism at work. This basal metabolic rate, as we are going to see, is related to weight by simple allometric laws. Part of the consumed energy is dissipated as heat through the organism's surface. But as heat dissipation is related to shape, the square-cube law has to be considered: as we have seen, the higher the gravitational field, the more pudgy and robust the shapes are, and therefore, less efficient the heat dissipation. On the contrary, in low gravitational fields, densely branched shapes with small sections are perfectly feasible. Therefore, maybe these allometric laws are not fulfilled with the same exponent under different gravitational circumstances. At least in animals under chronic acceleration conditions, it has been seen that the basal metabolic rate increases with gravity (Economos et al., 1982; Kelly et al., 1966; Catlett, 1973).

As early as in 1839, Sarrus and Rameaux proposed that metabolic rates might depend on heat dissipation and therefore increase with surface area, something originally checked in dogs by Rubner in 1889. In 1932, Max Kleiber empirically observed (Kleiber, 1932) that, indeed, a simple and robust allometric scaling between B and the animal mass M could account for most of the metabolic rate variability, $B \propto M^{\alpha}$. However, he found that $\alpha = 3/4$, instead of $\alpha = 2/3$, that would result of heat dissipation according to a simple surface-to-volume argument. Since then, extensive data have been collected, encompassing a fervent debate on the origin and concrete shape of the so-called Kleiber's law. While some of the empirical works seem to comply better to $\alpha = 2/3$ (White and Seymour, 2003), a great majority took for granted a 3/4 power law (West et al., 1997), raising it to the level of central paradigm in comparative physiology. This scaling was subsequently elegantly explained by space-filling fractal nutrient distribution network models (West et al., 1999), thus apparently closing the debate on its origin. However, additional statistical evidence challenges the validity of $\alpha = 3/4$. Thus, recently Kolokotrones et al. (2010), after fitting the encyclopedic data set of basal metabolic rates for mammalians compiled by McNab (2008), concluded that the scaling law was not after all a pure power law but had curvature in double logarithmic scale (see Fig. 5), giving an heuristic explanation as to why different exponents could be fitted depending on the range of masses considered. In the last decades, a large number of theories of different garment and degrees of formality have been proposed to justify the occurrence of particular scaling forms, organized into four major brands (by Glazier, 2014): surface area, resource transport, system composition, and resource demand models.

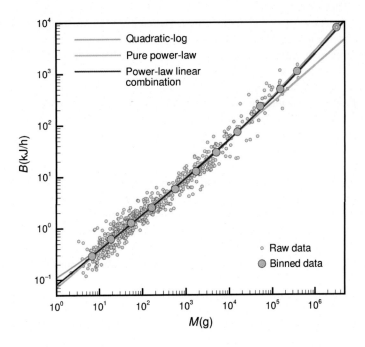

FIG. 5

Basal metabolic rates (BMR) for mammals. *Gray dots*: basal metabolic rate data for mammals compiled by McNab (2008). *Green line*: fitting to a pure power law. *Blue line*: Kolokotrones et al. (2010) statistical model. *Red line*: fitting of linear combination of two power laws. We also include a logarithmic binning of the data *(pink dots)* where the curvature is better appreciated (fits have been performed using the raw data).

Not all of the organism's energy income is wasted as heat. Recently, Ballesteros et al. (2016) proposed a simple thermodynamic balance that allows to explain in a quantitative way the correct allometric curves for mammals (in different environments), birds, and insects, to account for other biological features such as the relation between energy lost as heat and mass, as well as to extend the analysis to plants. Part of the energy incorporated to the organism from food is used for the synthesis of ATP and proteins, cellular division, etc... it keeps the animal alive. The second law of thermodynamics requires that such process is inefficient, in the sense that a significant part of energy consumption is always dissipated as heat. The trade-off between the energy dissipated as heat and the energy efficiently used by the organism to keep it alive results in a model for the dependence between B and M with an isometric (proportional to M) and an allometric (proportional to $M^{2/3}$) term, balanced respectively by coefficients a and b, which have clear biological meaning and can thus be estimated empirically:

$$B = aM + bM^{2/3} \qquad (4)$$

This balance, between work generated by metabolism and thermogenesis, complies with an effective (apparent) pure power law in a double logarithmic plot, with varying exponent in the range [2/3–1]. Note that this equation is not a pure power law, but the linear combination of two, with exponents 1 and 2/3, respectively. In a double logarithmic plot, this equation yields a curved graph with convex curvature, in good agreement with the findings of Kolokotrones et al. (2010). For small values of M, this approximates to a power law with exponent 2/3, whereas for a large range of masses, this equation approximates to an apparent pure power law with an effective exponent that can range between 2/3 and 1, in good agreement with empirical evidence. As can be seen in Fig. 5, this model fits exceptionally well the collection of close to 700 mammal basal metabolic rates recently compiled by McNab (2008).

To round it all off, we consider the case of plants. Now, the term associated to heat dissipation must take into account that plants have a branched fractal surface encompassing their volume. As the surface to volume ratio is higher, $S \propto V^{D/3}$ where $2 < D < 3$ is the surface fractal dimension, the risk of overheating is smaller, allowing much bigger sizes than in animals. According to West et al. (1997), $S \propto V^{3/4}$, yielding D equal approximately to 2.25, and thus the effective model reduces to $B = aM + bM^{3/4}$. As the exponents of the isometric and allometric parts are now closer, we expect a much less curved relationship with a higher effective slope ranging between 0.75 and 1. To test these predictions, we can see the database of basal metabolic rates compiled by Mori et al. (2010) that includes about 200 trees and seedlings. They showed measures of metabolic rate against both total mass (including the roots) and aboveground mass. To make the comparison with mammal data homogeneous, we have used metabolic rate against total mass. Fig. 6 shows these data, together with a fit to the effective model.

This effective thermodynamic model seems to result from the evolutionary trade-offs between the energy dissipated as heat and the energy efficiently used by the organism to keep it alive. Due to the generality of the underlying principles, thermodynamic laws, and evolution, it is expected that metabolic scaling will be present in other possible life forms in the universe. And the same consideration can be made about the other relationships between gravity, structure, and function in living beings. We do not know, even in the case of life on our planet, which are the absolute limits imposed by gravity. But since these relationships are based on universal physical laws, we hope to find them, conveniently scaled, in other past, present, and future life forms in environments with different gravitational intensities.

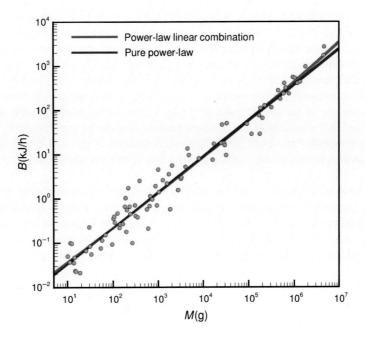

FIG. 6

Extension of the basal metabolic rates model for plant data. *Green dots*: metabolic rate data for plants compiled by Mori et al. (2010) for masses $M > 10$ g. *Green line*: fitting of the effective model. *Red line*: fitting of a pure power law.

7 THE PLURALITY OF EARTH-LIKE GRAVITIES

Although no gravitational limits to life have been discovered, it seems feasible that extreme gravity could be a serious problem. Leaving aside the case of the hypothetical inhabitants of a neutron star, as described in the novel "Dragon's Egg" (1980) by Robert L. Forward, monstrous tides caused by a nearby cosmic body with extreme gravitation force could turn a habitable planet into an uninhabitable one. Consider, for example, the giant tides of Miller's planet induced by the nearby black hole *Gargantua* in the film "Interstellar" (2014) directed by Christopher Nolan. Although it is not necessary to reach the extreme case of a black hole: Jupiter itself induces in the nearby Io an active volcanism that recycles its entire surface in a short time, making unlikely the existence of stable seas in the event that this satellite had possessed atmosphere. And it is almost certain that there will be much more extreme cases in the universe, although those extreme cases do not seem to be the norm but rather the exception, given the typology of words discovered by the Kepler space mission.

Excessive planetary surface gravity could also be a drawback. The atmospheric pressure is proportional to the planet's gravitational force, and a very high pressure could preclude the existence of liquid water, converting it into solid, into ice VII.[3] Although this would require really extreme values: it is

---

[3]Ice VII in the Bridgman nomenclature is one of the crystalline forms of ice that can exist at room temperature. It is a plausible constituent of giant planets.

necessary to reach 30,000 atmospheres of pressure (3 GPa), that is to say, gravities around thousands of G. Are there worlds like this? What gravity force can we expect on other planets? For transiting exoplanets, there is a quite straightforward way to estimate surface gravity: first, by analyzing the transit light curve, one can measure the amount of star light blocked by the planet, which, combined with a good model of the central star size, yields an estimate of the planet size (given by its upper opaque layer). Second, by measuring spectroscopically the radial velocity amplitude of its parent star, one can obtain an estimate of the planet mass M. In fact, this second technique sets a lower bound to the planetary mass, as it measures $M \sin(i)$, where i corresponds to the viewing angle, but for transiting planets, $\sin(i)$ is reasonably close to 1. Therefore, with estimates of size and mass and using Newton's gravity, $g_s = GM/R^2$, one can assess surface gravity. The results are shown in Fig. 7 (Ballesteros and Luque, 2016).

The figure shows the estimates of surface gravity versus mass (blue dots) using data from the aforementioned exoplanet.org database. Here we can see super-Earths, planets with masses ranging from 2 to 10 times greater than the Earth, such as GJ 1214b or gas giants as Kepler-7b, both discovered in 2009. As the exoplanets plotted in the figure are more massive than the Earth, we can also add data from Solar System bodies (red dots) to get a more complete picture of how surface gravity works. Scatter in

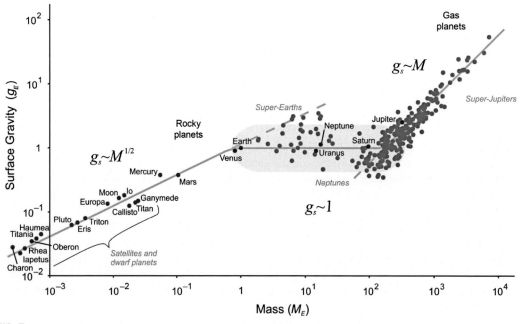

FIG. 7

Mass versus surface gravity. Mass is represented in Earth's units, M_E, and surface gravity - in units normalized by the terrestrial surface gravity, g_E. From left to right, three scaling regions can be clearly distinguished: (i) Rocky planets with $g_s \sim M^{1/2}$, (ii) the transition zone (green stripe) with $g_s \sim 1$ and (iii) gas planets with $g_s \sim M$. Solid line is a guide to the eye, not a fit. Red dots: Solar system objects. Blue dots: transiting exoplanets from The Exoplanet Orbit Database (Wright et al., 2011).

exoplanet data is much larger than in Solar System data, partly due to intrinsic causes (Howard, 2013), but also because gravity can be measured much more accurately in worlds near us, especially when they have satellites (by using Kepler's laws) or a space probe passes near. Still, both sets of data overlap quite well.

This representation has the advantage of classifying planets into three distinct regimes from left to right: (i) rocky bodies with masses below that of the Earth, (ii) the transition zone, with super-Earths, Neptunes, and some Solar System planets, with masses ranging from one to hundreds of Earth masses, and (iii) gas giants, with masses above hundreds of Earth masses. In the first regime, planet radius grows with mass as $R \sim M^{1/4}$ and therefore surface gravity grows as $g_s \sim M^{1/2}$ (faster than what would be expected for incompressible bodies, $g_s \sim M^{1/3}$). On the other hand, for gas worlds, planet radius remains roughly constant (i.e., gas giants with very different masses have similar sizes, due to electron degeneracy) and so surface gravity grows linearly with mass, $g_s \sim M$. But in the transition zone, we find some sort of plateau where planet radius has the fastest growth, as $R \sim M^{1/2}$, thereby yielding a constant surface gravity roughly similar to the Earth's one. This is especially evident in the Solar System: surface gravities for Venus, Uranus, Neptune, and Saturn, respectively, with 0.82, 14, 17, and 95 times the Earth's mass, are 0.91, 0.9, 1.14, and 1.06 G. This similarity with Earth's surface gravity is surprising, considering the difference of mass between the plateau's extremes and the contrasting chemical compositions and physical structures of the planets along this region (Rogers, 2015; Seager et al., 2007).

Competing planetary formation models still present several issues. There are worlds in principle not excluded by models that have never been observed, and worlds observed with characteristics unpredicted by models (Spiegel et al., 2014). It still remains unclear how to connect allowable masses and radii, as for a given mass one could expect a diversity of sizes depending on the planetary composition and atmospheric size. And we do not even know whether all what we call super-Earths have a solid surface (although if they are over the curve $g_s \sim M^{1/2}$, they will be rocky). This flat transition zone supposes another challenge for theorists. In this region, the contribution of the atmosphere to the total planetary mass plays an increasing role as one moves from rocky worlds to Neptunes, but curiously does so in a way that total mass and radius compensate each other. Although one could in principle propose a rocky planet as big and massive as one would wish, with no atmosphere at all, no natural process produces it. The accretion process and the competition for materials during planetary formation impose severe constraints to feasible planets. Current models of population synthesis (Mordasini et al., 2015) try to take this into account, being able to explain many of the observed features. However, they fail to explain this plateau, predicting instead a noticeable increasing trend in surface gravities in this region.

These kind of observational properties from real planetary systems are becoming a helpful input to improve models. Transiting planets, whose number grows daily, are a valuable resource to refine theoretical models, showing that what we see in our Solar System (five worlds with nearly the same surface gravity) is not a coincidence but a general trend. We still don't know the origin of this feature, but whatever the reason behind it, today several rocky super-Earths with surface gravities similar to that of our planet have already been discovered. This is bad news for Superman: Krypton, a rocky world inhabited by humanoids in orbit around the red star Rao, whose enormous surface gravity (source of the fantastic powers of the steel man) is estimated at about 30 G, does not exist. There are no rocky worlds with such high gravities, which we can only find in the gaseous super-Jupiters. But it is good news for future interstellar settlers: the gravity we live in is more common than we thought. Therefore, our descendants (as the crew of the U.S.S. Enterprise) will find it comfortable to walk on Kepler-48 b.

ACKNOWLEDGMENTS

This work has been supported by projects AYA2016-81065-C2-2-P, AYA2013-48623-C2-2 and FIS2013-41057-P from the Spanish Council.

REFERENCES

Adami, C., 1998. Introduction to Artificial Life. Springer-Verlag, New York, NY.

Alexander, R.M., 2004. Models and the scaling of energy costs for locomotion. J. Exp. Biol. 208, 1645–1652.

Ballesteros, F.J., Luque, B., 2016. Walking on exoplanets: is star wars right? Astrobiology 16 (5), 325–327.

Ballesteros, F.J., Vicent, J., Martinez, V.J., Luque, B., Lacasa, L., Valor, E., Moya, A., 2016. Kleiber's Law: A Thermal Origin After All? https://arxiv.org/abs/1407.3659.

Basri, G., Borucki, W.J., Koch, D., 2005. The Kepler mission: a wide-field transit search for terrestrial planets. New Astron. Rev. 49, 478–485.

Biewener, A.A., 1983. Allometry of quadrupedal locomotion: the scaling of duty factor, bone curvature and limb orientation to body size. J. Exp. Biol. 105, 147–171.

Blanckenhorn, W.U., 2000. The evolution of body size: what keeps organisms small? Q. Rev. Biol. 75 (4), 385–407.

Bräucker, R., Cogoli, A., Hemmersbach, R., 2002. Graviperception and graviresponse at the cellular level. In: Horneck, G., Baumstark-Khan, (Eds.), Astrobiology: The Quest for the Conditions of Life. Springer-Verlag, Berlin, pp. 287–296.

Carter, B., 1974. Large number coincidences and the anthropic principle in cosmology. In: IAU Symposium 63: Confrontation of Cosmological Theories with Observational Data. Dordrecht, Reidel, pp. 291–298. Republished in: General Relativity and Gravitation, 2011, 43, 11, 3225–3233.

Catlett, R.H., 1973. Readings in Animal Energetics. Ardent Media, London.

Charbonneau, D., et al., 2009. A super-Earth transiting a nearby low-mass star. Nature 462, 891–894.

Economos, A.C., Miquel, J., Ballard, R.C., Blunden, M., Lindseth, K.A., Fleming, J., Philpott, D.E., Oyama, J., 1982. Effects of simulated increased gravity on the rate of aging of rats: implications for the rate of living theory of aging. Arch. Gerontol. Geriatr. 1 (4), 349–363.

EO (2016). Available online at http://exoplanets.org.

EPE (2016). Available online at http://www.exoplanet.eu.

Feric, M., Brangwynne, C.P., 2013. A nuclear F-actin scaffold stabilizes ribonucleoprotein droplets against gravity in large cells. Nat. Cell Biol. 15, 1253–1259.

Feynman, R., 1967. The Character of Physical Law. MIT Press, Cambridge.

Galilei, G., 1914 [1638]. Dialogues Concerning Two New Sciences. The MacMillan Company, New York.

Glazier, D.S., 2014. Metabolic scaling in complex living systems. Systems 2, 451–540.

Glein, C.R., 2015. Noble gases, nitrogen, and methane from the deep interior to the atmosphere of Titan. Icarus 250, 570–586.

Häder, D.-P., Hemmersbach, R., Lebert, M., 2005. Gravity and the behavior of unicellular organisms. Developmental and Cell Biology Series, vol. 40. Cambridge University Press, Cambridge.

Haldane, J.B.S., 1926. On Being the Right Size. Harper's Magazine.

HSDB, 2016. Hazardous Substances Data Bank. U.S. National Library of Medicine. https://toxnet.nlm.nih.gov/newtoxnet/hsdb.htm.

Howard, A.W., 2013. Observed properties of extrasolar planets. Science 340, 572–576.

Huang, S., 1959. Occurrence of life in the Universe. Am. Sci. 47, 397–402.

Ida, S., Lin, D.N.C., 2004a. Towards a deterministic model of planetary formation. I. A desert in the mass and semimajor axis distribution of extrasolar planets. Astrophys. J. 604, 388–413.

Ida, S., Lin, D.N.C., 2004b. Towards a deterministic model of planetary formation. II. The formation and retention of gas giant planets around stars with a range of metallicities. Astrophys. J. 616, 567–572.

Kelly, C.F., Smith, A.H., Besch, E.L., Burton, R.R., Sluka, S.J., 1966. Chronic Acceleration Studies—Physiological Responses to Artificial Alterations in Weight. NASA Publications N66, Davis, p. 35168.

Kim, H.W., Matin, A., Rhee, M.S., 2014. Microgravity alters the physiological characteristics of *Escherichia coli* O157:H7 ATCC 35150, ATCC 43889, and ATCC 43895 under different nutrient conditions. Appl. Environ. Microbiol. 80 (7), 2270–2278.

Kleiber, M., 1932. Body size and metabolism. Hilgardia 6, 315–353.

Kolokotrones, T., Savage, V., Deeds, E.J., Fontana, W., 2010. Curvature in metabolic scaling. Nature 464, 753–756.

Lada, C.J., 2006. Stellar multiplicity and the initial mass function: most stars are single. Astrophys. J. 640, L63–L66.

Leconte, J., Wu, H., Menou, K., Murray, N., 2015. Asynchronous rotation of Earth-mass planets in the habitable zone of lower-mass stars. Science 347 (6222), 632–635.

Lillywhite, H.B., 1987. Circulatory adaptations of snakes to gravity. Am. Zool. 27, 81–93.

Luque, B., Ballesteros, F.J., Márquez, A., González, M., Agea, A., Lara, L., 2009. Astrobiología, un puente entre el Big Bang y la vida. Akal, Madrid.

Malhotra, R., 2015. The mass distribution function of planets. Astrophys. J. 808, 71.

McMahon, T.A., Bonner, J.T., 1983. On Size and Life. Scientific American Books-W. H. Freeman & Co, New York, NY.

McNab, B.K., 2008. An analysis of the factors that influence the level and scaling of mammalian BMR. Comp. Biochem. Physiol. A 151, 5–28.

Mordasini, C., Mollière, P., Dittkrist, K.M., Jin, S., Alibert, Y., 2015. Global models of planet formation and evolution. Int. J. Astrobiol. 14, 201–232.

Mori, S., et al., 2010. Mixed-power scaling of whole-plant respiration from seedlings to giant trees. Proc. Natl. Acad. Sci. U. S. A. 107, 1447–1451.

Mosterín, J., 2005. Anthropic explanations in cosmology. In: Háyek, P., Valdés, L., Westerstahl, D. (Eds.), Logic, Methodology and Philosophy of Science. Proceedings of the 12th International Congress of the LMPS, King's College Publications, London, pp. 441–473.

NEA (2016). Available online at http://exoplanetarchive.ipac.caltech.edu.

OEC (2016). Available online at http://www.openexoplanetcatalogue.com.

Penrose, R., 2005. The Road to Reality: A Complete Guide to the Laws of the Universe. A.A. Knopf, New York, NY.

Rees, M., 2001. Just Six Numbers: The Deep Forces That Shape The Universe. Basic Books, New York, NY.

Rogers, L.A., 2015. Most 1.6 Earth-radius planets are not rocky. Astrophys. J. 801 (1), 13.

Sagan, C., 1980. One Voice in the Cosmic Fugue in: Cosmos. Random House, New York, NY.

Sagan, C., Salpeter, E.E., 1976. Particles, environments, and possible ecologies in the Jovian atmosphere. Astrophys. J. Suppl. Ser. 32, 737–755.

Schmidt-Nielsen, K., 1975. Scaling in Biology: The Consequences of Size. J. Exp. Zool. 194, 287–307.

Seager, S., Kuchner, M., Hier-Majumder, C.A., Militzer, B., 2007. Mass-radius relationships for solid exoplanets. Astrophys. J. 669, 1279–1297.

Spiegel, D.S., Fortney, J.J., Sotin, C., 2014. Structure of exoplanets. Proc. Natl. Acad. Sci. 111 (35), 12622–12627.

Vazquez-Semadeni, E., Gonzalez-Samaniego, A., Colin, P., 2016. Hierarchical Cluster Assembly in Globally Collapsing Clouds. https://arxiv.org/abs/1611.00088.

Waldrop, M.M., 1992. Complexity: The Emerging Science at the Edge of Order and Chaos. Simon & Schuster, New York.

Weinberg, S., 2005. The Quantum Theory of Fields. Cambridge University Press, Cambridge.

West, G.B., Brown, J.H., Enquist, B.J., 1997. A general model for the origin of allometric scaling laws in biology. Science 276, 122–126.

West, G.B., Brown, J.H., Enquist, B.J., 1999. A general model for the structure and allometry of plant vascular systems. Nature 399 (6745), 664–667.

White, C.R., Seymour, R.S., 2003. Mammalian basal metabolic rate is proportional to body mass 2/3. Proc. Natl. Acad. Sci. U. S. A. 100, 4046–4049.

Wright, J.T., Fakhouri, O., Marcy, G.W., Han, E., Feng, Y., Johnson, J.A., Howard, A.W., Fischer, D.A., Valenti, J.A., Anderson, J., 2011. The Exoplanet Orbit Database. PASP 123, 412–422.

FURTHER READING

Anken, R.H., Rahmann, H., 2002. Gravitational zoology: how animals use and cope with gravity. In: Horneck, G., Baumstark-Khan (Eds.), Astrobiology: The Quest for the Conditions of Life. Springer-Verlag, Berlin, pp. 315–333.

Morey-Holton, E.R., 2003. The impact of gravity on life. In: Rothschild, L.J., Lister, M.L. (Eds.), Evolution on Planet Earth: The Impact of the Physical Environment. Elsevier Ltd, Amsterdam.

RADIATION AS A CONSTRAINT FOR LIFE IN THE UNIVERSE

Ximena C. Abrevaya, Brian C. Thomas

Instituto de Astronomía y Física del Espacio (UBA–CONICET), Buenos Aires, Argentina
Washburn University, Topeka, KS, United States

CHAPTER OUTLINE

1 INTRODUCTION

There are several factors that can constrain the existence of life on planetary bodies. To determine the possibility of existence and emergence of life, it is essential to consider astrophysical radiation which itself can be a constraint for the origin of life and its development. Additionally, the radiation received by the planetary body and the plasma environment provided by the parent star play a crucial role on the evolution of the planet and its atmosphere. Therefore, radiation can determine the conditions for the origin, evolution, and existence of life on planetary bodies.

Habitability of the Universe Before Earth, editors: Richard Gordon & Alexei Sharov, Volume 1 in the series:
Astrobiology: Exploring Life on Earth and Beyond, series editors: Pabulo Henrique Rampelotto,
Joseph Seckbach & Richard Gordon. ISSN 2468-6352. https://doi.org/10.1016/B978-0-12-811940-2.00002-2

2 TYPES OF RADIATION

Several types of radiation are relevant to life in the universe. The word "radiation" itself may first need some definition. We use this term very broadly, to cover both electromagnetic radiation and energetic particles. The electromagnetic radiation of interest to us is the higher energy end of the spectrum—gamma-ray, X-ray, and ultraviolet. Gamma-ray and X-ray forms of light are ionizing, and along with UV, have the potential to destroy or damage essential biological molecules of life "as we know it," including DNA and proteins. Energetic particles that may cause damage include electrons, protons, neutrons, and muons. The energy of these particles is determined mainly by their kinetic energy. A muon is an elementary particle similar to an electron, but with a greater mass. Muons are highly penetrating and ionizing, but not very much is known about their biological effects. Unlike gamma-rays or neutrons, or even electrons, muons are not produced in most artificial sources of radiation, and so their biological effects have not been studied (Atri and Melott, 2011; Rodriguez et al., 2013). They are normally assumed to have effects similar to electrons, but could be more severe due to greater penetration.

High-energy electromagnetic radiation is produced by many different processes. UV light is produced by blackbody emitters with sufficiently high temperatures, including our own Sun. In general, the ultraviolet region of the electromagnetic spectrum can be subdivided into bands. These subdivisions are arbitrary and can differ depending on the discipline involved (Diffey, 1991), but in biology it is possible to distinguish three main bands: UVA (400–315 nm), UVB (315–280 nm), and UVC (280–100 nm) and additional bands for shorter wavelengths as VUV (200–10 nm) or EUV (121–10 nm). Gamma- and X-rays can be produced by radioactive decay of certain elements, by electron-positron pair annihilation, inverse-Compton scattering, some intraatom electron transitions, and Bremsstrahlung radiation.

Charged particles (e.g., electrons, protons) can be accelerated to high energies by various processes, especially involving plasma shocks and interactions with magnetic fields. (For readers interested in more of the physics involved, we suggest Rieger et al. (2007) and Balogh and Treumann (2013).) Neutrons and muons are generated by nuclear reactions. Neutrons may be ejected from nuclei that radioactively decay (e.g., a Uranium nucleus). Both neutrons and muons can be generated by nuclear reactions of atomic nuclei with high-energy "primary" protons that enter a medium such as a planetary atmosphere. The primary protons induce a so-called air shower of secondary particles, which includes electromagnetic and particle constituents, including neutrons and muons. In addition, helium nuclei (termed "alpha particles") and electrons are produced in some radioactive decay processes. Electrons with relatively high energy are also found in the magnetospheres of planets with significant magnetic fields. This is the case for some terrestrial planets (e.g., Earth) as well as giant planets. The moons of giant planets may experience significant irradiation due to the planet's magnetospheric electrons; this is the case for Jupiter's four largest moons, for instance. In this case, however, the electron radiation is not very penetrating, so some shielding by ice/rock will prevent significant impacts below the surface.

3 SOURCES OF HIGH-ENERGY RADIATION

3.1 STELLAR EMISSIONS

Stars are of course sources of visible light, but they also emit UV, X-ray, and even gamma-ray light. The emission of stellar radiation depends on their surface temperature and also on their activity. Therefore, emission is related to the evolution and spectral type of the star and it can be highly variable.

A star's surface temperature determines its blackbody emission. A high-mass star on the main sequence will have a high surface temperature and emit a significant amount of UV. However, from the perspective of habitability, such high-mass stars are short-lived (as short as a few million years), probably too short-lived to host habitable planets (Turnbull and Tarter, 2003). Stars that are of most interest for habitable planets include those similar to the Sun's mass (types G and K), but also those of much lower mass (type M) (Tarter et al., 2007; Kopparapu et al., 2013).

The Sun emits enough UV light to be problematic for planets without a UV shield, such as ozone in the atmosphere. Lower mass stars, on the other hand, do not emit very much UV, but are more active, with frequent energetic flares. These flares themselves are sudden and energetic explosive events and emit UV and X-ray (and possibly some gamma-ray) light. They originate in magnetic processes affecting all the layers of the stellar atmosphere (photosphere, chromosphere, and corona), which heat the stellar plasma and accelerate its protons, electrons, and heavy ions, to velocities near the speed of light. Flare emissions (usually being several magnitudes higher compared to the quiescent state) undergo interactions with planets and it is not well-understood whether it could be lethal or unfavorable for life. In general, it is known that the strongest stellar flares exceed the strongest solar ones by a factor of 100 in X-ray and EUV flux. The quiescent X-ray and EUV radiation of young stars are up to a factor of 1000 higher than on the present-day Sun (Guinan and Ribas, 2002; Ribas et al., 2005). They also are likely to eject plasma through processes similar to those that produce Coronal Mass Ejections on our own Sun. CMEs and flares represent important sources of both electromagnetic and particle radiation. The intermittent nature of the emission from low-mass stars may present more of a hazard than a steady background emission (such as UV from a Sun-like star), since it may be harder for life to adapt to the varying levels of radiation, as opposed to a more constant value (e.g., Ayres, 1997; Gershberg, 2005; Scalo et al., 2007).

Over a star's lifetime, its radiation emission changes. A young star tends to be less luminous overall, but often more active, producing more frequent and more intense flare/CME events. As a star ages its luminosity slowly increases, but the activity tends to decrease; this decrease is more pronounced for stars of higher mass, while low-mass (e.g., M-dwarf) stars continue to be highly active.

3.2 STELLAR EXPLOSIONS

Explosions on the scale of whole stars fall into a couple of major categories: individual stars that explode and pairs of stars that interact leading to explosions. These events are usually categorized by how they are observed. A supernova is typically observed as a rapid brightening in visible light. Observations can also be made in UV and, for a few cases, there are observations in X-ray and gamma-ray; the data is limited in these wavebands due to the lower luminosity, but also the relative lack of observational equipment.

Supernovae are categorized by features in their light curve (the variation in luminosity with time) and the strength of the hydrogen absorption lines in their spectra. Type I events have a sharp increase in luminosity followed by a steady, gradual dimming, and show little to no H absorption, while Type II have a sharp increase in luminosity followed in most cases by a plateau lasting a few months and then a gradual dimming, and show stronger H lines; each type also has subtypes determined by other details in the spectrum. For a recent review see Hillebrandt (2011).

Broadly, Type II events are explosions of individual high-mass stars that undergo core collapse. This progenitor is also responsible for Type IB and Type IC supernovae, but in these cases

H absorption is weak. Type IIL, IIP, IIN, and IIB are defined by differences in the spectra, except that a Type IIL does not show the light curve plateau that Type IIP supernovae do.

A Type Ia supernova, on the other hand, is thought to be the explosion of a white dwarf that has accreted matter from a companion (larger, main sequence, or giant) star. In this model, the white dwarf is near the critical mass of 1.4 solar masses (the Chandrasekhar limit), which is the most mass that can be supported by the electron degenerate matter that makes up a white dwarf. When more mass is accreted, the star collapses and explodes.

All supernovae produce visible and UV light and likely all produce higher energy light as well, though observations are limited. Gamma-rays emitted from supernovae are the result of radioactive decay of certain elements that are synthesized in the explosion process (see, for instance, Karam, 2002a,b). Supernovae emit much of their energy in neutrinos, almost massless elementary particles, which interact so weakly as to pose no threat to organisms (Karam, 2002b).

Supernovae also produce an ejecta blast wave that propagates outward. These blast waves form "remnants" that are visible for some time after the explosion and inject the progenitor and synthesized material into the interstellar medium. In addition, the shock front accelerates protons to high energies, producing at least a portion of the cosmic rays observed on Earth, which would also be present for most other habitable planets. An exception may be moons of giant planets which could be shielded by their host planet's strong magnetic field (in which case, however, those moons would be subject to the magnetospheric radiation of the planet, as noted above).

Another category of stellar explosions, again defined by how they are observed, is gamma-ray bursts (GRBs). As the name implies, they are observed initially as a "burst" of gamma radiation, which is followed by emission in lower-energy wavebands, all the way through radio. For an excellent review of GRBs, see Gehrels et al. (2009). These bursts fall into two subcategories, "long" and "short," defined by the duration of the gamma-ray emission. Long GRBs are of order 10's of seconds, while short GRBs are about 1 s or less (in both cases referring only to the gamma emission; the "afterglow" in other wavebands may last much longer). The two types also show a difference in their spectra, with long GRBs having "softer" spectra, dominated by lower-energy gamma-rays (with a spectral peak around 100–200 keV), and short GRBs having "harder" spectra, with greater emission of high-energy gamma-rays (with a spectral peak closer to 1 MeV).

The progenitors of long GRBs are most likely individual stars that explode as core-collapse supernovae and are situated such that they launch an intense "jet" of material along their rotation axis which happens to be pointed at Earth, leading to the burst of high-energy light observed. The fact that the emission is strongly "beamed" allows for what may be a fairly normal supernova explosion to be observed as such an intense blast. While this scenario is the most widely accepted model, the full picture may be more complicated. (For a good review of GRBs, see Kouveliotou et al. (2012).) Short GRBs, while also thought to emit radiation along a jet, are most likely the result of the merger of two compact objects, such as neutron stars or black holes.

Other short-term stellar events also produce high-energy radiation, but are of low enough intensity as to not be significant on large scales. These include "soft gamma repeaters" thought to be powered by "magnetar" stars that periodically emit lower energy gamma-rays, but with relatively low luminosity.

Black holes are also a source of high-energy radiation, particularly X-rays and energetic protons, but only if they are actively accreting matter. This is most likely in the case of supermassive black holes associated with active galactic nuclei (AGNs). Emission from stellar mass black holes is rare enough and of small enough luminosity to not be significant from the point of view of habitability. AGN, on the other hand, may be significant, when the black hole is particularly active, and could have an effect on much of their host galaxy, primarily through accelerating particles to high energies, thereby increasing the background cosmic ray flux.

4 EFFECTS

In the previous sections, we described the main astrophysical sources of radiation in the universe and the different types of radiation that can be derived from them. Two main factors determine the effect of radiation on habitability: the total energy received by a given habitat and the "hardness" of the radiation (where hardness refers to the relative amount of higher- to lower-energy photons or particles received from the source). That means that the effects of radiation on life will depend in fact on the kind of radiation (electromagnetic or particle and their energy), the amounts of radiation (dose or fluence), aand the capability of the living beings to cope with radiation, as summarized in (Fig. 1).

Biologically, damaging radiation could reach the surface of the planet, depending on the existence of a magnetic field and the presence of an atmosphere. Magnetic fields can shield the surface from charged particles, depending on the strength of the field and "rigidity" (a combination of momentum and charge) of the particles. An atmosphere can protect from both particle and electromagnetic radiation depending on the energy of the radiation and thickness of the atmosphere (Dartnell, 2011).

We can consider effects on life as being either direct or indirect. Direct effects involve the interaction of radiation directly from the event with biological material (cells, prebiotic molecules);

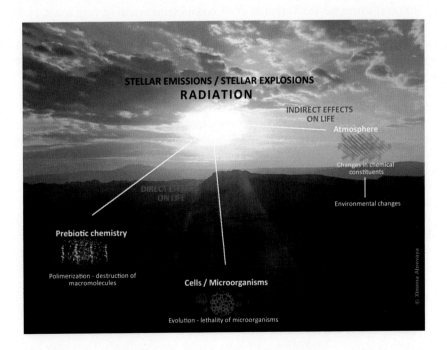

FIG. 1

Sources of radiation and its impact on life through direct and indirect effects.

(modified from Abrevaya X.C., 2013).

meanwhile indirect effects are those related to the interaction of the radiation with the environment (atmosphere), therefore favoring or limiting the possibility of life to arise and evolve (Abrevaya, 2013).

A fair amount of work has been done on the subject of astrophysical ionizing radiation and life. We cite much of that work below and also refer interested readers to the excellent reviews by Horneck et al. (2010), Olsson-Francis and Cockell (2010), and Dartnell (2011).

4.1 DIRECT EFFECTS

In general, radiation can be very harmful and even lethal to living beings, as it is capable of damaging DNA and other cellular components through different kinds of mechanisms. If we consider a planet with an atmosphere and magnetic field, UV radiation will be capable of reaching the surface, as well as muons and neutrons if sufficient energetic particles are incident at the top of the atmosphere.

In the case of UV, the most damaging effects are exhibited through direct interaction of UV photons with essential macromolecules such as DNA or proteins. As these molecules have a maximum of absorption of UV radiation at 260 nm and 280 nm, respectively, these effects are seen at UVC (100–280 nm) and UVB wavelengths (280–315 nm). The predominant kinds of damage on DNA are chemical modifications such as cyclobutane pyrimidine dimers (CPDs) and (6–4) photoproducts (6–4PPs). DNA single-strand and double-strand breaks can also be induced by UV, but these are produced as a consequence of failures during the DNA repair steps of CPDs and 6–4PPs, as was described in Bonura and Smith (1975a,b) and later by Bradley (1981).

Other kinds of damage are produced by indirect mechanisms, for example, at longer wavelengths as UVA (315–400 nm) where the absorption of DNA and proteins is null or very weak. In this case, free radicals such as reactive oxygen species are generated during the radiolysis of water molecules. The hydroxyl radical (OH) is the main damaging species producing a plethora of DNA lesions in the form of chemical modifications (e.g., 8-hydroxyguanine, DNA-protein cross-links) (for more details see Kielbassa et al., 1997 and references therein).

Other cellular components can also be damaged by UV, such as proteins. Oxidation of prokaryotic proteins during irradiation was documented for different microorganisms (Daly et al., 2007; Qiu et al., 2006). It was also suggested that UV radiation can damage membrane proteins with the concomitant leakiness of membranes (Koch et al., 1976). Membrane damage was also documented for microorganisms exposed to the 200–400 nm UV range (Fendrihan et al., 2009).

UV is also capable of inhibiting metabolism, enzymatic activity, and several cellular processes in general, such as photosynthesis (Sinha et al., 1995; Renger et al., 1989; Neale et al., 1998; Neale and Thomas, 2016).

From the experimental point of view, few studies analyzed the effects of stellar UV radiation on life considering planets orbiting habitable stars (G, F, K, and M-type stars). Nevertheless, several experimental simulations have been performed considering the solar radiation environment on Mars. Fendrihan et al. (2009) exposed halophilic archaea to several UV doses over a wavelength range of 200–400 nm to simulate the Martian UV flux. Cells that were embedded in halite showed survival under UV exposure doses as high as 10^4 kJ m^{-2} (exposure at Earth's surface today is around 3-4 kJ m^{-2}). Cockell et al. (2005) also exposed dried monolayers of *Chroococcidiopsis* sp. 029, a desiccation-tolerant, endolithic cyanobacterium, to a simulated Martian-surface UV flux (doses up to 720 kJ), also equivalent to the worst-case scenario for irradiation conditions on the Archean Earth. They have found loss of viability after 30 min of exposure.

The probability of survival of radiation-tolerant microorganisms (halophilic archaea) was evaluated considering flare activity from the dM star EV-Lacertae (EV Lac, Gliese 873, HIP 112460) taking the

UVC region (254 nm). Microorganisms survived the exposure to irradiation conditions (Abrevaya et al., 2011a). The same UV-resistant profiles were observed in experiments simulating radiation of the interplanetary environment or exposed in the low Earth orbit, where microorganisms have been exposed to EUV (e.g., Mancinelli et al., 1998; Abrevaya et al., 2011b; Mancinelli, 2015). Other studies have analyzed potential effects on life of stellar UV radiation, but they are only based on theoretical modeling and do not consider their effects on microorganisms but on isolated DNA molecules (Cockell, 1998, 1999; Cockell et al., 2005; Scalo and Wheeler, 2002; Rontó et al., 2003; Segura et al., 2003, 2010; Cockell and Raven, 2004; Buccino et al., 2007; Cuntz et al., 2010; Rugheimer et al., 2015).

At wavelengths shorter than UV, the effects of X-rays and gamma-rays are also well known. In general, direct action on the DNA molecule produces both DNA single-strand and double-strand breaks. Additionally, damage through indirect mechanisms as free radicals by radiolysis of water molecules is generated. There is no direct experimental data on the effects of this kind of radiation in the planetary context. Theoretical modeling has made predictions concerning the effects of radiation on the Earth's biosphere and revealed the biological importance of UV-flashes from GRBs delivered to the surface of the Earth, considering different present and prehistoric atmospheres (Galante and Horvath, 2007; Martín et al., 2009, 2010; Horvath and Galante, 2012).

On the other hand, since the "flash" from a GRB lasts at most 10s of seconds, this may have only a small impact on the biosphere. It is likely that the more important aspect, in the long run, is severe depletion of stratospheric O_3, caused by the formation of odd-nitrogen oxides after ionization induced by high-energy photons and cosmic rays (in the case of nearby supernovae). Thomas et al. (2005), for instance, estimated an increment in the DNA damage of up to 16 times the normal annual global average, which may be lethal for microorganisms such as phytoplankton. On the other hand, the biological impacts of increased UV following a GRB or similar event are complicated and depend on the particular organism or impact considered (Thomas et al., 2015). For two important (modern day) oceanic primary producers, Neale and Thomas (2016) found only a small impact on productivity. However, this study was limited in that it modeled only relatively short-term impacts and much remains to be learned about the long-term effects, including the level of mortality under post-GRB-type conditions.

Based on anticipated effects of reduced O_3, it has been argued that GRBs are likely to have impacted the Earth during the last billion years and could be responsible for mass extinctions (Melott et al., 2004; Melott and Thomas, 2009, 2011).

If we consider the space radiation environment, high-energy charged particles are present and they can interact at multiple scales with biological structures. Additionally, they can produce secondary particles capable of interacting with biological material. This kind of radiation should be distinguished from X-rays or gamma-rays as their deposition energy is done through a different mechanism along a "linear" track. Therefore, this produces distinguishable biological effects, different from those generated by other kinds of radiation, as particles can induce instantaneous damage, which is not compatible with repair mechanisms on cells, for example, when damaging molecules such as DNA. A detailed description of this phenomenon can be found in Nelson (2003). Some biological effects of low-energy particle radiation are also described in Yang et al. (1991).

Taking into account charged particles in an astrobiological context, Paulino-Lima et al. (2011) replicated charged particles under laboratory conditions to simulate solar wind. The radio-resistant microorganism *Deinococcus radiodurans* was exposed to electrons, protons, and ions to test its probability of survival. The results indicated that low-energy particle radiation (2–4 keV) had no significant effects on the survival of this microorganism, even if the microorganisms were irradiated with an equivalent fluence of 1000 years of exposure at 1 AU. However, as the authors mention, the effect of high-energy ions as those we could find in solar flares (200 keV) could have more deleterious effects on microbial cells, with estimated 90% cell inactivation, considering a distance of 1 AU and several flare events in one year.

It should be noted, however, that life on Earth evolved to cope with radiation as cells have developed different strategies that allow repair or prevent damage. Different DNA repair systems depending on specific enzymes exist in all life forms "as we know it" and are necessary to recognize and rebuild the injured sites, to prevent cell death. These processes are diverse from the point of view of mechanisms, but globally are highly conserved from prokaryotes to eukaryotes (and also including some viruses such as bacteriophages) (Cromie et al., 2001). One of the most unique and relevant features in the radiation-resistant microorganism *par excellence*, *D. radiodurans*, is its extremely powerful DNA repair mechanism (e.g., Cox and Battista, 2005). Several hypotheses have been suggested to explain the evolution of DNA repair and can be found in O'Brien (2006).

During biological evolution, living beings also developed other physiological strategies not only to repair DNA damage, but also to prevent it. Pigments, for example, such as melanin (Brenner and Hearing, 2008; Cordero and Casadevall, 2017), can act as a radiation shield, in particular for UV. Scytonemin, a sheath pigment in cyanobacteria, was found to protect these microorganisms against UVC radiation (Garcia Pichel et al., 1992; Dillon and Castenholz, 1999). Carotenoid pigments have also shown to protect microorganisms from UV. In fact, a positive correlation between the presence of carotenoids and resistance to radiation in bacteria was already documented several decades ago (Mathews and Krinsky, 1965). Moreover, carotenoids could have a role in DNA repair mechanisms such as photoreactivation or act as protective agents against the effects of free radicals such as hydrogen peroxide (Shahmohammadi et al., 1998). A detailed review of UV screening compounds and its relevance can be found in Cockell and Knowland (1999).

In haloarchaea, high intracellular concentrations of KCl seem to also provide protection against radiation through interaction with free radicals (Kish et al., 2009). Other radio-resistant microorganisms such as *D. radiodurans* showed that high intracellular Mn/Fe ratio combined with desiccation contributes to ionizing radiation resistance (Paulino-Lima et al., 2016). Also, physiological mechanisms such as polyploidy present in haloarchaea seem to provide advantages against radiation damage (Breuert et al., 2006).

Additionally, highly resistant structures such as bacterial spores (dormant structures produced by some bacteria that are formed in response to adverse environmental conditions) have also been shown to offer effective protection against the effects of UV radiation. Results obtained by Risenman and Nicholson (2000) indicate that the spore coat in *Bacillus subtilis* endospores is necessary for spore resistance to environmentally relevant UV wavelengths. Spores have also been shown to be 10- to 50-times more resistant to UV than growing cells and also more resistant to gamma radiation than cells during the growing state (Nicholson et al., 2000, 2005). Different kinds of photoproducts can be generated in spores by UV irradiation than those acquired when *B. subtillis* is in its growing state (Setlow, 2006). A summary can be found in Horneck et al. (2014).

In addition to physiological mechanisms that provide protection against radiation, the habitat where life forms exist and develop can be also particularly protective, for instance, in the cases of endolithic microorganisms living inside rocks (a detailed description of different kind of endoliths can be found in Golubic et al., 1981) and evaporitic environments. In addition to the obvious case of shielding from UV by opaque rock materials, haloarchaea inhabiting fluid inclusions of halite crystals have also been shown to be protected, as these crystals absorb short UV wavelengths and reemit them at longer, less damaging wavelengths (Fendrihan et al., 2009). In a series of papers, Horneck et al. (2001) and Rettberg et al. (2002, 2004) also showed that thin layers of clay, rock, or meteorite material are successful in UV-shielding.

Aquatic ecosystems can also provide shielding from UV depending on the optical properties of the water that control light penetration, which is influenced by dissolved and suspended organic material (Diffey, 1991). Vertical mixing has also been found to be an important factor (Huot et al., 2000).

This chapter is focused mainly on negative effects of radiation, but astrophysical radiation has likely had positive effects for life, especially in the context of prebiotic molecules. UV radiation, for instance, could have played an important role during the polymerization of the first prebiotic organic molecules (Dauvillier, 1947; Ponamperuma et al., 1963; Sagan and Khare, 1971; Sagan, 1973; Pestunova et al., 2005). Ranjan et al. (2017) determined the UV environment on prebiotic Earth-analog planets orbiting M-dwarfs as the recently discovered Proxima Centauri, TRAPPIST-1, and LHS 1140. They obtained dose rates to quantify the impact of different host stars on prebiotically important photoprocesses. According to the results obtained in this study, M-dwarf planets have access to 100–1000 times less bioactive UV fluence than the young Earth. Therefore it is unclear whether Earth-like planets orbiting M-dwarf could function considering UV-sensitive prebiotic chemistry that may have been important to abiogenesis on Earth (e.g., pyrimidine ribonucleotide synthesis). However, it is unknown if transient elevated UV irradiation due to flares may suffice. Experimental work under laboratory conditions is needed to constrain all these possibilities. Atri (2016b) has even proposed that cosmic rays could provide energy for existing subsurface radiolysis-powered life. In general, ionizing radiation could have had an important role in the origin of life and is relevant for the generation of habitable planetary environments (Dartnell, 2011).

In the context of biological evolution, other positive effects can be considered if the radiation doses are nonlethal for microorganisms. In this case, they could induce mutations increasing the genetic variability, thus providing new raw material for all sorts of selective pressure. For example, UV can act as a selective pressure itself, leading to the appearance of organisms adapted to live under UV stress, such as those with pigments (Scalo and Wheeler, 2002; Wynn-Williams et al., 2002). It is also postulated that UV radiation could have influenced protistan evolution (Rothschild, 1999).

4.2 INDIRECT EFFECTS

Indirect effects can be seen through the interaction of radiation with the atmosphere. Life on Earth is currently shielded from most ionizing radiation from space. The atmosphere of the present Earth (which started to increase its levels of oxygen around 2.5 Gyr ago) is thick enough to screen out high-energy photons (gamma- and X-rays), O_2 absorbs short-wavelength UV (UVC), and ozone in the middle atmosphere absorbs most UV between 200 and 350 nm (the biologically damaging UVB). In the opposite way, primitive Earth, which had a different atmospheric composition (anoxygenic atmosphere), was unable to shield the surface of the planet from the effects of UV radiation through O_3, but may have had instead "hazy" conditions that could have reduced the UV transmission (Wolf and Toon, 2010).

Stellar X-rays could affect the atmospheric evolution and the chances for life to emerge (Kulikov et al., 2007; Lammer et al., 2008). Theoretical modeling has shown that this radiation is capable of dissociating N_2 and O_2 in the atmosphere, releasing important quantities of very reactive species (atomic nitrogen and oxygen) which leads to the formation of nitrogen oxides that act as catalyzers of ozone dissociation, and therefore, increase the irradiation of the planet's surface with stellar UV radiation, among other important effects (Martín et al., 2010).

Similarly, high-energy charged particles (cosmic rays, mainly protons) interact with air molecules high in the atmosphere. On the other hand, those interactions lead to "showers" of secondary particles, some of which can be penetrating and damaging, in particular neutrons and muons, depending on the altitude considered. Charged particles with energy below about 10 GeV are deflected by Earth's large-scale magnetic field. Particles with energy below about 1 GeV are mostly deflected by the Sun's field, but that shielding varies with solar activity (more shielding when the Sun is more active). For a review of the effects on terrestrial life by cosmic rays, see Atri and Melott (2014).

On Earth then, life is mainly protected from direct effects of ionizing radiation by a thick atmosphere and large-scale magnetic field. In contrast, Mars has a thin atmosphere and no large-scale magnetic field. Smith et al. (2004a,b) and Smith and Scalo (2007) performed detailed computations of radiative transfer of high-energy photons and found that the surface of Mars would be exposed to a substantial fraction of any incident gamma radiation, while X-rays are effectively blocked. Due to the lack of a large-scale magnetic field, a planet like Mars is exposed to charged particle radiation of all energies. While the atmosphere will shield the surface to some extent, there will still be a significant flux of damaging primary and secondary charged particle radiation at the surface (Dartnell et al., 2007; Pavlov et al., 2012).

Moons around giant planets are generally too small to hold a significant atmosphere or have a large-scale magnetic field. The surfaces of Europa and Enceladus, for instance, are exposed to any incident photons. A moon's host planet can have a strong magnetic field, which, if the moon is sufficiently within that field, provides protection from high-energy cosmic rays, but will at the same time subject the moon to magnetospheric ions and electrons with energy up to tens of MeV (Cooper et al., 2001). However, a few hundred meters of ice and rock are effective shields against both high-energy photons and charged particles, so habitats existing sufficiently deep under the surface of such moons should not be affected even by the most intense irradiation from outside.

While thick, Earth-like atmospheres protect life on the surface from direct radiation effects, that life may still experience increased ionizing radiation during rare but intense high-energy astrophysical events. As noted above, high-energy protons (with energy above a few GeV) generate "showers" of secondary particles. For life around sea-level, energetic muons are the greatest threat. These "heavy electrons" can penetrate several hundred meters of rock, ice, or water and damage biological material. An enhancement of cosmic rays due to, for instance, a nearby supernova can increase the background muon radiation level by several times (depending on factors such as the distance to the supernova; Thomas et al. 2016), lasting hundreds to thousands of years, due to the slow diffusion of charged particles through interstellar space.

In addition, a thick atmosphere can "redistribute" the energy of high-energy photons (gamma- and X-rays) to UV photons, increasing the UV radiation at the surface for the duration of a gamma-ray event (Smith et al., 2004a,b; Smith and Scalo, 2007).

Finally, thick atmospheres can experience an increase in ionization due to both high-energy photons (gamma- and X-rays) and high-energy charged particles (above a few MeV), with higher-energy radiation affecting the atmosphere at lower altitudes. This ionization in an N_2-O_2 dominated atmosphere can lead to production of nitrogen oxides that catalytically destroy ozone, leading to increased penetration of UV from the host star (Thomas et al., 2005, 2015). This indirect irradiation, in fact, appears to be the most significant effect for Earth-like planets following short duration, high-energy ionizing photon events such as GRBs.

We now summarize what is known about the impacts of specific sources. In all cases, the severity of impacts depends on two main factors: (1) the total energy received, with more energy meaning greater impact and (2) the "hardness" of the radiation spectrum, with a "harder" spectrum having relatively greater flux of high-energy particles/photons, which tend to have a greater impact than lower energy particles/photons. Different types of event (SNe, GRBs, stellar activity) will have different spectra and total luminosity. The received energy depends on the intrinsic luminosity and the distance from the event (except in the case of a planet exposed to its host star's radiation, in which case the distance is negligible). The intensity of radiation decreases with the square of the distance in general, but the dependence may be more complicated for charged particles, which have significant interactions with magnetic fields in the Galaxy that cause diffusive instead of ballistic motion from the source.

GRBs are the simplest source to consider. All GRBs are relatively short in duration, ranging from tens of seconds to fractions of a second. They deliver a burst of high-energy photons, but do not appear to generate charged particle (cosmic ray) flux (Aartsen et al., 2016), at least at the highest energies (10^{18} eV or more). On the other hand, long duration GRBs are known to be associated with supernovae, which are sources of cosmic rays. For planets with thick atmospheres, the high-energy photons lead to redistributed UV radiation at the surface, but this persists only as long as the gamma- and X-rays are incident on the atmosphere, so the effect is quite short-lived (Martín et al., 2009; Peñate et al., 2010). Longer-term atmospheric chemistry effects occur following the ionization induced by the gamma-/X-rays. For planets with significant O_2, the chemistry changes lead to destruction of the ozone shield that is naturally present in the middle atmosphere of planets with O_2 and a stellar UV flux (Thomas et al., 2005). The destruction of O_3 then leads to unusual increases in stellar UV irradiance at the planet's surface and into the first 100 m or so of bodies of water, depending on their clarity (Peñate et al., 2010; Thomas et al., 2015). Overall, the depletion of O_3 can last for years to a decade. While there are two categories of GRB, both have essentially the same effect. Short GRBs have a harder spectrum but generally lower luminosity, while long GRBs have a softer spectrum but higher luminosity. Overall, they have very similar effects.

Supernovae are a more complicated source. First, they emit high-energy photons, which travel directly from the source with a $1/r^2$ intensity dependence. The photons are for the most part in the X-ray range and lower, with emission lasting on the order of months. The X-rays will have effects similar to the photon radiation from a GRB, with again the most important result being the depletion of O_3. The photons are not high enough in energy (above about 100 keV) to lead to redistributed UV as in the case of a GRB.

Supernovae also accelerate protons in the explosion blast wave. These protons travel outward from the supernova ahead of, with, and behind the ejected stellar material. Charged particles follow more complex paths in regions of space with magnetic fields present. Lower-energy particles are more strongly affected and may take many thousands of years longer than the photons to arrive. Higher-energy protons will take a more direct path. If the space in between the SN and the receiving planet is essentially empty of material and magnetic field, then the travel will be more direct and the protons may arrive within a few hundred years of the photons (Kachelrieß et al., 2015).

The accelerated protons will have two main impacts on a planet. First, they will cause ionization in a thick atmosphere, in essentially the same way as high-energy photons. This can lead to depletion of O_3, but depends strongly on the spectrum of the received protons. Harder spectra (with more of the higher energy particles) generate ionization closer to the ground and may therefore "miss" the ozone, which is concentrated in the middle atmosphere. However, high-energy protons generate showers of secondary particles, as discussed above, and these secondaries (especially muons) can be damaging at the surface and under hundreds of meters of water, ice, and rock. This is likely to be the most significant biological impact, since ozone depletion is likely to be associated mostly with the photons, which have a duration of months, while the high-energy proton flux will lead to increased biological damage for thousands of years.

For the case of a SN, the presence of a planetary magnetic field is generally not relevant, since the accelerated protons are of high enough energy to be only minimally affected (if at all) by the planet's magnetic field, unless it is much stronger than the present-day Earth's. This is true for isolated planets with their own magnetic fields as well as for moons of giant planets, which will be shielded by their host planet's field from most cosmic ray protons, but may not be shielded from the harder spectrum of protons received from a nearby supernova.

Stellar activity is most significant for close-in planets around lower mass (M type) stars (for an excellent collection of work on this topic see Lammer and Khodachenko, 2015). These stars are more active and the habitable zone is relatively close to the star (due to their low luminosity), meaning that a potentially habitable planet is more directly and more frequently exposed to radiation from stellar flares and CMEs. The relevant radiation in this case is mainly UV and protons. This radiation will mainly affect the atmosphere (see, e.g., Segura et al., 2010; Tabataba-Vakili et al., 2016). The protons will be of too low energy to generate significant showers of secondary particles, therefore not increasing the surface radiation significantly (Atri, 2016a). Another threat to habitability in this environment is atmospheric mass loss due to UV flux and the plasma stellar wind (see for instance Zendejas et al., 2010; See et al., 2014).

There may be some danger for planets located in a galaxy with an active supermassive black hole at its center (an AGN). AGN produce high-energy light (i.e., X-rays) and accelerate protons to very high energies. This may increase the background cosmic ray flux in a fairly steady way for as long as the black hole is active. This could put a constraint on habitability, but on the other hand, a steady enhancement could lead to greater radiation resistance adaptation.

5 RATES

The frequency of "dangerous" ionizing radiation events is relevant to their impact on habitability. Estimating such rates depends on a number of factors. First, as noted above, the most important parameters for determining impact are the total energy received and the hardness of the radiation spectrum. Details of the radiation spectrum depend on the particular type of event. For instance, short GRBs have very hard photon spectra, while supernovae tend to have softer photon spectra. However, the impact of SNe is also determined by the longer-lived and more spread out (in time) cosmic ray flux. For any event (except those of a host star on its planets), distance is the key factor in determining total energy received. The overall luminosity (total emitted energy) varies with event type. Short GRBs, for instance, are less luminous than long GRBs, but also have harder spectra.

Estimates of rates of "dangerous" events depends then on the basic rate of occurrence in some chosen volume (e.g., a single galaxy) as well as the distance at which that event may have a serious impact on a biosphere, which again is determined by the total energy received (in turn determined by event luminosity and distance) and spectral hardness. Existing estimates have mainly been made considering impacts on an Earth-like planet, with depletion of O_3 as the main "dangerous" effect. This is likely oversimplified. First, some recent work has indicated that O_3 depletion associated with a GRB (and events with similar total energy received and spectral hardness) may not be as disastrous as previously thought, at least on certain primary producers in the oceans (Thomas et al., 2015, 2016). If correct, this reduces the rate of dangerous events, since it requires either more energetic or closer events, both of which would be less common. On the other hand, very recent work has shown that SNe may be more damaging through the extended high-energy cosmic ray flux, not so much through O_3 depletion but through irradiation by secondary particles (muons), and possibly through increased atmospheric ionization at very low altitudes, which may impact global climate (Thomas et al., 2016).

Estimates for the frequency of severe effects, using O_3 depletion as a measure of "severe," arrive at one dangerous event every few hundred million years for Earth for SNe and both types of GRBs, with SNe and short GRBs being slightly more frequent than long GRBs (Melott and Thomas, 2011). One could extend that to any Earth-like planet with oxygen-containing atmospheres, but the rates vary through cosmic time, as discussed below.

Of particular interest is the recent discovery that at least one, and probably several, core-collapse SNe occurred relatively near Earth a few million years ago (Fry et al., 2015; Thomas et al., 2016; Wallner et al., 2016). This has been very well established by geochemical evidence, but the distance to the SNe is large enough so that terrestrial effects were not very severe.

In all cases, it should be noted that "sterilization" of a habitat is an extreme condition. For every realistic event, refugia would exist in the deep ocean and under at least 100s of meters of ice or rock. While surface life may be dramatically affected and mass extinction may result, it is likely that some life would persist. In Earth's history, at least five major mass extinctions have occurred, including one that wiped out some 90% of species on Earth at the end of the Permian period. At least, one of these is statistically likely to have been connected with an astrophysical ionizing radiation event (a specific proposal has been made regarding the late Ordovician mass extinction, see Melott and Thomas, 2009). But in every case, life has returned and flourished. Therefore, talk of "sterilization" of planets is likely overblown except in the most extreme and rare of events.

On the other hand, when considering conditions in the universe before Earth's formation, such sterilization may be more realistic. In galaxies with very high star formation rate, planets formed within dense stellar areas could be exposed to intense and repeated supernova and even GRB events. A long-term exposure to very closeby events could indeed knock back or delay the development of life.

In addition, planets in the liquid-water habitable zone around low-mass stars may experience so much bombardment from stellar activity as to be stripped of their atmosphere which is quite likely to spell the end of any complex life there.

When considering the threats over cosmic time (the last 13 billion years), rate estimates need to take into account various factors. In particular, estimates of the rates of GRBs and SNe depend on star formation rate histories. Long GRBs and core-collapse SNe result from high-mass stars that are relatively short-lived (a few million years or so) and so track regions and periods of active star formation. Short GRBs require pairs of evolved objects such as neutron stars. These objects are generally considered to be the remnant of high-mass stars, and so depend in a similar way on star formation. Type Ia supernovae require a white dwarf, which is the remnant of a star with mass similar to that of the Sun or a few times higher. Such objects, then, require longer time periods to form, since a Solar lifetime is roughly 10 billion years. These events, then, will not directly track with active star formation. Simulations that track star formation and metallicity have been used to investigate where, as well as when, different regions of our own galaxy may have been habitable, as controlled by SNe and GRBs (see Gowanlock et al., 2011; Morrison and Gowanlock, 2015; Gowanlock, 2016). In general, they find that the inner part of the galaxy is more dangerous.

The picture for GRBs is complicated by the observation that long GRBs tend to occur in lower metallicity environments. This means that the long GRB rate would have been higher earlier in the universe's history. On the other hand, short GRBs do not show such a metallicity dependence. Recently, two groups have examined the role of long GRBs in the history of life in the Universe. Piran and Jimenez (2014) find that, due mainly to the metallicity dependence, the inner part of our galaxy is most dangerous and that the existence of life in any galaxy would be severely constrained by GRBs before about 5 billion years ago. If this is correct, then habitability before the rise of life on Earth may have been significantly limited by this kind of stellar explosion.

However, Li and Zhang (2015) come to a more optimistic conclusion, that about 50% of galaxies would be hospitable (considering only effects of GRBs) at about 9 billion years ago and 10% at about 11 billion years ago, and that the most hospitable galaxies are those similar to the Milky Way. These results make the earlier universe look much more likely to have been habitable, at least from the perspective of GRB threats. Li and Zhang (2015) also note that their results should be similar for SNe, though may not track exactly, since SNe do not have the same metallicity dependence as long GRBs.

Since AGN are powered by supermassive black holes at the centers of galaxies, there will be a "sweet spot" in cosmic history where they will be most active. First, enough time must have passed for the galaxy and its central black hole to form. Second, AGN appear to be active for some time and then become less active. This is likely due to the black hole clearing out material in the central part of its galaxy. Once most of the accessible matter has been consumed, the activity is likely to cease or at least become less intense and less sustained. In general, AGN are not thought to be a major constraint on habitability, except within the central regions of galaxies, which are already dangerous due to higher rates of SNe (Dartnell, 2011; Gobat and Hong, 2016; Dayal et al., 2016).

---

6 CONCLUSIONS

Here we have presented an overview of sources of biologically relevant astrophysical radiation, and effects of that radiation on organisms and their habitats. This chapter was focused on radiation as a constraint for habitability, due to the potential harmful effects of radiation on life "as we know it." Some of these effects have been known for a long time from studies of photobiology and radiobiology. The impact of radiation on life can be varied and complicated, and in some cases, by no means fully understood. From the astrobiological point of view, it is necessary to consider these effects in the context of astrophysical scenarios, which significantly may differ from the conditions of the present Earth. Even though some limitations may arise in reproducing or simulating these environments from the experimental point of view, these kinds of studies may provide an approximation of a real case scenario to estimate the probability of a planetary body to be habitable. Additionally, of particular interest is the potential for radiation to have positive effects, either for individuals or for the development and evolution of life. Some of them were briefly described in this chapter. This is an active area of research and it may well be that a future review such as this will find that radiation is as helpful as harmful, from the broad perspective of life in the universe.

Of necessity, our review does not cover all the details of particular impacts or the responses available to organisms for dealing with radiation. We encourage the interested reader to follow-up with the sources cited for more details and to follow the continually changing landscape of this work. Surely, there is much more to be learned and we look forward to seeing what our community discovers over the next years and decades.

ACKNOWLEDGMENTS

BCT acknowledges support from NASA grant number NNX14AK22G under the Astrobiology: Exobiology and Evolutionary Biology program.
XCA acknowledges support from CONICET, Argentina.

REFERENCES

Aartsen, M.G., Abraham, K., Ackermann, M., Adams, J., Aguilar, J.A., Ahlers, M., Ahrens, M., Altmann, D., Anderson, T., Ansseau, I., et al., 2016. An all-sky search for three flavors of neutrinos from gamma-ray bursts with the icecube neutrino observatory. Astrophys. J. 824, 115. https://doi.org/10.3847/0004-637X/824/2/115.

Abrevaya, X.C., 2013. Astrobiology in Argentina and the study of stellar radiation on life. BAAA 56, 113–122.

Abrevaya, X.C., Cortón, E., Mauas, P.J.D., 2011a. Flares and habitability. Proceedings IAU Symposium 286. Comparative magnetic minima, characterizing quiet times in the Sun and stars, Cambridge University Press S286, 405–409.

Abrevaya, X.C., Paulino-Lima, I.G., Galante, D., Rodrigues, F., Cortón, E., Mauas, P.J.D., de Alencar Santos Lage, C., 2011b. Comparative survival analysis of *Deinococcus radiodurans* and the haloarchaea *Natrialba magadii* and *Haloferax volcanii*, exposed to vacuum ultraviolet irradiation. Astrobiology 11, 1034–1040.

Atri, D., 2016a. Modeling Stellar Proton Event-induced particle radiation dose on close-in exoplanets. MNRAS Lett. 465 (1), L34–L38. https://doi.org/10.1093/mnrasl/slw199.

Atri, D., 2016b. On the possibility of galactic cosmic ray-induced radiolysis-powered life in subsurface environments in the Universe. J. R. Soc. Int. 13, 20160459. https://doi.org/10.1098/rsif.2016.0459.

Atri, D., Melott, A.L., 2011. Biological implications of high-energy cosmic ray induced muon flux in the extragalactic shock model. Geophysical Research Letters 38, #L19203.

Atri, D., Melott, A.L., 2014. Cosmic rays and terrestrial life: a brief review. Astroparticle Physics. 53, 186. https://doi.org/10.1016/j.astropartphys.2013.03.001.

Ayres, T.R., 1997. Evolution of the solar ionizing flux. J. Geophys. Res. 102, 1641–1652.

Balogh, A., Treumann, R.A., 2013. Physics of Collisionless Socks: Space Plasma Shock Waves. Springer, New York.

Bonura, T., Smith, K.C., 1975a. Enzymatic production of deoxyribonucleic acid double-strand breaks after ultraviolet irradiation of *Escherichia coli* K-12. J. Bacteriol. 121, 511–517.

Bonura, T., Smith, K.C., 1975b. Quantitative evidence for enzymatically-induced DNA double-strand breaks as lethal lesions in UV-irradiated pol+ and polA1 strains of *E. coli* K-12. Photochem. Photobiol. 22, 243–248.

Bradley, M.O., 1981. Double-strand breaks in DNA caused by repair of damage due to ultraviolet light. J. Supramolec. Struct. Cell Biochem. 16, 337–343.

Brenner, M., Hearing, V.J., 2008. The protective role of melanin against UV damage in human skin. Photochemistry and Photobiology 84 (3), 539–549.

Breuert, S., Allers, T., Spohn, G., Soppa, J., 2006. Regulated polyploidy in halophilic archaea. PLoS One 1. e92. https://doi.org/10.1371/journal.pone.0000092.

Buccino, A.P., Lemarchand, G.A., Mauas, P.J.D., 2007. UV habitable zones around M stars. Icarus 192, 582–587.

Cockell, C.S., 1998. Biological effects of high ultraviolet radiation on early Earth—a theoretical evaluation. J. Theor. Biol. 193, 717–719.

Cockell, C.S., 1999. Carbon biochemistry and the ultraviolet radiation environments of F, G, and K main sequence stars. Icarus 141, 399–407.

Cockell, C.S., Knowland, J., 1999. Ultraviolet radiation screening compounds. Biol. Rev. 74, 311–345.

Cockell, C.S., Raven, J.A., 2004. Zones of photosynthetic potential on Mars and the early Earth. Icarus. 169, 300–310.

Cockell, C.S., Schuerger, A.C., Billi, D., Friedmann, E.I., Panitz, C., 2005. Effects of a simulated martian UV flux on the cyanobacterium *Chroococcidiopsis* sp. 029. Astrobiology 5, 127–140.

Cooper, J.F., Johnson, R.E., Mauk, B.H., Garrett, H.B., Gehrels, N., 2001. Energetic Ion and Electron Irradiation of the Icy Galilean Satellites. Icarus 149, 133–159. https://doi.org/10.1006/icar.2000.6498.

Cordero, R.J.B., Casadevall, A., 2017. Functions of fungal melanin beyond virulence. Fungal Biology Reviews 31 (2), 99–112.

Cox, M.M., Battista, J.R., 2005. *Deinococcus radiodurans*—the consummate survivor. Nat. Rev. Microbiol. 3, 882–892.

Cromie, G.A., Connelly, J.C., Leach, D.R., 2001. Recombination at double-strand breaks and DNA ends: conserved mechanisms from phage to humans. Mol. Cell. 8, 1163–1174.

Cuntz, M., Guinan, E.F., Kurucz, R.L., 2010. Biological damage due to photospheric, chromospheric and flare radiation in the environments of main-sequence stars. Planetary Systems as Potential Sites for Life. IAUS 264, 1–8.

Daly, M.J., Gaidamakova, E.K., Matrosova, V.Y., Vasilenko, A., Zhai, M., Leapman, R.D., Leapman, R.D., Lai, B., Ravel, B., Li, S.-H.W., Kemner, K.M., Fredrickson, J.K., 2007. Protein oxidation implicated as the primary determinant of bacterial radioresistance. PLoS Biol. 5, 0769–0779.

Dartnell, L.R., 2011. Ionizing radiation and life. Astrobiology 11, 551–582. https://doi.org/10.1089/ast.2010.0528.

Dartnell, L., Desorgher, L., Ward, J., Coates, A., 2007. Modelling the surface and subsurface martian radiation environment: implications for astrobiology. Geophys. Res. Lett. 34. L02207.

Dauvillier, A.,1947. Nature et evolution des planétes. Hermann, Paris.

Dayal, P., Ward, M., Cockell, C., 2016. The habitability of the Universe through 13 billion years of cosmic time, arXiv,1606.09224.

Diffey, B.L., 1991. Solar ultraviolet radiation effects on biological systems. Phys. Med. Biol. 36, 299–328.

Dillon, J.G., Castenholz, R.W., 1999. Scytonemin, a cyanobacterial sheath pigment protects against UVC radiation, implications for early photosynthetic life. J. Phycol. 35, 673–681.

Fendrihan, S., Berces, A., Lammer, H., Musso, M., Rontó, G., Polacsek, T.K., Holzinger, A., Kolb, C., Stan-Lotter, H., 2009. Investigating the effects of simulated Martian ultraviolet radiation on *Halococcus dombrowskii* and other extremely halophilic archaebacteria. Astrobiology 9, 104–112.

Fry, B.J., Fields, B.D., Ellis, J.L., 2015. Astrophysical shrapnel: discriminating among near-Earth stellar explosion sources of live radioactive isotopes. ApJ. 800, 71. https://doi.org/10.1088/0004-637X/800/1/71.

Galante, D., Horvath, J.E., 2007. Biological effects of gamma-ray bursts, distances for severe damage on the biota. Int. J. Astrobiol. 6, 19–26.

Garcia-Pichel, F., Sherry, N.D., Castenholz, R.W., 1992. Evidence for a UV sunscreen role of the extracellular pigment scytonemin in the terrestrial cyanobacterium *Chlorogloeopsis* sp. Photochem. Photobiol. 56, 17–23.

Gehrels, N., Ramirez-Ruiz, N.E., Fox, D.B., 2009. Gamma-ray bursts in the *Swift* Era. Annu. Rev. Astron. Astrophys. 47, 567–617. https://doi.org/10.1146/annurev.astro.46.060407.145147.

Gershberg, R.E., 2005. Solar-Type Activity in Main-Sequence Stars. Springer-Verlag, Berlin Heidelberg.

Gobat, R., Hong, S.E., 2016. Evolution of galaxy habitability. Astron. Astrophys. 592, A96. https://doi.org/10.1051/0004-6361/201628834.

Golubic, S., Friedmann, E.I., Schneider, J., 1981. The lithobiotic ecological niche, with special reference to microorganisms. J. Sediment. Res. 51, 475–478.

Gowanlock, M.G., 2016. Astrobiological effects of gamma-ray bursts in the Milky Way galaxy. Astrophys. J. 832, 38. https://doi.org/10.3847/0004-637X/832/1/38.

Gowanlock, M.G., Patton, D.R., McConnell, S.M., 2011. A model of habitability within the Milky Way Galaxy. Astrobiology 11, 855–873. https://doi.org/10.1089/ast.2010.0555.

Guinan, E.F., Ribas, I., 2002. Our Changing Sun: the role of solar nuclear evolution and magnetic activity on Earth's atmosphere and climate. ASPC 269, 85–106.

Hillebrandt, W., 2011. The physics and astrophysics of supernova explosions. In: von Berlepsch, R. (Ed.), Reviews in Modern Astronomy, Zooming in, The Cosmos at High Resolution, Volume 23, Wiley-VCH Verlag GmbH & Co. KGaA, Weinheim, Germany. https://doi.org/10.1002/9783527644384.ch4.

Horneck, G., Rettberg, P., Reitz, G., Wehner, J., Eschweiler, U., Strauch, K., Panitz, C., Starke, V., Baumstark-Khan, C., 2001. Protection of bacterial spores in space, a contribution to the discussion on panspermia. Orig. Life Evol. Biosph. 31, 527–547.

Horneck, G., Klaus, D., Mancinelli, R., 2010. Space microbiology. Microbiol. Mol. Biol. Rev. 74, 121–156.

Horneck, G., Douki, T., Cadet, J., Panitz, C., Rabbow, E., Moeller, R., Rettberg, P., 2014. UV-Photobiology of bacterial spores in space. 40th COSPAR Scientific Assembly. Held 2-10 August 2014, in Moscow, Russia, Abstract F3.1-4-14.

Horvath, J.E., Galante, D., 2012. Effects of high-energy astrophysical events on habitable planets. Int. J. Astrobiol. 11, 279–286.

Huot, Y., Jeffrey, W.H.R., Davis, F., Cullen, J.J., 2000. Damage to DNA in bacterioplankton: a model of damage by ultraviolet radiation and its repair as influenced by vertical mixing. Photochem. Photobiol. 72, 62–74.

Kachelrieß, M., Neronov, A., Semikoz, D.V., 2015. Signatures of a two million year old supernova in the spectra of cosmic ray protons, antiprotons, and positrons. Phys. Rev. Lett. 115, 181103. https://doi.org/10.1103/PhysRevLett.115.181103.

Karam, P.A., 2002a. Terrestrial radiation exposure from supernova-produced radioactivities. Radiat. Phys. Chem. 64, 77–87.

Karam, P.A., 2002b. Gamma and neutrino radiation dose from gamma ray bursts and nearby supernovae. Health Phys. 82, 491–499.

Kielbassa, C., Roza, L., Epe, B., 1997. Wavelength dependence of oxidative DNA damage induced by UV and visible light. Carcinogenesis 18, 811–816.

Kish, A., Kirkali, G., Robinson, C., Rosenblatt, R., Jaruga, P., Dizdaroglu, M., DiRuggiero, J., 2009. Salt shield, intracellular salts provide cellular protection against ionizing radiation in the halophilic archaeon, Halobacterium salinarum NRC-1. Environ. Microbiol. 11, 1066–1078.

Koch, A.L., Doyle, R.J., Kubitschek, H.E., 1976. Inactivation of membrane transport in *Escherichia coli* by near-ultraviolet light. J. Bacteriol. 126, 140–146.

Kopparapu, R.K., Ramirez, R., Kasting, J.F., Eymet, V., Robinson, T.D., Mahadevan, S., Terrien, R.C., Domagal-Goldman, S., Meadows, V., Deshpande, R., 2013. Habitable zones around main-sequence stars: new estimates. Astrophys. J. 765, 131. https://doi.org/10.1088/0004-637X/765/2/131.

Kouveliotou, C., Wijers, R.A.M.J., Woosley, S.E. (Eds.), 2012. Gamma-Ray Bursts, 51. Cambridge University Press, Cambridge, UK.

Kulikov, Y.N., Lammer, H., Lichtenegger, H.I.M., Penz, T., Breuer, D., Spohn, T., Lundin, R., Biernat, H.K., 2007. A comparative study of the influence of the active young Sun on the early atmospheres of Earth. Venus and Mars. Space Sci. Rev. 129, 207–243.

Lammer, H., Kasting, J.F., Chassefière, E., Johnson, R.E., Kulikov, Yu, N., Tian, F., 2008. Atmospheric escape and evolution of terrestrial planets and satellites. Space Sci. Rev. https://doi.org/10.1007/s11214-008-9413-5.

Lammer, H., Khodachenko, M., 2015. Characterizing Stellar and Exoplanetary Environments. Springer International Publishing, Switzerland.

Li, Y., Zhang, B., 2015. Can life survive gamma-ray bursts in the high-redshift universe? Astrophysical J. 810, 41. https://doi.org/10.1088/0004-637X/810/1/41.

Mancinelli, R., 2015. The affect of the space environment on the survival of *Halorubrum chaoviator* and *Synechococcu*s (Nägeli) data from the space experiment OSMO on EXPOSE-R. Int. J. Astrobiol. 14, 123–128.

Mancinelli, R., White, M., Rothschild, L., 1998. BIOPAN-SURVIVAL I, exposure of the osmophiles *Synechococcus* sp. (Nageli) and *Haloarcula* sp. to the space environment. Advances in space research 22, 327–334.

Martín, O., Galante, D., Cárdenas, R., Horvath, J.E., 2009. Short-term effects of gamma ray bursts on Earth. Astrophys. Sapce Sci. 321, 161–167. https://doi.org/10.1007/s10509-009-0037-3.

Martín, O., Cardenas, R., Guimarais, M., Peñate, L., Horvath, J., Galante, D., 2010. Effects of gamma ray bursts in Earth's biosphere. Astrophys. Space Sci. 326, 61–67.

Mathews, M.M., Krinsky, N.I., 1965. The relationship between carotenoid pigments and resistance to radiation in non-photosynthetic bacteria. Photochem. Photobiol. 4, 813–817.

Melott, A.L., Thomas, B.C., 2009. Late Ordovician geographic patterns of extinction compared with simulations of astrophysical ionizing radiation damage. Paleobiology 35, 311–320.

Melott, A.L., Thomas, B.C., 2011. Astrophysical ionizing radiation and Earth: a brief review and census of intermittent intense sources. Astrobiology 11, 343–361. https://doi.org/10.1089/ast.2010.0603.

Melott, A., Lieberman, B., Laird, C., Martin, L., Medvedev, M., Thomas, B., Cannizzo, J., Gehrels, N., Jackman, C., 2004. Did a gamma-ray burst initiate the late Ordovician mass extinction? Int. J. Astrobiol. 3, 55–61. https://doi.org/10.1017/S1473550404001910.

Morrison, I.S., Gowanlock, M.G., 2015. Extending the galactic habitable zone modeling to include the emergence of intelligent life. Astrobiology 15, 683–696. https://doi.org/10.1089/ast.2014.1192.

Neale, P.J., Thomas, B.C., 2016. Solar irradiance changes and phytoplankton productivity in Earth's ocean following astrophysical ionizing radiation events. Astrobiology 16, 245–258. https://doi.org/10.1089/ast.2015.1360.

Neale, P.J., Davis, R.A., Cullen, J.J., 1998. Interactive effects of ozone depletion and vertical mixing on photosynthesis of Antarctic phytoplankton. Nature 392, 585–589.

Nelson, G.A., 2003. Fundamental space radiobiology. Gravit. Space Biol. Bull. 16, 29–36.

Nicholson, W.L., Munakata, N., Horneck, G., Melosh, H.J., Setlow, P., 2000. Resistance of *Bacillus* endospores to extreme terrestrial and extraterrestrial environments. Microbiol. Mol. Biol. Rev. 64, 548–572.

Nicholson, W.L., Schuerger, A.C., Setlow, P., 2005. The solar UV environment and bacterial spore UV resistance, considerations for Earth-to-Mars transport by natural processes and human spaceflight. Mutat. Res. 571, 249–264.

O'Brien, P.J., 2006. Catalytic promiscuity and the divergent evolution of DNA repair enzymes. Chem. Rev. 106, 720–752.

Olsson-Francis, K., Cockell, C.S., 2010. Experimental methods for studying microbial survival in extraterrestrial environments. J. Microbiol. Methods 80, 1–13. https://doi.org/10.1016/j.mimet.2009.10.004.

Paulino-Lima, I.G., Janot-Pacheco, E., Galante, D., Cockell, C., Olsson-Francis, K., Brucato, J.R., Baratta, G.A., Strazzulla, G., Merrigan, T., McCullough, R., Mason, N., Lage, C., 2011. Survival of *Deinococcus radiodurans* against laboratory-simulated solar wind charged particles. Astrobiology 11, 875–882.

Paulino-Lima, I.G., Fujishima, K., Navarrete, J.U., Galante, D., Rodrigues, F., Azúa-Bustos, A., Rotschild, L.J., 2016. Extremely high UV-C radiation resistant microorganisms from desert environments with different manganese concentrations. J. Photochem. Photobiol. B Biol. 163, 327–336.

Pavlov, A.A., Vasiliev, G., Ostryakov, V.M., Pavlov, A.K., Mahaffy, P., 2012. Degradation of the organic molecules in the shallow subsurface of Mars due to irradiation by cosmic rays. GRL 39. L13202. https://doi.org/10.1029/2012GL052166.

Peñate, L., Martín, O., Cárdenas, R., Agustí, S., 2010. Short-term effects of gamma ray bursts on oceanic photosynthesis. Astrophys. Space Sci. 330, 211–217. https://doi.org/10.1007/s10509-010-0450-7.

Pestunova, O., Simonov, A., Snytnikov, V., Stoyanovsky, V., Parmon, V., 2005. Putative mechanism of the sugar formation on prebiotic Earth initiated by UV-radiation. Adv. Space Res. 36, 214–219.

Piran, T., Jimenez, R., 2014. Possible role of gamma ray bursts on life extinction in the Universe. Physical Rev. Lett. 113, 231102. https://doi.org/10.1103/PhysRevLett.113.231102.

Ponamperuma, C., Mariner, S., Sagan, C., 1963. Formation of adenosina by ultraviolet irradiation of a solution of adenine and ribose. Nature 198, 1199–1200.

Qiu, X., Daly, M.J., Vasilenko, A., Omelchenko, M.V., Gaidamakova, E.K., Wu, L., Zhou, J., Sundin, G.W., Tiedje, J.M., 2006. Transcriptome analysis applied to survival of *Shewanella oneidensis* MR-1 exposed to ionizing radiation. J. Bacteriol. 188, 1199–1204.

Ranjan, S., Wordsworth, R.D., Sasselov, D.D., 2017. The Surface UV Environment on Planets Orbiting M-Dwarfs: Implications for Prebiotic Chemistry & Need for Experimental Follow-Up. The Astrophysical Journal 843 (3), 11, Article ID 110.

Renger, G., Volker, M., Eckert, H.J., Fromme, R., Hohm-Veit, S., Graber, P., 1989. On the mechanism of photosystem II deterioration by UV-B irradiation. Photochem. Photobiol. 49, 97–105.

Rettberg, P., Eschweiler, U., Strauch, K., Reitz, G., Horneck, G., Wänke, H., Brack, A., Barbier, B., 2002. Survival of microorganisms in space protected by meteorite material: results of the experiment EXOBIOLOGIE of the PERSEUS mission. Adv. Space Res. 30, 1539–1545.

Rettberg, P., Rabbow, E., Panitz, C., Horneck, G., 2004. Biological space experiments for the simulation of Martian conditions: UV radiation and Martian soil analogues. Adv. Space Res. 3, 1294–1301.

Ribas, I., Guinan, E.F., Gudel, M., Audard, M., 2005. Evolution of the solar activity over time and effects on planetary atmospheres. I. High-energy irradiances, 1–1700 A°. ApJ. 622, 680–694.

Rieger, F.M., Bosch-Ramon, V., Duffy, P., 2007. Fermi acceleration in astrophysical jets. Astrophys. Space Sci. 309, 119–125.

Risenman, P.J., Nicholson, W.L., 2000. Role of the spore coat layers in *Bacillus subtilis* spore resistance to hydrogen peroxide, artificial UV-C, UV-B, and solar UV radiation. Appl. Environ. Microbiol. 66, 620–626.

Rodriguez, L., Cardenas, R., Rodriguez, O., 2013. Perturbations to aquatic photosynthesis due to high-energy cosmic ray induced muon flux in the extragalactic shock model. International Journal of Astrobiology 12 (4), 326–330.

Rontó, G., Bérces, A., Lammer, H., Cockell, C.S., Molina-Cuberos, G.J., Patel, M.R., Selsis, F., 2003. Solar UV irradiation conditions on the surface of Mars. Photochem. Photobiol. 77, 34–40.

Rothschild, L.J., 1999. The influence of UV radiation on protistan evolution. J. Euk. Micro. 46, 548–555.

Rugheimer, S., Kaltenegger, L., Segura, A., Linsky, J., Mohanty, S., 2015. Influence of UV activity on the spectral fingerprints of Earth-like planets around M dwarfs. ApJ 809, 57.

Sagan, C., 1973. Ultraviolet selection pressure on the earliest organisms. J. Theor. Biol. 39, 195–200.

Sagan, C., Khare, B.N., 1971. Long-wavelength ultraviolet photoproduction of aminoacids on the primitive Earth. Science 173, 417–420.

Scalo, J., Wheeler, J.C., 2002. Astrophysical and astrobiological implications of gamma-ray burst properties. Astrophys. J. 566, 723–737.

Scalo, J., Kaltenegger, L., Segura, A., Fridlund, M., Ribas, I., Kulikov, Y.N., Grenfell, J.L., Rauer, H., Odert, P., Leitzinger, M., Selsis, F., Khodachenko, M.L., Eiroa, C., Kasting, J., Lammer, H., 2007. M stars as targets for terrestrial exoplanet searches and biosignature detection. Astrobiology 7, 85–166.

See, V., Jardine, M., Vidotto, A.A., Petit, P., Marsden, S.C., Jeffers, S.V., do Nascimento, J.D., 2014. The effects of stellar winds on the magnetospheres and potential habitability of exoplanets. Astron. Astrophys. 570, A99. https://doi.org/10.1051/0004-6361/201424323.

Segura, A., Krelove, K., Kasting, J.F., Sommerlatt, D., Meadows, V., Crisp, D., Cohen, M., Mlawer, E., 2003. Ozone concentrations and ultraviolet fluxes on Earth-like planets around other stars. Astrobiology 3, 689–708.

Segura, A., Walkowicz, L., Meadows, V., Kasting, J., Hawley, S., 2010. The effect of a strong stellar flare on the atmospheric chemistry of an Earth-like planet orbiting an M dwarf. Astrobiology 10, 751–771.

Setlow, P., 2006. Spores of *Bacillus subtilis*, their resistance to and killing by radiation, heat and chemicals. J. Appl. Microbiol. 101, 514–525.

Shahmohammadi, H.R., Asgarini, E., Terat, H., Saito, T., Ohyama, Y., Gekko, K., Yamamoto, O., Ide, H., 1998. Protective roles of bacterioruberin and intracellular KCl in the resistance of Halobacterium salinarium against DNA-damaging agents. J. Radiat. Res. 39, 251–262.

Sinha, R.P., Kumar, H.D., Kumar, A., Hlider, D.-P., 1995. Effects of UV-B irradiation on growth, survival, pigmentation and nitrogen metabolism enzymes in cyanobacteria. Acta Protozool. 34, 187–192.

Smith, D.S., Scalo, J., 2007. Solar X-ray flare hazards on the surface of Mars. Planet. Space Sci. 55, 517–527. http://dx.doi.org/10.1016/j.pss.2006.10.001.

Smith, D.S., Scalo, J., Wheeler, J.C., 2004a. Transport of ionizing radiation in terrestrial-like exoplanet atmospheres. Icarus 171, 229–253. https://doi.org/10.1016/j.icarus.2004.04.009.

Smith, D.S., Scalo, J., Wheeler, J.C., 2004b. Importance of biologically active aurora-like ultraviolet emission, stochastic irradiation of Earth and Mars by flares and explosions. OLEB 34, 513–532. https://link.springer.com/article/10.1023/B%3AORIG.0000043120.28077.c9.

Tabataba-Vakili, F., Grenfell, J.L., Grießmeier, J.-M., Rauer, H., 2016. Atmospheric effects of stellar cosmic rays on Earth-like exoplanets orbiting M-dwarfs. Astron. Astrophys. 585, A96. https://doi.org/10.1051/0004-6361/201425602.

Tarter, J.C., Backus, P.R., Mancinelli, R.L., Aurnou, J.M., Backman, D.E., Basri, G.S., Boss, A.P., Clarke, A., Deming, D., Doyle, L.R., Feigelson, E.D., Freund, F., et al., 2007. A reappraisal of the habitability of planets around M dwarf stars. Astrobiology 7, 30–65.

Thomas, B.C., Melott, A., Jackman, C., Laird, C., Medvedev, M., Stolarski, R., Gehrels, N., Cannizzo, J., Hogan, D., Ejzak, L., 2005. Gamma-ray bursts and the Earth: exploration of atmospheric, biological, climatic, and biogeochemical effects. ApJ 634, 509–533.

Thomas, B.C., Neale, P.J., Snyder, B.R., 2015. Solar irradiance changes and photobiological effects at Earth's surface following astrophysical ionizing radiation events. Astrobiology 15, 207–220. https://doi.org/10.1089/ast.2014.1224.

Thomas, B.C., Engler, E.E., Kachelrieß, M., Melott, A.L., Overholt, A.C., Semikoz, D.V., 2016. Terrestrial effects of nearby supernovae in the early Pleistocene. Astrophys. J. Lett. 826, L3. https://doi.org/10.3847/2041-8205/826/1/L3.

Turnbull, M.C., Tarter, J.C., 2003. Target Selection for SETI. I. A catalog of nearby habitable stellar systems. Astrophys. J. 145, 181–198.

Wallner, A., Feige, J., Kinoshita, N., Paul, M., Fifield, L.K., Golser, R., Honda, M., Linnemann, U., Matsuzaki, H., Merchel, S., Rugel, G., Tims, S.G., Steier, P., Yamagata, T., Winkler, S.R., 2016. Recent near-Earth supernovae probed by global deposition of interstellar radioactive ^{60}Fe. Nature 532, 69–72. https://doi.org/10.1038/nature17196.

Wolf, E.T., Toon, O.B., 2010. Fractal organic hazes provided an ultraviolet shield for early Earth. Science 328, 1266–1268. https://doi.org/10.1126/science.1183260.

Wynn-Williams, D.D., Edwards, H.G.M., Newton, E.M., Holder, J.M., 2002. Pigmentation as a survival strategy for ancient and modern photosynthetic microbes under high ultraviolet stress on planetary surfaces. Int. J. Astrobiol. 1, 39–49.

Yang, Z., Gan, B., Lin, J., 1991. The mutagenic effect on plant growth by ion implantation on wheat. Anhui Agric. Univ. Acta 18, 282–288.

Zendejas, J., Raga, A., Segura, A., 2010. Atmospheric mass loss by stellar wind from planets around main sequence M stars. Icarus 210, 539–544. https://doi.org/10.1016/j.icarus.2010.07.013.

FURTHER READING

Chyba, C.F., Sagan, C., 1992. Endogenous production, exogenous delivery and impact-shock synthesis of organic molecules, an inventory for the origins of life. Nature 355, 125–132.

Cockell, C.S., 2000. Ultraviolet radiation and the photobiology of Earth's early oceans. OLEB 30, 467–499.

THE WHEN AND WHERE OF WATER IN THE HISTORY OF THE UNIVERSE

Karla de Souza Torres*, Othon C. Winter[†]

*CEFET-MG, Curvelo, Brazil
[†]FEG-UNESP, Guaratinguetá, Brazil

CHAPTER OUTLINE

Habitability of the Universe Before Earth, editors: Richard Gordon & Alexei Sharov, Volume 1 in the series:
Astrobiology: Exploring Life on Earth and Beyond, series editors: Pabulo Henrique Rampelotto,
Joseph Seckbach & Richard Gordon. ISSN 2468-6352. https://doi.org/10.1016/B978-0-12-811940-2.00003-4

Acronyms

HMSR	high mass star-forming regions
HST	Hubble Space Telescope
ISM	interstellar medium
LESIA	Linear Etalon Imaging Spectral Array
Masers	microwave amplification by stimulated emission of radiation
RSL	recurring slope lineae
SMOW	standard mean ocean water
SRVs	semiregular variable stars
SSSB	small solar system body
YSOs	young stellar objects

1 INTRODUCTION. WHY IS WATER ESSENTIAL FOR LIFE?

It is well known that liquid water has played the essential and undeniable role in the emergence, development, and maintenance of life on Earth. Two thirds of the Earth's surface is covered by water, however, fresh water is most valuable as a resource for animals and plants. Thus, sustainability of our planet's fresh water reserves is an important issue as population numbers increase. Water accounts for 75% of human body mass and is the major constituent of organism fluids. All these facts indicate that water is one of the most important compounds for life on Earth. Thus, "follow the water" has become a mantra of the science of astrobiology (Irion, 2002).

Water is present on the surface of our planet at ambient temperatures and pressures in three different states: liquid, vapor, and solid. Water is also found everywhere in the universe: in the most distant galaxies, among the stars, on the Sun, on planets and their satellites and ring systems, in asteroids, and in comets. Water exhibits unique properties that make it extremely important for life as known on Earth. First, it is the only substance on Earth that is abundant in liquid form at temperatures commonly found on the planet's surface. Second, it is a superb solvent, implying that other substances can easily dissolve in it. Thus, water carries nutrients to cells and is used to wash away the waste (Lynden-Bell et al., 2010).

Water is formed from two very abundant elements in the universe: hydrogen, the most abundant element, and oxygen, the third (helium is second) most abundant element in the universe.

In this chapter we discuss when did water appear after the Big Bang, where did it spread in the universe, what potential roles did it play in the emergence of life? First we discuss the physical and chemical characteristics, and then, turn to the history of water in the universe, its cosmic formation, and abundance. Next, we individually discuss the diverse cosmic sites, from distant galaxies to nearer stars and planets, where water has been discovered. Finally, we describe the strong relation between water and life, as we know it, and formulate conclusions about the great endeavor of determining the when and where of water in the universe's history.

2 WHAT IS WATER?

2.1 CHEMICAL PROPERTIES OF WATER

A water molecule includes two hydrogen atoms and one oxygen atom, connected with a strong polar covalent bond. Water molecules are stable and can last for millions and even billions of years. Water forms in a chemical reaction of two molecules of hydrogen (H_2) with one molecule of oxygen (O_2), as shown in Eq. (1):

$$H_2 + H_2 + O_2 = H_2O + H_2O \tag{1}$$

This process is one of the most exothermic among known chemical reactions, with released energy of 572 kJ/mol (Hanslmeier, 2010).

The water molecule is not linear but has a shape of a triangle (Fig. 1). At the apex is the oxygen atom, with two hydrogen atoms forming an angle of 104.5°. With six valence electrons, oxygen needs eight to fill its valence shell. Thus, it shares two electrons from the two hydrogen atoms, which become positively charged. Since the small hydrogen atoms have weaker affinity for electrons than the large oxygen atom, the molecule is bent, and the two hydrogen atoms appear on the same side. Water is thus classified as a polar molecule because of its polar covalent bonds and its bent shape (Encrenaz, 2007).

The polarity of water explains why it is an excellent solvent. Water can induce temporary dipoles in even nonpolar molecules, and interact differently with charged and polar substances. The polar molecules interact with the partially positive and negative ends of the water molecule, which results in the formation of a three-dimensional sphere of water molecules surrounding the solute. Water can thus dissolve and accumulate a variety of substances that are important for life. If the bonds were linear, then water would not be a strong solvent, which could affect the origin of life.

2.2 PHYSICAL PROPERTIES

On Earth, water is the only compound that exists naturally in all three phases: gas, liquid, and solid. Its unique properties have allowed life to be possible on Earth. Hanslmeier (Hanslmeier, 2010) wrote:

- Liquid water is more dense than water ice. This is essential for life because ice always forms at the surface, protecting life below it from freezing;
- The pH of pure water is neutral, value of 7, which is neither acidic nor basic;
- Water boils at 100°C and freezes at 0°C under standard pressure conditions[1].

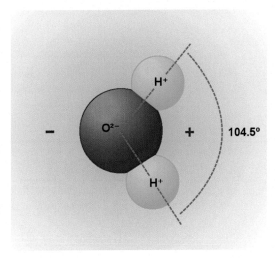

FIG. 1

Structure of the H_2O molecule, electrically polarized.

Liquid water boils at a temperature at which vapor pressure reaches the environmental pressure around the liquid. The higher is the environmental pressure, the higher is the temperature at which the liquid will boil. This temperature is known as the boiling point. At standard pressure,[1] water boils at 100°C. The pressure on the top of Mount Everest is 260 Pa, where the boiling point of water is 69°C.

The curves on the phase diagram shown in Fig. 2 correspond to the boundaries between the different phases of water, according to the temperature and pressure. The triple point, at the intersection of the

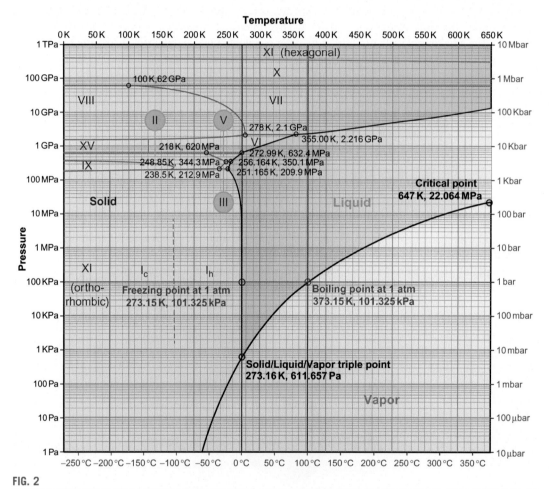

FIG. 2

Log-lin pressure-temperature phase diagram of water. The *Roman numerals* indicate various ice phases.

From: By Cmglee (original work), via Wikimedia Commons.

[1]The standard pressure is the pressure at 1 bar (100 kPa, the current IUPAC (International Union of Pure and Applied Chemistry) definition). The standard reference conditions are: temperature 0°C, pressure 100 kPa. The standard atmosphere (symbol: atm) is a unit of pressure defined as 101,325 Pa (1.01325 bar).

three curves, indicates the pressure and temperature where water can coexist in all three phases. It is at a temperature of 0.01°C (273.16 K) and a pressure of 611.657 Pa. At a low pressure of just 7.000 Pa, water boils at 38.5°C. This is about one order of magnitude higher than the atmospheric pressure on Mars. Therefore, liquid water cannot exist on the Martian surface at present. Despite that, salty liquid water seems to flow from some steep, relatively warm slopes on the surface of Mars (Gough et al., 2016). This is addressed in Section 4.5.3.

3 WHEN DID WATER APPEAR?

The raw materials for producing water molecules are hydrogen and oxygen atoms. Here we discuss where these elements come from, and how they were formed.

Spectroscopic analysis of sunlight indicates that the Sun's photosphere is composed of hydrogen (74.9%), helium (23.8%), and other heavier elements (Lodders, 2003b,a). Among the heavier elements, the most abundant is oxygen (around 1% of the solar mass) (Hansen et al., 2004). The abundance of the metals is usually estimated considering not just the spectroscopy of the Sun's photosphere but also the measures from meteorites that are believed to contain the original Solar composition (Piersanti et al., 2007).

However, our Sun was not created in the first generation of stars in the universe. It is only approximately 4.6 billion years old, while the universe is known to be approximately 13.7 billion years old. In fact, the chemical composition of our Sun was inherited from the interstellar medium, which was produced from previous generations of stars. Moreover the ingredients of the first stars were generated in the Big Bang nucleosynthesis.

The simplest chemical element in nature is hydrogen. Through stellar evolutions and nuclear fusion hydrogen leads to other elements, including oxygen. In the current section, we present how, when, and where these elements were and are being created in the universe.

3.1 PRIMORDIAL NUCLEOSYNTHESIS

After the Big Bang, the early universe was initially very dense and hot. It cooled down as it expanded, and the quark-gluon plasma gave origin to neutrons and protons (and other hadrons, but in very small quantities). The universe continued to cool, and quickly (after 15–30 min) the nucleosynthesis ceased because it became too cold (Rauscher and Patkós, 2011). The decay time of a free neutron is approximately 10 min. However, before their decay, neutrons may interact with protons forming deuterium nuclei. If the deuterium obtains another neutron it forms tritium, which in turn can absorb a proton to form a ^{4}helium. There is no stable element of mass 5 or 8. Therefore, it is generally not possible to have additional nucleosynthesis via $H + {}^4He = {}^5Li$ or ${}^4He + {}^4He = {}^8Be$; nevertheless, traces of one or two heavier elements, most notably ^{7}lithium, do form. Most of the matter was then hydrogen and ^{4}helium, with a small amount of deuterium, and just traces of ^{3}helium and ^{7}lithium. Neutrons and protons started to form only after the first 1/1000th of a second from the Big Bang, when the temperatures dropped low enough. From the first 1/100th of a second up to 3 or 4 min after the Big Bang, the abundances of the first very light atomic nuclei were defined. The ratio of cosmic abundance today, expressed in terms of mass, is approximately 71% of hydrogen, 28% of helium, and 1% for all the remaining elements.

However, at most 4% of this helium could be the result of burning hydrogen inside of stars since the beginning of the universe. Then, the initial ratio must have been approximately 24% of helium and 76% of hydrogen (Rauscher and Patkós, 2011).

Therefore, the primordial stars in the universe formed from a gaseous mixture of hydrogen and helium, as well as a very few traces of some rare light elements, such as ^{7}lithium, or isotopes such as deuterium or ^{3}helium (Karlsson et al., 2013). They did not have any oxygen.

3.2 ENERGY PRODUCTION IN STARS

One of the most intense research areas in the early 20th century was the source of stellar energy. A seminal paper on the subject was written by Hans Bethe in 1939, entitled "Energy Production in Stars," in which he presented two processes as being the main sources of stars' energy (Bethe, 1939). In 1967, Bethe received the Nobel Prize in Physics for his discovery. The first process is the proton-proton chain reaction (see Box 1), which is the main source of energy for stars with the same or smaller mass than that of the Sun. However, the carbon-nitrogen-oxygen (CNO) cycle (see Box 2), which was also considered by von Weizsacker (1938), is the one that dominates in more massive stars.

It is interesting to note that the initial goal was to explain the source of energy of the stars, but these studies also showed how some light chemical elements could have been generated. The studies of Bethe did not address the creation of heavy nuclei. This was studied later by Hoyle (1946, 1954). He showed that stars with advanced fusion stages were able to synthesize elements in the mass range from carbon and iron. The works of Hoyle are considered fundamental for the field of stellar nucleosynthesis (Clayton, 1968, 2008).

BOX 1 PROTON-PROTON CHAIN

The fusion of hydrogen occurs primarily following a chain of reactions called the proton-proton chain (Wallerstein et al., 1997):

$$4^1H \rightarrow 2^2H + 2e^+ + 2\nu_e$$
$$2^1H + 2^2H \rightarrow 2^3He + 2\gamma \quad (2)$$
$$2^3He \rightarrow ^4He + 2^1H$$

The overall reaction corresponds to the following equation:

$$4^1H \rightarrow ^4He + 2e^+ + 2\nu_e + 2\gamma \quad (3)$$

As illustrated in Fig. 3, four nuclei of hydrogen (i.e., protons) collide in pairs of two. Each collision results in a nucleus of deuterium, a positron, and a neutrino. The positrons collide with electrons and become annihilated, emitting gamma rays, whereas each nucleus of deuterium collides with a nucleus of hydrogen (proton) generating a nucleus of ^{3}He and emitting energy. In the last stage of the cycle, the two nuclei of ^{3}He are fused forming a nucleus of helium (^{4}He) and two nuclei of hydrogen. In stars of the mass our Sun or smaller, the proton-proton chain is the dominating reaction. In the core of the Sun, the proton-proton chain occurs approximately 9.2×10^{37} times per second, converting 3.7×10^{38} protons into helium nuclei (Phillips, 1995).

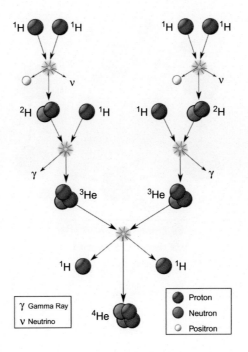

FIG. 3

The proton-proton chain reaction.

From: Wikipedia.

BOX 2 CNO CYCLE

The carbon-nitrogen-oxygen (CNO) cycle is the other set of fusion reactions that convert hydrogen into helium in the stars. Unlike the proton-proton chain reaction, CNO is a catalytic cycle. In stars with mass larger than 1.3 solar masses, the CNO cycle is the main source of energy according to theoretical models (Salaris and Cassisi, 2005). As illustrated in Fig. 4, four protons fuse, using isotopes of carbon, nitrogen, and oxygen as catalysts, producing one alpha particle, two electron neutrinos and two positrons. The same result is obtained in the CNO cycles, despite the different paths and catalysts involved.

3.3 STELLAR NUCLEOSYNTHESIS

In his Nobel Prize Lecture, Hans Bethe said "Stars have a life cycle much like animals. They are born, they grow, they go through a definite internal development, and finally die, to give back the material of which they are made so that new stars may live" (Bethe, 1968). Stars exist in the balance between two forces. On one hand, the star's gravity attempts to compress the stellar material into the smallest possible sphere, and on the other hand, enormous pressure and heat are produced by nuclear reactions at the center of the star pushing all the material outward. The outcome of this balance depends largely on the star's total mass. The stars are traditionally divided into three categories (Rauscher and Patkós, 2011):

1. stars less massive than the Sun;
2. stars with mass larger than the Sun and smaller than approximately eight solar masses; and
3. stars more massive than eight solar masses.

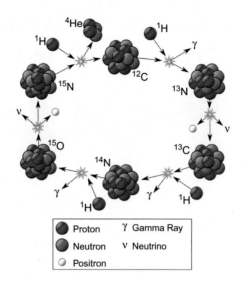

FIG. 4

The CNO cycle reaction.

From: Wikipedia.

The dominant reaction within small stars (in the first category) is the conversion of hydrogen into helium, whereas stars in the second mass category have further reactions that convert helium to carbon and oxygen (see Box 3). Only very massive stars in the third category support chain reactions that produce heavy elements up to the mass of iron.

BOX 3 NUCLEOSYNTHESIS OF CARBON AND OXYGEN

When the fusion of protons into helium continues until the star has exhausted its hydrogen, the temperature in its core rises to about a few times 10^8 K, allowing the fusion of helium into heavier nuclei. In the first reaction two nuclei of helium, ^{4}He, fuse with each other, creating the nucleus of beryllium, ^{8}Be. However, the ^{8}Be nucleus has an extremely short mean life of just 10^{16} s, before it decays back again to two ^{4}He nuclei. The rate of production equals the rate of destruction of ^{8}Be nucleus:

$$^4\text{He} + {}^4\text{He} \leftrightarrow {}^8\text{Be} \tag{4}$$

Nevertheless, the ^{8}Be can capture another ^{4}He nucleus producing the ^{12}C nucleus by the reaction:

$$^8\text{Be} + {}^4\text{He} \rightarrow {}^{12}\text{C} + \gamma \tag{5}$$

The reactions in Eqs. (4), (5) are called the triple-alpha reaction, because three ^{4}He nuclei or alpha particles are necessary for the creation of ^{12}C. This reaction can only create carbon in appreciable amounts because of the existence of a resonance in ^{12}C at the relevant energy for helium burning. Through this resonance the reaction in Eq. (5) is enhanced by many orders of magnitude.

The production of oxygen nuclei ^{16}O is the result of a capture of another ^{4}He nucleus by the carbon nuclei created in helium burning:

$$^{12}\text{C} + {}^4\text{He} \rightarrow {}^{16}\text{O} + \gamma \tag{6}$$

About half of the carbon nuclei produced are converted into oxygen.

It is interesting to note that the elements ^{12}C and ^{16}C are extremely fine-tuned with respect to the nuclear force. In the case of the strength of this force were just 0.5% different from their current values, the average abundance of carbon or oxygen in the universe would be more than two orders of magnitude smaller. That would make life based on carbon much more difficult to occur (Oberhummer et al., 2000; Schlattl et al., 2004).

3.4 WATER MOLECULE

Once a star like the Sun has exhausted its nuclear fuel, its core collapses and the outer layers are expelled as a planetary nebula, while the massive stars (more than eight solar masses) can explode in a supernova as their inert iron cores collapse. At these stages huge quantities of new nuclei are rapidly synthesized in the nuclear reactions triggered by the flood of neutrons. Most of the elements heavier than the iron group are generated either by nucleosynthesis, or by radioactive decay of unstable isotopes that were produced. This material ejected at the end of life of such stars resulted in huge interstellar clouds of gas and dust. In general, the gas is made of about 90% hydrogen, 9% helium, and 1% heavier atoms, while the dust is composed of silicates, carbon, iron, water ice, methane, ammonia, and some organic molecules (Dalgarno, 2006).

Therefore, the first water molecules of the universe might have emerged in interstellar clouds produced at the end of life of the first generation of massive stars. Interstellar clouds are abundant in our galaxy, and it is generally considered that all stars and planets have been formed from them.

4 DISTRIBUTION OF WATER IN THE UNIVERSE
4.1 WATER IN GALAXIES

Water vapor in galaxies is best detected by observing maser emissions. Masers (microwave amplification by stimulated emission of radiation) are similar to lasers, only emitting microwave radiation instead of visible light. Water molecules can absorb energy available around them in high mass star-forming regions (HMSR) or near dying stars, and re-emit it as microwave radiation. Several water masers were found in our Milky Way galaxy (Walsh et al., 2008). Mochizuki et al. (2009) studied water masers from young stellar objects (YSOs) in the outer regions of the galaxy.

Water masers were also found in nearby galaxies (Darling et al., 2008, 2016). Braatz and Gugliucci (2008) reported eight new sources of water maser emission in surveys conducted in nuclei of a hundred of galaxies. van der Tak (2015) reviewed the presence of water in nearby galactic nuclei and galactic interstellar clouds and concluded that the emission of radiation is necessary to detect water in those sites.

4.2 WATER IN STARS AND INTERSTELLAR SPACE

The space between stars, the interstellar medium (ISM), is permeated with dust, gas in atomic, molecular and ionic forms, and cosmic rays. Stars are born out of dense regions within molecular clouds in the interstellar space and, when they die, the interstellar medium is enriched with elements heavier than helium (see Section 3) (Hanslmeier, 2010). The Orion nebula is an example of star-forming regions (Fig. 5).

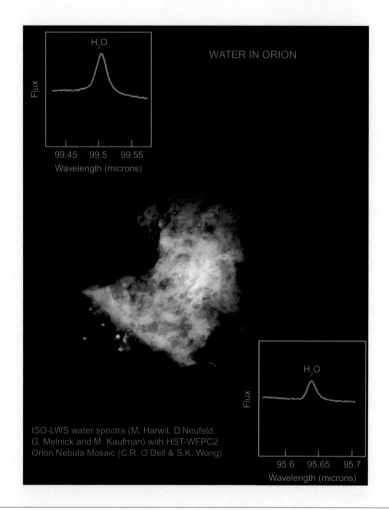

FIG. 5

Water signatures in M42, the Orion nebula. Mosaic picture made from more than 40 individual Hubble Space Telescope (HST) images.

From: ESA/NASA.

Water ice is abundant in the interior of molecular clouds (Allamandola and Sandford, 1988). Cheung et al. (1969) were the first to detect water molecules in the interstellar medium. Elitzur et al. (1989) proposed a comprehensive model of water masers in star-forming regions. Felli et al. (2007) monitored a sample of 43 masers within star-forming regions for 20 years and created a database of their variability. More than 1010 galactic water maser sources have been listed in the Arcetri catalogue (Brand et al., 1994; Valdettaro et al., 2001) and there is a distinction between water masers associated with star forming regions and late-type stars. Furuya et al. (2003) investigated water masers in low-mass YSOs.

Mira was the first star, where water was detected in its spectrum, in 1963 (Kuiper, 1963). Russell (1934) had already predicted the presence of water in the atmosphere of late type stars.[2] He showed theoretically that water should be the most abundant molecule beside atomic or molecular hydrogen in stars of approximately 2800 K.

Late type stars generally show strong mass loss and in many cases they form a circumstellar shell. Maercker et al. (2008) reported high water abundance in the circumstellar envelope of the star W Hya (an M-dwarf) and investigated water in the envelopes of six other M-type stars. They concluded that high amounts of water found in the majority of these sources may be explained by some kind of internal chemical processes. Updates on their research were reviewed by Maercker et al. (2009, 2016)

Jones et al. (2002) analyzed the spectra of a range of M stars and concluded that their observations match well with previous ground-based observations.

Tsuji (2000) reported the presence of water in the spectrum of the M2 supergiant star μ Cephei. Winnberg et al. (2008) investigated the variability of water masers in circumstellar shells of late-type stars, using RX Bootis and SV Pegasi as representatives of semiregular variable stars (SRVs).

4.3 WATER IN PLANETARY DISKS

Most emerging stars (protostars) have a protoplanetary disk that forms from a molecular cloud. Water was present in the molecular cloud that gave birth to our solar system. The presence of water played an important role as the cloud settled to form the protoplanetary disk from which the planets were formed (Mottl et al., 2007).

T Tauri are young stars that are often immersed in large molecular clouds and have accretion disks around them. Shiba et al. (1993) conducted a survey of 52 T Tauri stars and found water vapor in the disk of 17 of them. They measured the temperature of the water vapor, which appeared to be 2000 K. It is expected that the temperatures vary throughout the disk: water is likely solid in the outer regions of the disk and in gas phase in the inner regions where the temperature is higher than the evaporation point ($\sim$150 K).

The habitable zone can be predicted as the distance range from the central star, where water can be found in liquid phase on a planet surface. In the solar system, it has been calculated by Kasting et al. (1993) to be from 0.95 to 1.15 AU. Other estimates were reported by Rasool and de Bergh (1970), Hart (1979), Fogg (1992), Spiegel et al. (2010), Abe et al. (2011), Kopparapu et al. (2013), and Way et al. (2014). The range of all these estimates goes from 0.75 to 3.0 AU.

The snowline is the distance from the protostar where water (and other volatiles) condensates into ice. The amount of water on the surface of terrestrial planets in the habitable zone depends on the snowline location, which is determined by the temperature profile of the disk, properties of the star, mass accretion rate, and size distribution of dust grains (Mulders et al., 2015). Mulders et al. calculated the snowline location in disks around different stars using estimates of mass accretion rates as a function of stellar mass. They used N-body simulations to predict the amount of water on the surface of terrestrial planets within the habitable zone. In addition, they determined that the variation of the snowline locations strongly affects the range of the water availability on terrestrial planets. They showed that a significant fraction of terrestrial planets in the habitable zone around Sun-like stars remained dry

---

[2]Stars cooler than the Sun (less than 5200 K), with a yellow-orange-red color variation.

(assuming ISM-like dust sizes) and no water was predicted on planets within the habitable zones of low-mass M stars (<0.5 M$^\circledR$), where M$^\circledR$ is solar mass. When considering larger grains of dust, the snowline got closer to the star and that enabled water to be delivered to the habitable zone of a significant fraction of M stars and all FGK stars.

A protoplanetary disk has its temperature and density profiles defined by the mass fraction of micrometer-sized dust grains and on their chemical composition (Bitsch and Johansen, 2016). The larger is the abundance of micrometer-sized grains, the higher is the overall temperature of the disk, and the further away from the star is the snowline. If the dust abundance is kept the same, an increase in the water fraction inside the disk may lower the temperature in the inner regions and increase the temperature in the outer regions of the disk. Disks with a smaller water fraction have the opposite effect. Bitsch and Johansen (2016) studied the formation and migration of planets exploring the dust composition and its abundance in the disk. Their results imply that hot and warm super-Earths may contain a significant fraction of water, if they have formed beyond the snowline and migrated inwards.

Water vapor was found in circumstellar disks such as AA Tauri (Hanslmeier, 2010), TW Hydrae (Salinas et al., 2016), and APM 08279+5255 (Lis et al., 2011). Eisner (2007) reported water within 1 AU of the young star MWC 480, which is believed to have resulted from the sublimation of inwards migrating icy bodies, that provided water for potential terrestrial planets. Salyk et al. (2015) reported the presence of water vapor in the protoplanetary disk around DoAr 44.

4.4 WATER IN EXTRASOLAR PLANETS

By 2016, more than 3300 extrasolar planets have been confirmed, with more than 570 multiplanet systems reported (NASA, 2015). Detecting atmospheres on extrasolar planets is a challenging task.

The planet HD209458b was detected in 1999 and was the first extrasolar planet confirmed by the transit method. It is a giant planet with mass of approximately 63% the mass of Jupiter. It is 100 times closer to the central star than Jupiter. With this proximity to the star, it is assumed that the planet continuously looses volatiles, and the outflow is estimated to be 10^4 tons/s. Water is among these volatiles as revealed by the data from Hubble Space telescope (Rauer et al., 2004).

Planets with masses between 1 and 10 Earth masses are known as super-Earths. The habitability of super-Earth planets was discussed, for example, by Kaltenegger and Kasting (2008). It is likely that super-Earths have a wide range of atmospheres types. Miller-Ricci et al. (2009) argued that some of them may have hydrogen-rich atmospheres.

Dominguez (2016) analyzed the abundance of water and its dependency on stellar metallicity in extrasolar planetary systems. They found that the ratio of H_2O/SiO_2 produced in a molecular cloud of solar metallicity can account for the ratio of these compounds on Earth today, supporting the "wet" hypothesis that implies that Earth could have obtained enough water locally during its formation (Stimpfl et al., 2004). Bialy et al. (2015) studied the first step of H_2O formation in molecular clouds with extremely low metallicity, and showed that these clouds could have high abundances of water vapor. Some of this water may have contributed to forming planets if the cloud collapses into a protoplanetary disk.

Ehrenreich et al. (2007) searched for water in the transit exoplanet HD189733b by using the Spitzer telescope and showed that the observational capabilities in that moment were insufficient for detecting water vapor. Water vapor was confirmed in the atmospheres of extrasolar planets HD 189733 b

(Barman, 2008; McCullough et al., 2014), HD 209458 b (Beaulieu et al., 2010), Tau Boötis b (Lockwood et al., 2014), HAT-P-11b (Fraine et al., 2014; Hanslmeier, 2010), XO-1b, WASP-12b, WASP-17b, and WASP-19b (Mandell et al., 2013).

4.5 WATER IN THE SOLAR SYSTEM

Water is very abundant in the solar system. It is present even in the Sun, as confirmed, for example, by Wallace et al. (1995).

The solar system can be divided in two distinct group of planets by their position relative to the snowline: in the inner solar system, the volatiles are in a gaseous form and planets are relatively dry, small, and rocky—the terrestrial planets, whereas in the outer solar system, the volatiles are in a solid form and planets are big, gaseous with a solid rocky-ice inner core—the giant planets.

4.5.1 Water in the outer solar system

Water is an important constituent of the four giant planets: Jupiter, Saturn, Uranus, and Neptune (Stevenson and Fishbein, 1981). All these planets have similar structure, with a rocky-ice core and a huge gaseous layer consisting mainly of hydrogen and helium.

The abundance of water in Jupiter's atmosphere was studied by Roos-Serote et al. (2004). They found that the O/H ratio in the atmosphere of Jupiter was comparable with the one in the sun. Water in Jupiter was also reported by Bjoraker et al. (1986). (Hueso and Sánchez-Lavega, 2004) studied water storms in the atmosphere of Jupiter and concluded that they may develop velocities of up to 150 m/s.

Jupiter has four large satellites, known as the Galilean satellites: Io, Europa, Ganymede, and Callisto. Water keeps flowing away from Io (Pilcher, 1979), which may be explained by thermal escape. Kumar and Hunten (1982) investigated the atmospheres of Io and other Jupiter satellites and found that Europa, Ganymede, and Callisto may have oxygen atmospheres resulting from photolysis of water vapor. Moreover, these three satellites may have internal oceans more than a hundred kilometers thick (Spohn and Schubert, 2003). Leitner et al. (2014) developed a model for the analysis of oceans on Europa and Ganymede, and compared the results with the measured composition of brines on the surface of Europa. Vance et al. (2014) analyzed the influence of salinity on the internal structure of Ganymede and predicted that water ice may be present in the aqueous magnesium sulfate. They concluded that the stability of ice under high-pressure implies water-rock contact.

Water in the deep atmosphere of Saturn was studied by Visscher and Fegley (2005), who discussed chemical constraints for the abundance of water in the planet. Saturn has 60 confirmed moons. In 1997, water was detected in the atmosphere of Titan, the largest satellite of Saturn. The observed water abundance appeared to be four times lower than that in comets, suggesting that Titan's atmosphere was formed by outgassing from the interior rather than having a cometary origin (Coustenis et al., 1998). More recent data are in good agreement with these findings (Nixon et al., 2006; Bjoraker et al., 2008). Raulin (2008) describes Titan as "another world, with an active prebiotic-like chemistry, but in the absence of permanent liquid water on the surface: a natural laboratory for prebiotic-like chemistry." Dunaeva et al. (2013) built models of Titan's possible internal structure and predicted that Titan consists of the rock-iron core, rock-ice mantle, and outer water-ice shell.

There is a water influx from the Saturnian rings that is caused by its satellite Enceladus (Mueller-Wodarg et al., 2006); this has been discussed earlier by Connerney and Waite (1984). This sixth largest moon of Saturn is mostly covered by clean fresh ice, being one of the most reflective bodies of the solar system. In 2004, Cassini detected water vapor and complex hydrocarbons emerging from the geologically active south-polar region of Enceladus (Spencer et al., 2006). Tobie et al. (2008) showed that its particular location at the south pole and the magnitude of dissipation rate can only be explained by the models that assume a liquid water layer at a certain depth. Ingersoll and Pankine (2010) affirmed that the existence of liquid water on Enceladus depends on the efficiency of subsurface heat transfer. Iess et al. (2014) studied the interior structure of Enceladus and its gravity field; their results suggest that the body deviates mildly from the hydrostatic equilibrium.

Atreya et al. (2006) assessed the existence of an ocean of water-ammonia on Neptune and Uranus. They argued that the tropospheric cloud structure and the existence of a magnetic field must be maintained by a water-ammonia ionic ocean creating a dynamo action.

The five main satellites of Uranus, Miranda, Ariel, Umbriel, Titania, and Oberon, have weaker bands of water ice in their infrared spectra than those in the spectra of Saturn's icy moons and Galilean satellites. The difference may be explained by the presence of other ices (e.g., NH_3 and CH_4) besides water on their surfaces (Encrenaz, 2007).

So far, water has evaded detection on the dwarf planet Pluto based on Earth-bound observatories. Cook et al. (2015) analyzed all data on Pluto from the Linear Etalon Imaging Spectral Array (LESIA: a component of the New Horizons spacecraft) searching for the presence of water. Brown and Calvin (2000) presented evidences of water ice on Charon, Pluto's Satellite.

4.5.2 Water in small bodies

In 2006, the International Astronomical Union defined the term "small solar system body" (SSSB) as an object in the solar system that is not sufficiently massive to be a planet or a dwarf planet, and it is not a satellite. SSSBs are generally located in the main asteroid belt between Mars and Jupiter, in the Kuiper belt outside the orbit of Neptune, and in the Oort cloud extended as far as 50,000 AU from the Sun. Comets and asteroids consist mainly of pristine material and are remnants from the formation of the solar system about 4.6 billion years ago.

Comets

Comets are icy SSSBs that start a process of outgassing when approach the inner solar system. Comets expel vaporized volatile materials, carrying dust away with them. They may have been an important source of water on terrestrial planets, as well as on satellites, although some estimates limit their contribution up to 10% for Earth's water (Izidoro et al., 2013). Water is the main component of interstellar and cometary ices (Allamandola and Sandford, 1988).

Water in comets was first detected in 1970 from H and OH observations in comet Halley (Mumma et al., 1986; Combes et al., 1988). The comet Hale-Bopp (C/1995 O1) had its spectrum analyzed by Davies et al. (1997), who found that "some of the absorption features can be matched by an intimate mixture of water ice and a low-albedo material such as carbon on the nucleus." Cosmovici et al. (1998) detected water on comet Hyakutake (C/1996 B2). Bockelée-Morvan et al. (2009) used the Spitzer Space Telescope to detect water on the comets 71P/Clark and C/2004 B1 (Linear). Schulz et al. (2006) detected water ice grains on Comet 9P/Tempel 1 by analyzing the results of the DEEP IMPACT mission. de Bergh (2004) presented general remarks about water ice and organics in the Kuiper belt objects.

Hsieh and Jewitt (2006) discovered comets in the main asteroid belt, a new class of objects in the solar system. The activity of these comets is consistent with dust ejection driven by water-ice sublimation.

Asteroids

Asteroids and comets were previously thought to be of different origin: it was assumed that asteroids formed inside the orbit of Jupiter and comets originated from the outer solar system. However, the discovery of main-belt comets and recent findings like the returned sample of comet 81P/Wild 2 (Ishii et al., 2008) have blurred the distinction between comets and asteroids.

Every year, thousands of new asteroids are found and several thousands of asteroids have been studied. The first evidence of water in an asteroid was found in Ceres, now classified as a dwarf planet. Lebofsky (1978) estimated that Ceres's surface may contain 10%–15% water of hydration.[3] Küppers et al. (2014) indicated that at least 10^{26} water molecules per second are being evaporated from the dwarf planet, and this phenomenon could be due to "comet-like sublimation or cryo-volcanism, in which volcanoes erupt volatiles instead of molten rocks."

Fanale and Salvail (1989) analyzed the spectral signature of water in asteroids and confirmed that 66% of the C-class asteroids in the investigated sample have hydrated silicate surfaces. Although it was believed that D-type asteroids do not have water (Barucci et al., 1996), Kanno et al. (2003) suggested that these asteroids could contain water ice or hydrated minerals.

Yang and Jewitt (2007) investigated spectral signatures of water ice on Trojan asteroids[4] and their analysis showed no signs of water. Treiman et al. (2004) analyzed the meteorite Serra de MagÃ©, an eucrite believed to come from the asteroid 4 Vesta, and inferred that polar ice deposits in Vesta and similar asteroids are possible remainders from comet impacts, similar to water ice deposits on the Moon and Mercury. Campins et al. (2009) used IR observations to confirm water ice on the surface of asteroid 24 Themis.

Meteorites

A meteoroid is a small rocky or metallic body moving around the Sun or in the outer space. They are significantly smaller than asteroids or comets. Most meteoroids are fragments from asteroids or comets, other originated from debris ejected from impacts on bodies such as Mars or the Moon. A meteoroid that reaches the surface of the Earth without being completely vaporized is called meteorite.

Ashworth and Hutchison (1975) analyzed hydrous alteration products of olivine (a magnesium iron silicate) in an ordinary chondrite and an achondrite—two classes of meteorites in which hydrous minerals are rare. Their observations suggest that both meteorites have unusual volatile constituents, and they argue that the Nakhla achondrite contains water of extraterrestrial origin, and this may also be the case for the Weston chondrite.

The Shergotty-Nakhla-Chassigny meteorites, believed to be of martian origin, contain 0.04%–0.4% water by weight. Karlsson et al. (1991) used oxygen isotopic analysis to resolve whether this water was terrestrial or extraterrestrial. The results revealed that some of the water was extraterrestrial.

Carbonaceous chondrites are a class of chondritic meteorites that includes the most primitive known meteorites. They contain high percentages of water (3%–22%) (Norton, 2002) and organic compounds. Water and D/H ratios in the Chainpur (an ordinary chondrite) and Orgueil (a carbonaceous chondrite) meteorites were measured by Robert et al. (1978). Orgueil is one of the most studied

---

[3]Hydration implies that water molecules are components of the crystal structure of a mineral. In this case a new mineral is created, a hydrate.

[4]These asteroids are located near the equilibrium points L4 and L5 in the Sun-Jupiter system.

meteorites, owing to its unique primitive composition and relatively large mass (Gounelle and Zolensky, 2014).

4.5.3 Water on Earth and other terrestrial planets

Mercury, Venus, Earth, Mars, and their satellites, have very different histories. Water is present in all these terrestrial planets, but in very distinct patterns: Mercury has ice at the poles; Earth has very abundant water, in lakes, oceans, underneath the surface, and in icy continents; Venus has vapor in the atmosphere; and Mars has ice at the poles, liquid water in salty flows, and vapor in the atmosphere.

Mercury

Mercury, the smallest and closest planet to the Sun, has a temperature of 450°C on the dayside and −180°C on the nightside. This extreme contrast is a consequence of the lack of substantial atmosphere. Butler et al. (2001) used a radar system to analyze Mercury surface and found that water could persist near the poles of Mercury inside of deep craters. Wood et al. (1992) structured a thermal model that predicted the temperatures on the surface of Mercury and argued that, despite the proximity to the sun, temperatures at the poles could be low enough to permit water ice, as long as the albedo was high. Water was confirmed on Mercury by MESSENGER in 2012 (Lawrence et al., 2013).

Venus

Venus is similar to Earth in its size and has a dry surface hidden by dense clouds. Contrary to Mercury, it has a dense atmosphere mainly consisting of the greenhouse gas CO_2, with a surface pressure of 90 times that of Earth's. The mean surface temperature is approximately 460°C, and thus finding ice or liquid water near the poles of Venus is unlikely. Water vapor is an important component of the atmosphere and contributes to the global greenhouse effect. Water is mainly found below the cloud base at approximately 47 km above the surface (Hanslmeier, 2010). Fedorova et al. (2008) measured vertical distributions and mixing ratios of H_2O and HDO in Venus' mesosphere. They reported an increase of deuterium closer to the surface indicating a lower escape rate of D atoms comparing to H atoms or (and maybe also) a lower photodissociation of HDO comparing to H_2O. Water loss of Venus was measured by Delva et al. (2009) as 10^{24} molecules per second.

Earth

Earth is the largest terrestrial planet, and the oceans comprise 2/3 of its surface. Between a very hot and a very cold planet lies the Earth, where the average surface temperature of 288 K and pressure of 1 bar create a favorable environment for life, and where water can be found as vapor, ice, and liquid, simultaneously.

Water is present in large quantities in the Earth's atmosphere, along with 77% of N_2, 22% of O_2, and 1% of other gases. An estimate of the amount of water inside Earth points to values ranging from 1 $O_\oplus$[5] to 50 $O_\oplus$, with ~10 $O_\oplus$ being the most likely value (Drake and Campins, 2006). The origin of water on Earth is one of the most intriguing and debated issues in astronomy. One way to tackle the question is to analyze the proportion between the heavy water (HDO or D_2O) and the light water (H_2O), also known as the D:H ratio, in Earth's water and in different potential sources. This ratio in the present day Earth's mantle and oceans (standard mean ocean water, SMOW) is about 6 times higher than in the gas of the proto-planetary disk. Other bodies in the solar system present a great variation in their water D/H ratio,

---

[5]$O_\oplus$ = mass of Earths oceans = 1.4×10^{24} g.

as shown in Fig. 6. Although the comparison of this ratio on Earth with those of various meteorite types suggests that the water on Earth was derived mainly from asteroids, the remnants of the protosolar nebula are still present in the Earth mantle, presumably signing the sequestration of nebula gas at an early stage of planetary growth (Marty, 2012). Marty has proposed that a small (≤10%) fraction of the mantle volatiles might have been derived from the protosolar nebula during an early stage of the proto-Earth growth. A small contribution, up to 10%, may have come from comets (Morbidelli et al., 2000). Izidoro et al. (2013) developed a model that considers all main possible sources of water and uses the D:H ratio to evaluate potential relative contributions from each source.

Regarding our satellite, Sridharan et al. (2010) reported evidence of water ice at high latitudes on the moon's surface.

Mars

Mars, the outmost terrestrial planet, has a mass of 0.1 $M_\oplus$ and surface temperatures in the range from $-100°C$ to $0°C$. Water ice is present in the two polar caps of the planet: it forms almost the entire north polar cap and the bottom layer of the south polar cap (Bibring et al., 2004; Christensen, 2006). Water vapor was detected in the atmosphere of Mars by Owen and Mason (1969), and later quantified using Viking observations (Fedorova et al., 2010) and data from the Curiosity rover Mahaffy et al. (2013).

Masson et al. (2001) presented geomorphologic evidence that the planet underwent a hydrologic cycle with liquid water on its surface in the past. Ojha et al. (2015) presented new evidence that salty

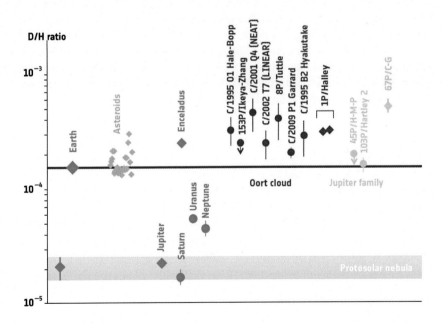

FIG. 6

The different values of the deuterium-to-hydrogen ratio (D/H ratio) in water, observed in various bodies of the solar system.

From: ESA, 2017. ESA's Rosetta Site 2015, Rosetta Fuels Debate on Origin of Earth's Oceans. http://www.esa.int/Our_Activities/Space_Science/Rosetta/Rosetta_fuels_debate_on_origin_of_Earth_s_oceans [Online] (Accessed 9 May 2017).

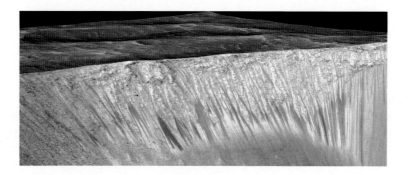

FIG. 7

Garni crater on Mars with recurring slope lineae (*dark narrow streaks*) in its walls. The *dark streaks* are believed to be the flow of salty liquid water on Mars.

From: NASA/JPL/University of Arizona.

liquid water flows sporadically on the present-day Mars despite the low atmospheric pressure (1% of that on Earth) and low temperature. Features known as recurring slope lineae (RSL), first identified in 2011, apparently resulted from the flows of salty liquid water on the surface of Mars. They look like dark streaks (Fig. 7) and appear seasonally. The water remains in a liquid phase at low temperatures due to the presence of salt, which also protects the water from boiling off in the thin atmosphere of Mars. Gough et al. (2016) analyzed the formation of liquid water via the deliquescence of calcium chloride at low temperatures and concluded that calcium chloride may help to form liquid water that could cause slope streaks on Mars. Pál and Kereszturi (2016) also predicted the appearance of microscopic amounts of liquid water on the hygroscopic mineral surfaces on Mars.

5 WATER AND LIFE

It is not easy to define life in the astrobiology context. First there is no consensus on the timing and mechanisms of the origin of life. Second, the distinction between living and nonliving systems can get vague. Are viruses forms of life? Is there any kind of artificial life?

Despite this uncertainty, the established features of living organisms include: an organized structure to perform specific functions, including cells as fundamental units of life; performing of anabolic and catabolic reactions to sustain life (metabolism); regulation of internal conditions to keep them stable even in unstable external environments (homeostasis); reaction to environmental stimuli or changes (response); growth; reproduction; and adaptation of organisms and populations to the environment (evolution) (Koshland, 2002).

Other aspects of life like being carbon-based or having its genetical information in the form of DNA, may also be considered. Even then, all these characteristics define a single model of life as we know it. It is an arduous task to envision different forms of life that may exist elsewhere in the universe. That is why, so far, we only seek for established aspects of extraterrestrial bodies that can provide conditions to support life as we know it.

Water is essential for life as we know it due to its unique properties. Because water can be found in all three phases on Earth, it allows an ample diversity of climates, habitats, and complex synergies between physical and chemical reactions (sch, 2004). It is an ideal solvent because of its polarity, and thus it can dissolve many different chemicals essential for metabolic reactions. Furthermore, the dipole character of water allows for hydrophobic organic molecules (e.g., lipids) to make cellular membranes. Alternate solvents may be possible, but there is a consensus that water is a prerequisite for life (Mottl et al., 2007).

However, the presence of water does not imply life. So far we know very little about the probability of the emergence and evolution of life in a cosmic body that contains water. This uncertainty stems from the fact we still do not know how life began on Earth. Was it brought from outside, or maybe Earth formation and evolution somehow made it possible for life to begin here? In this case, are those initial conditions replicable somewhere else in the universe? The necessary chemical reactions and environmental conditions that allowed the emergence of life on Earth are still debated (Line, 2002; Trevors and Abel, 2004; Pascal et al., 2006; Jortner, 2006; Benner and Kim, 2015). At the present date, it remains impossible to determine precisely all the circumstances that led to the emergence of first living cells on Earth (Pascal et al., 2006).

Nevertheless, one thing is certain: water must be in liquid state for all the living organisms we know. Although water is present everywhere in the universe (see Section 4), liquid water seems to be extraordinarily rare. The water we have found so far on the surface of other cosmic bodies is always in the solid or gaseous states, but not liquid. The only exception we know so far (besides Earth itself) is the potential presence of small amounts of surface water on Mars. Direct measurement of temperature and pressure that could favor liquid water on remote planets and satellites is generally not possible today. Therefore, we are still far from mapping where life could emerge and evolve outside our planet (Encrenaz, 2007).

In the last two decades, the astrochemical and astrophysical conditions for the emergence and evolution of life have been intensively debated (e.g., Ehrenfreund et al., 2002 and Chyba and Hand, 2005). Water on Earth, in the solar system and in the interstellar medium, and its strong association with life, has been reviewed by Mottl et al. (2007). Cottin et al. (2015) presented an interdisciplinary review of astrobiology, covering the most recent facts and hypotheses.

Even though we still have no evidence of extraterrestrial life, the resilience and presence of life in a wide variety of environments on Earth, even in very challenging niches, suggest that life may not be restricted to our planet. Since the discovery of extremophilic microbes (Rothschild and Mancinelli, 2001; Rampelotto, 2010), there has been less scepticism regarding the possibility of extraterrestrial life (Sagan, 1996; Chyba, 1997; Montmerle et al., 2006). Essentially, there are no chemical or physical barriers to extremophiles: where there is liquid water, there is life on Earth (Rothschild and Mancinelli, 2001). Therefore, all the recent discoveries, including liquid water on Mars (Ojha et al., 2015), the possibility of liquid oceans underneath the surface of Europa and Enceladus (Raulin, 2005; Tobie et al., 2008), and also the presence of organic molecules on Titan (Raulin, 2008), have fueled astrobiological interest in the solar system and beyond.

Finding extraterrestrial life seems to be only a question of time now. As we have seen in this chapter, water is a key aspect in this search. That is why understanding how water came to be and spread in the universe give us the first steps in this great endeavor of looking for life outside of our planet. Once we succeed, it will certainly expand our knowledge about what is to be a living organism in this vast universe. Once knowing we are not alone, human beings will never look at themselves at the same way again.

ACKNOWLEDGMENTS

This work was partially funded by CNPq and FAPESP (proc. 2016/24561-0). This support is gratefully acknowledged. The authors also would like to thank Tais Ribeiro for making Fig. 1.

REFERENCES

Abe, Y., Abe-Ouchi, A., Sleep, N.H., Zahnle, K.J., 2011. Habitable zone limits for dry planets. Astrobiology 11, 443–460. https://doi.org/10.1089/ast.2010.0545.

Allamandola, L.J., Sandford, S.A., 1988. Laboratory simulation of dust spectra. In: Bailey, M.E., Williams, D.A. (Eds.), Dust in the Universe, pp. 229–263.

Ashworth, J.R., Hutchison, R., 1975. Water in non-carbonaceous stony meteorites. Nature 256, 714. https://doi.org/10.1038/256714a0.

Atreya, S.K., Baines, K.H., Egeler, P.A., 2006. An ocean of water-ammonia on Neptune and Uranus: clues from tropospheric cloud structure. In: AAS/Division for Planetary Sciences Meeting Abstracts #38, Bulletin of the American Astronomical Society, vol. 38, p. 489.

Barman, T.S., 2008. On the presence of water and global circulation in the transiting planet HD 189733b. Astrophys. J. Lett. 676, L61. https://doi.org/10.1086/587056.

Barucci, M.A., Fulchignoni, M., Lazzarin, M., 1996. Water ice in primitive asteroids? Planet. Space Sci. 44, 1047–1049. https://doi.org/10.1016/0032-0633(96)00002-5.

Beaulieu, J.P., Kipping, D.M., Batista, V., Tinetti, G., Ribas, I., Carey, S., Noriega-Crespo, J.A., Griffith, C.A., Campanella, G., Dong, S., Tennyson, J., Barber, R.J., Deroo, P., Fossey, S.J., Liang, D., Swain, M.R., Yung, Y., Allard, N., 2010. Water in the atmosphere of HD 209458b from 3.6–8 μm IRAC photometric observations in primary transit. Mon. Not. R. Astron. Soc. 409, 963–974. https://doi.org/10.1111/j.1365-2966.2010.16516.x.

Benner, S.A., Kim, H.J., 2015. The case for a Martian origin for Earth life. In: Instruments, Methods, and Missions for Astrobiology XVII, Proceedings of the SPIE, vol. 9606, p. 96060C.

Bethe, H.A., 1939. Energy production in stars. Phys. Rev. 55, 434–456. https://doi.org/10.1103/PhysRev.55.434.

Bethe, H.A., 1968. Energy production in stars. Science 161, 541–547. https://doi.org/10.1126/science.161.3841.541.

Bialy, S., Sternberg, A., Loeb, A., 2015. Water formation during the epoch of first metal enrichment. Astrophys. J. Lett. 804, L29. https://doi.org/10.1088/2041-8205/804/2/L29.

Bibring, J.P., Langevin, Y., Poulet, F., Gendrin, A., Gondet, B., Berthé, M., Soufflot, A., Drossart, P., Combes, M., Bellucci, G., Moroz, V., Mangold, N., Schmitt, B., Team, O.M.E.G.A., Erard, S., Forni, O., Manaud, N., Poulleau, G., Encrenaz, T., Fouchet, T., Melchiorri, R., Altieri, F., Formisano, V., Bonello, G., Fonti, S., Capaccioni, F., Cerroni, P., Coradini, A., Kottsov, V., Ignatiev, N., Titov, D., Zasova, L., Pinet, P., Sotin, C., Hauber, E., Hoffman, H., Jaumann, R., Keller, U., Arvidson, R., Mustard, J., Duxbury, T., Forget, F., 2004. Perennial water ice identified in the south polar cap of Mars. Nature 428, 627–630. https://doi.org/10.1038/nature02461.

Bitsch, B., Johansen, A., 2016. Influence of the water content in protoplanetary discs on planet migration and formation. Astron. Astrophys. 590, A101. https://doi.org/10.1051/0004-6361/201527676.

Bjoraker, G.L., Larson, H.P., Kunde, V.G., 1986. The abundance and distribution of water vapor in Jupiter's atmosphere. Astrophys. J. 311, 1058–1072. https://doi.org/10.1086/164842.

Bjoraker, G., Achterberg, R., Anderson, C., Samuelson, R., Carlson, R., Jennings, D., 2008. Cassini/CIRS observations of water vapor in Titan's stratosphere. In: AAS/Division for Planetary Sciences Meeting Abstracts #40, Bulletin of the American Astronomical Society, vol. 40, p. 448.

Bockelée-Morvan, D., Woodward, C.E., Kelley, M.S., Wooden, D.H., 2009. Water in comets 71P/Clark and C/ 2004 B1 (linear) with Spitzer. Astrophys. J. 696, 1075–1083. https://doi.org/10.1088/0004-637X/696/2/1075.

Braatz, J.A., Gugliucci, N.E., 2008. The discovery of water maser emission from eight nearby galaxies. Astrophys. J. 678, 96–101. https://doi.org/10.1086/529538.

Brand, J., Cesaroni, R., Caselli, P., Catarzi, M., Codella, C., Comoretto, G., Curioni, G.P., Curioni, P., Di Franco, S., Felli, M., Giovanardi, C., Olmi, L., Palagi, F., Palla, F., Panella, D., Pareschi, G., Rossi, E., Speroni, N., Tofani, G., 1994. The Arcetri catalogue of H_2O maser sources update. Astron. Astrophys. Suppl. 103.

Brown, M.E., Calvin, W.M., 2000. Evidence for crystalline water and ammonia ices on Pluto's Satellite Charon. Science 287, 107–109. https://doi.org/10.1126/science.287.5450.107.

Butler, B.J., Slade, M.A., Muhleman, D.O., 2001. The nature of the Mercury polar radar features. In: Robinbson, M., Taylor, G.J. (Eds.), Workshop on Mercury: Space Environment, Surface, and Interior, vol. 1097, p. 9.

Campins, H., Hargrove, K., Howell, E.S., Kelley, M.S., Licandro, J., Mothé-Diniz, T., Ziffer, J., Fernandez, Y., Pinilla-Alonso, N., 2009. Confirming water ice on the surface of asteroid 24 Themis. In: AAS/Division for Planetary Sciences Meeting Abstracts, vol. 41, p. 32.05.

Cheung, A.C., Rank, D.M., Townes, C.H., Thornton, D.D., Welch, W.J., 1969. Detection of water in interstellar regions by its microwave radiation. Nature 221, 626–628. https://doi.org/10.1038/221626a0.

Christensen, P., 2006. Water at the poles and in permafrost regions of Mars. Elements 2 (3), 151–155. https://doi.org/10.2113/gselements.2.3.151.

Chyba, C.F., 1997. Life on other moons. Nature 385, 201. https://doi.org/10.1038/385201a0.

Chyba, C.F., Hand, K.P., 2005. Astrobiology: the study of the living universe. Ann. Rev. Astron. Astrophys. 43, 31–74. https://doi.org/10.1146/annurev.astro.43.051804.102202.

Clayton, D.D., 1968. Principles of Stellar Evolution and Nucleosynthesis. McGraw-Hill, New York.

Clayton, D.D., 2008. Fred Hoyle, primary nucleosynthesis and radioactivity. Nucleic Acids Res. 52, 360–363. https://doi.org/10.1016/j.newar.2008.05.007.

Combes, M., Crovisier, J., Encrenaz, T., Moroz, V.I., Bibring, J.P., 1988. The 2.5-12 micron spectrum of Comet Halley from the IKS-VEGA experiment. Icarus 76, 404–436. https://doi.org/10.1016/0019-1035(88)90013-9.

Connerney, J.E.P., Waite, J.H., 1984. New model of Saturn's ionosphere with an influx of water from the rings. Nature 312, 136–138. https://doi.org/10.1038/312136a0.

Cook, J.C., Cruikshank, D.P., Dalle Ore, C.M., Ennico, K., Grundy, W.M., Olkin, C.B., Protopapa, S., Stern, S.A., Weaver, H.A., Young, L.A., 2015. The search for Pluto water. In: AAS/Division for Planetary Sciences Meeting Abstracts, vol. 47, p. 200.02.

Cosmovici, C.B., Montebugnoli, S., Orfei, A., Pogrebenko, S., Cortiglioni, S., 1998. The puzzling detection of the 22 GHz water emission line in Comet Hyakutake at perihelion. Planet. Space Sci. 46, 467–470. https://doi.org/10.1016/S0032-0633(97)00230-4.

Cottin, H., Kotler, J.M., Bartik, K., Cleaves, H.J., Cockell, C.S., de Vera, J.P.P., Ehrenfreund, P., Leuko, S., Ten Kate, I.L., Martins, Z., Pascal, R., Quinn, R., Rettberg, P., Westall, F., 2015. Astrobiology and the possibility of life on Earth and elsewhere.... Space Sci. Rev. https://doi.org/10.1007/s11214-015-0196-1.

Coustenis, A., Salama, A., Lellouch, E., Encrenaz, T., Bjoraker, G.L., Samuelson, R.E., de Graauw, T., Feuchtgruber, H., Kessler, M.F., 1998. Evidence for water vapor in Titan's atmosphere from ISO/SWS data. Astron. Astrophys. 336, L85–L89.

Dalgarno, A., 2006. Interstellar chemistry special feature: the galactic cosmic ray ionization rate. Proc. Natl. Acad. Sci. U. S. A. 103, 12269–12273. https://doi.org/10.1073/pnas.0602117103.

Darling, J., Brogan, C., Johnson, K., 2008. Ubiquitous water Masers in nearby star-forming galaxies. Astrophys. J. Lett. 685, L39. https://doi.org/10.1086/592294.

Darling, J., Gerard, B., Amiri, N., Lawrence, K., 2016. Water masers in the Andromeda galaxy. I. A survey for water masers, ammonia, and hydrogen recombination lines. Astrophys. J. 826, 24. https://doi.org/10.3847/0004-637X/826/1/24.

Davies, J.K., Roush, T.L., Cruikshank, D.P., Bartholomew, M.J., Geballe, T.R., Owen, T., de Bergh, C., 1997. The detection of water ice in Comet Hale-Bopp. Icarus 127, 238–245. https://doi.org/10.1006/icar.1996.5673.

de Bergh, C., 2004. Kuiper belt: water and organics. In: Ehrenfreund, P., Irvine, W.M., Owen, T., Becker, L., Blank, J., Brucato, J.R., Colangeli, L., Derenne, S., Dutrey, A., Despois, D., Lazcano, A., Robert, F. (Eds.), Astrobiology: Future Perspectives, Astrophysics and Space Science Library, vol. 305, p. 205.

Delva, M., Volwerk, M., Mazelle, C., Chaufray, J.Y., Bertaux, J.L., Zhang, T.L., Vörös, Z., 2009. Hydrogen in the extended Venus exosphere. Geophys. Res. Lett. 36, L01203. https://doi.org/10.1029/2008GL036164.

Dominguez, G., 2016. On the abundance of water in extrasolar planetary systems as a function of stellar metallicity. In: American Astronomical Society Meeting Abstracts, vol. 228, p. 404.02.

Drake, M.J., Campins, H., 2006. Origin of water on the terrestrial planets. In: Daniela, L., Sylvio Ferraz, M., Angel, F.J. (Eds.), Asteroids, Comets, Meteors, IAU Symposium, vol. 229, pp. 381–394.

Dunaeva, A.N., Kronrod, V.A., Kuskov, O.L., 2013. Numerical models of Titan's interior with subsurface ocean. In: Lunar and Planetary Science Conference, vol. 44, p. 2454.

Ehrenfreund, P., Irvine, W., Becker, L., Blank, J., Brucato, J.R., Colangeli, L., Derenne, S., Despois, D., Dutrey, A., Fraaije, H., Lazcano, A., Owen, T., Robert, F., International Space Science Institute Issi-Team, 2002. Astrophysical and astrochemical insights into the origin of life. In: Lacoste, H. (Ed.), Exo-Astrobiology, ESA Special Publication, vol. 518, pp. 9–14.

Ehrenreich, D., Hébrard, G., Lecavelier des Etangs, A., Sing, D.K., Désert, J.M., Bouchy, F., Ferlet, R., Vidal-Madjar, A., 2007. A Spitzer search for water in the transiting exoplanet HD 189733b. Astrophys. J. Lett. 668, L179–L182. https://doi.org/10.1086/522792.

Eisner, J.A., 2007. Water vapour and hydrogen in the terrestrial-planet-forming region of a protoplanetary disk. Nature 447, 562–564. https://doi.org/10.1038/nature05867.

Elitzur, M., Hollenbach, D.J., McKee, C.F., 1989. H_2O masers in star-forming regions. Astrophys. J. 346, 983–990. https://doi.org/10.1086/168080.

Encrenaz, T., 2007. Searching for Water in the Universe, Springer Praxis Books. Springer, New York.

Fanale, F.P., Salvail, J.R., 1989. The water regime of asteroid (1) Ceres. Icarus 82, 97–110. https://doi.org/10.1016/0019-1035(89)90026-2.

Fedorova, A., Korablev, O., Vandaele, A.C., Bertaux, J.L., Belyaev, D., Mahieux, A., Neefs, E., Wilquet, W.V., Drummond, R., Montmessin, F., Villard, E., 2008. HDO and H_2O vertical distributions and isotopic ratio in the Venus mesosphere by Solar Occultation at Infrared spectrometer on board Venus Express. J. Geophys. Res. Planets 113, E00B22. https://doi.org/10.1029/2008JE003146.

Fedorova, A.A., Trokhimovsky, S., Korablev, O., Montmessin, F., 2010. Viking observation of water vapor on Mars: revision from up-to-date spectroscopy and atmospheric models. Icarus 208, 156–164. https://doi.org/10.1016/j.icarus.2010.01.018.

Felli, M., Brand, J., Cesaroni, R., Codella, C., Comoretto, G., Di Franco, S., Massi, F., Moscadelli, L., Nesti, R., Olmi, L., Palagi, F., Panella, D., Valdettaro, R., 2007. Water maser variability over 20 years in a large sample of star-forming regions: the complete database. Astron. Astrophys. 476, 373–664. https://doi.org/10.1051/0004-6361:20077804.

Fogg, M.J., 1992. An estimate of the prevalence of biocompatible and habitable planets. J. Br. Interplanet. Soc. 45, 3–12.

Fraine, J., Deming, D., Benneke, B., Knutson, H., Jordán, A., Espinoza, N., Madhusudhan, N., Wilkins, A., Todorov, K., 2014. Water vapour absorption in the clear atmosphere of a Neptune-sized exoplanet. Nature 513, 526–529. https://doi.org/10.1038/nature13785.

Furuya, R.S., Kitamura, Y., Wootten, A., Claussen, M.J., Kawabe, R., 2003. Water maser survey toward low-mass young stellar objects in the northern sky with the Nobeyama 45 meter telescope and the very large array. Astrophys. J. Suppl. 144, 71–134. https://doi.org/10.1086/342749.

Gough, R.V., Chevrier, V.F., Tolbert, M.A., 2016. Formation of liquid water at low temperatures via the deliquescence of calcium chloride: implications for Antarctica and Mars. Planet. Space Sci. 131, 79–87. https://doi.org/10.1016/j.pss.2016.07.006.

Gounelle, M., Zolensky, M.E., 2014. The Orgueil meteorite: 150 years of history. Meteorit. Planet. Sci. 49, 1769–1794. https://doi.org/10.1111/maps.12351.

Hansen, C.J., Kawaler, S.D., Trimble, V., 2004. Stellar Interiors: Physical Principles, Structure, and Evolution. Springer, New York.

Hanslmeier, A., 2010. Water in the Universe. Astrophysics and Space Science Library. Springer, Netherlands.

Hart, M.H., 1979. Habitable zones about main sequence stars. Icarus 37, 351–357. https://doi.org/10.1016/0019-1035(79)90141-6.

Hoyle, F., 1946. The synthesis of the elements from hydrogen. Mon. Not. R. Astron. Soc. 106, 343. https://doi.org/10.1093/mnras/106.5.343.

Hoyle, F., 1954. On nuclear reactions occurring in very hot stars. I. The Synthesis of elements from carbon to nickel. Astrophys. J. Suppl. 1, 121. https://doi.org/10.1086/190005.

Hsieh, H.H., Jewitt, D., 2006. A population of comets in the main asteroid belt. Science 312, 561–563. https://doi.org/10.1126/science.1125150.

Hueso, R., Sánchez-Lavega, A., 2004. A three-dimensional model of moist convection for the giant planets II: Saturn's water and ammonia moist convective storms. Icarus 172, 255–271. https://doi.org/10.1016/j.icarus.2004.06.010.

Iess, L., Stevenson, D.J., Parisi, M., Hemingway, D., Jacobson, R.A., Lunine, J.I., Nimmo, F., Armstrong, J.W., Asmar, S.W., Ducci, M., Tortora, P., 2014. The gravity field and interior structure of Enceladus. Science 344, 78–80. https://doi.org/10.1126/science.1250551.

Ingersoll, A.P., Pankine, A.A., 2010. Subsurface heat transfer on Enceladus: conditions under which melting occurs. Icarus 206, 594–607. https://doi.org/10.1016/j.icarus.2009.09.015.

Irion, R., 2002. Astrobiologists try to 'follow the water to life'. 296 (5568), 647–648. https://doi.org/10.1126/science.296.5568.647.

Ishii, H.A., Bradley, J.P., Dai, Z.R., Chi, M., Kearsley, A.T., Burchell, M.J., Browning, N.D., Molster, F., 2008. Comparison of comet 81P/Wild 2 dust with interplanetary dust from comets. Science 319, 447. https://doi.org/10.1126/science.1150683.

Izidoro, A., de Souza Torres, K., Winter, O.C., Haghighipour, N., 2013. A compound model for the origin of Earth's water. Astrophys. J. 767, 54. https://doi.org/10.1088/0004-637X/767/1/54.

Jones, H.R.A., Pavlenko, Y., Viti, S., Tennyson, J., 2002. Spectral analysis of water vapour in cool stars. Mon. Not. R. Astron. Soc. 330, 675–684. https://doi.org/10.1046/j.1365-8711.2002.05090.x.

Jortner, J., 2006. Conditions for the emergence of life on the early Earth: summary and reflections. Philos. Trans. R. Soc. Lond. B: Biol. Sci. 361, 1877.

Kaltenegger, L., Kasting, J., 2008. Session 22. Habitability of Super-Earths. Astrobiology 8, 394–396. https://doi.org/10.1089/ast.2008.1246.

Kanno, A., Hiroi, T., Nakamura, R., Abe, M., Ishiguro, M., Hasegawa, S., Miyasaka, S., Sekiguchi, T., Terada, H., Igarashi, G., 2003. The first detection of water absorption on a D type asteroid. Geophys. Res. Lett. 30, 1909. https://doi.org/10.1029/2003GL017907.

Karlsson, H.R., Clayton, R.N., Gibson, E.K., Mayeda, T.K., Socki, R.A., 1991. Extraterrestrial water of possible Martian origin in SNC meteorites: constraints from oxygen isotopes. In: 54th Annual Meeting of the Meteoritical Society, LPI Contributions, vol. 766.

Karlsson, T., Bromm, V., Bland-Hawthorn, J., 2013. Pregalactic metal enrichment: the chemical signatures of the first stars. Rev. Mod. Phys. 85, 809–848. https://doi.org/10.1103/RevModPhys.85.809.

Kasting, J.F., Whitmire, D.P., Reynolds, R.T., 1993. Habitable zones around main sequence stars. Icarus 101, 108–128. https://doi.org/10.1006/icar.1993.1010.

Kopparapu, R.K., Ramirez, R., Kasting, J.F., Eymet, V., Robinson, T.D., Mahadevan, S., Terrien, R.C., Domagal-Goldman, S., Meadows, V., Deshpande, R., 2013. Habitable zones around main-sequence stars: new estimates. Astrophys. J. 765, 131. https://doi.org/10.1088/0004-637X/765/2/131.

Koshland, D.E., 2002. The seven pillars of life. Science 295 (5563), 2215–2216. https://doi.org/10.1126/science.1068489.

Kuiper, G.P., 1963. Infrared spectra of stars and planets, II. Water vapor in Omicron Ceti. Commun. Lunar Planet. Lab. 1, 179–188.

Kumar, S., Hunten, D.M., 1982. The atmospheres of Io and other satellites. In: Morrison, D. (Ed.), Satellites of Jupiter, pp. 782–806.

Küppers, M., O'Rourke, L., Bockelée-Morvan, D., Zakharov, V., Lee, S., von Allmen, P., Carry, B., Teyssier, D., Marston, A., Müller, T., Crovisier, J., Barucci, M.A., Moreno, R., 2014. Localized sources of water vapour on the dwarf planet (1)Ceres. Nature 505, 525–527. https://doi.org/10.1038/nature12918.

Lawrence, D.J., Feldman, W.C., Goldsten, J.O., Maurice, S., Peplowski, P.N., Anderson, B.J., Bazell, D., McNutt, R.L., Nittler, L.R., Prettyman, T.H., Rodgers, D.J., Solomon, S.C., Weider, S.Z., 2013. Evidence for water ice near Mercury's North Pole from MESSENGER neutron spectrometer measurements. Science 339, 292. https://doi.org/10.1126/science.1229953.

Lebofsky, L.A., 1978. Asteroid 1 Ceres—evidence for water of hydration. Mon. Not. R. Astron. Soc. 182, 17P–21P. https://doi.org/10.1093/mnras/182.1.17P.

Leitner, M.A., Bothamy, N., Choukroun, M., Pappalardo, R.T., Vance, S., 2014. Ocean compositions on Europa and Ganymede. In: AGU Fall Meeting Abstracts.

Line, M.A., 2002. The enigma of the origin of life and its timing. Microbiology 148, 21–27.

Lis, D.C., Neufeld, D.A., Phillips, T.G., Gerin, M., Neri, R., 2011. Discovery of Water Vapor in the High-redshift Quasar APM 08279+5255 at $z = 3.91$. Astrophys. J. Lett. 738, L6. https://doi.org/10.1088/2041-8205/738/1/L6.

Lockwood, A.C., Johnson, J.A., Bender, C.F., Carr, J.S., Barman, T., Richert, A.J.W., Blake, G.A., 2014. Near-IR direct detection of water vapor in Tau Boötis b. Astrophys. J. Lett. 783, L29. https://doi.org/10.1088/2041-8205/783/2/L29.

Lodders, K., 2003. Abundances and condensation temperatures of the elements. Meteorit. Planet. Sci. Suppl. 38, .

Lodders, K., 2003. Solar system abundances and condensation temperatures of the elements. Astrophys. J. 591, 1220–1247. https://doi.org/10.1086/375492.

Lynden-Bell, R.M., Morris, S.C., Barrow, J.D., Finney, J.L., Harper, C., 2010. Water and Life: The Unique Properties of H2O. CRC Press, Boca Raton.

Maercker, M., Schöier, F.L., Olofsson, H., Bergman, P., Ramstedt, S., 2008. Circumstellar water vapour in M-type AGB stars: radiative transfer models, abundances, and predictions for HIFI. Astron. Astrophys. 479, 779–791. https://doi.org/10.1051/0004-6361:20078680.

Maercker, M., Schöier, F.L., Olofsson, H., Bergman, P., Frisk, U., Hjalmarson, Å., Justtanont, K., Kwok, S., Larsson, B., Olberg, M., Sandqvist, A., 2009. Circumstellar water vapour in M-type AGB stars: constraints from H{2}O(1{10}-1{01}) lines obtained with Odin. Astron. Astrophys. 494, 243–252. https://doi.org/10.1051/0004-6361:200810017.

Maercker, M., Danilovich, T., Olofsson, H., De Beck, E., Justtanont, K., Lombaert, R., Royer, P., 2016. A HIFI view on circumstellar H_2O in M-type AGB stars: radiative transfer, velocity profiles, and H_2O line cooling. Astron. Astrophys. 591, A44. https://doi.org/10.1051/0004-6361/201628310.

Mahaffy, P.R., Webster, C.R., Atreya, S.K., Franz, H., Wong, M., Conrad, P.G., Harpold, D., Jones, J.J., Leshin, L.A., Manning, H., Owen, T., Pepin, R.O., Squyres, S., Trainer, M., Team, M.S., 2013. Abundance and isotopic composition of gases in the martian atmosphere from the curiosity rover. Science 341, 263–266. https://doi.org/10.1126/science.1237966.

Mandell, A.M., Haynes, K., Sinukoff, E., Madhusudhan, N., Burrows, A., Deming, D., 2013. Exoplanet transit spectroscopy using WFC3: WASP-12 b, WASP-17 b, and WASP-19 b. Astrophys. J. 779, 128. https://doi.org/10.1088/0004-637X/779/2/128.

Marty, B., 2012. The origins and concentrations of water, carbon, nitrogen and noble gases on Earth. Earth Planet. Sci. Lett. 313, 56–66. https://doi.org/10.1016/j.epsl.2011.10.040.

Masson, P., Carr, M.H., Costard, F., Greeley, R., Hauber, E., Jaumann, R., 2001. Geomorphologic evidence for liquid water. Space Sci. Rev. 96, 333–364.

McCullough, P.R., Crouzet, N., Deming, D., Madhusudhan, N., 2014. Water vapor in the spectrum of the extrasolar planet HD 189733b. I. The transit. Astrophys. J. 791, 55. https://doi.org/10.1088/0004-637X/791/1/55.

Miller-Ricci, E., Seager, S., Sasselov, D., 2009. The atmospheric signatures of Super-Earths: how to distinguish between hydrogen-rich and hydrogen-poor atmospheres. Astrophys. J. 690, 1056–1067. https://doi.org/10.1088/0004-637X/690/2/1056.

Mochizuki, N., Hachisuka, K., Umemoto, T., 2009. Survey of outer galaxy molecular lines associated with water masers. In: Hagiwara, Y., Fomalont, E., Tsuboi, M., Yasuhiro, M. (Eds.), Approaching Micro-Arcsecond Resolution with VSOP-2: Astrophysics and Technologies, Astronomical Society of the Pacific Conference Series, vol. 402, p. 384.

Montmerle, T., Claeys, P., Gargaud, M., López-García, P., Martin, H., Pascal, R., Reisse, J., Selsis, F., 2006. From suns to life: a chronological approach to the history of life on Earth 9. Life on Earth.. And elsewhere? Earth Moon Planets 98, https://doi.org/10.1007/s11038-006-9093-7.

Morbidelli, A., Chambers, J., Lunine, J.I., Petit, J.M., Robert, F., Valsecchi, G.B., Cyr, K.E., 2000. Source regions and time scales for the delivery of water to Earth. Meteorit. Planet. Sci. 35, 1309–1320. https://doi.org/10.1111/j.1945-5100.2000.tb01518.x.

Mottl, M., Glazer, B., Kaiser, R., Meech, K., 2007. Water and astrobiology. Chem. Erde/Geochem. 67, 253–282. https://doi.org/10.1016/j.chemer.2007.09.002.

Mueller-Wodarg, I., Mendillo, M., Moore, L., 2006. Water on Saturn: global effects on ionospheric densities. In: AGU Fall Meeting Abstracts.

Mulders, G.D., Ciesla, F.J., Min, M., Pascucci, I., 2015. The snow line in viscous disks around low-mass stars: implications for water delivery to terrestrial planets in the habitable zone. Astrophys. J. 807, 9. https://doi.org/10.1088/0004-637X/807/1/9.

Mumma, M.J., Weaver, H.A., Larson, H.P., Williams, M., Davis, D.S., 1986. Detection of water vapor in Halley's comet. Science 232, 1523–1528. https://doi.org/10.1126/science.232.4757.1523.

NASA, 2015. NASA Exoplanet Archive. http://exoplanetarchive.ipac.caltech.edu/ [Online] (Accessed 28 September 2016).

Nixon, C.A., Jennings, D.E., de Kok, R., Coustenis, A., Flasar, F.M., 2006. Water in Titan's atmosphere from Cassini CIRS observations. In: AAS/Division for Planetary Sciences Meeting Abstracts #38, Bulletin of the American Astronomical Society, vol. 38, p. 529.

Norton, O.R., 2002. The Cambridge Encyclopedia of Meteorites. Cambridge University Press, Cambridge.

Oberhummer, H., Csótó, A., Schlattl, H., 2000. Stellar production rates of carbon and its abundance in the Universe. Science 289, 88–90. https://doi.org/10.1126/science.289.5476.88.

Ojha, L., Wilhelm, M.B., Murchie, S.L., McEwen, A.S., Wray, J.J., Hanley, J., Massé, M., Chojnacki, M., 2015. Spectral evidence for hydrated salts in seasonal brine flows on Mars. In: European Planetary Science Congress 2015, vol. 10. EPSC2015-838.

Owen, T., Mason, H.P., 1969. Mars: water vapor in its atmosphere. Science 165, 893–895. https://doi.org/10.1126/science.165.3896.893.

Pál, B., Kereszturi, Á., 2016. Possibility of microscopic liquid water formation at landing sites on Mars and their observational potential. ArXiv e-prints.

Pascal, R., Boiteau, L., Forterre, P., Gargaud, M., Lazcano, A., Lopez-Garcia, P., Maurel, M.C., Moreira, D., Pereto, J., Prieur, D., Reisse, J., 2006. Prebiotic chemistry biochemistry emergence of life (4.42 Ga). Earth Moon Planets 98, 153.

Phillips, K.J.H., 1995. Guide to the Sun. Cambridge University Press, Cambridge.

Piersanti, L., Straniero, O., Cristallo, S., 2007. A method to derive the absolute composition of the Sun, the solar system, and the stars. Astron. Astrophys. 462, 1051–1062. https://doi.org/10.1051/0004-6361:20054505.

Pilcher, C.B., 1979. The stability of water on Io. Icarus 37, 559–574. https://doi.org/10.1016/0019-1035(79)90014-9.

Rampelotto, P.H., 2010. Resistance of microorganisms to extreme environmental conditions and its contribution to astrobiology. Sustainability 2, 1602–1623. https://doi.org/10.3390/su2061602.

Rasool, S.I., de Bergh, C., 1970. The runaway greenhouse and the accumulation of CO_2 in the Venus atmosphere. Nature 226, 1037–1039. https://doi.org/10.1038/2261037a0.

Rauer, H., Collier-Cameron, A., Barnes, J., Harris, A.W., 2004. Search for signatures of an atmosphere of HD209458 b. In: Penny, A. (Ed.), Planetary Systems in the Universe, IAU Symposium, vol. 202, p. 109.

Raulin, F., 2005. Exo-astrobiological aspects of Europa and Titan: from observations to speculations. Space Sci. Rev. 116, 471–487. https://doi.org/10.1007/s11214-005-1967-x.

Raulin, F., 2008. Astrobiology and habitability of Titan. Space Sci. Rev. 135, 37–48. https://doi.org/10.1007/s11214-006-9133-7.

Rauscher, T., Patkós, A., 2011. Origin of the Chemical Elements. Springer, Netherlands.

Robert, F., Merlivat, L., Javoy, M., 1978. Water and deuterium content in the Chainpur and Orgueil meteorites. Meteoritics 13, 613.

Roos-Serote, M., Atreya, S.K., Wong, M.K., Drossart, P., 2004. On the water abundance in the atmosphere of Jupiter. Planet. Space Sci. 52, 397–414. https://doi.org/10.1016/j.pss.2003.06.007.

Rothschild, L.J., Mancinelli, R.L., 2001. Life in extreme environments. Nature 409, 1092–1101. https://doi.org/10.1038/35059215.

Russell, H.N., 1934. Molecules in the Sun and Stars. Astrophys. J. 79, 317. https://doi.org/10.1086/143539.

Sagan, C., 1996. Circumstellar habitable zones: an introduction. In: Doyle, L.R. (Ed.), Circumstellar Habitable Zones, p. 3.

Salaris, M., Cassisi, S., 2005. Evolution of Stars and Stellar Populations. Wiley-VCH, New Jersey.

Salinas, V.N., Hogerheijde, M.R., Bergin, E.A., Cleeves, L.I., Brinch, C., Blake, G.A., Lis, D.C., Melnick, G.J., Panić, O., Pearson, J.C., Kristensen, L., Yıldız, U.A., van Dishoeck, E.F., 2016. First detection of gas-phase ammonia in a planet-forming disk. NH_3, N_2H^+, and H_2O in the disk around TW Hydrae. Astron. Astrophys. 591, A122. https://doi.org/10.1051/0004-6361/201628172.

Salyk, C., Lacy, J.H., Richter, M.J., Zhang, K., Blake, G.A., Pontoppidan, K.M., 2015. Detection of water vapor in the terrestrial planet forming region of a transition disk. Astrophys. J. Lett. 810, L24. https://doi.org/10.1088/2041-8205/810/2/L24.

Schulze-Makuch, D., Irwin, L.N. (Eds.), 2004. Life in the Universe. Expectations and Constraints, vol. 3.

Schlattl, H., Heger, A., Oberhummer, H., Rauscher, T., Csótó, A., 2004. Sensitivity of the C and O production on the 3α rate. Astrophys. Space Sci. 291, 27–56. https://doi.org/10.1023/B:ASTR.0000029953.05806.47.

Schulz, R., Owens, A., Rodriguez-Pascual, P.M., Lumb, D., Erd, C., Stüwe, J.A., 2006. Detection of water ice grains after the DEEP IMPACT onto Comet 9P/Tempel 1. Astron. Astrophys. 448, L53–L56. https://doi.org/10.1051/0004-6361:200600003.

Shiba, H., Sato, S., Yamashita, T., Kobayashi, Y., Takami, H., 1993. Detection of water vapor in T Tauri stars. Astrophys. J. Suppl. 89, 299–319. https://doi.org/10.1086/191850.

Spencer, J.R., Pearl, J.C., Segura, M., Flasar, F.M., Mamoutkine, A., Romani, P., Buratti, B.J., Hendrix, A.R., Spilker, L.J., Lopes, R.M.C., 2006. Cassini encounters Enceladus: background and the discovery of a south polar hot spot. Science 311, 1401–1405. https://doi.org/10.1126/science.1121661.

Spiegel, D.S., Raymond, S.N., Dressing, C.D., Scharf, C.A., Mitchell, J.L., 2010. Generalized Milankovitch cycles and long-term climatic habitability. Astrophys. J. 721, 1308–1318. https://doi.org/10.1088/0004-637X/721/2/1308.

Spohn, T., Schubert, G., 2003. Oceans in the icy Galilean satellites of Jupiter? Icarus 161, 456–467. https://doi.org/10.1016/S0019-1035(02)00048-9.

Sridharan, R., Ahmed, S.M., Pratim Das, T., Sreelatha, P., Pradeepkumar, P., Naik, N., Supriya, G., 2010. 'Direct' evidence for water (H_2O) in the sunlit lunar ambience from CHACE on MIP of Chandrayaan I. Planet. Space Sci. 58, 947–950. https://doi.org/10.1016/j.pss.2010.02.013.

Stevenson, D.J., Fishbein, E., 1981. The behavior of water in the giant planets. In: Lunar and Planetary Science Conference, vol. 12, pp. 1040–1042.

Stimpfl, M., Lauretta, D.S., Drake, M.J., 2004. Adsorption as a mechanism to deliver water to the Earth. Meteorit. Planet. Sci. Suppl. 39.

Tobie, G., Čadek, O., Sotin, C., 2008. Solid tidal friction above a liquid water reservoir as the origin of the south pole hotspot on Enceladus. Icarus 196, 642–652. https://doi.org/10.1016/j.icarus.2008.03.008.

Treiman, A.H., Lanzirotti, A., Xirouchakis, D., 2004. Ancient water on asteroid 4 Vesta: evidence from a quartz veinlet in the Serra de Magé eucrite meteorite. Earth Planet. Sci. Lett. 219, 189–199. https://doi.org/10.1016/S0012-821X(04)00004-4.

Trevors, J.T., Abel, D.L., 2004. Chance and necessity do not explain the origin of life. Cell Biol. Int. 28, 729–739.

Tsuji, T., 2000. Water in emission in the infrared space observatory spectrum of the early M Supergiant Star μ Cephei. Astrophys. J. Lett. 540, L99–L102. https://doi.org/10.1086/312879.

Valdettaro, R., Palla, F., Brand, J., Cesaroni, R., Comoretto, G., Di Franco, S., Felli, M., Natale, E., Palagi, F., Panella, D., Tofani, G., 2001. The Arcetri catalog of H_2O maser sources: update 2000. Astron. Astrophys. 368, 845–865. httpss://doi.org/10.1051/0004-6361:20000526.https://arxiv.org/pdf/1511.09352.

van der Tak, F., 2015. Water in the interstellar media of galaxies. ArXiv e-prints https://arxiv.org/pdf/1511.09352.

Vance, S., Bouffard, M., Choukroun, M., Sotin, C., 2014. Ganymede's internal structure including thermodynamics of magnesium sulfate oceans in contact with ice. Planet. Space Sci. 96, 62–70. https://doi.org/10.1016/j.pss.2014.03.011.

Visscher, C., Fegley Jr., B., 2005. Chemical constraints on the water and total oxygen abundances in the deep atmosphere of Saturn. Astrophys. J. 623, 1221–1227. https://doi.org/10.1086/428493.

von Weizsacker, C.F., 1938. Uber Elementumwandlungen im Inneren der Sterne II. Phys. Z. 39, 633–646.

Wallace, L., Bernath, P., Livingston, W., Hinkle, K., Busler, J., Guo, B., Zhang, K., 1995. Water on the Sun. Science 268, 1155–1158. https://doi.org/10.1126/science.7761830.

Wallerstein, G., Iben, I., Parker, P., Boesgaard, A.M., Hale, G.M., Champagne, A.E., Barnes, C.A., Käppeler, F., Smith, V.V., Hoffman, R.D., Timmes, F.X., Sneden, C., Boyd, R.N., Meyer, B.S., Lambert, D.L., 1997. Synthesis of the elements in stars: forty years of progress. Rev. Mod. Phys. 69, 995–1084. https://doi.org/10.1103/RevModPhys.69.995.

Walsh, A.J., Lo, N., Burton, M.G., White, G.L., Purcell, C.R., Longmore, S.N., Phillips, C.J., Brooks, K.J., 2008. A pilot survey for the H_2O southern galactic plane survey. Publ. Astron. Soc. Aust. 25, 105–113. https://doi.org/10.1071/AS07053.

Way, M., Del Genio, A.D., Kelley, M., Aleinov, I.D., Clune, T., 2014. Exploring the inner edge of the habitable zone in the early Solar System. In: AGU Fall Meeting Abstracts.

Winnberg, A., Engels, D., Brand, J., Baldacci, L., Walmsley, C.M., 2008. Water vapour masers in long-period variable stars. I. RX Bootis and SV Pegasi. Astron. Astrophys. 482, 831–848. https://doi.org/10.1051/0004-6361:20078295.

Wood, S.E., Vasavada, A.R., Paige, D.A., 1992. Temperatures in the polar regions of Mercury: implications for water ice. In: AAS/Division for Planetary Sciences Meeting Abstracts #24, Bulletin of the American Astronomical Society, vol. 24, p. 957.

Yang, B., Jewitt, D., 2007. Spectroscopic search for water ice on Jovian Trojan asteroids. Astron. J. 134, 223–228. https://doi.org/10.1086/518368.

FURTHER READING

ESA, 2017. ESA's Rosetta Site 2015, Rosetta Fuels Debate on Origin of Earth's Oceans. http://www.esa.int/Our_Activities/Space:Science/Rosetta/Rosetta_fuels_debate_on_origin_of_Earth_s_oceans [Online].

THE COSMIC EVOLUTION OF BIOCHEMISTRY

Aditya Chopra, Charles H. Lineweaver
The Australian National University, Canberra, ACT, Australia

CHAPTER OUTLINE

Fundamental ingredients for the origin of life include specific chemical elements, water, rocky planets, and geochemistry. These ingredients were not present at the big bang $\sim$13.8 billion years ago. Nor were they present when the first stars formed $\sim$13.6 billion years ago. Here we review the element production and fractionation that led from the hot big bang, to the production of elements necessary for rocky planets (e.g., Fe, O, Si, Mg) and the production of elements necessary for life as we know it (e.g., C, O, N). Thus, we trace the potentially universal evolution from physics to chemistry to biochemistry and life. During the first $\sim$2 billion years of star formation, the abundance of rock-forming elements was probably not high enough to produce rocky planets that were both massive enough to hold an atmosphere able to maintain water at its surface and within the circumstellar habitable zone. This suggests that the earliest elemental bottleneck for the emergence of life was not the low abundance of the life forming elements (e.g., C, O, N), but rather the low abundance of Fe, Si, Mg and other metals that make up the rocky planets for life to live on.

After passing through this bottleneck $\sim$11 billion years ago, all the fundamental ingredients for life were present. There does not seem to be anything special about the Earth compared to other wet rocky planets. Thus, for 6 billion years preceding the formation of the Earth (from $\sim$11 to $\sim$5 billion years ago), wet rocky Earth-like planets in circumstellar habitable zones formed and evolved. Thus, it is plausible that some kind of life emerged many billions of years before the emergence of life on Earth.

Habitability of the Universe Before Earth, editors: Richard Gordon & Alexei Sharov, Volume 1 in the series:
Astrobiology: Exploring Life on Earth and Beyond, series editors: Pabulo Henrique Rampelotto,
Joseph Seckbach & Richard Gordon. ISSN 2468-6352. https://doi.org/10.1016/B978-0-12-811940-2.00004-6

1 BIG BANG TO PALE BLUE DOTS

Life as we know it is made of carbon, hydrogen, oxygen, and nitrogen (C, H, O, N). Rocky Earth-like planets are made of iron, oxygen, silicon, and magnesium (Fe, O, Si, Mg). Here we review the chronology and the processes that produced these fundamental ingredients for the origin of life. First, we discuss the relevant time scales. Then we discuss the ways in which the earliest stars increased the abundances of different elements and how the composition of rocky planets is critically dependent on the carbon-to-oxygen ratio. We review recent evidence for the ubiquity of planetary systems suggesting that Earth-like planets are common and have been common. We conclude with a summary of what the study of the origin of life on Earth can tell us about the origin of life elsewhere.

The Universe is 13.80 ± 0.02 billion years old (Planck, 2016). The Solar System formed 4.5673 ± 0.0002 billion years ago (Connelly et al., 2012). Thus, the universe existed 9.2 (=13.8–4.6) billion years before the Sun and Earth existed. Here, we argue that the prerequisites for the emergence of wet rocky planets and biochemistry were fulfilled in the first ~3 billion years of this ~9 billion year period, that is ~6 billion years before the existence of the Earth.

2 THE FIRST STARS: THE INCREASING METALLICITY OF POP III AND POP II STARS

The chemical context and the preconditions for the origin of life were set in the early universe (Kolb & Turner, 1994; Peacock, 1999). As the universe expanded and cooled, many things froze-out into metastable structures (Fig. 1). In the first few milliseconds after the big bang, quark soup cooled and froze into protons and neutrons. This produced six protons for every neutron due to the slightly lower mass of the proton. Ten minutes later, as a result of big bang nucleosynthesis, the protons and neutrons had frozen into the various light nuclei of hydrogen, helium, and lithium ($^1H^+$, $^2H^+$, $^3He^{++}$, $^4He^{++}$ and $^6Li^{+++}$ and $^7Li^{+++}$). As the universe cooled further, these nuclei and electrons froze into atoms; first He^{++} became He^+ (~30,000 years after the big bang), then He^+ became neutral He (~100,000 years after the big bang), then at 380,000 years after the big bang the dominant proton component (H^+) combined with electrons and became neutral hydrogen.

This period of recombination introduced a persistent transparency to the universe that now allows us to look back at the surface of last scattering—the photosphere of the universe—the source of the cosmic microwave background (Smoot et al., 1992). At recombination, when the universe became transparent, photon pressure could no longer prevent the collapse of baryonic over-densities into preexisting dark matter over-densities that had been able to start collapsing much earlier (~40,000 years after the big bang). The initially small baryonic over-densities collapsed slowly over the next few hundred million years until eventually they collapsed into the first stars (Loeb, 2010).

The exact timing of the formation of the first stars is debated. Some argue that the very first extremely rare stars could have begun shining only 10 million years after the big bang (Loeb, 2014). The cosmic microwave background gives us an estimate of when there were enough massive stars to re-ionize the neutral hydrogen in the universe. This happened between 500 and 600 million years after the big bang (Planck, 2016). Thus, the dominant population of first stars formed in the first few hundred million years. These first stars had negligible amounts of carbon, oxygen, or nitrogen for life

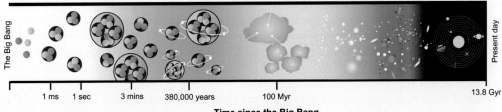

Time since the Big Bang

FIG. 1

Timeline of structure formation in the universe. As the universe expands and cools various structures freeze-out. Within 1 millisecond after the big bang, quarks (green and yellow balls) freeze into triplets called protons and neutrons. Approximately 3 minutes after the big bang, some of the neutrons and protons bound together to form helium nuclei. Approximately 380,000 years after the big bang, the dominant plasma of protons and electrons froze into neutral hydrogen atoms and began to collapse into dark matter haloes that had formed earlier. A few hundred million years later, the first high mass Pop III stars were formed. By ~2 billion years after the big bang, enough metals (e.g., oxygen, iron, silicon, magnesium) had been spread around the universe to allow wet rocky Earth-like planets to form. Our Solar System formed about 9 billion years after the big bang. In our ~12 billion year old galaxy, many billions of Earth-like planets formed before the Earth formed and life emerged on it ~4 billion years ago (Nutman et al., 2016).

Artwork by Aditya Chopra, modified from Domagal-Goldman, S.D., Wright, K.E., Adamala, K., Arina de la Rubia, L., Bond, J., Dartnell, L.R., Goldman, A.D., Lynch, K., Naud, M.-E., Paulino-Lima, I.G., Singer, K., Walter-Antonio, M., Abrevaya, X.C., Anderson, R., Arney, G., Atri, D., Azúa-Bustos, A., Bowman, J.S., Brazelton, W.J., Brennecka, G.A., Carns, R., Chopra, A., Colangelo-Lillis, J., Crockett, C.J., DeMarines, J., Frank, E.A., Frantz, C., de la Fuente, E., Galante, D., Glass, J., Gleeson, D., Glein, C.R., Goldblatt, C., Horak, R., Horodyskyj, L., Kaçar, B., Kereszturi, A., Knowles, E., Mayeur, P., McGlynn, S., Miguel, Y., Montgomery, M., Neish, C., Noack, L., Rugheimer, S., Stüeken, E.E., Tamez-Hidalgo, P., Imari Walker, S., Wong, T., 2016. The astrobiology primer v2.0. Astrobiology 16, 561–653.

and negligible amounts of oxygen, iron, silicon, or magnesium to produce rocky planets. Essentially, they were purely hydrogen and helium (Loeb, 2010).

With only tiny amounts of carbon, oxygen, and dust, clouds of molecular hydrogen do not cool quickly into the distribution of stellar masses that we see forming in the plane of the Milky Way today. Small amounts of carbon, oxygen, and/or dust (more than $\sim 10^{-5}$ of the Sun's amount) are needed to remove energy efficiently from a collapsing proto-star through gas-cooling and/or dust cooling (Frebel & Norris, 2013; Alexander et al., 2014). With less C, O, and dust than this threshold value, much more hydrogen and helium is needed before a blobby over-density of hydrogen and helium can collapse into a star. Thus, the first generation of stars were super-massive ultra-metal-poor stars. They are known as Pop III stars, but have not been directly observed. They have masses fifty to hundreds of times more massive than the Sun. Because they are so massive, they also have short lifetimes of only ~ 3 million years. And because they appeared only very early in the history of the universe, they would only be directly observable before the redshift of reionization, $z \sim 11$ (Planck, 2016).

Many generations of these Pop III stars are responsible for increasing the metallicity of the universe. The amount of material and the relative abundances of the elements in this material depends strongly on the mass of the star. For example, Pop III stars in the mass range $M > \sim 250\, M_{\mathrm{Sun}}$ collapse into black holes directly through photodisintegration. Their accretion-disk jets spray only a negligible fraction of their mass into the universe for recycling in the next generation of stars. Pop III stars in the mass range ($\sim$130, $\sim$250) M_{Sun} explode as pair instability supernovae. They disperse almost all of the heavy elements that they produce back into the interstellar medium, leaving no black hole behind. Pop III stars in the mass range ($\sim$50, $\sim$130) M_{Sun} produce type II supernovae which shed many of the lighter and medium weight elements in their outer layers. The heavier elements (e.g., Fe) in their cores are not shed for recycling, but collapse into central black holes.

Each of these three major classes of Pop III stars produced different relative abundances of the elements. The explosions of Pop III stars produced elements which were dispersed and recycled into the next generation of stars. The relative abundances of the elements in the most metal-poor Pop II stars are hints at what was produced by earlier Pop III stars (Norris et al., 2012; Karlsson et al., 2013; Frebel & Norris, 2015). When the metallicity of molecular clouds increases above the $\sim10^{-5}$ threshold, to values in the range $\sim10^{-4}$–10^{-2} of the metallicity of the Sun, the cooling mechanisms for molecular cloud collapse change. Higher metallicity Pop II stars begin to form. These have an initial mass function that extends to much lower masses (by comparison we say that the initial mass function of Pop III stars is "top heavy"). The highest redshift, damped Lyman-alpha systems in the line of sight to quasars, are observational probes of the elemental yields of Pop III stars (e.g., Wasserburg & Qian, 2000; Rafelski et al., 2014).

In contrast to Pop III stars, the earliest Pop II stars *have* been observed and the relative abundances of elements in these most metal-poor stars are therefore good tracers of the elemental yields of early Pop III stars. Some of the oldest Pop II stars have small masses and therefore have lifetimes long enough to still be shining locally (in our galaxy) and visible today.

Rocky terrestrial planets like the Earth are (at the 93% level by mass) made up of Fe, O, Si, and Mg (Wang et al., 2018). Life on Earth (at the $\sim$97% level by mass) is made up of C, H, O, and N (Lineweaver & Chopra, 2012b). Thus, to examine the earliest origins of terrestrial planets and the earliest biochemistry, we need to focus on the origin and evolution of these suites of elements: the rocky elements and the life elements.

When Pop III stars formed, the amount of metals in the universe was less than 100,000[th] of the current solar metallicity. Thus, if Pop III stars had planets, they were not terrestrial planets. However, if core-accretion is the dominant mode in which Jupiter-like planets form (Papaloizou & Terquem, 2006), even the formation of giant gaseous (predominantly H and He planets like Jupiter) was problematic. The core-accretion model of planet formation requires the initial formation of a dusty (nonhydrogen, nonhelium) core early enough in the evolution of a gas-rich proto-planetary disk to cause the exponential infall of hydrogen and helium onto the core. However, the abundances of the dominant metals (C, O, Si, Fe, Mg) in Pop III stars may have been too low to initialize the accretion of gaseous planets around them.

Early in the history of the universe, during the first few hundred million years when Pop III stars formed, there was lots of H and He but negligible amounts of other elements; negligible carbon, oxygen, water, and thus, probably no life. As the abundance of these ingredients for life increased, the potential for the emergence of life increased.

Traditionally, because of the many iron lines in the optical part of the spectrum (and because iron abundance has been a good rough proxy for the total amount of nonhydrogen and nonhelium elements), the iron abundance normalized to hydrogen (Fe/H) represents the metallicity of stars and of the universe in general. For example, panel B of Fig. 2 shows an estimate of the evolution of the metallicity of the universe. This panel can be improved by adding more recent metallicity estimates from high redshift damped Lyman-alpha systems (e.g., Rafelski et al. 2014). We need to augment this

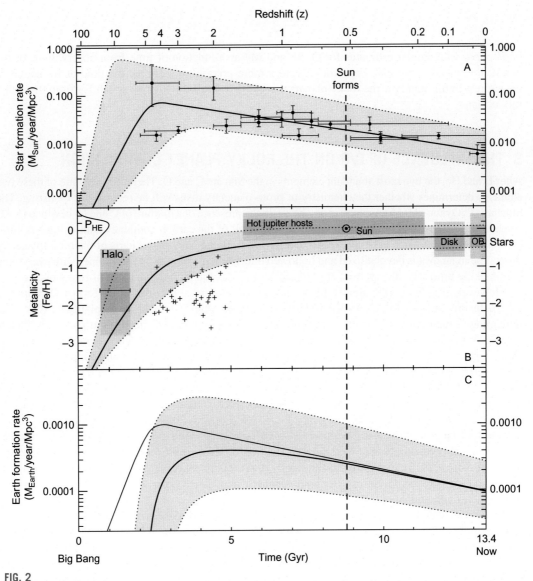

FIG. 2

The Age Distribution of Earth-like Planets in the Universe. The *x*-axis for all three panels is time from the big bang (left), until now (right). Panel A shows an estimate of the star formation rate in the universe. Since stars produce metallicity, the integral of the curve in Panel A produces the monotonically increasing metallicity curve in Panel B. In this integral, the conversion between star formation and metallicity depends on what types of stars are converting H and He into metals. This dependence is particularly important for the initial mass function of Pop III stars in the first 2 billion years. Recent estimates of star formation peak at redshifts between 1 and 3. We have made the plausible assumption that the probability of harboring rocky planets as a function of metallicity is the curve labeled "P_{HE}" in the upper left of Panel B. The probability distribution of harboring Earths, "P_{HE}", applied to the curve in Panel B tells us that during the first ~3 billion years, stars did not have enough metals to form rocky planets. Thus, we obtain the age distribution of terrestrial planets in Panel C.

Modified from Lineweaver, C.H., 2001. An estimate of the age distribution of terrestrial planets in the universe: quantifying metallicity as a selection effect. Icarus 151, 307–313.

Fe/H curve with similar estimates for O, Si, and Mg as a function of time from observations of the oldest, low metallicity stars. This would yield a more precise estimate for the time of formation of the earliest Earths and how the rocky planets that formed $\sim$11 billion years ago differ from those that formed later.

3 THE INFLUENCE OF C/O ON THE ROCKY PLANET COMPOSITION

After H and He, the two most abundant elements in the Sun are C and O. Therefore, the ratio of these two elements determines whether the chemistry of proto-planetary disks will be reducing or oxidizing. The Sun has a C/O ratio of 0.55 (Asplund et al., 2009). In the presence of a mixture of C and O, in which $O > C$ (such as was the case in our presolar nebula), almost all of the carbon combines with oxygen to form carbon monoxide (CO), a volatile gas that is removed from the inner hotter parts of the proto-planetary disks. The inner disk is devolatilized and hollowed out. The remaining O then combines with everything else, making primarily silicon, iron, and magnesium oxides found in the mantles of rocky planets.

However, if $C/O > \sim 0.8$, almost all of the *oxygen* combines with carbon to form CO (Larimer, 1975). In this volatile form, oxygen is largely removed from the inner proto-planetary disk. The remaining C then combines with everything else making silicon and magnesium carbides. Planets formed from this material are called carbon planets (Kuchner & Seager, 2005; Lodders & Fegley, 1997, Fig. 3). Very little is known about carbon planets and whether they could support life. When $C/O > \sim 0.8$, methane (CH_4) would be the dominant ice/liquid—with H_2O much less abundant than it is in $C/O < \sim 0.8$ systems. Life that depends on water as a solvent could be much less likely to evolve on carbon planets. Thus, the importance of carbon planets for our analysis depends on the frequency of stars with $C/O > \sim 0.8$, which currently may be $\sim 10\%$ (Fortney, 2012; Nakajima & Sorahana, 2016). We also need a plot of the distribution of C/O as a function of time to see if the very earliest and extremely metal-poor stars would have a large fraction of stars with $C/O > \sim 0.8$ (Frebel et al., 2007; Frebel & Norris, 2013). One caveat is that rocky planet composition should reflect the bulk composition of its host star and not necessarily a photospheric abundance superficially enriched in C through dredge-up or contamination of a thin convective layer.

In the stars with the lowest Fe/H, it is often the case that the abundances of the life elements (C, O, N) are not similarly low. For the earliest stars, iron is not as reliable a tracer of C, O, and N as it is currently. That is, (C+O+N)/Fe for the lowest Fe/H stars can be an order of magnitude higher than the solar value of (C+O+N)/Fe. This probably results from Pop III stars producing a higher value of (C+O+N)/Fe than did the subsequent supernovae explosions of massive Pop II stars. The relative abundances of recycled elements produced by type I, type II, and the many other subtypes of supernovae is also not constant over cosmic time. Thus, the abundances of the elements of life (C, O, N) increased earlier than the abundances of Fe, Si, and Mg that enabled the formation of rocky planets. This suggests that the earliest elemental bottleneck for the emergence of life was not the low abundance of the life forming elements (e.g., C, O, N), but rather the low abundance of Fe, Si, Mg and other metals that make up the rocky planets for life to live on.

The take-home message from the age distribution of rocky planets (panel C, Fig. 2) is that many rocky planets formed before the Earth formed. The average rocky planet is about 2 billion years older than Earth and the oldest formed $\sim$3 billion years after the big bang or $\sim$11 billion years ago. Although these earliest "Earths" might have had smaller iron cores, we could not identify any special or important geochemical feature of our Earth that would make it a more hospitable place for the origin and persistence of life than these early rocky planets.

4 THE UBIQUITY OF HABITABLE PLANETARY SYSTEMS

The universe is filled with stars similar to our Sun (Robles et al., 2008). Exoplanet statistics suggest that rocky planets similar to our Earth are common (Burke et al., 2015; Petigura et al., 2013; Bovaird et al., 2015). The Sun's metallicity seems to be common among the metallicity distribution of the stellar hosts of rocky planets (Robles et al., 2008). Or said in different way, the hosts of rocky planets (in contrast to hot Jupiter hosts) do not seem to have an anomalously high metallicity (Buchhave & Latham, 2015). Also, recently, the eccentricity distribution of Kepler exoplanets (in contrast to the high eccentricities of hot Jupiters) seems to be consistent with the low eccentricities of our planetary system (Xie et al., 2016).

Will the wet rocky planets that formed between 5 and 11 billion years ago in our galaxy (or other galaxies) be different from those that formed ~4.5 billion years ago? Did "life before Earth" have access to different resources (e.g., stellar spectra, mineral compositions, or abundances of radiogenic nuclides)? The answers to both these questions seem to be "No." The processes that led to our wet rocky planet orbiting a Sun-like star are likely to be old and universal (Chopra & Lineweaver, 2008; Lineweaver & Chopra, 2012a). Water, heat, chemical disequilibria, and energy sources would have been present on early wet rocky planets because of the universal nature of the processes that produced them. For example, the alkaline hydrothermal vent scenario for the origin of life on Earth could very well have taken place on other Earth-like planets billions of years before the emergence of life on Earth (Martin & Russell, 2007; Lane, 2009; Smith & Morowitz, 2016).

5 WHAT CAN TERRESTRIAL LIFE TELL US ABOUT EXTRATERRESTRIAL LIFE?

Terrestrial life emerged from nonlife. Quirky biology emerged from deterministic physics and chemistry. If this is correct, then the first steps of molecular evolution are deterministic or quasideterministic. DeDuve (1995) has argued that this initial determinism makes life a "cosmic imperative" built into the chemistry of the universe.

So far, without detection of extraterrestrial life, life on Earth provides the best basis for informed speculation about life elsewhere (see however, Lineweaver, 2006). The similarity in atomic and molecular composition of life on Earth suggests that the scope of variability in biochemistry of life is limited—e.g., the building block elements of life elsewhere will be H, O, C, N, P, and S and the solvent will be H_2O. The ubiquity and deep ancestry of the most fundamental features of life on Earth makes them the best candidates for being features of extraterrestrial life (Lineweaver & Chopra, 2012b; Cockell, 2016). See Weiss et al. (2016) for estimates of the first metabolisms on Earth.

Fundamental features of a species in a clade can provide information about the whole clade because the most fundamental features appear earliest in the embryonic development of individuals of a species.

These features are often phylogenetically the most deeply rooted. Thus, they are likely to be shared by ancient ancestors and their other descendants. For example, embryologists studying the larval forms of organisms often produce more accurate phylogenetic trees than a comparison of adult forms would. We suggest that this idea can be extrapolated to life elsewhere in the universe. In other words, the earliest forms of life near the root of the phylogenetic tree of terrestrial life should resemble the earliest forms of life that emerge on other planets. This resemblance would not be the result of a genetic connection, but is plausible if the conditions for the origin of life are specific enough (e.g., temperature, temperature gradients, redox potentials, hydrolytic cycling, relative elemental compositions, pH) (Chopra et al., 2010). This idea is reminiscent of Haekel's controversial "ontogeny recapitulates phylogeny" or "recapitulation theory" (Haekel, 1874). We hypothesize that the more reasonable, nuanced versions (DeBeer, 1940; Gould, 1977; Kalinka & Tomancak, 2012) of "ontogeny recapitulating phylogeny" can become "terrestrial phylogeny recapitulates extraterrestrial phylogeny." Another way to express this is to paraphrase the first two of von Baer's laws of embryology (von Baer, 1828; Huxley, 1853; Abzhanov, 2013) as laws of extraterrestrial phylogeny:

1. (The) more general characters of life appear earlier and closer to the roots of phylogenetic trees (on any planet) than the more special characters.
2. From the most general forms, the less general are developed, and so on, until finally the most special arises.

The earliest "embryonic" ontogeny of life on a planet—the earliest steps of development of the single example of terrestrial life we know—may recapitulate the paths that all life in the universe has to follow during its earliest evolution. The shortest branches in the phylogenetic tree of nonviral life (Fig. 3) are more likely to be representative of life elsewhere than will be the quirky longer branches. As an example, take the hyperthermophilia of the shortest branches of terrestrial life. The shortest branches of the 16S rRNA phylogenetic tree are hyperthermophilic (right side of Fig. 3, see also Lineweaver & Schwartzman, 2005; Wong et al., 2007). Extant organisms with the shortest branches are hyperthermophiles able to tolerate temperatures above 90°C. These organisms—Aquifex, Thermotoga, Nanoarchaeota, and Korarchaeota—seem to be the best representatives of the Last Universal Common Ancestor of all terrestrial life. We suggest that they may be our best guesses for the organisms at the roots of analogous phylogenies of extraterrestrial life (see however, Boussau et al., 2008).

In the book *Vital Dust*, deDuve (1995) presented the case that water and energy are common and that the emergence of life may be a cosmic imperative. However, in a recent paper (Chopra & Lineweaver, 2016), we propose that the most important constraint on the persistence of life in the Universe may be whether life, after emerging and evolving into a biosphere, can evolve global mechanisms rapidly enough to mediate the positive and negative feedbacks of abiotic atmospheric evolution. We hypothesize that the early evolution of biologically mediated negative feedback processes, or Gaian regulation as proposed by Lovelock and Margulis (1974), may be necessary to maintain habitability because of the strength, rapidity, and universality of abiotic positive feedbacks on the surfaces of rocky planets in traditional circumstellar habitable zones.

The almost inevitable nonevolution of biospheric regulation of surface volatiles, temperature, and albedo can become a Gaian bottleneck to the persistence of life. This would suppress the persistence (but not necessarily the emergence) of life elsewhere in the universe and make the 4-billion-year survival of our biosphere a lucky accident.

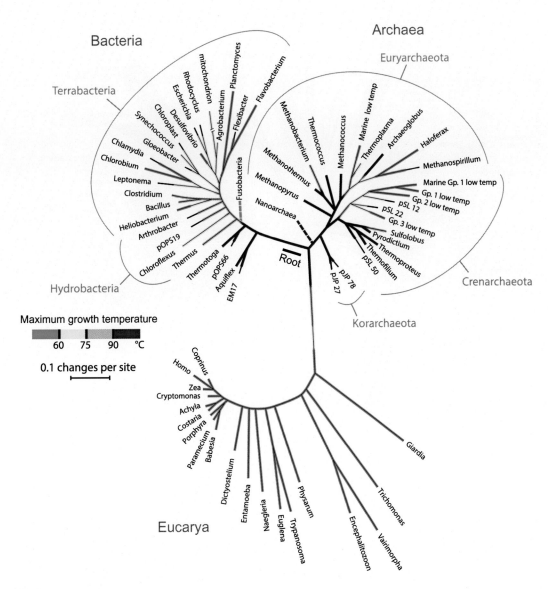

FIG. 3

Old and new trees of Life. The old diagram (*top*) is from Lineweaver and Chopra (2012b) and the new tree of all nonviral life (*bottom*) is from Hug et al. (2016). In the new tree, eukaryotes are well-established as a branch in the TACK region (Williams et al., 2013). Thus, the group labeled "Eucarya" in the lower part of the old tree (Fig. 3B) needs to be cut off and embedded further from the root among the branches of Archaea, closer to the Crenarchaeota and Korarchaeota. See also Zaremba-Niedzwiedzka et al. (2017).

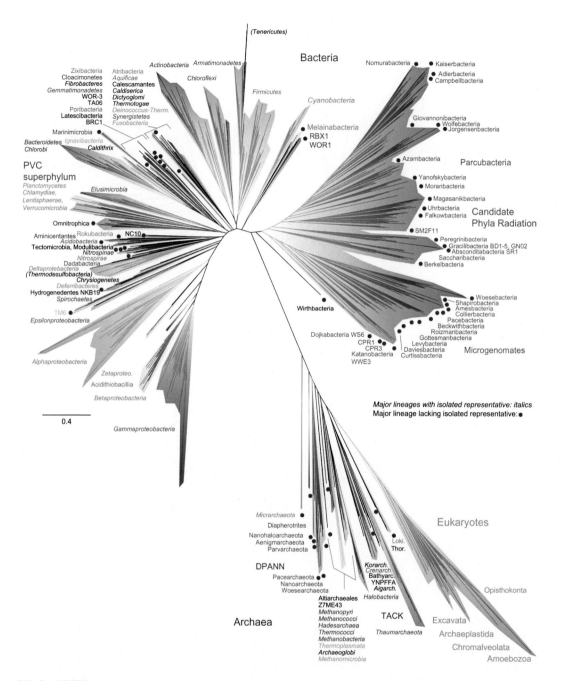

FIG. 3—CONT'D

6 CONCLUSION

The fundamental ingredients for the origin of life were not present ~13.8 billion years ago. During the first ~2 billion years of star formation, the abundance of rock-forming elements was probably not high enough to produce rocky planets in the circumstellar habitable zone, massive enough to hold an atmosphere and maintain water at its surface. This suggests that the earliest elemental bottleneck for the emergence of life was not the low abundance of the life elements (C, H, O, N), but rather the low abundance of the elements that make rocky planets (Fe, Si, Mg) for life to live on. However, after passing through this bottleneck ~11 billion years ago, all the fundamental ingredients for life were present. Thus, for ~6 billion years preceding the formation of the Earth (from ~11 to ~5 billion years ago), wet rocky Earth-like planets in circumstellar habitable zones formed and evolved. It is therefore reasonable to speculate that life may have emerged many billions of years before the emergence of life on Earth. As we learn more about the geochemistry of the early Earth and the origin of the first life on Earth, we are learning about the life which probably emerged before the existence of our Solar System.

REFERENCES

Abzhanov, A., 2013. von Baer's law for the ages: lost and found principles of developmental evolution. Trends Genet. 29, 712–722.

Alexander, P.J., Anna, F., Volker, B., 2014. The chemical imprint of silicate dust on the most metal-poor stars. Astrophys. J. 782, 95.

Asplund, M., Grevesse, N., Sauval, A.J., Scott, P., 2009. The chemical composition of the sun. Annu. Rev. Astron. Astrophys. 47, 481–522.

Boussau, B., Blanquart, S., Necsulea, A., Lartillot, N., Gouy, M., 2008. Parallel adaptations to high temperatures in the Archaean eon. Nature 456, 942–945.

Bovaird, T., Lineweaver, C.H., Jacobsen, S.K., 2015. Using the inclinations of Kepler systems to prioritize new Titius-Bode-based exoplanet predictions. Mon. Not. R. Astron. Soc. Lett. 448, 3608–3627.

Buchhave, L.A., Latham, D.W., 2015. The metallicities of stars with and without transiting planets. Astrophys. J. 808.

Burke, C.J., Christiansen, J.L., Mullally, F., Seader, S., Huber, D., Rowe, J.F., Coughlin, J.L., Thompson, S.E., Catanzarite, J., Clarke, B.D., Morton, T.D., Caldwell, D.A., Bryson, S.T., Haas, M.R., Batalha, N.M., Jenkins, J.M., Tenenbaum, P., Twicken, J.D., Li, J., Quintana, E., Barclay, T., Henze, C.E., Borucki, W.J., Howell, S.B., Still, M., 2015. Terrestrial planet occurrence rates for the Kepler Gk Dwarf sample. Astrophys. J. 809.

Chopra, A., Lineweaver, C.H., 2008. The major elemental abundance differences between life, the oceans and the Sun. In: Proceedings from 8th Australian Space Science Conference.

Chopra, A., Lineweaver, C.H., 2016. The case for a Gaian Bottleneck: the biology of habitability. Astrobiology 16, 7–22.

Chopra, A., Lineweaver, C.H., Brocks, J.J., Ireland, T.R., 2010. Palaeoecophylostoichiometrics: searching for the elemental composition of the last universal common ancestor. In: Proceedings from 9th Australian Space Science Conference.

Cockell, C.S., 2016. The similarity of life across the universe. Mol. Biol. Cell 27, 1553–1555.

Connelly, J.N., Bizzarro, M., Krot, A.N., Nordlund, A., Wielandt, D., Ivanova, M.A., 2012. The absolute chronology and thermal processing of solids in the solar protoplanetary disk. Science 338, 651–655.

Domagal-Goldman, S.D., Wright, K.E., Adamala, K., Arina de la Rubia, L., Bond, J., Dartnell, L.R., Goldman, A.D., Lynch, K., Naud, M.-E., Paulino-Lima, I.G., Singer, K., Walter-Antonio, M., Abrevaya, X.C., Anderson, R., Arney, G., Atri, D., Azúa-Bustos, A., Bowman, J.S., Brazelton, W.J., Brennecka, G.A., Carns, R., Chopra, A., Colangelo-Lillis, J., Crockett, C.J., DeMarines, J., Frank, E.A.,

Frantz, C., de la Fuente, E., Galante, D., Glass, J., Gleeson, D., Glein, C.R., Goldblatt, C., Horak, R., Horodyskyj, L., Kaçar, B., Kereszturi, A., Knowles, E., Mayeur, P., McGlynn, S., Miguel, Y., Montgomery, M., Neish, C., Noack, L., Rugheimer, S., Stüeken, E.E., Tamez-Hidalgo, P., Imari Walker, S., Wong, T., 2016. The astrobiology primer v2.0. Astrobiology 16, 561–653.

DeBeer, G.R., 1940. Embryos and Ancestors. University College London, Oxford, ISBN: 9781406700954.

DeDuve, C., 1995. Vital Dust. Basic books, New York, ISBN: 9780465090457.

Fortney, J.J., 2012. On the carbon-to-oxygen ratio measurement in nearby sun-like stars: implications for planet formation and the determination of stellar abundances. Astrophys. J. 747, L27.

Frebel, A., Norris, J.E., 2013. Metal-poor stars and the chemical enrichment of the universe. In: Oswalt, T.D., Gilmore, G. (Eds.), Planets, Stars and Stellar Systems. Galactic Structure and Stellar Populations, vol. 5. Springer, Dordrecht, Netherlands, pp. 55–114.

Frebel, A., Norris, J.E., 2015. Near-field cosmology with extremely metal-poor stars. Annu. Rev. Astron. Astrophys. 53, 631–688.

Frebel, A., Johnson, J.L., Bromm, V., 2007. Probing the formation of the first low-mass stars with stellar archaeology. Mon. Not. R. Astron. Soc. Lett. 380, L40–L44.

Gould, S.J., 1977. Ontogeny and Phylogeny. Harvard University Press, Cambridge, MA, ISBN: 0-674-63940-5.

Haekel, E., 1874. Anthropogenie: oder, Entwickelungsgeschichte des Menschen (Anthropogeny: Or, the Evolutionary History of Man). Wilhelm Engelmann, Leipzig. 5th and enlarged edition—1903.

Hug, L.A., Baker, B.J., Anantharaman, K., Brown, C.T., Probst, A.J., Castelle, C.J., Butterfield, C.N., Hernsdorf, A.W., Amano, Y., Ise, K., Suzuki, Y., Dudek, N., Relman, D.A., Finstad, K.M., Amundson, R., Thomas, B.C., Banfield, J.F., 2016. A new view of the tree of life. Nat. Microbiol. 1.

Huxley, T.H., 1853. Scientific memoirs, selected from the transactions of foreign academies of science, and from foreign journals. In: Henfrey, A. (Ed.), Natural History. Taylor and Francis, London.

Kalinka, A.T., Tomancak, P., 2012. The evolution of early animal embryos: conservation or divergence? Trends Ecol. Evol. 27, 385–393.

Karlsson, T., Bromm, V., Bland-Hawthorn, J., 2013. Pregalactic metal enrichment: the chemical signatures of the first stars. Rev. Mod. Phys. 85, 809–848.

Kolb, E.W., Turner, M.S., 1994. The Early Universe. Westview, New York.

Kuchner, M., Seager, S., 2005. Extrasolar Carbon Planets. https://arxiv.org/abs/astro-ph/0504214.

Lane, N., 2009. Life Ascending. Norton, NewYork (Chapter 1).

Larimer, J.W., 1975. The effect of CO ratio on the condensation of planetary material. Geochim. Cosmochim. Acta 39, 389–392.

Lineweaver, C.H., 2001. An estimate of the age distribution of terrestrial planets in the universe: quantifying metallicity as a selection effect. Icarus 151, 307–313.

Lineweaver, C.H., 2006. We have not detected extraterrestrials, or have we? In: Seckbach, J. (Ed.), Life as We Know It. Cellular Origins and Life in Extreme Habitats and Astrobiology, vol. 10. Springer, Dordrecht, pp. 445–457.

Lineweaver, C.H., Chopra, A., 2012a. The habitability of our earth and other earths: astrophysical, geochemical, geophysical, and biological limits on planet habitability. Annu. Rev. Earth Planet. Sci. 40 (40), 597–623.

Lineweaver, C.H., Chopra, A., 2012b. What can life on earth tell us about life in the universe? In: Seckbach, J. (Ed.), Genesis—In The Beginning: Precursors of Life Chemical Models and Early Biological Evolution. Springer, Dordrecht, Netherlands, pp. 799–815.

Lineweaver, C.H., Schwartzman, D., 2005. Cosmic thermobiology. In: Seckbach, J. (Ed.), Origins: Genesis Evolution and Diversity of Life. Springer, Dordrecht, Netherlands, pp. 233–248.

Lodders, K., Fegley, B., Jr., 1997. Condensation chemistry of carbon stars. AIP Conf. Proc. 402, 391–423.

Loeb, A., 2010. How did the first stars and galaxies form? Princeton University Press, Princeton, NJ, ISBN: 9780691145167.

Loeb, A., 2014. The habitable epoch of the early Universe. Int. J. Astrobiol. 13, 337–339.

Lovelock, J.E., Margulis, L., 1974. Atmospheric homeostasis by and for the biosphere: the gaia hypothesis. Tellus 26, 2–10.

Martin, W., Russell, M.J., 2007. On the origin of biochemistry at an alkaline hydrothermal vent. Philos. Trans. R. Soc. B 362, 1887–1925.

Nakajima, T., Sorahana, S., 2016. Carbon-to-oxygen ratios in M dwarfs and solar-type stars. Astrophys. J. 830, 159.

Norris, J.E., Bessell, M.S., Yong, D., Christlieb, N., Barklem, P.S., Asplund, M., Murphy, S.J., Beers, T.C., Frebel, A., Ryan, S.G., 2012. The most metal-poor stars. I. Discovery, data, and atmospheric parameters. Astrophys. J. 762 (1), 25.

Nutman, A.P., Bennett, V.C., Friend, C.R.L., Van Kranendonk, M.J., Chivas, A.R., 2016. Rapid emergence of life shown by discovery of 3,700-million-year-old microbial structures. Nature 537, 535–538.

Papaloizou, J.C.B., Terquem, C., 2006. Planet formation and migration. Rep. Prog. Phys. 69, 119–180.

Peacock, J., 1999. Cosmological Physics. Cambridge University Press, Cambridge, ISBN: 978-0521422703.

Petigura, E.A., Howard, A.W., Marcy, G.W., 2013. Prevalence of earth-size planets orbiting sun-like stars. Proc. Natl. Acad. Sci. 110, 19273–19278.

Planck Collaboration, 2016. XIII. Cosmological parameters. (Table 4). Astron. Astrophys. 594. A13.

Rafelski, M., Neeleman, M., Fumagalli, M., Wolfe, A.M., Prochaska, J.X., 2014. Ap. J. Lett. 782, L29.

Robles, J.A., Lineweaver, C.H., Grether, D., Flynn, C., Egan, C.A., Pracy, M.B., Holmberg, J., Gardner, E., 2008. A comprehensive comparison of the Sun to other stars: searching for self-selection effects. Astrophys. J. 684, 691–706.

Smith, E., Morowitz, H., 2016. The Origin and Nature of Life on Earth: The Emergence of the Fourth Geosphere. Cambridge University Press, Cambridge, ISBN: 978-1107121881.

Smoot, G.F., Bennett, C.L., Kogut, A., Wright, E.L., Aymon, J., Boggess, N.W., Cheng, E.S., De Amici, G., Gulkis, S., Hauser, M.G., Hinshaw, G., Lineweaver, C., Loewenstein, K., Jackson, P.D., Janssen, M., Kaita, E., Kelsall, T., Keegstra, P., Lubin, P., Mather, J.C., Meyer, S.S., Moseley, S.H., Murdock, T., Rokke, L., Silverberg, R.F., Tenorio, L., Weiss, R., Wilkinson, D.T., 1992. Structure in the COBE differential microwave radiometer first year maps. Astrophys. J. Lett. 396, L1–L5.

von Baer, K.E., 1828. Über Entwickelungsgeschichte der Thiere. Beobachtung und reflexion (On the Developmental History of the Animals. Observations and Reflections). Königsberg, Bei den Gebrüdern Bornträger.

Wang, H.S., Lineweaver, C.H., Ireland, T.R., 2018. The elemental abundances (with uncertainties) of the most Earth-like planet. Icarus 299, 460–474. ISSN 0019–1035, https://doi.org/10.1016/j.icarus.2017.08.024. (http://www.sciencedirect.com/science/article/pii/S0019103517302221).

Wasserburg, G.J., Qian, Y.Z., 2000. A model of metallicity evolution in the early universe. Astrophys. J. Lett. 538, L99.

Weiss, M.C., Sousa, F.L., Mrnjavac, N., Neukirchen, S., Roettger, M., Nelson-Sathi, S., Martin, W.F., 2016. The physiology and habitat of the last universal common ancestor. Nat. Microbiol. 1.

Williams, T.A., Foster, P.G., Cox, C.J., Embley, T.M., 2013. An archaeal origin of eukaryotes supports only two primary domains of life. Nature 504, 231–236.

Wong, J.T.F., Chen, J., Mat, W.K., Ng, S.K., Xue, H., 2007. Polyphasic evidence delineating the root of life and roots of biological domains. Gene 403, 39–52.

Xie, J.W., Dong, S.B., Zhu, Z.H., Huber, D., Zheng, Z., De Cat, P., Fu, J.N., Liu, H.G., Luo, A., Wu, Y., Zhang, H.T., Zhang, H., Zhou, J.L., Cao, Z.H., Hou, Y.H., Wang, Y.F., Zhang, Y., 2016. Exoplanet orbital eccentricities derived from LAMOST-Kepler analysis. Proc. Natl. Acad. Sci. U. S. A. 113, 11431–11435.

Zaremba-Niedzwiedzka, K., Caceres, E.F., Saw, J.H., Backstrom, D., Juzokaite, L., Vancaester, E., Seitz, K.W., Anantharaman, K., Starnawski, P., Kjeldsen, K.U., Stott, M.B., Nunoura, T., Banfield, J.F., Schramm, A., Baker, B.J., Spang, A., Ettema, T.J.G., 2017. Asgard archaea illuminate the origin of eukaryotic cellular complexity. Nature 541, 353–358.

ASTROPHYSICAL AND COSMOLOGICAL CONSTRAINTS ON LIFE

Paul A. Mason*,[1], Peter L. Biermann[†,‡,§,¶,2]
*New Mexico State University, Las Cruces, NM, United States
†Max Planck Institute for Radio Astronomy, Bonn, Germany
‡Inst. for Nucl. Phys., Karlsruhe Institute for Technology, Karlsruhe, Germany
§Dept. of Physics & Astronomy, University of Alabama, Tuscaloosa, AL, United States
¶Department of Physics and Astronomy, University of Bonn, Germany

CHAPTER OUTLINE

---

[1]P. A. Mason is a member of the Sloan Digital Sky Survey (SDSS) Collaboration.
[2]P. L. Biermann is a member of the Pierre Auger, JEM-EUSO, and LOPES Collaborations.

Habitability of the Universe Before Earth, editors: Richard Gordon & Alexei Sharov, Volume 1 in the series:
Astrobiology: Exploring Life on Earth and Beyond, series editors: Pabulo Henrique Rampelotto,
Joseph Seckbach & Richard Gordon. ISSN 2468-6352. https://doi.org/10.1016/B978-0-12-811940-2.00005-8

Acronyms

AGN	Active Galactic Nucleus/Nuclei
AU	Astronomical Unit, the average distance from Earth to Sun: 1 AU $= 1.496 \times 10^{11}$ m
CHNOPS	Carbon, Hydrogen, Nitrogen, Oxygen, Phosphorous, Sulfur
CNO	Carbon, Nitrogen and Oxygen
CR	Cosmic Ray—high energy particle from beyond Earth
EGCR	ExtraGalactic Cosmic Ray
eV	electron-Volt 1 eV $= 1.602 \times 10^{-19}$ J, 1keV $= 10^3$ eV, 1Mev $= 10^6$ eV, 1GeV $= 10^9$ eV, 1Tev $= 10^{12}$ eV, 1PeV $= 10^{15}$ eV, and 1EeV $= 10^{18}$ eV
far-IR	long wavelength Infrared Radiation
FR-I	massive galaxies often with jets and associated with X-ray emitting gas
GCR	Galactic Cosmic Ray
GHZ	Galactic Habitable Zone
GRB	Gamma-Ray Burst
Gyr	Gigayear: 1Gyr $= 10^9$ years
HMXB	High Mass X-ray Binary
ΛCDM	a specific dark energy plus Cold Dark Matter cosmology
LMXB	Low Mass X-ray Binary
$M_\odot$	the mass of the Sun: 1 $M_\odot = 1.989 \times 10^{30}$ kg
pc, kpc, Mpc	parsec: 1 pc $= 3.086 \times 10^{16}$ m, 1 kpc $= 10^3$ pc, 1 Mpc $= 10^6$ pc
SEP	Solar/Stellar Energetic Particle
SFR	Star Formation Rate
SN, plural SNe	SuperNova, a Generic Stellar Explosion
SN Ia	white dwarf explosion after exceeding the Chandrasekhar mass limit, no star remains
SN Ib/c	massive star explosion after nuclear fusion, a neutron star or a black hole is formed
SN II	core collapse SuperNova
SMBH	SuperMassive Black Hole
SGHZ	Super-Galactic Habitable Zone, See Fig. 4
UHECR	Ultra High Energy Cosmic Ray
ULX	Ultra-Luminous X-ray source
UVB	UltraViolet B-band, lower energy than UVA

1 INTRODUCTION

In this chapter, constraints placed on life in the ever evolving Universe are examined. A prerequisite for life is the availability of materials that form planets, support biospheres, and protect life from catastrophic threats. We focus especially on how those factors have changed over cosmic time. Also, we investigate how the location within a galaxy and the location of galaxies within galaxy clusters may affect the development and maintenance of life on planets. In this section, we outline the main factors affecting habitability of the Universe through cosmic time. Subsequent sections are focused on the constraints placed on life, as supported by recent observations and theoretical considerations.

Our analysis indicates that conditions for life on the surface of planets depend on the chance of avoiding catastrophic events described in Sections 2 and 3, as well as on the protecting role of the planetary and circumstellar environment, as discussed in Sections 4 and 5. Certain types of host stars, certain locations within galaxies, and certain galaxies within clusters, are more favorable for the origin and evolution of life than others. Habitability factors include being in a galaxy with a low, but nonzero, star formation rate (SFR) and no, or at least an inactive, central supermassive black hole (SMBH).

While some galaxies have very massive SMBHs, many if not most galaxies do not have a SMBH. The Milky Way has a low mass and nearly inactive SMBH (which certainly has been active in the past). In this context, habitability—the probability of developing complex life, as we know it, is severely compromised in the presence of ionizing radiation and high-energy particle flux. A decrease in the local SFR, the reduction of growth and activity of SMBHs, and the expansion of the Universe all generally enhance the probability of habitable planet development over cosmic time. However, habitability does not necessarily improve monotonically, as galaxy mergers increase the SFR and thereby elevate the supernova (SN) and gamma-ray burst (GRB) rates. SMBH mergers create havoc on a galactic scale and reduce habitability for extended periods of time in merging galaxies. In addition to the Milky Way, we discuss the potential for life in the nearby galaxies M31, M32, M33, M81, M82, M87, M90, M94, and Cen A.

1.1 FORMATION OF THE ELEMENTS OF LIFE

According to the standard cosmological model, the Universe in the first few minutes after the Big Bang went through phases having similar density and temperature as the center of the Sun. In these conditions, the fusion of protons into more massive nuclei occurred. This phase of nucleosynthesis briefly persisted until the Universe had cooled below threshold temperatures and pressures by expansion (Gamow, 1948; Alpher et al., 1948). Between about 10 and 2000 s after the Big Bang, roughly 25% of the hydrogen H was converted into helium He, mainly 4He. Some deuterium, lithium, beryllium, and possibly boron nuclei were also produced at this time. The main contribution of lithium, beryllium, and boron, in the interstellar medium today comes from cosmic ray spallation, which also produces trace amounts of some more massive elements. However, important elemental constituents for life were not yet available. It took quite some time for the heavier elements to emerge by nuclear fusion in the first stars. This ended a period known as the cosmic dark ages—a star-less era lasting until of order 100 million years after the expansion began. The primordial 1H and 4He nuclei became the seeds for the production of critical elements in the cores of stars (Burbidge et al., 1957) especially those elements that are essential for the formation of rocky planets and for life itself. See Longair (1994) for a general introduction to high-energy astrophysics.

The first stars apparently formed from pure H/He clouds (Loeb and Barkana, 2001). The interstellar clouds formed H_2 molecules allowing them to cool and facilitate contraction. These earliest stars likely formed before galaxies, in regions of collapsing gas possibly associated with clumps of dark matter. The first stars are thought to form primarily with high masses, consume most available fuel quickly, and to explode as a core-collapse supernova (SN-Type II), or collapse directly into black holes (collapsars) relatively soon after formation. Hence, they do not exist in the Universe today or at least not in normal galaxies. However, they played a critical role in synthesizing the first heavy nuclei and spreading their remnants back into the interstellar medium. The formation of heavier elements also facilitated cooling, and positive feedback may have increased the rate of star formation. In this way, the interstellar medium was slowly populated with the elements required for life; carbon, nitrogen, oxygen, phosphorus, and sulfur, which in cold environments combined with H to form the many molecules observed in molecular clouds today. This group of life elements is often abbreviated CHNOPS.

The first galaxies are assumed to form from these first stars, their black hole remnants, and massive gas clouds. Within the heart of these galaxies, active galactic nuclei (AGN) turned on, as SMBHs formed and grew in an ubiquitous, yet poorly understood process. SMBHs exist in the center of essentially all galaxies with a stellar bulge (Ferrarese and Merritt, 2000; Gebhardt et al., 2000). Within galaxies, H/He clouds slowly transformed into molecular clouds as newer generations of stars expelled elements generated in their cores during the final stages of stellar evolution. Metals, defined somewhat loosely as all elements with mass from carbon and up and produced in stellar nucleosynthesis, were necessary for the formation of lower mass stars. So later generations of short lived high mass stars as well as lower mass stars with a longer life span formed from increasingly metal-enriched molecular clouds. Steadily, the stellar population within each host galaxy increased in numbers and became enriched in elemental composition.

Over 100 years ago, Hess (1912) performed balloon experiments and discovered that electroscopes discharge more rapidly at high altitudes than on the ground. It became clear that this increase of discharge rate is caused by radiation coming from beyond Earth. We now know that these cosmic rays (CRs) are highly energetic atomic nuclei, electrons, antiprotons, and positrons impacting the atmosphere. The most massive stars are rare, but they play an important role in determining the habitability of their cosmic neighborhood as they form and accelerate CRs. Stars with mass above ca. 40 $M_\odot$ at the start of H fusion, ultimately eject a lot of mass including carbon and oxygen. If by chance carbon dominates, the outflows of these stars include soot (Woosley et al., 2002). Thus, these massive stars are the main suppliers of carbon, nitrogen, and oxygen (CNO).

Many stars are formed as binary star systems which provide additional channels for interstellar medium enrichment as galaxies evolve. For example, a single low mass star a bit more massive than the Sun produces a large amount of CNO locked in a white dwarf. However, in the right binary system, mass accretion onto a white dwarf may lead to its destruction as a SN-Type Ia if the Chandrasekhar limit is exceeded. SN Ia are driven by the explosive burning of the CNO to Ni, which gets ejected, and then decays into Fe. So SNe Ia supply lots of Fe. SN-Type Ia are not visible in CRs as a known distinct particle population, even though their frequency is relatively large. They may contribute to the same CR-population as do the stars formed with 8 to about 25 $M_\odot$, that explode into the interstellar medium and not into their previously ejected winds.

SN-Type Ia events occur generally later in the age of a galaxy compared to the SN-Type II events, which are explosions of high mass stars at the end of nuclear fusion, because SNe Ia involve the formation of white dwarfs which is the end point of the evolution of longer lived lower mass stars. Other important channels of element enrichment open as the Universe ages. For example, neutron star mergers may supply most of the r-process elements; i.e., those elements formed by rapid neutron capture during the explosion. Binary neutron stars or a neutron star plus black hole binary mergers are probably associated with observable events called kilonovae and/or short GRBs and cause gravitational wave emission (Abbott et al., 2016).

The formation of rocky/metal planets became possible only after the abundances of Si, Mg, and Fe increased enough to allow for their condensation into grains within protoplanetary disks. The formation of planets is a complex process. It involves circumstellar disk formation and subsequent feeding of protoplanets as well as stochastic and chaotic interactions within both nascent and aged planetary systems. Protoplanetary collisions and migration of planets ensure the wide variety of planetary systems observable today, see e.g., Seager (2013). On some of these planets conditions are suitable for life. Earth as the singular known location of life in the Universe remains the paragon of habitability, and indicates one of the things universal to all life as we know it, its dependence on water.

1.2 PROTECTION OF LIFE ON PLANETS

Habitability of a particular planet depends on the balance between the possibility of various threats that may temporarily, or even permanently, sterilize the surface and on the availability of protective mechanisms that reduce the impact and frequency of such events. Earth, for example, has undergone many mass extinctions with subsequent biosphere recovery. Just in the last 0.5 Gyr, the so called big five extinctions each eliminated from 75% to 96% of species on the Earth, see e.g., Raup and Sepkoski (1982). Some of these have been attributed to impacts, hypervolcanism, and global heating or cooling. Many lesser mass extinctions have also occurred. Such events are salient to the history of biological evolution on Earth itself. Indeed, mass extinctions likely have allowed the diverse array of biological changes to occur on Earth, eventually allowing one species of mammals to investigate astrobiology.

The availability of liquid water is by now a classical concept of habitability and it can be quantified by stellar luminosity and temperature. Its limits have been defined for single stars and extended to planets in binary star systems. However, being within the standard habitable zone does not imply habitability of a particular planet. In many cases, habitability is strongly affected by local conditions, such as stellar X-ray and ultraviolet radiation, stellar winds, as well as other catastrophic events (e.g., climate instability, volcanism, asteroid impacts, nearby supernovae, GRBs, microquasars, and local AGN). The ability for newly formed planets to support life depends on the abundance of heavier elements after generations of both high and lower mass stars. The sterilizing effect of accreting and exploding stars decreased as the SFR declined in the Milky Way and in similar galaxies (Madau and Dickinson, 2014). Over time, there has been a decrease in the number and activity of ionizing radiation sources in the local Universe. Planets with sufficiently thick atmospheres and strong magnetic fields provided protection against radiation from the host stars, background CRs, and some nearby radiation events. Formation of a protective atmosphere requires a significant enrichment of CNO elements. High abundance of Si, Ni, and Fe is required to form rocky planets capable of maintaining a strong magnetic field over astrobiologically significant (Gyr) time-scales. The geomagnetic field periodically reverses its polarity; and thus the magnetic protection may become weak during the transition period. If present, atmospheric ozone can provide a significant ultraviolet shield for the surface of planets. The interconnected features of planet habitability, such as the presence of liquid water, plate-tectonics, a magnetic dynamo, and atmospheric composition, are driven by the planet's internal thermal evolution. Habitability of planets orbiting long lived stars can be limited by the geophysical lifetime (Franck et al., 2000), rather than the stellar lifetime.

Astrospheres akin to the Solar heliosphere, formed by single and binary main sequence star winds, provide additional protection for habitable zone planets from CRs. Stellar winds are magnetized plasma flows that may remove atmospheric gas from inner planets, but then expand into the local interstellar medium providing a bubble of protection against the onslaught of CRs. If the stellar winds are too dense and fast, atmospheric erosion may occur (Vidotto et al., 2014). If protection mechanisms are not strong enough, then the habitability of the planet is compromised.

A magnetic field also permeates the disk of the Milky Way and other disk galaxies. This Galactic magnetic field originates from a relativistic flux of CRs accelerated by SN remnants mostly in the central plane of the disk. This CR flux drives a Galactic wind perpendicular to the disk (Hanasz et al., 2013). The SN rate per area must be sufficiently high to drive this wind (e.g., Rossa and Dettmar, 2003). The Galactic dynamo also needs the shear flow of differential rotation of the disk (Jokipii and Morfill, 1987). If this dynamo operates as expected, a galaxy's magnetic field may deflect many CRs from extragalactic sources (Axford, 1981; Hanasz et al., 2009a). However, when the SFR, and hence, the SN rate is too high then particles trapped by the Galactic magnetic field will be hazardous to life on planets. If the SFR is too low, then there would be no galactic wind, and presumably CRs just diffuse out of the disk rather than being convected out at an Alfvénic surface transition layer. In general, considering the gradual build up of the abundances of heavy elements, habitability has increased for planets formed after the bulk of star formation has occurred and most threats have subsided.

1.3 ASSUMPTIONS

To analyze astrophysical constraints on the occurrence of life in the evolving Universe we use the term *habitability* and the phrase *potentially habitable* for planets, or regions in stellar systems, galaxies, and galaxy clusters with a significant probability that life could originate and/or be sustained on suitable planets within those regions. Habitable planets may or may not be inhabited. The key is that they have a nonzero probability of being inhabited. In this context, habitability is not a binary concept. Planets throughout the Universe undoubtedly offer a wide multidimensional spectrum of both hospitable and inhospitable environments. However, some environments are better than others and some planets are more habitable than others. A major contributing factor is time. In many cases, simple or complex life may develop, but be extinguished by a catastrophic event. One assumption made here is that life as we know it requires suitable conditions that remain relatively stable over very long (Gyr) time-scales.

It is generally assumed that complex life is most probable on rocky/metal planets. Also, planets need to be large enough to support an atmosphere with CNO at sufficient surface pressure to maintain surface water (Kasting et al., 1993). Life flourished in Earth's oceans long before it spread over land; another role of oceans is to support the hydrologic cycle. Oxygen, the third most abundant element in the Universe and a constituent of water is assumed to be critical for life. The build up of O_2 in the atmosphere of planets, as a result of respiration of photo-synthetic life-forms or by other mechanisms, is assumed to be of great benefit (if not essential) for the development of complex life, both aquatic and on the land. Hence, we will regard a minimum oxygen abundance to be a necessary condition for habitable planets. The probability of complex life is enhanced in the case of a thick and oxygen-rich atmosphere. Thus, planets can be habitable for complex life only if they can retain surface water and that depends on the radiation and high-energy particle environment.

Extremophiles, by definition, are organisms that tolerate and even thrive in conditions that are harsh or even lethal for most life-forms on Earth. See the review of extremophiles by Rothschild and Mancinelli (2001) which motivates this discussion. We assume that environments on Earth and elsewhere in the Solar System may be studied and extrapolated to potentially habitable exoplanets (Preston and Dartnell, 2014). Consider hyperthermophiles living in the hot geyser pools of Yellowstone Park. These geothermal habitats are rare on Earth. One can imagine a planet with much more intense geothermal activity than Earth. These hyperthermophiles would be abundant on such an imagined world. Similarly, salt-loving halophiles might become abundant on a planet with a surface exposed to repeated evaporation. A lesson from extremophile studies on Earth is that these life-forms protect themselves by isolation from external conditions and by removing threats or repairing damage quickly in order to survive. Generalizing this example we thus consider physiological capacity of extremophiles as one among several layers of biological protection against environmental challenges. Rothschild and Mancinelli (2001) suggest that even humans are extremophiles in some sense as they thrive in an oxygenated atmosphere and experience damaging effects of free oxygen radicals. Some organisms, anaerobes, flourish in anoxic environments. However, life as we know it, transformed the environment, allowing oxygen consuming aerobes to thrive on the surface, in the air, and in shallow waters on Earth. At the same time, we do not assume that the Earth is the best of all possible worlds. For example, planets with reducing atmospheres may harbor a rich biosphere.

For simplicity, two types of habitability can be distinguished. The first is basic habitability supported by the existence of elements required to construct planets, their atmospheres, and organic compounds. Within this basic habitability condition, microbial life, extremophile and otherwise, may have existed in the past and may even exist today over vast areas of the Universe. There are likely even entire galaxies without complex life as we know it. Deep ocean life on planets may only be limited by the availability of water and atmospheric pressure to maintain the ocean. Complex life may arise on planets within lakes and oceans even if the environments above the water are not habitable. Thus, the overall habitability may depend on local niches on the surface, in shallow waters or in the air, and provide a diversity of environments comparable to or exceeding (Heller and Armstrong, 2014) that on Earth. Long-term environmental stability has likely been of great benefit to the life on Earth. Thus, the second more restrictive condition of habitability, is the capacity to support complex life as we know it here on Earth. In this case, biospheres can be challenged by factors such as atmosphere erosion, ozone removal, and long-term temperature regulation. Finally, we assume that the second condition of habitability includes atmospheric oxygen O_2, which is a powerful energy source for animals, and an ozone O_3 layer for ultraviolet protection of the surface of the planet and shallow waters. These conditions seem to be important mechanisms or at least catalysts for promoting the emergence of complex life.

2 HAZARDOUS RADIATION AND PARTICLES

In this section we discuss CRs, which are energetic particles of astrophysical origin. Most CRs are protons, but they also include a mixture of He nuclei and heavier element ions. At the highest energies, massive nuclei are much more common (Thoudam et al., 2016). These highest energy particles collide with particles in the upper atmosphere of planets, producing many secondary particles; especially muons, which are highly penetrating. Some secondary particles are absorbed in the air whereas others reach the planetary surface and contribute to the surface radiation dose of living organisms. Other particles penetrate the surface and may impact subsurface organic chemistry and life. CR particles may be classified into three types based on three different source categories, as follows from data on particles detected on Earth. This classification is related to protection mechanisms that prevent particles within certain energy ranges from reaching Earth (see next section for details).

The effect of radiation and particles on life is reviewed by Dartnell (2011), also see Abrevaya and Thomas (2017) in this volume. Here we address sources of high energy radiation and particles that are potentially hazardous to the biota. A plot of the CR particle flux as a function of energy is shown in Fig. 1. The lowest energy CRs (E up to a few GeV) include solar energetic particles (SEPs), which can occasionally be quite strong. This is because most of the lowest energy particles from nonsolar sources do not reach the surface of Earth. SEPs represent the high energy tail of the solar wind particle distribution and do not contribute to the CRs shown in Fig. 1. Higher energy particles (with E above a few

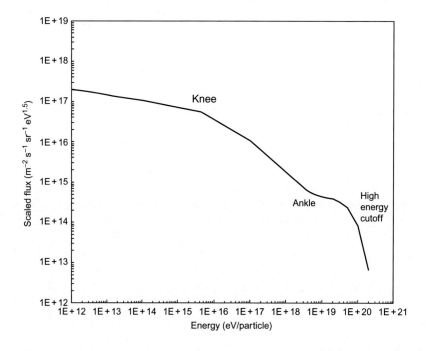

FIG. 1

The cosmic ray (CR) spectrum as detected on Earth. Galactic cosmic rays (GCRs) dominate at Earth from below 1 GeV all the way to about $10^{18.5}$ eV, the ankle, at which point they are taken over by extragalactic CRs, which reach the highest energies so far observed, the high-energy cutoff, about 10^{20} eV. SEPs (solar energetic particles) can contribute significantly, if you are outside Earth, because of their burst-like character, but on average they are low in flux. SEPs can dominate on another planet, but at Earth, the spectrum shown includes no SEPs whatsoever.

GeV) have a different origin and are called galactic CRs (GCRs) or extragalactic CRs (EGCRs), depending on their source. GCRs are the dominant population of CRs up to the so called ankle seen in Fig. 1, with EGCRs dominating above it.

A significant fraction of primary CRs likely originate from massive star SNe. However, this is probably not their only source. AGN and stellar mass black holes, neutron stars, and white dwarfs probably also produce some CRs detectable on Earth. Note that charged particles accelerated from galactic sources supply a galactic magnetosphere which traps CRs with lower energy. Planets in the disk of galaxies with ongoing star formation, like our galaxy, get bathed by GCRs. In the Milky Way GCRs stay diffusively in the disk until they convect out in a Galactic wind, which takes about ten million years at the location of the Sun and somewhat less in the inner Galaxy (Biermann et al., 2010, 2015; Wiegert et al., 2015). Some particles encounter the atmosphere of a planet, and those with the highest energy produce an air-shower. Information of their origin is usually lost as they follow trajectories deflected by the galactic magnetic field and thus are detected isotropically (Biermann et al., 2013). While GCRs trapped in the galactic magnetic field are a hazard to life imposed by SNe, the galactic magnetic field generated by GCRs accelerated in the remnants of SNe provides a shield against all but the highest energy EGCRs.

2.1 SOLAR/STELLAR ENERGETIC PARTICLES (SEPs)

The heliosphere extends out to a (time variable) distance of about 120 AU, depending on direction, from the Sun, where the solar wind and interstellar medium pressures balance. The solar wind velocity falls below the supersonic threshold and the density of particles increases. Magnetic fields within the termination shock provide a barrier to lower energy particles, mostly GCRs, with original energies less than 300 MeV. The strength of the solar wind is variable and there is an observed anticorrelation between solar activity and GCR flux, confirming this protection mechanism, see e.g., Shaviv (2003).

SEPs are responsible for mass loss of planetary atmospheres. However, planets may be magnetically protected. For example, the magnetic field of the Earth deflects or traps SEPs, enhancing the habitability of Earth. Mars currently does not have a significant global magnetic field, and its mass is about 10% of the mass of the Earth. Mars still experiences significant atmospheric mass loss due to solar activity as measured by instruments aboard the MAVEN (Mars Atmosphere and Volatile EvolutioN) spacecraft (Dong et al., 2015). In the early life of the Sun it was less luminous than it is now, however it was also much more active.

Considering stars of other types, the more massive upper main sequence stars consume available fuel and die before life can develop on their planets. Short stellar lifetimes effectively limit habitable planets to stars on the lower main sequence, below about 1.1 $M_\odot$. Stars are typically formed spinning much more rapidly than the present day Sun, e.g., Soderblom et al. (1993). As stars in this lower mass range have outer convection zones, a magnetic dynamo is generated and a stellar corona forms. As they age they gradually lose their angular momentum, thereby slowing their rotation as the result of winds driven by the corona. This model was introduced by Parker (1958) and studied further by Weber and Davis (1967). Planets in the habitable zone of magnetically active stars are subject to X-ray and ultraviolet radiation as well as stellar winds. Magnetic activity declines, with age, as the rotation period increases. Today, the activity of the Sun is lower by two to three orders of magnitude than it was during the formation of the Earth. Correspondingly, the SEP flux from the Sun has minimal effect on Earth today. However, the lowest mass stars remain active much longer.

2.2 GALACTIC COSMIC RAYS (GCRs)

GCRs have long been thought to originate from the gaseous remnants of SN explosions (Baade and Zwicky, 1934). Direct detection of pion-decay signatures in supernova remnants, IC 433 and W44, were made using the Fermi gamma-ray telescope (Ackermann et al., 2013). Specifically, they detected a cutoff in the gamma-ray spectrum due to the decay of neutral pions, produced when accelerated CRs interacted with the interstellar medium surrounding these SNe. Neutral pions quickly decay into gamma-rays, while charged pions become electrons, positrons, and neutrinos. SNe are probably not the only source of GCRs, but this detection suggests that shocks within SN remnants do accelerate particles. Benyamin et al. (2016) considered the spiral arms of galaxies as the main location of GCR sources. Aartsen et al. (2013) found a large-scale anisotropy in PeV emission and suggested that this result is consistent with the superposition of flux from a just few nearby sources, likely SN remnants. SN explosions of the most massive stars (above 25 $M_\odot$) are more important in many aspects, but the standard SNe (stars between 8 and 25 $M_\odot$) are the major contributors to CR-protons, and dominate the overall energetics.

Other sources of GCRs likely include white dwarf novae, low and high mass X-ray binaries (LMXB and HMXBs)—a low or high mass normal star transferring matter onto a neutron star or a black hole, and microquasars. Microquasars are scaled-down versions of AGN where matter accretes onto a stellar mass black hole, rather than a SMBH, and mildly relativistic collimated particle jets are ejected. Very high energy, short duration, bursts from a galactic black hole binary were first observed by Mason et al. (1997) with CGRO BATSE- Compton Gamma-Ray Observatory's Burst And Transient Source Experiment and later confirmed at even higher, 100 GeV, energies using MAGIC (Major Atmospheric Gamma Imaging Cherenkov Telescopes) (Albert et al., 2007) from the HMXB Cygnus X-1, the first confirmed Galactic black hole. See Mirabel et al. (2011) for an illuminating discussion on ionization due to stellar mass black holes.

2.3 EXTRAGALACTIC COSMIC RAYS (EGCRs)

The galactic magnetic field prevents many of the lower energy EGCRs from entering the Galactic disk; however the highest energy particles penetrate the magnetic field of the Galaxy and continue unimpeded through the heliosphere and the Earth's magnetic field to impact the atmosphere. The highest energy CRs detected, with energies up to 3×10^{20} eV are called ultra-high energy CRs (UHECRs), see Fig. 1. The Cosmic Microwave Background Radiation as well as the far-IR radiation fields interact with the UHECRs, limiting their travel distance, an effect known for protons as the GZK-cutoff (Greisen, 1966; Zatsepin and Kuz'min, 1966). At the upper end of the UHECR spectrum, massive nuclei dominate, so that the cutoff is modified, see the review by Olinto (2012).

However, the sources of UHECRs are naturally limited at the highest energies. Apparently, acceleration mechanisms are not able to accelerate CRs to the highest energies beyond the detection cutoff shown in Fig. 1. Theoretically, around 10^{21} eV (for protons, and without relativistic boosting) there seems to be a general limit. This combination derives firstly from the spatial limit (Lovelace, 1976) given by the magnetic field, which ties into the jet and is energetically limited by the Eddington power. So the more power, or luminosity of the jet, the higher is the possible particle energy, giving $E_{\max} \sim L_{jet}^{1/2}$. A second constraint comes from losses, such as synchrotron and photon interaction losses, giving a maximum energy that runs inversely with L_{jet}, as discussed in Hillas (1984) and Biermann and Strittmatter (1987).

Pierre Auger Collaboration et al. (2011) argues that the energy dependence of the composition of UHECRs supports the ankle interpretation of the transition from GCRs to EGCRs, see Fig. 1. The determination of EGCR sources is difficult mainly due to insufficient statistics, but also due to magnetic deflections of particle trajectories that are energy dependent. A hot spot of 19 UHECRs detected over 5 years by the Telescope Array (TA) collaboration is consistent with the starburst galaxy M82 (Abbasi et al., 2014). While the localization is consistent with other sources including a blazar, M82 remains the best candidate. Taken at face value, UHECRs from M82 suggest that planets in galaxies with rapid star formation are subject to high fluxes of UHECRs from local sources, like GRBs, microquasars, and relativistic SNe remnants. However, the incoming directions of UHECRs, which are not strongly deflected by magnetic fields, indicate that Milky Way sources do not significantly contribute to the UHECR flux currently observed impacting Earth. While the threat of UHECRs seems low because the flux is low now, see Fig. 1, many planets throughout the local Universe are exposed to higher fluxes of CRs at all energies, especially before the time Earth was formed.

Accretion onto SMBHs, associated with AGN, are also probably responsible, aided by magnetic fields, for accelerating UHECRs, see Biermann et al. (2016) for a review of the origin of UHECRs. An AGN likely inhibits life or even eliminates habitability within its host galaxy. AGN might even adversely affect other nearby galaxies as UHECRs are able to break free from the magnetic confines of the host galaxy. Only some AGN or some brief episodes of SMBH activity are significant CR sources, while many more are sources of beamed high energy electromagnetic radiation. As in the case of GRBs, the ionizing radiation from SMBHs is also a major threat to habitability.

Radio galaxies are effective particle accelerators. The Milky Way, and its cluster of galaxies known as the Local Group, are on the edge of the Virgo Supercluster of galaxies, which is centered on the rich Virgo Cluster of galaxies. In particular, at the heart of this supercluster with a mass 10^{15} $M_\odot$ lies the giant elliptical galaxy M87 and its radio source Vir A, with a SMBH powering an impressive jet. Other relatively near radio galaxies are Cen A and Cyg A. Benford and Protheroe (2008) suggested that the radio lobe structure in Cen A, shown in Fig. 2, might be a source of detected EGCRs, specifically the UHECRs ($E \geq 3$ EeV: above the "ankle" in Fig. 1). UHECRs may also include a proton component from many radio galaxies integrated over vast distances, visible already below 3 EeV (Biermann et al., 2015).

Becker and Biermann (2009) developed a detailed AGN model that suggests that neutrinos are produced near the launching region of the AGN jet, due to high optical depths for proton-photon interactions. Subsequently, protons escape from shocks where optically thin proton-photon interactions may occur. This model predicts CRs from FR-I galaxies (Fanaroff, 1974)—massive galaxies often with jets and associated with X-ray emitting gas as they move through a rich galaxy cluster. Importantly, this effect is independent of source orientation. Direction is important for neutrino detection as they are ejected only along the jet direction. BL Lac objects and flat spectrum radio sources are accreting SMBHs with jets directed toward Earth and they are significant particle accelerators. However, some AGN may be just in a spin-down mode, not in an accretion mode. The spin-down mode gives low luminosity, but for an extended duration, and thus it may also affect habitability.

FIG. 2

Centaurus A (Cen A) located 4 Mpc away in a composite X-ray, submillimeter, and optical image. It is by far the most violent nearby accreting supermassive black hole (SMBH) in observable display. The radio emission spans 20 times the size of the full Moon in the sky. A merger took place just a few tens of millions of years ago. Both old and new jets are seen pointing in different directions, indicating the merger a pair of SMBHs. As discussed in the text, ultra-high energy CRs (UHECRs) from Cen A may be among those detected on Earth.

From: NASA (https://www.nasa.gov/topics/universe/features/radio-particle-jets.html), X-ray: NASA/CXC/CfA/.; Submillimeter: MPIfR/ESO/APEX/A.; Optical: ESO/WFI.

2.4 THE STAR FORMATION RATE (SFR)

The SFR history plays an important role in the habitability of galaxies. The most massive stars do not last long and explode as SN-Type Ib/c, followed by the chemical enrichment of the interstellar medium. However, nearby SNe and GRBs probably have quite adverse effects on life (Gehrels et al., 2003; Thomas et al., 2005), causing mass extinctions, temporary sterilization, and even atmosphere removal if they happen very close. Images of two nearby SNe remnants, Cas A and the Crab Nebula, are shown in Fig. 3. Over billions of years, the occurrence of nearby SNe is unavoidable, however a lower SFR and associated lower SN frequency increases the probability that a planet will avoid the most catastrophic events. In Fig. 4, the SFR history of galaxies (Madau and Dickinson, 2014) is shown along with other trends discussed herein. The far-IR luminosity of galaxies is used as a proxy for SFR. Also, far-IR luminosity can be used as a proxy for gravitational wave events as essentially all massive stars are in binaries (Chini et al., 2012). However, gravitational wave events are not dangerous at the distances of concern here.

Planck observatory results (Planck Collaboration et al., 2016) suggest that the reionization of the Universe occurred at redshift value about $z = 8.8$, directly or indirectly as the result of star formation

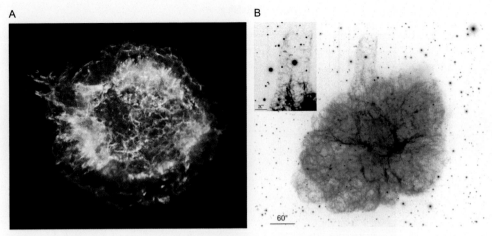

FIG. 3

Supernova remnants accelerate galactic cosmic rays. Left: A composite image of the roughly 300 year old supernova remnant Cas A, which is 3.4 kpc away. It has two oppositely opposed jets, the prominent one seen pointing toward the upper left. Image credit: NASA/CXC/SAO. Right: The 960 year old Crab Nebula supernova remnant which is located at a distance of 2 kpc. This Subaru telescope image shows remarkable details of its jet in this O III image that has been scaled logarithmically to show jet details. Inset: More detailed view of the jet.

From Rudie, G.C., Fesen, R.A., Yamada, T., 2008. The Crab Nebula's dynamical age as measured from its northern filamentary jet. Mon. Not. R. Astron. Soc. 384, 1200–1206. https://doi.org/10.1111/j.1365-2966.2007.12799.x, used by permission.

that began the process of heavy element fusion. Notably, a single epoch for ionization is extremely unlikely; it is suggested by all flat ΛCDM (dark energy + cold dark matter) models with the latest Planck cosmology parameters that this epoch could be quite protracted. Kogut et al. (2003) wrote that star formation started probably no later than $z = 20$, using a simple two-step function.

A promising reionization source was a flurry of star formation in galaxies that are similar to the local Green Pea galaxies, discovered by a citizen science program (Cardamone et al., 2009) of the Sloan Digital Sky Survey (SDSS). These compact low mass galaxies appear small and green because of a vast region of doubly ionized oxygen. They include galaxy scale star formation regions capable of ionizing the intergalactic medium with escaping Lyman continuum radiation. One local green pea, J0925+1403, at $z = 0.3$, is leaking ionizing radiation with an escape fraction of 8% (Izotov et al., 2016) and is a shining example of the ionizing power of high mass star formation. If the SFR can be used to trace the SN rate (i.e., due to the short lifetimes of stars that become SN-Type II), then it also likely tracks the GCR levels over cosmic time.

3 LOCAL ASTROPHYSICAL THREATS TO LIFE

Astrophysical sources of radiation often display dramatic variability over time. Occasionally, enormous bursts occur, which may be associated with relativistic particle ejection. Sometimes these are one-time events like SNe or GRBs. For accreting sources like microquasars and AGN, hazardous

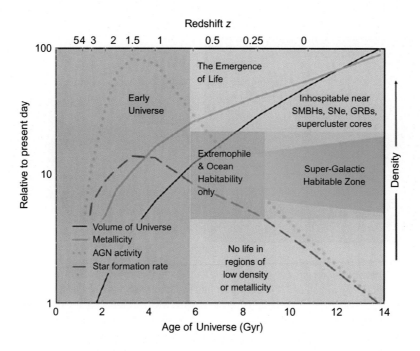

FIG. 4

Time evolution curves of star formation rate (SFR) (a proxy for supernovae (SNe) and gamma-ray burst (GRB) rates), Active galactic nuclei (AGN) activity, metallicity, and volume of the Universe. The elements of life slowly build up, while habitability threats decrease. On the right side is a schematic view of the development of a super-galactic habitable zone (SGHZ). Life is compromised near the center of rich superclusters and merging galaxies. Far from dense galaxy concentrations, habitability is possible in regions with sufficient metallicity. The *shaded region* around 6 Gyr and extending to later times in certain places corresponds to extremophile habitability without complex life on land as we know it.

events may occur repeatedly. Here we discuss such sources and their impact on planet habitability. The danger posed by individual objects is a strong function of distance and depends on the direction of beamed radiation sources. Planetary habitability is especially inhibited when high local SFRs are coupled with the growth of SMBHs.

The worst threats are those that cause complete planetary sterilization; these are most often local threats as a planet's location is a key to habitability. Life on Earth has avoided the worst catastrophes, since evidence shows the continuous presence of life for 3.5 Gyr. As discussed in Section 1, mass extinctions were quite common on Earth, yet there is no evidence of a complete planetary sterilization. Mass extinction events on Earth, have paved the way for an increase of evolutionary diversity as new niches became open. A key factor of habitability is the severity and frequency of sterilization and mass extinction events and how readily life is able to recover from them. The expansion of the Universe and the local reduction in the SFR improves the habitability of planets protected by magnetic fields and thick atmospheres.

Life depends on water, and therefore, a catastrophic or gradual loss of atmospheric water can result in a loss of habitability. Here we propose that a planet sufficiently supplied with a protective atmosphere and magnetic field in the mid-disk of the Milky Way is currently more habitable than similar planets in many other galaxies in the contemporary Universe. We argue this because, first, the Sun stays relatively far away from the SMBH in the center of the Milky Way, a major threat. Second, the Galactic SMBH has a relatively low mass and has been unusually inactive for some time, since only minor mergers have taken place in the Milky Way over the last 10–12 Gyr (Gilmore et al., 2002). Third, the galactic disk currently forms new stars at a rate which sustains a global magnetic disk wind, providing an efficient shield against EGCRs without supplying too many GCRs. It is difficult to evaluate the danger of radiation and particle dosage as we do not know what complex life is capable of withstanding. However, we speculate that life, and especially metazoan life on the surface, can be harmed by large doses of particles and radiation. Within this context, we conclude that life as we know it was rare or impossible in the disk of the Milky Way earlier than about 5–6 Gyr ago. An earlier limit, at about 8 Gyr ago, also exists for only microbial life. These temporal limits are illustrated in Fig. 4. Habitable conditions probably exist somewhat earlier in some disk galaxies, i.e., those without recent mergers or SMBH activity, like the Milky Way, compared to other more recently active galaxies.

3.1 SUPERNOVAE (SNe)

Single massive stars above about 8 $M_\odot$ explode as SN-Type Ib/c. Two examples of these massive star explosions, showing a stem-shaped breakout or jet, are shown in Fig. 3. Such SN remnants produce powerful shock-waves, which in turn accelerate CR particles. There are two types of precursors of these SNe: red super giant stars with slow dense winds, and blue super giant stars with fast low-density winds. Very massive stars above about 25 $M_\odot$, depending on their heavy element abundance, commonly produce stellar mass black holes (Heger et al., 2003; Woosley and Heger, 2015). These very high mass stars explode via an unknown mechanism. Models involving neutrinos (Bethe, 1990), or the magneto-rotational mechanism (Bisnovatyi-Kogan, 1970) have been proposed. All of them occur in binary star systems (Chini et al., 2012). A small fraction of them explode as GRBs or as relativistic SNe.

As a dying star ejects its envelope into space it encounters gas, either the stellar wind of the predecessor star, or the immediate environment of an OB-super-bubble, or directly the interstellar medium, forming a supersonic shock. A mechanism called diffusive shock acceleration proposed by Fermi (1949, 1954) and investigated by many others (Axford et al., 1977; Axford, 1981; Krymskii, 1977; Bell, 1978a,b; Blandford and Ostriker, 1978; Drury, 1983) is thought to accelerate particles to GCR energies.

The existence of expanding SNe remnants in the disk of the Milky Way and other disk galaxies provides a continuous supply of CRs. This occurs as long as the frequency of SNe events is high enough to replenish the GCR medium, which escapes the Galaxy on a time scale of 10^7 years in the solar neighborhood. At higher CR energies, that containment time is lower by $E^{-1/3}$. In the central region of the Galaxy the containment time may be much shorter (Biermann et al., 2010). The efficiency of the conversion of SNe kinetic energy to CRs is uncertain. However, Duric et al. (1995) performed an analysis of optical and radio observations of SNe in the Local Group spiral galaxy M33. They find that most (90%) of the SNe remnants are not significant CR accelerators, with the brightest 40 SNe remnants in M33 producing the bulk of the GCRs.

High mass stars, O and B types, explode either into the interstellar medium, or into their own wind. When the shock goes through the wind, it continues into the OB-super-bubble (Binns et al., 2007; Rauch et al., 2009; Murphy et al., 2016). It is only a statistical fluke that the Milky Way has only slow cases of interstellar medium-SNe, and not the fast cases as seen in M82 (Kronberg, 1998; Bartel et al., 1987), all of which are by several powers of ten more luminous than any SN remnant in our galaxy. We interpret these luminous SNe in M82 as explosions of stars into their previously ejected winds. Hence, some SNe may have more adverse effects on habitability than has been previously considered, using only Galactic SNe.

3.2 GAMMA-RAY BURSTS (GRBs)

GRBs are stellar explosions that generate enormous radiation beamed in two opposing directions. Local GRBs may occasionally sterilize large portions of the land-based life in the Galaxy. Annis (1999) argues that GRBs are so critical to astrobiology that the Universe is currently undergoing a phase transition from no intelligent life to intelligence as a result of the reduction of GRBs. While we generally agree that the reduction in number and luminosity of GRBs is critical to habitability, the timing of this transition is likely not universal. Galaxies with active SMBHs, especially those resulting from mergers, will remain uninhabitable for many Gyr into the future.

Notwithstanding atmospheric removal by a close and beamed GRB, the greatest impact of GRBs on surface or shallow marine life is the depletion of ozone and the subsequent exposure to stellar ultraviolet radiation. Melott et al. (2005) pointed out that DNA damage in organisms on Earth from GRBs is greatest at mid latitudes due to local ozone depletion due to enhanced ultraviolet B-band (UVB) radiation and the direct angle of GRB at these latitudes. Hence, in most cases (except the worst scenarios) GRBs would result in a mass extinction over large surface areas but not planetary sterilization. Galante and Horvath (2007) also examined the astrobiological impacts of direct GRBs and found that at distances up to 100 kpc there may be significant ozone depletion and thus more stellar UVB reaching the ground. Compounding the situation there would be less 350–450 nm stellar radiation reaching the ground resulting in a less efficient photo-repair of DNA damage. Also, the reduction of visible light, also known as photosynthetically active radiation (PAR), would slow down photosynthesis.

Direct CR flux from GRBs would be sterilizing only for a very close GRB source, less than about 10 pc. Since the integrated radiation is what drives their impact, GRBs are too brief to do extensive damage, unless they are quite close and/or if they are beamed directly at the planet. Using determinations of the luminosity functions and the rates of GRBs with consideration of host galaxy properties, Piran and Jimenez (2014) estimated the probability of a lethal GRB within a galaxy, which appeared to be much greater in the inner Milky Way, than locally. Specifically, that GRBs have a 95% probability of sterilizing a planet within 4 kpc of the galactic center. They further suggested that planets further out in the Galaxy play a game of "GRB Roulette" as the probability of mass extinction by GRB decreases to 50% beyond 10 kpc. Their work also suggests that the Earth, at the distance of 8 kpc from the Galactic center, has likely been exposed to one sterilizing GRB event.

There is an indication that at least one Galactic GRB has occurred in the last million years or so. The argument is as follows. The highest energy particles accelerated from a GRB are mostly neutrons. Protons remain trapped in the magnetic field and thus lose energy adiabatically during escape. A relativistic neutron travels through the Galactic disk and decays after some time into a proton, an electron, and an antineutrino. The proton is then caught by the Galactic magnetic field and with a small probability (about

0.05) interacts with the interstellar medium, again becoming a neutron traveling to us undeflected. Biermann et al. (2004) estimated that one to a few GRBs occurring about a million years ago within 3 kpc of the Galactic center can account for the excess of EV CRs by this process, detected by one instrument, AGASA- the Akeno Giant Air Shower Array (Hayashida et al., 1999; Teshima et al., 2001), but not confirmed by IceTop - the surface array of the IceCube neutrino telescope (Aartsen et al., 2016).

3.3 NEARBY SUPER-MASSIVE BLACK HOLES (SMBHs)

As we observe SMBHs in the local Universe, it is clear that many of them remain quiet for a long time, but these inactive SMBH have been active in the past and the right kind of accretion event will turn an inactive SMBH into an active one. AGN activity as a function of redshift is shown in Fig. 4. The peak around $z = 2$ indicates that AGN were most active 10 Gyr years ago. AGN activity dropped by an order of magnitude by the time the Earth formed, about 4.6 Gyr ago, corresponding to $z = 0.45$, and it has fallen by another order of magnitude since then.

The density of EGCRs was dramatically higher in the past not only because the density of CRs sources was orders of magnitude higher, but also due to an important cosmological effect. The co-moving volume of the Universe was smaller by a factor of $(1+z)^3$ in the past, exposing planets to more direct UHECRs from SMBHs as well as from extragalactic SNe. This means that at the peak of the cosmic SFR, the local GCR density was about 15 times higher than it is today, recall again Fig. 4. As mentioned, AGN activity peaked at roughly the same redshift, $z = 2$, when the Universe was a factor of 27 smaller. Thus the EGCR flux from AGN was about 2400 times greater than at present. The activity of SNRs counts with $(1+z)^3$ only in regards to other galaxies in the field, not SN remnants in our Galaxy nor inside of our own local co-moving group. As groups and clusters slowly grow by accretion, there is a weak countervailing effect. Comparing with Fig. 1, at $z = 2$ the entire CR spectrum was higher and the "ankle" flux was as high as the "knee" flux! These effects combine to suggest that the background level of CRs may strongly compromise or even prevent surface or shallow water life until about the formation of the Earth. Furthermore, short, very energetic energy bursts have probably adversely affected habitability in many locations, and many of them happened recently.

Planets that reside in large clusters of galaxies have larger ionizing radiation fluxes than planets in the Milky Way because they are likely to be near SN remnants and accreting SMBHs. In our cosmic neighborhood we have several SMBH. These are in M31, M32, M81, M94, NGC5128 (Cen A), and one in our own Galaxy, all within about 5 Mpc today. Within about 20 Mpc there are many other SMBHs. Having more local AGN would be a serious threat to life. M87, for example, has consumed roughly 100 Milky Way mass galaxies and has generated an enormous amount of radiation and relativistic particles in the process of growing its SMBH (Andrade-Santos et al., 2016). Several other giant elliptical galaxies in the heart of the Virgo supercluster probably have also accreted many galaxies, but are currently between merger events.

3.4 GALAXY MERGERS AND SMBH MERGERS

Galaxy mergers amplify the SN and GRB rates and SMBH activity. During a merger, giant molecular clouds collide, resulting in a considerable increase in the SFR and its associated SN-Type II rate. Considering the starburst galaxy M82, Kronberg et al. (1981, 1985) found that the radio, infrared, and X-ray luminosities from a region that is 600 pc in diameter within M82 is the result of an extreme burst of massive O and B type star formation. They detected many SN remnants and gave evidence for buried SNe exploding within giant molecular clouds and then breaking through the dense cloud to become visible. The SFR in that region is about 10 times that of a normal star forming galaxy like the Milky Way. The VERITAS - Very Energetic Radiation Imaging Telescope Array System Collaboration detected gamma radiation from M82 and estimated a CR density roughly 500 times greater than that of the Milky Way. M82 is likely experiencing a temporary enhanced star formation period due to a recent merger, as M82 did not have significant star formation outside the star forming nucleus in the last 300 million years. Based on observations of planetary nebulae, M82 has cannibalized a galaxy that was probably a $6 \times 10^9 \, M_\odot$ star forming galaxy within the last Gyr (Longobardi et al., 2015). Mergers may also result in accretion onto SMBHs. For example, the nearby galaxy M51 has a companion galaxy, NGC 5194, and Chandra X-ray observations show huge particle ejection events from the SMBH in the smaller companion (Schlegel et al., 2016).

Planets in rich clusters of galaxies were exposed to high energy radiation and particles each time their host galaxy merged. The fraction of merging galaxies is approximated as function of redshift, z:

$$F_M = F_{M_0}(1+z)^M$$

This equation is used by a number of investigators, see Khochfar and Burkert (2001) for a summary of results. F_{M_0} is the current fraction of galaxies undergoing mergers, estimated to be about 1%. The power M ranges from 6 in rich clusters of galaxies to 3–4 for field galaxies (Le Fèvre et al., 2000). At the conservative end, the fraction of merging field galaxies is $F_M = 27\%$ at $z = 2$. In a rich cluster of galaxies the majority of galaxies are merging at $z = 1$. Complex life as we know it is highly unlikely to occur in galaxies undergoing mergers, or in those that have merged recently.

When two big galaxies merge and both contain SMBHs, then their SMBHs also merge. Gergely and Biermann (2009) showed that the spin of the final merged black hole will be aligned with the direction of the angular momentum of the SMBH binary orbit, and so it is in a totally different direction than the original spins of the two SMBHs. This means that while this happens the two jets sweep through the sky, cleaning out a large conical solid angle (Gopal-Krishna et al., 2003, 2012).

Kun et al. (2017) presented a model of the merger of two SMBHs driving the consecutive emission of gravitational waves, high energy neutrinos, and UHECRs, followed by a luminous radio afterglow. This means that SMBH binary mergers involve interaction with matter on a galactic scale, so maximal injection and acceleration of UHECRs occurs. Also tremendous mutual interaction of the SMBHs leads to powerful beams not just of neutrinos, but also TeV photons, neutrons, and copious UHECRs. These beams could be devastating, even at great distances. In fact, this mechanism may explain the neutrino events detected by the IceCube Collaboration et al. (2016), showing that such beams are visible in neutrinos across the entire Universe.

The blazar PKS 0723-008 is a strong candidate with evidence for a spin-flip, high energy neutrino emission, and a recent increase in radio emission by a factor of about 5. Kun et al. (2017) also discussed four flat spectrum quasar identifications with IceCube track events, two of which have a flat spectrum to near THz. Radio interferometry evidence showed that seven others are undergoing mergers right now (± a few million years) and that all these have a flat spectrum to near THz, suggesting that a recent merger may have caused such extreme emission. A flat spectrum of a compact source implies a jet pointing at us, and thus extreme relativistic boosting may occur, and only about 10% of all GHz flat spectrum radio quasars have a spectrum extending flat to near THz.

3.5 AGN, SMBHs, AND ULTRA-LUMINOUS X-RAY SOURCES (ULXs)

An important quantity concerning accretion onto compact objects such as black holes is the Eddington luminosity—the radiation luminosity for which momentum transfer via radiation balances the momentum transfer from gravitational attraction, or in other words, accretion is stopped. Interestingly, van Paradijs (1981) showed how accretion can circumvent this effect in some cases that concern us here. Consider the filling of a bubble with relativistic gas by a radio galaxy, clearly visible in many cases like M87, NGC1275, Cen A (Fig. 2), Her A, etc. (e.g., Owen et al., 2000 for M87). All these radio galaxies are in a cluster of galaxies with its associated hot gas, at a temperature of about 10^8 K, and central density of order 10^{-3} cm^{-3}, so a pressure of about $10^{-10.5}$ dyn cm^{-2}. The radio galaxy fills up the bubble, of radius, say, 1 Mpc. This filling is highly nonsteady. Whenever the central SMBH merges with another one (e.g., Gergely and Biermann, 2009; Kun et al., 2017), a gigantic burst of new relativistic energy is released (Kronberg et al., 2001), possibly approaching a good fraction of $M_{BH}c^2$, say 1/2 (Hawking, 1971). In a fresh refilling, the energy may also go high enough so that the equation of state might get close to relativistic.

Consider a fresh outburst of energy spread around by a freshly merged central SMBH of 10^{10} M⊙ (Caramete and Biermann, 2010). This would add an energy of 10^{64} erg to the bubble, resulting in a pressure that exceeds the cluster pressure discussed above. The bubble would then be capable of bursting, as we have "light fluid"—the relativistic gas, being held in by a "heavy fluid"—the thermal gas. Such outbursts are directly visible in the SNR Cas A (Hwang et al., 2004; Fesen and Milisavljevic, 2016), and the Crab SN remnant = M1 (Black and Fesen, 2015), both showing stem-shaped jets in Fig. 3, and in some radio galaxies like the giants DA240 (Peng et al., 2004) and 4C+73.08 (Strom et al., 2013). Outburst flow has been modeled by Kompaneets (1960) and Moellenhoff (1976), and these models predicted the rather characteristic stem-like morphology clearly visible in Fig. 3. Outbursts from an overpressured gas result in flow velocities far above the speed of sound, as derived in Fluid Mechanics (Landau and Lifshitz, 1959), which can approach the speed of light often seen directly in relativistic flows. This outburst can then lead to an energy outflow of 10^{64} erg in possibly only 10^{14} s, to give a short-lived luminosity of 10^{50} erg/s, clearly above the Eddington limit for the central black hole. Therefore, it is quite plausible that temporarily particle population flows may significantly exceed the Eddington limit luminosity. The Eddington ratio deduced here can be written as:

$$f_{Edd} = \frac{1/2 \times M_{BH}c^2}{10^{38.3} \times M_{BH}/\text{M}_\odot \times R/c}, \tag{1}$$

where M_{BH} is the SMBH mass used (here 10^{10} M⊙) to produce a bubble of CRs (both hadronic and leptonic); $10^{38.3}$ erg/s is the Eddington luminosity of a SMBH per M⊙, and R is the radius of the bubble (here 1 Mpc is used). The precise mass of the SMBH actually drops out again in formulating the Eddington ratio. This specific example, patterned after M87 in SMBH mass, gives an Eddington ratio of 50. Even if we allow an efficiency far less than 1/2 to produce a relativistic fluid, the Eddington ratio might still exceed the value of unity, what is normally referred to as the Eddington limit. As linear scales of radio bubbles have an observed range from below 30 kpc to 4 Mpc, the Eddington ratio could occasionally also be far higher. The CR luminosity is likely to be more constrained, but clearly Eddington ratios far greater than unity are possible (Stollman and van Paradijs, 1985).

A relatively rare class of accreting objects, called Ultra-luminous X-ray sources (ULXs), are located, off-nuclear, in galaxies and are often found nearby star forming regions. Because of their very high luminosity (10^{40} erg/s) these are powered by black holes or neutron stars that are accreting at Eddington or super-Eddington rates. The detection of the spin period pulsations in M82X-2 securely establishes it as an accreting neutron star with a luminosity $L_X \sim 100\, L_{Edd}$ (Bachetti et al., 2014). In this case, a highly magnetized neutron star funnels accreting plasma from a mass losing companion star with a binary period of 2.5 days. The current high luminosity state of M82X-2 will be relatively short lived. Biological threats from neutron star or black hole ULXs thus likely follow the SFR, see Fig. 4.

Many ULXs are likely black holes with high masses that may have luminosities such that $L = L_{Edd}$, while others are neutron stars like M82X-2 (King and Lasota, 2016). The most energetic ULXs may involve black holes with $M_{BH} > 40$ M$_\odot$. Cygnus X-3 and XTE J1118+480 are Galactic microquasars with observed bipolar jets, likely to be similar to some ULXs. Accretion onto compact objects occurs over a wide range of masses and observed properties generally scale with the mass of the accretor, a white dwarf, a neutron star, or a black hole. For much higher mass black holes we need to consider the SMBHs powering the emission of AGN. As opposed to the off-center ULXs, AGN have black hole masses $M_{BH} = 10^{6-11}$ M$_\odot$.

By comparison, the current CR output of M87 is about 100 times higher than that of Cen A (see Fig. 2), but M87 is about 4 times as far as Cen A. So, M87 briefly emitting at L_{Edd} would also increase the flux of GCRs at Earth by a factor of 100 over current levels. Whysong and Antonucci (2003) show that the total power output of these two radio galaxies have this scaling. The claim that the CR output scales in the same manner is empirical, and not specific. We must conclude that past AGN activity of SMBHs, like that in M87, compromised life on a super-galactic scale. When considering the effects of nearby AGN, SMBH, or ULX activity, we must conclude that habitability is not a phase transition happening everywhere in the Universe at once. This point is illustrated in Fig. 4.

3.6 THE GALACTIC CENTER SMBH

The nucleus of the Milky Way emits high energy radiation and particles, potentially harmful to life. Extremely high energy gamma-rays have been detected using Cherenkov telescopes, including CANGAROO - Collaboration of Australia and Nippon for a Gamma-Ray Observatory in the Outback (Tsuchiya et al., 2004), VERITAS (Kosack et al., 2004), H.E.S.S. - High Energy Stereoscopic System, gamma-ray telescope (Aharonian et al., 2004, 2006, 2008), and MAGIC (Albert et al., 2006). An exponential cut-off above some 10 s of TeV is observed (Aharonian et al., 2009). Thus the maximum accelerated energy for a proton is 200 TeV (Guo et al., 2013). A diffuse gamma-ray component, with the same spectrum is observed from the disk surrounding the Galactic center (Aharonian et al., 2006; HESS Collaboration et al., 2016). The observed cutoff around 10 TeV results from a combination of interactions in the space between the source and us and interactions happening at the source. The maximum energy of protons in their synchrotron emission is about 10^{12} GeV (Biermann and Strittmatter, 1987; Biermann, 2006). The current radiation flux from the Galactic center is low and it is not a significant concern for life. However, this certainly has not always been the case. High levels of mass accretion onto the Galactic SMBH could have resulted in an X-ray flux at Earth comparable to that of the Sun, since the sun emits 10^{-7} of its bolometric luminosity in X-rays.

UHECRs detected at Earth are probably from accreting SMBHs at the centers of galaxies (see Section 2.3). The Galactic magnetic field helps to protect us from these EGCRs. Now let us consider CRs from the Galactic SMBH, which is a mere 8 kpc away. It is currently one of the least luminous of SMBHs observed. The gas accretion and luminosity of Sgr A* has certainly been much higher in the past. If it were to radiate at $10\, L_{Edd}$ even briefly, the CR flux would be $10^{7.6}$ times the current CR flux from Cen A and hence $10^{5.6}$ times the current GCR flux at Earth. These estimates are quite uncertain and very energy dependent. With the "ankle" energies used here, we see that the Galactic SMBH poses the greatest threat to global sterilization of planetary surfaces in the Galaxy. By analogy, the greatest threat to planets in external galaxies might arise from the presence of local SMBHs. Many if not most galaxies do not harbor SMBHs, so these may be more suitable galaxies for complex life.

4 PLANETARY PROTECTION

Many astrophysical events have a temporary or permanent adverse effects on planets. Long-term habitability requires a significant amount of luck. The circumstellar habitable zone is discussed in Section 5.3. However, many additional factors also affect so called habitable zone planets. Planetary system dynamics, host star activity, asteroid and comet impacts are important habitability factors. A planet's geomagnetic, atmospheric, hydrological, and biological history shapes its ability to support life. Shock waves from SNe, direct GRB exposure, and local AGN activity may stress and even exterminate life on many planets.

The question of whether life is prevalent in the Universe now and at times before Earth requires analysis of not only the threats to life on planets, but also on the development of planetary protection. Life on planets have several lines of protection from high energy particles. First, planets in the disk of a star forming galaxy are protected from all but the highest energy EGCRs because of shielding by the galactic magnetic field. Lower energy CRs accelerated from galactic sources, especially SN remnants, become deflected by magnetic fields to be contained in the Galactic magnetosphere. Those, mostly very high energy, EGCRs that penetrate the galactic magnetosphere. Both magnetically confined GCRs and the dangerous high energy GCRs with direct trajectories impact the astrosphere generated by the magnetized stellar wind. The astrosphere protects planets from the lower energy ($E < 300$ MeV) GCRs. A global planetary magnetic field provides the next layer of protection, not only from GCRs but also from the SEPs emitted by the host star(s). The last line of protection is the planetary atmosphere.

4.1 THE RISE OF THE ELEMENTS

Before the formation of the first stars in the Universe, hydrogen H and helium He were practically the only existing nuclei as there were mere traces of deuterium 2H and lithium Li, beryllium Be, and perhaps even smaller amounts of boron B. Today, life on Earth benefits from an atmosphere consisting mostly of N_2, O_2, H_2O, and CO_2. The availability of CNO atoms plays an important role in the potential for life on the surface, in the air, and water. For example, sufficient atmospheric pressure is needed to keep surface water from boiling. Planet formation also requires sufficient abundances of heavier elements including silicon Si, magnesium Mg, nickel Ni, iron Fe, and probably enough radioactive elements to maintain enough internal heat to drive magnetic field generation (see Section 4.4).

If oxygen is present in the atmosphere at sufficient density and atmospheric pressure, an ozone layer may be established. Atmospheric ozone O_3 provides a significant shield against ultraviolet radiation and CRs (see Section 4.5). It is thought that the complexity and diversity of life on Earth increased as a result of the increase of oxygen in the atmosphere produced by photosynthetic life. This increase of oxygen coincided with the Cambrian explosion in the geological record, 541 million years ago. CRs indirectly cause harm to potentially habitable planets by removing ozone. With the removal of ozone, X-ray and ultraviolet radiation from the host star may desiccate the planet's atmosphere even if it is in the traditional habitable zone by definition.

A high abundance of Fe is necessary to establish the planetary magnetic field. Planets without a sufficient Fe core probably cannot maintain a dynamo of sufficient strength to supply magnetic protection against CRs from a variety of sources over biologically relevant timescales. Planets with weak magnetic moments subject to high and intermittently intense fluxes of CRs are not expected to support life on the surface.

4.2 GALACTIC MAGNETIC FIELDS: PROTECTION FROM EGCRs

As discussed above, life on planets in the disks of star forming galaxies are protected from EGCRs by a magnetic galactopause generated by the galactic wind. This wind is analogous to the solar wind generating the heliosphere surrounding the solar system. This protective effect of the magnetic field in galaxies is closely connected to the spiral structure. It is driven by SNe explosions within the disk as well the shearing effect from differential rotation of the disk (Jokipii and Morfill, 1987; Hanasz et al., 2009b), see López-Cobá et al. (2017) as well as other recent Galactic wind investigations.

Consider star formation within the young Milky Way disk and the central role of SN explosions. When the stellar population was small and CNO elements were rare, a weak Galactic magnetic field likely provided little protection from EGCRs. Later, the SFR increased and the frequency of nearby SN became much higher as well. The resulting organization and strengthening of the Galactic magnetic field improved the protection of life from EGCRs. Later after CNO abundances had increased locally, the SFR slowed, and the associated threats to life decreased.

The Galactic halo wind has a k^{-2} (k wave-number: $k = 2\pi/\lambda$, $\lambda = $ wavelength), spectrum in magnetic irregularities (Biermann et al., 2015), which is standard for super-sonic flow (Federrath, 2013). The short wavelength for this spectrum is about 20 pc, and that corresponds to about $10^{16.6}$ Z eV in the disk. The energy scales linearly with charge Z. Now a k^{-2} spectrum implies that the scattering has no energy dependence and this means that the flux may be reduced inside the galaxy by forcing the CRs out.

The WMAP (Wilkinson Microwave Anisotropy Probe) haze from the Galactic center region is due to massive star explosions. The spectrum of magnetic irregularities is measured from radio scattering data, e.g., using LOFAR (Low Frequency Array Radio telescope), and direct solar wind data. In the Galactic center region these assumptions allow an interpretation of the haze and the 511 keV annihilation line, as well as for the size scales of the bubble (Biermann et al., 2010). The irregularity spectrum is found to be Kolmogorov ($k^{-5/3}$), in this region which means that the scattering is proportional to $E^{1/3}$ and the time-scale to scatter across any region of a specific scale is proportional to $E^{-1/3}$. This spectrum affects only particles with energy below about $10^{17.5}$ Z eV, as far as we know. CR transport occurs by scattering alone, so incoming particles may get moved around, stored for some time and then ejected again. This suggests that the Galactic halo wind transports all or most of the particles below about $10^{16.6}$ Z eV back out and protects life on planets in the Galactic disk.

There is some analogy between the Solar wind, which also has a Kolmogorov spectrum at small scales, and pushes out things below about a GeV, and the Galactic wind, which pushes out things below about $10^{16.6}$ eV. However, the time scale of driving the wind is of the order of 10 kpc/(500 km/s) = 2×10^7 years, not surprisingly of the same order of magnitude as the storage time of CRs in the CR disk. The wind might go much further out, but the interaction of the wind with the intra-cluster medium is not known.

Locally, we are just barely above the protective threshold of a strong galactic wind (Rossa and Dettmar, 2003; Biermann et al., 2015), and thus it is conceivable that potential life in a region not too far from the Galactic plane might not have enough protection from CRs, or suffer from a temporary weakening of that Galactic wind. The same holds true for a planet that is farther out in the disk, than about 10 kpc; out there the conditions to drive a galactic wind may no longer hold. We can see that effect in radio observations of external galaxies. So a sufficiently high level of local star formation per area in a galactic disk drives a super-sonic wind which is required for protection of life from EGCRs. Maybe today, galactic magnetic shielding is not essential for Earth. However, during a star-burst period, especially with associated SMBH accretion, in a neighbor galaxy like the Andromeda Galaxy (M31), it would be.

4.3 ASTROSPHERES: PROTECTION FROM GCRs

The circumstellar environment is dominated by electromagnetic radiation from the host star(s) and stellar winds launched by magnetic fields in the hot corona of cool main sequence stars (Biermann, 1951; Parker, 1958). As the wind escapes it fills an astrosphere—a bubble of outflowing plasma originating as a stellar wind and terminating at a dense hydrogen wall just outside the astropause. In the Solar System, the heliopause is about 120 AU from the Sun. Protection from host star winds is a prerequisite to planetary habitability, however these winds also protect life on planets from GCRs. Interaction of high energy charged particles with the dense layer of neutral atoms results in charge exchange interactions for CRs with energies below about 300 MeV. Dense stellar winds effectively reduce the flux of GCRs on the magnetosphere and atmosphere of planets. The protective value of the astrosphere against GCRs depends on the star type(s) and also on stellar activity cycles. A strong stellar wind significantly reduces the flux of GCRs, but the higher energy CRs penetrate the astrosphere. Smith and Scalo (2009) examined the effect of astrosphere collapse on planetary habitability during close encounters with nearby stars.

The Voyager spacecrafts provided the first opportunity to directly measure the interstellar GCR flux. They detected a sharp increase in GCRs as they passed through the heliopause on their trek from the heliosphere into interstellar space, where the plasma flow changes from the one dominated by the solar wind into one dominated by the GCR flux (Schlickeiser et al., 2014). The interstellar proton number flux per kinetic energy is 15 times higher than it is at 1 AU from the Sun (Cummings et al., 2016).

Planets in the habitable zone of the lowest mass M-type main sequence stars are subject to intense flares and SEPs, due to the combination of long lasting magnetic activity and close proximity of the habitable zone. However, in absolute terms, low-mass star winds are weaker and corresponding astrospheres are smaller and less effective at stopping GCRs than winds from higher mass stars. Having a sufficiently protective astrosphere is especially important during episodes of increased GCR flux and during times of planetary magnetic field reversals. Atri et al. (2013) pointed out that if such planets are rotating slowly, due to synchronization with its orbital motion, their habitability is likely compromised. On the other hand, a planet in the circumbinary habitable zone of a pair of solar like stars and with a binary period of say 25 days, will be exposed to reduced X-ray and ultraviolet radiation due to tidal deceleration of stellar rotation in binaries (Mason et al., 2015; Zuluaga et al., 2016). In addition, a circumbinary astrosphere is produced from a combined wind, thereby reducing GCRs compared to the single star case.

4.4 PLANETARY MAGNETIC FIELDS: PROTECTION FROM GCRs AND SEPs

To examine the importance of planetary magnetic fields in protecting potential life we have to look no further than at the nearest Solar System planets. Venus is too close to the Sun to have surface water, but may have previously been in the habitable zone when the Sun was younger and fainter. However, Venus rotates very slowly, once every 243 days, and as a result it does not have a global magnetic field. Venus lost its habitability as the result of its proximity to the Sun and it suffered a runaway greenhouse effect. Mars previously had surface water, but its low gravity is unable to prevent atmospheric mass loss. Mars lost its habitability for complex life once it cooled, lost its dynamo, and its atmosphere lost its protection capacity from SEPs.

For our purposes consider the stellar wind ram pressure $\rho_w v_{wp}^2$ where ρ_w is the stellar wind density and v_{wp} is the velocity of the wind relative to the planet and the magnetic pressure $B^2/8\pi$. The balance of these opposing pressures

$$\rho_w v_{wp}^2 = \frac{B^2}{8\pi} \qquad (2)$$

forms a magnetopause between the star and the planet located at a stand-off distance above the planet surface. Larger planets of similar composition will generate stronger magnetic fields. However, the strength of the magnetic field depends on its rotational frequency and on its heat content from sources such as gravitational contraction and radiogenic heating. Tidal forces provide additional heating in some cases. Atmospheric erosion and its protection by the magnetic field has complicated effects on habitability because a significant amount of the original H-rich atmosphere of Earth had to be removed. The loss of a significant amount of light gasses while retaining a relatively thick CNO atmosphere is critical for planetary habitability. See Güdel et al. (2014) for a review of H-rich proto-atmospheres.

4.5 THE ATMOSPHERE: A STRONG LAST LINE OF PROTECTION

Ruderman (1974) considered potential effects of a nearby supernova on the life on Earth, and found that hard X-ray pulses or increased CR flux may temporarily remove much of the Earth's protective ozone. High energy electromagnetic radiation from the formation of a black hole and heralded by a GRB can be devastating to life. Thomas et al. (2005) modeled several GRB effects including ozone depletion and associated UVB surface exposure resulting in DNA damage, as well as nitrous oxide (NO_2) production in the atmosphere resulting in acid rain and global cooling. For a typical burst, local ozone depletions were found to be highly dependent on geography and seasonal factors with up to 74% predicted drop in ozone. The resulting NO_2 opacity could provide strong local cooling for up to several months. In extreme cases, lethal UVB exposure of life forms at the base of the marine food chain, such as phytoplankton, are expected to be significant (Thomas et al., 2005).

SEP and GCR effects on atmospheres of Earth-like planets in the habitable zone around M-dwarf stars were studied by Grießmeier et al. (2005, 2016) and Grenfell et al. (2007). These studies showed that ozone levels are reduced, but some ozone remains, except in the highest SEP flux cases. Segura et al. (2010) modeled atmospheric effects of the ultraviolet and SEP flux from a large flare of the active M dwarf AD Leonis. These effects on an Earth-like planet without a magnetic field in the habitable zone were simulated. They found that the ultraviolet emission from such a flare should not have a significant impact on the atmospheric ozone, however, NO_2 produced by ionization from SEP protons may result in a greater than 90% ozone depletion about two years after the flare, with a predicted recovery time of 50 years.

A solar proton event in 1989 resulted in a 1%–2% drop in the column averaged ozone levels (Jackman et al., 2000). The most powerful solar flare ever observed is the Carrington event of 1859. Thomas et al. (2007) estimated that this high proton flux event may have caused up to 14% localized depletion of ozone for 4 years, causing an increase in nitrate deposition. The beryllium-10 isotope provides means to measure the CR history on Earth. 10Be is formed mainly by the CR spallation of oxygen and is preserved in ice. Thus, it is a direct proxy for CR flux, and ice core measurements in Greenland and Antarctica provide a record of CRs events.

High energy CRs, especially UHECRs, hit the atmosphere and create an air-shower (Thoudam et al., 2016), as the enormous energy is converted into potentially millions of secondary particles, including muons, neutrinos, electrons, positrons, neutrons, and protons, with muons having the greatest penetration capacity. If the secondary particles are energetic enough, and their flux is sufficiently high, muons can impact even subsurface life. If the radiation dose is too high, the chances of sustaining life as we know it on the planet are very low. Atri and Melott (2011) examined the dependence of the GCR-induced radiation dose on the strength of the planetary magnetic field and its atmospheric depth, and found that the latter is the decisive factor for the protection of a planetary biosphere. Thicker atmospheres provide longer path lengths for both primary and secondary CRs, especially lowering the flux of secondary particles at the surface. An atmosphere should be thick enough to provide surface pressure that is sufficient to maintain liquid water (or other solvent), which is required for surface life.

5 HABITABILITY IN SPACE AND IN TIME

Complex life requires a rather strict set of essential ingredients as well as a protected environment. In this section, we discuss limits on complex life based on our assumptions and the evidence presented in this review. We introduce the concept of a super-galactic habitable zone (SGHZ) and discuss the proposed galactic habitable zone (GHZ). Then we discuss the stellar, or radiative, habitable zone and stress that all of these must include the effects of high energy radiation and CRs. Finally, we discuss the cosmological implications for life over time, what we call cosmobiology.

5.1 THE SUPER-GALACTIC HABITABLE ZONE (SGHZ)

We propose that there are SGHZs. It is suggested that complex life is severely compromised on the surface of planets within the central region of large clusters of galaxies and superclusters. This hypothesis is based on several independent observations including (1) the ubiquity of galaxy mergers, (2) periods of intense star formation experienced by many galaxies either during natural spiral arm formation or as the result of a merger, and (3) activity associated with SMBH accretion and CR acceleration, especially as the result of a SMBH merger. The merger and star formation history is an important factor in the habitability of a galaxy. The occurrence of central black holes is tied to structural properties of galaxies. Many galaxies have no SMBH according to Spitzer observations (Buta et al., 2015). We may find, that only relatively small galaxies are able to maintain habitable environments over long times.

Our own galaxy is probably on the borderline. On the other hand, the metal abundance is correlated with galaxy size, and thus oxygen (e.g., in water) and other life-required elements may not be available in sufficient quantities in small galaxies. So between these two requirements, (a) no central SMBH and no extreme central star density (prone to make GRBs), and (b) sufficient amount of heavy elements, there may be only limited time in the Universe, and only a few galaxies, to allow life, especially complex life to develop. These scenarios are illustrated schematically on the right side of Fig. 4. The limit estimated for minimal or microbial life depends greatly on local conditions, but is estimated here to be about 8 Gyr ago or when the Universe was about 6 Gyr old (see Fig. 4). At this time, the elemental abundances were about 10% of the Solar value and AGN and SN rates had only just started to decline. Complex life was possible somewhat later. About 5–6 Gyr ago, when the Universe was 8–9 Gyr old, metal abundances were comparable to present day in some locations and AGN and the SFR had decreased to about 10 times present day levels. Where habitable conditions were met, life probably transitioned from simple to complex about that time.

5.2 THE GALACTIC HABITABLE ZONE (GHZ)

The GHZ introduced by Gonzalez et al. (2001) was originally based on the idea that habitability is rendered impossible unless heavy elements are widely available. While threats are common in the inner region of the Galaxy where gas is metal rich, threats are few but life supporting elements are sparse in the outer regions of the galactic disk. Thus, life exists only (or at least predominately) in a thin, metal rich annulus around the galactic center, beyond the galactic bulge.

GHZ modeling (Lineweaver et al., 2004; Prantzos, 2008; Gowanlock et al., 2011) has focused on the frequency of hazards and on the build-up of elements within the disk of the Milky Way. Where stars formed most rapidly first near the Galactic center and then continued more slowly further outwards, constructing a radial metallicity gradient. Models apply observed metallicity gradients in the our Galaxy and others including M31 (Carigi et al., 2013) and M33 (Forgan et al., 2017) and consider building up of elements as well as sterilization by exposure of planets to SN and GRBs. While predictions vary greatly, most argue that planets within 3–4 kpc of the Galactic center are subject to either too frequent or too violent events, thus forming the inner boundary of the GHZ.

The GHZ is represented here, in Fig. 5 (top view) and Fig. 6 (side view), showing the distribution of metallicity in the disk as derived from the SDSS APOGEE (APO Galactic Evolution Experiment) survey (Bovy et al., 2014; Hayden et al., 2015). An approximate GHZ is shown bounded by the box in Fig. 6. The inner bound of 4 kpc is based on age and sterilization threats. The outer and thin disk borders are based on a metallicity criterion for complex (metazoan) life of roughly 50% of the Solar value. Around Solar distances (8 kpc) planets are expected to have a moderate chance of avoiding a GRB sterilization event for at least 1–2 Gyr. However, evidence is increasing that GRB have a strong luminosity and a dependence on metallicity. Low metallicity GRBs are a rare class of highly luminous GRBs. Thus planets located nearby metal poor star forming regions will likely have a much higher probability of direct impact by a devastating GRB. Lower metallicity star formation is currently taking place in the outer spiral arms of the Milky Way, recall Figs. 5 and 6. We emphasize that a planet at 8 kpc from the galactic center, like Earth, is exposed to a significant risk from GRB sources located in the lower metallicity star forming regions at 10–12 kpc from the galactic center.

It likely that the GHZ in the disk of a spiral galaxy generally moves from an initial inner edge gradually outwards with time. For any localized region within a galaxy, the habitability increases only after

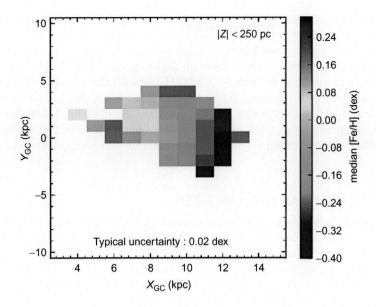

FIG. 5

The median Iron abundance [Fe/H] distribution near the mid-plane of the Milky-Way. The *square brackets* indicate dex units, meaning decimal exponent defined here with respect to the Sun, such that 0 equals the Solar Fe/H value and −1 is one-tenth of the Solar abundance. The plot includes 4330 stars in the SDSS APOGEE DR11 Red Clump sample, such that $Z \leq 250$ pc, where Z is the vertical distance from Galactic disk.

From Bovy, J., et al., 2014. The APOGEE red-clump catalog: precise distances, velocities, and high-resolution elemental abundances over a large area of the Milky Way's disk. Astrophys. J. 790, 127. https://doi.org/10.1088/0004-637X/790/2/127, used by permission.

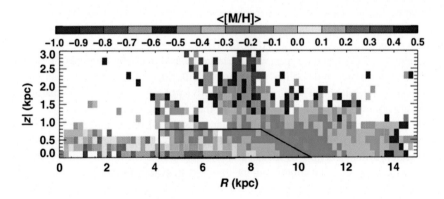

FIG. 6

SDSS APOGEE metallicity [M/H] distribution as a function of Galactocentric and mid-plane distances. The galactic habitable zone (GHZ), estimated by the box, is the most probable region of the Galaxy for complex life as we know it. The vertical distance from the galactic plane is the absolute value of the coordinate Z, as in Fig. 5. The *square brackets* indicate dex units, meaning decimal exponent defined here with respect to the Sun, such that 0 equals the Solar M/H value and −1 is one-tenth of Solar abundance. Metallicity refers to the sum of all of the elements other than hydrogen and helium.

From Hayden, M.R., et al., 2015. Chemical cartography with APOGEE: metallicity distribution functions and the chemical structure of the Milky Way disk. Astrophys. J. 808, 132. https://doi.org/10.1088/0004-637X/808/2/132, used by permission.

a significant number of stars have formed, substantial metal enrichment has occurred, and the SFR has declined. Only then the probability of sterilization events and CR density can decline below the harmful threshold. Thus, many stars formed before complex life can take hold on the planets.

If the argument made in the previous section for the existence of SGHZs is true, then the GHZ concept must also take into account the activity of SMBH. Consider two examples: NGC4258, where the central black hole is currently producing major outflows into the disk (Moran, 2008), and M33 = NGC 0598, where there is no known central black hole at all, and therefore no such activity. Even if major outflows are not currently taking place, they may occur from time to time from any SMBH. The inner radius of the GHZ in those examples may be quite different and complex life may not be possible galaxy wide.

5.3 THE CIRCUMSTELLAR HABITABLE ZONE

For more than a half a century, a cornerstone of astrobiology is that there exists a shell surrounding a star called the habitable zone, characterized by stellar luminosity and temperature within which an Earth-like planet could maintain surface water (Huang, 1959; Dole, 1964; Hart, 1979). The stability and habitability of planets in orbits around one or both stars in a close binary star system has also been investigated (Huang, 1960; Holman and Wiegert, 1999; Dvorak et al., 1989; Eggl et al., 2012; Pilat-Lohinger et al., 2012; Mason et al., 2013, 2015; Haghighipour and Kaltenegger, 2013; Zuluaga et al., 2016).

Kasting et al. (1993) derived the boundaries of the habitable zone for main sequence stars as a function of time. The boundaries are affected by the changes in the luminosity of the star that reduces the part of the habitable zone. Kopparapu et al. (2013, 2014) updated those calculations. This classical habitable zone concept is fundamental to assessment of the potential for liquid water on planetary surfaces. Many additional habitability factors must be considered, see the reviews by Lammer et al. (2009), Dartnell (2011), Forget (2013), France et al. (2014), Güdel et al. (2014), and Gonzalez (2014). For example, habitability within the habitable zone is strongly constrained by stellar X-ray and ultraviolet radiation, winds, and catastrophic events such as nearby SNe, asteroid impacts as well as other planetary dynamical issues. Especially deleterious situations occur if the planet is caught in the beam of a microquasar, ULX, GRB, or AGN (SMBH).

Main sequence stars O, B, and A types, with the largest mass have short lifetimes. So considering, times required for planet formation and cooling, only stars with lifetimes of greater than or about 2 Gyr are considered to have habitable planets. The chromospheric activity of low mass stars is comparatively high, with X-ray luminosity about $0.1 L_{bol}$. In addition, because they lose angular momentum more slowly, the high activity of these stars lasts much longer than for solar mass stars. Long lasting, high activity of low-mass stars suggests that planets orbiting these stars cannot be habitable unless they were formed with enormous quantities of water (Zuluaga et al., 2013; Tian and Ida, 2015). In order to fully address the habitability of planets over a wide range of environments, an integrated system approach of geodynamical, geochemical, climatological, and biochemical factors is needed (Franck et al., 2000; Franck et al., 2007).

6 LIFE AS WE KNOW IT IN THE UNIVERSE

Here we consider cosmological effects on life as we know it. The simplest cosmological model consistent with the four main observations, namely the accelerating expansion of the Universe, cosmic microwave background radiation, abundance of the primordial elements from big-bang nucleosynthesis, and the large scale galaxy distribution is called ΛCDM cosmology. The first Friedman-Robertson-Walker equation gives the expansion rate H, also known as the Hubble parameter, as a function of the distance scale $a(t)$ and in terms of the constant of gravitation G, the total matter plus radiation density ρ, the Friedman curvature of space k_F, and the Einstein cosmological constant Λ as follows:

$$H^2(a) = \frac{8\pi G}{3}\rho - \frac{k_F c^2}{a^2} + \frac{\Lambda c^2}{3}, \tag{3}$$

where the scale factor $a(t)$ and its time derivative are related by $H(a) = \dot{a}/a$. Presently, $a = 1$ and $H = H_0 = 67.74 \pm 0.46$ km/s/Mpc (Planck Collaboration et al., 2016). The total radiation plus matter density may be simplified by recognizing that the current radiation and neutrino densities are much smaller than the baryon and dark matter densities. Assuming a zero curvature, the time of transition from a dark matter dominated Universe to a dark energy Universe depends only on the respective densities, Ω_B, Ω_{DM}, Ω_Λ, as defined by Frieman et al. (2008), and the Hubble Constant H_0. Let $\Omega_M = \Omega_B + \Omega_{DM}$. The simplified Friedman-Robertson-Walker equation may be solved analytically for the scale factor and then setting $\ddot{a} = 0$. It follows that the transition to a dark energy dominated Universe takes place when

$$a = \left(\frac{\Omega_M}{2\Omega_\Lambda}\right)^{1/3}. \tag{4}$$

Evaluating with measured values for the Universe today: $\Omega_\Lambda = 0.6911 \pm 0.0062$, $\Omega_{DM} = 0.2603 \pm 0.0062$, and $\Omega_B = 0.04860 \pm 0.00031$ from the Planck Collaboration et al. (2016) yields $a = 0.61 \pm 0.01$, i.e., when the Universe was 61% of its current size. Along with $a = (1+z)^{-1}$ we find that the transition occurred near $z = 0.62$, 6–7 Gyr ago, see Fig. 4.

Let us consider a time-line of astrobiologically salient features of the evolving Universe. The formation and accumulation of planet and life building elements was made possible by the large quantity of dark matter in halos, binding galaxies and clusters together. In the Milky Way, significant abundances of C, N, O, Si, and Fe were not available much before 8 Gyr ago, recall Fig. 4. Planets formed after that time might develop oceans with deep sea life, but surface life was often compromised. From then until the formation of the Earth, 4.5 Gyr ago, and until today the GRBs possibly had a devastating effect on surface habitability, although the average luminosity of GRBs declined over time. The transition to a dark energy dominated Universe (and thereby expanding with acceleration) about 6.5 Gyr ago means that some denser regions of the Universe, like the Virgo and Perseus clusters with giant elliptical radio galaxies along with the numerous spiral galaxies they consume, remain inhospitable to life even today, well after the rise of elemental abundances. However, galaxies in other groups with less aggressive SMBHs and field galaxies are moving safely away owing to the transition to dark energy domination.

The Local Group galaxies M31 and M32 may have far fewer habitable planets that the Milky Way has today. The mass of the SMBH in M31 is 60 times the mass of the SMBH in the Milky Way. M32 appears to have been stripped of most of its stars, so the remaining stars are very close to the SMBH. The local starburst galaxy M82, the merging Cen A, and M87 appear to be inhospitable even toward neighboring galaxies, like M81 and M90 both of which also harbor SMBHs. The Local Group spiral galaxy M33 appears to be well suited for life as it has a significant galactic magnetic field and lacks a SMBH. To properly address the habitability over cosmic time in the Local Group, we must consider that the Andromeda Galaxy and others containing SMBH likely have been much closer to the Milky Way than they are today.

To consider now life after Earth, prospects for Milky Way planets to sustain complex life may worsen as the Andromeda Galaxy = M31 and our Galaxy are approaching each other and a major merger will take place in about 5 Gyr (Cox and Loeb, 2008). During the merger, Milky Way stars and their planets will become close to several SMBHs much more massive than our own. Especially the SMBH in the Andromeda Galaxy and in its current companion M32. It's worth noting that NGC185 and NGC205 are also close to Andromeda, but they do not have SMBHs.

Finally, our motion in space relative to the Virgo super-cluster is appreciably slowed down from cosmic expansion and it will reverse at some point in the future. Then we will start our descent into the neighborhood of the galaxy guzzling SMBH in M87 and others. In fact, the radial scale of reversal of relative motion between any two galaxies or galaxy groups/clusters increases with cosmic epoch, and so at times in the deep future, life in the Milky Way will likely be close to some monster black holes and will have to emigrate and/or use high level intelligence to deal with them.

7 SUMMARY OF CONCLUSIONS

Habitability of the Universe in general has improved due to progressively increasing protection of life from high energy radiation and CRs. The materials to form planets, and to protect life on their surfaces became possible as CHNOPS, Si, Mg, Ni, and Fe were forged by generations of stars. These elements allow the formation of a vigorous magnetic dynamo, protecting planets from atmospheric erosion and water loss by stellar winds. Long-term magnetic protection of life on rocky planets requires either enough mass to trap heat to keep the core cooling very slowly, or a significant supply of radioactive elements: K, Th, and U, with long decay lifetimes, or as in Earth's case, both. Far in the Galaxy's future, new habitable planets may form from recent SN Ia ejecta. New life will need planets formed from freshly produced Fe and radioactive nuclei, so as to restart the clock on the inner planet core heating necessary for the dynamo.

However, the presence of complex life is subject to conditions that vary dramatically, from place to place and over time, from the relatively calm present day Galactic disk with modest SFRs and an inactive SMBH to the ionizing environments of starburst galaxies or near accreting SMBH. Life on Earth and probably elsewhere benefits from four layers of protection: (1) the galactic magnetic field protects life from EGCRs, (2) the solar wind protects life from GCRs, (3) Earth's magnetic field protects life from SEPs, and (4) a thick atmosphere is the best protection against GCRs. However, UHECRs can reach the surface of Earth and even penetrate underground. While low now, UHECR fluxes have been orders of magnitude higher in the past. All protection mechanisms are dynamic. During magnetic field reversals, planetary magnetic fields are quite weak, and reversals might last thousands of years. Ozone provides the best protection against biologically damaging radiation, but it may be depleted by SNe, GRBs, or during outbursts of the local SMBH. Mergers of galaxies, both major and minor, have a profound effect on local habitability.

Many planets in the densest regions of the local Universe may remain uninhabitable due to intense particle fluxes and sterilizing events. We introduce the concept of a SGHZ to address this possibility. For the future, all this means that if humanity survives long enough, then our descendants may need to keep moving, first into space or to another planet with a different star (as the Sun dies), and then certainly to another galaxy.

REFERENCES

Aartsen, M.G., et al., 2013. Observation of cosmic-ray anisotropy with the IceTop air shower array. Astrophys. J. 765, 55. https://doi.org/10.1088/0004-637X/765/1/55.

Aartsen, M.G., et al., 2016. Search for sources of high-energy neutrons with four years of data from the IceTop detector. Astrophys. J. 830, 129. https://doi.org/10.3847/0004-637X/830/2/129.

Abbasi, R.U., et al., 2014. Indications of intermediate-scale anisotropy of cosmic rays with energy greater than 57 EeV in the northern sky measured with the surface detector of the telescope array experiment. Astrophys. J. Lett. 790, L21. https://doi.org/10.1088/2041-8205/790/2/L21.

Abbott, B.P., et al., 2016. Observation of gravitational waves from a binary black hole merger. Phys. Rev. Lett. 116 (6), 061102. https://doi.org/10.1103/PhysRevLett.116.061102.

Abrevaya, X., Thomas, B., 2017. Radiation as a constraint for life in the Universe. In: Gordon, R., Sharov, A.A. (Eds.), Habitability of the Universe before Earth. Elsevier B.V, Amsterdam (in series: Astrobiology: Exploring Life on Earth and Beyond, eds. P.H. Rampelott, J. Seckbach & R. Gordon), 27–46 (Chapter 2).

Ackermann, M., et al., 2013. Detection of the characteristic pion-decay signature in supernova remnants. Science 339, 807–811. https://doi.org/10.1126/science.1231160.

Aharonian, F., et al., 2004. Very high energy gamma rays from the direction of Sagittarius A*. Astron. Astrophys. 425, L13–L17. https://doi.org/10.1051/0004-6361:200400055.

Aharonian, F., et al., 2006. HESS observations of the galactic center region and their possible dark matter interpretation. Phys. Rev. Lett. 97 (22), 221102. https://doi.org/10.1103/PhysRevLett.97.221102.

Aharonian, F., et al., 2008. Simultaneous HESS and Chandra observations of Sagittarius A^{star} during an X-ray flare. Astron. Astrophys. 492, L25–L28. https://doi.org/10.1051/0004-6361:200810912.

Aharonian, F., et al., 2009. Spectrum and variability of the galactic center VHE γ-ray source HESS J1745-290. Astron. Astrophys. 503, 817–825. https://doi.org/10.1051/0004-6361/200811569.

Albert, J., et al., 2006. Observation of gamma rays from the galactic center with the MAGIC telescope. Astrophys. J. Lett. 638, L101–L104. https://doi.org/10.1086/501164.

Albert, J., et al., 2007. Very high energy gamma-ray radiation from the stellar mass black hole binary Cygnus X-1. Astrophys. J. Lett. 665, L51–L54. https://doi.org/10.1086/521145.

Alpher, R.A., Herman, R., Gamow, G.A., 1948. Thermonuclear reactions in the expanding Universe. Phys. Rev. 74, 1198–1199. https://doi.org/10.1103/PhysRev.74.1198.2.

Andrade-Santos, F., et al., 2016. Binary black holes, gas sloshing, and cold fronts in the X-ray halo hosting 4C +37.11. Astrophys. J. 826, 91. https://doi.org/10.3847/0004-637X/826/1/91.

Annis, J., 1999. An astrophysical explanation for the "great silence" J. Br. Interplanet. Soc. 52, 19–22.

Atri, D., Melott, A.L., 2011. Modeling high-energy cosmic ray induced terrestrial muon flux: a lookup table. Radiat. Phys. Chem. 80, 701–703. https://doi.org/10.1016/j.radphyschem.2011.02.020.

Atri, D., Hariharan, B., Grießmeier, J.M., 2013. Galactic cosmic ray-induced radiation dose on terrestrial exoplanets. Astrobiology 13, 910–919. https://doi.org/10.1089/ast.2013.1052.

Axford, W.I., 1981. Acceleration of cosmic rays by shock waves. In: International Cosmic Ray Conference, vol. 12, pp. 155–203.

Axford, W.I., Leer, E., Skadron, G., 1977. The acceleration of cosmic rays by shock waves. In: International Cosmic Ray Conference, vol. 11, pp. 132–137.

Baade, W., Zwicky, F., 1934. Cosmic rays from super-novae. Proc. Natl. Acad. Sci. U. S. A. 20, 259–263.

Bachetti, M., et al., 2014. An ultraluminous X-ray source powered by an accreting neutron star. Nature 514, 202–204. https://doi.org/10.1038/nature13791.

Bartel, N., et al., 1987. VLBI observations of 23 hot spots in the starburst galaxy M82. Astrophys. J. Lett. 323, 505–515. https://doi.org/10.1086/165847.

Becker, J.K., Biermann, P.L., 2009. Neutrinos from active black holes, sources of ultra high energy cosmic rays. Astropart. Phys. 31, 138–148. https://doi.org/10.1016/j.astropartphys.2008.12.006.

Bell, A.R., 1978. The acceleration of cosmic rays in shock fronts. I. Mon. Not. R. Astron. Soc. 182, 147–156. https://doi.org/10.1093/mnras/182.2.147.

Bell, A.R., 1978. The acceleration of cosmic rays in shock fronts. II. Mon. Not. R. Astron. Soc. 182, 443–455. https://doi.org/10.1093/mnras/182.3.443.

Benford, G., Protheroe, R.J., 2008. Fossil AGN jets as ultrahigh-energy particle accelerators. Mon. Not. R. Astron. Soc. 383, 663–672. https://doi.org/10.1111/j.1365-2966.2007.12565.x.

Benyamin, D., Nakar, E., Piran, T., Shaviv, N.J., 2016. The B/C and sub-iron/iron cosmic ray ratios—further evidence in favor of the spiral-arm diffusion model. Astrophys. J. 826, 47. https://doi.org/10.3847/0004-637X/826/1/47.

Bethe, H.A., 1990. Supernova mechanisms. Rev. Mod. Phys. 62, 801–866. https://doi.org/10.1103/RevModPhys.62.801.

Biermann, L., 1951. Kometenschweife und solare Korpuskularstrahlung. Z. Astrophys. 29, 274.

Biermann, P.L., 2006. Galactic cosmic rays. J. Phys. Conf. Ser. 47, 78–85.

Biermann, P.L., Strittmatter, P.A., 1987. Synchrotron emission from shock waves in active galactic nuclei. Astrophys. J. 322, 643–649. https://doi.org/10.1086/165759.

Biermann, P.L., Medina Tanco, G., Engel, R., Pugliese, G., 2004. The last gamma-ray burst in our galaxy? On the observed cosmic-ray excess at particle energy 10^{18} eV. Astrophys. J. Lett. 604, L29–L32. https://doi.org/10.1086/382072.

Biermann, P.L., Becker, J.K., Dreyer, J., Meli, A., Seo, E.S., Stanev, T., 2010. The origin of cosmic rays: explosions of massive stars with magnetic winds and their supernova mechanism. Astrophys. J. 725, 184–187. https://doi.org/10.1088/0004-637X/725/1/184.

Biermann, P.L., Becker Tjus, J., Seo, E.S., Mandelartz, M., 2013. Cosmic-ray transport and anisotropies. Astrophys. J. 768, 124. https://doi.org/10.1088/0004-637X/768/2/124.

Biermann, P.L., Caramete, L.I., Meli, A., Nath, B.N., Seo, E.S., de Souza, V., Becker Tjus, J., 2015. Cosmic ray transport and anisotropies to high energies. ASTRA Proc. 2, 39–44. https://doi.org/10.5194/ap-2-39-2015.

Biermann, P.L., et al., 2016. The nature and origin of ultra-high energy cosmic ray particles. ArXiv e-prints.

Binns, W.R., et al., 2007. OB associations, Wolf Rayet stars, and the origin of galactic cosmic rays. Space Sci. Rev. 130, 439–449. https://doi.org/10.1007/s11214-007-9195-1.

Bisnovatyi-Kogan, G.S., 1970. The explosion of a rotating star as a supernova mechanism. Astron. Zh. 47, 813.

Black, C.S., Fesen, R.A., 2015. A 3D kinematic study of the northern ejecta 'jet' of the Crab nebula. Mon. Not. R. Astron. Soc. 447, 2540–2550. https://doi.org/10.1093/mnras/stu2641.

Blandford, R.D., Ostriker, J.P., 1978. Particle acceleration by astrophysical shocks. Astrophys. J. Lett. 221, L29–L32. https://doi.org/10.1086/182658.

Bovy, J., et al., 2014. The APOGEE red-clump catalog: precise distances, velocities, and high-resolution elemental abundances over a large area of the Milky Way's disk. Astrophys. J. 790, 127. https://doi.org/10.1088/0004-637X/790/2/127.

Burbidge, E.M., Burbidge, G.R., Fowler, W.A., Hoyle, F., 1957. Synthesis of the elements in stars. Rev. Mod. Phys. 29, 547–650. https://doi.org/10.1103/RevModPhys.29.547.

Buta, R.J., et al., 2015. A classical morphological analysis of galaxies in the Spitzer survey of stellar structure in galaxies (S4G). Astrophys. J. Suppl. 217, 32. https://doi.org/10.1088/0067-0049/217/2/32.

Caramete, L.I., Biermann, P.L., 2010. The mass function of nearby black hole candidates. Astron. Astrophys. 521, A55. https://doi.org/10.1051/0004-6361/200913146.

Cardamone, C., et al., 2009. Galaxy Zoo Green Peas: discovery of a class of compact extremely star-forming galaxies. Mon. Not. R. Astron. Soc. 399, 1191–1205. https://doi.org/10.1111/j.1365-2966.2009.15383.x.

Carigi, L., García-Rojas, J., Meneses-Goytia, S., 2013. Chemical evolution and the galactic habitable zone of M31. Rev. Mex. Astron. Astrofis. 49, 253–273.

Chini, R., Hoffmeister, V.H., Nasseri, A., Stahl, O., Zinnecker, H., 2012. A spectroscopic survey on the multiplicity of high-mass stars. Mon. Not. R. Astron. Soc. 424, 1925–1929. https://doi.org/10.1111/j.1365-2966.2012.21317.x.

Cox, T.J., Loeb, A., 2008. The collision between the Milky Way and Andromeda. Mon. Not. R. Astron. Soc. 386, 461–474. https://doi.org/10.1111/j.1365-2966.2008.13048.x.

Cummings, A.C., et al., 2016. Galactic cosmic rays in the local interstellar medium: voyager 1 observations and model results. Astrophys. J. 831, 18. https://doi.org/10.3847/0004-637X/831/1/18.

Dartnell, L.R., 2011. Ionizing radiation and life. Astrobiology 11, 551–582. https://doi.org/10.1089/ast.2010.0528.

Dole, S.H., 1964. Habitable Planets for Man. RAND Corp., Santa Monica, CA.

Dong, Y., et al., 2015. Strong plume fluxes at Mars observed by MAVEN: an important planetary ion escape channel. Geophys. Res. Lett. 42, 8942–8950. https://doi.org/10.1002/2015GL065346.

Drury, L.O., 1983. An introduction to the theory of diffusive shock acceleration of energetic particles in tenuous plasmas. Rep. Prog. Phys. 46, 973–1027. https://doi.org/10.1088/0034-4885/46/8/002.

Duric, N., Gordon, S.M., Goss, W.M., Viallefond, F., Lacey, C., 1995. The relativistic ISM in M33: role of the supernova remnants. Astrophys. J. 445, 173–181. https://doi.org/10.1086/175683.

Dvorak, R., Froeschle, C., Froeschle, C., 1989. Stability of outer planetary orbits (P-types) in binaries. Astron. Astrophys. 226, 335–342.

Eggl, S., Pilat-Lohinger, E., Georgakarakos, N., Gyergyovits, M., Funk, B., 2012. An analytic method to determine habitable zones for S-type planetary orbits in binary star systems. Astrophys. J. 752, 74. https://doi.org/10.1088/0004-637X/752/1/74.

Fanaroff, B.L., 1974. A search for flux density variations in the central components of the extended extragalactic radio sources Virgo A and 3C III. Mon. Not. R. Astron. Soc. 166, 1P–8P. https://doi.org/10.1093/mnras/166.1.1P.

Federrath, C., 2013. On the universality of supersonic turbulence. Mon. Not. R. Astron. Soc. 436, 1245–1257. https://doi.org/10.1093/mnras/stt1644.

Fermi, E., 1949. On the origin of the cosmic radiation. Phys. Rev. 75, 1169–1174. https://doi.org/10.1103/PhysRev.75.1169.

Fermi, E., 1954. Galactic magnetic fields and the origin of cosmic radiation. Astrophys. J. 119, 1. https://doi.org/10.1086/145789.

Ferrarese, L., Merritt, D., 2000. A fundamental relation between supermassive black holes and their host galaxies. Astrophys. J. Lett. 539, L9–L12. https://doi.org/10.1086/312838.

Fesen, R.A., Milisavljevic, D., 2016. An HST survey of the highest-velocity ejecta in Cassiopeia A. Astrophys. J. 818, 17. https://doi.org/10.3847/0004-637X/818/1/17.

Forgan, D., Dayal, P., Cockell, C., Libeskind, N., 2017. Evaluating galactic habitability using high-resolution cosmological simulations of galaxy formation. Int. J. Astrobiol. 16, 60–73. https://doi.org/10.1017/S1473550415000518.

Forget, F., 2013. On the probability of habitable planets. Int. J. Astrobiol. 12, 177–185. https://doi.org/10.1017/S1473550413000128.

France, K., Linsky, J.L., Parke Loyd, R.O., 2014. The ultraviolet radiation environment in the habitable zones around low-mass exoplanet host stars. Astrophys. Space Sci. 354, 3–7. https://doi.org/10.1007/s10509-014-1947-2.

Franck, S., Block, A., von Bloh, W., Bounama, C., Scellnhuber, H.J., Svirezhev, Y., 2000. Reduction of biosphere life span as a consequence of geodynamics. Tellus B 52 (1), 94–107. https://doi.org/10.1034/j.1600-0889.2000.00898.x.

Franck, S., von Bloh, W., Bounama, C., 2007. Maximum number of habitable planets at the time of Earth's origin: new hints for panspermia and the mediocrity principle. Int. J. Astrobiol. 6, 153–157. https://doi.org/10.1017/S1473550407003680.

Frieman, J.A., Turner, M.S., Huterer, D., 2008. Dark energy and the accelerating Universe. Annu. Rev. Astron. Astrophys. 46, 385–432. https://doi.org/10.1146/annurev.astro.46.060407.145243.

Galante, D., Horvath, J.E., 2007. Biological effects of gamma-ray bursts: distances for severe damage on the biota. Int. J. Astrobiol. 6, https://doi.org/10.1017/S1473550406003545.

Gamow, G., 1948. The origin of elements and the separation of galaxies. Phys. Rev. 74, 505–506. https://doi.org/10.1103/PhysRev.74.505.2.

Gebhardt, K., et al., 2000. A relationship between nuclear black hole mass and galaxy velocity dispersion. Astrophys. J. Lett. 539, L13–L16. https://doi.org/10.1086/312840.

Gehrels, N., Laird, C.M., Jackman, C.H., Cannizzo, J.K., Mattson, B.J., Chen, W., 2003. Ozone depletion from nearby supernovae. Astrophys. J. 585, 1169–1176. https://doi.org/10.1086/346127.

Gergely, L.Á., Biermann, P.L., 2009. The spin-flip phenomenon in supermassive black hole binary mergers. Astrophys. J. 697, 1621–1633. https://doi.org/10.1088/0004-637X/697/2/1621.

Gilmore, G., Wyse, R.F.G., Norris, J.E., 2002. Deciphering the last major invasion of the Milky Way. Astrophys. J. Lett. 574, L39–L42. https://doi.org/10.1086/342363.

Gonzalez, G., 2014. Setting the stage for habitable planets. ArXiv e-prints.

Gonzalez, G., Brownlee, D., Ward, P., 2001. The galactic habitable zone: galactic chemical evolution. Icarus 152, 185–200. https://doi.org/10.1006/icar.2001.6617.

Gopal-Krishna, Gopal-Krishna, P.L., Wiita, P.J., 2003. The origin of X-shaped radio galaxies: clues from the Z-symmetric secondary lobes. Astrophys. J. Lett. 594, L103–L106. https://doi.org/10.1086/378766.

Gopal-Krishna, Biermann, P.L., Gergely, L.Á., Wiita, P.J., 2012. On the origin of X-shaped radio galaxies. Res. Astron. Astrophys. 12, 127–146. https://doi.org/10.1088/1674-4527/12/2/002.

Gowanlock, M.G., Patton, D.R., McConnell, S.M., 2011. A model of habitability within the Milky Way galaxy. Astrobiology 11, 855–873. https://doi.org/10.1089/ast.2010.0555.

Greisen, K., 1966. End to the cosmic-ray spectrum? Phys. Rev. Lett. 16, 748–750. https://doi.org/10.1103/PhysRevLett.16.748.

Grenfell, J.L., et al., 2007. Biomarker response to galactic cosmic ray-induced NO_x and the methane greenhouse effect in the atmosphere of an Earth-like planet orbiting an M dwarf star. Astrobiology 7, 208–221. https://doi.org/10.1089/ast.2006.0129.

Grießmeier, J.M., Stadelmann, A., Motschmann, U., Belisheva, N.K., Lammer, H., Biernat, H.K., 2005. Cosmic ray impact on extrasolar Earth-like planets in close-in habitable zones. Astrobiology 5, 587–603. https://doi.org/10.1089/ast.2005.5.587.

Grießmeier, J.M., Tabataba-Vakili, F., Stadelmann, A., Grenfell, J.L., Atri, D., 2016. Galactic cosmic rays on extrasolar Earth-like planets. II. Atmospheric implications. Astron. Astrophys. 587, A159. https://doi.org/10.1051/0004-6361/201425452.

Güdel, M., Dvorak, R., Erkaev, N., Kasting, J., Khodachenko, M., Lammer, H., Pilat-Lohinger, E., Rauer, H., Ribas, I., Wood, B.E., 2014. Astrophysical conditions for planetary habitability. In: Protostars and Planets VI, pp. 883–906. https://doi.org/10.2458/azu_uapress_9780816531240-ch038.

Guo, Y.Q., Yuan, Q., Liu, C., Li, A.F., 2013. A hybrid model of GeV-TeV gamma ray emission from the galactic center. J. Phys. G: Nucl. Phys. 40 (6), 065201. https://doi.org/10.1088/0954-3899/40/6/065201.

Haghighipour, N., Kaltenegger, L., 2013. Calculating the habitable zone of binary star systems. II. P-type binaries. Astrophys. J. 777, 166. https://doi.org/10.1088/0004-637X/777/2/166.

Hanasz, M., Otmianowska-Mazur, K., Kowal, G., Lesch, H., 2009a. Cosmic-ray-driven dynamo in galactic disks. A parameter study. Astron. Astrophys. 498, 335–346. https://doi.org/10.1051/0004-6361/200810279.

Hanasz, M., Wóltański, D., Kowalik, K., 2009b. Global galactic dynamo driven by cosmic rays and exploding magnetized stars. Astrophys. J. Lett. 706, L155–L159. https://doi.org/10.1088/0004-637X/706/1/L155.

Hanasz, M., Lesch, H., Naab, T., Gawryszczak, A., Kowalik, K., Wóltański, D., 2013. Cosmic rays can drive strong outflows from gas-rich high-redshift disk galaxies. Astrophys. J. Lett. 777, L38. https://doi.org/10.1088/2041-8205/777/2/L38.

Hart, M.H., 1979. Habitable zones about main sequence stars. Icarus 37, 351–357. https://doi.org/10.1016/0019-1035(79)90141-6.

Hawking, S.W., 1971. Stable and generic properties in general relativity. Gen. Rel. Gravit. 1, 393–400. https://doi.org/10.1007/BF00759218.

Hayashida, N., et al., 1999. The anisotropy of cosmic ray arrival directions around 10^{18} eV. Astropart. Phys. 10, 303–311. https://doi.org/10.1016/S0927-6505(98)00064-4.

Hayden, M.R., et al., 2015. Chemical cartography with APOGEE: metallicity distribution functions and the chemical structure of the Milky Way disk. Astrophys. J. 808, 132. https://doi.org/10.1088/0004-637X/808/2/132.

Heger, A., Fryer, C.L., Woosley, S.E., Langer, N., Hartmann, D.H., 2003. How massive single stars end their life. Astrophys. J. 591, 288–300. https://doi.org/10.1086/375341.

Heller, R., Armstrong, J., 2014. Superhabitable worlds. Astrobiology 14, 50–66. https://doi.org/10.1089/ast.2013.1088.

Hess, V.F., 1912. Über Beobachtungen der durchdringenden Strahlung bei sieben Freiballonfahrten. Phys. Z. 13, 1084.

HESS Collaboration, et al., 2016. Acceleration of petaelectronvolt protons in the galactic centre. Nature 531, 476–479. https://doi.org/10.1038/nature17147.

Hillas, A.M., 1984. The origin of ultra-high-energy cosmic rays. Annu. Rev. Astron. Astrophys. 22, 425–444. https://doi.org/10.1146/annurev.aa.22.090184.002233.

Holman, M.J., Wiegert, P.A., 1999. Long-term stability of planets in binary systems. Astron. J. 117, 621–628. https://doi.org/10.1086/300695.

Huang, S.S., 1959. The problem of life in the Universe and the mode of star formation. Publ. Astron. Soc. Pac. 71, 421. https://doi.org/10.1086/127417.

Huang, S.S., 1960. Life-supporting regions in the vicinity of binary systems. Publ. Astron. Soc. Pac. 72, 106. https://doi.org/10.1086/127489.

Hwang, U., et al., 2004. A million second Chandra view of Cassiopeia A. Astrophys. J. Lett. 615, L117–L120. https://doi.org/10.1086/426186.

IceCube Collaboration, Pierre Auger Collaboration, Telescope Array Collaboration, 2016. Search for correlations between the arrival directions of IceCube neutrino events and ultrahigh-energy cosmic rays detected by the Pierre Auger Observatory and the Telescope Array. J. Cosmol. Astropart. Phys. 1, 037.

Izotov, Y.I., Guseva, N.G., Fricke, K.J., Henkel, C., 2016. The bursting nature of star formation in compact star-forming galaxies from the Sloan Digital Sky Survey. Mon. Not. R. Astron. Soc. 462, 4427–4434. https://doi.org/10.1093/mnras/stw1973.

Jackman, C.H., Fleming, E.L., Vitt, F.M., 2000. Influence of extremely large solar proton events in a changing stratosphere. Geophys. Res. Lett. 105, 11659–11670. https://doi.org/10.1029/2000JD900010.

Jokipii, J.R., Morfill, G., 1987. Ultra-high-energy cosmic rays in a galactic wind and its termination shock. Astrophys. J. 312, 170–177. https://doi.org/10.1086/164857.

Kasting, J.F., Whitmire, D.P., Reynolds, R.T., 1993. Habitable zones around main sequence stars. Icarus 101, 108–128.

Khochfar, S., Burkert, A., 2001. Redshift evolution of the merger fraction of galaxies in cold dark matter cosmologies. Astrophys. J. 561, 517–520. https://doi.org/10.1086/323382.

King, A., Lasota, J.P., 2016. ULXs: neutron stars versus black holes. Mon. Not. R. Astron. Soc. 458, L10–L13. https://doi.org/10.1093/mnrasl/slw011.

Kogut, A., et al., 2003. First-year Wilkinson Microwave Anisotropy Probe (WMAP) observations: temperature-polarization correlation. Astrophys. J. Suppl. 148, 161–173. https://doi.org/10.1086/377219.

Kompaneets, A.S., 1960. A point explosion in an inhomogeneous atmosphere. Sov. Phys. Dokl. 5, 46.

Kopparapu, R.K., et al., 2013. Habitable zones around main-sequence stars: new estimates. Astrophys. J. 765, 131. https://doi.org/10.1088/0004-637X/765/2/131.

Kopparapu, R.K., Ramirez, R.M., SchottelKotte, J., Kasting, J.F., Domagal-Goldman, S., Eymet, V., 2014. Habitable zones around main-sequence stars: dependence on planetary mass. Astrophys. J. Lett. 787, L29. https://doi.org/10.1088/2041-8205/787/2/L29.

Kosack, K., et al., 2004. TeV gamma-ray observations of the galactic center. Astrophys. J. Lett. 608, L97–L100. https://doi.org/10.1086/422469.

Kronberg, P.P., 1998. Galactic and extragalactic magnetic fields in the local universe: an overview. In: Krizmanic, J.F., Ormes, J.F., Streitmatter, R.E. (Eds.), American Institute of Physics Conference SeriesWorkshop on Observing Giant Cosmic Ray Air Showers From >10(20) eV Particles From Space, vol. 433, pp. 196–211.

Kronberg, P.P., Biermann, P., Schwab, F.R., 1981. The continuum radio structure of the nucleus of M82. Astrophys. J. 246, 751–760. https://doi.org/10.1086/158970.

Kronberg, P.P., Biermann, P., Schwab, F.R., 1985. The nucleus of M82 at radio and X-ray bands—discovery of a new radio population of supernova candidates. Astrophys. J. 291, 693–707. https://doi.org/10.1086/163108.

Kronberg, P.P., Dufton, Q.W., Li, H., Colgate, S.A., 2001. Magnetic energy of the intergalactic medium from galactic black holes. Astrophys. J. 560, 178–186. https://doi.org/10.1086/322767.

Krymskii, G.F., 1977. A regular mechanism for the acceleration of charged particles on the front of a shock wave. Akademiia Nauk SSSR Doklady 234, 1306–1308.

Kun, E., Biermann, P.L., Gergely, L.Á., 2017. A flat spectrum candidate for a track-type high energy neutrino emission event, the case of blazar PKS 0723-008. MNRAS Lett. 466.

Lammer, H., et al., 2009. What makes a planet habitable? Astron. Astrophys. Rev. 17, 181–249. https://doi.org/10.1007/s00159-009-0019-z.

Landau, L.D., Lifshitz, E.M., 1959. Fluid Mechanics. Pergamon Press, Oxford.

Le Fèvre, O., et al., 2000. Hubble Space Telescope imaging of the CFRS and LDSS redshift surveys—IV. Influence of mergers in the evolution of faint field galaxies from $z \sim 1$. Mon. Not. R. Astron. Soc. 311, 565–575. https://doi.org/10.1046/j.1365-8711.2000.03083.x.

Lineweaver, C.H., Fenner, Y., Gibson, B.K., 2004. The galactic habitable zone and the age distribution of complex life in the Milky Way. Science 303, 59–62. https://doi.org/10.1126/science.1092322.

Loeb, A., Barkana, R., 2001. The reionization of the Universe by the first stars and quasars. Annu. Rev. Astron. Astrophys. 39, 19–66. https://doi.org/10.1146/annurev.astro.39.1.19.

Longair, M.S., 1994. High Energy Astrophysics. Volume 2. Stars, the Galaxy and the Interstellar Medium. Cambridge University Press, Cambridge.

Longobardi, A., Arnaboldi, M., Gerhard, O., Mihos, J.C., 2015. The build-up of the cD halo of M 87: evidence for accretion in the last Gyr. Astron. Astrophys. 579, L3. https://doi.org/10.1051/0004-6361/201526282.

López-Cobá, C., et al., 2017. Star formation driven galactic winds in UGC 10043. Mon. Not. R. Astron. Soc. 467, 4951–4964. https://doi.org/10.1093/mnras/stw3355.

Lovelace, R.V.E., 1976. Dynamo model of double radio sources. Nature 262, 649–652. https://doi.org/10.1038/262649a0.

Madau, P., Dickinson, M., 2014. Cosmic star-formation history. Annu. Rev. Astron. Astrophys. 52, 415–486. https://doi.org/10.1146/annurev-astro-081811-125615.

Mason, P.A., McNamara, B.J., Harrison, T.E., 1997. High energy transient events from Cygnus X-1: evidence for a source of galactic gamma-ray bursts. Astron. J. 114, 238. https://doi.org/10.1086/118468.

Mason, P.A., Zuluaga, J.I., Clark, J.M., Cuartas-Restrepo, P.A., 2013. Rotational synchronization may enhance habitability for circumbinary planets: Kepler binary case studies. Astrophys. J. Lett. 774, L26. https://doi.org/10.1088/2041-8205/774/2/L26.

Mason, P.A., Zuluaga, J.I., Cuartas-Restrepo, P.A., Clark, J.M., 2015. Circumbinary habitability niches. Int. J. Astrobiol. 14, 391–400. https://doi.org/10.1017/S1473550414000342.

Melott, A.L., Thomas, B.C., Hogan, D.P., Ejzak, L.M., Jackman, C.H., 2005. Climatic and biogeochemical effects of a galactic gamma ray burst. Geophys. Res. Lett. 32, L14808. https://doi.org/10.1029/2005GL023073.

Mirabel, I.F., Dijkstra, M., Laurent, P., Loeb, A., Pritchard, J.R., 2011. Stellar black holes at the dawn of the universe. Astron. Astrophys. 528, A149. https://doi.org/10.1051/0004-6361/201016357.

Moellenhoff, C., 1976. An explosion model for extragalactic double radio sources. Astron. Astrophys. 50, 105–112.

Moran, J.M., 2008. The black-hole accretion disk in NGC 4258: one of nature's most beautiful dynamical systems. In: Bridle, A.H., Condon, J.J., Hunt, G.C. (Eds.), Astronomical Society of the Pacific Conference SeriesFrontiers of Astrophysics: A Celebration of NRAO's 50th Anniversary, vol. 395, p. 87.

Murphy, R.P., et al., 2016. Galactic cosmic ray origins and OB associations: evidence from SuperTIGER observations of elements $_{26}$Fe through $_{40}$Zr. Astrophys. J. 831, 148. https://doi.org/10.3847/0004-637X/831/2/148.

Olinto, A.V., 2012. Cosmic rays at the highest energies. J. Phys. Conf. Ser. 375 (5), 052001. https://doi.org/10.1088/1742-6596/375/1/052001.

Owen, F.N., Eilek, J.A., Kassim, N.E., 2000. M87 at 90 centimeters: a different picture. Astrophys. J. 543, 611–619. https://doi.org/10.1086/317151.

Parker, E.N., 1958. Dynamics of the interplanetary gas and magnetic fields. Astrophys. J. 128, 664. https://doi.org/10.1086/146579.

Peng, B., Strom, R.G., Wei, J., Zhao, Y.H., 2004. Galaxies around the giant double radio source DA 240. Redshifts and the discovery of an unusual association. Astron. Astrophys. 415, 487–498. https://doi.org/10.1051/0004-6361:20034363.

Pierre Auger Collaboration, et al., 2011. Anisotropy and chemical composition of ultra-high energy cosmic rays using arrival directions measured by the Pierre Auger Observatory. J. Cosmol. Astropart. Phys. 6, 022. https://doi.org/10.1088/1475-7516/2011/06/022.

Pilat-Lohinger, E., Eggl, S., Gyergyovits, M., 2012. On the habitability of planets in binary star systems. In: Abbasi, A., Giesen, N. (Eds.), EGU General Assembly Conference Abstracts, vol. 14, p. 12406.

Piran, T., Jimenez, R., 2014. Physical Review Letters 113, 1102.

Planck Collaboration, et al., 2016. Planck 2015 results. XIII. Cosmological parameters. Astron. Astrophys. 594, A13. https://doi.org/10.1051/0004-6361/201525830.

Prantzos, N., 2008. On the "Galactic Habitable Zone" Space Sci. Rev. 135, 313–322. https://doi.org/10.1007/s11214-007-9236-9.

Preston, L.J., Dartnell, L.R., 2014. Planetary habitability: lessons learned from terrestrial analogues. Int. J. Astrobiol. 13, 81–98. https://doi.org/10.1017/S1473550413000396.

Rauch, B.F., et al., 2009. Cosmic ray origin in OB associations and preferential acceleration of refractory elements: evidence from abundances of elements $_{26}$Fe through $_{34}$Se. Astrophys. J. 697, 2083–2088. https://doi.org/10.1088/0004-637X/697/2/2083.

Raup, D.M., Sepkoski, J.J., 1982. Mass extinctions in the marine fossil record. Science 215, 1501–1503. https://doi.org/10.1126/science.215.4539.1501.

Rossa, J., Dettmar, R.J., 2003. An Hα survey aiming at the detection of extraplanar diffuse ionized gas in halos of edge-on spiral galaxies. II. The Hα survey atlas and catalog. Astron. Astrophys. 406, 505–525. https://doi.org/10.1051/0004-6361:20030698.

Rothschild, L.J., Mancinelli, R.L., 2001. Life in extreme environments. Nature 409, 1092–1101. https://doi.org/10.1038/35059215.

Ruderman, M.A., 1974. Possible consequences of nearby supernova explosions for atmospheric ozone and terrestrial life. Science 184, 1079–1081. https://doi.org/10.1126/science.184.4141.1079.

Schlegel, E.M., Jones, C., Machacek, M., Vega, L.D., 2016. NGC 5195 in M51: feedback 'Burps' after a massive meal? Astrophys. J. 823, 75. https://doi.org/10.3847/0004-637X/823/2/75.

Schlickeiser, R., Webber, W.R., Kempf, A., 2014. Explanation of the local galactic cosmic ray energy spectra measured by voyager 1. I. Protons. Astrophys. J. 787, 35. https://doi.org/10.1088/0004-637X/787/1/35.

Seager, S., 2013. Exoplanet habitability. Science 340, 577–581. https://doi.org/10.1126/science.1232226.

Segura, A., Walkowicz, L.M., Meadows, V., Kasting, J., Hawley, S., 2010. The effect of a strong stellar flare on the atmospheric chemistry of an Earth-like planet orbiting an M dwarf. Astrobiology 10, 751–771. https://doi.org/10.1089/ast.2009.0376.

Shaviv, N.J., 2003. Toward a solution to the early faint Sun paradox: a lower cosmic ray flux from a stronger solar wind. J. Geophys. Res. Space Phys. 108, 1437. https://doi.org/10.1029/2003JA009997.

Smith, D.S., Scalo, J.M., 2009. Habitable zones exposed: astrosphere collapse frequency as a function of stellar mass. Astrobiology 9, 673–681. https://doi.org/10.1089/ast.2009.0337.

Soderblom, D.R., Stauffer, J.R., MacGregor, K.B., Jones, B.F., 1993. The evolution of angular momentum among zero-age main-sequence solar-type stars. Astrophys. J. 409, 624–634. https://doi.org/10.1086/172694.

Stollman, G.M., van Paradijs, J., 1985. Super-Eddington fluxes in a quasi-hydrostatic, optically thick region of a neutron-star outer envelope. Astron. Astrophys. 153, 99–105.

Strom, R.G., Chen, R., Yang, J., Peng, B., 2013. Structure and environment of the giant radio galaxy 4C 73.08. Mon. Not. R. Astron. Soc. 430, 2090–2096. https://doi.org/10.1093/mnras/stt033.

Teshima, M., et al., 2001. Anisotropy of cosmic-ray arrival direction at 10^{18}eV observed by AGASA. In: International Cosmic Ray Conference, vol. 1, p. 337.

Thomas, B.C., et al., 2005. Gamma-ray bursts and the Earth: exploration of atmospheric, biological, climatic, and biogeochemical effects. Astrophys. J. 634, 509–533. https://doi.org/10.1086/496914.

Thomas, B.C., Jackman, C.H., Melott, A.L., 2007. Modeling atmospheric effects of the September 1859 solar flare. Geophys. Res. Lett. 34, L06810. https://doi.org/10.1029/2006GL029174.

Thoudam, S., Rachen, J.P., van Vliet, A., Achterberg, A., Buitink, S., Falcke, H., Hörandel, J.R., 2016. Cosmic-ray energy spectrum and composition up to the ankle: the case for a second galactic component. Astron. Astrophys. 595, A33. https://doi.org/10.1051/0004-6361/201628894.

Tian, F., Ida, S., 2015. Water contents of habitable zone rocky planets and biosignature detection around M dwarfs. In: Pathways Towards Habitable Planets, p. 20.

Tsuchiya, K., et al., 2004. Detection of sub-TeV gamma rays from the galactic center direction by CANGAROO-II. Astrophys. J. Lett. 606, L115–L118. https://doi.org/10.1086/421292.

van Paradijs, J., 1981. On the maximum luminosity in X-ray bursts. Astron. Astrophys. 101, 174.

Vidotto, A.A., Jardine, M., Morin, J., Donati, J.F., Opher, M., Gombosi, T.I., 2014. M-dwarf stellar winds: the effects of realistic magnetic geometry on rotational evolution and planets. Mon. Not. R. Astron. Soc. 438, 1162–1175. https://doi.org/10.1093/mnras/stt2265.

Weber, E.J., Davis Jr., L., 1967. The angular momentum of the solar wind. Astrophys. J. 148, 217–227. https://doi.org/10.1086/149138.

Whysong, D., Antonucci, R., 2003. New insights on selected radio galaxy nuclei. New Astron. Rev. 47, 219–223. https://doi.org/10.1016/S1387-6473(03)00029-0.

Wiegert, T., et al., 2015. CHANG-ES. IV. Radio continuum emission of 35 edge-on galaxies observed with the Karl G. Jansky very large array in D configuration-data release 1. Astron. J. 150, 81. https://doi.org/10.1088/0004-6256/150/3/81.

Woosley, S.E., Heger, A., 2015. The deaths of very massive stars. In: Vink, J.S. (Ed.), Astrophysics and Space Science LibraryVery Massive Stars in the Local Universe, vol. 412, p. 199.

Woosley, S.E., Heger, A., Weaver, T.A., 2002. The evolution and explosion of massive stars. Rev. Mod. Phys. 74, 1015–1071. https://doi.org/10.1103/RevModPhys.74.1015.

Zatsepin, G.T., Kuz'min, V.A., 1966. Upper limit of the spectrum of cosmic rays. Sov. J. Exp. Theor. Phys. Lett. 4, 78.

Zuluaga, J.I., Bustamante, S., Cuartas, P.A., Hoyos, J.H., 2013. The influence of thermal evolution in the magnetic protection of terrestrial planets. Astrophys. J. 770, 23. https://doi.org/10.1088/0004-637X/770/1/23.

Zuluaga, J.I., Mason, P.A., Cuartas-Restrepo, P.A., 2016. Constraining the radiation and plasma environment of the Kepler circumbinary habitable-zone planets. Astrophys. J. 818, 160. https://doi.org/10.3847/0004-637X/818/2/160.

FURTHER READING

Rudie, G.C., Fesen, R.A., Yamada, T., 2008. The Crab Nebula's dynamical age as measured from its northern filamentary jet. Mon. Not. R. Astron. Soc. 384, 1200–1206. https://doi.org/10.1111/j.1365-2966.2007.12799.x.

PRIMITIVE CARBON: BEFORE EARTH AND MUCH BEFORE ANY LIFE ON IT

Chaitanya Giri

Earth-Life Science Institute, Tokyo Institute of Technology, Tokyo, Japan
Geophysical Laboratory, Carnegie Institution of Washington, Washington, DC, United States

CHAPTER OUTLINE

1 INTRODUCTION: THE FOUNDATIONAL CARBON

The stunning quest to discern the "origin of life" has been pursued for ages through numerous disciplines in basic sciences, applied sciences, and philosophy. Despite these relentless pursuits, life as a form of physical matter continues to be undefined and its origin is a mystery. The consistent ineffectiveness in deciphering the origin of life is telling us that the answers are concealed in a multitude of locations across space and time. Moreover, the pursuit of finding these far-located answers is also triggering several *ad rem* questions pertaining to the operations, specificities, and evolution of

Habitability of the Universe Before Earth, editors: Richard Gordon & Alexei Sharov, Volume 1 in the series:
Astrobiology: Exploring Life on Earth and Beyond, series editors: Pabulo Henrique Rampelotto,
Joseph Seckbach & Richard Gordon. ISSN 2468-6352. https://doi.org/10.1016/B978-0-12-811940-2.00006-X

bio-geo-physico-chemical processes and phenomena occurring on the host Earth, and those pertaining to the far more complex cosmo-physico-chemical processes occurring in the universe. The quest for the origin of life does not have specific answers; The answers are many, they are scalable and variable. However, in all the ambiguity and complexity associated with the quest, what remains constant is the element that is fundamental to the biochemical architecture of life on Earth—carbon.

One of the major objectives of space exploration, which is now in its sixth decade, has been to find evidences of the factors that had assisted in the origin of carbon-based life on Earth. At the onset of space exploration, this evidence was cognized based on the concept of "cosmic panspermia" that purported the possible presence of extraterrestrial micro-organismic "seeds" of life. With growing capabilities to extensively characterize extraterrestrial materials from fallen meteorites, in situ analyses carried out on extraterrains by space probes, and analyses of extraterrestrial samples returned to Earth by sample-return space missions, the evidence is now pursued with a nuanced and scientific *modus operandi*. This has galvanized our readiness to further explore prominent as well as obscure astrobiochemical signatures from assorted locales of the cosmos.

The life-inhabited Earth is surrounded by organic-abundant solar system objects, such as comets (Elsila et al., 2009; Goesmann et al., 2015; Altwegg et al., 2016), carbonaceous chondrites (Kvenvolden et al., 1970; Bandurski and Nagy, 1976; Zenobi et al., 1989; Pizzarello et al., 2001), Saturn's moons Titan (Neish et al., 2008; Hörst et al., 2012) and Enceladus (Parkinson et al., 2007; McKay et al., 2014) and Jupiter's moon Europa (Lunine and Lorenz, 1997; Levy et al., 2000). These objects and their organic abundances are not serendipitous, but indicative of the vast and unquantified volume of carbon that could have existed in diverse conformations—ions, molecules, macromolecules and materials—in the protoplanetary nebula that preceded the solar system (Henning and Salama, 1998; Ehrenfreund and Charnley, 2000; Ehrenfreund and Cami, 2010). That being so, could there be any exchange of organics among these solar system objects?

The Earth is not a solitary planet. Over the course of its existence, it has accreted huge quantities of extraterrestrial materials, often benignly and at times violently. Even in the present Quaternary epoch, when there is no heavy bombardment of meter- and kilometer-sized asteroids, the Earth continues to accrete an estimated $40,000 \pm 20,000 \times 10^3$ kg of micrometeoritic dust annually (Love and Brownlee, 1993). The Southern Ocean, which is located far from any major continental land-mass, receives ~30%–300% more iron from meteoritic dust than from regular eolian dust (Johnson, 2001). This meteoritic iron is believed to be regulating the marine carbon cycle, which is important for sustaining life in the ocean (Johnson, 2001). This finding prompts two questions—whether life on Earth has emerged from the strategic supplies of prebiotic material of extraterrestrial origin? And is life on Earth yet dependent on extraterrestrial material for its long-term sustenance?

The quests for deciphering the origin of life and the place of life within physical universe go hand in hand. Life has not been comprehended yet, neither on the levels of biomolecules, cells, organisms, populations, or habitats and definitely nor in its entirety. It is not yet known if life exists elsewhere in the universe and if it is absolutely identical, roughly similar, or entirely different in terms structure, composition, and function to life on Earth. Therefore, deeper investigations into the quest of "origin of life" trigger enquiries from the accompanying quests of the origin of universe and of habitability.

Habitability has acquired several connotations that differ in their expanse—planetary, circumstellar, and galactic—since it was first etymologized (Strughold, 1953; Shapley, 1953; Huang, 1959, 1960; Hart, 1979; Kasting et al., 1993; Gonzalez et al., 2001; Ward and Brownlee, 2004;

Gowanlock et al., 2011). At the planetary scale, the origin, evolution, and sustainability of life are assumed to result from the geophysical and chemical composition of the planet, whereas the circumstellar habitability depends on the additional factor of orbital position of a planet in the star system. At larger galactic scales, the habitability could also be affected by the positioning of stars, their sizes, ages, stellar types, and cosmo-physico-chemical processes and events occurring in the sub-galactic interstellar space.

Potentially habitable exoplanets are being discovered steadily in our neighboring universe, but the approaches for determining if exoplanets are inhabited by life-forms are yet uncertain. Some of the exoplanets could be hosting no life-forms, some could be populated only in hyper-local environments, and some could carry life-forms proliferated all over the surface. Given this enigma, how should life on exoplanets be corroborated? Can this be solved simply by observing and characterizing the atmospheres and surfaces of planets at higher resolutions? Or should the exoplanetary sources of organics be determined by astronomically ascertaining direct and indirect biogeochemical signatures that could exist around the habitable planets on a large spatial scale? If yes, which are the signatures that should be examined? These and many other tactical questions need to be addressed.

If carbon-based life is searched for in both the local and faraway universe, then there is no better signature of potential life than the element carbon itself. The extraordinary characteristics of carbon, some of which are discussed in this chapter, should be used for connecting the dots between the origins of elemental carbon in the universe and that of carbon-based life.

2 VIEWING THE FIRST BILLION YEARS OF THE UNIVERSE

The primordial chemical elements, hydrogen-1 (^{1}H), helium-4 (^{4}He), and lithium-7 (^{7}Li) (Alpher and Herman, 1950), were nucleosynthesized to varying degree of abundances instantly after the Big Bang (Schramm and Turner, 1998; Olive et al., 2000). The universe, in its first billion years (Gyr), presumably transitioned from an opaque neutral to a translucent ionized state; thereby constructing the very first stars and galaxies (Davies, 1972). It is now becoming apparent that during this period, the universe also underwent significant chemical enrichment, which was possible only through the nucleosynthesis of several chemical elements.

In astronomy, chemical elements with atomic masses higher than the primordial elements are deemed as "metals." The dominant chemistry and mineralogy prevailing in the universe is largely due to metals. But their exact origin and the way in which they proliferated in the universe is yet to be discerned. Cosmological redshift (denoted by z) is an important measure of locating extremely distant luminous objects like galaxies and quasars in the universe. When objects move far away from the observer on Earth, due to the expansion of the universe, the light emanating from them is shifted to higher or redder wavelengths. In that respect, the higher the redshift numerically, the farther and earlier is its existence in the universe.

It is estimated that at $z = 10$, which corresponds to a period approximately 400 million years (Myr) after the Big Bang, methylylidene (CH) with a relative abundance of $\sim 6.2 \times 10^{-21}$ and hydroxyl (OH) with $\sim 1.2 \times 10^{-23}$ could have been the two most abundant radicals (Vonlanthen et al., 2009). This implies the presence of the metals, carbon and oxygen, in a very early epoch of the universe which for very long was perceived to only radiate the ^{1}H electromagnetic spectral line. In spite of that, the calculation does not yield a detailed perspective on the cosmochemical processes through which these metals were first generated. Such a perspective can only be garnered through characterization-driven astronomical surveys of the deep universe.

In the highly metal-rich extant universe in our neighborhood, the bulk of carbon-12 (^{12}C) is presumably generated by the triple-alpha (3 times ^{4}He nuclei) process that occurs during the He-burning phase in the last stages of stellar evolution of low- to medium-mass stars. As the ^{1}H begins to deplete in the stellar interiors, the low-temperature proton-proton chain is arrested, resulting in a rise in the temperature (up to $\sim 10^8$ K) and pressure. This triggers the fusion of two alpha (^{4}He) particles (Weizsäcker, 1938; Bethe, 1939) to synthesize a highly unstable ^{8}Be that has an extremely short lifetime of $\sim 10^{-18}$ s. Despite this instability, the fast fusion rate of ^{8}Be helps in maintaining a small equilibrium concentration in its favor (Salpeter, 1952). This minuscule ^{8}Be then fuses with another ^{4}He to form an excited ^{12}C. An excited ^{12}C is susceptible to rapid disintegration, but some of these isotopes attain a stable state by emitting gamma rays even more rapidly than their decay. This stable carbon is then expelled out of the stellar interiors by dredge-up processes or supernova explosions and it then attains several conformations—radicals like CH, simple molecules like CO, complex carbonaceous molecules, macromolecules, dust, or even carbon-based life, depending on the circumstellar environment it resides in.

In the recent past, considerable advances have been made towards detecting the brightest objects that existed during the universe's first Gyr. More than 6000 star-forming galaxies that are particularly luminous in the ultraviolet (UV) and lyman-α wavelengths and existed 0.9–2.0 Gyr after the Big Bang ($8 > z > 3$) have been discovered (Bouwens et al., 2007; Labbé et al., 2013). The technique of measurement of UV luminosity has garnered wide acceptance as a galaxy tracer due to its close correspondence with the star formation rates. Exhaustive survey of the Hubble Ultra-Deep Field, a spatial region in the Fornax constellation containing thousands of galaxies, led to the identification of a candidate galaxy UDFj-39546284, which existed merely within 500 Myr after the Big Bang. UDFj-39546284 was found to have lower star formation rate density than galaxies existing only 200 Myr later. This hints at a pronounced and rapid increase in the formation of stars only within a span of 200 Myr and well within the first Gyr after the Big Bang (Bouwens et al., 2011). UV luminosity also has its share of drawbacks. It under-performs as a tracer for galaxies with very high star-formation rates and those that are extremely dusty (Wong and Heckman, 1996; Adelberger and Steidel, 2000). Nonetheless, where the capabilities of UV luminosity measurements are inhibited, ultra-deep infrared (IR) astronomy has come to the rescue of astronomers.

The Infrared Array Camera onboard the Spitzer Space Telescope, the Wide Field Camera-3-Infrared (WFC-3/IR) onboard the Hubble Space Telescope, and the Multi-Object Spectrometer for Infrared Exploration (MOSFIRE) spectrograph on the WM Keck Observatory together have identified candidate objects existing at even farther redshifts ($11 > z > 9$). The WFC-3/IR recently confirmed an extremely luminous galaxy "GN-z11," which existed approximately 400 Myr after the Big Bang (Oesch et al., 2016). GN-z11 is the most distant galaxy confirmed to date and is also the most luminous galaxy identified at $z > 10$. Some other marginally younger galaxies detected are 330 EGSY8p7 ($\sim$600 Myr) (Zitrin et al., 2015), EGS-zs8-1 ($\sim$670 Myr) (Oesch et al., 2015), and z8_GND_5296 (700 Myr) (Finkelstein et al., 2013). These extremely old and massive galaxies possibly consisted of the now rare, supermassive, short-lived, and extremely metal poor (consisting only of ^{1}H and ^{4}He) population III (POP-III) stars, which are now estimated to have begun to exist as early as 100 Myr after the Big Bang (Tegmark et al., 1997; Barkana and Loeb, 2001; Yoshida et al., 2003; Loeb and Furlanetto, 2013).

Recently, a candidate galaxy, Cosmos Redshift 7 (CR7) with an extremely high lyman-α luminosity, was detected at $z = 6.604$. The high luminosity of CR7 is best explained by a mixed population of normal stars, similar to those present in the local universe, as well as the extremely metal poor POP-III stars (Sobral et al., 2015). Since POP-III stars have never been directly detected astronomically, several aspects of their existence have only been hypothesized via theoretical simulations (Tornatore et al., 2007). From the standpoint of provenance of metals, POP-III stars could have been ancestral to the metal-deficient population-II (POP-II) stars that nucleosynthesized lighter metals, especially carbon. It is important to comprehend that the genesis of carbon in the universe was the first step towards the formation of the CH radical (Vonlanthen et al., 2009), penultimately organic molecules, and ultimately towards the origin of carbon-based life.

3 THE ORIGIN OF METALLICITY
3.1 BRIEF OVERVIEW OF POP-II STARS

The POP-II stars nucleosynthesize metals in concentrations higher than POP-III stars, but less than the metal-rich population-I (POP-I) stars. Despite their metal deficiencies, POP-II stars surveyed to date have exhibited noticeably diverse chemical compositions. Close to one quarter of POP-II stars are observed to be "iron-poor," in comparison to the Sun with an iron-to-hydrogen abundance ratio $[Fe/H] < -2.0$ (Beers and Sommer-Larsen, 1995; Christlieb et al., 2008). Iron-poor stars with $[Fe/H] \ll -3.0$ apparently show over-abundance of lighter metals like carbon, nitrogen, and oxygen. It is hypothesized that these lighter elements in the circumstellar disks around the POP-II stars could have synthesized dust, organics, and water, which further could have played a major role in the formation of some of the earliest exoplanets in the universe (Kornet et al., 2005; Johansen et al., 2009; Yasui et al., 2009; Ercolano and Clarke, 2010; Mashian and Loeb, 2016).

Given the importance associated with POP-II stars, their identification in the local universe has been a major objective in astronomy. Galactic haloes, the extreme outer regions of galaxies located more than 18 kiloparsecs away from the galactic center, are widely presumed to be regions analogous to the early metal-deficient universe. The star formation processes in the galactic haloes are very different from that occurring in the neighborhood of our Sun (Ferguson et al., 1998; Kobayashi et al., 2008) and probably are analogous to those of the POP-II stars (Rudolph et al., 2006).

3.2 CARBON-ENHANCED METAL POOR STARS

Carbon-enhanced metal poor (CEMP) stars are an anomalous sub-class of POP-II stars often found in the galactic haloes. They comprise ~15%–20% of the POP-II star population with $[Fe/H] \approx -2.0$ (Lucatello et al., 2006, 2014; Frebel et al., 2007; Carollo et al., 2012; Lee et al., 2013). The carbon abundances in the CEMP stars are measured by the carbon-to-iron abundance ratio [C/Fe] ratios which typically are ≥ 0.7. The CEMP population is ~30% of the metal-poor stars with $[Fe/H] < -3.0$ and ~75% for those with $[Fe/H] < -4.0$ (Beers and Christlieb, 2005; Norris et al., 2013; Frebel and Norris, 2015). When [Fe/H] of POP-II stars is -1.5, the [C/Fe] is ~1.0, but as [Fe/H] plummets to -2.7, the [C/Fe] increases to 1.7 (Carollo et al., 2012).

CEMP stars are further distinguished into four different classes—CEMP-r, CEMP-s, CEMP-r/s, and CEMP-no class stars, based on their hypothesized neutron-capture elements (Beers and

Christlieb, 2005). The CEMP-*r* class stars are presumed to undergo rapid neutron-capture at extremely high temperatures and high neutron densities; these conditions are usually produced in supernovae (Pagel, 2009). The CEMP-*s* stars are hypothesized to be part of binary star systems where an asymptotic giant branch (AGB) star undergoes slow neutron-capture and transfers its innate carbonaceous material to a long-lived and low-mass companion star (McClure, 1985; Lucatello et al., 2005; Cohen et al., 2006; Starkenburg et al., 2014). The CEMP-*r/s* stars supposedly exhibit transfer of carbonaceous materials similar to CEMP-*s* stars, and in addition, they transfer elements heavier than carbon (Hill et al., 2000; Goswami et al., 2006). The final category, the CEMP-*no* stars do not show neutron-capture enhancement of elements as in the other three classes. Their carbon abundances are known to decrease with the decrease in the overall metallicity (Masseron et al., 2010).

CEMP stars are vital in the current models of the formation of earliest metals in the universe. Stable carbon could have been first synthesized from the earliest supernovae explosions of POP-II stars (Omukai, 2000; Bromm et al., 2001; Bromm and Loeb, 2003), probably from a stellar population akin to the CEMP-no stars (Aoki, 2010). The circumstellar environments around these earliest CEMP stars could have been the nurseries of the universe's earliest carbon-containing heteroatomic molecules and solid-state materials. When and how did carbon assume its molecular association with the lighter elements? What were the conditions then? When and how was carbon able to chemically evolve from an element to a material capable of forming planetesimals? All these and similar questions would be addressed once extensive chemical characterization of the comparatively nearby galactic haloes and the high-redshift POP-II stars become attainable.

4 CARBON: THE REACTANT AND SUBSTRATE IN THE EARLY UNIVERSE
4.1 CARBON MONOXIDE: THE REACTANT

Carbon monoxide (CO) is known to be the second-most abundant molecule in the universe after molecular hydrogen (H_2) (Ehrenfreund and Cami, 2010). But unlike H_2 whose precursor is the primordial H^+ ejected from the Big Bang, the carbon and oxygen of the heteroatomic CO could not have formed until the nucleosynthesis of metals in the universe. Cosmic CO has the propensity to transition to its first excited state even at an extremely low temperature of 5 K, thus making it a readily detectable and an efficient tracer for H_2, since the latter is difficult to detect in the cold ISM (Solomon et al., 1971; Neininger et al., 1998; Carilli and Walter, 2013). Having encountered this characteristic of CO in the local universe, attempts are being made to trace it in the high-redshift universe (Decarli et al., 2016).

Ionized carbon (C^+) emissions from two ultraluminous infrared/submillimeter galaxies (ULIRGs), located at z of 4.42 and 4.44, were observed using the Atacama Large Millimeter/Submillimeter Array (Swinbank et al., 2012). The C^+ emissions indicate extreme star formation and vast CO reservoirs in these ULIRGs that could have formed via the $C^+ \rightarrow C^0$ (neutral) $\rightarrow$ CO mechanism (Beuther et al., 2014). Similarly in a survey done by the Australia Telescope Compact Array, a high-redshift submillimeter galaxy ALESS65.1 ($z = 4.44$) was found to emit the ^{12}CO radio emission lines. The total volume of the cold gas (including H_2 and He) in ALESS65 was estimated to be $\sim 1.7 \times 10^{10}$ solar masses ($M_\odot$), its gas depletion rate to be ~ 25 Myr, and the galaxy also showed evidences of metallicity (Huynh et al., 2014). The presence of CO and H_2 in high-redshift objects raises the possibility of an organic cosmochemistry that could utilize them as reactants.

4.2 CARBONACEOUS DUST: THE SUBSTRATE

A significant number of the early universe galaxies have emissions due to heating of the circumstellar and interstellar dust by stars within them. With the backing of this astrophysical certitude, the cosmic infrared measurements carried out by the Infrared Astronomy Satellite, Infrared Space Observatory, Spitzer Space Telescope and the Cosmic Background Explorer, and several ground-based infrared spectrographs have led to direct detection of numerous high-redshift objects exclusively in the submillimeter/far-infrared (sub-mm/far-IR) wavelengths.

A high-redshift galaxy z8_GND_5296 that existed only 700 Myr after the Big Bang was identified using the MOSFIRE near-infrared spectrograph (Finkelstein et al., 2013). This galaxy demonstrated a star formation rate higher by a factor of 100 than that in the Milky Way and had colors consistent with high metallicity and dust (Finkelstein et al., 2013). In another observational campaign, a much older high-redshift galaxy A1689-zD1 was identified, which possibly existed well within the first 500 Myr after the Big Bang. A1689-zD1 showed an intense star-forming rate of $\sim 12\ M_\odot$/year, and dust-to-gas ratio roughly equivalent to the Milky Way galaxy (Watson et al., 2015). These two galaxies— z8_GND_5296 and the A1689-zD1—are representative of high-redshift dusty star formation galaxies (DSFGs) that had star formation rates higher than the relatively younger Milky Way galaxy (Robitaille and Whitney, 2010; Chomiuk and Povich, 2011). So what actually made these star-formation galaxies so dusty?

The POP-II stars in their terminal stages could have released massive volumes of both carbonaceous and non-carbonaceous dust. This ejected dust could have played out both as a substrate and a catalyst to a variety of cosmochemical processes, including the very crucial conversion of atomic to molecular hydrogen (Tegmark et al., 1997; Glover, 2003; Cazaux and Spaans, 2004; Mashian and Loeb, 2016). Shock-waves from the frequent supernovae explosions in the DSFGs are presumed to easily destroy hydrogenated amorphous carbon (a:C-H) dust (Panagia, 2005; Serra Diaz-Cano and Jones, 2008; Jones, 2009). However, in one particular astronomical observation carried out using the Herschel Space Observatory, the dust in the ejecta of supernova 1987A showed the presence of a huge volume of amorphous carbon (Matsuura et al., 2015). This observation points to the possibility that supernovae can dissociate a robust molecule like CO to form amorphous carbon (Clayton, 2011). This indicates that carbon, owing to its ability of catenation, is able to preserve itself in its most reactive and refractory form, even after being subjected to the energies unleashed by supernovae.

4.3 DUST-GRAIN INTERACTION: ESCALATING ORGANIC ENRICHMENT

Laboratory experiments simulating the inner warm solar nebula conditions (700 torr and 873 K) by exposing graphitic dust to a flowing gas mixture consisting of CO, H_2, and N_2 led to the synthesis of carbon nanotubes on the dust grain surfaces (Nuth et al., 2010). This experiment demonstrates that at higher temperatures (>800 K), in regions close to a stellar radiation source, carbonaceous dust can probably chemically reduce the gaseous CO into solid-state carbonaceous allotropes, whereas at lower temperatures, in regions slightly away from the radiation source, similar chemical reduction can probably synthesize a wide repertoire of organic molecules and macromolecules (Fries and Steele, 2008; Nuth et al., 2010; Giri et al., 2016).

Exoplanets entirely made of carbon have been theorized (Lodders, 2004; Kuchner and Seager, 2005; Bond et al., 2010; Madhusudan et al., 2011, 2012a; Öberg et al., 2011; Johnson et al., 2012; Mashian and Loeb, 2016) and tentatively observed in the local universe, especially around stars with high C/O and C/H ratios (Madhusudan et al., 2011; Madhusudan, 2012b). In the early universe and most probably in the circumstellar envelopes of CEMP stars, carbon dust might have accreted under the influence of gravitational hydrodynamics to grow into meter- and kilometer-scale planetesimals and further to carbonaceous exoplanets (Mashian and Loeb, 2016). The detection of carbon exoplanets at high-redshifts, whenever it happens, would be of immense scientific significance.

5 FINDING ORGANICS: ANALOGUES OF HIGH-REDSHIFT GALAXIES IN THE LOCAL UNIVERSE
5.1 SIGNATURES OF ORGANICS IN THE LOCAL UNIVERSE

The cosmosyntheses of organic chemical species from the abundant and chemically simpler precursors like CO and H_2 could have occurred via numerous chemical pathways. These pathways would differ from each other based on factors like the loci in the universe where the reactions occurred, the concentrations of precursors, substrates, and catalysts, interstellar and circumstellar thermodynamics, dust and gas hydrodynamics, and radiation fluxes. Regardless of the pathway and the status as reactant and product, organic chemical species could be identified by their typical spectral signatures.

Astronomical extinction features suggesting numerous complex organics and carbonaceous allotropes have been identified along several lines of sight within the Milky Way (van Dishoeck and Blake, 1998; Kwok, 2004, 2009). One such prominent interstellar extinction feature is the UV-extinction bump at 217.5 nm (hereafter UV-bump) (Stecher, 1965). UV-bump typically demonstrates an invariant central wavelength, but an irregular bandwidth; the latter is believed to be caused by the different interstellar environments existing in various lines of sight (Whittet, 2003; Ehrenfreund and Charnley, 2000; Fitzpatrick and Massa, 2007). Graphite grains (Stecher and Donn, 1965; Mathis et al., 1977; Ferrière, 2001), condensed-phase polycyclic aromatic hydrocarbon (PAH) clusters (Joblin et al., 1992; Li and Draine, 2001; Duley, 2006), polycrystalline graphite (Papoular and Papoular, 2009), hydrogenated fullerenes (Cataldo and Iglesias-Groth, 2009), and amorphous carbon (Mennella et al., 1998) are some of the candidate chemical materials that are conjectured to cause the UV-bump.

The UV-bump was recently identified for the first time outside the Local Group, a conglomerate of more than 54 galaxies, especially in the gamma-ray bursts GRB 080605 ($z = 1.64$), GRB 080805 ($z = 1.505$), and several star-forming regions and intermediate mass galaxies located at $1 < z < 2.5$ (Elíasdóttir et al., 2009; Noll et al., 2009; Zafar et al., 2012). Although the relative strength of the UV-bump is weaker in the Local Group observations in comparison to those emanating within the Milky Way, its central position and width remains consistent. Amorphous silicates and carbonaceous grains present in chondritic interplanetary dust particles were found to exhibit the UV-bump (Bradley et al., 2005), implying that these natural grains could be analogous to the interplanetary UV-bump carriers in terms of chemical and mineralogical compositions and structures.

Another set of absorption features at 3.4, 6.85, and 7.25 μm, possibly emanating from the aliphatic CH_3/CH_2 deformation modes, have been observed in the ULIRGs of the local universe (Wickramasinghe and Allen, 1980; Sandford et al., 1991; Bridger et al., 1994; Pendleton et al., 1994; Adamson et al., 1999; Imanishi, 2000; Mason et al., 2004; Spoon et al., 2004; Dartois and Muñoz-Caro, 2007; Sajina et al., 2009; Desai et al., 2009; Godard et al., 2011). Likewise, features at 3.3 and between 6.0 and 6.4 µm attributed to the aromatic C=C stretching modes and the 5.8 µm feature attributed to the carbonyl C=O vibration modes have also been identified (Harrington et al., 1998; Duley, 2000; Dartois et al., 2005; Zijlstra et al., 2006; Dartois and Muñoz-Caro, 2007). The mid-IR region (3–20 µm) particularly presents emission bands to which these materials are attributed: buck-minsterfullerenes (Sellgren et al., 2010; Kwok and Zhang, 2013; Campbell et al., 2015), PAHs (Leger and Puget, 1984; Galliano et al., 2008), and heavy petroleum fractions (Cataldo et al., 2004, 2013; Kwok and Zhang, 2013). PAH emission features are also found to correspond with the 158 µm line of ionized carbon, which in turn is often utilized as a tracer of star formation (Joblin et al., 2010). The high-redshift sub-mm galaxies are found to be comparable to ULIRGs of the local universe in terms of their IR, radio, and CO measurements (Tacconi et al., 2008, 2010), possibly indicating abundant organics.

5.2 AGB STARS: REFUGE FOR ORGANICS?

Not all stars meet their ultimate fate as supernovae; some in their last stages alternatively enter into a less destructive asymptotic giant branch (AGB) phase. Intermediate and low-mass stars ($1 < M < 8 \, M_\odot$) during their AGB phases can benevolently lose $\sim$80% of their original mass at loss rates typically around 10^{-6} to $10^{-4} \, M_\odot \, yr^{-1}$ to form enormous circumstellar disks (Zijlstra et al., 2006). The circumstellar disks of carbon-rich AGB stars show absorption features characteristic of aliphatic and aromatic macromolecular solids similar to coal and kerogen, amorphous carbon, and silicon carbide. On the other hand, the oxygen-rich AGB stars show features that could be attributed more so to amorphous and crystalline silicates. Certain AGB stars show features of both C-rich and O-rich stars, which is inferred by the presence of silicates and carbonaceous materials in substantial amounts (Waters et al., 1998; Kwok, 2004; Perea-Calderón et al., 2009). The circumstellar disk of the AGB-phase carbon star IRC+10216 shows the presence of more than 65 different organic species (Cernicharo et al., 2000, 2008; Ziurys, 2006; Thaddeus et al., 2008; Tenenbaum et al., 2010) as well as carbonaceous dust (Henning and Salama, 1998). The abundance of organics in the extremely large circumstellar disk of IRC+10216 is most probably due to the benevolent release of matter (Prantzos et al., 1996; Willson, 2000).

The prospects of detecting and characterizing the circumstellar disks of CEMP stars in the high-redshift universe therefore are exciting. Their disks could be profuse with carbonaceous materials and organic compounds. The processes and environments that led to the formation, preservation, and enrichment of organic molecules in the circumstellar disks of CEMP stars could have been indispensable for the formation of habitable zones in the early universe. The identification of the reactants CO, H_2, and the substrate dust in the DSFGs at $z \sim 7$ hints at the possible presence of a diverse organic galactic inventory within them. In this milieu, efforts should be invested in finding astronomical signatures of organics in the high-redshift universe.

6 CONCLUSION: WHERE DOES THE SCIENCE OF ORIGINS OF HABITABILITY GO FROM HERE?

The biochemical architecture prevalent on Earth, which first arose some ten billion years after the Big Bang, is laid largely on carbon, hydrogen, oxygen, and nitrogen followed by a string of other elements like sulfur, phosphorus, calcium, magnesium, iron, sodium, chlorine, potassium, boron, iodine, molybdenum, selenium, chromium, nickel, cobalt, copper, zinc, and manganese to name a few. For all its diversity, life utilizes these elements in different permutations and combinations. But is the origin of life a biochemical singularity? Was life, a higher order chemistry in itself, able to spin-off on a habitable planet like Earth from a long-running heritage of organic cosmochemistry occurring in the cosmos?

6.1 THE FIRST YARDSTICK OF FINDING HABITABILITY IN THE ANCIENT UNIVERSE

In the universe's primordial epochs when C, H, O, and N first co-existed, they could have together synthesized a significant catalog of chemical functional groups. This primitive organic cosmochemistry could have undergone enrichment with the amplification of metallicity in the universe. It is possible that the earliest organics might someday be astronomically detected in the high-redshift galaxies, but would such detections be a direct indication for the complex chemistry of life? The clues to the quest of the origin of carbon-based life are linked to the quest for the origin of carbon and origin of simple organic chemistry in the universe.

Life is an ultra-complex and adaptable supra-systemic *automaton* and demands the availability of diverse chemical functional groups working in concert to sustain its functions as well as the integrity of structures. The presence of complex life anywhere in the universe depends on the availability of the main chemical element—carbon—and other accompanying elements to form a vast diversity vast diversity of organic molecules. This points to a possibility that galaxies that host a substantial population of metal-rich stars and a diverse inventory of elements and physical conditions that permit carbon-based molecules to exist in both fluidic and solid states, stand a stand a greater chance of accommodating complex carbon-based life.

The quests of origins of habitability, life, and the universe have a long history in human thought. In antiquity, these questions, however, were largely pursued through metaphysics, philosophy, and abstract theory. Technologies capable of practically pursuing these quests have been realized only in the past few decades. The James Webb Space Telescope, the European Extremely Large Telescope, the Giant Magellan Telescope, the Primeval Structure Telescope, and many upcoming observatories will be pivotal in imaging and characterizing galactic and sub-galactic objects in the high-redshift universe. Apart from the attempts to identify known astronomical signatures of carbon-containing molecules and materials, greater efforts would be needed in deciphering organic signatures that could be archetypal to the high-redshift universe and which are yet unknown. The comparative surveys of POP-II stars, particularly of the CEMP-class, with POP-III will be crucial for determining the provenance of carbon and the provenance of metallicity in the universe. But the detection of potentially habitable exoplanets in the high-redshift universe would continue to be a massive technological challenge, given their smaller dimensions. This should nonetheless prompt a greater interest in determining the characteristics of galactic habitable zones, which due to their greater dimensions stand a better chance of being recognized.

Galactic habitable zones could be ear-marked by assessing factors such as the age of stars and their location within the zones, the metallicity in the zone, the pressure and temperature within the zone, the number of potential supernovae and evolved stars and their distances from nascent stars, and the star-formation rates. These and many such factors could be explored at much greater precisions in the coming years.

6.2 THE CUTTING-EDGE SCIENCE OF ORIGINS

In the twentieth century, the quest to decipher various processes and phenomena in the universe created new interdisciplinary domains of astrophysics, astrochemistry, astrobiology, cosmochemistry, and exogeology. This transpired only because new technologies were integrated with the classical scientific disciplines that were curated over centuries. Interdisciplinary studies now generate huge volumes of novel scientific data, which was not possible before. However, interdisciplinarity is not enough to address philosophically and technologically the questions of the origins of the universe, habitability, and life. Considering the epistemological formidability of these questions, interdisciplinarity should eventually turn into transdisciplinarity. In particular, the study of habitable zones in the high-redshift universe is likely to become a trans-disciplinary enterprise.

The habitability of the universe will be probed by the most pioneering methods in science and the most cutting-edge technologies. Precision "-scopy" and "metry" instrumentation (spatial, spectrometer, spectral, sensor, and optical), novel composites, ceramics, and metamaterials, next-generation semi- and superconductors, quantum computing, sequence mining, pattern recognition, deep-space and long-duration space exploration, and immersion will become integrated drivers for the three quests. So what can be anticipated from combining advanced technologies?

The quantifications necessary to model the highly complex and enormous processes and phenomena in the universe have been grasped only recently. It would become increasingly important to perceive matter and processes in scales beyond the current extent of modeling approaches, and this could become the vantage point from where the current data visualization is transformed into the futuristic data virtualization. Huge volumes of data gathered from astronomical observations, in situ extraterrestrial sample analyses, mathematical simulations, and laboratory simulations could be analyzed synchronously by creating an immersive environment, co-simulating some simple processes of the universe occurring at a particular location for a particular period of time. To support such research and development, there is a need to design analytical instruments that operate in the yotta (10^{24}) to yocto (10^{-24}) scalar ranges. Of the many needs of many faculties, here are the few needs pertaining to the focal points of this chapter.

As the quest to decipher the origin of carbon-based life progresses, *ad rem* attempts to quantify the total carbon in the universe, the volume of each conformations, the volume of carbon in a fixed spatial domains, the speed and distances covered by chemical species emitting from stellar outflows into the interstellar and intergalactic medium, and the concentration of desired chemical species that enhance the habitability of galactic regions would also be undertaken. These and similar phenomena cannot be studied in isolation as we do it today and would demand data virtualization.

It is increasingly becoming ostensible that life is not merely a "local" phenomena in the universe, but a manifested legacy of cosmic processes that occur in intricate ways over scales of space and time larger than the limits of the host planet. We certainly do not yet comprehend these processes, but we definitely know their common thread, which is carbon. "Follow the Carbon" on that account should be the core doctrine, if we are to truly search for life in the universe.

ACKNOWLEDGMENTS

The author is very grateful to the editors of this book and book series A. Sharov, R. Gordon, P.H. Rampelotto, and J. Seckbach. Special thanks to D. Giovannelli, E. Smith, H.J. Cleaves, J.A. Nuth, A. Steele, M. Fries, F. Fergusson,

N. Johnson, and K. Chandru for stimulating discussions. The author gratefully acknowledges the ELSI Origins Network (EON) which is supported by a grant from the John Templeton Foundation.

REFERENCES

Adamson, A.J., Whittet, D.C.B., Chrysostomou, A., Hough, J.H., Aitken, D.K., Wright, G.S., Roche, P.F., 1999. Spectropolarimetric constraints on the nature of the 3.4 micron absorber in the interstellar medium. Astrophys. J. 512, 224–229.

Adelberger, K.L., Steidel, C.C., 2000. Multi wavelength observation of dusty star formation at low and high-redshift. Astrophys. J. 544, 218–241.

Alpher, R.A., Herman, R.C., 1950. Theory of the origin and relative abundance distribution of the elements. Rev. Mod. Phys. 22, 153–212.

Altwegg, K., Balsiger, H., Bar-Nun, A., Berthelier, J.-J., Bieler, A., Bochsler, P., Briois, C., Calmonte, U., Combi, M.R., Cottin, H., De Keyser, J., Dhooghe, F., Fiethe, B., Fuselier, S.A., Gasc, S., Gombosi, T.I., Hansen, K.C., Haessig, M., Jäckel, A., Kopp, E., Korth, A., Le Roy, L., Mall, U., Marty, B., Mousis, O., Owen, T., Réme, H., Rubin, M., Sémon, T., Thou, C.-Y., Waite, J.H., Wurz, P., 2016. Prebiotic chemicals—amino acid and phosphorus—in the coma of comet 67P/Churyumov-Gerasimenko. Sci. Adv. 2, 1600285.

Aoki, W., 2010. Chemical abundances in the universe: Connecting first stars to planets. Proc. Int. Astron. Union, IAU Symp. 265, 111–116.

Bandurski, E.L., Nagy, B., 1976. The polymer-like organic material in the Orgueil meteorite. Geochim. Cosmochim. Acta 40, 1397–1406.

Barkana, R., Loeb, A., 2001. In the beginning: The first sources of light and the deionisation of the universe. Phys. Rep. 349, 125–238.

Beers, T.C., Sommer-Larsen, J., 1995. Kinematics of metal-poor stars in the galaxy. Astrophys. J. Suppl. Ser. 96, 175–221.

Beers, T.C., Christlieb, N., 2005. The discovery and analysis of very metal-poor stars in the galaxy. Ann. Rev. Astron. Astrophys. 43, 531–580.

Bethe, H.A., 1939. Energy production in stars. Phys. Rev. 55, 434–456.

Beuther, H., Ragan, S.E., Ossenkopf, V., Glover, S., Henning, Th., Linz, H., Nielbock, M., Krause, O., Stutzki, J., Schilke, P., Güsten, R., 2014. Carbon in different phases ([CII], [CI], and CO) in infrared dark clouds: Cloud formation signatures and carbon gas fractions. Astron. Astrophys. 571, A53.

Bond, J.C., O'Brien, D.P., Lauretta, D.S., 2010. The compositional diversity of extrasolar terrestrial planets. I. in situ simulations. Astrophys. J. 715, 1050.

Bouwens, R.J., Illingworth, G.D., Franx, M., Ford, H., 2007. UV luminosity functions at z ~4, 5, and 6 from the Hubble Ultra Deep Field and Other Deep Hubble Space Telescope ACS Fields: Evolution and star formation history*. Astrophys. J. 670, 928–958.

Bouwens, R.J., Illingworth, G.D., Labbé, I., Oesch, P.A., Trenti, M., Carollo, C.M., van Dokkum, P.G., Franx, M., Stiavelli, M., Gonzalez, V., Magee, D., Bradley, L., 2011. A candidate redshift z ≈ 10 galaxy and rapid changes in that population at an age of 500 Myr. Nature 469, 504–507.

Bradley, J., Dai, J.R., Emi, R., Browning, N., Graham, G., Weber, P., Smith, J., Hutcheon, I., Ishii, H., Bajt, S., Floss, C., Stadermann, F., Sandford, S., 2005. An astronomical 2175 angstrom feature in interplanetary dust particles. Science 307, 244–247.

Bridger, A., Wright, G.S., Geballe, T.R., 1994. Dust absorption in NGC1068. In: McLead, I.S. (Ed.), Infrared Astronomy with Arrays: The Next Generation. Astrophysics and Space Science Library 190, Springer, Netherlands.

Bromm, V., Ferrara, A., Coppi, P.S., Larson, R.B., 2001. The fragmentation of pre-enriched primordial objects. Mon. Not. Royal Astron. Soc. 328, 969–976.

Bromm, V., Loeb, A., 2003. The formation of the first low-mass stars from gas with low carbon and oxygen abundances. Nature 425, 812–814.

Campbell, E.K., Holz, M., Gerlich, D., Maier, J.P., 2015. Laboratory confirmation of C60+ as the carrier of two diffuse interstellar bands. Nature 523, 322–323.

Carilli, C.L., Walter, F., 2013. Cool gas in high-redshift galaxies. Ann. Rev. Astron. Astrophys. 51, 105–161.

Carollo, D., Beers, T.C., Bovy, J., Sivarani, T., Norris, J.E., Freeman, K.C., Aoki, W., Lee, Y.S., Kennedy, C.R., 2012. Carbon-enhanced metal-poor stars in the inner and outer halo components of the Milky Way. Astrophys. J. 744, 195.

Cataldo, F., Keheyen, Y., Heymann, D., 2004. Complex organic matter in space: About the chemical composition of carriers of the Unidentified Infrared Bands (UIBs) and protoplanetary emission spectra recorded from certain astrophysical objects. Origin Life Evol. Biospheres 34, 13–24.

Cataldo, F., Iglesias-Groth, S., 2009. On the action of UV photons on hydrogenated fullerenes C60H36 and C60D36. Monthly Notices of the Royal Astronomical Society 400, 291–298.

Cataldo, F., García-Hernández, D.A., Manchado, A., 2013. Far- and mid-infrared spectroscopy of complex organic matter of astrochemical interest: coal, heavy petroleum fractions and asphaltenes. Monthly Notices of the Royal Astronomical Society 429, 3025–3039.

Cazaux, S., Spaans, M., 2004. Molecular hydrogen formation on dust-grains in the high-redshift universe. Astrophys. J. 611, 40–51.

Cernicharo, J., Guélin, M., Kahane, C., 2000. A lambda 2 mm molecular line survey of the C-star envelope IRC +10216. Astron. Astrophys. Suppl. Ser. 142, 181–215.

Cernicharo, J., Guélin, M., Agúndez, M., McCarthy, M.C., Thaddeus, P., 2008. Detection of C5N- and vibrationally excited C6H in IRC+10216. Astrophys. J. 688, L83.

Cohen, J.G., McWilliam, A., Shectman, S., Thompson, I., Christlieb, N., Melendez, J., Ramirez, S., Swensson, A., Zickgraf, F.-Z., 2006. Carbon stars in the Hamburg/ESO survey: Abundances. Astron. J. 132, 137–160.

Chomiuk, L., Povich, M.S., 2011. Toward a unification of star formation rate determinations in the Milky Way and other galaxies. Astron. J. 142, 6.

Christlieb, N., Schörck, T., Frebel, A., Beers, T.C., Wisotzki, L., Riebers, D., 2008. The stellar content of the Hamburg/ESO survey. IV. Selection of candidate metal-poor stars. Astron. Astrophys. 484, 721–732.

Clayton, D.D., 2011. A new astronomy with radioactivity: Radiogenic carbon chemistry. New Astron. Rev. 55, 155–165.

Dartois, E., Muñoz-Caro, G.M., Deboffle, D., d'Hendecourt, L., 2005. Diffuse interstellar medium organic polymers: Photoproduction of the 3.4, 6.85 and 7.25 μm features. Astron. Astrophys. 423, 33–36.

Dartois, E., Muñoz-Caro, G.M., 2007. Carbonaceous dust grains in luminous infrared galaxies. Spitzer/IRS reveals a-C:H as an abundant and ubiquitous ISM component. Astron. Astrophys. 476, 1235–1242.

Davies, P.C.W., 1972. Is the universe transparent or opaque? J. Phys. A: Gen. Phys. 5, 1722–1737.

Desai, V., Soifer, B.T., Dey, A., Le Floc'h, E., Armus, L., Brand, K., Brown, M.J.I., Brodwin, M., Jannuzi, B.T., Houck, J.R., Weedman, D.W., Ashby, M.L.N., Gonzalez, A., Huang, J., Smith, H.A., Teplitz, H., Willner, S.P., Melbourne, J., 2009. Strong polyaromatic hydrocarbon emission from $z \approx 2$ ULIRGs*. Astrophys. J. 700, 1190.

Duley, W.W., 2000. Chemical evolution of carbonaceous material in interstellar clouds. Astrophys. J. 528, 841.

Duley, W.W., 2006. A plasmon resonance in dehydrogenated coronene ($C_{24}H_x$) and its cations and the origin of the interstellar extinction band at 217.5 nm. Astrophys. J. 639, 59–62.

Decarli, R., Walter, F., Aravena, M., Carilli, C., Bouwens, R., da Cunha, E., Daddi, E., Elbaz, D., Riechers, D., Smail, I., Swinbank, M., Weiss, A., Bacon, R., Bauer, F., Bell, E.F., Bertoldi, F., Chapman, S., Colina, L., Cortes, P.C., Cox, P., Gónzalez-López, J., Inami, H., Ivison, R., Hodge, J., Karim, A., Magnelli, B., Ota, K., Popping, G., Rix, H.W., Sargent, M., van der Wel, A., van der Werf, P., 2016. The ALMA spectroscopic survey in the Hubble ultra deep field: molecular gas reservoirs in high-redshift galaxies. Astrophys. J. 833, 70.

Ehrenfreund, P., Charnley, S., 2000. Organic molecules in the interstellar medium, comets and meteorites: A voyage from dark clouds to early Earth. Ann. Rev. Astron. Astrophys. 38, 427–483.

Ehrenfreund, P., Cami, J., 2010. Cosmic carbon chemistry: from the interstellar medium to the early earth. Cold Spring Harbor Perspect. Biol. 2. a002097.

Elíasdóttir, Á., Fynbo, J.P.U., Hjorth, J., Ledoux, C., Watson, D.J., Andersen, A.C., Malesani, D., Vreeswijk, P.M., Prochaska, J.X., Sollerman, J., Jaunsen, A.O., 2009. Dust extinction in high z-galaxies with gamma-ray burst afterglow spectroscopy: The 2175 Å feature at $z = 2.45$. Astrophys. J. 697, 1725–1740.

Elsila, J.E., Glavin, D.P., Dworkin, D.P., 2009. Cometary glycine detected in samples returned by Stardust. Meteorit. Planet. Sci. 44, 1323–1330.

Ercolano, B., Clarke, C.J., 2010. Metallicity, planet formation and disc lifetimes. Mon. Not. Royal Astron. Soc. 402, 2735–2743.

Ferguson, A.M.N., Gallagher, J.S., Wyse, R.F.G., 1998. The extreme outer regions of disk galaxies. Astron. J. 116, 673.

Ferrière, K.M., 2001. The interstellar environment of our galaxy. Rev. Mod. Phys. 73, 1031–1066.

Finkelstein, S.L., Papovich, C., Dickinson, M., Song, M., Tilvi, V., Koekemoer, A.M., Finkelstein, K.D., Mobasher, B., Ferguson, H.C., Giavalisco, M., Reddy, N., Ashby, M.L.N., Dekel, A., Fazio, G.G., Fontana, A., Grogin, N.A., Huang, J.-S., Kocevski, D., Rafelski, M., Weiner, B.J., Willner, S.P., 2013. A rapidly star-forming galaxy 700 million years after the Big Bang at $z = 7.51$. Nature 502, 524–527.

Fitzpatrick, E.L., Massa, D., 2007. An analysis of the shapes of interstellar extinction curves. V. The IR-through-UV curve morphology. Astrophys. J. 663, 320–341.

Frebel, A., Johnson, J.L., Bromm, V., 2007. Probing the formation of the first low-mass stars with spectral archeology. Mon. Not. Roy. Astron. Soc. 380, 40–44.

Frebel, A., Norris, J.E., 2015. Near-field cosmology with extremely metal-poor stars. Ann. Rev. Astron. Astrophys. 53, 631–688.

Fries, M., Steele, A., 2008. Graphite whiskers in CV3 meteorites. Science 320, 91–93.

Galliano, F., Madden, S.C., Tielens, A.G.G.M., Peeters, E., Jones, A.P., 2008. Variations of the mid-IR aromatic features inside and among galaxies. Astrophys. J. 679, 310.

Giri, C., McKay, C.P., Goesmann, F., Schäfer, N., Li, X., Steininger, H., Brinckerhoff, W.B., Gautier, T., Reitner, J., Meierhenrich, U.J., 2016. Carbonization in Titan Tholins: implication for low albedo on surfaces of Centaurs and trans-Neptunian objects. Int. J. Astrobiol. 15, 231–238.

Glover, S.C.O., 2003. Comparing gas-phase and grain-catalyzed H_2 formation. Astrophys. J. 584, 331–338.

Godard, M., Féraud, G., Chabot, M., Carpentier, Y., Pino, T., Brunetto, R., Duprat, J., Engrand, C., Bréchignac, P., d'Hendecourt, L., Dartois, E., 2011. Ion irradiation of carbonaceous interstellar analogues: Effects of cosmic rays on the 3.4 µm interstellar absorption band. Astron. Astrophys. 529, A146.

Goesmann, F., Rosenbauer, H., Bredehöft, J., Cabane, M., Ehrenfreund, P., Gautier, T., Giri, C., Krüger, H., Le Roy, L., MacDermott, A.J., McKenna-Lawlor, S., Meierhenrich, U.J., Muñoz Caro, G.M., Raulin, F., Roll, R., Steele, A., Steininger, H., Sternberg, R., Szopa, C., Thiemann, W., Ulamec, S., 2015. Organic compounds on comet 67P/Churyumov-Gerasimenko revealed by COSAC mass spectrometry. Science 349. aab0689.

Gonzalez, G., Brownlee, D., Ward, P., 2001. The galactic habitable zone I., galactic chemical evolution. Icarus 152, 185–200.

Goswami, A., Aoki, W., Beers, T.C., Christlieb, N., Norris, J.E., Ryan, S.G., Tsangarides, S., 2006. A high-resolution spectral analysis of three carbon-enhanced metal-poor stars. Mon. Not. Roy. Astron. Soc. 372, 343–356.

Gowanlock, M.G., Patton, D.R., McConnell, S.M., 2011. A model of habitability within the Milky Way galaxy. Astrobiology 11, 855–873.

Harrington, J.P., Lame, N.J., Borkowski, K.J., Bregman, J.D., Tsvetanov, Z.I., 1998. Discovery of a 6.4 micron dust feature in hydrogen-poor planetary nebulae. Astrophys. J. 501, 123–126.

Hart, M.H., 1979. Habitable zones about main sequence stars. Icarus 37, 351–357.

Henning, Th., Salama, F., 1998. Carbon in the universe. Science 282, 2204.

Hill, V., Barbuy, B., Spite, M., Spite, F., Cayrel, R., Plez, B., Beers, T.C., Nordstroem, B., Nissen, B., 2000. Heavy-element abundances in the CH/CN-strong very metal-poor stars CS 22948-27 and CS 29497-34. Astron. Astrophys. 353, 557–568.

Hörst, S.M., Yelle, R.V., Buch, A., Carrasco, N., Cernogora, G., Dutuit, O., Quirico, E., Sciamma-O'Brien, E., Smith, M.A., Somogyi, A., Szopa, C., Thissen, R., Vuitton, V., 2012. Formation of amino acids and nucleotide bases in a Titan atmosphere simulation experiment. Astrobiology 12, 809–817.

Huang, S.-S., 1959. Occurrence of life in the universe. Am. Sci. 47, 397–402.

Huang, S.-S., 1960. The sizes of habitable planets. Publ. Astron. Soc. Pacific 72, 489–493.

Huynh, M.T., Kimball, A.E., Norris, R.P., Smail, I., Chow, K.E., Coppin, K.E.K., Emonts, B.H.C., Ivision, R.J., Smolcic, V., Swinbank, A.M., 2014. Detection of molecular gas in an ALMA [CII]-identified submillimeter galaxy at $z = 4.44$. Mon. Not. Roy. Astron. Soc. 443, 54–58.

Imanishi, M., 2000. The 3.4-μm absorption feature towards three obscured active galactic nuclei. Mon. Not. Roy. Astron. Soc. 319, 331–336.

Joblin, C., Leger, A., Martin, P., 1992. Contribution of polycyclic aromatic hydrocarbon molecules to the interstellar extinction curve. Astrophys. J. 393, 79–82.

Joblin, C., Pilleri, P., Montillaud, J., Fuente, A., Gerin, M., Berné, O., Ossenkopf, V., Le Bourlet, J., Teyssier, D., Goicoechea, J.R., Le Petit, F., Röllig, M., Akyilmaz, M., Benz, A.O., Boulanger, F., Bruderer, S., Dedes, C., France, K., Güsten, R., Harris, A., Klein, T., Kramer, C., Lord, S.D., Martin, P.G., Martin-Pintado, J., Mookerjea, B., Okada, Y., Phillips, T.G., Rizzo, J.R., Simon, R., Stutzki, J., van der Tak, F., Yorke, H.W., Steinmetz, E., Jarchow, C., Hartogh, P., Honingh, C.E., Siebertz, O., Caux, E., Colin, B., 2010. Gas morphology and energetics at the surface of PDRs: New insights with *Herschel* observations of NGC 7023. Astron. Astrophys. 521, L25.

Johansen, A., Youdin, A., Mac Low, M.-M., 2009. Particle clumping and planetesimal formation depend strongly on metallicity. Astrophys. J. 704, L75.

Johnson, K.S., 2001. Iron supply and demand in the upper ocean: Is extraterrestrial dust a significant source of bioavailable iron? Global Biogeochem. Cycles 15, 61–63.

Johnson, T.V., Mousis, O., Lunine, J.I., Madhusudan, N., 2012. Planet. Compos. Exoplanet Syst. 757, 192.

Jones, A.P., 2009. The cycle of carbon dust in the ISM. In: Henning, T., Grün, E., Steinacker, J. (Eds.), Cosmic Dust-Near and Far ASP Series, San Francisco: Astronomical Society of the Pacific, September 8–12, 2008. Germany pp, Heidelberg, p. 473.

Kasting, J.F., Whitmire, D.P., Reynolds, R.T., 1993. Habitable zones around main sequence stars. Icarus 101, 108–128.

Kobayashi, N., Yasui, C., Tokunaga, A.T., Saito, M., 2008. Star formation in the most distant molecular cloud in extreme outer galaxy: A laboratory of star formation in an early epoch of the galaxy's formation. Astrophys. J. 683, 178.

Kornet, K., Bodenheimer, P., Rózyczka, M., Stepinski, T.F., 2005. Formation of giant planets in disks with different metallicities. Astron. Astrophys. 430, 1133–1138.

Kuchner, M.J., Seager, S., 2005. Extrasolar carbon planets. arXiv:astro-ph/0504214.

Kvenvolden, K., Lawless, J., Pering, K., Peterson, E., Flores, J., Ponnamperuma, C., Kaplan, I.R., Moore, C., 1970. Evidence for extraterrestrial amino-acids and hydrocarbons in the Murchison meteorite. Nature 228, 923–926.

Kwok, S., 2004. The synthesis of organic and inorganic compounds in evolved stars. Nature 430, 985–991.

Kwok, S., 2009. Organic matter in space: from star dust to the Solar System. Astrophys. Space Sci. 319, 5–21.

Kwok, S., Zhang, Y., 2013. Unidentified infrared emission bands: PAHs or MAONs? Astrophys. J. 771, 5.

Labbé, I., Oesch, P.A., Bouwens, R.J., Illingoworth, G.D., Magee, D., González, V., Carollo, C.M., Franx, M., Trenti, M., van Dokkum, P.G., Stiavelli, M., 2013. The spectral energy distributions of z∼8 galaxies from the IRAC Ultra Deep Fields: Emission lines, stellar masses, and specific star formation rates at 650 Myr*. Astrophys. J. Lett. 777, L19.

Lee, Y.S., Beers, T.C., Masseron, T., Plez, B., Rockosi, C.M., Sobeck, J., Yanny, B., Lucatello, S., Sivarani, T., Placco, V., 2013. Carbon-enhanced metal-poor stars in SDSS/SEGUE. I. Carbon abundance estimation and frequency of CEMP stars. Astron. J. 148, 132.

Leger, A., Puget, J.L., 1984. Identification of the 'unidentified' IR emission features of interstellar dust? Astron. Astrophys. 137, L5–L8.

Levy, M., Miller, S.L., Brinton, K., Bada, J.L., 2000. Prebiotic synthesis of adenine and amino acids under Europa-like conditions. Icarus 145, 609–613.

Li, A., Draine, B.T., 2001. Infrared emission from interstellar dust. II. The diffuse interstellar medium. Astrophys. J. 554, 778–802.

Lodders, K., 2004. Solar system abundances and condensation temperatures of the elements. Astrophys. J. 591, 1220–1247.

Loeb, A., Furlanetto, S.R., 2013. The First Galaxies in the Universe. Princeton University Press, Princeton and Oxford. pp. 133–170.

Love, S.G., Brownlee, D.E., 1993. A direct measurement of the terrestrial mass accretion rate of cosmic dust. Science 262, 550–553.

Lucatello, S., Tsangarides, S., Beers, T.C., Carretta, E., Gratton, R.G., Ryan, S.G., 2005. The binary frequency among carbon-enhanced, s-process rich, metal-poor stars. Astrophys. J. 625, 825.

Lucatello, S., Beers, T.C., Christlieb, N., Barklem, P.S., Rossi, S., Marsteller, B., Sivarani, T., Lee, Y.S., 2006. The frequency of carbon-enhanced metal-poor stars in the galaxy from the HERES sample. Astrophys. J. 652, 37–40.

Lucatello, D., Freeman, K., Beers, T.C., Placco, V.M., Tumlinson, J., Martell, S.L., 2014. Carbon-enhanced metal-poor stars: CEMP-s and CEMP-no subclasses in the halo system of the Milky Way. Astrophys. J. 788, 180.

Lunine, J.I., Lorenz, R.D., 1997. Light and heat in cracks of Europa: Implications for prebiotic synthesis. In: 28th Annual Lunar and Planetary Science Conference, March 17-21, 1997, Houston TX. pp. 855.

Madhusudan, N., Harrington, J., Stevenson, K.B., Nymeyer, S., Campo, C.J., Wheatley, P.J., Deming, D., Blecic, J., Hardy, R.A., Lust, N.B., Anderson, D.R., Collier-Cameron, A., Britt, C.B.T., Bowman, W.C., Hebb, L., Hellier, C., Maxted, P.F.L., Pollacco, D., West, R.G., 2011. A high C/O ratio and weak thermal inversion in the atmosphere of exoplanet WASP-12b. Nature 469, 64–67.

Madhusudan, N., Lee, K.K.M., Mousis, O., 2012. A possible carbon-rich interior in super-Earth 55 Cancri e. Astrophys. J. Lett. 759, L40.

Madhusudan, N., 2012. C/O ratio as a dimension for characterising exoplanetary atmospheres. Astrophys. J. 758, 36.

Mashian, N., Loeb, A., 2016. CEMP stars: possible hosts to carbon planets in the early Universe. Mon. Not. Roy. Astron. Soc. 460, 2482–2491.

Mason, R.E., Wright, G., Pendleton, Y., Adamson, A., 2004. Hydrocarbon dust absorption in Seyfert galaxies and ultraluminous infrared galaxies. Astrophys. J. 613, 770–780.

Masseron, T., Johnson, J.A., Plez, B., Van Eck, S., Primas, F., Goriely, S., Jorissen, A., 2010. A holistic approach to caroon-enhaned metal-poor stars. Astron. Astrophys. 509, A93.

Mathis, J.S., Rumpl, W., Nordsieck, K.H., 1977. The size distribution of interstellar grains. Astrophys. J. 217, 425–433.

Matsuura, M., Dwek, E., Barlow, M.J., Babler, B., Baes, M., Meixner, M., Cernicharo, J., Clayton, G.C., Dunne, L., Fransson, C., Fritz, J., Gear, W., Gomez, H.L., Groenewegen, M.A.T., Indebetouw, R., Ivison, R.J., Jerkstrand, A., Lebouteiller, V., Lim, T.L., Lundqvist, P., Pearson, C.P., Roman-Duval, J., Royer, P.,

Staveley-Smith, L., Swinyard, B.M., van Hoof, P.A.M., van Loon, J.Th., Verstappen, J., Wesson, R., Zanardo, G., Blommaert, J.A.D.L., Decin, L., Reach, W.T., Sonneborn, G., Van de Steene, G.C., Yates, J.A., 2015. A stubbornly large mass of cloud dust in the ejecta of Supernova 1987A. Astrophys. J. 800, 50.

McClure, R.D., 1985. The carbon and related stars. J. Roy. Astron. Soc. Canada 79, 277–293.

McKay, C.P., Anbar, A.D., Porco, C., Tsou, P., 2014. Follow the plume: The habitability of Enceladus. Astrobiology 14, 352–355.

Mennella, V., Colangeli, L., Bussoletti, E., Palumbo, P., Rotundi, A., 1998. A new approach to the puzzle of the ultraviolet interstellar extinction bump. Astrophys. J. 507, 177–180.

Neininger, N., Guélin, M., Ungerechts, H., Lucas, R., Wielbinski, R., 1998. Carbon monoxide emission as a precise tracer of molecular gas in the Andromeda galaxy. Nature 395, 871–873.

Neish, C.D., Somogyi, A., Imanaka, H., Lunine, J.I., Smith, M.A., 2008. Rate measurements of the hydrolysis of complex organic molecules in cold aqueous solutions: Implications for prebiotic chemistry on the early Earth and Titan. Astrobiology 8, 273–287.

Noll, S., Pierini, D., Cimatti, A., Daddi, E., Kurk, J.D., Bolzonella, M., Cassata, P., Halliday, C., Mignoli, M., Pozzetti, L., Renzini, A., Berta, S., Dickinson, M., Franceschini, A., Rodighiero, G., Rosati, P., Zamorani, G., 2009. GMASS ultradeep spectroscopy of galaxies at $z \sim 2$ IV. The variety of dust populations. Astron. Astrophys. 499, 69–85.

Norris, J.E., Yong, D., Bessell, M.S., Christlieb, N., Asplund, M., Gilmore, G., Wyse, R.F.G., Beers, T.C., Barkalem, P.S., Frebel, A., Ryan, S.G., 2013. The most metal-poor stars. IV. The two populations with [Fe/H] ≲ -3.0. Astrophys. J. 762, 28.

Nuth, J.A., Kimura, Y., Lucas, C., Ferguson, F., Johnson, N.M., 2010. The formation of graphite whiskers in the primitive solar nebula. Astrophys. J. Lett. 710, 98.

Öberg, K.I., Murray-Clay, R., Bergin, E.A., 2011. The effects of snow lines on C/O in planetary atmospheres. Astrophys. J. Lett. 714, L16.

Oesch, P.A., van Dokkum, P.G., Illingworth, G.D., Bouwens, R.J., Momcheva, I., Holden, B., Roberts-Borsani, G.W., Smit, R., Franx, M., Labbé, I., Gonzalez, V., Magee, D., 2015. A spectroscopic redshift measurement for a luminous lyman break galaxy at $z = 7.730$ using Keck/MOSFIRE. Astrophys. J. Lett. 804, L30.

Oesch, P.A., Brammer, G., van Dokkum, P.G., Illingworth, G.D., Bouwens, R.J., Labbé, I., Franx, M., Momcheva, I., Ashby, M.L.N., Fazio, G.G., Gonzalez, V., Holden, B., Magee, D., Skelton, R.E., Smit, R., Spitler, L.R., Trenti, M., Willner, S.P., 2016. A remarkably luminous galaxy at $z = 11.1$ measured with Hubble Space Telescope Grism Spectroscopy. Astrophys. J. 819, 129–140.

Olive, K.A., Steigman, G., Walker, T.P., 2000. Primordial nucleosynthesis: theory and observations. Phys. Rep. 333–334, 389–407.

Omukai, K., 2000. Protostellar collapse with various metallicities. Astrophys. J. 534, 809–824.

Pagel, B., 2009. Nucleosynthesis and Chemical Evolution of Galaxies. Cambridge University Press, Cambridge.

Panagia, N., 2005. High-redshift supernovae: Cosmological implications. arXiv:astro-ph/0502247.

Papoular, R.J., Papoular, R., 2009. A polycrystalline graphite model for the 2175 Å interstellar extinction band. Mon. Not. Roy. Astron. Soc. 394, 2175–2181.

Parkinson, C.D., Liang, M.-C., Hartman, H., Hansen, C.J., Tinetti, G., Meadows, V., Kirschvink, J.L., Yung, Y.L., 2007. Enceladus: Cassini observations and implications for the search for life. Astron. Astrophys. 463, 353–357.

Pendleton, Y.J., Sandford, S.A., Allamandola, L.J., Tielens, A.G.G.M., Sellgren, K., 1994. Near-infrared absorption spectroscopy of interstellar hydrocarbon grains. Astrophys. J. 437, 683–696.

Perea-Calderón, J.V., García-Hernández, D.A., García-Lario, P., Szczerba, R., Bobrowsky, M., 2009. The mixed chemistry phenomenon in Galactic Bulge PNe. Astron. Astrophys. 495, L5–L8.

Pizzarello, S., Huang, Y., Becker, L., Poreda, R.J., Nieman, R.A., Cooper, G., Williams, M., 2001. The organic content of the Tagish Lake meteorite. Science 293, 2236–2239.

Prantzos, N., Aubert, O., Audouze, J., 1996. Evolution of the carbon and oxygen isotopes in the Galaxy. Astron. Astrophys. 309, 760–774.

Robitaille, T.P., Whitney, B.A., 2010. The present-day star formation rate of the Milky Way determined from *Spitzer*-detected young stellar objects. Astrophys. J. Lett. 710, L11.

Rudolph, A.L., Fich, M., Bell, G.R., Norsen, T., Simpson, J.P., Haas, M.R., Erickson, E.F., 2006. Abundance gradients in the galaxy. Astrophys. J. Suppl. Ser. 162, 346–374.

Sajina, A., Spoon, H., Yan, L., Imanishi, M., Fadda, D., Elitzur, M., 2009. Detection of water ice, hydrocarbons, and 3.3 μm PAH in z∼2 ULIRGs. Astrophys. J. 703, 270–284.

Salpeter, E.E., 1952. Nuclear reactions in stars without hydrogen. Astrophys. J. 115, 326–328.

Sandford, S.A., Allamandola, L.J., Tielens, A.G.G.M., Sellgren, K., Tapia, M., Pendleton, Y., 1991. The interstellar C-H stretching band near 3.4 microns: Constraints on the composition of organic material in the diffuse interstellar medium. Astrophys. J. 371, 607–620.

Schramm, D.N., Turner, M.S., 1998. Big-Bang nucleosynthesis enters the precision era. Rev. Mod. Phys. 70, 303–318.

Sellgren, K., Werner, M.W., Ingalls, J.G., Smith, J.D.T., Carleton, T.M., Joblin, C., 2010. C60 in reflection nebulae. Astrophys. J. 722, 54–57.

Serra Diaz-Cano, L., Jones, A.P., 2008. Carbonaceous dust in interstellar shock waves: hydrogenated amorphous carbon (a:C-H) vs. graphite. Astron. Astrophys. 492, 127–133.

Shapley, H., 1953. Climatic Change—Evidence, Causes, and Effects. Harvard University Press, Cambridge. p. 318.

Sobral, D., Matthee, J., Darvish, B., Schaerer, D., Mobasher, B., Röttgering, H.J.A., Santos, S., Hemmati, S., 2015. Evidence for PopIII-like stellar populations in the most luminous Lyα emitters at the epoch of reionization: spectroscopic confirmation. Astrophys. J. 808, 139–153.

Solomon, P., Jefferts, K.B., Penzias, A.A., Wilson, R.W., 1971. Observation of CO emission at 2.6 millimeters from IRC+10216. Astrophys. J. 163, 53–56.

Spoon, H.W.W., Armus, L., Cami, J., Tielens, A.G.G.M., Chiar, J.E., Peeters, E., Keane, J.V., Charmandaris, V., Appleton, P.N., Teplitz, H.I., Burgdorf, M.J., 2004. Fire and ice: Spitzer infrared spectrograph (IRS) mid-infrared spectroscopy of IRAS F00183-7111. Astrophys. J. Suppl. Ser. 154, 184–187.

Starkenburg, E., Shetone, M.D., McConnachie, A.W., Venn, K.A., 2014. Binarity in carbon-enhanced metal-poor stars. Mon. Not. Roy. Astron. Soc. 441, 1217–1229.

Stecher, T.P., Donn, B., 1965. On graphite and interstellar extinction. Astrophys. J. 142, 1681–1682.

Stecher, T.P., 1965. Interstellar extinction in the ultraviolet. Astrophys. J. 142, 1683–1684.

Strughold, H., 1953. The green and red planet: a physiological study of the possibility of life on Mars. J. Am. Med. Assoc. 153, 1410.

Swinbank, A.M., Karim, A., Smail, I., Hodge, J., Walter, F., Bertoldi, F., Biggs, A.D., de Breuck, C., Chapman, S.C., Coppin, K.E.K., Cox, P., Danielson, A.L.R., Dannerbauer, H., Ivison, R.J., Greve, T.R., Knudsen, K.K., Menten, K.M., Simpson, J.M., Schinnerer, E., Wardlow, J.L., Weiss, A., van der Werf, P., 2012. Mon. Not. Roy. Astron. Soc. 427, 1066–1074.

Tacconi, L.J., Genzel, R., Smail, I., Neri, R., Chapman, S.C., Ivison, R.J., Blain, A., Cox, P., Omont, A., Bertoldi, F., Greve, T., Förster Schreiber, N.M., Genel, S., Lutz, D., Swinbank, A.M., Shapley, A.E., Erb, D.K., Cimatti, A., Dada, E., Baker, A.J., 2008. Submillimeter galaxies at z∼2: Evidence for major mergers and constraints on lifetimes, IMF, and CO-H_2 conversion factor. Astrophys. J. 680, 246–262.

Tacconi, L.J., Genzel, R., Neri, R., Cox, P., Cooper, M.C., Shapiro, K., Bolatto, A., Bouché, N., Bournaud, F., Burkert, A., Combes, F., Comerford, J., Davis, M., Förster Schreiber, N.M., Garcia-Burillo, S., Gracia-Carpio, J., Lutz, D., Naab, T., Omont, A., Shapley, A., Sternberg, A., Weiner, B., 2010. High molecular gas fractions in normal massive star-forming galaxies in the young universe. Nature 463, 781–784.

Tegmark, M., Silk, J., Rees, M.J., Blanchard, A., Abel, T., Palla, F., 1997. How small were the first cosmological objects? Astrophys. J. 474, 1–12.

Tenenbaum, E.D., Dodd, J.L., Milam, S.N., Woolf, N.J., Ziurys, L.M., 2010. The Arizona Radio Observatory 1 mm spectral survey of IRC+10216 and VY Canis Majoris (215–285 GHz). Astrophys. J. Suppl. Ser. 190, 348.

Thaddeus, P., Gottlieb, C.A., Brünken, S., McCarthy, M.C., Agúndez, M., Guélin, M., Cernicharo, J., 2008. Laboratory and astronomical detection of the negative molecular ion C_3N^-. Astrophys. J. 677, 1132.

Tornatore, L., Ferrara, A., Schneider, R., 2007. Population III stars: hidden or disappeared? Mon. Not. Roy. Astron. Soc. 382, 945–950.

van Dishoeck, E.F., Blake, G.A., 1998. Chemical evolution of star-forming regions. Ann. Rev. Astron. Astrophys. 36, 317–368.

Vonlanthen, P., Rauscher, T., Winteler, C., Puy, D., Signore, M., Dubrovich, V., 2009. Chemistry of heavy elements in the dark ages. Astron. Astrophys. 503, 47–59.

Ward, P., Brownlee, D., 2004. Rare Earth: Why Complex Life is Uncommon in the Universe. Copernicus, New York.

Waters, L.B.F.M., Waelkens, C., van Winckel, H., Molster, F.J., Tielens, A.G.G.M., van Loon, J.Th., Morris, P.W., Cami, J., Bouwman, J., de Koter, A., de Tong, T., de Graauw, Th., 1998. An oxygen-rich dust disk surrounding and evolved star in the Red Rectangle. Nature 391, 868–871.

Watson, D., Christensen, L., Knudsen, K.K., Richard, J., Gallazzi, A., Michalowski, M.J., 2015. A dusty, normal galaxy in the epoch of reionization. Nature 519, 327–330.

Weizsäcker, C.F.V., 1938. Neuere Modellvorstellungen über den Bau der Atomkerne. Naturwissenschaften 26, 209–217.

Whittet, D.C.B., 2003. Dust in the galactic environment. In: Whittet, D.C.B. (Ed.), 2003 Series in Astronomy and Astrophysics. Institute of Physics Publishing, Bristol.

Wickramasinghe, D.T., Allen, D.A., 1980. The 3.4-micron interstellar absorption feature. Nature 287, 518–519.

Willson, L.A., 2000. Mass loss from cool stars: Impact on the evolution of stars and stellar populations. Ann. Rev. Astron. Astrophys. 38, 573–611.

Wong, B., Heckman, T.M., 1996. Internal absorption and the luminosity of disk galaxies. Astrophys. J. 457, 645–657.

Yasui, C., Kobayashi, N., Tokunaga, A.T., Saito, M., Tokoku, C., 2009. The lifetime of protoplanetary disks in a low-metallicity environment. Astrophys. J. 705, 54.

Yoshida, N., Abel, T., Hernquist, L., Sugiyama, N., 2003. Simulations of early structure formation: Primordial gas clouds. Astrophys. J. 592, 645–663.

Zafar, T., Watson, D., Elíasdóttir, Á., Fynbo, J.P.U., Kruhler, T., Schady, P., Leloudas, G., Jakobsson, P., Thone, C.C., Perley, D.A., Morgon, A.N., Bloom, J., Greiner, J., 2012. The properties of the 2175AA extinction feature discovered in GRB afterglows. Astrophys. J. 753, 82.

Zenobi, R., Philippoz, J.-M., Buseck, P.R., Zare, R.N., 1989. Spatially resolved organic analysis of the Allende meteorite. Science 246, 1026–1029.

Zijlstra, A.A., Matsuura, M., Wood, P.R., Sloan, G.C., Lagadec, E.T., van Loon, J., Groenewegen, M.A.T., Feast, M.W., Menzies, J.W., Whitelock, P.A., Blommaert, J.A.D.L., Cioni, M., Habing, H.J., Hony, S., Loup, C., Waters, L.B.F.M., 2006. A Spitzer mid-infrared spectral survey of mass-losing carbon stars in the Large Magellanic Cloud. Mon. Not. Roy. Astron. Soc. 370, 1961–1978.

Zitrin, A., Labbé, I., Belli, S., Bouwens, R., Ellis, R.S., Roberts-Borsani, G., Stark, D.P., Oesch, P.A., Smit, R., 2015. Lyα emission from a luminous $z = 8.68$ galaxy: implications for galaxies as tracers of cosmic reionization. Astrophys. J. 810, L12.

Ziurys, L., 2006. The chemistry in circumstellar envelopes of evolved stars: Following the origin of the elements to the origin of life. Proc. Natl. Acad. Sci. U. S. A. 103, 12274–12279.

PREDICTING HABITABILITY

THE HABITABILITY OF OUR EVOLVING GALAXY

Michael G. Gowanlock*,†, Ian S. Morrison‡
*Massachusetts Institute of Technology, Haystack Observatory, Westford, MA, United States
†Northern Arizona University, Flagstaff, AZ, United States
‡Swinburne University of Technology, Hawthorn, VIC, Australia

CHAPTER OUTLINE

1 INTRODUCTION

Astrobiology has developed significantly over the past decades through the advances made in multiple fields. The search for and detection of extrasolar planets has made possible estimates of the number of planets in the Milky Way (Borucki et al., 2010; Petigura et al., 2013). New observational constraints have permitted modeling efforts to constrain the conditions necessary for planets to host a biosphere (Godolt et al., 2016). The Solar System has been used as an analog to understand the possible conditions for life on other planets. And beyond planets, other bodies such as moons (Williams et al., 1997; Kaltenegger, 2010; Quarles et al., 2012) have been suggested as potentially able to host a biosphere. Furthermore, the life sciences have made inroads into understanding the origin and constraints for life on Earth (Martin et al., 2008; Ehrenfreund et al., 2006). Combining the findings in these fields and others, new assessments of habitability on larger spatial and temporal scales are now possible.

The Earth hosts many habitable environments, but the current conditions are possibly due to the presence of liquid water on the planet's surface. The circumstellar habitable zone (HZ) (Kasting et al., 1993; Kopparapu et al., 2013) constrains the distance from a host star such that water remains in a

Habitability of the Universe Before Earth, editors: Richard Gordon & Alexei Sharov, Volume 1 in the series:
Astrobiology: Exploring Life on Earth and Beyond, series editors: Pabulo Henrique Rampelotto,
Joseph Seckbach & Richard Gordon. ISSN 2468-6352. https://doi.org/10.1016/B978-0-12-811940-2.00007-1

liquid state on the surface of a planet having sufficient atmospheric pressure. Indeed, our Solar System is a prototype for an exoplanetary system, where the Earth is found to be roughly within the middle of the HZ. Beyond the Solar System, exoplanet detections suggest that terrestrial planets in the HZ are common within the Milky Way (Petigura et al., 2013).

Although the study of habitability beyond the Earth is still in its early stages, there are astrophysical constraints on the suitability of planets to have the carrying capacity to host life over time within the Milky Way. Studies have examined various aspects of the habitability of the Milky Way (Gonzalez et al., 2001; Lineweaver et al., 2004; Prantzos, 2008; Gowanlock et al., 2011; Spitoni et al., 2014; Dayal et al., 2015; Morrison and Gowanlock, 2015; Forgan et al., 2017; Thomas et al., 2016; Vukotić et al., 2016; Gobat and Hong, 2016) as increasing consensus on astronomical observational constraints have been realized over the past few decades. Some of these works define the habitability of the Milky Way by spatial locations suitable for life over time, thus giving rise to the term the galactic habitable zone. Some of the aforementioned references will be reviewed within this chapter.

Largely due to the many uncertainties involved in assessing the habitability of the Milky Way, the astrophysical processes that may lead to or are detrimental to life necessitate a coarse-grained view of the factors that make the Milky Way a habitable environment. Furthermore, the discussion of habitability often assumes an implicit definition of what constitutes an environment suitable for life. In terms of the GHZ, a habitable environment is considered one that is not frequently exposed to transient radiation events, where these events can deplete ozone in planetary atmospheres, exposing life to harmful radiation from its host star. However, this implies a habitable environment for surface dwelling life that would be particularly exposed to radiation from its host star. Thus, existing life in other potentially protected environments (e.g., the ocean floor) are exempt from any analysis of the effects on their environments. Furthermore, the study of the GHZ makes no claims on where preexisting life may be within the Galaxy; rather it is an assessment of the suitability for land-based life given the astrophysical mechanisms known to limit or bolster what are believed to be habitable environments. Given the uncertainties involved in this field of study, the habitability of the Milky Way is still largely an open question.

In this chapter, we review the concept of habitability on large spatial and temporal scales, we illuminate some of the key advancements made by the exoplanet community, we give an overview of the differing predictions and models of the habitability of the Milky Way, discuss the habitability of other Galaxies, revisit types of transient radiation events and their implications for the Milky Way, extrapolate and speculate on the habitability of the Galaxy before the formation of the Earth and finally conclude the chapter and discuss future research directions in the emergent field of habitability on large spatiotemporal scales.

2 HABITABILITY

Before discussing the details of the habitability of our evolving galaxy, the Milky Way, we will discuss what is meant by a habitable environment. The term habitability is widely used in differing contexts across the constituent research communities that make up the field of astrobiology. The study of the habitability of the Earth starts with determining a list of the items that are common to all life on our planet. There are broadly three major requirements for life as we know it: (1) an energy source (McKay, 2004), such as chemical energy or the Sun (for life that requires photosynthesis); (2) a range of elements, C, H, N, O, P, S (Falkowski, 2006), as all life on Earth is composed of these elements as they make up biomolecules; and (3) water to facilitate chemical interactions (Mottl et al., 2007). These appear to be necessary requirements for all life on Earth and are an accepted part of the definition of habitable environments on Earth. However, life on other bodies may be able to thrive without these specific ingredients. For instance, it is possible that a solvent other than water may be used to facilitate chemical interactions. Despite the many uncertainties inherent in applying our knowledge of the habitability of the Earth to other environments, having a minimal list of requirements for life on Earth forms the basis for understanding the conditions for life in other environments.

If we take a closer look at some of these requirements, it becomes clear that there are many prerequisites for these conditions. For instance, having a solvent such as water implies that the temperature on a planet must be within a range such that water does not freeze or evaporate. This means that the planet must have a suitable average orbital distance from its host star, and an atmosphere that can provide those conditions. The atmospheres of Venus and Mars are two cases that cannot provide liquid water on their respective surfaces, where the Venusian atmosphere is too dense and traps heat, and the Martian atmosphere is too depleted and cannot retain enough heat at the surface causing freezing, or the water rapidly evaporates. Thus, there are many interesting questions regarding the combination of the orbital radius and atmospheric composition required for planets to retain liquid water on their surfaces (Kopparapu et al., 2013). Exoplanets likely have similar requirements to be habitable, and to remain so over long timescales.

From the previous discussion, we have made the leap from a rudimentary list of requirements for life on Earth, to speculating about how these requirements may lead to defining particular *habitable environments*. This opens up many questions about the limits to habitability on Earth, the formation of planets and their configurations in planetary systems, and the galactic environments that may bolster or limit the carrying capacity for life. Exoplanets and their configurations have been detected through radial velocity and planetary transit methods (Saar et al., 1998; Charbonneau et al., 2000). These observations are the outcome of planet formation which occurs in protoplanetary disks (Pollack et al., 1996; Boss, 1997; Matsuo et al., 2007). Modeling the formation of planets from protoplanetary disks attempts to reconcile fundamental physical processes with those planetary configurations that are similar to the Solar System and exoplanetary systems. In combination with observations, they have shown that the (previously thought) exotic exoplanetary systems must be common throughout the Milky Way, and furthermore, that there are a wide range of possible planetary configurations that are significantly different than the Solar System.

By shifting the focus from the requirements for life on Earth to those environments that can support life, we implicitly ignore that the origin of life must occur, and that life itself influences its environment. For instance, the first forms of life on Earth used an anaerobic instead of aerobic metabolism. However, once oxygen was plentiful in the atmosphere as a by-product of photosynthesis, new aerobic forms of life began to develop and thrive. Utilizing oxygen yields more energy from the metabolic process, and as a result, life is more efficient in its homeostasis. This suggests that biology is rooted in physical processes, and that life having a tendency to change its environment is not a random process. The build up of oxygen in the Earth's atmosphere is accounted for by photosynthesizing cyanobacteria (Catling et al., 2005). Without an increase in atmospheric oxygen, it is possible that only anaerobic organisms would still be in existence, and the evolutionary paths of ancient organisms would have been significantly different. Therefore, it is very likely that animal life, particularly land-based animal life, would not exist today, had it not been for life affecting its environment. This necessarily presents a problem for the study of habitability. If habitable conditions are required for life, but life is required for maintaining long term habitable conditions, then how can we study habitability without first understanding the conditions for the origin of life? Other works discuss this problem (Catling et al., 2005; Lineweaver and Chopra, 2012; Chopra and Lineweaver, 2016), and we refer the reader to those references and other chapters in this book, such as that by Chopra and Lineweaver, for additional discussion. When it comes to the study of habitability, we must assume that we are interested in maintaining the conditions for life for a sufficient timescale. Furthermore, it is necessary to be specific about the type of life and requisite environment that is of interest. For instance, the Galilean moon Europa that orbits Jupiter is a target of interest because it may contain liquid water under its frozen water ice crust. Its source of heat is thought to be tidal heating as a result of its orbit around its parent body. Constraining habitable conditions on Europa equates to understanding subsurface life. The study of habitability needs to be mindful of these constraints, the influence that life has on its environment, and the numerous uncertainties that result when applying our knowledge of life on Earth to other bodies in the Solar System or Galaxy. With that in mind, we now move on to the habitability of our evolving Galaxy.

We may think that there are many requirements for life on Earth because we can study the numerous environments in which life is found in the biosphere. However, when assessing habitability at larger scales, many of these prerequisites cannot be quantified or observed. For instance, exoplanet searches may claim a habitable planet detection based on its location in the HZ, without knowledge of its atmospheric characteristics or other factors that may make the planet habitable. Unfortunately, until future missions can make measurements that detect biosignatures or other factors that constrain habitability, we are limited to a coarse-grained view of habitability throughout the Milky Way. The GHZ is an idea that is similar to the HZ, initially defined by an annular region of the Milky Way that can host habitable planets that are not too close to the center of the Galaxy or the outskirts. The inner edge has been thought to be constrained by a high frequency of transient radiation events, and the outer edge constrained by insufficient metallicity for planet formation (Gonzalez et al., 2001) and these factors will be discussed in greater detail in upcoming sections. By these constraints, a habitable planet is one that is considered to maintain an environment free from transient radiation events for timescales commensurate with complex life, or more particularly land-based surface dwelling life. Therefore, the definition of the GHZ specifically considers a particular environment for life (the surface). This is because it is not clear that a transient radiation event would have a catastrophic impact on life found in the subsurface or oceans. In all that follows, when we discuss the habitability of a planet in the context of the Milky Way, we are referring to those conditions that may support land-based complex life. The region of habitability, or "galactic habitable zone" refers to the regions in the Galaxy that have the highest carrying capacity for complex life over time. Examining the effects of astrophysical events on the habitability of planets contributes a small part to our understanding of the habitability of the Milky Way throughout cosmic time, and our precarious situation as a species on our planet.

3 THE EXOPLANET ERA

The first exoplanet was detected in 1992 orbiting a pulsar (Wolszczan and Frail, 1992). Since then, at the time of writing, there have been over 3000 additional exoplanets confirmed.[1] The *Kepler* mission (Borucki et al., 2010) has detected the majority of these exoplanets through the photometric transit

---

[1]http://www.exoplanet.eu, 3610 planets as of May 17, 2017.

technique. Combining transit and radial velocity measurements, the orbital period, planet radius and mass can be constrained, thereby providing insight into the configurations of other planetary systems. These detections have given rise to an intense period of research on exoplanets. Statistical constraints across the sample of detected planets have been conducted to estimate the fraction of HZ planets around Sun-like stars (Petigura et al., 2013). The mass, radius, and orbital period have been used as parameters into models of a planet's atmospheric composition to determine whether it can retain liquid water on its surface (Shields et al., 2016). Furthermore, the study of atmospheric biosignatures may yield evidence for the existence of life on a planet (Pilcher, 2003; Segura et al., 2005; Kaltenegger et al., 2007). These avenues of research have combined the efforts of astronomy, planetary science, geology, biology, and other fields, to determine the conditions for potential life on other planets. Future missions and observing campaigns will provide much needed insight into whether some of the detected HZ exoplanets are in fact habitable, and will guide models that constrain the habitability of planets around their host stars.

The two techniques that have indirectly detected the majority of the exoplanets are the transit and radial velocity techniques. When a planet's orbit passes through the line of sight between the observer and its host star, the transit of the planet will cause a small drop in the observed brightness of the star that can be used to infer the presence of the planet. This technique yields the planet's radius and orbital period. With regard to the radial velocity technique, as a planet orbits its host star, the gravitational effect on the star can be measured. By examining Doppler variations in the stellar spectrum, the star will appear to be periodically moving toward and away from the Earth. This method is sensitive to the relative masses of the planet and host star, favoring larger planets orbiting lower mass stars. Combining findings from both the transit and radial velocity methods, a planet's orbital period, size, and upper limit on mass can be obtained. With the mass estimate and size, the bulk densities can be calculated, and chemical compositions inferred. From the study of large numbers of exoplanets, statistics can be gathered that can be directly applied to understanding the habitability of the Galaxy. We now review some of these key findings.

We now know that the majority of stars are likely to host planets. Mayor et al. (2011) finds that 75% of solar type stars host a detectable planet with an orbital period less than 10 years. This estimate does not extrapolate to those planets that are not detectable due to observational constraints; therefore, this may be a lower limit on the fraction of solar type stars with planets.

Properties of host stars are correlated with the type of planets that are likely to orbit them. Before large scale transit searches, the radial velocity method was the primary means for detecting exoplanets, and this method has those aforementioned observation biases that mean it is able to more easily detect large planets. Fischer and Valenti (2005) found a planet-metallicity correlation that shows gas giant planets are strongly correlated with host star metallicity, where metallicity refers to the abundance of elements in the star that are not hydrogen or helium. From the *Kepler* mission, Buchhave et al. (2012) find that the correlation between small planets and metallicity is flat, which means that small planets form from protoplanetary disks that have a range of chemical abundances, and do not require a significant abundance of metals for small planet formation. Therefore, small Earth-sized planet formation is moderately independent of chemical evolution, thus Earth-sized planets will be common in the Galaxy. Fig. 1 shows the host star metallicity as a function of planet radius from Buchhave et al. (2012). From the figure, it can be seen that small planets form around stars with a wide range of metallicities, and larger planets are correlated with higher metallicities, consistent with Fischer and Valenti (2005) and Mayor et al. (2011).

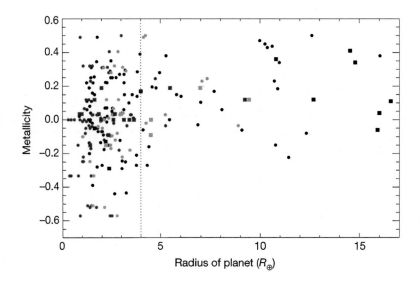

FIG. 1

Figure 3 from Buchhave et al. (2012) that plots host star metallicity vs. planet radius in Earth radii ($R_\oplus$). Small planets form around stars with a range of metallicity values, and large planets are strongly correlated with high metallicity. The mean metallicity for small size planets is near the solar value. The *vertical dotted line* separates the two planet sizes in the sample at $4.0R_\oplus$. For more information on the sample of planets and the representation of figure markers that refer to single planet systems, and planet sizes in planetary systems with multiple planets, see Buchhave et al. (2012).

The observations of planet size and metallicity are consistent with the core accretion model of planet formation. Gas giant formation first needs to produce a large rocky core in the protoplanetary disk that then has a runaway gas accretion period (Pollack et al., 1996). This explains why gas giants are primarily found around metal-rich host stars. In comparison, smaller rocky planets do not need a large abundance of planetary building blocks (high metallicity) in the protoplanetary disk. These findings have direct implications for the habitability of the Milky Way, as the metallicity of the Galaxy has evolved with time, and this evolution will influence how planets form over time.

To constrain the habitability of planets, and not just their abundance around stars of various masses and chemical compositions, determining the fraction of Earth-size planets that orbit within the HZ of their host stars is required. Petigura et al. (2013) finds that 22% of Sun-like stars have Earth-size planets orbiting in their HZ. This suggests that, of the many billions of Earth-sized planets predicted in the Milky Way, a significant fraction may have conditions favorable for life.

4 THE HABITABILITY OF THE MILKY WAY

In this chapter, we have discussed habitable zones on the Earth, Solar System, and the Galaxy. The GHZ originated as an idea that was analogous to the HZ (Gonzalez et al., 2001), and thus it was assumed to be annular in nature, with sharp inner and outer limits to habitability. However, as will be examined in this section, an annular perspective may not necessarily be reasonable for the Milky Way. Therefore, when discussing the morphology of the GHZ, we do not mean for the reader to assume an annular region, such that we are not predisposed to a goldilocks view of habitability.

When considering the habitability of the Galaxy, a coarse-grained view of habitability is assumed, as astrophysical factors that may inhibit or advance life are considered. Some astrophysical prerequisites for life include a star, with a rocky planet in the HZ, that can withstand transient radiation events, such as supernovae (SNe), gamma-ray bursts (GRBs), or active galactic nuclei (AGN) over a sufficient time period that allows for land-based complex life to evolve and survive. As mentioned in the discussion regarding habitability, an implicit assumption must be made that the conditions that allow complex life to thrive must exist, even though those environments may have only been made possible through life transforming its environment. However, when future observations are better able to constrain the fraction of truly habitable planets, those statistics will be able to be used as input into newer models of the habitability of the Milky Way.

Through stellar nucleosynthesis, the generations of (mostly high mass) stars produce the materials needed for planet formation. The initial generations of stars would have formed nearly entirely out of hydrogen (and helium) without any of the heavier elements to produce planets in protoplanetary disks. The metallicity of the Milky Way evolves with time; therefore, the abundance of planet-building material has increased over time in the Galaxy. Gonzalez et al. (2001) proposed quantifying the metallicity selection effect to constrain the GHZ. Using constraints on the chemical evolution of the Galaxy, and an estimate of the abundance of metals needed to form a habitable planet (half the Solar value), the habitability of the Milky Way can be assessed. Gonzalez et al. (2001) report that metal-poor regions of the Milky Way, such as the halo, thick disk, and outer thin disk (at higher galactocentric radii), may not have enough metals for planet formation. Furthermore, they compare the metallicity of the Milky Way to other galaxies. Since higher luminosity galaxies are likely to be more metal-rich (luminosity traces star formation, and star formation traces metal abundance), they find that the Milky Way, being more luminous than other galaxies in the local Universe, is likely to host more planets. Lineweaver (2001) uses the metallicity evolution and star formation rate of the Universe to quantify the age distribution of planets in the Universe over cosmic timescales. Lineweaver (2001) finds that the majority of Earth-like planets in the Universe are 1.8 ± 0.9 Gyr older than the Earth. Furthermore, he suggests that Hot Jupiters (gas giants orbiting near their host stars) would inhibit the formation of Earth-like planets, as they may accrete the material required for terrestrial planet formation. Since these Hot Jupiters are found around metal-rich host stars, then there may be a selection effect, where high metallicity inhibits rocky planet formation, and low metallicity inhibits planets from forming entirely.

Lineweaver et al. (2004) constrained the habitability of the Milky Way as a function of both metallicity and catastrophic transient radiation events. They find that the habitability of the Milky Way is described in terms of an annular region, where at the present day, the inner edge of the GHZ is defined by too high of a frequency of lethal SN events, and too high metallicity that produces Hot Jupiters (as discussed above in Lineweaver, 2001), and the outer edge is defined by insufficient metallicity for planet formation. Lineweaver et al. (2004) finds that the GHZ is between 7 and 9 kpc in galactocentric radii that widens with time (the radius of the disk of the Milky Way is approximately 15 kpc). Additionally they find that 75% of the stars in this region are older than the Sun. Gonzalez et al. (2001) and Lineweaver et al. (2004) set the stage for constraining habitability on larger spatial and temporal scales, and carefully considered these calculations largely before many exoplanets had been detected.

With the concept of the GHZ described as an extension to the HZ, Prantzos (2008) notes that the question of habitability on galactic scales may be too difficult to quantify, because the prerequisites for habitability are not well established. Despite this perspective, Prantzos (2008) models the star formation history, rocky planet formation, presence of Hot Jupiters, and SN events, and finds that the entire disk of the Milky Way may be habitable. Furthermore, they find that the inner Galaxy is likely more habitable than the outer Galaxy at later epochs. Early epochs would have had too many SNe, but after these lethal events become less frequent, there are likely to be more Earth-like planets in the region than at the solar neighborhood or outskirts. Furthermore, our take-away from Prantzos (2008) is that the framework of a GHZ may be insignificant because the entire disk may be habitable, the physical processes underlying habitability are not well constrained, and the results of such studies are largely dependent upon model assumptions.

Our previous work, Gowanlock et al. (2011) used the same prerequisites for complex life as Lineweaver et al. (2004). We considered the formation of rocky planets in the HZ of their host stars that are free from SN events over a timescale sufficient for the rise and long-term stability of land-based complex life. In particular, using a Monte Carlo approach, we populated stars using the three-dimensional stellar number density of the thin and thick disks (Jurić et al., 2008), then for each star we applied the following properties: a mass from the initial mass function (IMF) (Salpeter, 1955; Kroupa, 2001), a stellar lifetime (from the IMF), a birth date from the inside-out star formation history, and a metallicity, as a function of radius and time (Naab and Ostriker, 2006). With this model of the Milky Way, we then assigned planets to stars as a function of metallicity. Then, in this population of disk stars, we assigned them a probability of becoming a SN, either type II (SNII) or type Ia (SNIa). SNII occur from massive stars; therefore, they occur primarily in star-forming regions and are young when they occur, and single degenerate SNIa are thought to occur in binary star systems that span a distribution of ages (Pritchet et al., 2008). Therefore, star forming environments are likely to host SNII, and SNIa can occur independently of recent star formation. Using the inside-out star formation history from Naab and Ostriker (2006), stars initially form in the inner Galaxy and star formation propagates outwards over time. Therefore, the average age of stars (and planets) in the inner Galaxy is greater than the outer Galaxy. Fig. 2 shows the star formation history and metallicity evolution of Model 4 in Gowanlock et al. (2011). From the figure, there is a burst of star formation that declines with time. The inner Galaxy hosts more stars and has more chemical evolution than at the solar radius ($R_\odot = 8$ kpc) or the outskirts.

Fig. 3 from Gowanlock et al. (2011) plots the area density (planets per pc^2) of the birth dates of habitable planets at the present day. We found that the inner Galaxy hosts the greatest density of habitable planets, and that the Earth is not in the most habitable region of the Milky Way. Furthermore, examining the fraction of stars with a habitable planet over all epochs, the region most favorable for life is the inner Galaxy (Fig. 4). However, the fraction of stars with a habitable planet is only a factor of ~ 2 higher in the inner Galaxy at recent epochs than the solar radius at 8 kpc (for instance, see the last 1 Gyr in Fig. 4). Stars form earlier in the inner Galaxy, which in turn build up a large number of planets through increased metallicity abundances before planet formation occurs at higher galactocentric radii (see the metallicity distribution in Fig. 2). The high planet density and older mean planet age in the inner Galaxy outweighs the negative impact of SNe. High metallicity abundances can hinder terrestrial planet formation through the formation of gas giants. Over time, the inner Galaxy builds up a large number of terrestrial planets, and likely only hinders the formation of terrestrial planets due to high

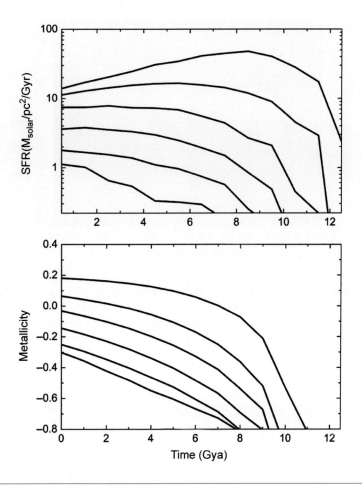

FIG. 2

The star formation rate (upper) and metallicity (lower) as a function of time in Model 4 from Gowanlock et al. (2011). The *curves* represent the values at six galactocentric regions as ordered from top to bottom: 2.5, 5, 7.5, 10, 12.5, 15 kpc. There is a burst of star formation in the inner Galaxy that declines with time (upper). Due to the inside-out formation history, the inner Galaxy produces metals earlier than the outskirts (lower). Note that the present day is 0 Gya.

metallicity abundances at later epochs. In summary, Gowanlock et al. (2011) find that the inner Galaxy is favorable to both metrics of habitability: (1) the number density, and (2) the fraction of stars that host a habitable planet.

The Monte Carlo modeling exercise of Gowanlock et al. (2011) was extended in Morrison and Gowanlock (2015) to consider not only habitability for complex land-based life, but also habitability over the extended timescales thought to be required for the emergence of intelligence. Depending on

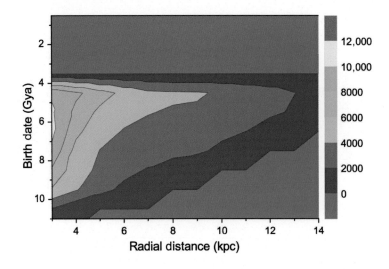

FIG. 3

The distribution of birth dates of the number of habitable planets per pc^2 (*color scale*) at the present day (Model 4 from Gowanlock et al., 2011) as a function of radial distance. Planets are not considered habitable if they have ages <4 Gyr. The inner Galaxy attains habitable planets earlier than at the solar radius ($R_\odot = 8$ kpc), and at the present day, the inner Galaxy at $R = 2.5$ kpc has the greatest area density of habitable planets.

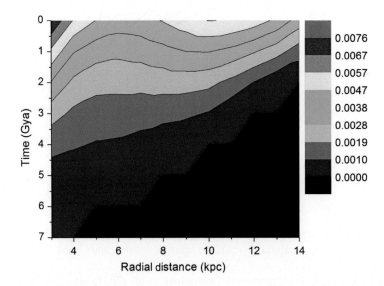

FIG. 4

The fraction of stars with a habitable planet (*color scale*) as a function of time and radial distance over all epochs (Model 4 from Gowanlock et al., 2011).

the epoch time and location within the Galaxy, habitable planets are at greater or lesser risk of experiencing sterilizing SNe events. Gaps between such events provide windows of opportunity for complex/intelligent life to evolve. Morrison and Gowanlock (2015) investigates the spatial and temporal distribution of these opportunities to gain an understanding of the way the likelihood of complex/intelligent life arising may have varied over cosmic timescales in different regions of the Galaxy.

In terms of the spatial distribution of opportunities, this was found to mirror the findings of Gowanlock et al. (2011). Regardless of the duration of undisturbed habitability presumed for an opportunity, the greatest number of opportunities of that duration are always expected in the inner Galaxy. This can be seen in Fig. 5 taken from Morrison and Gowanlock (2015), which shows histograms of the number of opportunities of different durations for a range of different distances from the Galactic center. For all duration lengths, the number of opportunities is at its maximum in the inner Galaxy and decreases monotonically with increasing galactocentric radius. Whether habitability is defined over relatively short or long timeframes, the significantly higher number density of habitable planets outweighs the impact of more frequent SNe events and the inner Galaxy is always favored.

Due to the inside-out star formation history of the Galaxy, the first habitable planets appear in the inner Galaxy and at progressively later epochs with increasing galactocentric radius. At all radii, the number density of habitable planets climbs steadily over time. However, since the inner Galaxy has a

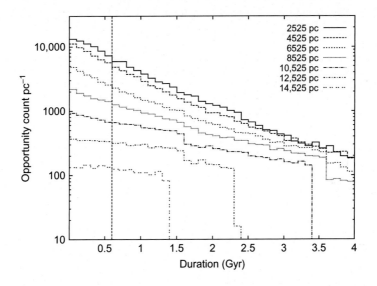

FIG. 5

Figure 9 from Morrison and Gowanlock (2015) that plots histograms of the number of opportunities for the emergence of complex/intelligent life of different durations, summed over all epochs, for a range of different galactocentric radii. The *dashed vertical line* crossing the horizontal axis at 0.6 Gyr corresponds to the opportunity duration experienced when intelligence arose on Earth. The opportunity count pc^{-1} is calculated by summing the total number of opportunities of a given duration in a radial bin (an annulus) and then dividing by the bin width (50 pc). This yields the absolute opportunity counts within each annular region.

"head start" due to the inside-out formation history, it is found that there are always a greater number of habitable planets toward the inner Galaxy, in all epochs. Even accounting for the higher rates of sterilizing SNe events in the inner Galaxy, across all epochs there are more windows of opportunity for complex life to emerge in the inner Galaxy than at higher radii. This can be seen in Fig. 6 taken from Morrison and Gowanlock (2015), which plots a count of the total number of active opportunities (defined as habitable planets experiencing a time gap between SNe of at least 2.15 Gyr) as a function of time since the formation of the Galaxy, for various galactocentric radii. At higher radii, opportunities do not begin to occur until later epochs, due to insufficient planetary age. At all radii the number of opportunities increases steadily over time, reaching its maximum today. This trend can be expected to continue for several billion years into the future as the metallicity increases throughout the Galactic disk, thus supporting higher planet formation rates.

A key point to note from Fig. 6 is that the level of opportunity for the emergence of complex life at the present time at Earth's galactocentric radius was matched in the distant past in regions closer to the Galactic center. Specifically the modeling is suggestive of there existing a level of habitability in the inner Galaxy around 2 Gyr ago that was similar to that in our local neighborhood today. Furthermore, the very first opportunities for complex life began to appear in the Galaxy more than 5 Gyr ago, *before the formation of the Earth*. We discuss these findings further in Section 7.

The chemical evolution of the Milky Way determines the regions of the Galaxy that host habitable planets over time. Spitoni et al. (2014) model the habitability of both the Milky Way and Andromeda (M31). In their work, the authors compare the effects of radial gas flows on the number of stars that host a habitable planet. By including radial gas flows, the number of habitable planets increases by 38% in

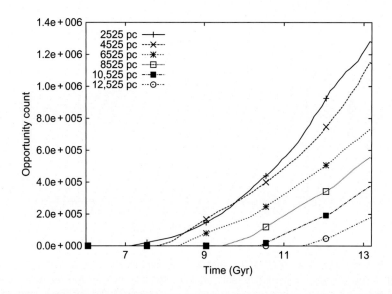

FIG. 6

Figure 10 from Morrison and Gowanlock (2015) that plots the number of active opportunities for the emergence of complex/intelligent life vs. time since the formation of the Galaxy, for a range of different galactocentric radii.

comparison to modeling the chemical evolution without radial gas flows, and they find that the location with the highest number of stars with planets is at a galactocentric radius of $R = 8$ kpc. This result is similar to that found in Lineweaver et al. (2004), which may be due to the method used to model the dangers of SNe events to planets in the Milky Way.

All of the GHZ models described above (Gonzalez et al., 2001; Lineweaver et al., 2004; Prantzos, 2008; Gowanlock et al., 2011; Spitoni et al., 2014; Morrison and Gowanlock, 2015) have employed major observational properties to model the GHZ; however, they have excluded the effects of stellar kinematics. Forgan et al. (2017) advanced a model of the habitability of the Milky Way using high resolution N-Body cosmological simulations to understand the effects of the evolution of the Local Group on the Milky Way. Thus, the work is able to give an account of the mass assembly history of the Milky Way in comparison to the other models that largely assume azimuthal symmetry.[2]Forgan et al. (2017) report that the GHZ of the Milky Way is between 2 and 13 kpc and that there are a large number of habitable planets in the inner Galaxy. The probability of a star hosting a habitable planet increases with galactocentric radius, whereas the number density of habitable planets decreases with radius. Another interesting paper by Vukotić et al. (2016) uses an N-Body simulation of an isolated Milky Way-like galaxy to study dynamical effects on the GHZ. They trace the locations of the stellar mass particles (roughly the mass of a star cluster) over time to determine the timespan within which they reside in a habitable region, as defined by a stellar number density and star formation rate below a particular threshold. Vukotić et al. (2016) find that the most habitable region of the Milky Way at the present day is toward the outskirts, and less than 1% of habitable planets are found near the solar radius at 8–10 kpc (see also Vukotić, 2017). This result suggests that the Earth's location in the Galaxy is atypical, and it is inconsistent with the 7–9 kpc GHZ found in Lineweaver et al. (2004) or Spitoni et al. (2014). As noted in Vukotić et al. (2016), this is likely due to the thresholded stellar density and star formation rate assumed to be required for a habitable environment, and in addition, a possible cause might be the dynamical effects that cause radial mixing of stellar populations with more metal rich systems migrating outwards from the inner regions of the Galactic disk.

Table 1 compares the GHZ literature discussed above, which is not a complete summary of all of the papers on the habitability of the Milky Way. In the table, we show selected literature, the type of model, what lethal events to habitability were considered, and the region that is predicted to be the most habitable at the present day. Most of the studies list numerous caveats and have differing metrics of habitability (e.g., the location of the greatest number density of habitable planets vs. the greatest fraction of stars with habitable planets). For instance, Lineweaver et al. (2004) find the location with the greatest probability of a star hosting a habitable planet to be between $R = 7$–9 kpc, Prantzos (2008) finds a maximum probability at $R = 10$ kpc, and Gowanlock et al. (2011) find this probability to be highest in the inner Galaxy (but only by a factor of $\sim$2 higher than at $R = 8$ kpc). However, Prantzos (2008) and Gowanlock et al. (2011) find that the number density of habitable planets is greatest in the inner Galaxy, and Lineweaver et al. (2004) does not calculate this number density. Given the different metrics and methods, we refer the reader to the papers in Table 1, and note that we have endeavored to interpret the results of these studies as accurately as possible.

---

[2]As an example of azimuthal symmetry, if you were to stand in the center of the Galaxy and studied an observable property, such as the number of stars in a region at a fixed distance, you would obtain the same number of stars if you were to rotate your body and look at the same sized region at the same fixed distance.

Table 1 Comparison of Selected Literature Sources

Source	Model Type	Lethal Events	Assessment at $z \sim 0$
Gonzalez et al. (2001)	Major Observables	N/A	Thin disk near the Sun for formation of Earth-like planets
Lineweaver et al. (2004)	Major Observables	SNe and Hot Jupiters	$R = 7$–9 kpc
Prantzos (2008)	Major Observables	SNe and Hot Jupiters	Entire disk
Gowanlock et al. (2011)	Major Observables	SNII, SNIa, and Hot Jupiters	Highest density at $R \approx 2.5$ kpc
Spitoni et al. (2014)	Major Observables	SNII, SNIa, and Hot Jupiters	$R = 8$ kpc
Morrison and Gowanlock (2015)	Major Observables	SNII, SNIa, and Hot Jupiters	$R \approx 2.5$ kpc
Forgan et al. (2017)	N-Body	SNII, SNIa, and Hot Jupiters	$R = 2$–13 kpc
Vukotić et al. (2016)	N-Body	SNe approximated by stellar number density and star formation rate thresholds	Predominantly the outskirts, $R \sim 16$ kpc

Major observables may include: the star formation history, metallicity evolution, stellar number density distribution, and other properties. Lethal events refer to those dangers to habitable planets in the models. Assessment of most habitable region at the present day (z ~ 0) refers to quoted values in the studies.

From Table 1 there are numerous predictions regarding the most habitable region of the Milky Way. Excluding Gonzalez et al. (2001) (because dangers to planets were not modeled) and Morrison and Gowanlock (2015) (because it is an extension of Gowanlock et al., 2011, and has a similar methodology), three papers suggest that nearly the entire disk is habitable, two papers suggest that the region encompassing the solar radius ($R_\odot = 8$ kpc) is the most habitable, and one paper suggests that the outer disk is the most habitable region at the present day. Thus, there is a divergence in the literature, which we believe to be due to the fact that model outcomes are heavily reliant on model assumptions and methodologies (as was pointed out in Prantzos, 2008). In what follows, we describe some of the reasons for the divergence in the literature.

Lineweaver (2001) informed the probability of forming an Earth-like planet in Lineweaver et al. (2004), where Earth-like planet formation was strongly correlated with metallicity (but declines when high metallicity environments inhibit Earth-like planet formation, due to Hot Jupiters). From simulations of planet formation, Prantzos (2008) argued for a constant probability of forming Earth-like planets as a function of metallicity. Gowanlock et al. (2011) combined the planet-metallicity correlation for Hot Jupiters (Fischer and Valenti, 2005) with statistics from the formation of Earth-mass planets in models of planet formation. Spitoni et al. (2014) and Forgan et al. (2017) use similar assumptions as Prantzos (2008). Vukotić et al. (2016) assign star particles a metallicity abundance from a Gaussian distribution and then determine if the particle is habitable based on whether the value is above that of a sample of detected Earth-like exoplanet host stars. With the exception of Lineweaver et al. (2004), the studies generally assume that small planet formation is weakly dependent on metallicity (as shown in Fig. 1). Although these methods slightly differ, over recent epochs, the models produce a large number of planets throughout the Galactic disk that can be examined to determine whether they are habitable. Thus, it is unlikely that the relationship between metallicity and terrestrial planet formation is responsible for the divergence in the model predictions.

Another difference in the literature is the metric that is used to determine whether a planet is habitable. Most of the papers define the GHZ as a probability function that takes as input the radial distance, and time. With this probabilistic formulation, it is straightforward to assume that the danger to planets due to proximity to SNe can be modeled as a function of the SN rate in the region. In particular, all of Lineweaver et al. (2004), Prantzos (2008), Spitoni et al. (2014), Forgan et al. (2017), and Vukotić et al. (2016) assume that the danger due to SNe is normalized to a particular SNe rate (typically that of the solar neighborhood). For instance, if we let RSN be the SN rate integrated over the past 4 Gyr at the solar neighborhood, then Lineweaver et al. (2004) assigns a probability value of surviving a SN, $P_{SN} = 1$ when the rate is $\leq 0.5RSN$, and $P_{SN} = 0$ when the rate is $\geq 4RSN$. As mentioned in Lineweaver et al. (2004), this normalization is somewhat arbitrary, as our knowledge of the effects of transient radiation events on the atmosphere and biosphere are not well established. This formulation is conducive to producing hard inner boundaries on the GHZ, as regions will not permit any habitable planets when the SN rate is more than $4\times$ the local neighborhood rate. In contrast to these methods, the dangers due to SNe in Gowanlock et al. (2011) were calculated by examining the times at which individual stars were sufficiently nearby a SN event, as determined by a distribution of sterilization distances for SNII and SNIa, in combination with using the Earth's history as a template for when SN events will negatively influence the biosphere. This may more carefully quantify the dangers to planets as a function of SN event rate, and more easily allows for studying habitability above and below the midplane, and not only as a function of galactocentric radius (and time). Additionally, it is less conducive to producing hard boundaries on the habitable regions of the Galaxy, although it still suffers from our incomplete understanding of the effects of transient radiation events on habitability. But it is a key factor in explaining why Gowanlock et al. (2011) find that the inner Galaxy is the most habitable region at the present day; even though the habitability of many planets does not survive SN explosions, a large number of planets still withstand these lethal events. A study that compares the differences between SN lethality modeling methods would help illuminate why the models give differing predictions.

5 THE HABITABILITY OF OTHER GALAXIES

The habitability of the Milky Way has understandably received greater attention than other galaxies, partly because it is our home, but predominantly because we have much more observational data at our disposal with which to calibrate our models of habitability. However, consideration of other galaxies is essential to characterizing the habitability of the wider Universe. Furthermore, observational data from distant galaxies may provide a valuable window into the past, and thus help to improve our understanding of the history of the Milky Way. The habitability of other galaxies is an emerging area of research that has generated a number of recent publications including Suthar and McKay (2012), Carigi et al. (2013), Spitoni et al. (2014), Forgan et al. (2017), and Dayal et al. (2015).

Carigi et al. (2013) and Spitoni et al. (2014) consider galaxy M31 (Andromeda); the nearest spiral galaxy to the Milky Way. The presumed similarities in structure and formation history of all spiral galaxies allows similar methodologies to be applied as discussed previously for the Milky Way, but modified to take account of M31's different mass; nearly twice that of the Milky Way. M31 makes an ideal test-bed for understanding the habitability of even larger spiral galaxies. Carigi et al. (2013) apply a similar methodology as Lineweaver et al. (2004) and estimate M31's GHZ. They find that when considering the highest surface density of habitable planets, the GHZ lies between 3 and 7 kpc. The region with the greatest number of habitable planets, however, is found between 12 and 14 kpc.

Spitoni et al. (2014) extend the work of Carigi et al. (2013) to include radial gas flows. They apply their model to both M31 and the Milky Way, predicting similar radii of maximum habitable planet density as Carigi et al. (2013) but slightly higher total counts for habitable planets. The fact that they applied precisely the same methodology to both M31 and the Milky Way allows for a direct comparison of the two galaxies. They estimate that the maximum habitable planet density occurs at 8 kpc radius for the Milky Way and 16 kpc for the more massive M31.

One may wonder about the near-future habitability of M31 and the Milky Way. The potential merger of M31 and the Milky Way in $\sim$4 Gyr may result in the formation of a single elliptical galaxy with a very different habitability distribution and trajectory than either of its constituent progenitor spiral galaxies. For this reason we cannot extrapolate the habitability of the Milky Way or M31 too far into the future.

The habitability of elliptical galaxies has been considered by Suthar and McKay (2012). This work is more speculative because it is necessary to make assumptions about star formation rates, metallicity distributions and Earth-like exoplanet formation that are less constrained by observational data than our Galaxy or M31. Based primarily on statistical considerations of metallicity distributions, Suthar and McKay (2012) conclude that the elliptical galaxies M87 and M32 may host a multitude of habitable Earth-like planets. However, this is based on numerous assumptions, so a more reasonable conclusion may be that certain regions of these elliptical galaxies can be expected to contain stars following metallicity distributions that are not completely disjoint from our region of the Milky Way, so we may reliably infer that small rocky planets will be present—but in what fraction of planetary systems is uncertain. Nevertheless, the fact that we expect the existence of Earth-like planets in elliptical galaxies is important because such galaxies are abundant in the Universe and are believed to contain generally older stars. Therefore they potentially host a great number of habitable planets on which there have been long durations undisturbed by sterilization events, offering many opportunities for the emergence of complex life.

Consideration of other galaxies and galaxy types has been discussed in the work of Forgan et al. (2017) and Dayal et al. (2015). In Forgan et al. (2017), high resolution cosmological simulations of galaxy formation in the Local Group are used to map out the spatial and temporal behavior of the Milky Way and M33 GHZs, in addition to tidal streams and satellite galaxies, allowing for more generalized three-dimensional structures that may not possess azimuthal symmetry. They conclude that each galaxy's GHZ depends critically on its evolutionary history, in particular on its accretion history. Dayal et al. (2015) propose a framework that allows the consideration of all galaxy types, with a view to understanding which types offer the greatest degree of habitability. Studying $\sim$100,000 galaxies from the Sloan Digital Sky Survey (Eisenstein et al., 2011), their work suggests that giant metal-rich elliptical galaxies at least twice as massive as the Milky Way could potentially host many thousands of times more habitable planets than the Milky Way. Due to the large sample of galaxies studied, they primarily rely on fewer observable properties than used in studies of the Milky Way to derive their estimates of habitability. Their approach makes a number of simplifying astrophysical assumptions and approximations, such as a homogeneous distribution in galaxies of stars/SNe and that the volume scales with the total stellar mass, together with simplistic assumptions about the effects of sterilizing radiation events on the evolution of complex life. Nonetheless, their work is a good step toward understanding habitability from a cosmological context. Future refinements to simulation methodologies and astrophysical assumptions in star/galaxy formation processes can be expected to yield steady improvements to our understanding of GHZ evolution in all galaxy types and hence the Universe at large.

6 TRANSIENT RADIATION EVENTS

Research on the GHZ has focused on SNe events. However, there are other transient radiation events that may inhibit life on planets in the Milky Way. In particular, GRBs have a beamed emission that can affect planets on the order of 1 kpc, potentially causing mass extinction events to many planets in the Milky Way (Thorsett, 1995; Thomas et al., 2005a,b; Melott and Thomas, 2011; Piran and Jimenez, 2014; Li and Zhang, 2015; Gowanlock, 2016). For overviews on the origin of GRBs, see Mészáros (2002); Woosley and Bloom (2006), and Gehrels et al. (2009), and references in the above mentioned articles. Furthermore, AGN may be dangerous to planets in the inner Galaxy and bulge (Clarke, 1981). Both AGN and GRBs have not been modeled with SNe in studies of the GHZ; rather, these dangers have often been modeled separately, for instance to study the dangers of GRBs to the Earth (Melott et al., 2004; Thomas et al., 2005a,b; Melott and Thomas, 2009; Thomas et al., 2015). AGN are likely not very dangerous to planets in the Milky Way, as the range in which their radiation would impact the habitability of planets is limited to the innermost regions of the Galaxy. Here, we briefly summarize recent findings on the effect of GRBs to the habitability of planets in the Milky Way.

Piran and Jimenez (2014) modeled the effects of GRBs to planets within the disk of the Milky Way. They find that long GRBs are the most dangerous GRB type, and that short GRBs are a fairly negligible source of transient radiation. Long GRBs have a metallicity dependence, as they are predominantly found in low metallicity galaxies (Fruchter et al., 2006), and thus are typically likely to occur at high redshift. Using the metallicity-dependent GRB rate and moderate GRB fluence lethality threshold values, Piran and Jimenez (2014) find that over the past 1 Gyr leading up to the present day, there is a 60% chance of a planet at the Earth's galactocentric radius to be irradiated by a long GRB. Li and Zhang (2015) find that there is roughly one long GRB at Earth's radius every 500 Myr, and Gowanlock (2016) reports that ~35% of planets at the solar radius are within the beam of a long GRB over the past 1 Gyr.

Without comparing the individual model assumptions of these works we simply note that the chemical evolution of the Milky Way may be sufficiently advanced to quench long GRB formation, as these GRBs are found in low-metallicity environments (Fruchter et al., 2006), and have primarily been found in galaxies less massive than the Milky Way (Jimenez and Piran, 2013). Thus, it may be the case that long GRBs may not strongly contribute to the sources of transient radiation hazards in the Galaxy. Furthermore, assuming a linear dependence between metallicity and GRB formation (low metallicity produces more GRBs), Gowanlock (2016) reports that the only environment that would produce GRBs at the present day is the galactic outskirts. However, at higher redshift, there are likely to be more long GRBs at lower galactocentric radii that may reduce the habitability of planets at that time. Fig. 7 plots

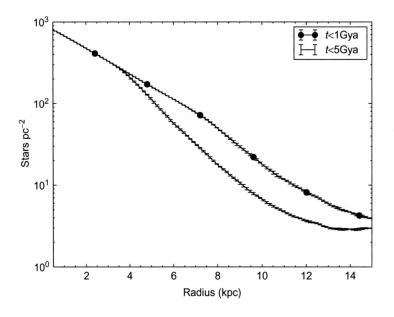

FIG. 7

The surface density of stars that are not within the beam of a GRB over the past 1 Gyr and 5 Gyr, as a function of galactocentric radius (Model 2 from Gowanlock, 2016).

the area density of stars that survive a long GRB over the past 1 Gyr and 5 Gyr from Model 2 in Gowanlock (2016). From the plot, the inner Galaxy has the highest number density of stars that survive a long GRB event, in part because the chemical evolution has quenched GRB formation in the region. The overlap in the 1 Gyr and 5 Gyr curves at $R \lesssim 4$ kpc, indicates that no GRB formation has occurred in that region over the past 5 Gyr. GRB events should be incorporated into calculations of the habitability of the Milky Way; however, SNe may still be the dominant source of transient sterilizing radiation in the Galaxy.

7 THE HABITABILITY OF THE GALAXY BEFORE THE EARTH

The majority of the studies of habitability on galactic scales focus on the present day. This is likely due to a few reasons. First, astronomy and astrophysics have intensely studied the structure and properties of the Milky Way. Therefore, the morphology and characteristics of the Milky Way at the present day are well known. In contrast, at earlier epochs, the overall characteristics of the Milky Way are less well constrained, and we rely on models that reproduce the major observable properties we see at present. Additionally, as our civilization exists today, notions of habitability are most relevant at the present day (such as estimating the distribution of the birth dates of planets that are habitable now in Fig. 3). Lastly, due to the many uncertainties in assessing habitability described heretofore, it is difficult to validate the habitability of the Galaxy in recent epochs, and even more challenging (or potentially impossible) at earlier epochs. In what follows, we appraise the habitability of the Galaxy before the Earth formed in the Milky Way 4.5 billion years ago.

From Fig. 2 (upper panel), it is clear that the inner Galaxy had a burst of star formation that has declined with time. Furthermore, the stars initially form in the inner Galaxy and the stellar mass builds outwards with time. Since the Milky Way would have been a more compact object and thus more dense on average at earlier epochs, SNII that explode shortly after they form, and prompt SNIa, would have had a catastrophic impact on the habitability of the inner Galaxy. However, with the decline in the star formation rate over time, the planets that formed in the inner Galaxy are eventually able to sustain complex life (according to Gowanlock et al., 2011; Morrison and Gowanlock, 2015). If we examine earlier epochs in Fig. 3, there are many planets in the inner Galaxy at $R \sim 2.5$ kpc that formed 9 Gyr ago that are habitable today. That is roughly double the lifetime of the Earth. Therefore, given the age distribution of stars in the Milky Way, it is likely that the first habitable planets began to appear $\gtrsim 10$ Gyr ago. In fact, Kapteyn's star is thought to host two super-Earth planets where the star is estimated to be between 10 and 12 Gyr old (Anglada-Escudé et al., 2014). Therefore, if we only consider the lethality of SNe to biospheres, then the Galaxy is likely to have hosted many habitable environments before the Earth formed.

This conclusion is supported by the findings of Morrison and Gowanlock (2015), as already discussed in Section 4. We saw in Fig. 6 that the number of opportunities for the emergence of complex life in the Milky Way has been steadily increasing over time, at all galactocentric radii. We might broadly equate the likelihood of complex life emerging at any given epoch with the number of active opportunities in that epoch. Given that complex life is known to have emerged at least once near the present time at Earth's galactocentric radius, it is worthwhile considering in what epoch a similar number of opportunities were available in other regions of the Galaxy. Fig. 6 suggests a similar likelihood for the emergence of complex life existed in the inner Galaxy around 2 Gyr ago to that seen at Earth's galactocentric radius today. We should not infer from this that complex life *did* in fact emerge previously within the inner Galaxy, since Earth might represent a statistical outlier event. But if it did, we might expect such life to be considerably older than ourselves. Indeed, it may predate the very formation of the Earth. As explained above, with the first habitable planets appearing more than 10 Gyr ago, there was a subsequent duration of more than 5.5 Gyr on those planets before the time that the Earth formed. This is a substantially longer duration than the ~ 4 Gyr that it took on Earth for complex life to appear after the planet's formation.

If we include the dangers of long GRBs to planets before the Earth formed (Fig. 7), then 5 Gyr ago, there were numerous stars that survive GRB events in the inner Galaxy. And even if we consider Model 1 in Gowanlock (2016) that favors hosting GRBs primarily in the inner Galaxy over all epochs, the number density of stars that survive a GRB event over the past 5 Gyr is roughly the same between $2.5 < R < 8$ kpc. Assuming that a planet is not deemed uninhabitable if it is within the beam of a GRB once every 5 Gyr, then the entire disk of the Galaxy would have hosted a large number of habitable planets before the Earth formed. This is consistent with Li and Zhang (2015) that report that even though the GRB rate is higher at $z > 0.5$ (roughly 5 Gyr ago), many planets are expected to survive long GRBs in the Milky Way and in other galaxies examined in samples from the Sloan Digital Sky Survey. If we combine the dangers of long GRBs and SNe, the Galaxy is likely to have hosted habitable planets throughout the Galactic disk, well before the Earth formed, with the exception of the very early Galaxy that would have hosted too many SNe, and likely long GRBs that would not have been quenched by sufficiently high metallicity abundances at that time.

8 CONCLUSIONS AND FUTURE OUTLOOK

The habitability zone of the Milky Way is still an open question. The study of habitability on large spatial scales necessitates that we define a habitable environment as one that would be endangered by a transient radiation event that would deplete protective ozone in a planetary atmosphere. With our limited knowledge of the consequences of such an event, we must presume that this would likely cause a mass extinction to surface dwelling complex life. From the exoplanet era, we have learned that a substantial fraction of stars host Earth-sized planets. These planets do not require a large abundance of metals to form, suggesting that terrestrial planets are located throughout the entire Galactic disk and are of a wide range of ages. The Milky Way may indeed host a large number of planets that have not been sterilized by transient radiation events, such as SNe and long GRBs; however, predictions vary in the literature and as we are in the early period of the development of this field, it is unlikely that consensus will be reached in the near term. Future observations that derive a large sample of the compositions of planetary atmospheres and possible signs of biosignatures will help determine a sample of truly habitable planets that can be used to better predict the distribution of habitable environments throughout the Galaxy. Galactic archaeology studies, starting with reconstructing the history of the solar neighborhood, may lead to a better understanding of the frequency that planets will survive lethal astrophysical events (Thomas et al., 2016; Breitschwerdt et al., 2016; Wallner et al., 2016) and emerging studies that incorporate galactic dynamics will illustrate that the galactic environments in which habitable planets are found may vary widely with time (Roškar et al., 2008). These new and future discoveries will further complicate our ideas regarding habitability on large spatio-temporal scales, but will surely help us better understand the prospects for life in our Galaxy and the Universe as a whole.

REFERENCES

Anglada-Escudé, G., et al., 2014. Two planets around Kapteyn's star: a cold and a temperate Super-Earth orbiting the nearest halo red dwarf. Mon. Not. R. Astron. Soc. 443, L89–L93.

Borucki, W.J., et al., 2010. Kepler planet-detection mission: introduction and first results. Science 327, 977.

Boss, A.P., 1997. Giant planet formation by gravitational instability. Science 276 (5320), 1836–1839.

Breitschwerdt, D., Feige, J., Schulreich, M., de Avillez, M., Dettbarn, C., Fuchs, B., 2016. The locations of recent supernovae near the Sun from modelling 60Fe transport. Nature 532 (7597), 73–76.

Buchhave, L.A., et al., 2012. An abundance of small exoplanets around stars with a wide range of metallicities. Nature 486, 375–377.

Carigi, L., García-Rojas, J., Meneses-Goytia, S., 2013. Chemical evolution and the galactic habitable zone of M31. Rev. Mex. Astron. Astrofis. 49, 253–273.

Catling, D.C., Glein, C.R., Zahnle, K.J., McKay, C.P., 2005. Why O2 is required by complex life on habitable planets and the concept of planetary "oxygenation time". Astrobiology 5 (3), 415–438.

Charbonneau, D., Brown, T.M., Latham, D.W., Mayor, M., 2000. Detection of planetary transits across a Sun-like star. Astrophys. J. Lett. 529 (1), L45.

Chopra, A., Lineweaver, C.H., 2016. The case for a Gaian bottleneck: the biology of habitability. Astrobiology 16 (1), 7–22.

Clarke, J.N., 1981. Extraterrestrial intelligence and galactic nuclear activity. Icarus 46 (1), 94–96.

Dayal, P., Cockell, C., Rice, K., Mazumdar, A., 2015. The quest for cradles of life: using the fundamental metallicity relation to hunt for the most habitable type of galaxy. Astrophys. J. Lett. 810, L2.

Ehrenfreund, P., Rasmussen, S., Cleaves, J., Chen, L., 2006. Experimentally tracing the key steps in the origin of life: the aromatic world. Astrobiology 6 (3), 490–520.

Eisenstein, D.J., et al., 2011. SDSS-III: massive spectroscopic surveys of the distant universe, the Milky Way, and extra-solar planetary systems. Astron. J. 142 (3), 72.

Falkowski, P.G., 2006. Tracing oxygen's imprint on Earth's metabolic evolution. Science 311 (5768), 1724–1725.

Fischer, D.A., Valenti, J., 2005. The planet-metallicity correlation. Astrophys. J. 622, 1102–1117.

Forgan, D., Dayal, P., Cockell, C., Libeskind, N., 2017. Evaluating galactic habitability using high resolution cosmological simulations of galaxy formation. Int. J. Astrobiol. 16 (1), 60–73.

Fruchter, A.S., et al., 2006. Long γ-ray bursts and core-collapse supernovae have different environments. Nature 441, 463–468.

Gehrels, N., Ramirez-Ruiz, E., Fox, D.B., 2009. Gamma-ray bursts in the swift era. Annu. Rev. Astron. Astrophys. 47, 567–617. https://doi.org/10.1146/annurev.astro.46.060407.145147.

Gobat, R., Hong, S.E., 2016. Evolution of galaxy habitability. Astron. Astrophys. 592, A96.

Godolt, M., Grenfell, J.L., Kitzmann, D., Kunze, M., Langematz, U., Patzer, A.B.C., Rauer, H., Stracke, B., 2016. Assessing the habitability of planets with Earth-like atmospheres with 1d and 3d climate modeling. Astron. Astrophys. 592, A36.

Gonzalez, G., Brownlee, D., Ward, P., 2001. The galactic habitable zone: galactic chemical evolution. Icarus 152, 185–200.

Gowanlock, M.G., 2016. Astrobiological effects of gamma-ray bursts in the Milky Way galaxy. Astrophys. J. 832, 38.

Gowanlock, M.G., Patton, D.R., McConnell, S.M., 2011. A model of habitability within the Milky Way galaxy. Astrobiology 11, 855–873.

Jimenez, R., Piran, T., 2013. Reconciling the gamma-ray burst rate and star formation histories. Astrophys. J. 773, 126.

Jurić, M., et al., 2008. The Milky Way tomography with SDSS. I. Stellar number density distribution. Astrophys. J. 673, 864–914.

Kaltenegger, L., 2010. Characterizing habitable exomoons. Astrophys. J. Lett. 712 (2), L125.

Kaltenegger, L., Traub, W.A., Jucks, K.W., 2007. Spectral evolution of an Earth-like planet. Astrophys. J. 658 (1), 598.

Kasting, J.F., Whitmire, D.P., Reynolds, R.T., 1993. Habitable zones around main sequence stars. Icarus 101 (1), 108–128.

Kopparapu, R.K., et al., 2013. Habitable zones around main-sequence stars: new estimates. Astrophys. J. 765 (2), 131.

Kroupa, P., 2001. On the variation of the initial mass function. Mon. Not. R. Astron. Soc. 322, 231–246.

Li, Y., Zhang, B., 2015. Can life survive gamma-ray bursts in the high-redshift universe? Astrophys. J. 810, 41.

Lineweaver, C.H., 2001. An estimate of the age distribution of terrestrial planets in the universe: quantifying metallicity as a selection effect. Icarus 151, 307–313.

Lineweaver, C.H., Chopra, A., 2012. The habitability of our Earth and other Earths: astrophysical, geochemical, geophysical, and biological limits on planet habitability. Annu. Rev. Earth Planet. Sci. 40 (1), 597.

Lineweaver, C.H., Fenner, Y., Gibson, B.K., 2004. The galactic habitable zone and the age distribution of complex life in the Milky Way. Science 303, 59–62.

Martin, W., Baross, J., Kelley, D., Russell, M.J., 2008. Hydrothermal vents and the origin of life. Nat. Rev. Microbiol. 6 (11), 805–814.

Matsuo, T., Shibai, H., Ootsubo, T., Tamura, M., 2007. Planetary formation scenarios revisited: core-accretion versus disk instability. Astrophys. J. 662 (2), 1282.

Mayor, M., et al., 2011. The harps search for southern extra-solar planets XXXIV. Occurrence, mass distribution and orbital properties of super-Earths and Neptune-mass planets. ArXiv e-prints.

McKay, C.P., 2004. What is life-and how do we search for it in other worlds? PLoS Biol. 2 (9).

Melott, A.L., Thomas, B.C., 2009. Late Ordovician geographic patterns of extinction compared with simulations of astrophysical ionizing radiation damage. Paleobiology 35 (3), 311–320.

Melott, A.L., Thomas, B.C., 2011. Astrophysical ionizing radiation and Earth: a brief review and census of intermittent intense sources. Astrobiology 11, 343–361.

Melott, A.L., et al., 2004. Did a gamma-ray burst initiate the late Ordovician mass extinction? Int. J. Astrobiol. 3, 55–61.

Mészáros, P., 2002. Theories of gamma-ray bursts. Annu. Rev. Astron. Astrophys. 40, 137–169. https://doi.org/10.1146/annurev.astro.40.060401.093821.

Morrison, I.S., Gowanlock, M.G., 2015. Extending galactic habitable zone modeling to include the emergence of intelligent life. Astrobiology 15, 683–696.

Mottl, M.J., Glazer, B.T., Kaiser, R.I., Meech, K.J., 2007. Water and astrobiology. Chem. Erde Geochem. 67 (4), 253–282.

Naab, T., Ostriker, J.P., 2006. A simple model for the evolution of disc galaxies: the Milky Way. Mon. Not. R. Astron. Soc. 366, 899–917.

Petigura, E.A., Howard, A.W., Marcy, G.W., 2013. Prevalence of Earth-size planets orbiting Sun-like stars. Proc. Natl. Acad. Sci. U. S. A. 110 (48), 19273–19278.

Pilcher, C.B., 2003. Biosignatures of early Earths. Astrobiology 3, 471–486.

Piran, T., Jimenez, R., 2014. Possible role of gamma ray bursts on life extinction in the universe. Phys. Rev. Lett. 113 (23), 231102.

Pollack, J.B., Hubickyj, O., Bodenheimer, P., Lissauer, J.J., Podolak, M., Greenzweig, Y., 1996. Formation of the giant planets by concurrent accretion of solids and gas. Icarus 124 (1), 62–85.

Pollack, J.B., Hubickyj, O., Bodenheimer, P., Lissauer, J.J., Podolak, M., Greenzweig, Y., 1996. Formation of the giant planets by concurrent accretion of solids and gas. Icarus 124, 62–85.

Prantzos, N., 2008. On the "galactic habitable zone". Space Sci. Rev. 135, 313–322.

Pritchet, C.J., Howell, D.A., Sullivan, M., 2008. The progenitors of type Ia supernovae. Astrophys. J. Lett. 683, L25.

Quarles, B., Musielak, Z.E., Cuntz, M., 2012. Habitability of Earth-mass planets and Moons in the Kepler-16 system. Astrophys. J. 750 (1), 14.

Roškar, R., Debattista, V.P., Quinn, T.R., Stinson, G.S., Wadsley, J., 2008. Riding the spiral waves: implications of stellar migration for the properties of galactic disks. Astrophys. J. Lett. 684, L79.

Saar, S.H., Butler, R.P., Marcy, G.W., 1998. Magnetic activity-related radial velocity variations in cool stars: first results from the lick extrasolar planet survey. Astrophys. J. Lett. 498 (2), L153.

Salpeter, E.E., 1955. The luminosity function and stellar evolution. Astrophys. J. 121, 161.

Segura, A., Kasting, J.F., Meadows, V., Cohen, M., Scalo, J., Crisp, D., Butler, R.A.H., Tinetti, G., 2005. Biosignatures from Earth-like planets around M dwarfs. Astrobiology 5, 706–725.

Shields, A.L., Barnes, R., Agol, E., Charnay, B., Bitz, C., Meadows, V.S., 2016. The effect of orbital configuration on the possible climates and habitability of Kepler-62f. Astrobiology 16, 443–464.

Spitoni, E., Matteucci, F., Sozzetti, A., 2014. The galactic habitable zone of the Milky Way and M31 from chemical evolution models with gas radial flows. Mon. Not. R. Astron. Soc. 440, 2588–2598.

Suthar, F., McKay, C.P., 2012. The galactic habitable zone in elliptical galaxies. Int. J. Astrobiol. 11 (3), 157–161.

Thomas, B.C., Jackman, C.H., Melott, A.L., Laird, C.M., Stolarski, R.S., Gehrels, N., Cannizzo, J.K., Hogan, D.P., 2005. Terrestrial ozone depletion due to a Milky Way gamma-ray burst. Astrophys. J. Lett. 622, L153–L156.

Thomas, B.C., et al., 2005. Gamma-ray bursts and the Earth: exploration of atmospheric, biological, climatic, and biogeochemical effects. Astrophys. J. 634, 509–533.

Thomas, B.C., Neale, P.J., Snyder II, B.R., 2015. Solar irradiance changes and photobiological effects at Earth's surface following astrophysical ionizing radiation events. Astrobiology 15, 207–220.

Thomas, B.C., Engler, E.E., Kachelrieß, M., Melott, A.L., Overholt, A.C., Semikoz, D.V., 2016. Terrestrial effects of nearby supernovae in the early pleistocene. Astrophys. J. Lett. 826, L3.

Thorsett, S.E., 1995. Terrestrial implications of cosmological gamma-ray burst models. Astrophys. J. Lett. 444, L53–L55.

Vukotić, B., 2017. N-body simulations and galactic habitability. In: Gordon, R., Sharov, A. (Eds.), Habitability of the Universe Before Earth. Elsevier, Amsterdam.

Vukotić, B., Steinhauser, D., Martinez-Aviles, G., Ćirković, M.M., Micic, M., Schindler, S., 2016. 'Grandeur in this view of life': N-body simulation models of the galactic habitable zone. Mon. Not. R. Astron. Soc. 459, 3512–3524.

Wallner, A., et al., 2016. Recent near-Earth supernovae probed by global deposition of interstellar radioactive 60Fe. Nature 532 (7597), 69–72.

Williams, D.M., Kasting, J.F., Wade, R.A., 1997. Habitable moons around extrasolar giant planets. Nature 385 (6613), 234–236.

Wolszczan, A., Frail, D.A., 1992. A planetary system around the millisecond pulsar PSR1257 + 12. Nature 355, 145–147.

Woosley, S.E., Bloom, J.S., 2006. The supernova gamma-ray burst connection. Annu. Rev. Astron. Astrophys. 44, 507–556. https://doi.org/10.1146/annurev.astro.43.072103.150558.

N-BODY SIMULATIONS AND GALACTIC HABITABILITY

Branislav Vukotić*

*Astronomical Observatory, Belgrade, Serbia

CHAPTER OUTLINE

Habitability of the Universe Before Earth, editors: Richard Gordon & Alexei Sharov, Volume 1 in the series:
Astrobiology: Exploring Life on Earth and Beyond, series editors: Pabulo Henrique Rampelotto,
Joseph Seckbach & Richard Gordon. ISSN 2468-6352. https://doi.org/10.1016/B978-0-12-811940-2.00008-3

1 FRAMING THE BIG QUESTION: WHERE ARE WE?

The first exoplanets were discovered at the turn of the past century. With the present advance of the detection techniques new discoveries are happening on a daily basis (Gillon et al., 2017). It is very likely that there is an abundance of the worlds similar to our Earth beyond the Solar system in the vast spaces of our Galaxy. A question of finding habitable places in the Universe was never more pronounced in mainstream scientific research. The same discoveries motivated the contextual research on development of life on Earth as an integral part of the evolution of the Universe. At the beginning of the space age, the geo-strategic political goals of the time have had an enormous cultural influence on the public over the following years. Since the same public is the one that churns out the scientists of the future, the discourse on the abundance of life in the Universe was seen as an easy research project that can be done with simple equations on the back of an envelope. The actual important discoveries of that time started to get credit very recently through the serious work on evolutionary time-scales, discovery of exoplanets and models of galactic habitability. In addition, any serious research in this area should discuss the future of humanity on Earth.

It became clear that the task is of great complexity and that the joint effort from various scientific disciplines, under the banner of an *astrobiological synthesis* is necessary for productive research. Simultaneously, the comprehension of such a complex phenomenon requires novel approaches in scientific modeling. Models based on complex system platforms became increasingly popular. The basic characteristic of these kinds of models is that the studied system is comprised from large numbers of smaller parts contributing individually or through their mutual interactions to the properties of the system as a whole. In this manner, the models based on probabilistic cellular automata (PCA) showed very promising in predicting the outcomes of fire spreading, traffic jams, urban development, etc. (see Ilachinski, 2001). PCA is essentially a grid of cells allowed to take different states at usually discrete time steps. The state of each cell in the next time step is determined with probabilistic rules applied to the present state of the cell and the present state of the surrounding cells. In this way, the simple rules of local evolution can generate a great amount of complex behavior on a global scale. Vukotić and Ćirković (2012) showed that a PCA platform can test various scenarios for assessing Contact chances by simply changing the input probabilities once a PCA kernel is developed.

Apart from obvious practicality, the motivation for using this powerful class of models can be sought in the significant advancement of complex systems understanding, from studies of mathematical complexity to complex nonlinear systems in physics, especially statistical physics. This opened new possibilities for this class of advanced models to proliferate to modern studies of life-related phenomena; from the physiological complexity of single living systems to, potentially the most interesting from the astrobiological point of view, the complexity of the biospheres. Since both structural and functional complexity are nowadays regarded as central (or even defining) properties of living systems, this direction of research is also relevant for the perennial problem of the definition of life.

The results of these modern studies indicate that evolution of life on Earth is a piece of a great puzzle and an integral part of the evolution of matter in the Universe where the matter is primarily organized into the structures called galaxies. The studies of galactic habitability are therefore crucial for understanding the evolution of life in the Universe, evolution of life on Earth, contact prospects and future of life in the Solar system.

Contemporary main stream research should be emancipated from the pitfalls and drawbacks of the past SETI projects that found their way into a scientific discourse masked as the legitimate science. Compared to these early SETI studies, there was an obvious need for the philosophical re-assessment of foundations of scientific method.

Without going into too much detail, it is obvious that in the spirit of Occam's razor it is reasonably simplistic to adopt the continuity thesis (Fry, 2000) as an axiom. The thesis states that we should expect the continuity in the process of matter formation (for an elaborate explanation of continuity thesis in astrobiological context see Ćirković, 2012). As the first atoms condensed from a primordial soup of elementary particles we should expect, on the other end of the scale, that large and complex organic molecules formed from inorganic matter. This can be further extended to involve metabolism and cell formation, etc. A simplistic and yet very powerful axiom! It eliminates the need for extra-scientific assumptions such as that life on Earth was formed by the intervention of some supreme intelligence factor. Now we have this continuous cosmic life development flow in the space-time vastness of our Universe, but it is still a far cry from producing some scientific predictions about the phenomenon of life.

As with other fields of research in science, in order for this framework to have a real predictive value, it needs to be constrained by boundary conditions. The only one we can invoke so far is that there are only us out there (in the present epoch and at our actual position). So all our models need to satisfy, within reasonable probability margins, that at the present Galactic epoch we have not yet discovered life outside Earth. This seems to contradict the fact that Milky Way is comprised of $\sim 10^{11}$ stars. Soaring numbers of newly discovered exoplanets makes this paradox even more pronounced. So far, many solutions have been proposed to this, often called Fermi paradox (or Contact absence), but only few of them were scientifically scrutinized (see Ćirković, 2017). Regardless of their individual scientific scrutiny, the large number of proposed solutions highlights the need for using the paradox as a boundary condition.

Depending on the adopted framework, the boundary condition can be viewed as exclusive (soft) or nonexclusive (hard). The exclusive nature means that there is something special about Earth and the Solar system, as well as with other habitats in the Universe. For example, civilizations may reach a sufficient level of maturity to self-destruct with nuclear warfare or misuse of biotechnology, artificial intelligence, etc. But such solutions are an extrapolation of our current knowledge and such predictions are less scientific than the ones achieved within the deductive framework discussed above. Hence, the solutions that are in the best accordance with scientific method have to be nonexclusive (see also Brin, 1983 arguing in the favor of hard solutions). The Universe is filled with giant gas clouds, stars (their remnants) and planets that all together dwell in a rare hot gas (and various photon and particle radiation) of vast interstellar and intergalactic space. Stars and planets have much different density and composition than the average values in the Universe. However, this is not the reason to consider them as an exclusive parts of that same Universe. The stars and planets formed and evolved according to the same laws like the rest of the matter in the Universe and for this reason they should be studied in a nonexclusive manner. Until we can confirm that, i.e., the pinnacle of matter evolution is self induced nuclear destruction of intelligence (if there is anybody left to make such an observation) it is more in-line with the philosophy of science and scientific method to first explore the nonexclusive branch.

2 HABITABILITY PROPERTIES

The continuity thesis implies that our knowledge of the history of life outside the Solar system should follow from the basic principles of Milky Way evolution. Our Galaxy with its satellite galaxies is located in the Local group of galaxies comprised of other galaxies and their satellites. All of these galaxies have a complex mutual gravitational interaction. It has become increasingly clear that galactic collisions significantly influence the evolution of a particular galaxy (apart from obvious dynamical effects, the tidal shockwaves can produce harmful amount of cosmic rays, Atri, 2011). Now we will introduce the basic properties that are relevant for the habitability of a single galaxy (Milky Way or any other and later these properties will be discussed in the context of galactic interactions).

2.1 METALLICITY

With respect to the continuity thesis, in a wider astrobiological sense, we can term the level of matter evolution as astrobiological complexity. This term can encompass all stages of matter evolution, from elementary particles to intelligent being condensates, and is of great importance for astrobiological related studies (Chaisson, 2003; Vukotić and Ćirković, 2012). For the inorganic part of the scale, highly sufficient for standard astrophysical studies, it is usually (and traditionally) called metallicity. Simply, metallicity represents the amount of elements heavier than helium expressed in some convenient form. As a good tracer of the amount of heavy elements a ratio of Fe relative to hydrogen is usually used, designated as [Fe/H]. It is expressed on a logarithmic scale where the zero value represents the Fe/H ratio for the Sun. Also, a usual designation for the abundance of heavy elements is Z and it represents the mass fraction of a star that is not hydrogen or helium.

Given the fine tuning of our Universe, the optimal conditions for biochemical reactions are, in a very rough approximation, in 0–100°C temperature interval (under a realistic range of atmospheric pressures). This temperature range is found on the surface on Earth. Among all other astronomical objects, planets in general are the ones that are the most likely to have the similar temperatures on their surfaces. Of course, we should not neglect the complex chemistry taking place in zero gravity conditions of the giant molecular clouds (and possible panspermia implications) but as far as complex and functional biospheres are concerned, planetary environments are the best suited for life.

Planets condense as a by-product of stellar formation from the same proto-stellar gas cloud. The cores of the planets are formed from heavy elements, such as iron. While the planet is cooling, heavy elements sink down and the crust is formed from lighter elements such as silica. There are also other elements that are important for organic life chemistry: carbon, hydrogen, oxygen, nitrogen, phosphorus, and sulfur (all of them usually designated as CHONPS). The larger and heavier the planetary core is, there are more chances that the planet will form and retain an atmosphere. Giant Jupiter-like planets have formed from heavy cores that enable them to accrete much of the hydrogen and helium gas from the proto-stellar cloud (Pollack et al., 1996; Lissauer, 1987; Stevenson, 1982). Because of their large mass their solid cores are enveloped with metallic hydrogen and helium (high pressure compression may degenerate electron clouds of nonmetals so that they show a metal-like properties such as lattice structure, conductivity, etc.) choking the biochemistry. On the other hand, if the abundance of heavy elements is sufficiently high, it is likely that the giant planets will form with large satellites (some of Jupiter's or Saturn's satellites are comparable in size to Earth and have a solid surface), opening the speculative possibility that some of those moons can be habitable. Unlike these giants, smaller planets like Earth have their solid surfaces protected by a thin layer of atmosphere where temperature and pressure are just right for various chemical reactions to take place and form complex biochemical structures. Whether the small, solid surface planets, or gas giants with satellites are concerned, the existence of the heavy elements as building blocks is necessary in order to form all of these potential habitats.

The heavy elements, with the number of possible electron transitions in their cumbersome electron clouds, are very efficient cooling channels for collapsing gas clouds. It is then to be expected that clouds with larger [Fe/H] should form giant planets with higher probability, since the more efficient cooling leads to a faster gas cloud collapse and larger bodies can be condensed before the radiation from the newborn star evaporates the cloud. However, a large sample of detected exoplanets did not show any significant correlation between the metallicity of the host stars and the mass of their largest planets once we have corrected for the selection effects inherent in spectroscopic methods. The discoveries of exoplanets are most frequent around stars of near Solar metallicity and Earth-sized planets are found around stars with a wide range of metallicity (Buchhave et al., 2012). Several discovered Earth-sized exoplanets have stars with $-0.32 <$ [Fe/H] < -0.13 (Schuler et al., 2015). The lower end can be attributed to a metallicity threshold for Earth-sized rocky planets formation. For the high end of the interval it is not clear whether the increased probability of the gas giants' formation is relevant for the probability of the same star to host the habitable Earth-size planets as well. Gas giants might just migrate toward their host star and absorb their smaller counterparts in the process. Together with the possibility of having the habitable moons around Jupiter-like giant planets, stars with $-0.32 <$ [Fe/H] should be considered as candidates for hosting habitable Earth-size rocky planets.

Lineweaver (2001) calculated the probability of hosting the Earth-sized exoplanets from the metallicity of the galactic gas. The results highlighted that an average Earth-like planet in the Universe should be ~ 2 Gyr older than Earth. Later study of Behroozi and Peeples (2015) found that Solar system formed after 80% of existing extrasolar planets while in the most recent work Zackrisson et al. (2016) estimate that the mean age of terrestrial planets in the local Universe is 7 ± 1 Gyr. This emphasizes the need for seeking other habitability relevant properties and mechanisms that can give satisfactory explanation for the absence of exo-life detection so far.

2.2 STAR FORMATION RATE

Stars predominantly form simultaneously in large numbers from a single giant gas cloud where initial density irregularities collapse to form individual stellar systems. Stars of smaller masses form more frequently, but although smaller in numbers, stars of large masses have a far more intense nuclear fusion in their cores during their lifetime. When massive stars spend their nuclear fuel there is nothing to counter the gravity collapse of the stellar material and stars die-out with violent explosions called supernovae and their matter collapses to form exotic objects called stellar remnants (neutron stars or black holes). The collapse is followed by the very energetic explosion (readily observed from other galaxies) spawning out life hazardous radiation in the form of γ and X-rays and various species of subatomic particles. At the same time these explosions seed the heavy elements (that the star has produced during its lifetime) into the surrounding space. This enriches the gas clouds in heavy elements and increases the probability of planetary core production for the next generation of stars. The more massive the star, the shorter it lives and the more energetic its terminal explosion is. For a review of contemporary knowledge of star formation, see e.g., Kennicutt and Evans (2012).

The places of intense star formation are not life friendly environments. Occasional devastating explosions from rare short lived massive stars are very hazardous for life on nearby newly formed exoplanets and in addition these stellar maternity places have a very high background UV flux. UV radiation can easily break protein bonds and is considered dangerous to life. A newly formed star needs to take its time to drift away from the places of intense star formation or until the described effects fade away in order to be in a friendly environment for its planets to host life. Also, passage into places of high star formation might be a considerable threat to life in a habitable stellar system.

On much larger temporal scales, a global galactic *climate* of star formation might just be one of the main reasons we have not yet perceived extrasolar life. In the paradigm of an astrobiological phase transition (see Annis, 1999; Ćirković and Vukotić, 2008; Ćirković et al., 2009), the development of life in the galaxy is severely constrain until the star formation rate (SFR) drops below a certain threshold. The supplies of hydrogen gas (the main fuel for stellar fusion) in the Universe are depleted over time as more new stars are produced to fuse hydrogen into heavier elements. The supply of gas in our own galaxy is now smaller than it was after the time of its formation $\approx$10 Gyr ago. Reduced gas content lowers the SFR and this makes the galaxy much more life friendly in the present epoch than in past times.

SFR surface density can be expressed in Solar masses (of gas cloud mass turned to stars) per year per galactic disk unit surface area. For the Solar neighborhood it is estimated to be $\approx$0.003 $M_\odot$ yr^{-1} kpc^{-2} and is observed ranging from 0.002 to 0.01 $M_\odot$ yr^{-1} kpc^{-2} (De Donder and Vanbeveren, 2002). Since there are no reasons to consider that the Solar system is currently in the Milky Way region with life hazardous high SFR it is to be expected that the habitable stellar systems predominantly can be found in the regions with SFR surface density <0.01 $M_\odot$ yr^{-1} kpc^{-2}.

It is important to note here that at the current level of our astrobiological knowledge we cannot hope to calibrate models with high precision. There are a number of studies that estimate the effects of the hazardous γ, X-ray, and particle radiation (possibly coming from the energetic supernovae explosions, γ-ray bursts, and supernova remnants) on the Earth's biosphere (for some of them see Thomas et al., 2008; Thomas, 2009; Beech, 2011; Brakenridge, 2011; Korschinek, 2016) but their results are dependent on a large number of calibrating parameters (thickness and composition of the atmosphere, incoming radiation incident angle, Earth's magnetic field, Solar wind, galactic magnetic field, etc.). In addition, some of these parameters are vaguely known. Also, when estimating the threat from a potential explosion progenitor, parameters of the explosion cannot be estimated with high certainty. With all these difficulties it is not possible to say whether the explosion of, e.g., an 80 $M_\odot$ star will affect the habitability of an Earth-like planet at a distance of 10 pc. At this point we can only strive to capture the mechanics and global trends in galactic habitability from some rather arbitrarily estimated values of SFR thresholds implemented in our models.

2.3 DYNAMICAL PROPERTIES

Galaxies like the Milky Way look like spirally structured thin disks with a bulge in the center of the disc. These two components are very different in their dynamical properties. The bulge is smaller in volume (typical size being less then 3 kpc in diameter) then the disk (around 40 kpc in diameter and about 1 kpc thick), but it has a larger density of stars. It is considered that the SFR in the bulge has ceased in the past because the high density environment with a larger SFR has depleted the gas much faster than in the disk. In the present epoch, the bulge is a swarm like structure mainly composed of older stars that orbit around the dynamical center of the Galaxy. A frequent nearby passages of stars can perturb their planetary orbits, or even cut the planets lose from the gravitational pull of their host star, releasing them into the interstellar space. On the other hand, this should give a significant number of rogue planets that can be recaptured by other stars. Such frequent changes of planet-star distance are not suitable for habitability, at least as we know it here on Earth. Even the Earth, with its almost circular orbit, constantly in the Sun's habitable zone, exhibits periodic severe glaciations that influence its biodiversity. These periodic climate changes called Milanković's cycles (Milanković, 1941) are in part caused by the existing small eccentricity of the Earth's orbit. Orbital stability is one of the imperatives of planetary habitability and it is not expected that in the crowded environment of a galactic bulge, where planetary systems frequently are perturbed by nearby passages, orbital stability can be maintained for a considerable time-span (Forgan, 2016).

In the Galactic disk, with orders of magnitude smaller density of stars, nearby passages are far less frequent and there is a higher probability for planets to maintain stable orbits around their host stars. Given this argument alone, life as we know it should favor the disk of the galaxy and possibly the outer parts of the bulge. However, it is interesting to note here that a flyby at the solar radius might cause a stronger gravitational perturbation than at the bulge. The strength of the average gravitational perturbation caused by a nearby passages depends on the velocity dispersion of stars in the considered part of the galaxy (Jiménez-Torres et al., 2013). According to Jiménez-Torres et al. (2013) the smaller velocity dispersion allows for a stronger gravitational interaction. The velocity dispersion is lower at the solar radius, therefore the gravitational interaction from a nearby passage is likely to be stronger than in the bulge. Apart from these local dynamical effects, the understanding of galactic habitability should also consider the importance of mixing between the disk and the bulge stellar populations.

Just like in other similar galaxies that are directly observed, indirect observations of the Milky Way show the existence of a spiral pattern in the disk. The spiral arms are denser parts of the disc with increased concentration of gas and stars and are the places of more intense star formation. This spiral pattern rotates with a frequency different from the orbital periods of most of the disk stars, so that stars encounter periodic spiral arms crossings. Like our Sun, the disk stars also oscillate perpendicular to the disk plane. The density of matter in the disk is the highest at the middle and decreases exponentially with distance from the disk plane. Similar as with the spiral arms, the higher density environment may affect the habitability. The complex dynamics of the disk and distribution of matter make the habitability assessment far from trivial.

Apart from circular motion around the center of the Galaxy and disk perpendicular oscillations, the galactocentric distance of disk stellar systems also oscillates. In addition, the effects of gravitational energy dissipation (dynamical friction) cause the stars to slowly drift toward the out(inn)er parts of the disk (Roškar et al., 2011). The same mechanism also leads to population mixing of disk and bulge stars. With the dynamical friction being the dissipation process, the flux of stars in the outward direction is larger than the inward flux. This leads to increase of the stellar disk radius. It is evident that there is a large degree of stellar migration in the disk of the galaxy. On such migrations stars experience significant changes in habitability relevant galactic parameters of their immediate surroundings. Given that life sustainability requires long-term stable favorable conditions, the influence of dynamics is very important for tracing parts of the galaxy that have long-term chances of hosting stars in habitable conditions.

2.4 THE GALACTIC HABITABLE ZONE

Is there an analogue of stellar habitable zone on a galactic scale? It is far more complex to constrain the zone of enhanced galactic habitability than to constrain the habitable zone of a Main Sequence star. The task is to pinpoint the parts of the Galaxy that support habitable environments over longer periods of time. After the galactic habitable zone (GHZ) was proposed by Gonzalez et al. (2001), there were several works on the concept over the previous decade (Peña-Cabrera and Durand-Manterola, 2004; Lineweaver et al., 2004; Ćirković, 2005). Apart from these pioneering works the constraints on Galactic habitability have been addressed with various studies in the last few years (Gowanlock et al., 2011; Spitoni et al., 2014; Forgan et al., 2016; Legassick, 2015; Vukotić et al., 2016).

Cylindrical symmetry of the Galactic disk implies that GHZ is in shape of an annular ring aligned with the disk plane. Depending on the modeling approach the position of the inner boundary of the ring varies in its range from the outer part of the central bulge to around 10 kpc of galactocentric distance, whereas the outer boundary is sometimes estimated to be at the outskirts of the galaxy at the present epoch. The estimates of the galactocentric distances that are the most suitable in habitability terms range from significantly inwards to significantly outwards the Solar circle. This indicates that we still do not well understand the structure of the GHZ (for a review on the extent of GHZ see also Gowanlock and Morrison, 2017).

The inner boundary should primarily depend on the density of the stellar component and dynamical stability: closer to the bulge the gas is depleted and SFR is very low whereas heavy elements are abundant due to their built up in previous epochs before the gas was depleted. However, in models where the inner boundary lies further away from the bulge, the SFR might play a significant role. Since the outer boundary usually lies at the peripheral areas of the disk, metallicity is also of key significance. The build-up of sufficient amount of heavy elements took a much longer time since a lower density of matter and depletion of gas content led to a very small SFR that kicked in only at latter epochs. A comparable amount of metallicity might arrive to the outskirts of the disk by the means of population mixing with stars that have migrated from the inner disk areas. But if the metallicity threshold required for planets to form is significantly high in the model, then the outer boundary might be positioned more inwards and overlap with the areas of higher star formation, and greater stellar density.

Usually, following on a secular increase in metallicity and decrease of SFR and stellar density distribution, the GHZ tends to drift outwards and widen with time, but the results of some recent works are somewhat different from this simplified picture. The studies that include dynamical models of Galactic habitability (Forgan et al., 2016; Vukotić et al., 2016) are not conclusive about where to expect the highest probability for habitability, or if even something like that is possible in the dynamically and chemically active environment of the galactic disk. The indisputable fact is that the concept of GHZ and the process of its confinement have served well its purpose in provoking advanced studies. This firmly established the galactic habitability branch of astrobiological studies as a well founded and justified scientific enterprise. At present we might not think that a simplified picture of an expanding annular ring is justified for describing and grasping the habitability on the galactic scales. It is the goal of our present efforts to further advance the field by, in the spirit of good science, *making things as simple as possible, but not simpler than that* (Einstein). To facilitate advancement in this manner in further text we describe and discuss N-body simulations as a tool to study galactic habitability phenomena. This tool is likely to best capture all of the complex galactic dynamics and consequential habitability implications.

3 N-BODY SIMULATIONS: GALACTIC HABITABILITY IN DYNAMICAL PERSPECTIVE

3.1 DESCRIPTION

With the advent of sufficient computing power, around two decades ago, N-body simulation entered the scene and became an indispensable tool in modeling of galactic related phenomena. At present, these models are standard in studies ranging from the formation of cosmic structure and galaxies, to studies of the development of individual stellar systems. Wherever there is gravity as a dominant driving force, N-body models are the best choice for simulation. A number of available code kernels and visualization tools can facilitate the research in this area.

Similar to complex systems models like PCA, an N-body simulation traces simple gravitational interactions between small parts of the modeled system (usually called particles), and analyses the emerging behavior on larger scales. This formalism is particularly convenient for investigating how the overall habitability of the galaxy depends on the complex dynamics of individual stellar systems. The results of such studies can give a better understanding of the spatial and temporal distribution of galactic habitats. Unlike the PCA where each cell interacts only with its immediate surroundings the N-body simulations can potentially account for gravitational interaction between all particles. However, modeling of all interactions is computationally impossible, and various approximations are implemented in N-body kernels. The practice shows that even these approximate models are very good in capturing complex phenomena related to gravitationally driven dynamics.

In the ideal case, the galaxy should be modeled as a composition of gravitationally attracted particles, where each particle has the mass of an individual star. For a realistic model of the Milky Way this translates to simulating the mutual interactions between $\sim 10^{11}$ particles which implies calculating of $\sim 10^{22}$ interactions. This is not possible even with the state of the art kernels and computing resources. Today's standards for a typical parallel computing facilities found at scientific departments are of the order $\sim 10^6$ particles for a simulation of the Milky Way-like galaxy. Each individual particle then has the mass of the order of a typical star cluster. For a successful astrobiological use of N-body simulations the habitability related characteristics of individual stars have to be deduced from the parameters of much more massive particles found in simulations.

Apart from dynamics of individual stars, chemical evolution and SFR-relevant phenomena are related to gas content of the galaxy. For more realistic models of galactic evolution the N-body simulations have to capture the physics of gaseous matter in addition to gas dynamics. Typically, there are several types of particles in N-body simulations of galaxies. For example, the GADGET-2 code for modeling of the N-body interactions (Springel, 2005a), which is a standard in the field, processes particles of 6 types: 0—Gas, 1—Halo, 2—Disk, 3—Bulge, 4—Stars, and 5—Boundary (for specific details see also Springel, 2005b). All particle types have the same gravitational interactions, whereas gas particles experience hydrodynamical acceleration in addition to gravity. Other codes may use different approaches but the different treatment of gas and stars is always essential. The habitability-related phenomena addressed in this chapter are specifically related to gas and stellar particles.

Gas fluid content in a galaxy is presented at discrete points called particles—each described by standard gas properties (pressure, density, temperature, chemical composition, mass, speed, position, and possibly even a viscosity). These properties are related not only to a singularity of a particle position (similar to delta functions) but rather smeared out over the surroundings with some form of the smoothing kernel function. The characteristic length for dispersing of the properties of a single particle in this manner is called a smoothing length. In this way it is possible to calculate the relevant gas property at any position in space by simply adding up contributions from all the particles in range of their smoothing kernel from the given position in space. Hydrodynamic acceleration of gas particles is then calculated from the pressure gradients exhibited on a given gas particle. This approach is called smoothed particle hydrodynamics (SPH) and it is an efficient way of reconciling the need to calculate the relevant properties over the entire space with increased efficiency of movement modeling. The best results for simulating spiral Milky Way sized galaxies with $\sim 10^5$ particles per species are achieved with smoothing lengths of the order of 1 kpc. The statistical formalism of kernel density smoothing states that smoothing lengths should depend on the density of the data points (e.g., for larger data densities the smoothing lengths should be smaller). In some SPH implementations the smoothing length is not fixed but is a function of the density of particles.

As the motion is implemented with particles that attract each other with gravitational force, there is a singularity of the force when the distance between two particles approaches zero. This problem is resolved by introducing a softening length, which is usually somewhat smaller than the smoothing length and no gravitational force calculations are performed if the distance between the particles is smaller than the softening length. A typical N-body simulation has the softening length of $\sim 10^2$–10^3 pc (while the smoothing lengths are usually a few times larger). These approximations enable the capturing of the relevant dynamical phenomena on the galactic scale without a significant computational burden.

3.2 METALLICITY AND SFR

Together with the available gas supplies, metallicity and SFR are properties that intrinsically depend on one another. As mentioned earlier, most of the heavy elements are produced in short lived massive stars and released after their terminal explosion. Intense stellar winds from massive stars and their terminal explosions transferred the energy and heavy elements to the interstellar gas, which later collapsed to produce a new generation of stars and the cycle repeated but this time with more heavier elements in the gas. Since the habitability models are all about timing and stability, the accurate modeling of chemical composition and SFR are crucial, and the adjustment of parameters usually done "by hand". The underlying kernel, that calculates gravitational and hydrodynamical acceleration, is combined with equations for calculating SFR, stellar feedback and gas cooling through radiative processes. The simulations usually start with far more gas than stellar particles. Over the simulated time span, SFR builds up stellar content at the expense of gas particles. When the implemented SFR laws allow for star formation to occur at some gas particle, the particle is either reduced in mass or removed from the simulation with newly formed stellar particle(s) added to the simulation. The newly formed stellar particles inherit parameters of motion, metallicity, and others from their gas progenitors.

3.3 MODEL ACCURACY AND LIMITATIONS

At a big city, living conditions of an average inhabitant are the result of overall conditions in the city (local climate, air and water quality, infrastructure level, population density, etc.) and conditions in the actual living quarter of the individual (size of the living space, number of flat mates, income, and other factors). In the same fashion, we can decompose basic habitability for planets into global and local conditions. The global ones reflect the influence of relevant galactic parameters while the parameters of the host stellar system (locally influencing the planet) are indicative of the local conditions. Part of the local conditions that is specifically related to a planet is described with the parameters of the planet itself (e.g., atmosphere, interior heating, magnetic field, plate tectonics, water content, etc.). Here we, however, address specifically the global conditions that define the environment where particular stellar systems might dwell. Until the appearance of Kardashev's Type 3 civilizations, capable of actively changing their galactic environment, we can consider the evolution of habitability relevant galactic parameters independent of various biological, cultural and technological time-scales. At least for the time being.

The overall movement of stellar particles in a simulation should approximate the motion of particular stars. For more accurate prediction of the habitability of the galactic environment imposed on a single stellar system the model should consider the trajectory of the system and environmental parameters at each position along the trajectory. Even with the approximation that trajectories of the single stellar systems are well represented by simulation particles the finite time resolution might still be an issue. Since the time sampling cannot be performed beyond the minimal time step within the simulation, the particles can only be sampled for environmental conditions at discrete steps along their trajectories. This modeling technology has issues of a rather epistemological nature. The more times the particle is sampled along its trajectory, the higher is the probability that it will be found in harsh conditions. A brute force resolution to this problem is to sample the particle on a time scale comparable to the duration of hazardous events (e.g., supernovae explosions, stellar collisions, etc.). Since supernovae explosions duration is rather a matter of days and not years this is likely to be computationally costly for a large number of particles and possible workarounds should be considered and developed. One way of resolving the sampling issue is to consider the average habitability conditions along the particle trajectory. A possible drawback to this solution is that the averaged habitability parameters might not be a good tracer of the largest continuous time interval that particles have spent in the habitable friendly conditions. For a more robust insight, this approach should be complemented with actual sampling at the best time resolution available. Another solution is to calculate the (possibly randomly chosen) time to the next hazardous event and use that as the time step (an analogous method for Ising models was formulated in Gordon, 1980).

With the state-of-the-art computing power of today it is possible to simulate a Milky Way with particles of $\sim$100 $M_\odot$. This is still far from an ideal case scenario of one simulation particle to represent one stellar system. With present standards, the mass of a common stellar particle is of the order of the mass of a stellar cluster. Although stars are predominantly formed in stellar clusters, the clusters are likely to disperse on time scales that are 1–2 orders of magnitude shorter than, e.g., the life-time of the Sun (10 Gyr). With stars drifting away from their birth clusters it is not expected that the environment experienced by the position of the simulated particle is accurate for considering individual stars. In addition to time resolution arguments, the low mass resolution also highlights the need for averaging instead of using an accurate approach. Similar considerations apply in favor of smoothing the other relevant properties.

4 SIMULATIONS
4.1 GENERAL DESCRIPTION

To our knowledge, there are only two N-body simulations where data were analyzed in terms of habitability (although the research in this area progresses fast). The work of Forgan et al. (2016) examines the habitability of the galaxies in the Local group, specifically the Milky Way and M33, while the work of Vukotić et al. (2016) deals with the habitability of an isolated Milky Way-like spiral galaxy. Both simulations were realized with the GADGET-2 SPH code.

Forgan et al. (2016) utilized the results of the CLUES simulation (for details see references in Forgan et al., 2016). The CLUES simulation was designed for the purpose of cosmological studies but it turned out to be also of great practical use for habitability studies of the Milky Way, M33 and other galaxies from the Local group. Around 200 runs of the simulation at low-resolution

(256^3 particles) were performed until a suitable candidate for the Local group was found (with good resemblance to the Milky Way and M31 as the two major galaxies in the group). Then the particles resembling the group of interest were tracked back to their original positions and new high-resolution runs (the equivalent of 4096^3 particles) were performed to obtain the required data.

The single run of an isolated galaxy's evolution from Vukotić et al. (2016) covers 10 Gyr of simulated time span and ends up with 1.055×10^6 stellar particles and 4.881×10^5 gas particles. Both works use the feedback implementation of Springel and Hernquist (2003). The common characteristic was also that each gas particle was allowed to produce up to two stellar particles with each one of them being half of the mass of the progenitor gas particle. The stellar particles have the constant mass of $3.1612 \times 10^4 \, M_\odot$ and $5 \times 10^4 \, M_\odot$ in Forgan et al. (2016) and Vukotić et al. (2016), respectively. The star particles in the CLUES simulation have the effective radius of 150 pc while simulation from Vukotić et al. (2016) has the softening length of 0.8 kpc (independent of stellar density).

4.2 HABITABILITY CALCULATIONS

Here we describe the habitability models from Forgan et al. (2016) termed Model 1 and Vukotić et al. (2016) referred to as Model 2.

4.2.1 Model 1

This model calculates the number of habitable planets in favorable habitable conditions as a product of the number of habitable planets hosted by a stellar particle (N_{hab}) and the probability of those planets to be in habitable conditions ($P_{survive}$):

$$N_{survive} = N_{hab} \, P_{survive}. \tag{1}$$

The probability $P_{survive}$ is equal to one when the local supernova rate (SNR) is zero and vanishes if SNR exceeds twice the Solar neighborhood supernovae rate $SNR_\odot$:

$$P_{survive} = 1.0 - \frac{SNR}{2 SNR_\odot}. \tag{2}$$

For

$$N_{hab} = N_* \, P_{planet,t} \, (1 - P_{planet,g}), \tag{3}$$

with N_* being the number of stars, it is assumed that the probability of forming terrestrial-like habitable planets is dependent on a metallicity (Z) threshold:

$$P_{planet,t} = \begin{cases} 0.4, & \dfrac{Z}{Z_\odot} > 0.1 \\[2ex] 0.0, & \dfrac{Z}{Z_\odot} \leq 0.1 \end{cases} \tag{4}$$

and that the probability of hosting close-in giant planets is also metallicity dependent:

$$P_{planet,g} = \begin{cases} 0.03 \times 10^{\log \frac{Z}{Z_\odot}}, & \dfrac{Z}{Z_\odot} > 1 \\[2ex] 0.03, & \dfrac{Z}{Z_\odot} \leq 1 \end{cases} \tag{5}$$

N_* is calculated from the mass of a stellar particle integrating the Salpeter initial mass function $M_*^{-2.35}$ in the $0.1 < M_* < 100$ M$_\odot$ mass range (M_* represents star mass). The numbers of stars per stellar particle, in the mass ranges relevant for calculation of supernova rates, were also obtained from the Salpeter law. Supernovae explosions occurring at one stellar particle are assumed not to influence other stellar particles since the stellar particles represent parts of the "stellar fluid" smoothed to an effective radius of $\approx$500 pc while supernovae are taken to be hazardous at distances of a few pc (see Gehrels et al., 2003).

4.2.2 Model 2

The relevant quantities are averaged from the properties of gas and stellar particles in a 2D grid aligned with the plane of the Galactic disk defined in respect to the Galactic angular momentum value. Averaging was also applied in the time domain (as a workaround to the previously described finite resolution and sampling problem). The fraction of sampling times in habitable-friendly grid cells is calculated for each stellar particle. The grid cell is considered to be in favorable (or habitable) conditions if the total SFR from gas particles (with supernovae rate being directly proportional to SFR) in the cell and the density from stellar particles in the cell are both below Solar neighborhood values of 3.0×10^{-3} M$_\odot$ kpc^{-2} yr^{-1} and 6.1×10^7 M$_\odot$ kpc^{-2}, respectively. The fraction of habitable time for stellar particles is calculated as:

$$f_{ht} = \frac{t_h}{a}, \tag{6}$$

where t_h is the number of times the particle was sampled to be in a habitable favorable grid cell and a is the total number of samplings.

The stellar particle is considered to be a candidate habitable particle (CHP) if it satisfies the following conditions:

- according to particle's [Fe/H] metallicity value the particle is randomly selected with the Gaussian probability (mean at $= 0.075$ and standard deviation of $\sigma = 0.1125$), and
- is older than 1 Gyr.

For a CHP to be considered as a habitable particle (HP) in addition to above conditions it needs to satisfy $f_{ht} > 0.5$.

4.3 COMPARISON OF MODELS

The two presented models are complementary to each other in the sense that Model 1 is geared toward precision calculations, while Model 2 emphasizes the averaging approach in calculating the degree of habitability. Since Model 2 results from a cosmological simulation, it is not trivial to establish a dominant plane for each of the considered galaxies in order to average the relevant properties. This is not like the case of an isolated galaxy where angular momentum is preserved and it is a natural choice for defining the dominant plane (aligned with the plane of the galactic disk). A possible solution is a 3D analysis, but this method is likely to be computationally intensive and may require parallel computing.

Model 2, with softening length of 0.8 kpc and a somewhat larger smoothing length, does not provide much gain in performing 3D analysis since the disk thickness is of the order of 1 kpc, even for the Galactic bulge which is up to 3 kpc in size. Smaller effective length for stellar particles in Model 1 (500 pc) is also a better option for high-precision calculation than Model 2.

Based on both models, the following factors are likely to have significant effect on the habitability calculations:

  (i) The metallicity thresholds.
 (ii) Stability of habitability conditions over longer time.
(iii) Mutual particle interaction in relation to the habitability relevant properties.

We discuss each of these factors in more detail below.

  (i) Common to both models is the stochastic "goldilocks" approach in implementing metallicity thresholds. In Model 1, the lower metallicity bound (Eq. 4) is more stringent than the upper (Eq. 5) bound. This should be in agreement with epistemological principles since the abundance threshold is likely to be more precisely defined than the threshold for a more complex process of giant planet migration. In contrast, Model 2 uses the symmetrical Gaussian function. Reliable empirical confirmations are not available at present since the sample of Earth-like exoplanets is rather small. On the contrary, the overall set of known exoplanets is dominated by giant planets that are close to their host stars (because they are easy to detect). The distribution of metallicity of the exoplanets hosts is steeper at the upper end than at the lower end. Although the implementations of models are not exactly the same, there is no apparent reason to expect that the obtained results will show a significant difference in the effect of metallicity thresholds.

 (ii) Model 1 does not explicitly account for the longevity of habitable conditions, and thus no averaging in time is done in this respect, in contrast to Model 2. The continuity of habitable conditions is implicit in the adopted stochastic approach that yields the number of habitable planets per stellar particle. A stellar particle with high supernova rate has on average fewer habitable planets that the stellar particle with lower supernova rate. On the conceptual level, this is similar to explicit averaging in Model 2 where it is required that life habitats are sampled more frequently (by 50%) in habitable environments. Occasional disturbances of lower magnitude might not have a severe impact in systems with high average habitability, and hence, the sampling problem is alleviated. As with the Earth climate feedback (carbon cycle), where the changes in parameters such as insolation can be amortized to some degree and enable the terrestrial biosphere to survive, similar effects might happen outside of the stellar system. For example, the intense radiation from a nearby supernova, or an impact of the asteroid from a disturbed Oort cloud, might just boost biological evolution by changing its direction, as happened with dinosaurs and mammals at Cretaceous-Paleogene great extinction event. Ćirković (2007) argued that conditional (Bayesian) probability analysis should be used to estimate the rates of catastrophic events. Standard (i.e., frequentist) probability inference has a tendency to underestimate the actual danger because of the observation selection effect, which favors chains of events ultimately supporting the persistence of observers who make the inference. This implies that simple averaging of the environmental galactic parameters might overestimate the longevity of habitable conditions. At present, this is the only feasible approach since the application of conditional probability analysis requires better understanding of galactic habitability processes. As with the item (i), the underlying similarity in implementation of models 1 and 2 should not yield conceptually different results with respect to averaging habitability conditions over time.

(iii) Except for gravitational interaction, Model 1 considers stellar particles as isolated entities that are not influenced by their environment at later times. The habitability of stellar particles is determined in a stochastic manner at their birth to be distributed throughout the galaxy by their motion. The authors justify their approach by poor mass resolution that gives an effective radius of stellar particles much greater than the typical supernova hazardous distance and also leaves much to be speculated about the close encounters of individual stars. Even with a lower mass resolution, Model 2 implements feedback from the environment in an explicitly averaged way. The averaged SFR is calculated from parameters of gas particles and stellar density is calculated from stellar particles in each individual grid cell, which together determine the habitability of the stellar particles found in the cell. These differences are likely to affect the predicted level of habitability but the results are not qualitatively different and allow cross-model comparison. The results of Model 1 may represent better the overall motion of potentially habitable stellar systems, whereas Model 2 is likely to provide more realistic understanding of the overall habitability trends.

In this respect, Model 1 gives a better idea of how interactions between galaxies can influence their habitability, whereas Model 2 is more helpful in grasping the basic mechanics of galactic habitability processes in isolation. Both models do not account for long-distance hazards such as influence of gamma-ray bursts over galactic-scale distances, possible radiation bursts from galactic centers, or correlated effects of the galactic-arm crossings. For better quantification of galactic habitability, future models should consider a synthesis of these approaches.

4.4 RESULTS

4.4.1 Model 1

Figures 3, 4, 6, and 7 in Forgan et al. (2016) show the number of biologically survived planets as the outcome of the implemented stochastic model. For both examined major galaxies, M33 and Milky Way (MW) similar patterns are predicted. A significant number of N_{survive} emerges after $\sim$5 Gyr of the simulation time ($\sim 10^7$ for M33 and $\sim 10^8$ for MW). In later epochs the numbers are even higher with $\sim 10^9$ for both galaxies. At all times the maximum average number of N_{survive} is at the center of the galaxies and it decreases with increasing distance from the galactic center. As the simulation time progresses from $\sim$5 to 13 Gyr, N_{survive} increases on average for all galactocentric distances and the plot becomes smoother. The increase of smoothness is evident on smaller scales as well as on the larger scales with peaks in N_{survive}. These peaks can be contributed to smaller satellite galaxies and are less pronounced than the signal from the central spiral galaxy. The spatial distribution of N_{survive} is close to spherical at the galactic bulge but becomes flatter as the galactocentric distance increase, similar as the overall distribution of visible matter in the galactic disk. High local values of N_{survive} appear throughout the whole galaxy.

As a measure of the GHZ extent, the overall spread of N_{survive}, is given as the interval between the first ($Q1$) and the third ($Q3$) quartile range at 4.505, 9.51, and 13.65 Gyr. In the case of M33, the ($Q1$, $Q3$) intervals are (1.6, 13.4), (2.78, 9.95), and (2.7, 10.05), respectively at stated times. The interval position and spread does not seem to change significantly indicating a rather well established and bounded GHZ. In the group of galaxies selected to resemble MW and its satellites the GHZ is much less constrained. At $\sim$5 Gyr it has a dual core like structure. High SFR at all times causes this GHZ to spread much more outwards before settling at the values similar as the ones of M33 at 4.505 Gyr. The respective Q pairs are (16.7, 18.1), (2.8, 66.1), and (1.9, 13.4). Overall, the distribution of N_{survive} looks more dispersed for MW than in the case of M33.

4.4.2 Model 2

Similar to Model 1, the simulation of an isolated MW like galaxy from Vukotić et al. (2016) shows a significant increase in habitability as late as $\sim$5 Gyr. The habitability signal also increases at later times. The analysis was performed up to 20 kpc galactocentric distance since even at the late stages there are very few stellar particles residing at galactic outskirts. The GHZ appears as a well defined entity at times >5 Gyr. It is located between $\sim$10 and $\sim$15 kpc galactocentric distance with the peak of the habitability distribution gradually moving closer to 15 kpc as time progresses.

The number of habitable particles grows with time, and by the end of the simulation the tails of related distributions reach below 5 kpc and up to 20 kpc. At the 20 kpc end the distribution has larger values and a slightly shallower slope than at the 5 kpc end. Also, a small peak appears at later times at $\sim$5 kpc. The presented results are the averages of the five runs of habitability analysis. Despite the additional averaging performed over galactocentric distance, the distributions do not appear smooth on smaller length scales. The distributions of HPs and distributions of the number of grid cells that host HPs appear similar in shape. This implies that HPs are evenly distributed in space.

4.5 HABITABILITY BEFORE THE EARTH WAS FORMED

In the times before the Earth was formed, the sufficient amount of metallicity for habitable systems formation was built up only in the areas close to the galactic centers. The SFR was higher and available stellar mass was smaller than at present. In the areas close to the galactic centers, this resulted in a smaller number (by an order of magnitude) of habitable systems than today. During the lifetime of the Sun, the increase in metallicity and radial migrations populated habitable systems well into the outskirts of today's galactic disks. A "wave" of migration and habitable systems formation made its way outwards sweeping across Solar galactocentric radius at $\sim$10 kpc. However, the interaction of the Milky Way with satellite galaxies makes this simplified picture much more complicated (as evident from the results of Model 1). This interaction produced large peaks in the overall habitability distribution of the Local group. This implies that, when compared to present times, earlier epochs had a greater fraction of habitable systems that had originated in the satellite galaxies rather than in the Milky Way itself. What are the chances that a typical Solar-like system in our Galaxy had such an origin remains to be investigated. As the production of habitable planets gained momentum in a much more massive Milky Way, the contribution from smaller satellites was much less evident and primarily could be observed in the regions outside the Milky Way.

4.6 DISCUSSION

According to the dark matter paradigm, a visible galaxy is a part of much larger halo comprised of dark matter. These haloes, although invisible, are very massive and steer the dynamics of the central galaxy and its satellites. Unlike Model 2, Model 1 is designed for studies at large spatial scales of up to 100 kpc that are comparable in size to the dark matter haloes. This is convenient for simulating the dynamical phenomena and corresponding habitability features at larger scales. However, it is less convenient for understanding the basic habitability features on galactic scale, since the effects of dark matter haloes might distort the formation of relevant habitability features that are common for galaxies of a certain type. In this respect, the N_{survive} signal from satellite galaxies might mask the signal produced by the central galaxy. A possible solution to this problem is to track the relevant particles in the simulations by their identification numbers. With particle tracking it is possible to separate the particles originated from the central galaxy from the particles originated in satellite galaxies. This makes it possible to evaluate the contribution of a particular galaxy to the overall signal at N_{survive} peaks. In addition, this approach provides insights on how the habitability features are affected by the interactions of galaxies. In Forgan et al. (2016) it is noted that, "[S]treams of stars between satellites and galaxies produce relatively large numbers of habitable planets." The appearance of habitable signal peaks from Model 1, dispersed around the galaxy, might be attributed to the lack of averaging of the stochastic calculations (although a significant averaging in Model 2 does not give smooth results). Other possible causes of such signal appearance might stem from the dynamical effects of galactic interactions that produce dissipating streams of matter in addition to dynamical diffusion in galaxies alone. In each case, it is expected that the actual habitability signal should be smoother, since the particles in simulations can exchange between the individual stellar systems.

To fully understand the effect of galactic collisions on habitability, the studies of galactic interactions need to be coupled with the studies of isolated galaxies. Such an approach can further advance the field toward understanding of the habitability features in large groups of galaxies (see Dayal et al., 2015 and Section 5.1) and better comprehension of the phenomenon of habitability in the cosmological context.

As discussed in Section 4.3 (iii), the difference in the implementation of the density effect on stellar systems produced an apparent discrepancy between the two models. Forgan et al. (2016) argued that one of the general trends in their results was that the central regions were more habitable (due to high number of stellar particles) but that the outer regions had a greater habitability potential (since the survival probabilities for individual stellar particles were higher). The later effect is apparent in the results of Model 2. This example shows the importance of dynamical effects on the density distributions of matter. Apart from the dense central region, the dynamics of particles might also be important for accurate habitability assessment of the middle and outer parts of a galactic disk.

Both models yield no sharp boundaries of the GHZ and authors argued that the outer parts of the disk were likely to have a larger habitability potential. In a strict sense, their results on assessing habitability differ from ours but can be understood as compatible and complementary. Both studies show no distinct inner boundary, and the outer boundary is likely to extend to the edge of the galaxy. Emerging habitability peaks in Model 1, positioned well outside the central galaxy, are likely to be a consequence of the interaction with satellite galaxies and structures emerging from galactic collisions. Based on the N-body models presented here, no sharp GHZ boundaries should be expected, in contrast to the original concept of Gonzalez et al. (2001), supported by the first-generation of semiquantitative studies. Instead, a comprehensive understanding of habitability factors (dominant in a single galaxy, as well as in galactic interactions) should give the most realistic quantification of this central astrobiological variable. Apart from the quantitative advancements, the described models provide significant gains in the understanding of habitability on galactic scales as compared to the simplified annulus models that opened this area of research more than a decade ago.

5 COMPARISON WITH OTHER STUDIES

Initial studies on Galactic habitability were mainly focused on predicting the inner and outer boundaries of the GHZ annulus. The estimated boundaries highly depended on the adopted habitability model. Nevertheless, all studies found the inner boundary to be inside the Solar circle while the outer boundary was found to be outside of this circle, i.e., toward the periphery of the Galactic disk.

The pioneering work of Gonzalez et al. (2001) and later work of Lineweaver et al. (2004) considered habitability mainly in the context of Galactic chemical evolution. Although Lineweaver et al. additionally accounted for the danger of supernovae radiation, both studies placed the highest probability of having the habitable planets in the GHZ annulus near the Solar circle. Lineweaver et al. (2004) conservatively placed the GHZ at 7–9 kpc from the Galactic center, whereas the calculations by Peña-Cabrera and Durand-Manterola (2004), which were based only on the metallicity requirements, yielded a much wider GHZ at 4–17.5 kpc from the Galactic center. In another model, Ćirković (2005) defined the inner boundary based on the criterion of dynamical stability and the outer boundary based on the build up of necessary heavy elements. This predicted the GHZ of similar width as the model from Peña-Cabrera and Durand-Manterola (2004) but shifted slightly closer to the Galactic center. Vukotić et al. (2016) developed a stochastic habitability model, where the metallicity distribution was derived from the observed samples of exoplanet hosts. The center of the derived distribution is at the metallicity value that is similar to the metallicity of the Sun. This model predicted the highest probability for habitable planets to be outside of the Solar circle at galactocentric distances larger than 10 kpc. This result is not in agreement with earlier findings discussed above. Vukotić et al. (2016) adopted the habitability threshold for SFR and stellar density at the values characteristic for the Solar circle neighborhood, which moved the peaks of habitability distribution outwards. This implies that, in a Milky Way (MW)-like galaxy, the SFR and stellar density might have a higher leverage on habitability than the abundance of heavy elements. Stellar density appeared particularly important, since in the simulation from Vukotić et al. (2016) the initial gas supply is not replenished in an isolated Galaxy and the SFRs decreased significantly after 3 Gyr, which reduced the habitability hazards.

In the work of Prantzos (2008), the GHZ concept was criticized on the grounds of incomplete understanding of the threats from energetic explosions and metallicity concerns. The author argued that the whole Galactic disk might be suitable for hosting planets with life. This argument is certainly appealing because it is consistent with the broad habitability distributions generated by the N-body models presented here. Also these models are in agreement with the studies from Roškar et al. (2011) since the radial population mixing may redistribute the heavy elements more evenly in the Galactic disk, especially at later times. All these arguments are in favor of the assumption that the habitability is primarily determined by the dynamical effects. In contrast, the metallicity based studies lose their weight and the emphasis is moved toward the examination of dynamics-related phenomena, such as the stellar density, dynamic stability on galactic scales, and interaction with other galaxies on even larger scales.

The latter studies are more diverse and sophisticated. The study of Gowanlock et al. (2011) does not include dynamics, but combines precise modeling of stellar distributions, supernovae influence, metallicity, the effects of sterilization distances, and tidal locking, which is the loss of habitability potential in planets that slow down their rotation and eventually face their host star with only one side. Similar to the results of Model 1, the model of Gowanlock et al. predicts the greatest habitability potential near the bulge of the galaxy, decreasing at larger galactocentric distances. The metallicity in the inner part of the Galaxy causes the increased abundance of planets which, on long time scales overcomes the negative effects of supernovae radiation. The authors note that the metallicity gradient should be less pronounced if radial mixing is taken into account but argue that this effect has a weaker effect on habitability.

The study of Spitoni et al. (2014), which also does not include dynamics, but models radial mixing with radial gas inflow, predicted that gas inflow and resulting metallicity increase may boost the habitability by 38% for MW and 10% for the Andromeda galaxy. They found the maximum of habitability distribution at 8 kpc for MW and 16 kpc galactocentric distance for Andromeda. The latest of the nondynamical approaches is a sophisticated model by Legassick (2015). The results of this model are similar to Model 2. The Sun is positioned on the inside of the galactocentric distances with the highest habitability, and older habitable planets are located mostly farther outside in the Galactic disk. In agreement with Model 2, this model predicts that the oldest (~6 Gyr) planets with the highest f_{ht} are located at 17 kpc galactocentric distance (see Vukotić et al., 2016 the lowest panel in Figure 9). The model of Legassick deals exclusively with G-type stars (like our Sun). It is very interesting that the results of this study do not consider the Sun as a typical habitable G-type star but rather an outlier, an assumption that may help to resolve the Contact paradox. Even more appealing is that the study finds a typical habitable planet to be 3.3 Gyr older than the Sun, similar to the findings of the later study by Zackrisson et al. (2016) where a terrestrial planet orbiting an FGK-type host star is up to 3.5 Gyr older than Earth. Future dynamical models with a higher mass resolution capable of modeling individual stars should be able to address this issue in more detail.

The model of Robles et al. (2008), which is based on the χ^2 analysis with 11 habitability-relevant parameters, predicted that the Sun is a typical star in the MW. Apart from the Sun's high mass they found another most atypical property—the Solar galactic orbit, which is less eccentric than the orbits of 93% of the stars. In addition, there are the studies that relate the Earth's geological mass extinctions record with the Solar galactic orbit (Feng and Bailer-Jones, 2013; Clube and Napier, 1982; Filipović et al., 2013). These studies argue that the spiral arms and galactic mid-plane crossings might increase the habitability influence from supernovae and stellar collisions. In addition to the average motion on galactic orbits, the effects of stellar velocity dispersion were investigated by Jiménez-Torres et al. (2013). Using the dynamical simulations on the scales of individual stellar systems they implied that recent perturbations of the Solar system by nearby passages were highly unlikely but cannot be ruled out.

Another dynamics factor, which is speculative but possibly of equal importance for habitability prediction and origin of life on Earth, is the long ago argued panspermia hypothesis (for some studies see, Lin and Loeb, 2015; McNichol and Gordon, 2012; Wallis and Wickramasinghe, 2004; Napier, 2004; Wickramasinghe et al., 2003; Melosh, 1988). With the work of Afanasiev et al. (2007) reporting the possible detection of an extragalactic meteorite, this leaves an impact even on the scales that outsize individual galaxies.

5.1 HABITABILITY OF OTHER GALAXIES IN THE DYNAMICAL PERSPECTIVE

The most distinctive feature in the appearance of galaxies is their shape. Most Galaxies are elliptical or spiral. The more numerous elliptical galaxies are thought to have resulted from numerous galactic collisions. A likely outcome of the collision of two spiral galaxies of similar size is an elliptical galaxy. In general, elliptical galaxies grow to larger sizes than their spiral counterparts and have lower gas content. With the gas-starved blunted SFR the ellipticals are primarily composed of older, less massive stars. More massive, short-lived stellar giants are predominantly found in the vicinity of the regions with intense SFR, usually in the disks of spiral galaxies. The bulges of spiral galaxies are very similar in their properties to elliptical galaxies. The stellar orbits within the ellipticals are far more chaotic as compared to the predominantly circular orbits in the disk planes of the spirals. Stars in the elliptical galaxies are observed to have mainly radial motion in respect to the observers line of sight. This means that they are in high eccentricity orbits.

Using the metallicity distribution as the only criterion of habitability, Suthar and McKay (2012) predicted that the majority of elliptical galaxies should support habitable zones but their dynamics is quite different from that of spiral galaxies. It is yet to be examined how the dynamics of elliptical galaxies may influence the overall habitability distribution derived solely from the metallicity criteria.

Carigi et al. (2013) did not include the radial flows of gas and stars, but only used the metallicity based habitable criterion corrected for supernova influence. They found that the GHZ for the spiral M31 (Andromeda) galaxy extends from 3 to 14 kpc of galactocentric distance for planets that are 3–9 Gyr old. The highest number of stars that may host habitable planets was found in 12–14 kpc galactocentric distance range with average stellar age of 7 Gyr. From the SDSS data findings, both spiral and elliptical galaxies make a tight surface in a 3D virtual space with coordinates of stellar mass, metal mass and SFR, dubbed Fundamental Metallicity Relation (Mannucci et al., 2010). In a recent work of Dayal et al. (2015) the authors utilize the Fundamental Metallicity Relation to argue, that shapeless elliptical galaxies, twice the size of the MW, have the highest habitability potential. They explain this by high stellar mass and low SFRs in those massive ellipticals, assuming no connection between the habitable planet formation rates and gas metallicity. They also add that the low mass spirals ($<10^9$ $M_\odot$) of any SFR, are far less likely to have planets with life since the small number of planets yields low habitability potential even for the small values of SFR. However the study did not provide any estimate of probability density for habitability, e.g., per Solar mass of total stellar galactic component. Such an estimate should be more conclusive for understanding habitability relevant processes in individual galaxies of any type, even when the density of habitable planets within individual galaxies is not quantified. It follows from their Figure 2 (lower panel), that the density of habitability has a rather uniform pattern over the plotted galactic mass range. This pattern is likely the result of an oversimplified model that did not include the dynamical components. Different dynamical properties of elliptical galaxies compared to spirals should give different habitability distributions.

Gobat and Hong (2016) quantified habitability as the number of habitable systems per total number of stellar systems. In addition to hazardous supernovae distance of <8 kpc, estimated in Gehrels et al. (2003) and used in Gowanlock et al. (2011) and Forgan et al. (2016), they also used a much restrictive case of <0.5 kpc. Their results imply a greater dependence of habitability on supernovae lethality distance as the mass of the galaxy increases.

6 CONCLUSIONS AND FUTURE PROSPECTS

In this chapter we have reviewed the models of habitability at the level of galaxies and highlighted the importance of dynamical effects. Some of the discrepancies in the presented results could not be explained at the moment. Future simulation projects (with increased computing power) should put emphasis on habitability models that are more detailed and stringent in the dynamical sense. A viable starting point would be to trace the individual particles and characterize the distribution of parameters of their galactic orbits.

Due to the complexity of the task and large size of the parameter space, most of the Milky Way habitability studies lack dynamical properties. The importance of dynamical effects is evident in the studies of galactic evolution. Hence, the dynamical effects are of great importance for models of galactic habitability. One of the most important effects is the mixing of matter from inner and outer parts of the galactic disks which is crucial when estimating the galactic distribution of habitable planets. On larger scales, the collisions of galaxies and tidal effects within the galactic groups and clusters influence the evolution of galactic habitability and other galactic properties.

The studies of galactic interaction should be coupled with the studies of isolated galaxies in order to better understand the influence of galactic interactions on habitability. This can be especially relevant for the collisions of galaxies that are comparable in size.

A primary concern of pioneering GHZ studies was to determine the GHZ boundaries. Unlike the rather strict boundaries of the stellar habitable zone, most of the GHZ studies (especially the recent ones) were rather inconclusive in this respect. Before we are able to be more accurate in this form of Galactic geodesy the emphasis must be put on thorough understanding of habitability relevant properties on galactic scales.

The distribution of heavy elements might be a good proxy for habitability relevant galactic properties, especially in the light of the new SDSS findings concerning the Fundamental Metallicity Plain of local galaxies. The stellar habitable zone studies advanced from simple energy conservation derived boundaries to the climate models of increasing complexity. In a similar manner, the time has come for more complex approaches in modeling habitability on galactic scales. The models including complex dynamics of stars within a galaxy (and dynamics of groups of galaxies on even larger scales), as a largely metallicity-independent feature, have a great potential to improve our understanding of the distribution of life.

Current dynamical models of galactic habitability have low resolution and further development is required to overcome this limitation. The publicly available data on high resolution cosmological simulation projects (such as Illustris, see Nelson et al., 2015) might give more insights even for the current habitability models by sheer increase in resolution.

So far, most of the studies concerning the galactic habitability were based on naive inductive approaches. These studies lead to unproductive discussions of the abundance of life in the Universe that were based on only one observed biosphere. As a result, they failed to explain the distribution of life in the Universe. A deductive approach based on cosmological and galactic evolution models might be more fruitful. With the steadily increasing number of newly discovered exoplanets, both inductive and deductive approaches should be used to evaluate habitability. The lucky accident is that the inductive line of research is already active for more than a century. From the vast volume of astronomical data we have learned a great deal about the Universe and evolution of space and matter. The process of evolution has culminated in having us here at present to observe it. Indeed, with our present astronomical knowledge it is very reasonable to construct a simulation study where star-forged particles of matter can eventually condense into conscious beings. The Big Question is, obviously, why we do not observe a similar phenomena on some other potentially habitable places? To address the chances of detection of extrasolar life, we still have not fully utilized the deductive line of thought founded in the simple but powerful continuity assumption and our present empirical knowledge.

It is a long road ahead to develop and refine tools presented in this chapter, but we are more serious than ever; instead of just asking "Are we alone?" we can add a prior probability to it.

ACKNOWLEDGMENTS

Milan M. Ćirković is acknowledged for his help in various stages of preparation of this manuscript and for many valuable discussions and tutoring that have shaped my comprehension of the Milky Way habitability over the past years. The financial support was provided by the Ministry of Education, Science and Technological Development of the Republic of Serbia through the project #176021 Visible and invisible matter in nearby galaxies: theory and observations. Many thanks to two referees and editors for their careful reading and comments that have improved the manuscript.

REFERENCES

Afanasiev, V.L., Kalenichenko, V.V., Karachentsev, I.D., 2007. Detection of an intergalactic meteor particle with the 6-m telescope. Astrophys. Bull. 62, 301–310. https://doi.org/10.1134/S1990341307040013.

Annis, J., 1999. An astrophysical explanation for the "great silence" J. Br. Interplanet. Soc. 52, 19–22.

Atri, D., 2011. Terrestrial effects of high energy cosmic rays. In: American Astronomical Society Meeting Abstracts #217, Bulletin of the American Astronomical Society, vol. 43, p. 319.05.

Beech, M., 2011. The past, present and future supernova threat to Earth's biosphere. Astrophys. Space Sci. 336, 287–302. https://doi.org/10.1007/s10509-011-0873-9.

Behroozi, P., Peeples, M.S., 2015. On the history and future of cosmic planet formation. Mon. Not. R. Astron. Soc. 454, 1811–1817. https://doi.org/10.1093/mnras/stv1817.

Brakenridge, G.R., 2011. Core-collapse supernovae and the Younger Dryas/terminal Rancholabrean extinctions. Icarus 215, 101–106. https://doi.org/10.1016/j.icarus.2011.06.043.

Brin, G.D., 1983. The great silence—the controversy concerning extraterrestrial intelligent life. Q. J. R. Astron. Soc. 24, 283–309.

Buchhave, L.A., et al., 2012. An abundance of small exoplanets around stars with a wide range of metallicities. Nature 486, 375–377. https://doi.org/10.1038/nature11121.

Carigi, L., García-Rojas, J., Meneses-Goytia, S., 2013. Chemical evolution and the galactic habitable zone of M31. Rev. Mex. Astron. Astrofis. 49, 253–273.

Chaisson, E.J., 2003. A unifying concept for astrobiology. Int. J. Astrobiol. 2, 91–101. https://doi.org/10.1017/S1473550403001484.

Ćirković, M.M., 2005. Boundaries of the habitable zone: unifying dynamics, astrophysics, and astrobiology. In: Knežević, Z., Milani, A. (Eds.), IAU Colloq. 197: Dynamics of Populations of Planetary Systems, pp. 113–118. https://doi.org/10.1017/S1743921304008579.

Ćirković, M.M., 2007. Evolutionary catastrophes and the Goldilocks problem. Int. J. Astrobiol. 6, 325–329. https://doi.org/10.1017/S1473550407003916.

Ćirković, M.M., 2012. The Astrobiological Landscape. Cambridge University Press, Cambridge, UK.

Ćirković, M.M., 2017. The Great Silence: Science and Philosophy of Fermi's Paradox. Oxford University Press, Oxford (in press).

Ćirković, M.M., Vukotić, B., 2008. Astrobiological phase transition: towards resolution of Fermi's paradox. Orig. Life Evol. Biosph. 38, 535–547. https://doi.org/10.1007/s11084-008-9149-y.

Ćirković, M.M., Vukotić, B., Dragićević, I., 2009. Galactic punctuated equilibrium: how to undermine carter's anthropic argument in astrobiology. Astrobiology 9, 491–501. https://doi.org/10.1089/ast.2007.0200.

Clube, S.V.M., Napier, W.M., 1982. Spiral arms, comets and terrestrial catastrophism. Q. J. R. Astron. Soc. 23, 45–66.

Dayal, P., Cockell, C., Rice, K., Mazumdar, A., 2015. The quest for cradles of life: using the fundamental metallicity relation to hunt for the most habitable type of galaxy. Astrophys. J. 810, L2. https://doi.org/10.1088/2041-8205/810/1/L2.

De Donder, E., Vanbeveren, D., 2002. The chemical evolution of the solar neighbourhood: the effect of binaries. New Astron. 7, 55–84. https://doi.org/10.1016/S1384-1076(01)00090-2.

Feng, F., Bailer-Jones, C.A.L., 2013. Assessing the influence of the solar orbit on terrestrial biodiversity. Astrophys. J. 768, 152. https://doi.org/10.1088/0004-637X/768/2/152.

Filipović, M.D., Horner, J., Crawford, E.J., Tothill, N.F.H., White, G.L., 2013. Mass extinction and the structure of the Milky Way. Serbian Astronomical Journal 187, 43–52. https://doi.org/10.2298/SAJ130819005F.

Forgan, D., 2016. Milankovitch cycles of terrestrial planets in binary star systems. Mon. Not. R. Astron. Soc. 463, 2768–2780. https://doi.org/10.1093/mnras/stw2098.

Forgan, D., Dayal, P., Cockell, C., Libeskind, N., 2016. Evaluating galactic habitability using high-resolution cosmological simulations of galaxy formation. Int. J. Astrobiol. 1–14. https://doi.org/10.1017/S1473550415000518.

Fry, I., 2000. The Emergence of Life on Earth: A Historical and Scientific Overview. Rutgers University Press, Brunswick, NJ.

Gehrels, N., Laird, C.M., Jackman, C.H., Cannizzo, J.K., Mattson, B.J., Chen, W., 2003. Ozone depletion from nearby supernovae. Astrophys. J. 585, 1169–1176. https://doi.org/10.1086/346127.

Gillon, M., et al., 2017. Seven temperate terrestrial planets around the nearby ultracool dwarf star TRAPPIST-1. Nature 542, 456–460. https://doi.org/10.1038/nature21360.

Gobat, R., Hong, S.E., 2016. Evolution of galaxy habitability. Astron. Astrophys. 592, A96. https://doi.org/10.1051/0004-6361/201628834.

Gonzalez, G., Brownlee, D., Ward, P., 2001. The galactic habitable zone: galactic chemical evolution. Icarus 152, 185–200. https://doi.org/10.1006/icar.2001.6617.

Gordon, R., 1980. Monte Carlo methods for cooperative Ising models. In: Karreman, G. (Ed.), Cooperative Phenomena in Biology. Pergamon Press, New York, pp. 189–241.

Gowanlock, M.G., Morrison, I.S., 2017. The habitability of our evolving galaxy. In: Gordon, R., Sharov, A.A., (Eds.), Habitability of the Universe Before Earth. In: Rampelotto, P.H., Seckbach, J., Gordon, R. (Eds.), Astrobiology: Exploring Life on Earth and Beyond. Elsevier B.V., Amsterdam, pp. 149–171. (Chapter 7).

Gowanlock, M.G., Patton, D.R., McConnell, S.M., 2011. A model of habitability within the Milky Way galaxy. Astrobiology 11, 855–873. https://doi.org/10.1089/ast.2010.0555.

Ilachinski, A., 2001. CELLULAR AUTOMATA: A Discrete Universe. World Scientific Publishing, Singapore.

Jiménez-Torres, J.J., Pichardo, B., Lake, G., Segura, A., 2013. Habitability in different Milky Way stellar environments: a stellar interaction dynamical approach. Astrobiology 13, 491–509. https://doi.org/10.1089/ast.2012.0842.

Kennicutt, R.C., Evans, N.J., 2012. Star formation in the Milky Way and nearby galaxies. Annu. Rev. Astron. Astrophys. 50, 531–608. https://doi.org/10.1146/annurev-astro-081811-125610.

Korschinek, G., 2016. Mass extinctions and supernova explosions. In: Murdin, P., Alsabeti, A. (Eds.), Handbook of Supernovae. Springer International Publishing, Cham, Switzerland.

Legassick, D., 2015. The Age Distribution of Potential Intelligent Life in the Milky Way. Master's thesis, University of Exeter, UK. ArXiv e-prints: 1509.02832.

Lin, H.W., Loeb, A., 2015. Statistical signatures of panspermia in exoplanet surveys. Astrophys. J. 810, L3. https://doi.org/10.1088/2041-8205/810/1/L3.

Lineweaver, C.H., 2001. An estimate of the age distribution of terrestrial planets in the universe: quantifying metallicity as a selection effect. Icarus 151, 307–313. https://doi.org/10.1006/icar.2001.6607.

Lineweaver, C.H., Fenner, Y., Gibson, B.K., 2004. The galactic habitable zone and the age distribution of complex life in the Milky Way. Science 303, 59–62. https://doi.org/10.1126/science.1092322.

Lissauer, J.J., 1987. Timescales for planetary accretion and the structure of the protoplanetary disk. Icarus 69, 249–265. https://doi.org/10.1016/0019-1035(87)90104-7.

Mannucci, F., Cresci, G., Maiolino, R., Marconi, A., Gnerucci, A., 2010. A fundamental relation between mass, star formation rate and metallicity in local and high-redshift galaxies. Mon. Not. R. Astron. Soc. 408, 2115–2127. https://doi.org/10.1111/j.1365-2966.2010.17291.x.

McNichol, J., Gordon, R., 2012. Are we from outer space? A critical review of the panspermia hypothesis. In: Seckbach, J. (Ed.), Genesis—In the Beginning: Precursors of Life, Chemical Models and Early Biological Evolution. Springer, Dordrecht, pp. 591–620.

Melosh, H.J., 1988. The rocky road to panspermia. Nature 332, 687–688. https://doi.org/10.1038/332687a0.

Milanković, M., 1941. Kanon der Erdbestrahlung und seine Anwendung auf das Eiszeitenproblem, Posebna izdanja, Srpska akademija nauka. Königlich Serbische Akademie. https://books.google.rs/books?id=oN0iQgAACAAJ.

Napier, W.M., 2004. A mechanism for interstellar panspermia. Mon. Not. R. Astron. Soc. 348, 46–51. https://doi.org/10.1111/j.1365-2966.2004.07287.x.

Nelson, D., et al., 2015. The Illustris simulation: public data release. Astron. Comput. 13, 12–37. https://doi.org/10.1016/j.ascom.2015.09.003.

Peña-Cabrera, G.V.Y., Durand-Manterola, H.J., 2004. Possible biotic distribution in our galaxy. Adv. Space Res. 33, 114–117. https://doi.org/10.1016/j.asr.2003.07.016.

Pollack, J.B., Hubickyj, O., Bodenheimer, P., Lissauer, J.J., Podolak, M., Greenzweig, Y., 1996. Formation of the giant planets by concurrent accretion of solids and gas. Icarus 124, 62–85. https://doi.org/10.1006/icar.1996.0190.

Prantzos, N., 2008. On the "Galactic Habitable Zone". Space Sci. Rev. 135, 313–322. https://doi.org/10.1007/s11214-007-9236-9.

Robles, J.A., Lineweaver, C.H., Grether, D., Flynn, C., Egan, C.A., Pracy, M.B., Holmberg, J., Gardner, E., 2008. A comprehensive comparison of the Sun to other stars: searching for self-selection effects. Astrophys. J. 684, 691–706. https://doi.org/10.1086/589985.

Roškar, R., Debattista, V.P., Loebman, S.R., Ivezić, Z., Quinn, T.R., 2011. Implications of radial migration for stellar population studies. In: Johns-Krull, C., Browning, M.K., West, A.A. (Eds.), 16th Cambridge Workshop on Cool Stars, Stellar Systems, and the Sun, Astronomical Society of the Pacic Conference Series, vol. 448, p. 371.

Schuler, S.C., et al., 2015. Detailed abundances of stars with small planets discovered by Kepler. I. The first sample. Astrophys. J. 815, 5. https://doi.org/10.1088/0004-637X/815/1/5.

Spitoni, E., Matteucci, F., Sozzetti, A., 2014. The galactic habitable zone of the Milky Way and M31 from chemical evolution models with gas radial flows. Mon. Not. R. Astron. Soc. 440, 2588–2598. https://doi.org/10.1093/mnras/stu484.

Springel, V., 2005a. The cosmological simulation code GADGET-2. Mon. Not. R. Astron. Soc. 364, 1105–1134. https://doi.org/10.1111/j.1365-2966.2005.09655.x.

Springel, V., 2005b. User Guide for GADGET-2. https://wwwmpa.mpa-garching.mpg.de/gadget/users-guide.pdf (Accessed 22 November 2016).

Springel, V., Hernquist, L., 2003. Cosmological smoothed particle hydrodynamics simulations: a hybrid multiphase model for star formation. Mon. Not. R. Astron. Soc. 339, 289–311. https://doi.org/10.1046/j.1365-8711.2003.06206.x.

Stevenson, D.J., 1982. Formation of the giant planets. Planet. Space Sci. 30, 755–764. https://doi.org/10.1016/0032-0633(82)90108-8.

Suthar, F., McKay, C.P., 2012. The galactic habitable zone in elliptical galaxies. Int. J. Astrobiol. 11, 157–161. https://doi.org/10.1017/S1473550412000055.

Thomas, B.C., 2009. Gamma-ray bursts as a threat to life on Earth. Int. J. Astrobiol. 8, 183–186. https://doi.org/10.1017/S1473550409004509.

Thomas, B.C., Melott, A.L., Field, B.D., Anthony-Twarog, B.J., 2008. Superluminous supernovae: no threat from η carinae. Astrobiology 8, 9–16. https://doi.org/10.1089/ast.2007.0181.

Vukotić, B., Ćirković, M.M., 2012. Astrobiological complexity with probabilistic cellular automata. Orig. Life Evol. Biosph. 42, 347–371. https://doi.org/10.1007/s11084-012-9293-2.

Vukotić, B., Steinhauser, D., Martinez-Aviles, G., Ćirković, M.M., Micic, M., Schindler, S., 2016. 'Grandeur in this view of life': N-body simulation models of the galactic habitable zone. Mon. Not. R. Astron. Soc. 459, 3512–3524. https://doi.org/10.1093/mnras/stw829.

Wallis, M.K., Wickramasinghe, N.C., 2004. Interstellar transfer of planetary microbiota. Mon. Not. R. Astron. Soc. 348, 52–61. https://doi.org/10.1111/j.1365-2966.2004.07355.x.

Wickramasinghe, N.C., Wainwright, M., Narlikar, J.V., Rajaratnam, P., Harris, M.J., Lloyd, D., 2003. Progress towards the vindication of panspermia. Astrophys. Space Sci. 283, 403–413. https://doi.org/10.1023/A:1021677122937.

Zackrisson, E., Calissendorff, P., González, J., Benson, A., Johansen, A., Janson, M., 2016. Terrestrial planets across space and time. Astrophys. J. 833, 214. https://doi.org/10.3847/1538-4357/833/2/214.

OCCUPIED AND EMPTY REGIONS OF THE SPACE OF EXTREMOPHILE PARAMETERS

Jeffrey M. Robinson*,†, Jill A. Mikucki‡
*Howard University, Washington, DC, United States
†National Institutes of Health, Bethesda, MD, United States
‡University of Tennessee, Knoxville, TN, United States

CHAPTER OUTLINE

Habitability of the Universe Before Earth, editors: Richard Gordon & Alexei Sharov, Volume 1 in the series:
Astrobiology: Exploring Life on Earth and Beyond, series editors: Pabulo Henrique Rampelotto,
Joseph Seckbach & Richard Gordon. ISSN 2468-6352. https://doi.org/10.1016/B978-0-12-811940-2.00009-5

1 INTRODUCTION

Microbial ecosystems have been detected in many extreme habitats on Earth, aided by advances in molecular tools and their application to microbial ecology (Pace, 1997). Discovery of novel microorganisms had remained limited by the inability to culture extremophilic microorganisms and poor accessibility to extreme environments for sample collection. Modern exploration using deep-sea submarine vehicles and deep subsurface drilling systems, for example, has provided scientists with unprecedented samples (e.g., Colwell et al., 1992; Van Dover et al., 1996; Tulaczyk et al., 2014) to analyze with new molecular tools. These advancements have facilitated the investigation of increasingly extreme microbial ecosystems (DeLong, 1998; Madigan & Mars, 1997). Some discoveries include chemosynthetic, thermophilic bacteria and archaea forming the basis of complex ecosystems at temperatures >100°C and mega-pascal (MPa) pressures at deep-sea hydrothermal vents (e.g., Van Dover, 2000). Others include microbial communities in acid-mine drainages at <pH 2 (Baker & Banfield, 2003), viable salt-tolerant archaeal cells surviving millennia within salt deposit fluid inclusions (McGenity et al., 2000; Vreeland et al., 2000), and cold-tolerant microbes occupying lakes below thick ice sheets or hypersaline brine channels within polar glaciers and sea ice (Boetius et al., 2015). In addition to the discovery of these extremophilic microbial communities, several complex animals including vertebrates can survive extreme conditions such as being frozen solid by using endogenously produced anti-freeze agents (Clarke et al., 2013). Some unicellular protists and multicellular meiofauna can even survive physiochemical parameters found only in laboratory conditions or in the interstellar space, such as extreme radiation, giga-pascal (GPa) high pressure, or zero gravity (Sharma et al., 2002; Moissl-Eichinger et al., 2016).

In 1974, MacElroy proposed the term *extremophiles* to describe the recent discoveries of microorganisms that existed beyond what he called "intermediate conditions," or parameters beyond that of mesophile limits of temperature, circumneutral pH, nominal salinity, and available, but nontoxic levels of oxygen (MacElroy, 1974; Horikoshi & Grant, 1998; Madigan & Mars, 1997; Rothschild & Mancinelli, 2001). Even the most extreme survivors, however, require the basic biochemical conditions that support metabolic biochemistry and thermodynamic disequilibria. As far as we know, these processes depend on liquid water as the universal solvent (Cleland & Chyba, 2007).

The discovery of extreme microbial ecosystems on Earth, coupled with the detection of liquid water on extraterrestrial bodies in our Solar System, has expanded the previously accepted limits of the habitable zone, which was originally determined based on the presence of liquid water at a planet's surface and its distance from the sun (Kasting et al., 1993). These discoveries supported the possibility that extraterrestrial life may be found within our Solar System (Vogel, 1999). Remote exploration of Mars and several ice moons of Jupiter (i.e., Europa, Ganymede, Callisto) and of Saturn (i.e., Titan, Enceladus) has provided strong evidence of the presence of either liquid water or other aqueous solutions. These bodies may be heated by radioactive decay within the interior mantle, gravitational friction, and/or seasonal changes. On Mars, seasonal surface brine seeps have been observed in regions where groundwater may exist (McEwen et al., 2011; Jakosky et al., 2007). Europa's geologically young surface reflects active subsurface dynamics of a putative liquid water ocean (Chyba & Phillips, 2007). Enceladus displays active cryovolcanism which is predicted to result from subsurface liquid water dynamics (Porco et al., 2006; Nimmo & Pappalardo, 2016), while Titan is inferred to have a deep subsurface ocean and possesses crater-like terrain associated with warmer surface temperatures consistent with cryovolcanic activity (Lopes et al., 2013; Lunine & Rizk, 2007). Less is known about the gas giants; however, they may harbor atmospheric layers with significant water vapor at high density, providing speculative possibilities for aerial life (Ingersoll et al., 2004; Libby, 1974).

Plausible scenarios have been proposed for extraterrestrial life that utilizes a solvent other than water, or alternative biochemistry (Bains, 2004; Benner et al., 2004). Proposed alternative solvents include methane, ammonia, hydrocyanic and hydrofluoric acids, hydrogen sulfide and sulfur dioxide, methanol, hydrazine, and others (Schulz-Makuch & Irwin, 2006). In fact, liquid methane lakes have been found on the surface of Titan (Stofan et al., 2007) and while surface temperatures on Titan are too cold for liquid water or Earth-like cellular membranes, hypothetical methane-based life could be supported (Schulz-Makuch & Grinspoon, 2005). It has been suggested that life forms in these nonpolar methane seas on Titan could form cellular membranes comprised of nitrogen head groups, which form azotomes rather than phospholipids-based liposomes (Stevenson et al., 2015b). Various hypotheses have been proposed for the origin of life on Earth, which are not the subject of this review; however, the deep conservation of biochemical pathways and phylogeny (Pace, 2001; Cleland & Chyba, 2007) supports the notion of a single origin event for Earth. Other hypotheses invoke multiple origins (Raup & Valentine, 1983; Forterre et al., 2004) and life may also have originated remotely, such as on Mars or elsewhere, then subsequently transferred to Earth via a meteor in a process called panspermia (reviewed in Kamminga, 1982; McNichol & Gordon, 2012). This possibility is relevant to our discussion below of extremophiles surviving interstellar space conditions.

Here we aim to review the parameter space of known Earth extremophiles and make comparisons to potential extraterrestrial habitats in a semiquantitative manner. Parameter space is derived from published observations of extremophile biology and data from direct observation of space probes. Our discussion of hypothetical biochemistries and origin of life questions are therefore limited. In Section 2, we review the known limits of temperature, pressure, salinity, water activity, and pH that may support life and present several multidimensional phase diagrams for visualization. In Section 3, we review putative extraterrestrial habitats of Mars, Europa, Titan, and Enceladus, in order to compare these parameter spaces with those found in analogous Earth environments supporting extremophilic life, and evaluate the extent of overlapping parameter space in these potential extraterrestrial habitats.

2 PARAMETER SPACE OF EXTREMOPHILIC ORGANISMS ON EARTH

Liquid water is a critical requirement for many biochemical processes. Temperatures outside the range for liquid water at surface pressure set a starting point for defining occupied physiochemical parameter space, with pressure and salinity exerting significant effects. Increased pressure raises the boiling point temperature (Fig. 1A; Chaplin, 2016), while increased salinity expands the thermal capacity of water, both lowering the melting point and increasing the boiling point (Fig. 1B, Brady, 1992; Driesner, 2007). At an average estimated seafloor depth of 3682 m (Charette & Smith, 2010), pressure is ~368 atm (1 atm/10 m depth), or ~37 MPa. Heated water does not become steam at these pressures, rather it becomes a supercritical fluid above ~22 MPa and at a temperature of ~374°C (Wagner & Kretzschmar, 2008; Fig. 1A). Microenvironments in otherwise frozen or solid substrates such as ice, halite, and sediments may contain pockets of liquid brine (Fig. 1B), and liquid water is also found in high porosity regions of deep subsurface strata (i.e., aquifers).

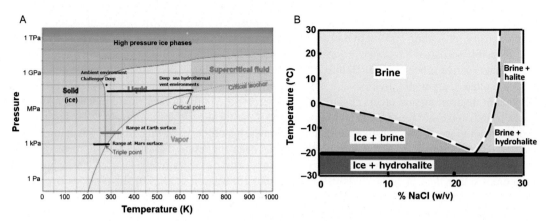

FIG. 1

Phase diagrams describe physiochemical parameter space. (A) Temperature-pressure ranges for Earth's surface, Mars' surface, and Earth's deep-sea are indicated on a two-dimensional phase space for temperature and pressure. The range of temperatures found in proximity to deep-sea hydrothermal vents is contrasted with the stable temperatures found at nongeothermal areas. At the Martian surface, liquid water can exist at the upper limit of the temperature range. (B) The Temperature-salinity phase diagram shows how increased salinity lowers the H_2O melting point; even above the saturation point ($\sim$24% NaCl), and down to ~ -20, fluid inclusions persist and can provide habitats for halophiles and halophilic psychrophiles, respectively.

(A) Modified from Chaplin (2016); (B) Modified from Brady, J., 1992. Does ice dissolve or does Halite melt?
J. Geol. Educ. 40(2), 116–118.

Observed natural temperatures on Earth's continental and oceanic crust span a large range. Fluid emerging from hydrothermal vents may exceed 400°C, but rapidly cools when mixed with the surrounding seawater of 1–4°C (Haymon et al., 1993, Fig. 1A). Volcanic magma can range in temperature from 650°C to over 1200°C depending on magma type (Larsen, 1929). At the cold end of the spectrum, −89.2°C was recorded at Vostok Station, Antarctica, and −93.2°C was measured over Vostok Station via NASA satellites (Turner et al. 2009).

2.1 HYPERTHERMOPHILES

To date, the highest temperature at which microbial reproduction has been observed ($\sim$122°C) is well below the critical point of water. At higher temperatures, even highly adapted thermophilic proteins and nucleic acids quickly degrade (Cowan, 2004). Organisms living in habitats with temperatures >60°C are considered hyperthermophilic, but even environments with temperatures reaching 80–100°C maintain relatively diverse microbial communities (Miroschnichenko & Bonch-Osmolovskaya, 2006). The maximum survival temperature predicted for hyperthermophiles is around 150°C based on the theoretically rapid denaturation of important biomolecules (i.e., DNA and proteins) (Madigan & Oren, 1999). Somewhere the range of 150–250°C might constitute a "hard" thermodynamic constraint on carbon-based life forms due to denaturation (Cowan, 2004).

Hot springs and geysers, such as geothermal features found in Yellowstone National Park (United States), provide hyperthermophile habitats at or near the surface, while the hottest known habitats are near hydrothermal vents at seafloor spreading centers. Microbial cultures have not been isolated directly from supercritical hydrothermal vent fluids (which can reach 400°C); however, abundant microbes were found immediately adjacent to vents. These microbes utilize minerals from the vent fluid for chemolithotrophic metabolisms and form the base of the food web which supports diverse marine ecosystems with crabs, clams, tube worms, and fish (Bonch-Osmolovskaya, 2012). Obtaining uncontaminated samples from hydrothermal fluid remains difficult and likely limits our ability to identify and isolate hyperthermophilic extremophiles from higher temperatures. Perhaps, novel "record-breakers" will be identified in the future by leveraging the ability to replicate high-temperature and high-pressure conditions in the laboratory.

The current champion of high-temperature survival is the hyperthermophilic archaeon *Pyrolobus fumarii*, originally isolated from a hydrothermal vent "black smoker" at a depth of 3650 m. Its cardinal growth rate in culture occurred at 106°C, with cell division occurring at 113°C under pressures of 25 MPa. A small percentage of these cultured cells even survived autoclaving, which in effect increases their survival temperature limit to 121°C (Blochl et al., 1997). More recently, *Geogemma barossii* "Strain 121" was collected from a black smoker along the Juan De Fuca Ridge hydrothermal vent field, which demonstrated growth at temperatures between 85°C and 121°C and continued to survive exposure up to 130°C (Kashefi & Lovley, 2003). The most extreme hyperthermophile currently known is *Methanopyrus kandleri* st. 116, a hydrothermal vent archaeon collected from the Gulf of California at 2450 m depth. This methanogen was originally reported to grow at temperatures of 85–116°C and slightly above atmospheric pressure, 0.4 MPa (Kurr et al., 1991). However, when cultivated at a pressure of 40 MPa, cells were capable of division at temperatures up to 122°C (Takai et al., 2008), a reminder that our understanding of extremophilic life may be hindered by our laboratory capabilities and our imagination.

2.2 PSYCHROPHILES

Psychrophiles are organisms capable of growth and reproduction in cold temperatures and are often xerophilic and halophilic as well. In polar regions, water exists mostly as ice with liquid water available intermittently, seasonally, or in insulated microenvironments. For example, hundreds of subglacial lakes have been detected under the Antarctic ice sheet (Wright & Siegert, 2012), which have been shown to host diverse chemosynthetic microbial communities (Christner et al., 2014; Mikucki et al., 2015). In permafrost regions, the majority of porewater is permanently frozen and therefore unavailable for life, except for the upper "active" layer that melts seasonally and in thin films around sediment grains. Also, below the permafrost layer, cold and salty groundwater can sustain liquid and support chemosynthetic microbial communities (i.e., Mikucki et al., 2015; Amato et al., 2010). The two-dimensional phase diagram of salt concentration and temperature shows that liquid habitat is available as fluid inclusions, even below the point of ice formation and salt saturation. (Fig. 1B). Psychrophiles are often adapted to life as xerophilic (organisms that survive in extremely dry environments) or halophilic extremophiles. These adaptions likely result from analogous conditions where availability of liquid water is also limited in frozen environments, or because high salinity is required to maintain liquid water at subzero temperatures. Phase equilibrium data indicates that H_2O ice above $\sim -21°C$ can contain fluid inclusions of brines at saturation point (Brady, 1992; Driesner, 2007; Fig. 1B) and microorganisms have been found within these briny pockets (Lowenstein et al., 2011).

In cryobiology, vitrification is the transformation of cells or tissues from the liquid phase into a "glass phase." Vitrification prevents cellular damage resulting from ice crystal formation allowing for cryopreservation. This process is facilitated by various cryoprotectant compounds, for example, dimethyl sulfoxide (DMSO) or glycerol are commonly used for long-term storage of bacterial strains, cell lines, and embryos in liquid nitrogen ($-196°C$) (Fahy & Wowk, 2014). Tardigrades (Hengerr et al., 2009) and some vertebrates can survive seasonal freezing due to endogenously produced cryoprotectants (Costanzo et al., 1995). Seeds and spores of plants and fungi and many types of resistant invertebrate eggs are viable after years and even decades of frozen storage, such as those stored at the Svalbard Global Seed Vault (Charles, 2006). At the very low temperatures of cryopreservation, active cellular processes such as cell division and metabolism cease. Clarke et al. (2013) predict the lower temperature limit for growth (cell division) of psychrophilic organisms to be approximately $-26°C$ for microbes and $-50°C$ for multicellular, thermoregulating organisms, which is based on the temperature which cells would vitrify under exposure to external temperature. Many physiological and behavioral adaptations allow vertebrates to survive seasonal freezing conditions; these include hibernation, shelter construction and burrowing, homeothermy, and insulating tissue layers such as fat, feathers, or fur (De Maayer et al., 2014).

Cell division and metabolism are halted in cryopreservation conditions, but various microbial psychrophiles actually carry on biological activity at subzero temperatures. A range of genomic adaptations and physiological modifications has been described (e.g., Andersson, 2011; Ayala del Rio et al., 2010; Bergholz et al., 2009). Cold-adapted enzymes are efficient at low temperatures and employ a more flexible structure to ensure biocatalysis at low temperatures (Siddiqui et al., 2013). Psychrophiles tend to alter the structure of their cell membranes to maintain fluidity and allow for the transport of molecules despite the onset of gelling from freezing (Feller & Gerday, 2003). Some bacteria can also produce extracellular ice-binding particles (Raymond et al., 2007) and other ice-active substances (Davies, 2014) that help maintain a liquid environment at low temperatures. Diatoms may do the same (Raymond & Knight, 2003).

The Gram-negative bacterium *Psychrobacter arcticus* str273-4 was isolated from 20,000 to 30,000 year-old Siberian permafrost and has been shown to perform selective metabolic processes including DNA synthesis at $-15°C$ (Youle et al., 2012; Amato et al., 2010). Multicellular eukaryotes such as the yeast *Rhodotolura glutinis* and the lichen *Umbilicaria* can grow at $-18°C$ and $-17°C$, respectively (De Maayer et al., 2014). Microbial metabolism at $-33°C$, including DNA repair, protein synthesis, and respiration, but not cell division, has been confirmed in laboratory studies of *Paenisporosarcina* sp. B5 and *Chryseobacterium* sp. V3519-10, isolates originally obtained from the base of an Antarctic glacier (Bakermans & Skidmore, 2011). These bacteria may be the current low temperature champions.

2.3 EXTREME HALOPHILES

Above $0°C$, the concentration of saturated NaCl-H_2O brine is ~5.5 molar (~26%); above this concentration, NaCl precipitates into halite (Fig. 1B). For reference, present-day seawater has 3.5% NaCl. On Earth, extremely saline habitats are found on the surface in artificial salt-evaporation ponds and closed-basin lakes such as Mono Lake in California or Utah's Great Salt Lake, as well as ancient seas, including the Dead Sea in Israel and Jordan. Subsurface salt deposits may occur in salt-dome geological formations and rock strata representing ancient, dried saltwater habitats. In cold marine regions, freezing seawater concentrates salt into hypersaline brine pockets within sea ice and sediments. Processes of evaporation and liquid-ice phase transition often create poly-extremophilic conditions. As described above, many halophiles share adaptations for cold, dry, and oligotrophic conditions (Stan-Lotter & Fendrihan, 2015). The high concentration of solutes in these environments may also create acidic or alkaline conditions resulting in adaptations for pH extremes (Banciu & Sorokin, 2013; Capece et al., 2013). High intracellular salt concentrations cause protein aggregation and molecular damage, which result from both desiccation and ionic imbalance. Cells maintain osmotic balance by concentrating non-ionic solutes, or actively pumping protons, preventing water loss and providing a neutral solvent environment. Additionally, modifications to protein structures have also been observed (Graziano & Merlino, 2014; Santos & da Costa, 2002; Plemenitas et al., 2014). Extreme halophiles include Eukaryotic fungal species and the unicellular green algae *Dunaliella salina*, which has been the subject of scientific research for over 100 years (Gunde-Cimerman et al., 2009; Oren, 2005). While there are extreme halophilic bacteria, the greatest diversity of halophiles is found among the Archaea (Kim et al., 2007; Oren, 2002) and fall primarily in the Order Halobacteriales (Grant & Larsen, 1989).

On geological timescales, bacterial and archaeal species can survive inside microscopic fluid inclusions in salt crystals. For example, microbes were isolated from salt deposits that were between 10,000 and 34,000 years old. These microbes were trapped along with the nonviable algae species *Dunaliella*. One hypothesis is that glycerol from these algal cells provide metabolic substrate for the bacteria and archaea (Schubert et al., 2010; Lowenstein et al., 2011). Several earlier reports described haloarchaea isolated from samples of a large Permo-Triassic era salt deposit that is ~240 million-years old (Fendrihan et al., 2006). While these microbes appear to represent an *in situ* community, the dynamics of their survival and ecology over geological time remain uncertain (McGenity et al., 2000; Jaakkola et al., 2016). It is nonetheless intriguing to consider viable microbial communities living in deep subsurface salt deposits over hundred-million-year geological time scales, with relevance to life surviving in extraterrestrial habitats.

2.4 TOLERANCE FOR LOW WATER ACTIVITY

Life processes require water as a solvent; therefore sufficient water, measured in an environment as water activity (a_w), is considered a critical parameter for life as we know it. No other liquid has been shown to function as a solvent for organic life on Earth. In general, most microbial life requires a water activity of 0.900 or higher. Diverse halophilic Archaea and Bacteria are capable of growth in saturated NaCl solutions (i.e., 5.2 M), which is 0.755 a_w at room temperature (Stevenson et al., 2015a). Fungi and molds have been shown to withstand lower water activities than prokaryotes. For example, the fungus *Xeromomyces bisporus* was capable of cell division and sporulation in a concentrated sugar solution with water activity (a_w) = 0.605 (Pitt & Christian, 1968; Williams & Hallsworth, 2009). Two strains of haloarchaea GN-2 and GN-5, originally isolated from a Mexican solar saltern (Javor, 1984) showed growth in a brine solution with water activity of 0.635. These authors then extrapolated their results to obtain a theoretical water activity minimum of 0.615–0.611, concluding that the lower limit for water activity that allows cell division in all three domains of life may be ~0.61 (Stevenson et al., 2014, 2015a). However, there are environmental conditions that might challenge this finding. Microbial communities within microliter-sized oil droplets were metabolically active, despite extremely low water content (only ~13.5%) (Meckenstock et al., 2014).

Don Juan Pond, a saturated $CaCl_2$ brine in Antarctica, is a possible example of an environment where the water activity is too low for life to persist ($a_w = 0.45$; Cameron et al., 1972). Despite numerous attempts to demonstrate life can grow in this extreme brine (i.e., Cameron et al., 1972; Siegel et al., 1979), there have been no definitive reports validating the presence of active, *in situ* life (Samarkin et al., 2010). Thus, Don Juan Pond provides an important research site for our understanding of extremophile parameter space in regard to low temperature and low water activity.

2.5 pH EXTREMOPHILES

Microbial communities are found across a wide range of the pH scale including the most acidic and alkaline environments known on Earth, both natural and artificial. Acidophiles are organisms that are found in acidic environments and grow optimally at pH < 6 (De Saro et al., 2013). Acid-mine drainages occur at abandoned metal mining sites or naturally at some geothermal sites. These habitats can be at pH ~0.08–2 and still harbor diverse microbial biofilm communities (Ram et al., 2005; Baker & Banfield, 2003). Sulfur-rich hydrothermal acid pools, such as those found in the Norris Geyser Basin area of Yellowstone National Park, United States, host diverse anaerobic, chemolithoautotrophic microbial communities at ~70°C and pH 2.5–3.0. Many acidophilic Eukaryotes also grow at similar pH ranges, for example the green algae *Dunaliella acidophila* photosynthesizes at pH values of ~1, while other eukaryotes including amoebas, ciliates, and fungi are found in habitats of pH 2–4 (Anguilera, 2013).

Alkaliphiles grow optimally at pH >9 (Horikoshi, 1999) and can grow at pH as high as 12–13 (Preiss et al., 2015). Natural alkaline environments include soda lakes, such as Lakes Natron, Bogoria, and Magadi in the Kenya rift valley or Mono Lake in California. These lakes are at near-saturation salinity and alkaline (with pH ~8.5–10.5), and yet harbor highly diverse microbial communities (Banciu & Sorokin, 2013). Hyper-alkaline freshwater environments (pH values as high as 13.2) exist where industrial iron-slag waste has been dumped into wetland areas. Diverse microbial communities, similar in composition to those found in naturally occurring alkaline environments, are found in these systems (Roadcap et al., 2006; Tiago et al., 2004). Alkaliphiles have also been found growing near alkaline hydrothermal vent fields and continental serpentinization sites (Preiss et al., 2015). Many extreme alkaliphiles are among the *Bacillus* lineage and include the well-studied *B. pseudofirmus* OF4, (Grant et al., 1990; Hicks & Krulwich, 1990). Recently, a novel Betaproteobacterial genus, with the proposed name *Serpentinomonas*, was isolated from a serpentinizing spring at The Cedars, California (Suzuki et al., 2014).

Acidophiles and alkaliphiles exhibit molecular, physiological, and biomechanical strategies for maintaining a neutral cytoplasmic pH as well as adaptations for dealing with high metal or salt concentrations (Krulwich et al., 2011). A common adaptation for acidophiles is active proton pumping to maintain neutral intracellular pH values (Baker-Austin & Dopson, 2007). These adaptations also translate into a significant energy cost associated with living in low-pH habitats (Messerli et al., 2005). The most extreme known acidophile is the archaea *Picrophilus oshimae*, also a poly-extremophile, which tolerates a pH of 0 and exhibits growth at temperatures up to 65°C (Futterer et al., 2004). In alkaliphiles, the pH gradient across the cytoplasmic membrane creates a more acidic environment inside rather than outside the cell, which can significantly reduce the proton motive force that is necessary for ATP production (Preiss et al., 2015). Still, nonfermentative alkaliphiles have been shown to produce ATP via oxidative phosphorylation despite this challenge. Research suggests these organisms have mechanisms for maintaining protons close to their ATP-synthase in order to overcome these thermodynamic challenges (Heberle, 2000).

2.6 MISSING LIFE IN POLY-EXTREMOPHILIC PARAMETER SPACES

Unoccupied regions of parameter space are assumed to be at temperatures above +150°C and below −50°C. Parameter models within multidimensional phase diagrams of temperature, pressure, and salinity can provide a framework for understanding these regions. Fig. 2A provides a graphical visualization of such "parameter space" and allows the relatively small volume of "occupied" parameter space to be visualized within the larger regions of unknown, or unavailable, parameter space.

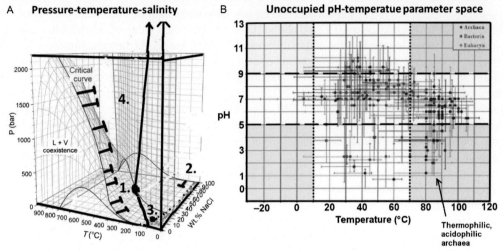

FIG. 2

Three-dimensional phase space identifies limitations in parameter space of poly-extremophiles. (A) This three-dimensional graph of temperature-pressure-salinity parameter space illustrates a specific volume inhabited by Earth extremophiles. The gray shades along the X([NaCl]) and Y(temperature) axes represent the two-dimensional space of known extremophiles. (1) Red blocks along the critical curve indicate that this is a likely physical boundary for complex biochemical molecules; (2) indicates 100% NaCl (halite). Extreme halophiles can be found living in fluid inclusions in saturated halite; however, these are not known to be hyperthermophiles. The red dot at (3) shows the approximate known parameters of a piezophilic hyperthermophile at ~120°C at deep-sea pressures. Tolerance for high temperature decreases as tolerance for high salinity increases, along the X-Y curve of the gray-shaded area. The area around (4) inhabits a large volume of parameter space, higher than 150°C, but somewhat distant from the critical curve, which is presumably uninhabited by Earth extremophiles. (B) pH-temperature two-dimensional parameter space reveals low poly-extremophile diversity under combinations of high/low temperatures and high/low pH, particularly in "hyper-extremophilic" ranges. Green indicates an empty area in high temperature, low pH. Yellow indicates how low temperature space is essentially empty of acidophiles or alkaliphiles. Red indicates a group of acidophilic, thermophilic Archaea, which do not inhabit the highest temperature ranges.

(A) Modified from Driesner, T., 2007. The system H_2O-NaCl I: correlation formulae for phase relations in temperature-pressure-composition space from 0 to 1000° C, 0 to 5000 bar, and 0 to 1 X_{NaCl}. Geochim. Cosmochim. Acta 71(20), 4880–4901 using the SoWat app, http://www.orefluids.ethz.ch/software/sowat.html; (B) Modified from Capece, M.C., Clark, E., Saleh, J.K., Halford, D., Heinl, N., Hoskins, S., Rothschild, L.J., 2013. Polyextremophiles and the constraints for terrestrial habitability. In: Seckbach, J., Oren, J., Stan-Lotter, H. (Eds.), Polyextremophiles: Life Under Multiple Forms of Stress. Springer, Dordrecht, pp. 3–59.

Parameter space occupation for combinations of giga-Pascal pressures and high temperatures (>150°C) (Fig. 2A-1) has not been studied. On the salinity axis, microbes may survive at the saturation point, but the temperature range of halophilic thermophiles is reduced as salinity approaches saturation (Fig. 2A-3). Even beyond the saturation point, microbes may survive within brine inclusions up to the 100% halite boundary (Fig. 2A-2), possibly for extended time periods as described above. Heat appears to be the most significant barrier for occupied parameter space; regions of high-temperature/high pressure (Fig. 2A-4) represent unknown territory. These conditions do not exist on Earth to the best of our knowledge, but may exist elsewhere.

The analysis by Capece et al. (2013) shows the range of growth-supporting conditions for species in a two-dimensional space of pH and temperature ranges. This study indicates that acidophiles and alkaliphiles have little diversity below 0°C and above 100°C. A significant group of thermophilic acidophilic Archaea is observed in the pH = 0–4 and temperature 60–100°C range, yet their alkaliphilic counterparts in the >pH 10 range are notably absent (Capece et al., 2013). Acidophiles (pH < 5) are lacking at temperatures <10°C (Fig. 2B). A three-dimensional representation of the occupancy of temperature-pH-salinity parameter space by Harrison et al. (2012) reflects this same pattern. Additionally, data from both Harrison et al. (2012) and Capece et al. (2013) show that extreme halophiles most commonly occupy a mesophilic to moderately thermophilic temperature range (<=60°C), and that pH extremophiles are also more abundant at milder temperature ranges. Even within the more moderate ranges of 0–100°C, thermophilic alkaliphiles and psychrophilic acidophiles are notably absent outside the pH range of 4–9 (Harrison et al., 2012). For example, Mesbah and Wiegel (2005) report on a group of halophilic thermophilic archaea; all of which are intermediate thermophiles (in the range of 40–60°C). Bowers et al. (2009) performed a three-dimensional analysis investigating the range of the anaerobic bacterium *Natranaerobius jonesii,* which perhaps occupies the most poly-extremophilic range for an alkaliphilic, halophilic thermophile, with optimal parameter ranges of 3.8–3.6 M (~%20) NaCl, pH >10, >50°C up to 66°C.

Physiological constraints arising from combinations of extreme parameter space may therefore constitute limiting factors. For example, at low temperatures, adaptations such as active proton pumping to maintain neutral cytoplasmic pH in highly acidic environments may not function due to thermodynamic constraints. High salinity may also adversely affect the ability of organisms to mitigate the effects of temperatures in the hyperthermophilic range (i.e., 80–120°C) (Chin et al., 2010). To date, only four taxa of poly-extremophiles that were extreme halophiles (>20% NaCl), alkaliphilic (>pH 8.5), and thermophilic (50–55°C cardinal growth temperatures) have been reported on including *Haloarcula quadrata, Haloferax elongans* and *Haloferax mediterranei*, and *Natronolimnoiaus aegyptiacus* (Bowers & Weigel, 2011).

Poly-extremophilic psychrophiles include the halophilic *Planococcus halocryophilus* Or1, a permafrost Actinobacterium collected from the Canadian High Arctic which is capable of division in an 18% NaCl medium at −15°C; respiration was detected as low as −25°C (Mykytczuk et al., 2012, 2013). Another species *Psychromonas ingrahamii* can grow at temperatures of −12°C and higher salinity 20% NaCl (Goordial et al., 2015; Kashefi & Lovley, 2003). The dearth of information on psychrophilic poly-extremophiles could be due to their slow growth rates, which make physiological characterization challenging (Baross et al., 2007). Observed patterns may not only reflect limits of genetic adaptation to physiochemical constraints, but may also be a result of missing habitat space, for example, there may be very few habitats with extremes of pH, temperature, and salinity in a single location. Still, the ability to isolate and study organisms experiencing multiple extremes remains technically difficult.

2.7 RADIATION- AND PRESSURE-RESISTANT EXTREMOPHILES: PARAMETER SPACES ANALOGOUS TO THE INTERSTELLAR MEDIUM

Some organisms exhibit resistance to high dosages of radiation under high and low-pressure scenarios including vacuum conditions. These conditions far exceed naturally occurring parameters on Earth, but may be found in space environments. The survival of organisms under laboratory-generated space conditions or exposure at low Earth orbit (Cockell et al., 2011) indicates the possibility that microbial life can survive exposure to interstellar space. Even multicellular animals like tardigrades have survived under vacuum conditions (Jonsson et al., 2008).

2.7.1 Radiation

Resistance to high ionizing radiation has been observed in various microbes as well as multicellular eukaryotes. The SI unit for quantifying ionizing radiation dosage is the gray (Gy), which is equivalent to one absorbed joule of radiation energy per kilogram of matter (BIPM, 2006). Average estimated yearly human radiation dosage is .0024 Gy (UN Scientific Committe on the Effects of Atomic Radiation, 2008). The common proteobacterium *Escherichia coli* is able to survive significantly higher dosages of 60 Gy. The tardigrade *Milnesium tardigradum* survives 5000 Gy. Various microbes can survive orders of magnitude higher levels of radiation than what naturally occurs on Earth, for example, *Deinococcus radiodurans* survived 15,000 Gy and *Thermococcus gammatolarans* has survived a dosage of 30,000 Gy (Hirsch et al., 2004; White et al., 1999). Experimental evidence shows that halophilic adaptations may also provide tolerance to space radiation (Leuko et al., 2015).

2.7.2 High Pressure: Mega-Pascal and Giga-Pascal Ranges

Organisms that tolerate high pressure (i.e., pressures $> \sim 116$ MPa) are considered piezophiles or barophiles (Kato, 1999). These organisms are typically found living in deep-sea habitats and range from microbes to large multicellular organisms. Many are obligate piezophiles and cannot survive the transition from pressures at the seafloor to surface pressure. Some examples include species of giant foraminiferous amoebae of the class *Xenophyophorea* (Gooday et al., 2008).

Laboratory-based high-pressure experiments show that some organisms, such as the non-obligate piezophile *Halobacterium salinarium* NRC-1, can survive exposure to 400 MPa (Kish et al., 2012); pressures higher than those that exist on Earth. More dramatically, recent experiments show that microbes including *Shewanella odenensis*, *Escherichia coli,* and others, not known to be piezophilic, can survive and even adapt to pressures in the *Giga-Pascal* range (Sharma et al., 2002; Hazael et al., 2014; Vanlint et al., 2011), although some changes in morphology and physiology have been noted (Marietou et al., 2014). Above pressures of ~ 1 GPa, high-density ice phases exist; and the performance of biological systems under such conditions is not well-known (Fig. 1A; Chaplin, 2016).

2.7.3 Vacuum and Low Pressure

Humans and other mammals exhibit physiological acclimation to low pressure in high altitude environments, except on the highest mountains. The "death-zone" for mammals is ~ 8000 m, above which death occurs even after long-term acclimatization due to low oxygen and severe physiological impacts resulting in cerebral and/or pulmonary edema (for reference, Mt. Everest peaks at 8,848 meters). On Mars, surface pressure is much lower (0.6 kPa) than on Earth (101 kPa), and experimental exposure of bacteria to simulated Martian conditions shows that this low pressure is well within the parameter ranges for many species, even species not known as "hypo-barophiles" (Schuerger & Nicholson, 2016).

Mammals are able to survive several minutes when exposed to vacuum (Gosline, 2008). Remarkably, many common and extremophilic microbes tested exhibit resistance to simulated or real space vacuum (Olsson-Francis & Cockell, 2010). Numerous organisms including viruses, bacteria, fungi, and nematodes were tested in space during the Apollo 16 mission (Taylor et al., 1975). Fungi (Novikova et al., 2015), lichens (Sancho et al., 2007), and animals such as tardigrades and nematodes (Jonsson et al., 2008) have also been tested. Many are able to survive medium- to long-term space exposure, and some are shown to survive even re-entry and impact (Pasini & Price, 2015).

2.7.4 Microbial Metabolism

Terrestrial microbes are known to obtain energy from the chemical bonds in organic or inorganic molecules and light. Deep below the surface or below a thick ice cover, sunlight is not available, and thus, life must rely on chemical energy. Organisms conserve the energy released from chemical oxidation-reduction reactions (i.e., redox reactions) by forming energy-rich molecules such as ATP. Microorganisms can harness energy from a wide variety of electron acceptors and donors. For example, CO_2 can be reduced to methane with H_2 as a form of anaerobic respiration to generate energy (known as methanogenesis). Sulfate $\left(SO_4{}^{2-}\right)$ reduction to H_2S can be performed with H_2 or organic matter as the reductant. Fermentation reactions are also performed where the substrate serves as both electron acceptor and donor.

Although never observed, it has been theorized that life could also grow by harnessing free energy from thermal gradients in a process called thermosynthesis (Muller, 1985; Muller & Schulze-Makuch, 2006). Perhaps, temperature gradients are important when other sources of energy become limited in subsurface or subice environments like Mars, Enceladus, or Titan (Muller, 2003). Since most predicted extraterrestrial conditions lack light, the presence and availability of thermodynamically favorable substrates for both metabolism and construction of macromolecules and molecular complexes appear to be major factors of habitability.

3 SETTINGS FOR LIFE IN OUR SOLAR SYSTEM: PHYSIOCHEMICAL PARAMETER SPACE ON MARS, EUROPA, TITAN, AND ENCELADUS

The Solar System contains a variety of planetary habitats, which could accommodate extremophilic organisms. The most obvious of these habitats includes Mars, a cold desert that once hosted surface water and today has thick H_2O-ice polar ice sheets. The icy worlds of Europa, Titan, Enceladus, and other ice moons likely possess oceans and seafloor sediments, deep-sea hydrothermal vent environments, and brine veins within their ices, all of which offer possible extremophile niche space. Analyses of active cryovolcanism on several icy moons have provided a sample of interior content showing the presence of many organic building blocks or potential substrates for metabolism (Solar System Exploration Survey, 2003; Committee on the Limits of Organic Life in Planetary Systems, 2007; Prockter, 2005; Deming & Eichen, 2007; Mendez, 2001).

3.1 MARS

Mars is the closest planet to Earth in terms of its geophysical settings. Although the low temperatures and pressure at the Martian surface make liquid water very tenuous, daytime temperatures often climb above 0°C. Because Mars hosts relatively mild conditions compared to the outer Solar System and other inner rocky planets, it is considered the most likely to host organic carbon-based life beyond Earth, while its proximity has made it accessible to an armada of robotic explorers.

Mars is just within the outer edge of the Solar System's traditional "habitable zone" and shows strong evidence for once hosting liquid water at its surface (Baker, 2001). The surface is painted with ancient geomorphological evidence of large lakes and channels from megafloods as well as smaller crater lakes and fluvial features (Balme et al., 2013). Mars was inferred to have had a thicker atmosphere and surface seas in its early geological epochs as evidenced from sedimentary rock formations, hydrological terrain features (drainage systems), and the presence of minerals that require liquid water for formation (Baker, 2001; Boynton et al., 2008). On modern Mars, much of the H_2O is now frozen in the northern and southern polar ice caps or as subsurface permafrost. The north polar ice cap is primarily water ice, while the south polar ice cap is a layered mixture of CO_2 and H_2O ices.

More recent observations have detected ground ice, glacial features, and seasonal liquid brine seeps (Balme et al., 2013) with *circum-global* distribution (Fig. 3B). These findings suggest subsurface liquid groundwater may exist seasonally, relatively close to the surface. Alternatively, the observed brine seeps could form as a result of deliquescence (Zorzano et al., 2009), where salts at the surface absorb water, forming liquid that is intermittently available (Chevrier et al., 2009). These various areas where water may exist close to the surface have been collectively classified as "Special Regions" (Rummel et al., 2014). An extensive analysis of the physiochemical parameters for Mars showed that such Special Regions might, even perhaps seasonally, support parameter space where terrestrial microbial life could potentially survive and reproduce (Rummel et al., 2014). Laboratory experiments showed that bacteria collected from a wide range of environments were able to survive and adapt to simulated Mars pressure and temperature conditions (Schuerger & Nicholson, 2016). Psychrophilic, xerotolerant, and halophilic extremophiles are of particular relevance to these Special Regions because they already have adaptations for conditions similar to those on Mars (Amato et al., 2010; Chin et al., 2010). Research on temperature-salinity phase space, described above, shows that even at post-precipitation salinities, fluid inclusions can provide long-term microbial habitat. Collectively, the data presented in the temperature-salinity phase diagrams (Figs. 1B and 2A), the ability of microbes to survive in ancient salt deposits, and experimental evidence of microbial viability under Martian conditions indicate a high potential that life could have existed on Mars and might have survived in subsurface locations (Fig. 3C). By the same token, Mars could easily be contaminated by our own exploration efforts (Debus, 2005).

3.2 EUROPA: "EARTH-LIKE" SUBSURFACE OCEAN

Europa is one of the four Galilean moons of Jupiter that gained significant astrobiological interest after the Voyager 1 and 2 flybys and subsequent Galileo orbiter mission because of the likelihood of an ocean of liquid water beneath its surface. The moon's density indicates high water content, while a lack of cratering and the relative youthful surface features indicate geological activity driven by dynamic fluid processes beneath the crust (Fig. 4A). Current models predict that tidal heating maintains enough heat to drive geothermal processes (Chyba & Phillips, 2007; Committee on Planetary & Lunar Exploration, 1999). Recently, active cryovolcanism was observed at Europa (Nimmo & Pappalardo, 2016).

Two other Galilean moons, Ganymede and Callisto, are also likely to contain layers of subsurface liquid water; however, their orbits are further away from Jupiter and experience less tidal heating, which is reflected in Callisto's ancient surface features. Modeling of Callisto and Ganymede also indicates that layers of liquid water are likely present, but probably not as close to the surface as on Europa (Solomonidou et al., 2011; Schenk et al., 2004). Europa's ocean is estimated to be 100–150 km deep or slightly deeper (168 km); the variation depending on the model and data used (Melosh et al., 2003;

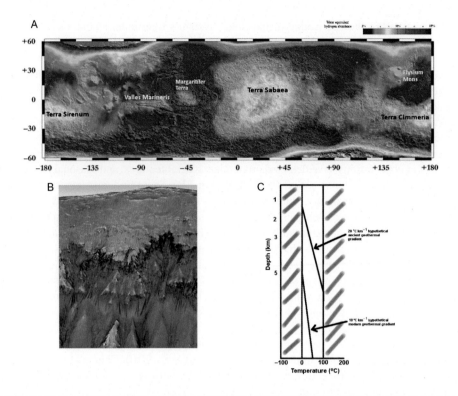

FIG. 3

Mars special regions and the deep subsurface: liquid water environments. (A) Global Mars water distribution shows a belt of H_2O (and mineral hydrates). (B) Recurring slope linae (RSL) is interpreted as liquid brine flows on Mars' surface during warm seasonal periods. (C) A geothermal gradient estimate of Mars shows that warm, liquid subsurface habitat may be likely on Mars and indicates a range of depths where temperature and pore size would allow microbial habitat.

(A) Modified NASA/JPL image, Feldman, W.C., Prettyman, T.H., Maurice, S., Plaut, J.J., Bish, D.L., Vaniman, D.T., Mellon, M.T., Metzger, A.E., Squyres, S.W., Karunatillake, S., Boynton, W.V., Elphic, R.C., Funsten, H.O., Lawrence, D.J., Tokar, R.L., 2004. Global distribution of near-surface hydrogen on Mars. J. Geophys. Res. Planets 109(E9); (B) Courtesy of NASA; (C) Modified from Michalski, J., Cuadros, J., Niles, P., Parnell, J., Rogers, A., Wright, S., 2013. Groundwater activity on Mars and implications for a deep biosphere. Nat. Geosci. 6, 133–138.

Pappalardo et al., 1998, 1999; Schubert et al., 2009). This is far deeper than the deepest point in Earth's ocean; the bottom of the Mariana Trench is measured at 10.99 km (Kato et al., 1998). The thickness of the ice crust and detailed understanding of the underlying layers remain unknown; estimates suggest the ice crust ranges from ~1 to 30 km thick. On Earth, the Antarctic ice sheet is ~4 km at its thickest point and harbors active aquatic environments at its base, including the large Lake Vostok, which is about the size of Lake Ontario. A current hypothesis is that Europa has a hard ice shell that covers a warmer, convective "stratosphere" of water near its freezing point or warm (i.e., slushy) ice (Fig. 4B). Significant heterogeneity between regions is likely, for example, some of the most active geological features occur in locations where warmer brine from the interior ocean may rise closer to the surface

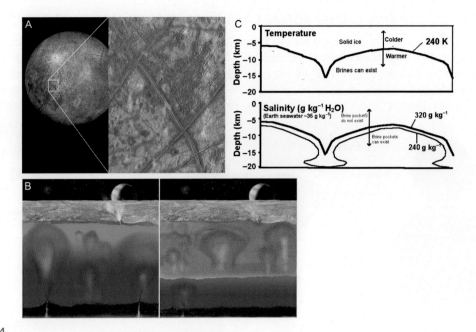

FIG. 4

Europa: Earth-like Ocean on a Moon of Jupiter. (A) Europa's surface is geologically young, with many features indicating shallow subsurface fluid dynamics; brown coloration is hypothesized to be salt originating from the subsurface ocean, discolored by exposure to solar radiation. (B) Models for the Europa subsurface with a thicker ice crust and water extending to the seafloor (left), a "stratosphere" layer of slushy ice below a thinner ice crust. (C) Plumes of warmer water originating from geothermal regions may approach within 5 km of the surface (or nearer, see text), driving observed active surface features.

(A) and (B) Courtesy of NASA/JPL; (C) Modified from Zolotov, M., Shock, E.L., 2004. Brine pockets in the icy shell on Europa: distribution, chemistry, and habitbility. In: Workshop on Europa's Icy Shell: Past, Present, and Future. Houston, TX.

(Billings & Kattenhorn, 2005; Melosh et al., 2003; Ruiz, 1999; Sonderlund et al., 2014). Measurements of electrical current from the NASA Galileo mission magnetometer indicated near-saturation salt content, while near-infrared absorption spectra indicate high magnesium sulfate and sodium chloride content (Hand & Chyba, 2007; Zolotov & Shock, 2001). The brown coloration seen on Europa's surface (Fig. 4A) is interpreted as irradiated salts from ocean water expunged during cryovolcanic activity (Hand & Carlson, 2015). Modeling of the interior thermodynamics and salinity suggests that plumes of geothermal heat would melt ice in dome-like formations reaching to 5–20 km of the ice surface (Fig. 4C; Zolotov & Shock, 2004; Zolotov & Kargel, 2009).

The latest evidence suggests that Europa's ocean is saline and contains various metabolic substrates, thus it may be the most Earth-like liquid ocean in the solar system, providing a compelling astrobiological target. Continued exploration of Europa is in the planning and execution stages. The NASA Juno mission is currently in-orbit collecting data on the Jupiter system, while the European Space Agency (ESA's) Jupiter Icy Moons Explorer (JUICE) is planned to orbit Ganymede and study subsurface liquid dynamics on Europa, Ganymede, and Callisto. NASA has also proposed continuing Europa exploration that includes a lander and drilling mission (Europa Study Team, 2012).

3.3 TITAN AND ENCELADUS: ACTIVE CRYOVOLCANISM ON MOONS OF SATURN

Despite their similar size, Titan and Europa have quite distinct compositions. Titan has greater water content, with a significant amount of ammonia (NH_3) and various 1-3 hydrocarbon molecules (Fig. 5A). The surface of Titan is a thick crust of H_2O-ice, with hydrocarbon lakes on the surface. There is also evidence for hydrocarbon-based cycling between these surface lakes and the atmosphere (Tobie et al., 2006). Models further predict the internal structure of Titan contains a ~400 km deep liquid

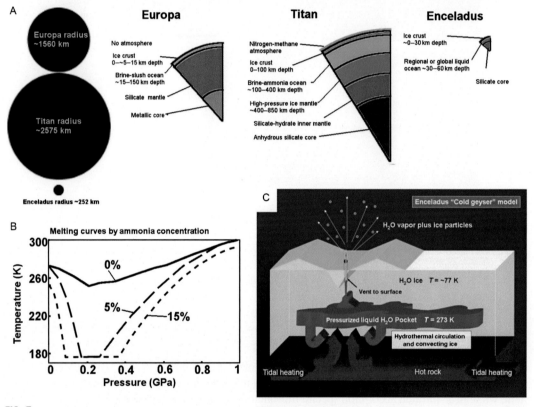

FIG. 5

Titan and Enceladus: Parameter Space in Cryovolcanic Magmas. (A) Europa, Titan, and Enceladus have different sizes, composition, and surface processes. Europa with a thin icy shell over a salty liquid water ocean. Titan with a thicker ice crust over a deep, ammonia-water mantle, which includes a lower layer high-pressure-phase ice, also with a thick atmosphere and surface liquid Methane lakes. Enceladus, a comparatively tiny ice moon with thick, frozen ice crust; nevertheless with pockets of water or cryo-fluid producing geyser plumes at the surface. (B) Phase equilibria show that 5%–15% aqueous ammonia solution may exist as liquid down to 180 K (−93°C), providing significant subsurface, liquid water habitat on Titan. (C) "Cold Geyser" model for cryovolcanic processes on Enceladus; heat from the deep interior may provide energy to melt ice and create liquid water, potentially ~273 K range (~0°C).

(B) Modified from Sohl et al. (2003); (C) Courtesy of NASA/JPL.

water ocean rich in NH_3 and CH_4 approximately 50–100 km below the ice crust. Below this water ocean, pressure in the giga-Pascal to tera-Pascal ranges compress water into high-pressure states of ice, extending another 400 km down to a rocky submantle (Fortes, 2012; Mousis et al., 2002; Sohl et al., 2003; Tobie et al., 2005).

Just as the Voyager and follow-up Galileo missions showed that Europa appears to have an active hydrologic system, the NASA Voyager missions also discovered that Saturn's moon, Titan, had a thick, hazy atmosphere of N_2 and CH_4. The following mission to Saturn was the NASA/ESA Cassini-Huygens probe, which explored Titan's surface directly, observing abundant great lake-sized bodies of liquid hydrocarbon (methane and ethane) and active cryovolcanism associated with warm regions near Titan's surface (Bouquet et al., 2015; Lopes et al., 2013). Based on these data, a complex, seasonal hydrological cycle for Titan was predicted (Lunine & Atreya, 2008; Lunine & Rizk, 2007; Le Gall et al., 2016).

Enceladus, another moon of Saturn and much smaller in diameter than Titan, exhibited active cryo-volcanic plumes. These plumes jetted thousands of kilometers from crack-like geological formations in the south polar region (Fig. 5C; Porco et al., 2006; Spencer et al., 2009). The observation of these active plumes provided significant data on the composition of Enceladus's interior. The Cassini orbiter was equipped with a mass spectrometer, which analyzed gas and particles in these plumes including H_2O, CO_2, H_2S, NH_3, Methanol, 2-carbon ethers, HCN, and a variety of 2-6 carbon hydrocarbons (Hansen et al., 2011; Waite et al., 2009). These cryo-magmas are reflective of the subsurface ocean composition, which is predicted to be a mixture of water, ammonia, nitrogen, and hydrocarbon at temperatures near 0°C (Tobie et al., 2009; Zolotov, 2007). The surfaces of these moons remain too cold for liquid water and no known terrestrial habitats are cold enough nor are psychrophiles known to be viable within this parameter space. In the subsurface, however, heat and melting do take place at the lower end of the H_2O melting point (Hsu et al., 2015). High salinity and NH_3 content lower water's melting point to predicted Titan and Enceladus' subsurface temperatures, and metabolic substrates such as ammonia and methane are abundant (Fig. 5B, Arjmandi et al., 2007). Many bacteria can survive in high ammonia environments (Kelly et al., 2012), suggesting these icy moons might harbor habitable parameter space (McKay et al., 2008).

3.4 SETTINGS FOR LIFE IN OUR SOLAR SYSTEM: PLAUSIBLE ECOSYSTEMS BASED ON ANALOG NICHES

Martian rovers have found evidence of ancient sedimentary basins; however, the most hospitable putative habitats on Mars today are likely deep within the subsurface (Fig. 3D). Briny subsurface groundwater systems may support a deep Martian biosphere either now or in the past (Michalski et al., 2013; Mikucki et al., 2015). Gale Crater on Mars provides an example of a habitat where a large body of liquid surface water could have once supported a chemosynthetic microbial ecosystem (Grotzinger et al., 2014) that since retreated into the subsurface. On Earth, deep subsurface communities are found in sediments, fracture fluids, and groundwater systems. In Antarctica, life survives in deep, ice-covered lakes and fluid inclusions in ice, analog environments approaching Europa, Titan, or Enceladus conditions. Microorganisms detected in these dark habitats are chemosynthetic, garnering energy from the oxidation of reduced S, Fe, and NH_4^+ (Christner et al., 2014; Purcell et al., 2014).

Estimates of subsurface microbial abundances are difficult to assess. However, some calculations suggest that the Earth's subsurface harbors 2%–19% of the Earth's total biomass (McMahon & Parnell, 2014), and this life can exist under increasingly reducing conditions (Reith, 2011). Marine sediments show microbial activity at hundreds of meters in depth. Deep mines in Africa (2.8 km) provide access to fracture fluids which support a single chemosynthetic, thermophilic species, *Candidatus Desulforudis audaxviato*, growing via sulfate reduction using radiolytically generated H_2 (Lin et al., 2005, 2006; Chivian et al., 2008). Subpermafrost brines below 100 m depth in Antarctica's McMurdo Dry Valleys provide another analog for a possible Martian subsurface ecosystem. A small volume of this deep brine leaks to the surface at a feature known as Blood Falls (Mikucki et al., 2015) and analyses of this surface-released brine revealed a halotolerant, psychrophilic chemosynthetic community that uses reduced minerals to sustain redox reactions such as sulfur oxidation and reduction (Mikucki & Priscu, 2007; Mikucki et al., 2009). However, modeling results on the potential primary productivity of Blood Falls is quite low relative to terrestrial ecosystems (i.e., only 6.4×10^{-5} g C l^{-1} year^{-1}, where C = carbon), suggesting biomass in subglacial aquifers might be difficult to detect (Fisher & Schulze-Makuch, 2013).

Europa, Enceladus, and other icy moons may support possible microbial habitats within a range of parameter space (Parkinson et al., 2008; Hsu et al., 2015). Infrared measurements of Europa's ice surface show that radiolysis of surface-exposed material produces a complex mixture of CO_2, CO, carbonate, methanol, ethanol, methane, H_2O_2, O_2, and SO_2 (Hand et al., 2007). It is hypothesized that these molecules could provide a large flux of material into the subsurface liquid layers through impact gardening and subduction (Chyba & Phillips, 2001), thereby delivering substantial organics and oxidants for redox-based metabolism (Chyba, 2000; Vance et al., 2016). Carbon monoxide (CO) serves as a carbon and energy source (electron donor) for some autotrophic microorganisms. Carboxydotrophic organisms are found in many bacterial lineages (King & Weber, 2007) and include extremophiles such as the thermophile, *Carboxydocella thermautotrophica* (Svetlichny et al., 1991). Nitrogen, a fundamental component of organic macromolecules, is also observed in abundance, often as ammonia or ammonium (NH_3/NH_4^+). Biologically available forms of nitrogen have been detected on Mars (as nitrate, NO_2; Stern et al., 2015) and on Titan and Enceladus (as NH_3; Waite et al., 2009). These findings show that extraterrestrial environments contain common organic building blocks and metabolic substrate for life.

Near and within the seafloor on Europa or Enceladus, serpentinization reactions and radiolytic dissociation of water may generate substrates that drive planetary energetics for extended periods of time. Iron-rich mineral oxides in Earth's mantle can interact with water from the overlying ocean resulting in the generation of a variety of reduced and oxidized species via serpentinization or other rock-water interactions (Fig. 6A; Sekine et al., 2015). Serpentinization is the aqueous alteration of minerals (such as olivine and pyroxenes) in ultramafic rocks; this process creates fluids rich in H_2 and CH_4 depending on pressure and temperature conditions (McCollom & Seewald, 2013). These fluids can support microbial communities. For example, H_2, generated through serpentinization, could fuel H_2-oxidation-based metabolisms (Parkes et al., 2011; Taubner et al., 2015). Depending on the oxidation state of the sub-ice ocean, methanogens, if present, may be an important consumer of generated H_2 (McCollum, 1999; Taubner et al., 2015). Beta-proteobacteria of the *Hydrogenophaga* genus grow by oxidizing H_2 with water as a metabolic byproduct and this group has been found associated with serpentinization incubation experiments (Morrill et al., 2014). Poly-extremophilic, H_2-oxidizing members of new *Serpentinomonas* genus have been found in natural serpentine environments and are also tolerant of extremely high pH. While electron acceptors (oxidants) may be limiting in oceans-sealed below ice, microbes are also capable of direct or indirect transfer of electrons to solid mineral oxides (e.g., Edwards et al., 2005). The radiolytic dissociation of water can generate strong oxidants including

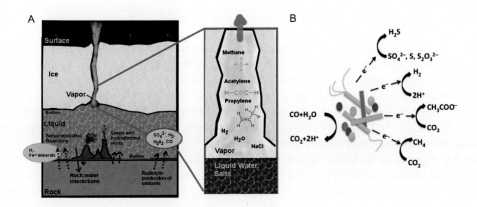

FIG. 6

Metabolic substrates in Ice Moon analogs. (A) Energy released from the decay of radioactive elements (e.g., U, Th, K) causes dissociation of water molecules, which produces hydrogen and reactive oxygen species (ROS). These can then react with other aqueous species including O_2 (Lin et al., 2005) and SO_4 (Li et al., 2016). Serpentinization reactions may provide additional H_2 along with iron oxides. Reduction of N_2 to NH_3 is also possible (Sekine et al., 2015). These molecules can all potentially serve as substrate for microbial chemolithotrophic metabolism. Biofilms may form near hydrothermal vents and above cold seeps, both of which emit important redox chemicals from the subsurface. Accretion processes at the base of an ice crust could result in the exclusion of substrates from the ice matrix, similar to the formation of sea ice, which encourages biofilms formation. Evidence of such putative energetic substrates has been detected in Enceladus' cryovolcanic plumes. (B) Molecules produced from radiolysis of H_2 and serpentinization provide a variety of oxidized and reduced substrates for microbial metabolism. Possible redox reaction energetics in an icy world ecosystem includes hydrogenotrophic methanogenesis, sulfate oxidation and reduction, iron oxidation and reduction, knallgass (hydrogen oxidation), and reduction of carbon monoxide. CO_2 and CO could serve as sole carbon sources for these consortia.

(A) Adapted from Postberg et al. (2009); (B) Adapted from Thauer et al. (1977).

hydroxyl radicals and H_2O_2 as well as additional reductant substrate (H_2). Li et al. (2016) have demonstrated that in deep subsurface ecosystems, sulfate can also be generated from oxidation of sulfide minerals with radiolytic oxidants. Together, serpentinization and radiolytic dissociation of water could provide oxidants and reductants that sustain redox reactions for chemolithoautotrophic metabolic processes (McCollum, 1999) for extended periods of time.

Physiochemical parameter space on Titan and Enceladus is colder than Mars or Europa, but cryovolcanic pockets of liquid and semiliquid pockets could fall within the lower temperature limit of terrestrial psychrophiles, within the salinity range tolerable for halophiles and within the laboratory-tested range of microbial survival in solutions containing high concentrations of ammonia. Low-temperature, high-pressure hydrocarbon seeps on Earth host microbial ecosystems that form based on the metabolism of methane-clathrates; such gas structures may be common on Titan and Enceladus (Choukroun et al., 2010; Arjmandi et al., 2007; Kieffer et al., 2006). The components of the Titan system, i.e., the raw materials for organic chemistry with a heat source and mixing processes, are similar to models of Earth's prebiotic chemistry (Raulin et al., 2012).

4 CONCLUSION

Extremes in physiochemical space do appear to present constraints for Earth life. Temperature emerges as a parameter of principle importance: above 150°C, even the hardiest hyperthermophilic archaea and dormant tardigrades rapidly become non-viable (Madigan & Oren, 1999). On the other hand, while active biochemical processes are completely, or very nearly, dormant below $\sim -30°C$ to $-50°C$, many microbes, invertebrates, and vertebrates can survive long-term cryopreservation through vitrification (Fahy & Wowk, 2014). Dormancy or sporulation during periods of stress, such as low temperature and desiccation, when considered with the fact that some organisms can survive in this dormant/sporulated state while exposed to the harsh vacuum, radiation, and desiccation of space and also survive subsequent impact stresses (Cockell et al., 2011; Olsson-Francis & Cockell, 2010), show that microbial life can survive interplanetary travel on ejected substrate as proposed by the panspermia hypothesis (Kamminga, 1982).

Within the parameter space of Earth extremophiles, combinations of extremes of temperature, pH, and salinity appear to provide additional limits to life. For example, organisms shown to be the most tolerant extremophiles in any given parameter invariably fall within the "normal" range of other parameters, aside from several low-diversity groups which are able to occupy multiple, moderately extreme parameters (Capece et al., 2013; Harrison et al., 2012).

Interestingly, long-term survival of halophiles surviving in fluid inclusions in halite deposits may indicate that low-temperature, high-salinity regions of ice moons may be habitable (Lowenstein et al., 2011). Resistance and adaptation of even common species of bacteria to Giga-Pascal pressures argue that high pressure alone may not provide a limitation to life, rather most parameter space, up to the high-pressure phases, may be habitable, though some organisms may become obligate at specific pressures (Sharma et al., 2002).

Particularly tolerant representatives of the Animal kingdom are members of the meiofauna, barely microscopic animals adapted to periodic and long-term desiccation in soil environments. Hardy multicellular Eukaryotes including lichens and algae survive simulated exposure to the dry, cold, low-pressure desert of Mars (Schuerger & Nicholson, 2016). The deep Martian subsurface could contain significant regions warm enough for liquid water to persist in aquifers analogous to Earth (Michalski et al., 2013).

The issue of dissolved solute composition and availability of metabolic substrate are additional constraints providing limitations to habitat on Earth and are likely to be components determining the habitability of extraterrestrial liquid-water habitat. High concentrations of chaotropic substances in solution can disrupt biological macromolecules and membranes (Ball & Hallsworth, 2015); while absence of important substrates can prevent or stunt growth.

Several of the ice moons of Jupiter and Saturn may contain deep-sea hydrothermal vent-like environments. In Europa (Chyba & Phillips, 2007), Titan (Lunine & Rizk, 2007), and Enceladus (Parkinson et al., 2008), interior heating processes may create both localized areas of thermophilic and hyperthermophilic habitat, suitable for thermophilic, chemolithotrophic microbes, while large regions on these moons could be cold and hypersaline, similar to habitat where halophilic psychrophiles are found on Earth. Data collected from the spectra of recently observed active cryo-geysers on Europa and that of Enceladus indicate a diverse mixture of simple and complex organic molecules, consistent with chemical processes potentially important for driving metabolic reactions (Hansen et al., 2011).

Titan's subsurface ocean is estimated to be colder and far deeper than Europa. Its significant ammonia composition and presence of methane-clathrates provide an exceptional array of potential prebiotic building blocks (Fortes, 2012; Tobie et al., 2005). Many mesophilic microbial species are able to survive concentrated ammonia solutions (Kelly et al., 2012), while methane clathrate seeps on Earth support diverse communities (Reed et al., 2002). Methane, as a nonpolar solvent for extraterrestrial biochemistry, has been proposed and, in this scenario, liquid hydrocarbon lakes on Titan's surface could provide abundant habitat in the absence of liquid water (McKay & Smith, 2005).

The "time" parameter is likely to play a key role in habitability as well. The rapidity of life's apparent origin on Earth (Schopf et al., 2007) suggests that any locale with liquid water and abundant organics could lead to an origin of life event. It is also possible that in the icy outer moons, like Titan, and the dwarf planets that long periods of prebiotic-like chemistry may take place under cold, aqueous conditions, but that life itself would originate during warmer, "red-giant" conditions of the late solar system (Iben, 1965). Multiple origins of life could therefore be "prebaked" into the prebiotic chemistries of the outer Solar System. Exchange of viable microbial life between planets also seems plausible based on the observed survival of diverse organisms under space conditions. These "space travelers" could also have a time limit, while microbial survival of several millions of years would allow time for interplanetary travel space exposure over larger geological timescales seems improbable for Earth-like life.

REFERENCES

Amato, P., Doyle, S.M., Battista, J.R., Christner, B.C., 2010. Implications of subzero metabolic activity on long-term microbial survival in terrestrial and extraterrestrial permafrost. Astrobiology 10 (8), 789–798.

Andersson, O., 2011. Glass-liquid transition of water at high pressure. Proc. Natl. Acad. Sci. U. S. A. 108 (27), 11013–11016.

Anguilera, A., 2013. Eukaryotic organisms in extreme acidic environments, the Rio Tinto case. Life (Basel) 3 (3), 363–374.

Arjmandi, M., Chapoy, A., Tohidi, B., 2007. Equilibrium data of hydrogen, methane, nitrogen, carbon dioxide, and natural gas in semi-clathrate hydrates of tetrabutyl ammonium bromide. J. Chem. Eng. Data 52 (6), 2153–2158.

Ayala del Rio, H., Chain, P.S., Gryzmski, J.J., Ponder, M.A., Ivanova, N., Bergholz, P.W., Di Bartolo, G., Hauser, L., Land, M., Bakermans, C., Rodrigues, D., Klappenbach, J., Zarka, D., Larimer, F., Richardson, P., Murray, A., Thomashow, M., Tiedje, J.M., 2010. The genome sequence of *psychrobacter arctitcus* 273-4, a psychroactive siberian permafrost bacterium, reveals mechanisms for adaptation to low-temperature growth. Appl. Environ. Microbiol. 76 (7), 2304–2312.

Bains, W., 2004. Many chemistries could be used to build living systems. Astrobiology 4, 13016.

Baker, V.R., 2001. Water and the Martian landscape. Nature 412, 228–236.

Baker, B., Banfield, J., 2003. Microbial communities in acid mine drainage. FEMS Microbiol. Ecol. 44 (2), 139–152.

Baker-Austin, C., Dopson, M., 2007. Life in acid: pH homeostatis in acidophiles. Trends Microbiol. 15 (4), 165–171.

Bakermans, C., Skidmore, M., 2011. Microbial respiration in ice at subzero temperatures (-4C to -33C). Environ. Microbiol. Rep. 3 (6), 774–782.

Ball, P., Hallsworth, J.E., 2015. Water structure and chaotropicity: their uses, abuses and biological implications. Phys. Chem. Chem. Phys. 17 (13), 8297–8305.

Balme, M.R., Gallagher, C.J., Hauber, E., 2013. Morphological evidence for geologically young thaw of ice on Mars: a review of recent studies using high-resolution imaging data. Prog. Phys. Geogr. 37 (3), 289–324.

Banciu, H., Sorokin, D., 2013. Adaptation in Haloalkaliphiles and Natronophilic Bacteria. In: Seckbach, J., Oren, J., Stan-Lotter, H. (Eds.), Polyextremophiles: Life Under Multiple Forms of Stress. Springer, Dordrecht, pp. 121–178.

Baross, J.A., Schrenk, M.O., Huber, J.A., 2007. Limits of carbon life on Earth and elsewhere. In: Sullivan III, W.T., Baross, J.A. (Eds.), Planets and Life: The Emerging Science of Astrobiology. Cambridge University Press, New York, pp. 275–291.

Benner, S.A., Ricardo, A., Carrigan, M., 2004. Is there a common chemical model for life in the universe? Curr. Opin. Chem. Biol. 8, 672–689.

Bergholz, P.W., Bakermans, C., Tiedge, J.M., 2009. *Pyschrobacter arcticus* 273-4 uses resource efficiency and molecular motion adaptations for subzero temperature growth. J. Bacteriol. 191 (7), 2340–2352.

Billings, S.E., Kattenhorn, S.A., 2005. The great thickness debate: ice shell thickness models for Europa and comparisons with estimates based on flexure at ridges. Icarus 177 (2), 397–412.

Blochl, E., Rachel, R., Burggraf, S., Hafenbradl, D., Jannasch, H.W., Stetter, K.O., 1997. Pyrolobus fumarii, gen. and sp. nov., represents a novel group of archaea, extending the upper temperature limit for life to 113 degrees C. Extremophiles 1 (1), 14–21.

Boetius, A., Anesio, A.M., Deming, J.W., Mikucki, J.A., Rapp, J.Z., 2015. Microbial ecology of the cryosphere: sea ice and glacial habitats. Nat. Rev. Microbiol. 13 (11), 677–690.

Bonch-Osmolovskaya, E., 2012. Metabolic diversity of thermophilic prokaryotes—what's new. In: Anitori, R.P. (Ed.), Extremophiles: Microbiology and Biotechnology. Caister Academic Press, Norfolk, pp. 109–131.

Bouquet, A., Mousis, O., Waite, J.H., Picaud, S., 2015. Possible evidence for a methane source in Enceladus' ocean. Geophys. Res. Lett. 42 (5), 1334–1339.

Bowers, K.J., Weigel, J., 2011. Temperature and pH optima of extremely halophilic archaea: mini-review. Extremophiles 15 (2), 119–128.

Bowers, K.J., Mesbah, N.M., Wiegel, J., 2009. Biodiversity of poly-extremophilic bacteria: does combining the extremes of high salt, alkaline pH and elevated temperature approach a physico-chemical boundary for life? Saline Syst. 5 (9).

Boynton, W.V., Taylor, G.J., Karunatillake, S., Reedy, R.C., Keller, J.M., 2008. Elemental abundances determined by Mars Odyssey GRS. In: Bell, J. (Ed.), The Martian Surface: Composition, Mineralogy, and Physical Properties. Cambridge University Press, New York, pp. 103–124.

Brady, J., 1992. Does ice dissolve or does Halite melt? J. Geol. Educ. 40 (2), 116–118.

Bureau International des Poids et Mesures (BIPM), 2006. The International System of Units (SI), eighth ed. Stedi Media, Paris.

Cameron, R.E., Morelli, F.A., Randall, L.P., 1972. Aerial, aquatic, and soil microbiology of Don Juan Pond Antarctica. Antarct. J. US, 254–258.

Capece, M.C., Clark, E., Saleh, J.K., Halford, D., Heinl, N., Hoskins, S., Rothschild, L.J., 2013. Polyextremophiles and the constraints for terrestrial habitability. In: Seckbach, J., Oren, J., Stan-Lotter, H. (Eds.), Polyextremophiles: Life Under Multiple Forms of Stress. Springer, Dordrecht, pp. 3–59.

Chaplin, M., 2016. Water Structure and Science. http://www1.lsbu.ac.uk/water/.

Charette, M., Smith, W., 2010. The volume of earth's ocean. Oceanography 23 (2), 112–114.

Charles, D., 2006. Species conservation. a "forever" seed bank takes root in the Arctic. Science 312 (5781), 1730–1731.

Chevrier, V.F., Hanley, J., Altheide, T.S., 2009. Stability of perchlorate hydrates and their liquid solutions at the Phoenix landing site, Mars. Geophys. Res. Lett. 36 (10).

Chin, J.P., Megaw, J., Magill, C.L., Nowotarski, K., Williams, J.P., Bhaganna, P., Linton, M., Patterson, M.F., Underwood, G.J.C., Mswaka, A.Y., Hallsworth, J.E., 2010. Solutes determine the temperature windows for microbial survival and growth. PNAS 107 (17), 7835–7840.

Chivian, D., Brodie, E.L., Alm, E.J., Culley, D.E., Dehal, P.S., DeSantis, T.Z., Gihring, T.M., Lapidus, A., Lin, L.H., Lowry, S.R., Moser, D.P., Richardson, P.M., Southam, G., Wanger, G., Pratt, L.M., Andersen, G.L., Hazen, T.C., Brockman, F.J., Arkin, A.P., Onstott, T.C., 2008. Environmental genomics reveals a single-species ecosystem deep within Earth. Science 322 (5899), 275–278.

Choukroun, M., Grasset, O., Tobie, G., Sotin, C., 2010. Stability of methane clathrate hydrates under pressure: influence on outgassing processes of methane on Titan. Icarus 205 (2), 581–593.

Christner, B.C., Priscu, J.C., Achberger, A.M., Barbante, C., Carter, S.P., Christianson, K., Michaud, A.B., Mikucki, J.A., Mitchell, A.C., Skidmore, M.L., Vick-Majors, T.J., 2014. A microbial ecosystem beneath the West Antarctic ice sheet. Nature 512 (7514), 310–313.

Chyba, C.F., 2000. Energy for microbial life on Europa. Nature 403, 381–382.

Chyba, C.F., Phillips, C.B., 2001. Possible ecosystems and the search for life on Europa. PNAS 98 (3), 801–804.

Chyba, C.F., Phillips, C.B., 2007. Europa. In: Sullivan III, W.T., Baross, J.A. (Eds.), Planets and Life: The Emerging Science of Astrobiology. Cambridge University Press, New York, pp. 388–423.

Clarke, A., Morris, G.J., Fonseca, F., Murray, B.J., Acton, E., Price, H.C., 2013. A low temperature limit for life on earth. PLoS One 8 (6). e66207.

Cleland, C.E., Chyba, C.F., 2007. Does "life" have a definition? In: Sullivan III, W.T., Baross, J.A. (Eds.), Planets and Life: The Emerging Science of Astrobiology. Cambridge University Press, New York, pp. 119–131.

Cockell, C.S., Rettbery, P., Rabbow, E., Olsson-Francis, K., 2011. Exposure of phototrophs to 548 days in low Earth orbit: microbial selection pressures in outer space and on early earth. ISME J. 5 (10), 1671–1682.

Colwell, F.S., Stormberg, G.J., Phelps, T.J., Birnbaum, S.A., McKinley, J., Rawson, S.A., Veverka, C., Goodwin, S., Long, P.E., Russell, B.F., Garland, T., Thompson, D., Skinner, P., Grover, S., 1992. Innovative techniques for collection of saturated and unsaturated subsurface basalts and sediments for microbiological characterization. J. Microbiol. Methods 15 (4), 279–292.

Committee on the Limits of Organic Life in Planetary Systems, Committee on the Origins and Evolution of Life, Space Studies Board Division on Engineering and Physical Sciences, Board on Life Sciences Division on Earth and Life Studies, National Research Council of the National Academies, 2007. The Limits of Organic Life in Planetary Systems. National Academies Press, Washington, DC.

Committee on Planetary and Lunar Exploration, Space Studies Board, Commission on Physical Sciences, Mathematics, and Applications, 1999. A Science Strategy for the Exploration of Europa. National Academies Press, Washington, DC.

Costanzo, J.P., Lee Jr., R.E., DeVries, A.L., Wang, T., Layne Jr., J.R., 1995. Survival mechanisms ov vertebrate ectotherms at subfreezing temperatures: applications in cryomedicine. FASEB J. 9, 351–358.

Cowan, D.A., 2004. The upper temperature for life—where do we draw the line? Trends Microbiol. 12 (2), 58–60.

Davies, P.L., 2014. Ice-binding proteins: a remarkable diversity of structures for stopping and starting ice growth. Trends Biochem. Sci. 39 (11), 548–555.

Debus, A., 2005. Estimation and assessment of Mars contamination. Adv. Space Res. 35 (9), 1648–1653.

De Maayer, P., Anderson, D., Cary, C., Cowan, D.A., 2014. Some like it cold: understanding the survival strategies of psychrophiles. EMBO Rep. 15 (5), 508–517.

De Saro, F.J.L., Gomez, M.J., Gonzalez-Tortuero, E., Parro, V., 2013. The dynamic genomes of acidophiles. In: Seckbach, J., Oren, J., Stan-Lotter, H. (Eds.), Polyextremophiles: Life Under Multiple Forms of Stress. Springer, Dordrecht, pp. 81–97.

Deming, J.W., Eichen, H., 2007. Life in ice. In: Sullivan III, W.T., Baross, J.A. (Eds.), Planets and Life: The Emerging Science of Astrobiology. Cambridge University Press, New York, pp. 292–312.

Driesner, T., 2007. The System H2O-NaCl I: correlation formulae for phase relations in temperature-pressure-composition space from 0 to 1000°C, 0 to 5000 bar, and 0 to 1 X_{NaCl}. Geochim. Cosmochim. Acta 71 (20), 4880–4901.

DeLong, E., 1998. Archaeal means and extremes. Science 280 (5363), 542–543.

Edwards, K.J., Bach, W., McCollom, T.M., 2005. Geomicrobiology in oceanography: microbe–mineral interactions at and below the seafloor. Trends Microbiol. 13 (9), 449–456.

Europa Study Team (JPL D-71990), 2012. Europa Study 2012 Report: Europa Lander Mission. NASA Task Order NMO711062, Outer Planets Flagship Mission.

Fahy, G.M., Wowk, B., 2014. Principles of cryopreservation by vitrification. In: Wolkers, W.F., Oldenhof, H. (Eds.), Methods in Molecular Biology, vol. 1257. Humana Press, New York, pp. 21–82.

Feller, G., Gerday, C., 2003. Psychrophilic enzymes: hot topics in cold adaptation. Nat. Rev. Microbiol. 1 (3), 200–208.

Fendrihan, S., Legat, A., Pfaffenhuemer, M., Gruber, C., Weidler, G., Gerbl, F., Stan-Lotter, H., 2006. Extremely halophilic archaea and the issue of long-term microbial survival. Rev. Environ. Sci. Biotechnol. 5 (2–3), 203–218.

Fisher, T.M., Schulze-Makuch, D., 2013. Nutrient and population dynamics in a subglacial reservoir: a simulation case study of the Blood Falls ecosystem with implications for astrobiology. Int. J. Astrobiol. 12 (04), 304–311.

Forterre, P., Filee, J., Myllykallio, H., 2004. Origin and evolution of DNA and DNA replication machineries. In: de Pouplana, L.R. (Ed.), The Genetic Code and the Origin of Life. Kluwer Academic/Plenum Publishers, Landes Bioscience, New York, pp. 145–168.

Fortes, A.D., 2012. Titan's internal structure and the evolutionary consequences. Planet. Space Sci. 60 (1), 10–17.

Futterer, O., Angelov, A., Liesegang, H., Gottschalk, G., Schleper, C., Schepers, B., Dock, C., Antranikian, G., Liebl, W., 2004. Genome sequence of Picrophilius torridus and its implications for life around pH 0. PNAS 101 (24), 9091–9096.

Gooday, A.J., Todo, Y., Uematsu, K., Kitazato, H., 2008. New organic-walled Foraminifera (Protista) from the ocean's deepest point, the Challenger Deep (western Pacific Ocean). Zool. J. Linn. Soc. 153 (3), 399–423.

Goordial, J., Raymond-Bouchard, I., Ronholm, J., Shapiro, N., Woyke, T., Whyte, L., Bakermans, C., 2015. Improved-high-quality draft genome sequence of *Rhodococcus sp.* JG-3, a eurypsychrophilic *Actinobacteria* from Antarctic dry valley permafrost. Stand. Genomic Sci. 10, 61.

Gosline, A., 2008. Survival in space unprotected is possible—briefly. Sci. Am. .

Grant, W.D., Larsen, H., 1989. Group III. Extremely halophilic archaeobacteria, order Halobacteriales ord. nov. In: Staley, J.T., Bryant, M., Pfennig, N., Holt, J.G. (Eds.), first ed. Bergey's Manual of Systematic Bacteriology, vol. 3.

Grant, W.D., Mwatha, W.E., Jones, B.E., 1990. Alkaliphiles: ecology, diversity and applications. FEMS Microbiol. Lett. 75 (2-3), 255–269.

Graziano, G., Merlino, A., 2014. Molecular bases of protein halotolerance. Biochim. Biophys. Acta 1844.

Grotzinger, J., Sumner, D., Kah, L., Stack, K., Gupta, S., Edgar, L., Rubin, D., Lewis, K., Schieber, J., Mangold, N., Milliken, R., 2014. A habitable fluvio-lacustrine environment at Yellowknife Bay, Gale Crater Mars. Science 343.

Gunde-Cimerman, N., Ramos, J., Plemenitas, A., 2009. Halotolerant and Halophilic fungi. Mycol. Res. 113 (Pt. 11), 1231–1241.

Hand, K.P., Chyba, C.F., 2007. Empirical constraints on the salinity of the europan oceans and implications for a thin ice shell. Icarus 189, 424–438.

Hand, K.P., Carlson, R.W., 2015. Europa's surface color suggests an ocean rich with sodium chloride. Geophys. Res. Lett. 42 (9), 3174–3178.

Hand, K.P., Carlson, R.W., Chyba, C.F., 2007. Energy, chemical disequilibrium, and geological constraints on Europa. Astrobiology 7 (6), 1006–1022.

Hansen, C.J., Shemansky, D.E., Esposito, L.W., Stewart, A.I.F., Lewis, B.R., Colwell, J.E., Hendrix, A.R., West, R.A., Waite Jr., J.H., Teolis, B., Magee, B.A., 2011. The composition and structure of the Enceladus plume. Geophys. Res. Lett. 38 (11). L11202.

Harrison, J.P., Gheeraert, N., Tsigelnitskiy, D., Cockell, C.S., 2012. The limits for life under multiple extremes. Trends Microbiol. 21 (4), 204–212.

Haymon, R.M., Fornari, D.J., Von Damm, K.L., Lilley, M.D., Perfit, M.R., Edmond, J.M., Shanks III, W.C., Lutz, R.A., Grebmeier, J.M., Carbotte, S., Wright, D., McLaughlin, E., Smith, M., Beedle, N., Olson, E.,

1993. Volcanic eruption of the mid-ocean ridge along the east pacific rise crest at 9°45-52′N: direct submersible observations of seafloor phenomena associated with an eruption event in april, 1991. Earth Planet. Sci. Lett. 119 (1-2), 85–101.

Hazael, R., Foglia, F., Kardzhaliyska, L., Daniel, I., Meersman, F., McMillan, P., 2014. Laboratory investigation of high pressure survival in Shewanella oneidensis MR-1 into the gigapascal pressure range. Front. Microbiol. 5, 612.

Heberle, J., 2000. Proton transfer reactions across bacteriorhodopsin and along the membrane. Biochim. Biophys. Acta, Bioenerg. 1458 (1), 135–147.

Hengerr, S., Worland, M.R., Reuner, A., Brummer, F., Schill, R.O., 2009. Freeze tolerance, supercooling points and ice formation: comparative studies on the subzero temperature survival of limno-terrestrial tardigrades. J. Exp. Biol. 212 (Pt. 6), 802–807.

Hicks, D.B., Krulwich, T.A., 1990. Purification and reconstitution of the F1F0-ATP synthase from alkaliphilic Bacillus firmus OF4. evidence that the enzyme translocates H+ but not Na+. J. Biol. Chem. 265 (33), 20547–20554.

Hirsch, P., Gallikowski, C.A., Siebert, J., Peissl, K., Kroppenstedt, R., Schumann, P., Stackebrandt, E., Anderson, R., 2004. Deinococcus frigens sp. nov., Deinococcus saxicola sp. nov., and Deinococcus marmoris sp. nov., low temperature and draught-tolerating, UV-resistant bacteria from continental Antarctica. Syst. Appl. Microbiol. 27 (6), 636–645.

Horikoshi, K., Grant, W. (Eds.), 1998. Extremophiles: Microbial Life in Extreme Environments. Wiley-Liss, New York, p. 322.

Horikoshi, K., 1999. Alkaliphiles: some applications of their products for biotechnology. Microbiol. Mol. Biol. Rev. 63 (4), 735–775.

Hsu, H.W., Postberg, F., Sekine, Y., Shibuya, T., Kempf, S., Horanyi, M., Juhasz, A., Altobelli, N., Suzuki, K., Masaki, Y., Kuwatani, T., Tachibana, S., Sirono, S.I., Moragas-Klostermeyer, G., Srama, R., 2015. Ongoing hydrothermal activities within Enceladus. Nature 519 (7543), 207–210.

Iben Jr., I., 1965. Stellar evolution II. The evolution of a 3 M Star from the main sequence through core helium burning. Astrophys. J. 142, 1447.

Ingersoll, A.P., Dowling, T.E., Gierasch, P.J., Orton, G.S., Read, P.L., Sanchez-Lavega, A., Showman, A.P., Simon-Miller, A.A., Vasavada, A.R., 2004. Dynamics of Jupiter's atmosphere. In: Bagenal, F., Dowling, T.E., McKinnon, W.B. (Eds.), Jupiter: The Planet, Satellites and Magnetosphere. Cambridge University Press, Cambridge, pp. 105–128.

Jaakkola, S.T., Ravantti, J.J., Oksanen, H.M., Bamford, D.H., 2016. Buried alive: microbes from ancient Halite. Trends Microbiol. 24 (2), 148–160.

Jakosky, B.M., Westall, F., Brack, A., 2007. Mars. In: Sullivan III, W.T., Baross, J.A. (Eds.), Planets and Life: The Emerging Science of Astrobiology. Cambridge University Press, New York, pp. 357–384.

Javor, B.J., 1984. Growth potential of halophilic bacteria isolated from solar salt environments: carbon sources and salt requirements. Appl. Environ. Microbiol. 48 (2), 352–360.

Jonsson, K., Rabbow, E., Schill, R.O., Harms-Ringdahl, M., Rettberg, P., 2008. Tardigrades survive exposure to space in low earth orbit. Curr. Biol. 18 (17), R729–R731.

Kamminga, H., 1982. Life from space—a history of panspermia. Vistas Astron. 26 (2), 67–86.

Kashefi, K., Lovley, D.R., 2003. Extending the upper temperature limit for life. Science 301 (5635), 934.

Kasting, J.F., Whitmire, D.P., Reynolds, R.T., 1993. Habitable zones around main sequence stars. Icarus 101 (1), 108–128.

Kato, C., 1999. Barophiles (Piezophiles). In: Horikoshi, K., Tsujii, K. (Eds.), Extremophiles in Deep-Sea Environments. Springer, Toyko, pp. 91–112.

Kato, C., Li, L., Nogi, Y., Nakamura, Y., Tamaoka, J., Horikoshi, K., 1998. Extremely barophilic bacteria isolated from the Mariana Trench, challenger Deep, at a depth of 11,000 meters. Appl. Environ. Microbiol. 64 (4), 1510–1513.

Kelly, L.C., Cockell, C.S., Summers, S., 2012. Diverse microbial species survive high ammonia concentrations. Int. J. Astrobiol. 11 (2), 125–131.

Kieffer, S.W., Lu, X., Bethke, C.M., Spencer, J.R., Marshak, S., Navrotsky, A., 2006. A Clathrate reservoir hypothesis for Enceladus' South Polar Plume. Science 314 (5806), 1764–1766.

Kim, K.K., Jin, L., Yang, H.C., Lee, S.T., 2007. Halomonas gomseomensis sp. nov., Halomonas janggokensis sp. nov., Halomonas salaria sp. nov. and Halomonas deitrificans sp. nov., moderately halophilic bacteria isolated from saline water. Int. J. Syst. Evol. Microbiol. 57 (Pt. 4), 675–681.

King, G.M., Weber, C.F., 2007. Distribution, diversity and ecology of aerobic CO-oxidizing bacteria. Nat. Rev. Microbiol. 5 (2), 107–118.

Kish, A., Griffin, P.L., Rogers, K.L., Fogel, M.L., Hemley, R.J., Steele, A., 2012. High-pressure tolerance in Halobacterium salinarum NRC-1 and other non-piezophilic prokaryotes. Extremophiles 16 (2), 355–361.

Krulwich, T.A., Liu, J., Morino, M., Fujisawa, M., Ito, M., Hicks, D.B., 2011. Adaptive mechanisms of extreme Alkaliphiles. In: Horikoshi, K. (Ed.), Extremophiles Handbook. Springer, New York, pp. 119–139.

Kurr, M., Huber, R., Konig, H., Jannasch, H.W., Fricke, H., Trincone, A., Krisjansson, J.K., Stetter, K.O., 1991. Methanopyrus kandleri, gen. and sp. nov. represents a novel group of hyperthermophilic methanogens, growing at 110°C. Arch. Microbiol. 156 (4), 239–247.

Larsen, E.S., 1929. The temperatures of Magmas. Am. Mineral. 14, 81–94.

Le Gall, A., Malaska, M.J., Lorenz, R.D., Janssen, M.A., Tokano, T., Hayes, A.G., Mastrogiuseppe, M., Lunine, J.I., Veyssiere, G., Encrenaz, P., Karatekin, O., 2016. Composition, seasonal change, and bathymetry of Ligeia Mare, Titan, derived from its microwave thermal emission. J. Geophys. Res. Planets 121, 233–251.

Leuko, S., Domingos, C., Parpart, A., Reitz, G., Rettberg, P., 2015. The survival and resistance of Halobacterium salinarum NRC-1, Halococcus hamelinensis, and Halococcus morrhuae to simulated outer space solar radiation. Astrobiology 15 (11), 987–997.

Li, L., Wing, B.A., Bui, T.H., McDermott, J.M., Slater, G.F., Wei, S., Lacrampe-Couloume, G., Lollar, B.S., 2016. Sulfur mass-independent fractionation in subsurface fracture waters indicates a long-standing sulfur cycle in Precambrian rocks. Nat. Commun. 7, 13252.

Libby, W.F., 1974. Life on Jupiter? Orig. Life 5 (3), 483–486.

Lin, L., Wang, P., Rumble, D., Lippmann-Pipke, J., Boice, E., Pratt, L., Lollar, B., Brodie, E., Hazen, T., Andersen, G., DeSantis, T., 2006. Long-term sustainability of a high-energy, low-diversity crustal biome. Science 314, 479–482.

Lin, L.H., Hall, J., Lippmann-Pipke, J., Ward, J.A., Lollar, B.S., DeFlaun, M., Rothmel, R., Moser, D., Gihring, T.M., Mislowack, B., Onstott, T.C., 2005. Radiolytic H2 in continental crust: nuclear power for deep subsurface microbial communities. Geochem. Geophys. Geosyst. 6 (7). Q07003.

Lopes, R.M.C., Kirk, R.L., Mitchell, K.L., LeGall, A., Barnes, J.W., Hayes, A., Kargel, J., Wye, L., Radebaugh, J., Stofan, E.R., Janssen, M.A., Neish, C.D., Wall, S.D., Wood, C.A., Lunine, J.I., Malaska, M.J., 2013. Cryovolcanism on Titan: new results from Cassini RADAR and VIMS. J. Geophys. Res. Planets 118 (3), 416–435.

Lowenstein, T.K., Schubert, B.A., Timofeeff, M.N., 2011. Microbial communities in fluid inclusions and long-term survival in halite. GSA Today 21 (1), 4–9.

Lunine, J.I., Atreya, S., 2008. The methane cycle on Titan. Nat. Geosci. 1, 159–164.

Lunine, J.I., Rizk, B., 2007. Titan. In: Sullivan III, W.T., Baross, J.A. (Eds.), Planets and Life: The Emerging Science of Astrobiology. Cambridge University Press, New York, pp. 424–443.

Madigan, M., Oren, A., 1999. Thermophilic and halophilic extremophiles. Curr. Opin. Microbiol. 2 (3), 265–269.

Madigan, M.T., Mars, B.L., 1997. Extremophiles. Sci. Am. 276 (4), 82–87.

Marietou, A., Nguyen, A.T.T., Allen, E.E., Bartlett, D.H., 2014. Adaptive laboratory evolution of *Escherichia coli* K-12 MG1655 for growth at high hydrostatic pressure. Front. Microbiol. 5, 749.

McCollum, T.M., 1999. Methanogenesis as a potential source of chemical energy for primary biomass production by autotrophic organisms in hydrothermal systems on Europa. J. Geophys. Res. Planets 104 (E12), 30729–30742.

McCollom, T.M., Seewald, J.S., 2013. Serpentinites, hydrogen, and life. Elements 9 (2), 129–134.

McEwen, A.S., Ojha, L., Dundas, C.M., Mattson, S.S., Byrne, S., Wray, J.J., Cull, S.C., Murchie, S.L., Thomas, N., Gulick, V.C., 2011. Seasonal flows on warm Martian slopes. Science 333 (6043), 740–743.

McGenity, T.J., Gemmell, R.T., Grant, W.D., Stan-Lotter, H., 2000. Origins of halophilic microorganisms in ancient salt deposits. Environ. Microbiol. 2 (3), 243–250.

McKay, C.P., Smith, H.D., 2005. Possibilities for methanogenic life in liquid methane on the surface of Titan. Icarus 178 (1), 274–276.

McKay, C.P., Porco, C.C., Altheide, T., Davis, W.L., Kral, T.A., 2008. The possible origin and persistence of life on Enceladus and detection of biomarkers in the plume. Astrobiology 8 (5), 909–919.

McMahon, S., Parnell, J., 2014. Weighing the deep continental biosphere. FEMS Microbiol. Ecol. 87 (1), 113–120.

McNichol, J., Gordon, R., 2012. Are we from outer space? A critical review of the panspermia hypothesis. In: Sechbach, J. (Ed.), Genesis—In the Beginning: Precursors of Life. Chemical Models and Early Biological Evolution. Springer, Dordrecht, pp. 591–620.

Meckenstock, R.U., von Netzer, F., Stumpp, C., Lueders, T., Himmelberg, A.M., Hertkorn, N., Schmitt-Kopplin, P., Harir, M., Hosein, R., Haque, S., Schulze-Makuch, D., 2014. Water droplets in oil are microhabitats for microbial life. Science 345 (6197), 673–676.

Melosh, H.J., Ekholm, A.G., Showman, A.P., Lornz, R.D., 2003. The temperature of Europa's subsurface water ocean. Icarus 168 (2), 498–502.

Mendez, A., 2001. Planetary habitable zones: the spatial distribution of life on planetary bodies. In: Chela-Flores, J., Tobias, O., Raulin, F. (Eds.), First Steps in the Origin of Life in the Universe. Springer Science+Business Media, Dordrecht, pp. 211–214.

Mesbah, N., Wiegel, J., 2005. Halophilic thermophiles: a novel group of extremophiles. In: Satyanarayana, T., Johri, B.N. (Eds.), Microbial Diversity: Current Perspectives and Potential Applications. I.K. International Publishing House Pvt. Ltd, New Delhi, pp. 91–118.

Messerli, M.A., Amaral-Zettler, L.A., Zettler, E., Jung, S.K., Smith, P.J.S., Sogin, M.L., 2005. Life at acidic pH imposes an increased energetic cost for a eukaryotic acidophile. J. Exp. Biol. 208 (Pt. 13), 2569–2579.

Michalski, J., Cuadros, J., Niles, P., Parnell, J., Rogers, A., Wright, S., 2013. Groundwater activity on Mars and implications for a deep biosphere. Nat. Geosci. 6, 133–138.

Mikucki, J.A., Pearson, A., Johnston, D.T., Turchyn, A.V., Farquhar, J., Schrag, D.P., Anbar, A.D., Priscu, J.C., Lee, P.A., 2009. A contemporary microbially maintained subglacial ferrous "Ocean" Science 324 (5925), 397–400.

Mikucki, J.A., Auken, E., Tulaczyk, S., Virginia, R.A., Schamper, C., Sorensen, K.I., Doran, P.T., Dugan, H., Foley, N., 2015. Deep groundwater and potential subsurface habitats beneath an Antarctic dry valley. Nat. Commun. 6, 6831.

Mikucki, J.A., Priscu, J.C., 2007. Bacterial diversity associated with blood falls, a subglacial outflow from the Taylor Glacier, Antarctica. Appl. Environ. Microbiol. 73 (12), 4029–4039.

Miroschnichenko, M.L., Bonch-Osmolovskaya, E.A., 2006. Recent developments in the thermophilic microbiology of deep-sea hydrothermal vents. Extremophiles 10 (2), 85–96.

Moissl-Eichinger, C., Cockell, C., Rettberg, P., 2016. Venturing into new realms? microorganisms in space. FEMS Microbiol. Rev. 40 (5), 722–737.

Morrill, P.L., Brazelton, W.J., Kohl, L., Rietze, A., Miles, S.M., Kavanagh, H., Schrenk, M.O., Ziegler, S.E., Lang, S.Q., 2014. Investigations of potential microbial methanogenic and carbon monoxide utilization pathways in ultra-basic reducing springs associated with present-day continental serpentinization: the Tablelands, NL, CAN. Front. Microbiol. 5, 613.

Mousis, O., Pargamin, J., Grasset, O., Sotin, C., 2002. Experiments in the NH_3-H_2O system in the [0, 1 GPa] pressure range – implications for the deep liquid layer of large icy satellites. Geophys. Res. Lett. 29 (24), 45-1–45-4.

Muller, A.W., 1985. Thermosynthesis by biomembranes: energy gain from cyclic temperature changes. J. Theor. Biol. 115 (3), 429–453.

Muller, A.W., 2003. Finding extraterrestrial organisms living on thermosynthesis. Astrobiology 3 (3), 555–564.

Muller, A.W., Schulze-Makuch, D., 2006. Thermal energy and the origin of life. Orig. Life Evol. Biosph. 36 (2), 177–189.

Mykytczuk, N.C., Wilhelm, R.C., Whyte, L.G., 2012. Planococcus halocryophilus sp. nov., an extreme sub-zero species from high Arctic permafrost. Int. J. Syst. Evol. Microbiol. 62 (Pt. 8), 1937–1944.

Mykytczuk, N.C., Foote, S.J., Omelon, C.R., Southam, G., Greer, C.W., Whyte, L.G., 2013. Bacterial growth at -15°C; molecular insights from the permafrost bacterium Planococcus halocryophilus Or1. ISME J. 7 (6), 1211–1226.

Nimmo, F., Pappalardo, R.T., 2016. Ocean worlds in the outer solar system. J. Geophys. Res. Planets 121 (8), 1378–1399.

Novikova, N., Deshevaya, E., Levinshkikh, M., Polikarpov, N., 2015. Study of the effects of the outer space environment on dormant forms of microorganisms, fungi and plants in the 'Expose-R' experiment. Int. J. Astrobiol. 14 (1), 137–142.

Olsson-Francis, K., Cockell, C.S., 2010. Experimental methods for studying microbial survival in extraterrestrial environments. J. Microbiol. Methods 80 (1), 1–13.

Oren, A., 2002. Molecular ecology of extremely Halophilc Archaea and bacteria. FEMS Microbiol. Ecol. 39 (1), 1–7.

Oren, A., 2005. A hundred years of Dunaliella research: 1905–2005. Saline Syst. 1, 2.

Pace, N.R., 1997. A molecular view of microbial diversity and the biosphere. Science 276 (5313), 734–740.

Pace, N.R., 2001. The universal nature of biochemistry. PNAS 98 (3), 805–808.

Pappalardo, R.T., Head, J.W., Greeley, R., Sullivan, R.J., Pilcher, C., Schubert, G., Moore, W.B., Carr, M.H., Moore, J.M., Belton, M.J., Goldsby, D.L., 1998. Geological evidence for solid-state convection in Europa's ice shell. Nature 391, 365–368.

Pappalardo, R.T., et al., 1999. Does Europa have a subsurface ocean? evaluation of the geological evidence. J. Geophys. Res. 104 (E10), 24015–24055.

Parkes, R.J., Linnane, C.D., Webster, G., Sass, H., Weightman, A.J., Hornibrook, E.R., Horsfield, B., 2011. Prokaryotes stimulate mineral H2 formation for the deep biosphere and subsequent thermogenic activity. Geology 39 (3), 219–222.

Parkinson, C.D., Liang, M.C., Yung, Y.L., Kirschivnk, J.L., 2008. Habitability of Enceladus: planetary conditions for life. Orig. Life Evol. Biosph. 38 (4), 355–369.

Pasini, D.L.S., Price, M.C., 2015. Panspermia survival scenarios for organisms that survive typical hypervelocity solar system impact events. In: EPSC Abstracts, vol. 9, EPSC2014-68.

Pitt, J.I., Christian, J.H., 1968. Water relations of xerophilic fungi isolated from prunes. Appl. Microbiol. 16 (12), 1853–1858.

Plemenitas, A., Lenassi, M., Konte, T., Kejzar, A., Zajc, J., Gostincar, C., Gunde-Cimerman, N., 2014. Adaptation to high salt concentrations in halotolerant/halophilic fungi: a molecular perspective. Front. Microbiol. 5, 199.

Porco, C.C., et al., 2006. Cassini observed the active south pole of Enceladus. Science 311 (5766), 1393–1401.

Postberg, F., Kempf, S., Schmidt, J., Brilliantov, N., Beinsen, A., Abel, B., Buck, U., Srama, R., 2009. Sodium salts in E-ring ice grains from an ocean below the surface of Enceladus. Nature 459, 1098–1101.

Preiss, L., Hicks, D., Suzuki, S., Meier, T., Krulwich, T., 2015. Alkaliphilic bacteria with impact on industrial applications, concepts of early life forms, and bioenergetics of ATP synthesis. Front. Bioeng. Biotechnol. 3, 75.

Prockter, L.M., 2005. Ice in the solar system. J. Hopkins APL Tech. Dig. 26 (2).

Purcell, A.M., Mikucki, J.A., Achberger, A.M., Alekhina, I.A., Barbante, C., Christner, B.C., Ghosh, D., Michaud, A.B., Mitchell, A.C., Priscu, J.C., Scherer, R., 2014. Microbial sulfur transformations in sediments from Subglacial Lake Whillans. Front. Microbiol. 5, 594.

Ram, M., Verberkmoes, N.C., Thelen, M.P., Tyson, G.W., Baker, B.J., Blake 2nd, R.C., Shah, M., Hettich, R.L., Banfield, J.F., 2005. Community proteomics of a natural microbial biofilm. Science 308 (5730), 1915–1920.

Raulin, F., Brasse, C., Poch, O., Coll, P., 2012. Prebiotic-like chemistry on Titan. Chem. Soc. Rev. 41, 5380–5393.

Raup, D.M., Valentine, J.W., 1983. Multiple origins of life. PNAS 80, 2981–2984.

Reed, D.W., Fujita, Y., Delwiche, M.E., Blackwelder, D.B., Sheridan, P.P., Uchida, T., Colwel, F.S., 2002. Microbial communities from methane hydrate-bearing deep marine sediments in a forearc basin. Appl. Environ. Microbiol. 68 (8), 3759–3770.

Reith, F., 2011. Life in the deep subsurface. Geology 39 (3), 287–288.

Raymond, J.A., Fritsen, C., Shen, K., 2007. An ice-binding protein from an Antarctic sea ice bacterium. FEMS Microbiol. Ecol. 61 (2), 214–221.

Raymond, J.A., Knight, C.A., 2003. Ice binding, recrystallization inhibition, and cryoprotective properties of ice-active substances associated with Antarctic sea ice diatoms. Cryobiology 46 (2), 174–181.

Roadcap, G.S., Sanford, R.A., Jin, Q., Pardinas, J.R., Bethke, C.M., 2006. Extremely alkaline (pH > 12) ground water hosts diverse microbial community. Groundwater 44 (4), 511–517.

Rothschild, L.J., Mancinelli, R.L., 2001. Life in extreme environments. Nature 409 (6823), 1092–1101.

Ruiz, J., 1999. Onset of convection, heat flow and thickness of the Europa's ice shell. Earth Moon Planet. 77, 99–104.

Rummel, J.D., et al., 2014. A new analysis of Mars "special regions": findings of the second MEPAG special regions science analysis group (SR-SAG2). Astrobiology 14 (11), 887–968.

Samarkin, V.A., Madigan, M.T., Bowles, M.W., Casciotti, K.L., Priscu, J.C., McKay, C.P., Joye, S.B., 2010. Abiotic nitrous oxide emission from the hypersaline Don Juan Pond in Antarctica. Nat. Geosci. 3 (5), 341–344.

Sancho, L.G., de la Torre, R., Horneck, G., Ascaso, C., de Los Rios, A., Pintado, A., Wierzchos, J., Schuster, M., 2007. Lichens survive in space: results from the 2005 LICHENS experiment. Astrobiology 7 (3), 443–454.

Santos, H., da Costa, M.S., 2002. Compatible solutes of organisms that live in hot saline environments. Environ. Microbiol. 4, 501–509.

Schenk, P.M., Chapman, C.R., Zahnle, K., Moore, J.M., 2004. Ages and interiors: the cratering record of the Galilean satellites. In: Bagenal, F., Dowling, T.E., McKinnon, W.B. (Eds.), Jupiter The Planet, Satellites and Magnetosphere. Cambridge University Press, Cambridge, pp. 427–456.

Schopf, J.W., Kudryavtsev, A.B., Czaja, A.D., Tripathi, A.B., 2007. Evidence of archean life: stromatolites and microfossils. Precambrian Res. 158 (3-4), 141–155.

Schubert, B.A., Timofeeff, M.N., Lowenstein, T.K., Polle, J.E.W., 2010. *Dunaliella* cells in fluid inclusions in halite: significance for long-term survival of prokaryotes. Geomicrobiol J. 27 (1), 61–95.

Schubert, G., Sohl, F., Hussmann, H., 2009. Interior of Europa. In: Pappalardo, R., McKinnon, W.B., Khurana, K. (Eds.), Europa. The University of Arizona Press/Lunar and Planetary Institute, Tuscon, AZ/Houston, TX.

Schuerger, A.C., Nicholson, W., 2016. Twenty-three species of hypobarophilic bacteria recovered from diverse ecosystems exhibit growth under simulated martian conditions at 0.7 kPa. Astrobiology 16 (5), 335–347.

Schulz-Makuch, D., Irwin, L.N., 2006. Life in the Universe: Expectations and Constraints. Springer, New York.

Schulz-Makuch, D., Grinspoon, D.H., 2005. Biologically enhanced energy and carbon cycling on titan? Astrobiology 5 (4), 560–567.

Sekine, Y., Shibuya, T., Postberg, F., Hsu, H.W., Suzuki, K., Masaki, Y., Kuwatani, T., Mori, M., Hong, P.K., Yoshizaki, M., Tachibana, S., Sirono, S.I., 2015. High-temperature water-rock interactions and hydrothermal environments in the chondrite-like core of Enceladus. Nat. Commun. 6, 8604.

Sharma, A., Scott, J.H., Cody, G.D., Fogel, M.L., Hazen, R.M., Hemley, R.J., Huntress, W.T., 2002. Microbial activity at gigapascal pressures. Science 295 (559), 1514–1516.

Siddiqui, K.S., Williams, T.J., Wilkins, D., Yau, S., Allen, M.A., Brown, M.V., Lauro, F.M., Cavicchioli, R., 2013. Psychrophiles. Annu. Rev. Earth Planet. Sci. 41, 87–115.

Siegel, B.Z., McMurty, G., Siegel, S.M., Chen, J., LaRock, P., 1979. Life in the calcium chloride environment of Don Juan Pond, Antarctica. Nature 280 (5725), 828–829.

Sohl, F., Hussmann, H., Schwentker, B., Spohn, T., Lorenz, R.D., 2003. Interior structure models and tidal Love numbers of Titan. J. Geophys. Res. Planets 108 (E12). 4-1.

Solar System Exploration Survey, National Research Council, 2003. Frontiers in Solar System Exploration. National Academies Press, Washington, D.C. https://doi.org/10.17226/10898.

Solomonidou, A., Coustenis, A., Bampasidis, G., Kyriakopoulos, K., Moussas, X., Bratsolis, E., Hirtzig, M., 2011. Water oceans of Europa and other moons: implications for life in other solar systems. J. Cosmol. 13, 4191–4211.

Sonderlund, K.M., Schmidt, B.E., Wicht, J., Blankenship, D.D., 2014. Ocean-driven heating of Europa's icy shell at low latitudes. Nat. Geosci. 7, 16–19.

Spencer, J.R., Barr, A.C., Esposito, L.W., Helfenstein, P., Ingersoll, A.P., Jaumann, R., McKay, C.P., Nimmo, F., Waite, J.H., 2009. Enceladus: an active cryovolcanic satellite. In: Dougherty, M.K., Esposito, L.W., Krimigis, S.M. (Eds.), Saturn from Cassini-Huygens. Springer, New York, pp. 683–724.

Stan-Lotter, H., Fendrihan, S., 2015. Halophilic archaea: life with desicccation, raditation, and oligotrophy over geological times. Live 5 (3), 1487–1496.

Stern, J., Sutter, B., Freissinet, C., Navarro-González, R., McKay, C., Archer, P., Buch, A., Brunner, A., Coll, P., Eigenbrode, J., Fairen, A., 2015. Evidence for indigenous nitrogen in sedimentary and aeolian deposits from the Curiosity rover investigations at Gale crater, Mars. Pnas 112, 4245–4250.

Stevenson, A., Burkhardt, J., Cockell, C., Cray, J., Dijksterhuis, J., Fox-Powell, M., Kee, T., Kminek, G., McGenity, T., Timmis, K., Timson, D., 2014. Multiplication of microbes below 0.690 water activity: implications for terrestrial and extraterrestrial life. Environ. Microbiol. 17 (2), 257–277.

Stevenson, A., et al., 2015a. Is there a common water-activity limit for the three domains of life? ISME J. 9 (6), 1333–1351.

Stevenson, J., Lunine, J., Clancy, P., 2015b. Membrane alternatives in worlds without oxygen: creation of an azotosome. Sci. Adv. 1 (1). p. e1400067.

Stofan, E.R., Elachi, C., Lunine, J.I., Lorenz, R.D., Stiles, B., Mitchell, K.L., Ostro, S., Soderblom, L., Wood, C., Zebker, H., Wall, S., 2007. The lakes of Titan. Nature 445 (7123), 61–64.

Suzuki, S., Kuenen, J., Schipper, K., van der Velde, S., Ishii, S., Wu, A., Sorokin, D., Tenney, A., Meng, X., Morrill, P., Kamagata, Y., 2014. Physiological and genomic features of highly alkaliphilic hydrogen-utilizing Betaproteobacteria from a continental serpentinizing site. Nat. Commun. 5, 3900.

Svetlichny, V.A., Sokolova, T.G., Gerhardt, M., Ringpfeil, M., Kostrikina, N.A., Zavarzin, G.A., 1991. Carboxydothermus hydrogenoformans gen. nov., sp. nov., a CO-utilizing thermophilic anaerobic bacterium from hydrothermal environments of Kunashir Island. Syst. Appl. Microbiol. 14 (3), 254–260.

Takai, K., Nakamura, K., Toki, T., Tsunogai, U., Miyazaki, M., Miyazaki, J., Hirayama, H., Nakagawa, S., Nunoura, T., Horikoshi, K., 2008. Cell proliferation at 122oC and isotopically heavy CH4 production by a hyperthermophilic methanogen under high-pressure cultivation. Proc. Natl. Acad. Sci. U. S. A. 105 (31), 10949–10954.

Taubner, R.-S., Schleper, C., Firneis, M.G., Rittman, S.K.-M., 2015. Assessing the ecophysiology of methanogens in the context of recent astrobiological and planetological studies. Life 5 (4), 1652–1686.

Taylor, G.R., Bailey, J.V., Benton, E.V., 1975. Physical dosimetric evaluations in the Apollo 16 microbial response experiment. Life Sci. Space Res. 13, 135–141.

Thauer, R.K., Jungermann, K., Kecker, K., 1977. Energy conservation in chemotrophic anaerobic bacteria. Bacteriol. Rev. 41 (1), 100–180.

Tiago, I., Chung, A.P., Verissimo, A., 2004. Bacterial diversity in a nonsaline alkaline environment: heterotrophic aerobic populations. Appl. Environ. Microbiol. 70 (12), 7378–7387.

Tobie, G., Grasset, O., Lunine, J.I., Mocquet, A., Sotin, C., 2005. Titan's internal structure inferred from a coupled thermal-orbital model. Icarus 175 (2), 496–502.

Tobie, G., Lunine, J.I., Sotin, C., 2006. Episodic outgassing as the origin of atmospheric methane on Titan. Nature 440, 61–64.

Tobie, G., Giese, B., Hurford, T., Lopes, R., Nimmo, F., Postberg, F., Retherfore, K., Schmidt, J., Spencer, J., Tokano, T., Turtle, E., 2009. Surface, subsurface and atmosphere exchanges on icy moons. Space Sci. Rev. 153 (1-4), 375–410.

Tulaczyk, S., Mikucki, J.A., Siegfried, M.R., Priscu, J.C., Barcheck, C.G., Beem, L.H., Behar, A., Burnett, J., Christner, B.C., Fisher, A.T., Fricker, H.A., Mankoff, K.D., Powell, R.D., Rack, F., Sampson, D., Scherer, R.P., Schwartz, S.Y., Team Wissard Sci, 2014. WISSARD at Subglacial Lake Whillans, West Antarctica: scientific operations and initial observations. Ann. Glaciol. 55, 51–58.

Turner, J., Anderson, P., Lachlan-Cope, T., Colwell, S., Phillips, T., Kirchgaessner, A., Marshall, G.J., King, J.C., Bracegirdle, T., Vaughan, D.G., Lagun, V., Orr, A., 2009. Record low surface air temperature at Vostok station, Antarctica. J. Geophys. Res. Atmos. 114. D24102. https://doi.org/10.1029/2009JD012104.

United Nations Scientific Committee on the Effects of Atomic Radiation, 2008. Sources and Effects of Ionizing Radiation. United Nations, New York.

Van Dover, C.L., Desbruyeres, D., Segonzac, M., Comtet, T., Saldanha, L., Fiala-Medioni, A., Langmuir, C., 1996. Biology of the lucky strike hydrothermal field. Deep-Sea Res. I Oceanogr. Res. Pap. 43 (9), 1509–1529.

Vance, S.D., Hand, K.P., Pappalardo, R.T., 2016. Geophysical controls of chemical disequilibria in Europa. Geophys. Res. Lett. 43 (10), 4871–4879.

Vanlint, D., Mitchell, R., Bailey, E., Meersman, F., McMillan, P.F., Michiels, C.W., Aertsen, A., 2011. Rapid acquisition of Gigapascal-high-pressure resistance by Escherichia coli. MBio 2 (1), e00130–00110.

Van Dover, C.L., 2000. The ecology of deep-sea hydrothermal vents. Princeton University Press, Princeton, NJ.

Vogel, G., 1999. Expanding the habitable zone. Science 286 (5437), 70–71.

Vreeland, R.H., Rosenzweig, W.D., Powers, D.W., 2000. Isolation of a 250 million-year-old halotolerant bacterium from a primary salt crystal. Nature 407, 897–900.

Wagner, W., Kretzschmar, H.-J., 2008. International steam tables: properties of water and steam based on the industrial formulation IAPWS-IF97, second ed. Springer-Verlag, Berlin.

Waite Jr., J.H., Lewis, W.S., Magee, B.A., Lunine, J.I., McKinnon, W.B., Glein, C.R., Mousis, O., Young, D.T., Brockwell, T., Westlake, J., Nguyen, M.J., Teolis, B.D., Niemann, N.B., McNutt Jr., R.L., Perry, M., Ip, W.H., 2009. Liquid water on Enceladus from observations of ammonia and 40Ar in the plume. Nature 460, 487–490.

White, O., et al., 1999. Genome sequence of the radioresistant bacterium Deinococcus radiodurans R1. Science 286 (5444), 1571–1577.

Williams, J.P., Hallsworth, J.E., 2009. Limits of life in hostile environments: no barriers to biosphere function? Environ. Microbiol. 11 (12), 3292–3308.

Wright, A., Siegert, M., 2012. A fourth inventory of Antarctic subglacial lakes. Antarct. Sci. 24 (6), 659–664.

Youle, M., Rohwer, F., Stacy, A., Whiteley, M., Steel, B.C., Delalez, N.J., Nord, A.L., Berry, R.M., Armitage, J.P., Kamoun, S., Hougenhout, S., Diggle, S.P., Gurney, J., Pollitt, E.J.G., Boetius, A., Cary, S.C., 2012. The microbial olympics. Nat. Rev. Microbiol. 10, 583–588.

Zolotov, M., Shock, E.L., 2001. Composition and stability of salts on the surface of Europa and their oceanic origin. J. Geophys. Res. 106 (E12), 32815–32827.

Zolotov, M., Shock, E.L., 2004. Brine pockets in the icy shell on Europa: distribution, chemistry, and habitbility. In: Workshop on Europa's Icy Shell: Past, Present, and Future. Houston, TX.

Zolotov, M.Y., Kargel, J.S., 2009. On the chemical composition of Europa's icy shell, ocean and underlying rocks. In: Pappalardo, R., McKinnon, W.B., Khurana, K. (Eds.), Europa'. The University of Arizona Press/Lunar and Planetary Institute, Tuscon, AZ/Houston, TX.

Zolotov, M.Y., 2007. An oceanic composition on early and today's Enceladus. Geophys. Res. Lett. 34 (23). https://doi.org/10.1029/2007GL031234.

Zorzano, M.P., Mateo-Marti, E., Prieto-Ballesteros, O., Osuna, S., Renno, N., 2009. Stability of liquid saline water on present day Mars. Geophys. Res. Lett. 36 (20). https://doi.org/10.1029/2009GL040315.

FURTHER READING

Angelov, A., Liebl, W., 2006. Insights into extreme thermoacidophily based on genome analysis of Picrophilus torridus and other thermoacidophilic archaea. J. Biotechnol. 126 (1), 3–10.

Feldman, W.C., Prettyman, T.H., Maurice, S., Plaut, J.J., Bish, D.L., Vaniman, D.T., Mellon, M.T., Metzger, A.E., Squyres, S.W., Karunatillake, S., Boynton, W.V., Elphic, R.C., Funsten, H.O., Lawrence, D.J., Tokar, R.L., 2004. Global distribution of near-surface hydrogen on Mars. J. Geophys. Res. Planets 109 (E9).

Lowrie, W., 2007. Fundamentals of Geophysics, second ed. Cambridge University Press, New York.

MacElroy, R.D., 1974. Some comments on the evolution of extremophiles. Biosystems 6 (1), 74–75.

Mikucki, J.A., Lee, P.A., Ghosh, D., Purcell, A.M., Mitchell, A.C., Mankoff, K.D., Fisher, A.T., Tulaczyk, S., Carter, S., Siegfried, M.R., Fricker, H.A., Hodson, T., Coenen, J., Powell, R., Scherer, R., Vick-Majors, T., Achberger, A.A., Christner, B.C., Tranter, M., Team, W.S., 2016. Subglacial Lake Whillans microbial biogeochemistry: a synthesis of current knowledge. Philos. Transact. A Math. Phys. Eng. Sci. 374 (2059).

Moeller, R., Raguse, M., Leuko, S., Berger, T., Hellweg, C.E., Fujimori, A., Okayasu, R., Homeck, G., Grp, S.R., 2017. STARLIFE: an international campaign to study the role of galactic cosmic radiation in astrobiological model systems. Astrobiology 17 (2), 101–109.

THE EMERGENCE OF STRUCTURED, LIVING, AND CONSCIOUS MATTER IN THE EVOLUTION OF THE UNIVERSE: A THEORY OF STRUCTURAL EVOLUTION AND INTERACTION OF MATTER

Dorian Aur*, Jack A. Tuszynski[†]

*University of Victoria, Victoria, BC, Canada
[†]University of Alberta, Edmonton, AB, Canada

CHAPTER OUTLINE

Habitability of the Universe Before Earth, editors: Richard Gordon & Alexei Sharov, Volume 1 in the series:
Astrobiology: Exploring Life on Earth and Beyond, series editors: Pabulo Henrique Rampelotto,
Joseph Seckbach & Richard Gordon. ISSN 2468-6352. https://doi.org/10.1016/B978-0-12-811940-2.00010-1

1 INTRODUCTION

At the beginning of the last century, Oparin (2003) made an attempt to unveil the mystery of life. He identified organic chemicals, which may have spontaneously led to primitive living organisms in the Earth's primordial ocean. Thirty years later, Stanley Miller performed the first experiment aimed at validating Oparin's hypothesis of primordial soup. He showed that complex organic molecules, amino acids, can be synthesized from inorganic substances in a short period of time. The experiment partially validated Oparin's hypothesis; however, the main issue remained unresolved.

Many think that our Earth is the sole place in the whole universe where a few billion years ago from inert matter organic molecules were built. This naive but very common perspective is perceived as the anthropic principle (see, Kane et al., 2002; Hetesi & Balázs, 2006). Within such a paradigm, the Earth is the center of life in the entire universe.

At least two essential steps of the evolution still remain a mystery. The first one is the creation of simple organisms. Contrary to common belief, Hoyle and Wickramasinghe (1981) did not concur with Oparin's idea that primordial soup would be the place for the emergence of the first life forms. Instead, they pioneered a different idea of cometary panspermia. Apparently in the 1970s, this was a radical idea and the concept of cometary panspermia was strongly rejected by the main stream of research.

Clear evidence points to terrestrial bacteria living in outer space (Johnson et al., 2015) and that extremophile tardigrades can survive in these harsh environmental conditions. They can withstand loss of water, extremely low temperatures below freezing, high pressures, or even radiation since they can efficiently repair damage to their DNA (Persson et al., 2011). Life may be more abundant than we imagine and different forms of life may exist in the Universe. Ehrenfreund and Charnley have not only shown the existence of organic molecules in the interstellar medium, they presented a possible path from molecular clouds to the early solar system and Earth (Ehrenfreund, & Charnley, 2000). Organic compounds found in meteorites have their origins in the interstellar medium. During the formation of the solar system, these compounds were incorporated into newly shaped planets.

Another evidence for Hoyle-Wickramasinghe's hypothesis came from the measurements of genetic complexity. It is often assumed that the complexity of the genome has influenced all functional levels and that potentially such information can be extracted from comparative genomic analysis. The genome constitutes the major source of heritable information and its complexity can be used to estimate the age of life. Hypothesizing an exponential growth of biological complexity, Sharov and Gordon have shown that the beginning of life in the Universe may have occurred about 10 billion years ago. In this scenario, life could not have originated on Earth since the Earth would have been too young (<5 billion years). To reach through evolution to the complexity of bacteria, about 5 billion years would have been required after the Big Bang (Sharov & Gordon, 2013). A fresh critical review of the panspermia hypothesis was recently published (see McNichol & Gordon, 2012).

We already know that life on Earth has astrophysical significance. Every living thing ever present on Earth has incorporated chemical components from extraterrestrial space (supernovae explosions). However, the presence of life in space does not solve the fundamental issue. The entire story just moves the origin of life to a different place in the Universe.

In addition, there is a certain degree of confusion regarding the role of dissipation in biological structures. The prospects of explaining organized life solely based on dissipation principles are disappointing. Biological organisms are not so open; they are not gasoline engines. Recently, England made another attempt (England, 2015), showing that in a simulation group of atoms can restructure their position to dissipate more energy if an external source of energy drives the process and they are surrounded by a heat bath (e.g., ocean, atmosphere). Indeed, it may take millions of years for life to occur from nonliving matter and it will be a slow process of evolution. The exchange of energy and matter in biological systems with the environment is slow and highly constrained, therefore one can consider in a theoretical framework slight variations of entropy ($dS \approx 0$).

Can England replicate in his lab the entire process? No, he cannot. In this sense, the remarkable work on a cellular automaton of Christopher Langton (1989) at the Santa Fe Institute is not a failure, it is a step forward in showing that diverse, complex structures can be formed based on simple fixed rules.

It is often assumed that there is little or no fundamental difference between lifeless matter and living organisms. This is untrue. Many attempts to generate "simple" life forms from scratch have failed because creating simple, self-replicating cells from organic or nonorganic matter does not seem to be an easy step. We are still far from making new cells entirely from nonliving materials. The transplantation of the genome into an already existing cell and the development of the first cell from synthetic DNA (Jeon et al., 1970; Gibson et al., 2010) did not solve the fundamental problem. Therefore, it would be important to identify the process that can transform organic (inorganic) matter into simple living forms and their further evolution.

Biological evolution can lead to adaptation by generating "useful structures and functions." Overall, there is a conceptual confusion that physical evolution tends to follow the classical principle of minimum energy towards stability and equilibrium. It would be wrong to believe that this is always the case for biological processes. Our view is that nonequilibrium physics can better explain the presence of oscillations, quasiperiodic, and chaotic regimes that accompany structural changes in many biological systems. In addition, we are less able to recognize structural changes that shape biological adaptation from a nanoscale level. We are more inclined to acknowledge observable differences in "physiological activity." We are not ignoring biological adaptation. Importantly, the creative power of biological evolution appears to be triggered by physical evolution from a microscopic scale. The presence of chaotic events may significantly change local conditions and accelerate the entire process of "evolution" of biological structures. From our perspective, biological evolution and physical "evolution" do not appear to be different things. In fact what is natural selection? As an example, consider that human skin pigmentation is regarded as a product of natural selection. The pigmentation is a process that can be described by changes in molecular dynamics (tyrosine oxidation) under the effect of UV radiation. The sensitivity to UV radiation, hypopigmentation, and a predisposition to structural changes can lead to skin cancer. Hidden molecular dynamics and what happens at the nanoscale levels can change biological evolution. With this simple example, we show that structural evolution of matter can put a different light on natural selection.

Biological adaptation has a clear functional role in the life of every organism; however, changes in biological activity can be expressed in general as alterations in the concentration of substances are required to give a certain biological response (Yousefinejad & Hemmateenejad, 2015). In addition, the electronic distribution and quantum-chemical descriptors of molecular constituents are very important and determinant in changing the trajectory of biological evolution (Fransson, 2010).

The second gap seems to be related to the evolution of the human race, the so-called puzzling theory of human/ape divergence. For decades, the theory of multiregional evolution placed Neanderthals as direct ancestors of humans. Based on a mitochondrial "DNA signature," Allan Wilson advanced a different theory regarding the origins of the human race (Cann et al., 1987). The evolution of humans was driven by a pure matrilineal progression involving people who lived in Africa a few hundred thousand years ago (Walker et al., 1987). In addition, explaining the loss of fur, or in other words, how intelligence and complex behavior appeared during evolution is still a challenge (Dennett, 2017). While the aquatic ape hypothesis was strongly rejected by many researchers, the presence of water could have played a certain role in natural selection. The loss of fur may be regarded as an adaptation to environmental conditions where the presence of water has generated selective pressure for human evolution (Trauth et al., 2010).

An investigation of the origin of intelligence must necessarily face the relationship between the mind and the biological organization (Piaget, 1951). Is intelligence a genetic accident? We know that the human brain is very different: some specific structures are not present in the cerebral cortex of monkeys and the representation of tool use in humans is completely different than in monkeys (Joly et al., 2009). Has such a process of evolution provided early humans with unique brain structures? Some explain these changes as due to natural, environmental fluctuations and epigenetic plasticity (Carja & Feldman, 2012). All life on Earth appears to have a single origin, the last universal common ancestor with identifiable ancestral traits (Di Giulio, 2011). We already know that genomes are constantly changing. Their rate of change can be used as a molecular clock to understand the tree of life. Human similarity with other apes can be explained by analyzing the hominid lineage. The fact that each lineage has its own evolution may explain why the evolution from a different lineage (e.g., monkeys) did not lead to hominids.

Encouraged by unexplained megalithic constructions such as Pumapunku, Stonehenge, and the Great Pyramid of Giza, von Däniken advanced a different hypothesis of possible extraterrestrial visitations (von Däniken, 2002). Not only ancient astronauts may have helped to build these monumental constructions, they may have also genetically engineered the human race. Since natural selection usually generates a more robust outcome, von Däniken claims that this origin of human race may explain why there are so many genetic defects in humans and only a few genetic defects in animals. Therefore, the origin of human intelligence and mental illness could be linked to a genetic accident, a proof of aliens' mistakes in engineering the human race (Forsdyke, 2009; Nithianantharajah et al., 2013).

Again, placing the origins of life and intelligence out in space does not seem to solve the fundamental issue of how intelligent life has appeared in the Universe, which is one of the greatest scientific problems. The truth may lie somewhere between these two extremes. In this chapter, we explore the evolution of matter from a purely physical unstructured form, to a structured form culminating in animate matter eventually acquiring intelligent qualities. The present chapter also tries to provide answers to several intriguing questions that arise in this context. What is the source of intelligent behavior? Why are EEG oscillations correlated with cognitive abilities? Is intelligence an outcome of neuroanatomical measures, brain connectivity, or neurochemistry? And also a more general question: what is the relation between matter and intelligence?

2 THE PHYSICS OF MATTER AND STRUCTURAL EVOLUTION

Classical physics represents the first attempt to decipher the relationship between matter and motion. Space, time, motion, matter, and energy are regarded as independent concepts. For example, the motion of the objects forming the solar system can be mathematically described as the so-called n-body problem where n-material mass points interact due to gravitational forces between them according to

Newtonian mechanics which is represented using a function H, the Hamiltonian of the system (Hamilton, 1833; Meyer et al., 2008). The real mathematical difficulty of solving the n-body problem ($n \geq 3$) was perceived only late at the end of the 19th century and has attracted the interest of many mathematicians ever since (Sundman, 1913; Qiu-Dong, 1990).

Well-known equations are explicitly displayed below in this paper to stress the importance of simple rules (e.g., electronic distribution, quantum descriptors, and chaotic regimes) that can reshape complex structures in living or nonliving systems. In biological organisms, similar equations that include concentration variables can describe metabolites and biochemical complexes and their relation to observed physiological rhythms is a critical aspect (see also, Heinrich & Schuster, 1996; Berthoumieux et al., 2011).

Many physical systems can be modeled as quasi-Hamiltonian systems (Deng & Zhu, 2009). In order to analyze the problem of solving equations of motion for systems, which are not always integrable, Arnold et al. have approximated solutions using perturbation expansions in terms of the small parameter ε (Arnold et al., 1985). The Lagrangian equations of motion (Lagrange, 1772) can be expressed in the form of action angle variables:

$$H = \sum_{i=1,N} H_0(I_i) + \varepsilon V(I_1, I_2, \ldots, I_N, \theta_1, \theta_2, \ldots, \theta_N) \tag{1}$$

where $I_1, I_2, \ldots, I_N, \theta_1, \theta_2, \ldots, \theta_N$ are action angle variables, H_0 is the unperturbed Hamiltonian, N is the number of degrees of freedom, and V is the perturbation term (Reichl, 2004).

The presence of perturbations can lead to internal nonlinear resonances (Chirikov, 1979; Reichl, 2004) with oscillations $\omega_1, \ldots \omega_N$ that satisfy the condition:

$$n_1 \omega_1(I_{10}) + \ldots, + n_N \omega_N(I_{N0}) = 0 \tag{2}$$

where $n_1, \ldots n_N$ are natural numbers. If the perturbation energy is higher than the energy difference between the two closest unperturbed resonant orbits:

$$\Delta H_i > E_{i+1} - E_i \tag{3}$$

chaotic dynamics will develop (Chirikov, 1979) and the origin of chaotic dynamics and fractional dynamics can be explained by the resonance-overlap criterion (Chirikov, 1979; Zaslavsky, 2005).

Within the solar system, over time, the presence of resonant frequencies has shifted the position of Mars and Jupiter, changed the arrangement of asteroids, and altered the entire dynamics (Morbidelli & Henrard, 1991; Guzzo & Morbidelli, 1996). Once the structure of the solar system was reshaped, the gravitational field and its frequency of vibration have also changed. Captured by a frequency analysis (Laskar, 1993, 2005), this shift in gravitational field provided the "first convincing evidence that the long time motion of the solar system is chaotic and to some extent unpredictable" (Reichl, 2004).

Over time, asymmetry and heterogeneity were naturally generated (Wyatt, 2008), providing a set of unique properties within every planetary system. Mathematically, the nonlinear Hamiltonian dynamics describes the formation of a planetary system and is an expression of structural "evolution" in cosmology. The planetary n-body problem represents an example of an n-body interaction, which is an exceedingly common situation in the physical world. On *smaller scales, n*-charge dynamics and the *electric field* become dominant components. The Coulomb interaction depends on the value of charges

e_i and e_j and their positions $\vec{r}_i$ and $\vec{r}_j$ in physical space. For n interacting charges, similar equations can be written as:

$$m_i \ddot{r}_i = \sum_j k_e \frac{e_i e_j \left(\vec{r}_i - \vec{r}_j \right)}{|\vec{r}_i - \vec{r}_j|^3}, \quad i = 1, 2, \ldots, n \tag{4}$$

where $r_1, r_2, \ldots, r_N$ are the position vectors of the point charges and $k_e = 1/4\varepsilon_0 \varepsilon_r$ where $\varepsilon_0 = 8.854 \times 10^{-12}\,\mathrm{F/m}$, and ε_r is the relative permittivity. At a quantum level, the interaction involving a large number of charges becomes the n-body problem and is described by the Schrödinger equation:

$$\frac{\partial \psi(t, x)}{\partial t} = \frac{i}{h} H(E) \psi(t, x) \tag{5}$$

where x is the position coordinate, ψ is the system's wavefunction, H is the Hamiltonian of the system, and E is the total energy. Instead of an n-body problem involving point particles with distinct positions in three-dimensional space, the solution of the Schrödinger equation gives a probability distribution for the quantum states of the system and the quantum approach can be further developed to take the form of a quantum field. As presented above in classical models, the occurrence of perturbations leads to internal nonlinear resonances and chaotic dynamics (Dixon et al., 1997).

The Hamiltonian equations of motion represent a bridge between nonlinear dynamics and quantum mechanics (Yang & Weng, 2012), which can include the dynamics of chemical reactions, the generation of chaos, and regularity (Komatsuzaki & Berry, 2002). The mathematical theory that underlies physical interaction of many-body dynamics calculations is far from being complete. Other attempts tried to provide an adequate description of the many-body problem. The path integral (Feynman & Hibbs, 1965), the density matrix theory (Fano, 1957), the second quantization approach (Dirac, 1927), and the variational pilot wave theory (Styer et al., 2002) are but a few examples of such attempts.

Either in a quantum approach of many-body physics (wave function) or in a classical framework of n-body dynamics (Greengard, 1994), the difficulties of solving the problem of interacting particles are well-known. Max Born made seminal contributions to the development of quantum mechanics and suggested that "Nature fortified herself against further advances behind the analytical difficulties of the many-body problem" (Born, 1960). A similar process of reorganization and "evolution" of matter occurs at a quantum level; however, our ability to perceive the particle positions and their dynamics in space is extremely limited as stated in the Heisenberg uncertainty principle. According to quantum mechanics and both special and general theories of relativity, space, time, motion, and matter are not as independent of each other as once thought in classical physics. Higher frequency of "vibrations" can be generated at a quantum scale. Classical models describe well the system's behavior, provided the average thermal energy per particle is $k_B T > hf$ (k_B is the Boltzmann constant, and h is Planck's constant). Therefore, the n-body (many-body) problems are in fact mathematical models that describe a dynamic process of structural change. A small structure can evolve faster than a large one as it operates at shorter time scales and has fewer degrees of freedom.

Einstein challenged the main stream of thought in classical physics by demonstrating that a fragment of matter with mass m can be transformed into energy according to the famous formula $E = mc^2$. Inspired by Einstein's idea, Louis de Broglie suggested that matter can be represented by waves (Broglie, 1923). With this view, a novel interpretation has emerged in physics and Einstein wrote "what we have called matter is energy, whose vibration has been so lowered as to be perceptible to the senses. There is no matter." However, neither classical physics nor quantum mechanics have fully resolved the mystery of matter. The greatest show about matter can be observed on Earth (Dawkins, 2009).

3 BUILDING THE BIOSTRUCTURE: THE MYSTERY OF LIFE

The existence of molecular asymmetry and micro-heterogeneity within the biological substrate remained largely unexplained. Schrödinger introduced the idea that biological information is kept at a molecular level in "aperiodic crystals" (Schrödinger, 1944) and, a few years later, the double helix structure of DNA was discovered (Watson & Crick, 1953), which revolutionized the world of biology.

All forms of life are shaped by molecular structures, which transport energy and process information inside living systems. Fundamental characteristics of living systems such as molecular asymmetry and micro-heterogeneity of biological "matter" were recognized since Pasteur's time (Robert et al., 2007). Chemical composition defines just a basic model required to build chunks of living matter. However, this does not seem to describe sufficiently the complexity of structured matter in a biostructure (Drochioiu, 2006; Murariu & Drochioiu, 2012). Notably, a simple change in spatial arrangement of chemical constituents can significantly alter emerging properties. Determined by structural changes, asymmetry and nonhomogeneity are universal features of biological matter. Stereochemical relevance of chiral molecules and the resulting change of olfactory characteristics are well-known examples of this mechanism (Zaslavsky, 2005).

Turning our attention to the brain, it is well-known that all neurons acquire their shape by spreading in the three-dimensional space of the brain (Bras et al., 1993; Scott & Luo, 2001; Cuntz et al., 2010; Packer & Yuste, 2011). Axonal and dendritic growths of neuronal cytoskeleton provide a clear basis for cell interactions, tissue development, and neuronal function (Ng, 2012). The structure of neurons consists of various protein polymers (e.g., microtubules, intermediate filaments, and actin filaments), which are spatially organized. Quantitatively, slow growing phenomena, which refine spatial development of cells, can be simply described using physical principles (Kollmannsberger et al., 2011). In addition, on a millisecond time scale, a functional protein can be generated by polymerizing several hundred amino acids. Made from a set of amino acids, newly synthesized proteins require efficient folding to enable their function (Hartl & Hayer-Hartl, 2002). Determined by amino-acid sequences, the conformation of various proteins can be spatially reshaped by many factors (Gething & Sambrook, 1992). Posttranslational modifications, transcriptional and translational control, and chaperones are involved in this process.

In addition, complex regulatory mechanisms are programmed to preserve the soluble state of proteins (Vendruscolo, 2012). Created and regulated by protein polymerization, protein folding, and protein phosphorylation, the newly generated structures can easily acquire complex functional roles. Many factors including neurotransmitters, hormones, and electrical fluctuations can highly influence spatial arrangement of proteins and therefore change functional characteristics of neurons or synapses (Pirooznia et al., 2012). This places strong emphasis on the mechanisms of transformation, which can dynamically occur inside neurons, but can be sensitive to external influences.

A constant supply of biochemical energy in the form of ATP and GTP molecules is required to maintain the functional state of proteins (Grzybowski et al., 2009). The spatial rearrangement of charges of atoms at the molecular level is highly regulated by changes in gene expression, protein folding, alterations in electric fields, polarizations, the effect of hormones, and neurotransmitters (Aur, 2012a,b; Plankar et al., 2012).

Since information is embedded in the structure by a transient effect of electric polarization, the generated electric field carries structural information. Therefore, the generation of action potentials (APs) and synaptic activities allows the system to "read" (decode) fast information stored within a protein structure (e.g., microtubule) (Aur & Jog, 2010; Aur et al., 2011). However, specific proprieties of biological material can be significantly changed from a nanometer-scale (Woolf & Priel, 2009) since the process of structural change continues inside neurons and synapses (Scheeff & Bourne, 2005). Therefore, it is equally important to include the relationship between the structure of matter and the generated rhythms in living systems, notably in the brain.

New theories are typically triggered by experimental observations that challenge old paradigms of science. For almost a century, the neuroscience community has believed that APs are stereotyped events which provide the basis for theories of brain function, integrating such concepts as the firing rate, temporal patterns, and connectivity. However, a growing body of experimental data points in a different direction. Only recently have we found that a fast change of structure happens within neurons in addition to synapses, and very recently, a reliable way has been found to predict seizures (Aur et al., 2013). These experimental observations show that we need to extend our conceptual framework beyond old theories to understand the brain and to better design intelligent systems. We have artificially established boundaries between important research fields, which have created many gaps in our view. Essentially, we need a physical framework to understand cognitive phenomena (Tuszynski et al., 2006; Tuszynski, 2013).

4 THE RHYTHMS IN THE DYNAMICS OF STRUCTURED MATTER

While the fundamental relationship between aggregated matter and intrinsic oscillations is well understood in physics, this association has remained elusive in biology. Various electric "vibrations" can be recorded in all living organisms and can be related to significant changes in their behavior. In addition, an adaptive behavior is present in plants, which develop minimal forms of cognition (Garzón & Keijzer, 2011). Structural evidence and electromagnetic characteristics of plant morphogenesis have been recently demonstrated (Pietak, 2012). The presence of electric fields can change the distribution of proteins in plants and determine modifications of the cytoskeleton (Antov et al., 2004; Schenk & Seabloom, 2010). In particular, the maximum intensity of electric fields in space points to the direction of root growth (Masi et al., 2009; Ciszak et al., 2012).

The smaller the structure, the faster the generated rhythm of transformation can be. This simple relationship applies to heterogeneous macromolecular structures (Sekulić et al., 2011) and their vibrations (natural frequency), which extends to neurons or larger ensembles of neurons.

At the early dawn of the genomic era, Burr's research at Yale provided evidence for the existence of electrodynamic fields of living organisms. His pioneering work on salamander embryo showed that recorded electric fields are stronger in the location where the salamander's head is shaped. A strong increase of the electric field in the brain region is not a random event (see neuroelectrodynamics). Ahead of his time, Harold Burr recognized the recurring electric vibrations as the fields of life, which "preserve the living tissue from falling into a chaotic state" (Burr & Northrop, 1935; Burr, 1972).

More recently, a similar observation has been made in relation to seizure generation (Aur et al., 2011; Aur, 2012b). A subject of intense theoretical analysis, the generated rhythm is in fact a characteristic of the biostructure and represents a direct link to subcellular morphology (Kučera & Havelka, 2012; Sekulić et al., 2011; Havelka et al., 2011). Energy released from ATP's phosphate bonds is converted into nonlinear vibrations of proteins (Sekulić et al., 2011; Pang, 2011) and fuels the entire change. At the nanoscale level, the amazing variety of protein structures and their dynamics have direct correspondence to the generated nonlinear oscillations, and the presence of nonlinear dynamics was theoretically demonstrated to lead to soliton generation (Sekulić et al., 2011; Pang, 2011; Scott, 1991).

Proteins being electrically polar structures can generate nonlinear oscillations and variable electric fields in the kHz to GHz frequency bandwidth (Cifra et al., 2010; Rahnama et al., 2011). These nonlinear vibrations intrinsically characterize the structure and are often presented as resonators of electromagnetic field (Cosic, 1994; Cifra, 2012). In addition, electromagnetic waves, e.g., infrared light, can induce nonthermal biological effects which can change the conformational states of proteins (Pang, 2012). The occurring molecular resonances play a stabilizing role that can reshape the entire molecular system (Mo, 2009; Kučera, & Havelka, 2012).

A small change in the free-energy range of several $k_B T$ can significantly affect the protein dynamics (Henzler-Wildman & Kern, 2007). Regulated by complex biochemical reactions, the presence of ions, neurotransmitters, and metabolic substances leads to a continuous generation of electrical field patterns inside neurons and in the brain. The simplest mathematical model, which describes a nonequilibrium system of this kind, can take the form of reaction-diffusion equations of the type:

$$\frac{\partial c_\sigma}{\partial t} = D_\sigma \nabla^2 c_\sigma + R_\sigma, \quad \sigma < m, \tag{6}$$

where c_σ is the concentration of species σ, D_σ is the diffusion coefficient, ∇^2 is the Laplacian operator and R_σ accounts for local reactions, and m is the number of species (Sivakumaran et al., 2003; Ayodele et al., 2011).

For more than two decades, the development of reaction-diffusion models was separated from Hamiltonian-based dynamical models of conservative systems. Elgart and Kamenev have recently established the correspondence between reaction-diffusion systems and their Hamiltonians (Elgart & Kamenev, 2006). Assuming similar local reactions and diffusion coefficients for all species, the reaction-diffusion system can be written using the corresponding Hamiltonians:

$$\frac{dq}{dt} = D\nabla^2 q - \frac{\partial H}{\partial p} \quad \text{and} \quad \frac{dp}{dt} = -D\nabla^2 q + \frac{\partial H}{\partial q} \tag{7}$$

where $p \doteq \dot{q}$ represents momentum and the resulting system can be modeled and analyzed based on corresponding, nonintegrable Hamiltonians (Elgart & Kamenev, 2006).

In addition, the evolution of electrical patterns and chemical kinetic equations is strongly related. The nonintegrable Hamiltonians can approximate the system dynamics. In addition, the angle action approximation can generate extreme examples of the kind of behavior the brain exhibits, which include the presence of resonant regimes and chaotic dynamics. Conceptually, nonlinear resonances are analogous to coupled oscillators (Scholes, 2003).

Chaotic dynamics can naturally develop in Hamiltonian systems with many degrees of freedom, which describe the motion of charged particles in an electric field (e.g., Arnold diffusion, resonance-overlap). The Hamiltonian structure extended to an infinitely dimensional case can model the interaction of electric fields with charges. The chaotic regime can be maintained only if neurons do not fire or have a very low firing rate (e.g., focal epileptogenic region (Aur et al., 2011). A quasiequilibrium state corresponds to neuronal resting phase, while the generation of an AP is a rare event and represents the moment when internal nonlinear resonances develop inside the neuron. Either the presence of electrochemical "perturbations" or the effect of chaotic dynamics causes protein channels to change their configuration. In order to transition to a lower energy of vibrational states, the energy of the excited vibrational mode is transferred to the kinetic modes and the flow of ions leads to a large diffusion at a synaptic level. If the neuron fires APs, the existent intracellular structural order is "exported" in a generated electric field (with no associated chaos).

Another point of misunderstanding comes from the difference between chaotic systems and stochastic systems. While in mathematics this difference is clear, distinguishing between stochasticity and chaotic behavior in any biological system is extremely difficult. We cannot rule out the presence of stochastic behavior in the brain; however, we can make short-term predictions of brain seizures since there is deterministic-chaotic behavior in the brain (Aur et al., 2011).

From a nanoscale level, the transfer of information is performed during electron conduction within proteins, vibrational signaling, and the resonant interaction between macromolecules (Cosic, 1994). The presence of various neurotransmitters and hormones can enhance or inhibit structural changes. A nonstereotyped vibration (e.g., an AP) generates an exchange of information between the electric field and protein structures, which embed fragments of information inside neurons and synapses. The structure and dynamics of proteins can be reshaped during resonant regimes (conceptually similar to those in a planetary system) and are enhanced during an increase of activity (firing rate). In addition, spatial propagation of APs makes synaptic "connectivity" between neurons highly variable.

The atomic-level characterization of protein dynamics provides an n-body level description specific to a free-energy perturbation model (Shaw et al., 2010). These transitions between structurally distinct states, protein folding and unfolding, can take place on time scales between 10 µs and a few ms (Kubelka et al., 2004; Anandakrishnan & Onufriev, 2010; Oladepo et al., 2011). One can readily find that the distribution of atoms and electric charges within a particular protein structure is not random (Brenner et al., 2000; Koehl & Levitt, 1999). Therefore, inside cells, proteins can embed fragments of information within an evolved nonnative structure. This process has been explained recently (Craddock et al., 2012a,b). The information embedded in proteins can be rapidly exchanged through specific "vibrations," which characterize the protein structure. Therefore, structural organization within neurons and synapses at a protein level is an important basis for information storage, while electric fields carry this specific information between various intracellular locations. In fact, we found that a generated electric field reflects essential structural properties of the matter genetically shaped inside neurons.

Both in the case of the solar system and in the brain, the respective equations of motion, the principle of least action, and the second law of thermodynamics represent physical foundations for a theory of structured matter. The energy constraints drive the system on the path of least action (Maupertuis, 1746), which follows the principle of increasing entropy and is reflected in the equations of motion. The use of a nonequilibrium thermodynamic treatment either to examine the generalized theories of gravitation (Chattopadhyay & Ghosh, 2012; Bamba, & Geng, 2009) or transformations that occur at a protein level (Annila & Salthe, 2010) describes the structural change of matter. The validity of the laws of thermodynamics has been affirmed beyond doubt and this area of physics represents an essential tool that can be used to determine the evolution of system.

In the physical realm, the transformation of the structure has always a one-way direction which is equivalent to "evolution." Darwin observed that natural selection determines gradual changes in organic structures; therefore, by definition, we can consider structural changes that follow the arrow of time as "evolution."

At different scales, the set of equations (Eqs. 1–5) shows the relationship between the evolution of the structure (self-assembly) and generated rhythms that develop from chaotic behavior to nonlinear resonances in a broad class of many-body system. These equations describe the transformation of the structure either in nonliving or within biological matter. Our recent work describes in detail structural transformations and changes in nonlinear dynamics in biological systems (Chan & Tuszynski, 2016; Aur & Vila-Rodriguez, 2017). That's not just a matter of language.

The ensemble of neurons in the brain can be considered to be surrounded by a similar bath (extracellular space) and information transfer can be presented as changes in the thermodynamic entropy (Aur & Jog, 2010). However, the human brain is extremely efficient; it needs <20 kcal each hour, which is around 23 W. This shows that biophysical processes required for cognition, thinking, memory, problem solving, and consciousness are almost reversible ($dS \approx 0$). In fact, difficulties in mimicking brain power in self-driving cars and robotics with so-called brain-like chips are well-known. Can we replicate the process of generating conscious states? We can, by designing a system that "evolves" in a similar way our brains do. Building conscious machines should not be science fiction since we already have the required technology. Briefly, to make progress on this specific problem, a two-step approach was proposed. First, a brain can be grown from cerebral organoids (see Lancaster et al., 2013). Second, the brain will be digitally integrated to supervise its transformation into a functioning brain. Conscious experience should then emerge in this hybrid system during a gradual process of training (Aur, 2014).

5 THE EMERGENCE OF INTELLIGENCE

Intelligence is often associated with intellectual abilities required to perform various cognitive tasks. Different researchers have hypothesized that intelligence is determined by particularities of synaptic "connectivity," while others have highlighted neuroanatomical characteristics (Li et al., 2009). The importance of neurochemistry (Thagard, 2002; Changeux & Edelstein, 2012) or of the brain size was frequently reinforced (Jerison, 1973; Willerman et al., 1991; Andreasen et al., 1993).

What caused Albert Einstein to move beyond conventional wisdom in physics? What made him a genius? In an attempt to discover the mystery of intelligence, Thomas Harvey has surgically removed, sliced, and performed neuroanatomical studies on Einstein's brain. Indeed, different details regarding brain parameters can be obtained by an anatomical assessment. A relatively wider parietal region was observed to be present in Einstein's brain; however, it was not clear how such a change can be linked with other brain characteristics (Anderson & Harvey, 1996; Witelson et al., 1999). Whether neuroanatomical measures (Witelson et al., 1999) correspond to creative abilities, intelligence, or cognitive skills has largely remained an open issue (Cairó, 2011). (Not surprisingly, because if it were explained this way, phrenology would be an accepted field, Bookstein, 1996.)

Quantitative genetic studies have shown that intelligence is inheritable (Murray & Herrnstein, 1994; Jensen, 1998; Deary et al., 2010; Davies et al., 2011; Chakravarty, 2010). In addition, recorded electrical activity provides meaningful details regarding the brain function and, in particular, EEG characteristics reflect genetic traits (Smit et al., 2012). Particular characteristics of EEG oscillations are strongly correlated with cognitive abilities (Anokhin & Vogel, 1996; Thatcher et al., 2008) and "fluid" intelligence (Silvia & Beaty, 2012).

In spite of the progress made in this field, the validity of many theories regarding intelligence continues to be disputed today. In fact, rudimentary forms of self-awareness, cognition, and intelligence as biological features did not originate with *Homo sapiens* (Trewavas, 2003; Barlow, 2010). In order to survive, cells rely on sensing and signaling environmental changes. Basic forms of intelligent behavior are known to have already appeared in touch-sensitive plants such as *Mimosa pudica*, which generates APs that trigger rapid movements (Sibaoka, 1991; Chamovitz, 2012). For several decades, the main thesis was that cells perform "rudimentary processing of sensed information." This view regarding the abilities of cells has radically changed lately. Environmental stress can activate DNA elements and slightly modify biological structure. In this sense, plants are "minimally cognitive" (Garzón & Keijzer, 2011). While in the past many researches would hesitate to call *Amoeba* intelligent, this view has drastically changed (Nakagaki et al., 2000). Subject to ongoing criticism, the hypothesis that cells do "rudimentary information processing" has already become a problem. *Amoebas* and *Paramecium* cells "have remarkable abilities to make decisions and take constructive action, which correlates with standard definitions of intelligence" (Albrecht-Buehler, 1985; Ford, 2009, 2010).

6 MICROSTRUCTURAL EVOLUTION, LEARNING, SELF-ORGANIZATION, AND SEMANTICS

The development and plasticity of the brain are well-recognized. A fast reorganization of temporal patterns is triggered during T-maze procedural tasks (Jog et al., 1999; Barnes et al., 2005) or repetitive presentation of images (Quiroga et al., 2005; Aur, 2012b). The fundamental "vibration" of neurons, i.e., the generation of APs, is considered to be a stereotyped event in spite of recent experimental observations (Aur et al., 2005, 2010; Sasaki et al., 2011; Aur, 2012b). However, in neural computation all previous models (temporal coding) rely on erroneous interpretation of experimental observations. The stereotyped AP is just an appearance. Since single-electrode recordings are inadequate to capture spatial propagation of the spike, a digital uniformity was attributed. Occasionally, a few electrophysiologists have observed changes in AP waveforms. While they recognized these changes, unfortunately they failed almost completely to understand their deeper meaning (Quirk et al., 2001). This simple observation (nonstereotyped spikes) has powerful implications since changes that occur within the molecular structure inside neurons can be related to electrical patterns generated during spiking activity and preferential propagation of information either synaptically or nonsynaptically (by an electric field). From a computational point of view, information can be quickly "read out" (Aur & Jog, 2010; Aur et al., 2011; Beggs & Tucker, 2007) or "written in" (encoded) during electric interactions within proteins (Craddock et al., 2012a,b). In an intact brain, temporal coding models are extremely limited and lack generality (e.g., spike-timing-dependent plasticity, Schulz, 2010; Spruston & Cang, 2010; Aur et al., 2007; Feldman, 2010). The rules that "are much more complex" describe the development of molecular structure at a neuronal and synaptic level. The evolution of the structure of matter provides a more general model.

In addition, our understanding of reality is necessarily based upon incomplete information extracted from experimental observations. Since APs are recorded with low sampling frequencies (<50 kHz), nonlinear phenomena that occur at a molecular level within neurons during AP generation can be hardly evidenced. The presence of electrical patterns is a direct effect of fast changes that can occur at a protein level. If one cannot record high frequencies, it does not necessarily mean that such frequencies do not exist or they cannot alter the evolution of the system. Recently, it has been shown that infrared neural stimulation can be used to stimulate the somatosensory cortex in rats (Albrecht-Buehler, 1985; Cayce et al., 2011). However, to be able to "see" fast molecular changes that can occur at a protein level inside the neuron during an AP, one would need high resolution infrared optical imaging of the neuron. Unfortunately, the required temporal and spatial resolution to obtain such images is still unattainable using current technology.

The change observed in temporal patterns reflects the reorganization that occurs within the molecular structure at a neuronal and synaptic level (Aur & Jog, 2010). Nature is far more subtle than we have imagined. Every AP can be seen as a brief moment (1 ms in duration) when the neuron "solves" at least a classical n-body problem rather than exclusively computes functions (Aur & Jog, 2006, 2010; Aur, 2012c). The limits generated by computing functions were largely presented by Goldin and Wegner as an introductory topic in interactive computing (Goldin & Wegner, 2008). Current ideas in computational neuroscience regarding the interspike interval and spike-timing-dependent plasticity are just extensions and misunderstandings of the all-or-none principle (Adrian, 1914). At this point, it is becoming abundantly clear that the mainstream in computational neuroscience adheres to a dogma older than Turing's paradigm (Turing, 1936). There is little understanding or appreciation for the hypothesis that information can be processed in a non-Turing manner at nanoscale levels within neurons and synapses. Detailed explanations regarding this issue are provided in (Aur et al., 2011; Aur & Jog, 2010) and in a more recent book (Cicurel & Nicolelis, 2015).

Known as habituation, the repetitive presentation of a stimulus leads to a decreased response of neurons (Thompson & Spencer, 1966). The development of habituation in neurons can be simply explained. A fast process of structural transformation at a protein level is triggered in specific neurons and synapses during a repeated presentation of images (Aur & Jog, 2010; Aur, 2012b) or repetitive T-maze behavioral tasks (Barnes et al., 2005; Aur & Jog, 2007, 2010). Less energy is required to identify the image or provide an adequate behavioral response required to get the reward. This repetitive presentation leads to an efficient response. With few generated spikes, the information regarding the semantics of presented images or behavioral meaning are efficiently processed and electrically transmitted. The entire process that *decreases the firing rate can be simply explained as a process of reorganization determined by internal structural changes, which* can be strongly triggered within specific neurons by a repetitive presentation of images or by a rewarding mechanism.

Structural changes inside the cell can change the expression of genes into proteins, since the structure of neuron is driven from a DNA level. It is now well-recognized that an increased expression of misfolded proteins, giving rise to pathological protein aggregation, is age-dependent. Macromolecular degeneration, reduced neurotrophic support, cytoskeletal abnormalities, posttranslational modifications of proteins, and protein denaturation are changes of structure, which develop over time (Orpiszewski et al., 2000). The presence of regulatory interactions that operate at gene and protein levels indicates a more complex mechanism for information flow than earlier proposed by the central dogma of molecular biology (Crick, 1970). The existence of epigenetic effects of chemical signals at several levels (e.g., chemical modifications to DNA) and the presence of complex regulatory networks which determines which genes are expressed and alter the genome function challenges the main dogma and limits genetic-causal explanations (Noble, 2006, 2011; Aur & Jog, 2010). Targeting the genome stability, DNA/RNA structure and artificially increasing the length of telomeres can provide solutions useful in the search for effective therapeutic interventions for cancer or various aging-related diseases (de Jesus et al., 2012a,b).

In addition, every process needs sources of energy to develop and expand. A new generation of synthesized proteins can bring new features and it is not an accident that protein turnover plays a key role in aging (Ryazanov & Nefsky, 2002). Caloric restriction (less energy consumed) can slow down the entire transformation and holds back the dynamics of the aging processes (Lee et al., 1999; Colman et al., 2009; Fontana et al., 2010).

A balance between protein synthesis and protein degradation is strongly regulated in healthy organisms through fine-tuned mechanisms of homeostasis. However, the regulation of protein stability is a small part of a general framework of changes in structure. While the generated structure is not random ("God doesn't play dice" as famously stated by Einstein), the fast changing structure of many proteins remained largely unnoticed. In addition, the degradation of proteins increases entropy and energy expenditure. It has been clearly stated recently that it "becomes evident that protein folding is an evolutionary process among many others" (Sharma et al., 2009). Being highly regulated, the hierarchical structure of proteins evolves from the primary structure, to the secondary structure to tertiary and finally the quaternary structure. In addition, the denaturation of the protein structure is a natural process of "evolution," which can be triggered by aging or epigenetic factors. While many generated proteins have a limited life-time, some structures have a long life and slowly degenerate in time over a period of years (Su et al., 2012).

7 WHAT IS BALANCED EXCITATION AND INHIBITION?

The balance between excitation and inhibition is extensively presented in many papers as a "basic functional principle" underlying neural activity in the brain (Zheng et al., 2012). Classified as excitatory or inhibitory, chemical synapses can generate various modulatory signals. A number of controversies regarding the nature of synaptic transmission (Isaacson & Walmsley, 1995), in addition to a dynamic shift between inhibition and excitation, have strongly limited the applicability of this principle (Heiss et al., 2008; Zuo et al., 1999). The inhibitory neurotransmitter $GABA_A$ (Luscher et al., 2011) can become excitatory in case of an increased Cl^- concentration inside the cell (Taketo & Yoshioka, 2000). In addition, $GABA_A$ receptors, the main ionotropic receptors for fast inhibitory transmission, are expressed by 19 distinct genes, which can generate at least 26 different receptors (Luscher et al., 2011; O'Rourke et al., 2012).

However, many scientists felt that this principle of inhibition and excitation can be extended beyond conventional neurotransmitter type classification (e.g., GABAergic, dopaminergic, and serotonergic) to explain the occurrence of neurological disorders, the effect of treatment (DBS = deep brain stimulation), or even the presence of consciousness. Extended to large brain areas, the idea of excitation and inhibition engendered controversies instead of explanations (e.g., the mechanism of action of DBS, Montgomery & Gale, 2008).

Inside neurons and synapses, various proteins are expressed. Understanding structural changes of synapses provides insights into the organization of matter inside the brain. Our convenience (inhibitory, excitatory connections) has hidden the fundamental role of proteins and their homeostatic contribution. The dense and rich protein network structure can change fast and generate the required plasticity in the brain. In addition, proteomic analysis has shown that protein phosphorylation can significantly change intracellular signaling with effects on synaptic plasticity (Craddock et al., 2012a,b; Tweedie-Cullen et al., 2012). A deep molecular diversity (O'Rourke et al., 2012) and the "balance between inhibition and excitation" are in fact determined by structural changes from a nanoscale (quantum) level. Within a systemic approach, the entire dynamics in the brain can be schematically represented by three inter-acting regulatory loops that include molecular computations, neurotransmitter release, and electric fields (Aur & Jog, 2010; Aur, 2012a).

8 THE GENETIC BASIS OF BRAIN DISORDERS AND AGING

The topic of age-related structural changes can be best discussed within this framework. Dissecting the genetic basis of common brain disorders provides an intrinsic relationship, between structural alter-ations determined by genetic changes which occur within neurons in epilepsy (Helbig et al., 2008; Poza, 2012), schizophrenia (Friedman et al., 2007) or Alzheimer's disease (Bertram & Tanzi, 2004), and visible changes of brain rhythms. The altered genetic structure within neurons and synapses leads to similar nonlinear dynamic regimes (chaotic dynamics, synchrony), which are always detected in Alzheimer's disease, Parkinson's disease, and epilepsy (Jeong, 2004; Dauwels et al., 2010; Uhlhaas & Singer, 2006). An abnormal change of the electric field precedes the generation of seizures in epilepsy (for a physical model, nonlinear dynamics, extended Kolmogorov- Arnold-Moser theory, transition to chaos, the reader is referred to Aur et al., 2011) and the onset of cancer (Assenheimer et al., 2001). Mathematically, persistent chaotic dynamics in the brain, which precedes a seizure, is no different than the chaotic dynamics in the solar system. Unfortunately, after 60 years of research in epilepsy, we still do not understand the "big picture" and we treat the generation of seizures by "fighting fire with fire" (Gwinn & Spencer, 2004). This metaphor is somewhat misleading regarding the role of therapy and the specific moment when stimulation needs to be provided (Aur et al., 2013).

The formation and evolution of solar cycles depend on energy resources. Aging is an irreversible phenomenon that severely affects the structure of matter. Over the course of millions of years, the solar material is converted into energy, while the entire structural evolution is shaped from a "quantum" level of nuclear fusion. The depletion of energy (matter) marks the end of an evolutionary state. Either in a star or in the brain, driven by evolution, this change of structured matter leads to aging. A similar pro-cess of structural evolution is present in every living organism. Maintained by a continuous influx of chemical energy, the process of evolution shapes the structure of matter from a nanoscale (quantum) level of proteins.

While information stored in the brain increases with time, the number of synapses and the volume of gray matter reduce with age (Taki et al., 2011, p. 9). The absolute number of synapses per neuron is reached by the age of 1 year and strongly decreases during the preschool years. Inside neurons and synapses, proteins can maintain their structural order within arrays of amino acids, by preferred spatial orientation of covalent bonds (Scheiner, 2011), specific spatial arrangement of electric charges, and intraprotein electric fields. The development of structural order at a molecular level is required to pre-serve and accumulate new information in neurons (structural plasticity of neurons and synapses). The loss of information can be generated by mechanisms that alter the life cycle of proteins and local aberrant protein degradation in neurons (Steward & Schuman, 2003). Therefore, the control of protein degradation and protein synthesis is highly regulated within every neuron. An entire cycle from DNA to

RNA through appropriate gene activation for protein synthesis is maintained in response to a wide variety of extracellular signals and electrical events (Pozo & Goda, 2010). The process of biophysical interaction and new protein synthesis is built to preserve previous fragments of information and to accumulate new information over time. Correct protein folding and binding depends on the local environment where proteins are synthesized. These processes are critically important for faithful information preservation within molecular structures. As always, information is retained and transferred to newly formed structures and different electric rhythms, and various neurotransmitters and hormones are involved in this process. Incompletely folded, misfolded proteins are required to establish a correct conformation or initiate the protein degradation process (Malgaroli et al., 2006). During sleep phases, different categories of genes provide the needed support for new protein synthesis (Cirelli et al., 2004). The internal molecular structure in neurons is continuously reshaped; the regulation and maintenance of the life cycle of proteins is a vital process required to preserve and incorporate new information.

Alzheimer's Disease (AD) is systemically induced by metabolic or genetic risk factors (APP, PS1, and PS2) that in different forms alter the life cycle of proteins, disrupt the unique repertoire of proteins secreted within neural cells, and lead to distinct pathological features (e.g., intracellular neurofibrillary tangles of MAP tau and extracellular beta amyloid plaques). Since similar fragments of information are distributed and stored within a large number of cells, few changes in a small number of neurons or the death of some cells do not necessarily generate memory loss or cognitive impairments. Therefore, in order to determine significant effects, a large number of neurons have to be affected by the progression of the disease. Inside the brain, there is no shield for generated electric events (APs, synaptic activities) and changes that occur in neurons and synapses transform the brain rhythms. The rhythm is critical to the brain's ability to "read" or integrate specific fragments of information.

At a biological level, a progressive deterioration of physiological function can occur in time triggered by unexpected events (e.g., brain trauma Moeller et al, 2011; pilocarpine treatment, Buckmaster & Haney, 2012), which lead to changes in protein structure at synapse and neuron levels. "Unfortunately, nature seems unaware of our intellectual need for convenience and unity and very often takes delight in complication and diversity," (Cajal, 1906). The expression of genes and environmental factors plays a significant role in shaping structural changes in a variety of ways and only a certain percentage of treated rats (34%) develop *status epilepticus* (Buckmaster & Haney, 2012).

Either due to aging or self-destruction, the senescence of cells appears to be programmed by gene expression and influenced by environmental factors and/or DNA damage. With the progression of aging, additional factors accumulate and may trigger cellular dysfunctions and influence the life cycle of proteins. Even in the absence of a specific disease, aging is associated with an increased aggregation of a large number of proteins (David et al., 2010). Since aggregated proteins are insoluble, they determine changes in electrical characteristics in intraprotein electric fields that can alter how information is stored or transferred between the flow of charges and the biological substrate during AP generation or synaptic activities. In addition, with aging, reduced periods of sleep may affect the process of "writing" fragments of information within molecular structures. Periods of sleep (REM = rapid eye movement) required to reshape the internal structure within neurons significantly decrease in Alzheimer's disease compared to age-matched control subjects (Onen & Onen, 2003). Exaggerated amyloid-β (Aβ) production is correlated with aneuploidy induced by different mutant forms (e.g., amyl, PS proteins).

The buildup of Aβ protein and MAP tau disrupts axonal transport and can limit the propagation of the signal during AP generation only in few axonal branches.

Synaptic dysfunction can be determined by a local buildup of amyloid-β protein at synapses or related to accumulation of intraneuronal amyloid-β and disrupts AP propagation. In addition, perturbation of Zn homeostasis can have profound effects on a system scale in the pathogenesis of Alzheimer's disease (Maret, 2005; Craddock et al., 2012a,b; Liu et al., 2008).

The availability of transgenic models in the case of Alzheimer's disease (Irizarry et al., 1997), Huntington's disease (Mangiarini et al., 1996), schizophrenia (Crawley, 1999), and even epilepsy (Croll et al., 1999) shows that neuropathological states can be initiated at a genetic level that controls the structure and function of proteins. A similar aspect can be observed in traumatic brain injuries. Biochemical evidence of DNA fragmentation shows that the expression of proteins is altered from a genetic level, which can lead to fast apoptosis and seizure generation (Lewen et al., 2000; Dubreuil et al., 2006). In addition to various neurotransmitters and hormones, the generated electric field can change the structure of matter within neurons and synapses. An aberrant "evolution" can be controlled and reshaped by changing the biostructure from a nanoscale level of proteins by electrical (DBS) or magnetic stimulation (May & Gaser, 2006; Dragicevic et al., 2011). In addition to many other effects, the stimulation alters relevant genes that are expressed, which explains why electrical (or magnetic) stimulation can provide better therapy in the case of major neurological disorders such as Parkinson's disease, Alzheimer's disease, or epilepsy.

9 ON THE ORIGIN OF TIME, MATTER, AND INTELLIGENCE OF LIFE

Physical systems often self-aggregate into complex structures under specific conditions, e.g., in the process of polymerization. Organized groups of bacteria acquire unimaginable capabilities for information processing, which reflect "cognitive, computational and evolutionary capabilities" (Shapiro, 2005, 2007). This simple organization of bacteria and their abilities in mobilizing and engineering DNA molecules show that the origins of intelligent behavior may actually precede cellular development (Shapiro, 2005). Either in plants or within the animal kingdom, the creation of cells and their interactions lies at the heart of intelligent behavior (Ford, 2009).

The presence of APs in "sensitive" plants, which trigger physiological responses and rapid movements in, e.g., *Mimosa pudica* (Fromm & Lautner, 2006), is a relevant aspect. Electrical excitability and signaling allows for a fast exchange of information, which sustains the bio-field hypothesis and its role beyond the "brain universe" (Burr, 1972; Rubik, 2002). A fast response of rats to X-rays (Garcia et al., 1964; Brust-Carmona et al., 1966) or the presence of blue light photoreceptors (e.g., cryptochrome proteins) in plants or animals can provide information regarding magnetic fields or circadian rhythms. The "vibration" of bio-structured matter at a protein level can respond to a large spectrum of signals generated by another source of energy. Thus, while we are unable to "see" directly a certain phenomenon, we might indirectly "feel" its effects.

In addition, the change of structure inside living organisms typically follows a nonlinear path (e.g., involving nonlinear resonances, chaotic dynamics, and fractal structure formation) that shapes the structured matter from a nanoscale level up. Even when biological organisms share identical genes (twins), such nonlinear interactions can create huge differences between them. This natural variability explains why certain phenomena such as remote perception (Puthoff & Targ, 1976), telepathy (Ritchie et al., 2012; Bem & Honorton, 1994), or retrocausal influence (May et al., 2005) occur preferentially only in certain human subjects.

10 WHAT IS HOLDING US BACK IN ARTIFICIAL INTELLIGENCE?

At different scales, the properties of "evolved" matter can be perceived to be similar. Theoretically, an adaptive evolution by (natural) selection can be extrapolated to a population of neurons. Coined by Edelman, neural Darwinism and neuronal group selection as a morphological argument (Edelman, 1987) reflect in fact a structural change of matter which can be either adaptive (Nei & Kumar, 2000; Fay, 2011) or neutral (Schiffels et al., 2011).

Having water as the highly preferred environment for the evolution of life is not a random phenomenon. Water molecules represent an active player in major biological processes. Fast building or regenerating flexible structures in water solutions is an essential event in all living systems. The brain and many other organs can function as a whole, with no separateness since in liquid water all components can strongly interact. Recently, a group led by Nobel laureate L. Montagnier has detected low-frequency electromagnetic waves emitted by the DNA of bacteria and viruses in low dilution solutions (Montagnier et al., 2011). It seems, therefore, that the DNA structure can send "spooky electromagnetic imprints of itself into distant cells and fluids." Such properties are largely attributed to the "quantum-coherent domains." The liquid water and internal components cannot be described as noninteracting point particles; the regulation of complex electrochemical solutions or fluctuating hydrodynamics is modeled using variational principles (Eisenberg et al., 2010).

The change of the structure inside neurons and synapses is the most powerful tool used to reshape the matter and generate intelligent behavior in living organisms. The increase of the electric field in the brain region (e.g., salamander) is not a random event. The presence of the electric field provides the required interaction. It is a very fast way to transfer information and increase the number of cells (neurons), which process information in the brain. Since every neuron functions (communicates, processes information) using an internal (variable) structure, it does not make sense to generate stereotyped ("digital") APs. Why would a complex biological neuron with an evolved structure behave as a metronome? However, only a few researchers understood that while it is convenient, the "digital" AP is a vastly oversimplified hypothesis.

Does the neuron doctrine (e.g., the firing rate) provide the required explanatory level to understand the mind-brain relationship? We do not believe so. Temporal coding, spiking models, balanced inhibition and excitation (instead of regulation), and Hebbian law ("fire together—wire together," instead of a change in the structure from a nanoscale-quantum level) represent our "intellectual need for convenience." Simplistic theoretical ideas about "neural coding," "connectivity," and "Hebbian laws" have distorted the general image regarding a fast internal evolution of matter in living organisms (especially in the brain). Indeed, there is no separation between a nanoscale level of information processing and the macroscopic observable events. However, the process of computation cannot be described by fitting the moments when the peaks of amplitudes are detected (APs) to a particular function since APs are nonstereotype events. Temporal patterns do not provide an approximation of relevant structural changes. Neither the degree of fidelity nor the accuracy of modeling such parameters has much value. Since the structure evolves, the function of proteins, synapses, and neurons can be totally transformed.

The generation of a stronger electric field in the brain is not a random event, the process of computation by physical interaction directly involves molecular structures from many cells (Aur & Jog, 2010; Aur, 2012b). It is a completely different process, which can change fast with structural evolution of matter.

In addition, current models patterned on Turing Machines have a limited ability to acquire properties, which are specific to evolved, structured matter. The well-known connectionist models and parallel computation represent attempts to describe structural changes at a network level (Rumelhart et al., 1986; Kohonen, 1988; McClelland & Rumelhart, 1986; Carpenter & Grossberg, 1988). They may in fact simulate changes that occur within a fast evolution of biological structure. Bayesianism, connectionist models, temporal coding models, statistical analysis, or searching databases are not "physically grounded" and do not express similar properties. In our opinion, the objective of generating "artificial consciousness" using digital computers is closer to science fiction (Wikipedia, 2017) than to science. Only with a paradigm shift that uses a physical, nonalgorithmic approach to build intelligent machines (Aur, 2012a), artificial consciousness can become "an engineered artifact."

11 INCOMPLETE MODELS, THE THEORY OF EVERYTHING

The human mind has developed the model of the world we know based on sensorial and experimental observations. Imperfect, flawed experimental observations (e.g., stereotyped action potentials) have led to a series of distorted, incomplete models. We are not only "made of star-stuff." Inside our brains, the creation of structured matter follows the same physical principles. In an attempt to build a theory of everything, string theory was proposed in different formulations following Einstein's legacy ("matter is an energy vibration"). However, without including how life and mind originate from the evolution and interaction of structured matter, we will always have an incomplete, elusive theory (Deacon, 2011; Hawking & Mlodinow, 2010; Copeland, 2002; Chalmers, 1995; Penrose, 1994; Scruton, 2005).

Different states of the same matter (e.g., water) can express various physical characteristics. Under a slight change of environmental factors, the properties of matter can and do change. A simple example can be the transition of water from the liquid state to the solid state. In philosophy, one may interpret this change as a "subjective" response to environmental factors (Chalmers, 1996; MacGregor & Vimal, 2008). Others will strongly oppose it (Dennett, 1971; Pinker & Bloom, 1990; Searle, 2000; Dawkins, 2009). They might state that it doesn't matter how many times this transition occurs, to their knowledge the water doesn't develop any subjective experience. The same interpretation is attributed to computers and smart phones where the outcome of algorithms neither changes their "subjective state" nor differentiates between self and nonself (Searle, 2000, 2013).

At this point in time, we cannot overcome our ignorance and fully explain how consciousness arises from brain processes. As well, we may not fully agree with a dual approach and its conceptual confusion that separates the mind and body (see Searle, 2002). Higher level features (e.g., solid, liquid) can be explained by the behavior of the lower level constituents (e.g., atoms, molecules). A similar phenomenon can be observed in the brain, where the lower level of cellular processes should generate consciousness. However, since consciousness occurs at the brain scales and it is unlikely to be present at the cellular level, "emergence" should be the most appropriate physical term that characterizes this specific aspect. Regarding the origins of consciousness, we already recognized that "we cannot overcome our ignorance and fully explain how consciousness arises from brain processes." Importantly, Hameroff and Penrose (1996) have changed our perception regarding consciousness, so that the entire field is now under the scrutiny of scientific investigation. That constitutes a big step in the right direction. However, a full validation/refutation of any theory (including Orch OR) should come from experimentation. This rule should be equally applied for any vocal critic of any theory advanced in the scientific community.

In this chapter, we try to avoid such "paralysis of thought" (Feynman et al., 1963); by hypothesizing that solely within structurally "evolved" matter mental, subjective aspects can be expressed. From fetal to neonatal life, clear markers of the emergence of human consciousness can be measured and quantified. The emergence of consciousness may be faster within biological structures where a rapid development from molecular (protein) level is enhanced and constrained by an evolutionary path (Sharma & Annila, 2007; Shaw et al., 2010). An effective detection of different states of consciousness is observer-independent (Welberg, 2012). Such detectable changes (e.g., EEG rhythms, functional MRI) are correlated with the evolution of the brain's structure at all levels and various states of consciousness. The emergence of consciousness can be viewed as a system property. Alone, different parts of the central nervous system (CNS) do not seem to inherit this characteristic. The intact brain behaves differently than the sliced brain (Steriade, 2001; Kohl et al., 2010).

The interaction with the external environment gradually changes the structure of matter inside the brain from a molecular level (e.g., protein) and captures the essential nature of experience. This hypothesis is in agreement with experimental observations, which show a gradual emergence of consciousness correlated with structural (functional) development of human brain from early fetal stages up to the adult state (Lagercrantz & Changeux, 2009). We propose that consciousness becomes an emergent property of "evolved" matter (Searle, 2002).

12 SUMMARY OF THEORETICAL CONCEPTS—NEW PREDICTIONS

Most previous research has followed an anthropomorphic approach. Sense or semantics were subjectively attributed to characterize different forms of evolution. Here, the term "evolution" is used in a broader, physical sense, an impersonal way to transmit information regarding scientific knowledge. Having a theoretical understanding of overarching evolution and interaction of matter from a nanoscale (quantum) level becomes increasingly important in life sciences and medical research. Clear possibilities for experimental verification and reliable predictions with direct applications in various fields are key elements for any theoretical model. Therapeutic advances in neurological disorders, cancer therapy, delaying the aging process, or even modeling natural computing systems all may require a deep understanding of the evolution of matter.

On a millisecond time scale, a functional protein can be generated by joining a number of amino acids together. The structure of proteins can evolve from primary structure to secondary structure and reach a functional tertiary structure. This process is reflected in changes generated in nonstereotyped "vibrations" (APs, synaptic activities) in the brain. Everything can be seen as a result of biophysical interactions. The change of temporal patterns with "learning" reflects the evolution of the structure inside synapses and neurons. Since a change in the structure at a protein level can be rapidly triggered by a repetitive presentation of stimuli, many recording techniques (e.g., single electrodes) are inadequate to detect a small variability and changes of APs remained largely unperceived.

Inadequate observations (stereotyped APs) have led to sophisticated "digital" models simulated on Turing Machines. However, these models are of limited scope and use; they do not provide or express essential attributes achieved through physical interaction within organized, structured matter (e.g., self-awareness, consciousness, and emotion). The entire gimmick to "read" out memories at the brain scale based on "connectivity" and temporal patterns is an impractical journey. As a particular property of organized, living matter, the electric field within a generated rhythm does not only transfer information, it can reshape the biostructure (the brain) from a nanoscale (quantum) level over the entire lifespan. The neuroelectrodynamic theory highlights the limits of connectionist models and temporal coding principles.

An aberrant evolution of the structure can be controlled, from a nanoscale level of proteins by electrical or magnetic stimulation (DBS, transcranial magnetic stimulation = TMS). Also, infrared neural stimulation can be used for therapy and does not require genetic manipulation as in optogenetics (LaLumiere, 2011). In addition, the evolutionary theory of aging has its intrinsic origins in a more general model of structural evolution of matter. Either in stars or in cells, driven from a nanoscale (quantum) level, the evolution shapes the structure of matter and leads to aging. Self-destruction, aging, or the senescence of cells appear to be programmed by gene expression and influenced by environmental factors and/or DNA damage. Inside the animal kingdom, caloric restriction (less energy) slows down the process of "evolution" and holds back the dynamics of aging. The process that shapes the structure of matter can be slowed down, but cannot be stopped. Aging is an irreversible phenomenon that severely affects the structure of matter. While targeting the DNA/RNA structure, the length of telomeres can increase longevity; it cannot eliminate the ultimate demise of any living organism.

Either in the brain or in the solar system, the presence of chaotic events may significantly change the conditions and accelerate the entire process of "evolution." After the transitory behavior ends, new different assemblies can be naturally born from chaos. Instead of a rock that is slowly rolling downhill based on dissipation, the presence of chaos in the many-body systems may be that stroke of luck required to trigger a different evolutionary path which in a long-term perspective can lead to a new route for life.

Biological "evolution" and physical "evolution" are not so different beasts. The perspective of the physics in biology provides a more general phenomenon. Many hypotheses advanced in biology including Darwin's theory seem to be far more restrictive and they can be considered special cases. Biologists have argued about natural selection since Darwin has put forward his theory. Our goal in this paper has not been to prove that biologists are wrong, we are just trying to show that their view regarding biological evolution is incomplete. We strongly feel that structural evolution of matter can provide a different perspective on our entire understanding of the living, nonliving universe.

13 CONCLUSION

From Darwin to Dawkins, the theory of evolution is presented as a gradual process of natural selection driven by genetic (heritable) features. Biological evolution, natural selection on Earth represents but a small step that started at least 3.7 billion years ago, while the age of the Universe is estimated to be about 13.8 billion years (Dodd et al., 2017). Darwin's theory represents the basis for adaptation and specialization to changes in the environment of any living organism.

Contrary to common wisdom, which has restricted the theory of life's origin to natural selection, in this chapter we have presented an argument that structural evolution of matter has shaped the entire universe, from solar systems to human brains. Over billions of years fashioned by gravitational interaction, the development of the universe is a clear expression of the "evolution" of matter in cosmology. Built from "stardust," we born, evolve, and die as any star in the Universe. The process of natural selection appears to be another form of the evolution of matter.

Initiated before birth in the mother's womb, the evolution of the child's brain follows a similar process and shares similar physical laws. From a nanoscale (*quantum*) level of electrical interaction, the structure of matter inside neurons and synapses is shaped based on genetic traits, presence of neurotransmitters, hormones, and environmental factors (Hameroff & Penrose, 1996). What is commonly referred to as "learning" and self-organization is a change of the structure inside neurons and synapses at a protein level. There is no "neural code," instead fast evolution of matter takes place triggered by external events (e.g., reward, image presentation).

From a nanoscale level to cosmic scales, energy constraints govern over hierarchical structural changes. Therefore, we do not need to invent new principles other than classical and quantum laws of physics at appropriate space and time scales to describe the evolution of matter. The smaller the structure under consideration, the faster the evolution can be. Specific properties of proteins and neurons depend on their refined structure (asymmetry, heterogeneity). Having a clear model of overarching evolution of matter can help us to provide effective therapies for various diseases and prolong human life spans. Environmental factors can alter the expression of genes, transcription, and translation of proteins. The biologically functional soluble state of proteins can be rapidly achieved either inside bacteria or in neurons. Natural intelligence is the outcome of this process, which restructures the matter in living organisms in a selective manner. Once structured, the matter acquires characteristics, which are not common for the inert, unstructured matter. Cognitive abilities, self-awareness, consciousness, and intelligent behavior become emergent features of structured matter inside the brain.

Therefore, the existence of natural intelligence is embedded within a refined structure and intrinsic interactions. Maintaining a functional state of proteins and their heterogeneous, asymmetric structure inside neurons and synapses requires a continuous influx of energy. Outside an intact, living brain, all proteins may rapidly lose their structure and function and irreversibly transform into abnormal forms so that the entire brain can become a mass of amorphous organic matter. *It is worth keeping in mind that there is a long history of testing wrong hypotheses in investigating biological intelligence.* Following the main stream of thought, Thomas Harvey and other scientists did not realize that a lifeless, sliced (aged) brain is no longer the exceptional brain that once epitomized the genius of Albert Einstein. It is always important to understand the big picture to avoid the expensive, vicious cycle of testing so many unreliable hypotheses. Natural selection is just a small step on the path of general evolution of matter. In this sense, the structural evolution of matter may represent the theory of everything and the ultimate quest to discover a final theory.

ACKNOWLEDGMENTS

The authors would like to thank Jay McClelland for excellent comments which helped improve the manuscript. JAT acknowledges funding received from the Natural Sciences and Engineering Council of Canada (NSERC).

REFERENCES

Adrian, E.D., 1914. The all-or-none principle in nerve. J. Physiol. 47 (6), 460–474.

Albrecht-Buehler, G., 1985. Is cytoplasm intelligent too? Cell Muscle Motil. 6, 1–21.

Anandakrishnan, R., Onufriev, A.V., 2010. An N log N approximation based on the natural organization of biomolecules for speeding up the computation of long range interactions. J. Comput. Chem. 31 (4), 691–706.

Anderson, B., Harvey, T., 1996. Alterations in cortical thickness and neuronal density in the frontal cortex of Albert Einstein. Neurosci. Lett. 210 (3), 161–164.

Andreasen, N.C., Flaum, M., Swayze, V.D., O'Leary, D.S., Alliger, R., Cohen, G., Yuh, W.T., 1993. Intelligence and brain structure in normal individuals. Am. J. Psychiatr. 150, 130.

Annila, A., Salthe, S., 2010. Physical foundations of evolutionary theory. J. Non-Equilib. Thermodyn. 35 (3), 301.

Anokhin, A., Vogel, F., 1996. EEG alpha rhythm frequency and intelligence in normal adults. Intelligence 23 (1), 1–14.

Antov, Y., Barbul, A., Korenstein, R., 2004. Electroendocytosis: stimulation of adsorptive and fluid-phase uptake by pulsed low electric fields. Exp. Cell Res. 297 (2), 348–362.

Arnold, V.I., Kozlov, V.V., Neištadt, A.I., 1985. Mathematical Aspects of Classical and Celestial Mechanics. Springer-Verlag, Berlin.

Assenheimer, M., Laver-Moskovitz, O., Malonek, D., 2001. The T-SCAN (TM) technology: electrical impedance as a diagnostic tool for breast cancer detection. Physiol. Meas. 22 (1), 1–8.

Aur, D., 2012a. From neuroelectrodynamics to thinking machines. Cogn. Comput. 4 (1), 4–12.

Aur, D., 2012b. A comparative analysis of integrating visual information in local neuronal ensembles. J. Neurosci. Methods 207 (1), 23–30.

Aur, D., 2012c. Reply to Comments on Neuroelectrodynamics: Where are the Real Conceptual Pitfalls? arXiv. preprint arXiv:1210.1983.

Aur, D., 2014. Can We Build a Conscious Machine? arXiv. preprint arXiv:1411.5224.

Aur, D., Jog, M.S., 2006. Building spike representation in tetrodes. J. Neurosci. Methods 157 (2), 364–373.

Aur, D., Jog, M.S., 2007. Reading the neural code: what do spikes mean for behavior? Nat. Proc. https://doi.org/10.1038/npre.2007.61.1.

Aur, D., Jog, M.S., 2010. Neuroelectrodynamics: Understanding the Brain Language, vol. 74. IOS Press.

Aur, D., Vila-Rodriguez, F., 2017. Dynamic cross-entropy. J. Neurosci. Methods 275, 10–18.

Aur, D., Connolly, C.I., Jog, M.S., 2005. Computing spike directivity with tetrodes. J. Neurosci. Methods 149 (1), 57–63.

Aur, D., Connolly, C.I., Jog, M.S., 2007. Spike timing–an incomplete description of neural code. BMC Neurosci. 8 (Suppl 2), 149.

Aur, D., Jog, M.S., Poznanski, R., 2011. Computing by physical interaction in neurons. J. Integr. Neurosci. 10 (4), 413–422.

Aur, D., Toyoda, I., Bower, M. R., Buckmaster, P., 2013. Seizure Prediction and Neurological Disorder Treatment. U.S. Patent No. 8,600,513. Washington, DC: U.S. Patent and Trademark Office.

Ayodele, S.G., Varnik, F., Raabe, D., 2011. Lattice Boltzmann study of pattern formation in reaction-diffusion systems. Phys. Rev. E 83 (1). 016702.

Bamba, K., Geng, C.-Q., 2009. Thermodynamics in F (R) gravity with phantom crossing. Phys. Lett. 679 (3), 282–287.

Barlow, P.W., 2010. Plastic, inquisitive roots and intelligent plants in the light of some new vistas in plant biology. Plant BioSyst. 144 (2), 396–407.

Barnes, T.D., Kubota, Y., Hu, D., Jin, D.Z., Graybiel, A.M., 2005. Activity of striatal neurons reflects dynamic encoding and recoding of procedural memories. Nature 437 (7062), 1158–1161.

Beggs, E.J., Tucker, J.V., 2007. Can Newtonian systems, bounded in space, time, mass and energy compute all functions? Theor. Comput. Sci. 371 (1), 4–19.

Bem, D.J., Honorton, C., 1994. Does psi exist? Replicable evidence for an anomalous process of information transfer. Psychol. Bull. 115 (1), 4.

Berthoumieux, S., Brilli, M., de Jong, H., Kahn, D., Cinquemani, E., 2011. Identification of metabolic network models from incomplete high-throughput datasets. Bioinformatics 27 (13), 86–95.

Bertram, L., Tanzi, R.E., 2004. The current status of Alzheimer's disease genetics: what do we tell the patients? Pharmacol. Res. 50 (4), 385.

Bookstein, F.L., 1996. Endophrenology: new statistical techniques for studies of brain form: life on the hyphen in neuro-informatics. Neuroimage 4 (3), S36–S38.

Born, M., 1960. The Classical Mechanics of the Atom. Ungar, New York.

Bras, H., Korogod, S., Driencourt, Y., Gogan, P., Tyc-Dumont, S., 1993. Stochastic geometry and electrotonic architecture of dendritic arborization of brain stem motoneuron. Eur. J. Neurosci. 5 (11), 1485–1493.

Brenner, S.E., Koehl, P., Levitt, M., 2000. The ASTRAL compendium for protein structure and sequence analysis. Nucleic Acids Res. 28 (1), 254–256.

Broglie, L., 1923. Waves and quanta. Nature 112 (2815), 540.

Brust-Carmona, H., Kasprzak, H., Gasteiger, E.L., 1966. Role of the olfactory bulbs in x-ray detection. Radiat. Res. 29 (3), 354–361.

Buckmaster, P.S., Haney, M.M., 2012. Factors affecting outcomes of pilocarpine treatment in a mouse model of temporal lobe epilepsy. Epilepsy Res. 102 (3), 153–159.

Burr, H.S., 1972. The Fields of Life. Ballantine, New York.

Burr, H.S., Northrop, F.S.C., 1935. The electro-dynamic theory of life. Q. Rev. Biol. 10 (3), 322–333.

Cairó, O., 2011. External measures of cognition. Front. Hum. Neurosci. 5.

Cajal, S.R., 1906. The Structure and Connexions of Neurons. Nobel Lecture, December 12.

Cann, R.L., Stoneking, M., Wilson, A.C., 1987. Mitochondrial DNA and human evolution. Nature 325 (6099), 31–36.

Carja, O., Feldman, M.W., 2012. An equilibrium for phenotypic variance in fluctuating environments owing to epigenetics. J. R. Soc. Interface 9 (69), 613–623.

Carpenter, G.A., Grossberg, S., 1988. The ART of adaptive pattern recognition by a self-organizing neural network. Computer 21 (3), 77–88.

Cayce, J.M., Friedman, R.M., Jansen, E.D., Mahavaden-Jansen, A., Roe, A.W., 2011. Pulsed infrared light alters neural activity in rat somatosensory cortex in vivo. NeuroImage 57 (1), 155–166.

Chakravarty, A., 2010. The creative brain—revisiting concepts. Med. Hypotheses 74 (3), 606–612.

Chalmers, D.J., 1995. Minds, machines, and mathematics. Psyche 2 (9), 117–118.

Chalmers, D.J., 1996. The Conscious Mind: In Search of a Fundamental Theory. Oxford University Press, Oxford.

Chamovitz, D., 2012. What a Plant Knows: A Field Guide to the Senses. Oxford Publications, Oxford.

Chan, A., Tuszynski, J.A., 2016. Automatic prediction of tumour malignancy in breast cancer with fractal dimension. R. Soc. Open Sci. 3 (12), 160558.

Changeux, J.-P., Edelstein, S.J., 2012. The Brain as a Chemical Machine. Odile Jacob, New York.

Chattopadhyay, S., Ghosh, R., 2012. A study of generalized second law of thermodynamics in modified f (R) Horava–Lifshitz gravity. Astrophys. Space Sci. 341 (2), 669–674.

Chirikov, B.V., 1979. A universal instability of many-dimensional oscillator systems. Phys. Rep. 52 (5), 263–379.

Cicurel, R., Nicolelis, M.A., 2015. The Relativistic Brain: How It Works and Why It Cannot Be Simulated By a Turing Machine. Kios Press.

Cifra, M., 2012. Electrodynamic eigenmodes in cellular morphology. Biosystems 109 (3), 356–366.

Cifra, M., Pokorný, J., Havelka, D., Kučera, O., 2010. Electric field generated by axial longitudinal vibration modes of microtubule. Biosystems 100 (2), 122–131.

Cirelli, C., Gutierrez, C.M., Tononi, G., 2004. Extensive and divergent effects of sleep and wakefulness on brain gene expression. Neuron 41 (1), 35–43.

Ciszak, M., Comparini, D., Mazzolai, B., Baluska, F., Arecchi, F.T., Vicsek, T., Mancuso, S., 2012. Swarming behavior in plant roots. PLoS One 7 (1). e29759.

Colman, R.J., Anderson, R.M., Johnson, S.C., Kastman, E.K., Kosmatka, K.J., Beasley, T.M., Allison, D.B., Weindruch, R., 2009. Caloric restriction delays disease onset and mortality in rhesus monkeys. Science 325 (5937), 201–204.

Copeland, B.J., 2002. Hypercomputation. Mind. Mach. 12 (4), 461–502.

Cosic, I., 1994. Macromolecular bioactivity: is it resonant interaction between macromolecules?-theory and applications. IEEE Trans. Biomed. Eng. 41 (12), 1101–1114.

Craddock, T.J., Tuszynski, J.A., Hameroff, S., 2012a. Cytoskeletal signaling: is memory encoded in microtubule lattices by CaMKII phosphorylation? PLoS Comput. Biol. 8 (3). e1002421.

Craddock, T.J., Tuszynski, J.A., Chopra, D., Casey, N., Goldstein, L.E., Hameroff, S.R., Tanzi, R.E., 2012b. The zinc dyshomeostasis hypothesis of Alzheimer's Disease. PLoS One 7 (3). e33552.

Crawley, J.N., 1999. Behavioral phenotyping of transgenic and knockout mice: experimental design and evalua- tion of general health, sensory functions, motor abilities, and specific behavioral tests. Brain Res. 835 (1), 18–26.

Crick, F., 1970. Central dogma of molecular biology. Nature 227 (5258), 561–563.

Croll, S.D., Suri, C., Compton, D.L., Simmons, M.V., Yancopoulos, G.D., Lindsay, R.M., Scharfman, H.E., 1999. Brain-derived neurotrophic factor transgenic mice exhibit passive avoidance deficits, increased seizure sever- ity and in vitro hyperexcitability in the hippocampus and entorhinal cortex. Neuroscience 93 (4), 1491–1506.

Cuntz, H., Forstner, F., Borst, A., Häusser, M., 2010. One rule to grow them all: a general theory of neuronal branching and its practical application. PLoS Comput. Biol. 6 (8). e1000877.

Dauwels, J., Vialatte, F., Cichocki, A., 2010. Diagnosis of Alzheimers disease from EEG signals: where are we standing? Curr. Alzheimer Res. 7 (6), 487–505.

David, D.C., Ollikainen, N., Trinidad, J.C., Cary, M.P., Burlingame, A.L., Kenyon, C., 2010. Widespread protein aggregation as an inherent part of aging in C. elegans. PLoS Biol. 8 (8). e1000450.

Davies, G., Tenesa, A., Payton, A., Yang, J., Harris, S.E., Liewald, D., Deary, I.J., 2011. Genome-wide association studies establish that human intelligence is highly heritable and polygenic. Mol. Psychiatry 16 (10), 996–1005.

Dawkins, R., 2009. The Greatest Show on Earth, the Evidence for Evolution. Free Press, New York.

de Jesus, B.B., Vera, E., Schneeberger, K., Tejera, A.M., Ayuso, E., Bosch, F., Blasco, M.A., 2012a. Telomerase gene therapy in adult and old mice delays aging and increases longevity without increasing cancer. EMBO Mol. Med. 4, 1–14.

de Jesus, B.B., Vera, E., Schneeberger, K., Tejera, A.M., Ayuso, E., Bosch, F., Blasco, M.A., 2012b. Telomerase gene therapy in adult and old mice delays aging and increases longevity without increasing cancer. EMBO Mol. Med. 4 (8), 691–704.

Deacon, T.W., 2011. Incomplete Nature: How Mind Emerged From Matter. W.W. Norton, New York.

Deary, I.J., Penke, L., Johnson, W., 2010. The neuroscience of human intelligence differences. Nat. Rev. Neurosci. 11 (3), 201–211.

Deng, M., Zhu, W., 2009. Some applications of stochastic averaging method for quasi Hamiltonian systems in physics. Sci. China, Ser. G 52 (8), 1213–1222.

Dennett, D.C., 1971. Intentional systems. J. Philos. 68 (4), 87–106.

Dennett, D.C., 2017. From Bacteria to Bach and Back: The Evolution of Minds. W.W Norton, New York.

Di Giulio, M., 2011. The last universal common ancestor (LUCA) and the ancestors of archaea and bacteria were progenotes. J. Mol. Evol. 72 (1), 119–126.

Dirac, P.A.M., 1927. The quantum theory of the emission and absorption of radiation. Proc. R. Soc. London, Ser. A 114.

Dixon, J.M., Tuszynski, J.A., Clarkson, P.L., 1997. From Nonlinearity to Coherence: Universal Features of Non- linear Behaviour in Many-Body Physics. Oxford University Press, Oxford.

Dodd, M.S., Papineau, D., Grenne, T., Slack, J.F., Rittner, M., Pirajno, F., O'Neil, J., Little, C.T.S., 2017. Evidence for early life in Earth's oldest hydrothermal vent precipitates. Nature 543 (7643), 60–64.

Dragicevic, N., Bradshaw, P.C., Mamcarz, M., Lin, X., Wang, L., Cao, C., Arendash, G.W., 2011. Long-term electromagnetic field treatment enhances brain mitochondrial function of both Alzheimer's transgenic mice and normal mice: a mechanism for electromagnetic field-induced cognitive benefit? Neuroscience 185, 135–149.

Drochioiu, G., 2006. Eugen Macovschi's concept of biostructure and its current development. In: Life and Mind in Search of the Physical Basis, pp. 43–60. Trafford Publ., Canada, USA, Ireland & UK.

Dubreuil, C.I., Marklund, N., Deschamps, K., McIntosh, T.K., McKerracher, L., 2006. Activation of Rho after traumatic brain injury and seizure in rats. Exp. Neurol. 198 (2), 361–369.

Edelman, G.M., 1987. Neural Darwinism: The Theory of Neuronal Group Selection. Basic Books, New York.

Ehrenfreund, P., Charnley, S.B., 2000. Organic molecules in the interstellar medium, comets, and meteorites: a voyage from dark clouds to the early Earth. Annu. Rev. Astron. Astrophys. 38 (1), 427–483.

Eisenberg, B., Hyon, Y., Liu, C., 2010. Energy variational analysis of ions in water and channels: field theory for primitive models of complex ionic fluids. J. Chem. Phys. 133 (10), 104104.

Elgart, V., Kamenev, A., 2006. Classification of phase transitions in reaction-diffusion models. Phys. Rev. E 74 (4). 041101.

England, J.L., 2015. Dissipative adaptation in driven self-assembly. Nat. Nanotechnol. 10 (11), 919–923.

Fano, U., 1957. Description of states in quantum mechanics by density matrix and operator techniques. Rev. Mod. Phys. 29 (1), 74–93.

Fay, J.C., 2011. Weighing the evidence for adaptation at the molecular level. Trends Genet. 27 (9), 343–349.

Feldman, J., 2010. Ecological expected utility and the mythical neural code. Cogn. Neurodyn. 4 (1), 25–35.

Feynman, R.P., Hibbs, A.R., 1965. Quantum Physics and Path Integrals. McGraw-Hill, New York.

Feynman, R.P., Leighton, R.B., Sands, M.L., 1963. The Feynman Lectures on Physics. Addison-Wesley, Reading, MA.

Fontana, L., Partridge, L., Longo, V.D., 2010. Extending healthy life span-from yeast to humans. Science 328 (5976), 321–326.

Ford, B., 2009. On intelligence in cells: the case for whole cell biology. Interdiscip. Sci. Rev. 34 (4), 350–365.

Ford, B.J., 2010. The secret power of the single cell. New Sci. 206 (2757), 26–27.

Forsdyke, D.R., 2009. Samuel Butler and human long term memory: is the cupboard bare? J. Theor. Biol. 258 (1), 156–164.

Fransson, J., 2010. Non-Equilibrium Nano-Physics: A Many-Body Approach, vol. 809. Springer, Berlin.

Friedman, J.I., Vrijenhoek, T., Markx, S., Janssen, I.M., Van der Vliet, W.A., Faas, B.H.W., Veltman, J.A., 2007. CNTNAP2 gene dosage variation is associated with schizophrenia and epilepsy. Mol. Psychiatry 13 (3), 261–266.

Fromm, J., Lautner, S., 2006. Electrical signals and their physiological significance in plants. Plant Cell Environ. 30 (3), 249–257.

Garcia, J., Buchwald, N.A., Feder, B.H., Koelling, R.A., Tedrow, L., 1964. Sensitivity of head to x-ray. Science 144 (3625), 1470–1472.

Garzón, P.C., Keijzer, F., 2011. Plants: adaptive behavior, root-brains, and minimal cognition. Adapt. Behav. 19 (3), 155–171.

Gething, M.J., Sambrook, J., 1992. Protein folding in the cell. Nature 355 (6355), 33.

Gibson, D.G., Glass, J.I., Lartigue, C., Noskov, V.N., Chuang, R.Y., Algire, M.A., … Merryman, C., 2010. Creation of a bacterial cell controlled by a chemically synthesized genome. Science 329 (5987), 52–56.

Goldin, D., Wegner, P., 2008. The interactive nature of computing: refuting the strong Church–Turing thesis. Mind. Mach. 18 (1), 17–38.

Greengard, L., 1994. Fast algorithms for classical physics. Science 265 (5174), 909.

Grzybowski, B.A., Wilmer, C.E., Kim, J., Browne, K.P., Bishop, K.J., 2009. Self-assembly: from crystals to cells. Soft Matter 5 (6), 1110–1128.

Guzzo, M., Morbidelli, A., 1996. Construction of a Nekhoroshev like result for the asteroid belt dynamical system. Celest. Mech. Dyn. Astron. 66 (3), 255–292.

Gwinn, R.P., Spencer, D.D., 2004. Fighting fire with fire: brain stimulation for the treatment of epilepsy. Clin. Neurosci. Res. 4 (1), 95–105.

Hameroff, S., Penrose, R., 1996. Orchestrated reduction of quantum coherence in brain microtubules: A model for consciousness. Mathematics and computers in simulation 40 (3–4), 453–480.

Hamilton, S.W.R., 1833. On a general method of expressing the paths of light, & of the planets, by the coefficients of a characteristic function. PD Hardy, Dublin.

Hartl, F.U., Hayer-Hartl, M., 2002. Molecular chaperones in the cytosol: from nascent chain to folded protein. Science 295 (5561), 1852–1858.

Havelka, D., Cifra, M., Kučera, O., Pokorný, J., Vrba, J., 2011. High-frequency electric field and radiation characteristics of cellular microtubule network. J. Theor. Biol. 286 (1), 31–40.

Hawking, S., Mlodinow, L., 2010. The (elusive) theory of everything. Sci. Am. 303 (4), 68–71.

Heinrich, R., Schuster, S., 1996. The Regulation of Cellular Systems. Chapman & Hall, New York.

Heiss, J.E., Katz, Y., Ganmor, E., Lampl, I., 2008. Shift in the balance between excitation and inhibition during sensory adaptation of S1 neurons. J. Neurosci. 28 (49), 13320–13330.

Helbig, I., Scheffer, I.E., Mulley, J.C., Berkovic, S.F., 2008. Navigating the channels and beyond: unravelling the genetics of the epilepsies. Lancet Neurol. 7 (3), 231–245.

Henzler-Wildman, K., Kern, D., 2007. Dynamic personalities of proteins. Nature 450 (7172), 964–972.

Hetesi, Z., Balázs, B., 2006. On the question of validity of the anthropic principles. Acta Phys. Pol. B 37 (9), 2729–2739.

Hoyle, F., Wickramasinghe, C., 1981. Evolution From Space. JM Dent, London. 176 p.

Irizarry, M.C., McNamara, M., Fedorchak, K., Hsiao, K., Hyman, B.T., 1997. APPSw transgenic mice develop age-related A beta deposits and neuropil abnormalities, but no neuronal loss in CA1. J. Neuropathol. Exp. Neurol. 56 (9), 965.

Isaacson, J.S., Walmsley, B., 1995. Counting quanta: direct measurements of transmitter release at a central synapse. Neuron 15 (4), 875–884.

Jensen, A.R., 1998. The g factor: the science of mental ability. Politics Life Sci. 17 (2), 230–232.

Jeon, K.W., Lorch, I.J., Danielli, J.F., 1970. Reassembly of living cells from dissociated components. Science 167, 1626–1627.

Jeong, J., 2004. EEG dynamics in patients with Alzheimer's disease. Clin. Neurophysiol. 115 (7), 1490–1505.

Jerison, H.J., 1973. Evolution of the Brain and Intelligence. Academic Press, New York.

Jog, M.S., Kubota, Y., Connolly, C.I., Hillegaart, V., Graybiel, A.M., 1999. Building neural representations of habits. Science 286 (5445), 1745–1749.

Johnson, N.F., Manrique, P.D., Mendoza, A.D., Caycedo, F.D., Rodríguez, F., Quiroga, L., 2015. Survivability of photosynthetic bacteria in non-terrestrial light. J. Astrobiol. Outreach, 1–4.

Joly, O., Vanduffel, W., Orban, G.A., 2009. The monkey ventral premotor cortex processes 3D shape from disparity. Neuroimage 47 (1), 262–272.

Kane, G.L., Perry, M.J., Zytkow, A.N., 2002. The beginning of the end of the anthropic principle. New Astron. 7 (1), 45–53.

Koehl, P., Levitt, M., 1999. A brighter future for protein structure prediction. Nat. Struct. Biol. 6, 108–111.

Kohl, P., Crampin, E.J., Quinn, T.A., Noble, D., 2010. Systems biology: an approach. Clin. Pharmacol. Ther. 88 (1), 25–33.

Kohonen, T., 1988. An introduction to neural computing. Neural Netw. 1 (1), 3–16.

Kollmannsberger, P., Bidan, C.M., Dunlop, J.W.C., Fratzl, P., 2011. The physics of tissue patterning and extracellular matrix organisation: how cells join forces. Soft Matter 7 (20), 9549–9560.

Komatsuzaki, T., Berry, R.S., 2002. Chemical reaction dynamics: many-body chaos and regularity. Adv. Chem. Phys. 123, 79–152.

Kubelka, J., Hofrichter, J., Eaton, W.A., 2004. The protein folding "speed limit". Curr. Opin. Struct. Biol. 14 (1), 76–88.

Kučera, O., Havelka, D., 2012. Mechano-electrical vibrations of microtubules–link to subcellular morphology. Biosystems 109 (3), 346–355.

Lagercrantz, H., Changeux, J.P., 2009. The emergence of human consciousness: from fetal to neonatal life. Pediatr. Res. 65 (3), 255–260.

Lagrange, J.L., 1772. Essay on the problem of the three bodies. Complete Works 6, 229–324.

LaLumiere, R.T., 2011. A new technique for controlling the brain: optogenetics and its potential for use in research and the clinic. Brain Stimul. 4 (1), 1–6.

Lancaster, M.A., Renner, M., Martin, C.A., Wenzel, D., Bicknell, L.S., Hurles, M.E., … Knoblich, J.A., 2013. Cerebral organoids model human brain development and microcephaly. Nature 501 (7467), 373–379.

Langton, C.G., 1989. Artificial Life. Addison-Wesley Publishing Company, Redwood City, CA. pp. 1–48.

Laskar, J., 1993. Frequency analysis for multi-dimensional systems. Global dynamics and diffusion. Physica D 67 (1), 257–281.

Laskar, J., 2005. Frequency Map Analysis and Quasiperiodic Decomposition, in Hamiltonian Systems and Fourier Analysis: New Prospects for Gravitational Dynamics. In: Benest, D., Froeschlé, C., Lega, E. (Eds.), Advances in astronomy and astrophysics. Cambridge Scientific Publishers, Cambridge, UK, pp. 99–129.

Lee, C.-K., Klopp, R.G., Weindruch, R., Prolla, T.A., 1999. Gene expression profile of aging and its retardation by caloric restriction. Science 285 (5432), 1390–1393.

Lewen, A., Matz, P., Chan, P.H., 2000. Free radical pathways in CNS injury. J. Neurotrauma 17 (10), 871–890.

Li, Y., Liu, Y., Li, J., Qin, W., Li, K., Yu, C., Jiang, T., 2009. Brain anatomical network and intelligence. PLoS Comput. Biol. 5 (5). e1000395.

Liu, J., Jiang, Y.G., Huang, C.Y., Fang, H.Y., Fang, H.T., Pang, W., 2008. Depletion of intracellular zinc downregulates expression of Uch-L1 mRNA and protein, and CREB mRNA in cultured hippocampal neurons. Nutr. Neurosci. 11 (3), 96–102.

Luscher, B., Fuchs, T., Kilpatrick, C.L., 2011. GABAA receptor trafficking-mediated plasticity of inhibitory synapses. Neuron 70, 385–409.

MacGregor, R.J., Vimal, R.L.P., 2008. Consciousness and the structure of matter. J. Integr. Neurosci. 7 (1), 75–116.

Malgaroli, A., Vallar, L., Zimarino, V., 2006. Protein homeostasis in neurons and its pathological alterations. Curr. Opin. Neurobiol. 16 (3), 270–274.

Mangiarini, L., Sathasivam, K., Seller, M., Cozens, B., Harper, A., Hetherington, C., … Bates, G.P., 1996. Exon 1 of the HD gene with an expanded CAG repeat is sufficient to cause a progressive neurological phenotype in transgenic mice. Cell 87 (3), 493–506.

Maret, W., 2005. Zinc coordination environments in proteins determine zinc functions. J. Trace Elem. Med. Biol. 19 (1), 7–12.

Masi, E., Ciszak, M., Stefano, G., Renna, L., Azzarello, E., Pandolfi, C., Mancuso, S., 2009. Spatiotemporal dynamics of the electrical network activity in the root apex. Proc. Natl. Acad. Sci. 106 (10), 4048–4053.

Maupertuis, P.D., 1746. The laws of motion and rest deduced from a metaphysical principle. Mémoires de l'académie des sciences de Berlin 2, 1746–1748.

May, A., Gaser, C., 2006. Magnetic resonance-based morphometry: a window into structural plasticity of the brain. Curr. Opin. Neurol. 19 (4), 407–411.

May, E.C., Paulinyi, T., Vassy, Z., 2005. Anomalous anticipatory skin conductance response to acoustic stimuli: experimental results and speculation about a mechanism. J. Altern. Complement. Med. 11 (4), 695–702.

McClelland, J.L., Rumelhart, D.E., & the PDP Research Group. 1986. Parallel distributed processing: Explorations in the microstructure of cognition: Vol. 2. Psychological and biological models, MIT Press, Cambridge, MA.

McNichol, J., Gordon, R., 2012. Are we from outer space? A critical review of the panspermia hypothesis. In: Seckbach, J. (Ed.), Genesis—In the Beginning: Precursors of Life, Chemical Models and Early Biological Evolution. Springer, Dordrecht, pp. 591–620.

Meyer, Y., Hall, G., Offin, D., 2008. Introduction to Hamiltonian Dynamical Systems and the N-Body Problem, vol. 90. Springer Science & Business Media.

Mo, Y., 2009. The resonance energy of benzene: a revisit. J. Phys. Chem. A 113 (17), 5163–5169.

Moeller, J.J., Tu, B., Bazil, C.W., 2011. Quantitative and qualitative analysis of ambulatory electroencephalography during mild traumatic brain injury. Arch. Neurol. 68 (12), 1595.

Montagnier, L., Aissa, J., Del Giudice, E., Lavallee, C., Tedeschi, A., Vitiello, G., 2011. DNA waves and water. J. Phys. Conf. Ser. 306 (1), 012007. IOP Publishing.

Montgomery Jr., E.B., Gale, J.T., 2008. Mechanisms of action of deep brain stimulation (DBS). Neurosci. Biobehav. Rev. 32 (3), 388–407.

Morbidelli, A., Henrard, J., 1991. Secular resonances in the asteroid belt: theoretical perturbation approach and the problem of their location. Celest. Mech. Dyn. Astron. 51 (2), 131–167.

Murariu, M., Drochioiu, G., 2012. Biostructural theory of the living systems. Biosystems 109 (2), 126–132.

Murray, C., Herrnstein, R., 1994. Bell curve: Intelligence and class structure in American life. Free Press.

Nakagaki, T., Yamada, H., Tóth, Á., 2000. Intelligence: Maze-solving by an amoeboid organism. Nature 407 (6803), 470.

Nei, M., Kumar, S., 2000. Molecular Evolution and Phylogenetics. Oxford University Press, Oxford.

Ng, J., 2012. Wnt/PCP proteins regulate stereotyped axon branch extension in Drosophila. Development 139 (1), 165–177.

Nithianantharajah, J., Komiyama, N.H., McKechanie, A., Johnstone, M., Blackwood, D.H., St Clair, D., … Grant, S.G., 2013. Synaptic scaffold evolution generated components of vertebrate cognitive complexity. Nat. Neurosci. 16 (1), 16–24.

Noble, D., 2006. The Music of Life: Biology Beyond the Genome. Oxford University Press, Oxford.

Noble, D., 2011. Neo-Darwinism, the modern synthesis and selfish genes: are they of use in physiology? J. Physiol. 589 (5), 1007–1015.

Oladepo, S.A., Xiong, K., Hong, Z., Asher, S.A., 2011. Elucidating peptide and protein structure and dynamics: UV resonance Raman spectroscopy. J. Phys. Chem. Lett. 2 (4), 334.

Onen, F., Onen, S.-H., 2003. Sleep rhythm disturbances in Alzheimer's disease. Altérations des rythmes du sommeil dans la maladie d'Alzheimer. Rev. Med. Interne 24 (3), 165–171.

Oparin, A.I., 2003. The Origin of Life. Courier Corporation, New York.

O'Rourke, N.A., Weiler, N.C., Micheva, K.D., Smith, S.J., 2012. Deep molecular diversity of mammalian synapses: why it matters and how to measure it. Nat. Rev. Neurosci. 13 (6), 365–379.

Orpiszewski, J., Schormann, N., Kluve-Beckerman, B., Liepnieks, J.J., Benson, M.D., 2000. Protein aging hypothesis of Alzheimer disease. FASEB J. 14 (9), 1255–1263.

Packer, A.M., Yuste, R., 2011. Dense, unspecific connectivity of neocortical parvalbumin-positive interneurons: a canonical microcircuit for inhibition? J. Neurosci. 31 (37), 13260–13271.

Pang, X.-F., 2011. The theory of bio-energy transport in the protein molecules and its properties. Phys Life Rev 8 (3), 264–286.

Pang, X.-F., 2012. The mechanism and properties of bio-photon emission and absorption in protein molecules in living systems. J. Appl. Phys. 111 (9). 093519.

Penrose, R., 1994. Shadows of the Mind, vol. 52. Oxford University Press, Oxford.

Persson, D., Halberg, K.A., Jørgensen, A., Ricci, C., Møbjerg, N., Kristensen, R.M., 2011. Extreme stress tolerance in tardigrades: surviving space conditions in low earth orbit. J. Zool. Syst. Evol. Res. 49, 90–97.

Piaget, J., 1951. The child's conception of the world (No. 213). Rowman & Littlefield.

Pietak, A.M., 2012. Structural evidence for electromagnetic resonance in plant morphogenesis. Biosystems 109 (3), 367–380.

Pinker, S., Bloom, P., 1990. Natural language and natural selection. Behav. Brain Sci. 13, 707–784.

Pirooznia, S.K., Sarthi, J., Johnson, A.A., Toth, M.S., Chiu, K., Koduri, S., Elefant, F., 2012. Tip60 HAT activity mediates APP induced lethality and apoptotic cell death in the CNS of a drosophila Alzheimer's disease model. PLoS One 7 (7). e41776.

Plankar, M., Brežan, S., Jerman, I., 2013. The principle of coherence in multi-level brain information processing. Progress in biophysics and molecular biology 111 (1), 8–29.

Poza, J.J., 2012. The genetics of focal epilepsies. Handbook of Clinical Neurology 107, 153–161.

Pozo, K., Goda, Y., 2010. Unraveling mechanisms of homeostatic synaptic plasticity. Neuron 66 (3), 337.

Puthoff, H.E., Targ, R., 1976. A perceptual channel for information transfer over kilometer distances: historical perspective and recent research. Proc. IEEE 64 (3), 329–354.

Qiu-Dong, W., 1990. The global solution of the n-body problem. Celest. Mech. Dyn. Astron. 50 (1), 73–88.

Quirk, M.C., Blum, K.I., Wilson, M.A., 2001. Experience-dependent changes in extracellular spike amplitude may reflect regulation of dendritic action potential back-propagation in rat hippocampal pyramidal cells. J. Neurosci. 21 (1), 240–248.

Quiroga, R.Q., Reddy, L., Kreiman, G., Koch, C., Fried, I., 2005. Invariant visual representation by single neurons in the human brain. Nature 435 (7045), 1102–1107.

Rahnama, M., Tuszynski, J.A., Bókkon, I., Cifra, M., Sardar, P., Salari, V., 2011. Emission of mitochondrial biophotons and their effect on electrical activity of membrane via microtubules. J. Integr. Neurosci. 10 (01), 65–88.

Reichl, L.E., 2004. The Transition to Chaos: Conservative Classical Systems and Quantum Manifestations. Springer, Berlin.

Ritchie, S.J., Wiseman, R., French, C.C., 2012. Failing the future: three unsuccessful attempts to replicate Bem's "retroactive facilitation of Recall" effect. PLoS One 7 (3). e33423.

Robert, A.M., Robert, C.S., Robert, L., 2007. Symmetry breaking in biological systems. From molecules to tissues. Struct. Chem. 18 (6), 899–907.

Rubik, B., 2002. The biofield hypothesis: its biophysical basis and role in medicine. J. Altern. Complement. Med. 8 (6), 703–717.

Rumelhart, D.E., Hinton, G.E., McClelland, J.L., 1986. A general framework for parallel distributed processing. Parallel distributed processing: Explorations in the microstructure of cognition, vol. 1. 45–76.

Ryazanov, A.G., Nefsky, B.S., 2002. Protein turnover plays a key role in aging. Mech. Ageing Dev. 123 (2), 207–213.

Sasaki, T., Matsuki, N., Ikegaya, Y., 2011. Action-potential modulation during axonal conduction. Sci. Signal. 331 (6017), 599.

Scheeff, E.D., Bourne, P.E., 2005. Structural evolution of the protein kinase–like superfamily. PLoS Comput. Biol. 1 (5). e49.

Scheiner, S., 2011. Weak H-bonds. Comparisons of CHO to NHO in proteins and PHN to direct PN interactions. Phys. Chem. Chem. Phys. 13 (31), 13860–13872.

Schenk, H.J., Seabloom, E.W., 2010. Evolutionary ecology of plant signals and toxins: a conceptual framework. In: Plant Communication from an Ecological Perspective. Springer Berlin Heidelberg, New York, pp. 1–19.

Schiffels, S., Szöllősi, G.J., Mustonen, V., Lässig, M., 2011. Emergent neutrality in adaptive asexual evolution. Genetics 189 (4), 1361–1375.

Scholes, G.D., 2003. Long-range resonance energy transfer in molecular systems. Annu. Rev. Phys. Chem. 54 (1), 57–87.

Schrödinger, E., 1944. What is Life? With Mind and Matter and Autobiographical Sketches. Cambridge University Press, Cambridge.

Schulz, J.M., 2010. Synaptic plasticity in vivo: more than just spike-timing? Front. Syn. Neurosci. 2, 150.

Scott, A.C., 1991. Davydov's soliton revisited. Physica D 51 (1), 333–342.

Scott, E.K., Luo, L., 2001. How do dendrites take their shape? Nat. Neurosci. 4, 359–366.

Scruton, R., 2005. The unobservable mind. Technol. Rev. 108 (2), 72–77.

Searle, J.R., 2000. Consciousness. Annu. Rev. Neurosci. 23, 557–578.

Searle, J.R., 2002. Why I am not a property dualist. J. Conscious. Stud. 9 (12), 57–64.

Searle, J.R., 2013. Can a photodiode be conscious? New York Rev. Books 60 (4), 43–44. Reply.

Sekulić, D.L., Satarić, B.M., Tuszynski, J.A., Satarić, M.V., 2011. Nonlinear ionic pulses along microtubules. Eur. Phys. J. E: Soft Matter Biol. Phys. 34 (5), 1–11.

Shapiro, J.A., 2005. A 21st century view of evolution: genome system architecture, repetitive DNA, and natural genetic engineering. Gene 345 (1), 91–100.

Shapiro, J.A., 2007. Bacteria are small but not stupid: cognition, natural genetic engineering and socio-bacteriology. Stud. Hist. Phil. Biol. Biomed. Sci. 38 (4), 807–819.

Sharma, V., Annila, A., 2007. Natural process—natural selection. Biophys. Chem. 127 (1-2), 123–128.

Sharma, V., Kaila, V.R.I., Annila, A., 2009. Protein folding as an evolutionary process. Physica A: Statistical Mechanics and its Applications 388 (6), 851–862.

Sharov, A.A., Gordon, R., 2013. Life Before Earth. arXiv preprint arXiv:1304.3381.

Shaw, D.E., Maragakis, P., Lindorff-Larsen, K., Piana, S., Dror, R.O., Eastwood, M.P., Wriggers, W., 2010. Atomic-level characterization of the structural dynamics of proteins. Science 330 (6002), 341–346.

Sibaoka, T., 1991. Rapid plant movements triggered by action potentials. Bot. Mag. (Tokyo) 104 (1), 73–95.

Silvia, P.J., Beaty, R.E., 2012. Making creative metaphors: the importance of fluid intelligence for creative thought. Intelligence 40 (4), 343–351.

Sivakumaran, S., Hariharaputran, S., Mishra, J., Bhalla, U.S., 2003. The database of quantitative cellular signaling: management and analysis of chemical kinetic models of signaling networks. Bioinformatics 19 (3), 408–415.

Smit, D.J., Boomsma, D.I., Schnack, H.G., Hulshoff Pol, H.E., de Geus, E.J., 2012. Individual differences in EEG spectral power reflect genetic variance in gray and white matter volumes. Twin Res. Hum. Genet. 15 (3), 384.

Spruston, N., Cang, J., 2010. Timing isn't everything. Nat. Neurosci. 13 (3), 277.

Steriade, M., 2001. Intact & Sliced Brain. MIT press.

Steward, O., Schuman, E.M., 2003. Compartmentalized synthesis and degradation of proteins in neurons. Neuron 40 (2), 347–359.

Styer, D.F., Balkin, M.S., Becker, K.M., Burns, M.R., Dudley, C.E., Forth, S.T., … Wotherspoon, T.D., 2002. Nine formulations of quantum mechanics. Am. J. Phys. 70, 288.

Su, S.P., Lyons, B., Friedrich, M., McArthur, J.D., Song, X., Xavier, D., … Aquilina, J.A., 2012. Molecular signatures of long-lived proteins: autolytic cleavage adjacent to serine residues. Aging Cell 11 (6), 1125–1127.

Sundman, K.F., 1913. Memory on the problem of the three bodies. Acta Mathematica 36 (1), 105–179.

Taketo, M., Yoshioka, T., 2000. Developmental change of GABA(A) receptor-mediated current in rat hippocampus. Neuroscience 96 (3), 507–514.

Taki, Y., Thyreau, B., Kinomura, S., Sato, K., Goto, R., Kawashima, R., Fukuda, H., 2011. Correlations among brain gray matter volumes, age, gender, and hemisphere in healthy individuals. PLoS One 6 (7). e22734.

Thagard, P., 2002. How molecules matter to mental computation. Philos. Sci. 69 (3), 429–446.

Thatcher, R.W., North, D.M., Biver, C.J., 2008. Intelligence and EEG phase reset: a two compartment model of phase shift and lock. NeuroImage 42 (4), 1639–1653.

Thompson, R.F., Spencer, W.A., 1966. Habituation: a model phenomenon for the study of neuronal substrates of behavior. Psychol. Rev. 73 (1), 16.

Trauth, M.H., Maslin, M.A., Deino, A.L., Junginger, A., Lesoloyia, M., Odada, E.O., Tiedemann, R., 2010. Human evolution in a variable environment: the amplifier lakes of Eastern Africa. Quat. Sci. Rev. 29 (23), 2981–2988.

Trewavas, A., 2003. Aspects of plant intelligence. Ann. Bot. 92 (1), 1–20.

Turing, A.M., 1936. On computable numbers, with an application to the Entscheidungsproblem. J. Math. 58, 345–363.

Tuszynski, J.A., 2013. The need for a physical basis of cognitive process: Comment on "Consciousness in the universe. A review of the 'Orch OR' theory" by Hameroff and Penrose. Phys Life Rev 11 (1), 79–80.

Tuszynski, J.A., Carpenter, E.J., Huzil, J.T., Malinski, W., Luchko, T., Ludueña, R.F., 2006. The evolution of the structure of tubulin and its potential consequences for the role and function of microtubules in cells and embryos. Int. J. Dev. Biol. 50 (2), 341–358.

Tweedie-Cullen, Y., Brunner, A.R., Mansuy, I.M., 2012. Proteomic analysis of phosphorylation in the brain. Curr. Proteomics 9 (3), 167–185.

Uhlhaas, P.J., Singer, W., 2006. Neural synchrony in brain disorders: relevance for cognitive dysfunctions and pathophysiology. Neuron 52 (1), 155–168.

Vendruscolo, M., 2012. Proteome folding and aggregation. Curr. Opin. Struct. Biol. 22 (2), 138–143.

von Däniken, E., 2002. The Return of the Gods: Evidence of Extraterrestrial Visitations. Tantor eBooks.

Walker, A.C., Smith, S.D., Smith, S.D., 1987. Mitochondrial DNA and human evolution. Nature 325, 1–5.

Watson, J.D., Crick, F.H.C., 1953. A structure for deoxyribose nucleic acid. Nature 171, 737–738.

Welberg, L., 2012. Consciousness: effective detection. Nat. Rev. Neurosci. 13 (3), 155.

Wikipedia, 2017. Artificial Consciousness. http://en.wikipedia.org/wiki/Artificial_consciousness.

Willerman, L., Schultz, R., Neal Rutledge, J., Bigler, E.D., 1991. In vivo brain size and intelligence. Intelligence 15 (2), 223–228.

Witelson, S.F., Kigar, D.L., Harvey, T., 1999. The exceptional brain of Albert Einstein. Lancet 353 (9170), 2149–2153.

Woolf, N.J., Priel, A., 2009. Nanoneuroscience: Structural and Functional Roles of the Neuronal Cytoskeleton in Health and Disease. Springer Science & Business Media.

Wyatt, M.C., 2008. Resonant trapping of planetesimals by planet migration: debris disk clumps and Vega's similarity to the solar system. Astrophys. J. 598 (2), 1321.

Yang, C.D., Weng, H.J., 2012. Nonlinear quantum dynamics in diatomic molecules: vibration, rotation and spin. Chaos, Solitons Fractals 45 (4), 402–415.

Yousefinejad, S., Hemmateenejad, B., 2015. Chemometrics tools in QSAR/QSPR studies: a historical perspective. Chemom. Intell. Lab. Syst. 149, 177–204.

Zaslavsky, G.M., 2005. Hamiltonian Chaos and Fractional Dynamics, vol. 1. Oxford University Press, London.

Zheng, Y., Luo, J.J., Harris, S., Kennerley, A., Berwick, J., Billings, S.A., Mayhew, J., 2012. Balanced excitation and inhibition: model based analysis of local field potentials. NeuroImage 63 (1), 81–94.

Zuo, Z., Tichotsky, A., Johns, R.A., 1999. Inhibition of excitatory neurotransmitter–nitric oxide signaling pathway by inhalational anesthetics. Neuroscience 93 (3), 1167–1172.

FURTHER READING

Brádler, K., Adami, C., 2014. The capacity of black holes to transmit quantum information. J. High Energy Phys. 5, 1–26.

Feynman, R.P., 1986. Quantum-mechanical computers. Found. Phys. 16 (6), 507–531.

Kimura, M., 1983. The Neutral Theory of Molecular Evolution. Cambridge University Press, Cambridge.

Shapiro, J.A., 2011. Evolution—A New View From the 21st Century. FT Press Science, New Jersey.

Tuszynski, J.A. (Ed.), 2006. The Emerging Physics of Consciousness. Heidelberg, Springer Verlag.

Zehnacker, A., Suhm, M.A., 2008. Chirality recognition between neutral molecules in the gas phase. Angew. Chem. Int. Ed. 47 (37), 6970–6992.

LIFE IN THE COSMIC SCALE

LIFE BEFORE EARTH

Alexei A. Sharov*, Richard Gordon[†,‡]

*National Institute on Aging (NIA/NIH), Baltimore, MD, United States
†Gulf Specimen Marine Laboratories, Panacea, FL, United States
‡Wayne State University, Detroit, MI, United States

CHAPTER OUTLINE

1 THE INCREASE OF GENETIC COMPLEXITY FOLLOWS MOORE'S LAW

Biological evolution is traditionally studied in two aspects. First, paleontological records show astonishing changes in the composition of major taxonomic groups of animals and plants deposited in sedimentary rocks of various ages (Cowen, 2009; Valentine, 2004). Aquatic life forms give rise to the first terrestrial plants and animals, amphibians lead to reptiles including dinosaurs, ferns lead to gymnosperms, and then to flowering plants. Extinction of most dinosaurs is followed by the spread of mammals and flying descendants of dinosaurs called birds. Second, Darwin's theory augmented with statistical genetics demonstrated that heritable changes may accumulate in populations and result in replacement of gene variants (Mayr, 2002). This process drives microevolution, which helps species

Habitability of the Universe Before Earth, editors: Richard Gordon & Alexei Sharov, Volume 1 in the series:
Astrobiology: Exploring Life on Earth and Beyond, series editors: Pabulo Henrique Rampelotto,
Joseph Seckbach & Richard Gordon. ISSN 2468-6352. https://doi.org/10.1016/B978-0-12-811940-2.00011-3

to improve their functions and adjust to changing environments. But despite the importance of these two aspects of evolution, they do not capture the core of the macroevolutionary process, which is the increase of functional complexity of organisms.

Function can be defined as a reproducible sequence of actions of organisms that satisfies specific needs or helps to achieve vital goals (e.g., capturing a resource or reproduction) (Sharov, 2010). To be passed on from one generation to the next, functions must be encoded within the genome or other information carriers. The genome plays the role of intergeneration memory that ensures the preservation of various functions. Other components of the cell (e.g., stable chromatin modifications, gene imprinting, and assembly of the outer membrane) (Frankel, 1989; Grimes and Aufderheide, 1991), and cultural transmission (Dennett, 2017; Frohoff and Oriel, 2016; Lefebvre, 2013), may also contribute to the intergeneration memory; however, their informational role is minor compared to the genome for most organisms. Considering that the increase of functional complexity is the major trend in macroevolution, which seems applicable to all kinds of organisms from bacteria to mammals, it can be used as a generic scale to measure the level of organization. Because functions are transferred to new generations in the form of genetic memory, it makes sense to consider genetic complexity as a reasonable representation of the functional complexity of organisms (Luo, 2009; Sharov, 2006). The mechanism by which the genome becomes more complex probably relies heavily on duplication of portions of DNA ranging from parts of genes to gene cascades to polyploidy (Gordon and Gordon, 2016; Gordon, 1999; Ohno, 1970), followed by divergence of function of the copies. Developmental plasticity and subsequent genetic assimilation also play a role (West-Eberhard, 2002).

We then have to ask what might be a suitable parameter, measurable from a genome, that reflects its functional complexity? Early studies of the genomes of various organisms showed little correlation between genome length and the level of organization. For example, the total amount of DNA in some single-cell organisms is several orders of magnitude greater than in human cells, a phenomenon known as C-value paradox (Patrushev and Minkevich, 2008). Sequencing of full genomes of eukaryotic organisms showed that the total amount of DNA per cell is not a good measure of information encoded by the genome. The genome includes numerous repetitive elements (e.g., LINE, LTR, and SINE transposons), which have no direct cellular functions; also, some portions of the genome may be represented by multiple copies. Large single-cell organisms (e.g., amoeba) need multiple copies of the same chromosome to produce the necessary amount of mRNA. In eukaryotes, DNA has additional functions besides carrying genes and regulating their expression. These non-informational functions include structural support of nuclear matrix and nuclear lamina, chromosome condensation, regulation of cell division and homologous recombination, maintenance and regulation of telomeres and centromeres (Cavalier-Smith, 2005; Patrushev and Minkevich, 2008; Rollins et al., 2006), and rate of metabolism (Gregory, 2001; Martin and Gordon, 1995). Segments of DNA with non-genetic functions are mostly not conserved and include various transposing elements as well as tandem repeats. While the ENCODE project has uncovered many functions for 80% of the noncoding DNA in humans (Pennisi, 2012), it has not yet solved the C-value paradox (Doolittle, 2013; Eddy, 2012, 2013; Germain et al., 2014; Niu and Jiang, 2013). Thus, we stick to the suggestion to measure genetic complexity by the length of functional and nonredundant DNA sequence rather than by total DNA length (Adami et al., 2000; Sharov, 2006).

Functionality of genome regions can be inferred from their conservation in evolution. Conserved sequences (5% in the human genome) include the majority of genes (both protein-coding and noncoding) and regulatory modules such as promoters and enhancers (Kellis et al., 2014). Regulatory regions are less stable than genes, and thus, they are not always highly conserved in evolution. An additional

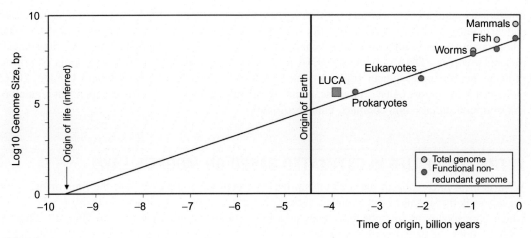

FIG. 1

On this semilog plot, the complexity of organisms, as measured by the length of functional nonredundant DNA in the genome counted by nucleotide base pairs (bp), increases linearly with time (Sharov and Gordon, 2013). Time is counted backwards in billions of years before the present (time 0). Modified from Fig. 1 in Sharov (2006). The genome size and age of LUCA are based on evolutionary reconstruction (Weiss et al., 2016) and molecular clock (Fournier, 2015; Hedges, 2002). As LUCA is a hypothetical construct, we have not included it as data for estimating this exponential curve.

criterion of functionality is the absence (or strong depletion) of transposable repeats (Simons et al., 2006). Based on our estimate, ca. 15% of mammalian genome is transposon-free, and therefore, likely functional (Sharov, 2006).

The time of origin of phylogenetic groups is based mostly on paleontological records, although the origin of eukaryotes is based on protein homology and the molecular clock (Sharov, 2006). Currently, more estimates of the origin of the major lineages are available, but not all of them agree with each other. For consistency, here we used the same dates of origin as in our original paper (Sharov and Gordon, 2013), although some of them have been updated recently.

If we plot genome complexity of major phylogenetic lineages on a logarithmic scale against the time of origin, the points appear to fit well to a straight line (Sharov, 2006) (Fig. 1). This indicates that genome complexity increased exponentially and doubled about every 340 million years. Similar estimates (360 million years) were obtained independently by Markov et al. (Markov et al., 2010). Such a relationship reminds us of the exponential increase of computer complexity known as a "Moore's law" (Lundstrom, 2003; Moore, 1965). Moore's law is an empirical observation that the number of transistors in integrated computer chips doubles every 18–24 months, which is much shorter than the doubling time of genome complexity.

Naturally, there is a question: What is common between computers and living organisms so that complexity measures of both follow an exponential curve? First, the evolution of human technology is similar to biological evolution because both processes are driven by goal-seeking agents (i.e., humans and other organisms (Sinnott, 1962)) that improve their functions in terms of reliability, speed, and extent. Second, both processes involve learning via trial and error because natural selection is functionally equivalent to learning (Hoffmeyer, 2011; p. 48). Finally, life is based on molecular-scale nanotechnology, which is even more advanced than human computer technology. Living cells and whole organisms use such molecular mechanisms as programmable construction devices (ribosomes), molecular copiers (DNA polymerase, RNA polymerase), communication via transporters (shuffling proteins, nuclear pore, dynein complex) and mechanochemical waves (Gordon and Gordon, 2016; Gordon, 1999), motors (ATP synthetase, actin-myosin interactions), and context-dependent regulators (transcription preinitiation complex, enhanceosome).

2 THE AGE OF LIFE IS ESTIMATED BASED ON MOORE'S LAW

What is most interesting in the relationship between genome complexity and the time of origin of organisms is that it can be extrapolated back to the origin of life. Genome complexity reaches zero, which corresponds to just one base pair, at a time about 9.7 billion years ago (Fig. 1) (Sharov, 2006). Among the data used for regression, the two earliest points are most uncertain, and thus, have major contribution to the variability of regression-based predictions. A sensitivity analysis that assumes variations of two earliest points in the range of ±0.3 billion years and ±0.3 log bp genome length gives a range for the regression of ca. ±2 billion years (Sharov, 2006). Based on this model, it is unlikely that life originated on Earth because the age of Earth is only 4.5 billion years. Combining our analysis with proposed scenarios of the evolution of the Universe, we come to a hypothesis that life is rather a cosmic-scale phenomenon that started many billion years before the formation of Solar system and planet Earth (Fig. 2).

The growth of organism complexity in evolution was later independently quantified using eco-exergy density (kJ/g) (Jørgensen, 2007). The conclusions of this study were similar to ours. Others have also concluded (qualitatively) that early life on Earth was too complex to have had time to evolve from an abiotic condition on Earth (Line, 2002, 2007; McNichol and Gordon, 2012). Jørgensen wrote: "Self-organizing systems as the organisms are expected to grow in complexity exponentially according to Moore's law, which is also the case for exergy density with good approximations.... Extrapolations of the graphs seem to indicate that the first primitive life form maybe of cells with less complexity than the prokaryote cells has been brought to us by the open space" (p. 490). His estimate for the origin of life was from 5 to 6 billion years ago. Another approach, applicable to multicellular organisms, would be to plot complexity measures of binary differentiation trees of organisms (Alicea and Gordon, 2014; Gordon and Gordon, 2016; Gordon, 1992, 1999) versus time of origin, but this will have to wait until enough differentiation trees have been constructed from 4D time-lapse observations of embryo development (Alicea and Gordon, 2016). Cavalier-Smith used a linear extrapolation of rate of evolution to roughly estimate that life on Earth required 18 billion years, coming to "the logically inescapable conclusion" that the rate of evolution sped up (Cavalier-Smith, 2014), rather than our conclusion that life may indeed have needed most of the age of the universe to evolve. The linearity of his extrapolation may account for his estimate exceeding the age of the universe.

Our approach to sensitivity analysis has been challenged by Marzban et al. (Marzban et al., 2014, 2017), who proposed to use prediction bands/intervals to evaluate sampling variability (their Fig. 2, black dashed lines, page 303 in this book). This method yielded an interval estimate for the age of life >7 billion years. In addition, Marzban et al. estimated prediction bands using measurement error models (Buonaccorsi, 2010), which yielded a much broader interval (Fig. 2, *red* dashed lines, page

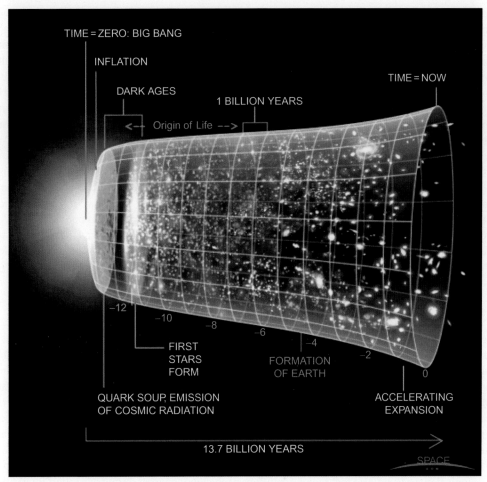

TIME = ZERO: BIG BANG

INFLATION

DARK AGES

TIME = NOW

1 BILLION YEARS

<-- Origin of Life -->

−12

−10

−8

−6

−4

−2

0

FIRST
STARS
FORM

FORMATION
OF EARTH

QUARK SOUP, EMISSION
OF COSMIC RADIATION

ACCELERATING
EXPANSION

13.7 BILLION YEARS

SPACE

FIG. 2

A schematic view of the development of the universe since the Big Bang, courtesy of the Hubble Space Telescope Science Institute, on which we have superimposed our estimate for the origin of life (Gordon and Gordon, 2016; Sharov and Gordon, 2013). Here we assume that life appeared >7 billion years ago based on the sensitivity analysis and prediction interval for individual points in the regression (see Fig. 2, black dashed lines, page 303 in this book). Note that the "Dark Ages" may have ended at −13.55 billion years (Zheng et al., 2012) (with the Big Bang more precisely placed at −13.75 billion years (Jarosik et al., 2011)), rather than an end of that period at −11.5 billion years, as depicted. Recent work suggests that acceleration of the expansion may not be occurring (Nielsen et al., 2016).

303 in this book). The interval includes 4.3 billion years, which they interpreted as being consistent with the hypothesis that life is not older than Earth.

We greatly appreciate that Marzban et al. applied a relatively new statistical approach to our data. However, the measurement error models assume that the frequency distribution of errors is uniform for all data points, which is not true for our data because some data points are much more accurate than others. Thus, the application of Eq. (9) in Marzban et al. (2014) is not justified. Also, individual points in Fig. 1 are not single measurements as assumed in statistics; instead each point represents our knowledge about large groups of organisms. It would be easy to replace one point for mammals by multiple points for each order of mammals. Various phyla and classes of invertebrates can be added also to increase the number of points. Larger number of points may yield narrower prediction intervals. However, these points are not fully independent as assumed in the model of Marzban et al. Thus, more complex models are needed that take into account (1) non-uniform measurement errors and (2) autocorrelation between points as a function of the evolutionary distance between lineages. Until such models are developed, the uncertainty in the predicted age of life cannot be estimated reliably.

Can we take these estimates as an approximate age of life in the universe? Answering this question is not easy because several other problems have to be addressed. First, why does the increase of genome complexity follow an exponential law instead of fluctuating erratically? Second, is it reasonable to expect that biological evolution started from something equivalent in complexity to one nucleotide? And third, if life is older than the Earth and the Solar System, then how can organisms survive interstellar transfer? These problems as well as consequences of the exponential increase of genome complexity are discussed below.

3 HOW VARIABLE ARE THE RATES OF EVOLUTION?

To extrapolate the rate of biological evolution into the past, we need to provide arguments why this rate is stable enough. There is no consensus among biologists on the question how variable are the rates of evolution. Darwin thought that in general evolutionary changes accumulate gradually through a series of small steps, rather than by sudden leaps (Darwin, 1866). However, he also pointed out that the rate of evolution is not uniform: "But I must here remark that I do not suppose that the process ever goes on so regularly as is represented in the diagram, though in itself made somewhat irregular, nor that it goes on continuously; it is far more probable that each form remains for long periods unaltered, and then again undergoes modification" (Darwin, 1866; p. 132). This concept, rediscovered (Eldredge and Gould, 1972) and now called punctuated equilibrium, assumes a high variation in the rates of evolution. Paleontological records indicate that major evolutionary changes occurred during very short intervals that separated long epochs of relative stability. If the concept of punctuated equilibrium is applied to the global trend of the increase of functional complexity of organisms (Fig. 1), then it may be argued that rates of evolution are so unstable that any extrapolation of them into the past is meaningless.

The hypothesis that life originated on Earth requires very fast increase of genome complexity in primordial living systems. Geological data suggests that bacteria-like organisms flourished as early as 3.7 billion years ago (Nutman et al., 2016). This finding is consistent with molecular clock studies that place the Last Universal Common Ancestor (LUCA), which is reconstructed from gene sequences of major domains of life, at ca. 4.0 billion years ago (Fournier, 2015; Fournier and Alm, 2015; Fournier et al., 2015; Hedges, 2002). Based on the composition of genes, LUCA was most similar to methanogenic archaea and acetogenic clostridia (Weiss et al., 2016). Organic molecules were found in 4.4 billion years old rocks (Trail et al., 2015); however, there is no direct proof of their biogenic origin. Thus, we consider the age of oldest bacteria-like organisms on Earth from 4.0 to 3.7 billion years. Papers reviewed in Mikhailovsky and Gordon (2017) suggest fossil and LUCA date back to 4.3 billion years ago. The study of zircons indicates that liquid water appeared at 4.4 billion years ago (Valley, 2005;

Valley et al., 2002), which leaves from 100 to 700 million years for the origin of prokaryote-like organisms.

The number of protein-coding genes in LUCA has been estimated from 404 to 742 (Tuller et al., 2010). Recent analysis indicates that LUCA likely had 355 protein-coding genes (Weiss et al., 2016). The total genome size of LUCA is difficult to estimate because ancient proteins may have been shorter than their extant homologs. It is assumed that the genome of LUCA was comparable in size to the simplest organisms on Earth today (McDonald, 2014). Taking parasitic microbe *Mycoplasma genitalium* as an example of such organisms with genome of 590 Kb long, we can estimate the rate of genome complexity increase that is consistent with the hypothesis that life originated on Earth. On this condition, primordial genomes were doubling in complexity between a presumed origin of life on Earth and LUCA every 21 to 36 million years, which is 10 to 16 times faster than our estimated rate of increase. While one may quibble about the reliability of any given "earliest" evidence of life or the size of genome in LUCA, we are not in the situation in which a star was found (by original estimates) to be slightly older than the universe (Bond et al., 2013; Creevey et al., 2015), as the age of the Earth is precisely known, and there is plenty of time between the origin of the Earth and the origin of the universe to be considered. The possible sterilization of the Earth after the impact that created our Moon, reviewed in Mikhailovsky and Gordon (2017), narrows the window for an origin of life on Earth even more. The usually unjustified assumption in so many books and articles on the origin of life, that it began on Earth, may be one of our last anthropocentrisms.

The notion of unusually rapid primordial evolution was discussed in earlier studies (Davies, 2003; Lineweaver and Davis, 2003). Koonin and Galperin (2003) explained it by the following factors: (1) early evolution "happened within the framework of the RNA world," (2) high error rate of RNA replication (comparable to viruses), (3) major protein folds have "crystallized" almost immediately after the emergence of protein synthesis, (4) high rate of horizontal gene transfer (HGT), and (5) strong positive selection.

Let us evaluate the potential contribution of each of these factors to the rate of increase in genome complexity:

- First, the three domains of life have very few noncoding genes in common (Hoeppner et al., 2012), and the majority of them are related to protein synthesis (tRNA, ribosomal subunits), which indicates that they did not exist in the RNA world. Thus, the contribution of the RNA world to the genome of LUCA is very small (not more than a few hundred bases of nonredundant sequence).
- Second, high error rates of RNA replication are more likely to have a negative than positive effect on the increase of genome complexity because longer RNA sequences experience more damage from mutations. According to the hypercycle theory, the genome size is inversely related to the mutation rate (Eigen and Schuster, 1979). For example, viruses with high mutation rate have relatively short genomes (from 5 to 250 Kb) as compared to prokaryotes, and therefore, their rates of increase in genome complexity over evolutionary time cannot be high. There are a few viral species with large genomes (Abergel et al., 2015; Colson et al., 2012; Sharma et al., 2016) that include many genes homologous to cellular organisms. We do not discuss these species because their origin and mutation rates are not sufficiently studied (Abrescia et al., 2012; Suzan-Monti et al., 2006).
- Third, the presumably fast emergence of major protein folds via "crystallization" is not supported by facts. The structure of ribosomal RNA indicates that the first kinds of proteins synthesized in living systems were predecessors of ribosomal proteins, and ribosomes at some stage of

evolution were independent replicating agents that used ribosomal proteins to enhance their self-replication (Root-Bernstein and Root-Bernstein, 2015). This scenario of the origin of protein synthesis requires only a few types of proteins. The additional types of proteins probably emerged in evolution billions of years later, when the organisms became complex enough to utilize these new protein folds.

- Fourth, the argument that HGT (horizontal gene transfer) accelerated the increase of genome complexity is also not convincing. All prokaryotes have frequent HGTs and yet their rates of genome complexity increase are lower than in eukaryotes (see Section 4).

- Finally, there is no reason to expect that the ratio of positive selection to purifying selection had been higher in primordial systems than in extant organisms. Stronger positive selection can be associated with projected low competition in primordial systems. However, it is questionable that life ever experienced much lower competition than now. Self-reproducing systems rapidly saturate any available ecological niche, and thus, competition is not only inevitable but starts quickly. Of course, there is a possibility of spread into a new niche, but the number of accessible new niches is always limited. Evolving organisms not only occupy niches, they create new niches (Beckage et al., 2011; Grosch and Hazen, 2015). Thus, organisms can more easily avoid competition in complex contemporary ecosystems than in bare rock primordial ecosystems. Moreover, even if the competition was indeed low, there is still no explainable connection between the rates of competition and increase of genome complexity. In fact, it is more likely that low mortality associated with low competition would make evolution slower.

Thus, there is no plausible mechanism that could explain the possibility of extremely fast increase of genome complexity in primordial evolution. These attempts to explain the presumed origin of life on Earth are strikingly similar to stretching and shrinking of time scales in Biblical Genesis to fit preconceptions (Schroeder, 1990).

Although we fully agree that evolutionary rates fluctuate in time as in punctuated equilibrium, and that catastrophic changes of the environment are followed by mass extinction (Hull, 2015; Jablonski, 2005; Keller, 2005; Lipowski, 2005; Newman, 2000) and provide a boost of novel adaptations to survived lineages, we strongly disagree that the concept of punctuated equilibrium is applicable to the general trend of the increase of functional complexity of organisms (Fig. 1). First, there is a difference in time scales: punctuated equilibrium refers to relatively short periods of evolutionary change (millions of years), whereas the global growth of functional complexity becomes apparent at the time scale of billions of years. Second, adaptive radiation of lineages observed during periods of rapid evolutionary change has nothing to do with the increase of functional complexity. Multicellular organisms have enough functional plasticity to produce a large variety of morphologies based on already existing molecular and cellular mechanisms. Thus, the differences in morphology of animals and plants may have stronger connections with epigenetic factors than with the increase of genome complexity. Third, many rapid changes in the composition of animal and plant communities resulted from migration and propagation of already existing species (Dawkins, 1986), a mechanism that does not require an increase in functional complexity. Finally, there is no reason to expect that functional complexity of organisms did not increase during long "equilibrium" periods with no dramatic change in the morphology of organisms. Morphology is the tip of the evolutionary iceberg as the greatest changes occur at the molecular level. The common idea that stabilizing selection simply preserves the status quo in evolution is based on the misunderstanding of the original theory of stabilizing selection (Schmalhausen, 1949).

Stabilizing selection leads to increased plasticity of organisms (West-Eberhard, 2002) which is achieved via novel signaling pathways that replace less reliable older pathways. All these adaptive changes increase the robustness of living functions, but may have no immediate effect on morphology. Nevertheless, these changes lead to the increase of functional complexity. Schmalhausen developed his theory without knowledge of molecular biology, which was not available at that time. But he managed to capture the idea on how phenotypic plasticity reshaped evolution.

The reason why living organisms cannot increase their functional complexity instantly may be that it takes a long time to develop each new function via trial and error. Thus, simultaneous and fast emergence of numerous new functions is very unlikely. In particular, the origin of life was then not a single lucky event, but a gradual increase of functional complexity in evolving primordial systems. Similarly, the emergence of eukaryotes from prokaryotes was not the result of one successful symbiosis, but may have involved as many as 100 discrete innovative steps (Cavalier-Smith, 2010). This view is consistent with Darwin's insight that early evolution was slow and gradual:

> During early periods of the earth's history, when the forms of life were probably fewer and simpler, the rate of change was probably slower; and at the first dawn of life, when very few forms of the simplest structure existed, the rate of change may have been slow in an extreme degree.
>
> **Darwin (1866), p. 214.**

Another factor that may have reduced the rates of primordial evolution was the absence of well-tuned molecular mechanisms, which are now present in every cell. In particular, there was no basic metabolism to produce a large set of simple organic molecules (e.g., sugars, amino acids, nucleic bases) and no template-based replication of polymers (see Section 5). These two obstacles substantially reduced the frequency of successful "mutations" and, thus, the initial rate of complexity increase was likely even slower than shown in Fig. 1. Thus, there is no basis to assume that there was enough time between cooling of the Earth and the appearance of LUCA for organisms of the complexity of LUCA to have emerged.

---

4 WHY DID GENOME COMPLEXITY INCREASE EXPONENTIALLY?

The increase of functional complexity in evolution can be modeled on the basis of known mechanisms, which appear to act as positive feedbacks (Sharov, 2006). First, the model of a hypercycle considers a genome as a community of mutually beneficial (i.e., cross-catalytic) self-replicating elements (Eigen and Schuster, 1979). For example, a mutated gene that improves the accuracy of DNA replication is beneficial not only for itself but also for all other genes. Moreover, these benefits are applied to genes that may appear in the future. Thus, already existing genes can help new genes to become established, and as a result, bigger genomes grow faster than small ones. Second, new genes usually originate via duplication and recombination of already existing genes in the genome (Massingham et al., 2001; Ohno, 1970; Patthy, 1999). Thus, larger genomes provide more diverse initial material for the emergence of new genes. Third, large genomes support more diverse metabolic networks and morphological elements (at various scales from cell components to tissues and organs) than small genomes, which, in turn, may provide new functional niches for novel genes. For example, genes in multicellular organisms operate in highly diverse environments represented by various types of cells and tissues. Progressive differentiation of cells supports the emergence of gene variants that either perform the same function in specific cell types or modify the original function for specific needs of some cells. Replication of branches of differentiation trees followed by divergence allows duplication of whole cell types, followed by them assuming different functions within an organism (Gordon and Gordon, 2016; Gordon, 1999, 2013). Large-scale and whole genome duplication has also occurred from time to time (DePamphilis and Bell, 2010; Holland, 2013; Isambert and Singh, 2016). In the Red Queen hypothesis (Van Valen, 1973), coevolving organisms have to "run" faster and faster just to stay in the same place (p. 42 in Carroll, 1875) (Benton, 2010; Dasgupta et al., 2005; Markov, 2000; Voje et al., 2015), implying increases in function and thus increases in the genome (Edger et al., 2015). These mechanisms of positive feedback may be sufficient to cause an exponential growth in the size of functional nonredundant genome.

Existing data also indicates that genetic complexity may have increased a little faster than exponentially (i.e., hyperexponentially), which may be explained by phase transitions to higher levels of functional organization (Markov et al., 2010; Sharov, 2006). For example, the time of genome doubling in Archae and Eubacteria was 1080 and 756 million years, respectively (these estimates are based on the largest known archaeal genome, 5 Mb, in *Methanogenium frigidum* and bacterial genome, 13 Mb, in *Sorangium cellulosum*) (Bernal et al., 2001). These estimates are 3.2- and 2.2-fold longer than the genome doubling time in eukaryotes. The difference between the rates of increase of genome complexity between most successful and lagging lineages can be explained by evolutionary constraints of the latter ones (e.g., inefficient DNA proofreading and absence of mitosis in prokaryotes). Thus, the rate of the "complexity clock" may have increased with the emergence of eukaryotes, and therefore, life may have originated even earlier than expected from the regression in Fig. 1. That would push the projected origin of life close to the origin of our galaxy and the universe itself (see Fig. 2). Thus, life may have originated shortly after parts of the universe cooled down from the Big Bang (Bialy et al., 2015; Gordon and Hoover, 2007; Loeb, 2014). For the sake of this chapter, we are assuming that the Big Bang model for the universe and its age of 13.75 ± 0.11 billion years is correct (Jarosik et al., 2011), although some evidence suggests that our universe is substantially older (Kazan, 2010).

The exponential increase of functional complexity is consistent with Reid's view of evolution as cascading emergences:

> As evolution progresses, the freedom of choice increases exponentially.... intrinsic complexification of differentiated cell types, is overall an exponential function of reproduction and time-quite a simple equation.... the historical curve of some lineages, especially that of hominids, fits the simple exponential equation, its logarithmic slope theoretically determined by the fact that the acceleration of complexification is virtually equivalent to increasing adaptability and freedom to explore unexploited environments.
>
> **Reid (2007).**

---

5 COULD LIFE HAVE STARTED FROM THE EQUIVALENT OF ONE NUCLEOTIDE?

Most popular theories of the origin of life imply a negative answer to this question. The RNA-world model assumes that life started from self-replicating RNA sequences (Gilbert, 1986) or other kinds of similar heteropolymers (TNA, PNA) (Nelson et al., 2000; Orgel, 2000), whereas the model of autocatalytic sets assumes that life started from a set of short peptides, where the synthesis of each peptide is catalyzed by some other peptides in the same set (Kauffman, 1986; Kauffman et al., 1986). Both models assume that life requires heteropolymers (i.e., either nucleic acids or peptides) to support heritable self-replication. Indeed, some RNA molecules can catalyze the polymerization of other RNAs (Johnston et al., 2001), which is perceived as a proof of the RNA-world hypothesis. These theories also assume that diverse monomers (nucleotides or amino acids) are abundant enough to support the synthesis of polymers. However, the natural concentration of nucleotides and amino acids is negligibly small without life and not sufficient to make polymers (Klevenz et al., 2010; Luisi, 2015). Nucleotides can be synthesized abiogenically (Benner et al., 2012; Powner et al., 2009), but special mechanisms, such as thermal gradients across microporous minerals (Copley et al., 2007; Deamer and Georgiou, 2015) or cycles of evaporation and rewetting (Damer, 2016), have to be invoked to produce significant concentrations from dilute sources. By chance, a few nucleotides may appear in a close proximity and become utilized for RNA synthesis, but there would be no monomers left for the next round of replication. RNA-mediated catalysis was shown to yield nucleotides from bases and sugars (Unrau and Bartel, 1998), but these precursors are not supplied in sufficient quantities in nature (Sharov, 2016, 2017). Polymers like nucleic acids and peptides may persist only on condition of an unlimited supply of monomers, and this requires a heritable mechanism for their synthesis from simple and abundant organic and nonorganic resources (Copley et al., 2007; Sharov, 2009). Thus, neither chemical evolution nor primordial soup can provide resources of self-replication of polymers (Lane et al., 2013). The RNA-world and autocatalytic set models have other weak points (e.g., no explanation for the origin of membranes), which are discussed by Sharov (Sharov, 2016). This review (Sharov, 2016) also discusses other models of the origin of life, such as lipid-world and GARD (Bar-Even et al., 2004; Segré et al., 2001; Segré and Lancet, 1999) and autocell (Deacon, 2006, 2011). The oil droplet first model for the origin of life is reviewed in this volume (Gordon et al., 2017).

One alternative to these theories is the *coenzyme-world* scenario of the origin of life, which assumes that first heritable primordial systems were composed of relatively simple monomers that reproduced via indirect autocatalysis (Sharov, 2009, 2016). Presumably, these molecules ware catalytically active and thus may have resembled existing coenzymes. Because many coenzymes (e.g., ATP, NAD, and CoA) are similar to nucleotides, they can be viewed as evolutionary predecessors of RNA and DNA nucleotides (Kritsky and Telegina, 2004). Of these coenzymes, the AMP-containing cofactors seem to be most ancient (Yarus, 2011), but they are certainly not the first coenzymes that existed in early primordial life. This observation is important because it justifies the use of the term "coenzyme" for primordial hereditary molecules.

These systems evolved via natural selection and eventually developed metabolic pathways for internal production of sugars, nucleotides, and amino acids that later were utilized for making heteropolymers. A primitive type of heredity can be supported by autonomous self-production, which is a generalization of autocatalytic synthesis. The general notion of self-reproduction has been defined using the formalism of Petri nets (Sharov, 1991). In short, a chemical system is self-reproducing if there is a finite sequence of transitions (i.e., reactions) that results in the increase of the numbers of all components within the system. For example, the formose reaction is autocatalytic and makes sugars from formaldehyde (Huskey and Epstein, 1989). Such reactions can propagate in space, which is similar to the growth and expansion of populations of living organisms (Gray, 1994; Tilman and Kareiva, 1997). Autocatalytic reactions have two alternative steady states: "on" and "off" (the "on" state is stabilized via a limited supply of resources). Thus, they represent the simplest hereditary system or memory unit (Jablonka and Szathmáry, 1995; Lisman and Fallon, 1999). For example, a reverse citric acid cycle, which captures carbon dioxide and converts it into sugars, may become self-sustainable, at least theoretically (Morowitz et al., 2000). Prions are examples of autocatalytic reproduction (Griffith, 1967; Laurent, 1997; Watzky et al., 2008), and indeed, have been invoked in various ways in speculations on the origin of life (Hu et al., 2010; Maury, 2009; Steele and Baross, 2006). However, they cannot support the synthesis of the primary (i.e., unfolded) polypeptide.

Autocatalysis is necessary for the origin of life, but not sufficient. The specific feature of autocatalysis in living systems is that it is linked functionally with a local environment (e.g., oil vesicle) (Sharov, 2009), giving it a spatial structure. In particular, the autocatalytic system has evolutionary potential only if it modifies (encodes) the features of its local environment, and this modification increases the rate of autocatalysis because this functional linkage is a necessary condition for cooperation between multiple autocatalytic components if they happen to share their local environment (Sharov, 2009, 2016). In economic terms, the system invests in the modification of its environment, and therefore, cannot leave its investment. This can also be viewed as a "property" relation at the molecular level. The autocatalytic system is the owner of its local environment, which plays the role of "home" or "body." Because the system is attached to its home, it is forced to cooperate with other autocatalytic systems that may appear in the same local environment. The local environment can be represented by either enclosure or attachment to a surface. Although all known free-living organisms have enclosures (cell membranes), life may have started from surface metabolism because autocatalysis has a much higher rate on a two-dimensional surface than in three-dimensional space (Wächtershäuser, 1988), an example of dimension reduction (Adam and Delbrück, 1968).

Let us consider a particular kind of coenzyme-world system represented by coenzyme-like molecules (referred to below as "coenzymes") that colonize tiny hydrocarbon droplets (referred below as "oil") in water (Sharov, 2009, 2016, 2017). We chose oil droplets/vesicles as the most likely environments for a population of coenzymes because (1) hydrocarbons are the most abundant organic molecules in the universe (Deamer, 2011) and are expected to exist on early terrestrial planets (Marcano et al., 2003), (2) oil droplets self-assemble in water, (3) some ancient organic molecules such as carotinoids and vitamin K are oleophilic and may be relics of the primordial metabolism on the surface of oil droplets, (4) oil can be used as a source of carbon for self-reproducing coenzymes, and (5) it is logical to project the evolutionary transformation of oil droplets into a lipid membrane (see below). Coenzymes can colonize oil droplets in water as follows. Assume that precursors of coenzymes cannot anchor to the hydrophobic oil surface; however, they can get attached to rare fatty acids with hydrophilic ends (Fig. 1B in Sharov, 2017). Once attached, precursors combine to make a coenzyme (synthesis is facilitated by attachment) and catalyze the oxidation of outer ends of nearby hydrocarbons in the oil droplet, thus providing the substrate for the synthesis of additional coenzymes. Accumulation of fatty acids reduces the surface tension, and thus, increases the chance of a droplet to split into smaller ones, and small droplets can infect other oil droplets, i.e., capture new oil resource (Fig. 1B in Sharov, 2017).

Some precursors of coenzymes (e.g., aromatic hydrocarbons) may come from the oil phase rather than from the water. Furthermore, organic molecules could have been derived from degradation of fatty acids. The idea that hydrocarbons were used by primordial systems as nutrients was proposed by

Oparin (1936), who assumed that hydrocarbons together with proteins and other organic molecules formed colloidal systems with spontaneous metabolism. The emergence of self-reproducing coenzymes could be facilitated by interaction with metals and other nonorganic compounds. Simple organic molecules are not likely to act as strong catalysts, but this limitation can be removed if these small molecules interacted with nonorganic compounds. For example, amino acids readily interact with cobalt, nickel, or zinc, and some of these complexes are catalysts (Dutta et al., 2014); vanadium has effects in carbohydrate metabolism (Gruzewska et al., 2014), whereas iron and sulfur facilitate electron transfer (Kümmerle et al., 2000). Oxidation of hydrocarbons can be facilitated by iron oxide. These inorganic compounds are abundant in petroleum (Nadkarni, 2009) and could have been present in abiogenic oil droplets.

Oil droplets populated with coenzymes represent a two-level system that includes self-reproduction at both levels: coenzymes on the surface and whole oil droplets. Moreover, coenzymes support self-reproduction at both levels because they change surface properties of oil droplets and this change enables them to reproduce, capture resources, and propagate to other oil droplets (Fig. 1B in Sharov, 2017). As a result, coenzymes play the role of coding molecules because they encode the surface properties of oil droplets (phenotype) and transfer the code to the next generation of colonized droplets with copies of the same molecular type (Sharov, 2009).

Several kinds of autocatalytic coenzymes may coexist on the same oil droplet, creating a system with compositional (or combinatorial) heredity (Sharov, 2009, 2016). Each kind of these coenzymes may perform a specific function (e.g., capturing resource, storing energy, or catalysis of a reaction) and ensure the persistence of this function (i.e., play a role of coding molecule). Not generally being bound to one another, they can be transferred to offspring systems in different combinations. But despite random transfer, such compositional heredity can be stable because (1) coding molecules are present in multiple copies and therefore each offspring has a high probability of getting the full set, and (2) natural selection preserves preferentially systems with a full set of coding molecules. The efficiency of the latter mechanism was shown in a "stochastic corrector model" (Szathmáry, 1999). New types of coding molecules can be added by (1) acquisition of entirely new coenzymes from the environment, (2) modification of existing coenzymes, and (3) polymerization of coenzymes. Compositional heredity can eventually lead to the emergence of synthetic polymers (Sharov, 2009). For example, if a new coenzyme, B, can catalyze the polymerization of another coenzyme, A, then together they encode long polymers AAAAA..., which may cover the surface of the oil droplet and substantially modify its physical properties.

This hypothesis of compositional heredity is supported by the GARD (Graded Autocatalysis Replication Domain) model, which assumes that life started with lipid-like vesicles/assemblies that control the exchange of their components with the environment (Segré et al., 2000). This model, also known as the "lipid-world" (Segré et al., 2001), assumes no covalent bond formation, and thus, vesicles do not include any novel kinds of molecules not present in the environment. A more advanced GARD model (called P-GARD) assumes polymerization of simple molecules into short oligomers via covalent bonds catalyzed by other components of the assembly (Shenhav et al., 2005). Thus, new molecules can be synthesized from simple resources absorbed from the environment. Compositional lipid assemblies can evolve via "mutations" (e.g., deletion of components) followed by natural selection (Inger et al., 2009). They can further modify their environment and make it more favorable for survival and reproduction (Shenhav et al., 2007).

6 HOW HERITABLE SURFACE METABOLISM MAY HAVE EVOLVED INTO AN RNA-WORLD CELL?

The surfaces of oil droplets provide discrete and abundant local environments for emerging primordial life. However, further evolution of life is constrained by (1) limited amount of abiotic liquid hydrocarbons, and (2) limited dimensionality (i.e., 2D) of the functional system. Limited abundance of hydrocarbons prevented primordial organisms from increasing in size, which is a serious constraint on the surface area and, consequently, the total rate of metabolism. Although surface metabolism is beneficial for the emergence of primordial life, it cannot support the storage of a large amount of resources. Also, the movement of long polymers on the surface is highly restricted in a 2D-space. Both problems could have been solved with one invention: a transition from surface metabolism to a membrane-enclosed cell (Sharov, 2016). From the topological point of view, an oil droplet can be converted into a membrane-enclosed cell via engulfing water. Such "double" droplets are easily generated by agitating an emulsion of liquid hydrocarbons in water, but they are not stable unless surfactants are present (Chong et al., 2015) or the internal aqueous phase is stabilized as a gel (Perez-Moral et al., 2014). Cell-like organization yields little functional advantage if the membrane breaks before the cell divides. Thus, the membrane has to be strong enough to sustain mechanical disturbances, and the volume of a cell has to be controlled to prevent bursting. Hypothetically, primordial systems passed through a transitional period with relatively unstable membranes, but managed to secure short-term benefits from temporary cell-like organization (e.g., via storage of accumulated energy at the membrane). Apparently, the major selection pressure during this period was to develop more stable membranes, which required the synthesis of glycerol that is the backbone of lipids and phospholipids. Later in evolution, glycerol could have been reused to make sugars, which are suitable for storing energy and regulating osmosis. This scenario of the origin of cells from oil droplets is supported by the predominantly oleophylic ancestral proteomes (Mannige et al., 2012).

Initial steps of primordial evolution were likely slow and inefficient because there was no universal rule for producing new coding molecules. Some improvement was likely achieved by transformation of old coding molecules into novel ones via modification of functional groups or polymerization (Sharov, 2009). However, there was no streamlined procedure for making new coding molecules until the invention of template-based (or digital) replication. Replication is a special case of digital autocatalysis, where each coding molecule is a linear sequence of a few kinds of monomers (digits), and copying is done sequentially via predefined actions applied to each monomer (Szathmáry, 1999). Digital replication makes the coding system universal because it works for any sequence. Thus, there is no need to invent individual recipes for copying modified coding molecules. In a like manner, the later invention of continuing differentiation (Gordon, 1999) in multicellular organisms allowed multiple cell types to be created without requiring a unique mechanism for each new cell type.

The starting point for the origin of template-based replication was the existence of polymeric coding molecules with either random or repetitive sequence (Sharov, 2009). Polymers may initially stick to each other to perform some other functions (e.g., increasing the stability of the membrane or facilitating polymerization (Gordon et al., 2017)). The shorter strand of the paired sequence can then become elongated by adding monomers that weakly match to the overhanging longer strand. Then the specificity of this process increased due to natural selection because it helped to produce better copies of existing polymers. Invention of digital replication, therefore, may have been the turning point in the origin of life that substantially increased the hereditary potential of primordial living systems (Jablonka and Szathmáry, 1995; Sharov, 2009). It is not clear if template-based replication of polymers appeared before or after the formation of a stable cell membrane. But in any case, these two evolutionary events produced cells with ribozyme-based catalysts, and this stage of evolution is known as the "RNA-world" (Gilbert, 1986). These organisms still had no tools for protein synthesis, but they likely were able to make simple peptides with the help of ribozymes.

Similar mechanisms for the emergence of cell membrane and template-based replication have been proposed within the paradigm of lipid-world (Hunding et al., 2006). In particular, it was assumed that membranes can be formed by establishment of strong bonds between lipid molecules at the surface of protocells, whereas the selection for polymer-polymer complementarity can enhance the encoding of polymer constituent information.

Despite the apparent similarity between the lipid-world and coenzyme-world scenarios, there is an important difference. Lipid-world assemblies utilize various kinds of lipids (and/or other molecules such as amino acids, nucleobases, and carbohydrates) as resources, and thus their existence is possible only in a water-based soup with high concentrations of various organic molecules. So far, there is no evidence that such soup ever existed on any planet devoid of life. Lipids and other organic molecules important for living cells are unstable and cannot be generated effectively without life. Thus, the production and enrichment processes cannot compensate for the loss of these molecules due to decay and diffusion. Heterotrophic life requires high concentrations of nutrients in the medium (Eagle et al., 1961), which are several orders of magnitude higher than those reported or predicted in the absence of living cells. For example, the concentration of dissolved free amino acids near hydrothermal vents is only 0.143 µM (Lang et al., 2013), which is <1% of the concentration required for protein synthesis (Eagle et al., 1961). Considering that life is present at hydrothermal vents, the concentration of amino acids in habitats devoid of life should be much lower. Concentration of formaldehyde in experimental simulator of hydrothermal vent was "in low nanomolar quantities," which is one million fold lower than required for the autocatalytic formose reaction (Herschy et al., 2014). In contrast to the lipid-world, the coenzyme-world scenario of the origin of life does not require a primordial soup enriched with organic molecules. Instead, it is assumed that hydrocarbon (oil) is used as the main organic resource. Oil vesicles are colonized with more complex but rare organic molecules, which play the role of seeds rather than resources. These more complex molecules are assumed to be catalytically active. Thus, they establish autocatalytic and cross-catalytic production networks together with other components recruited from the environment, including inorganic molecules and crystals (see Sharov, 2017).

The hypothetical LUCA was far more complex than the RNA-world organisms described above (Doolittle, 2000; Theobald, 2010). Its heredity was likely based on DNA (de Farias et al., 2016), as follows from the presence of DNA-binding proteins that regulated transcription in the genome of LUCA (Weiss et al., 2016). However, the nature of LUCA's hereditary molecules is still debated (Di Giulio, 2011; Glansdorff et al., 2008). LUCA had fully developed protein synthesis programmed by genes. The integrity of the DNA was maintained by a group of "maintenance" enzymes. The RNA-polymerase complex was used to synthesize mRNA copies of each gene, and then each mRNA was translated into a protein by the action of ribosomes and tRNAs. LUCA was able to make fatty acids from sugars, and therefore, it was no longer dependent on the supply of abiotically synthesized hydrocarbons. Although the genome of LUCA is a product of phylogenetic reconstruction, it is possible that organisms that arrived to the primordial Earth from cosmic sources had functional complexity similar to LUCA. Based on the rate of complexity increase (Fig. 1), we expect that RNA-world cells emerged ca. 2 billion years after the origin of life, and LUCA-like organisms appeared ca. 3 billion years later. The origin of LUCA may or may not have been on Earth.

7 HOW CAN ORGANISMS SURVIVE INTERSTELLAR TRANSFER?

Bacterial spores have unusually high survival rates even in very harsh conditions, and therefore, they are likely candidates for interstellar transfer. Contaminated material can be ejected into space from a planet via collision with comets or asteroids (Ehrlich and Newman, 2008). Then bacterial spores may remain alive in a deep frozen state for a long time that may be sufficient for interstellar transfer. Bacterial spores were reported revived after 25–35 million years of dormancy (Lambert et al., 1998). A more recent discovery of viable bacteria trapped in the 0.75 million year old ice (Katz, 2012) suggests that bacteria may be preserved in ice for long times. One of the scenarios of life's arrival to new planets is the capture of contaminated material by a protoplanetary disc before planet formation (Wallis and Wickramasinghe, 2004).

Survival of active bacteria during their cosmic travel is an alternative attractive hypothesis because living bacteria are able to repair their DNA and withstand the damaging effect of radiation (Johnson et al., 2007). In comparison, dormant spores cannot repair their DNA, and therefore, damage may only accumulate during long travel. Active bacteria can be present only on large cosmic bodies such as rogue planets and possibly asteroids. Prediction and discovery of rogue planets (Debes and Sigurdsson, 2002; Gibson, 2000; Gladman and Chan, 2006; Samuel, 2001; Vanhamäki, 2011) that could harbor life (Abbot and Switzer, 2011; Badescu, 2011) strengthen the panspermia hypothesis which can no longer be dismissed on the basis of disbelief (McNichol and Gordon, 2012). The hypothesis of panspermia becomes even more plausible if the Solar System originated from the remnants of the exploded parental star (Joseph, 2009). Remnants of planets from exploded supernovae (Gordon and Hoover, 2007) can carry billions of bacterial spores and maybe even active chemosynthetic bacteria deep beneath the surface. If the Earth incorporated some of these planet fragments, it could have been seeded by a diverse community of bacterial species and their viruses.

8 IMPLICATIONS OF THE COSMIC ORIGIN OF LIFE ON EARTH

The first implication of our study of genomic complexity is that the early appearance of life on Earth most likely stemmed from contamination with prokaryotes (bacteria or archaea or their predecessors, such as LUCA) from space. Thus, despite the fact that we don't have a final answer, it makes sense to explore the implications of a cosmic origin of life, before the Earth existed. The idea that life was transferred to Earth by intelligent beings (i.e., "directed panspermia") (Crick and Orgel, 1973) is unlikely because there was no intelligent life in our universe at the time of the origin of Earth by Fig. 1. The universe was only 8 billion years old at that time, whereas the development of intelligent life seems to require ca. 10 billion years of evolution.

Second, our analysis indicates that life took a long time, perhaps 5 billion years, to reach the complexity of prokaryotes. Thus, the possibility of repeated and independent origins of life of this complexity on other planets in our Solar System can be ruled out. Extrasolar life is likely to be present at least on some planets or satellites within our Solar System, because (1) all planets had comparable chances of being contaminated with microbial life, and (2) some planets and satellites (e.g., Mars, Europa, and Enceladus) provide niches where certain prokaryotes may survive and reproduce. If extraterrestrial life is present in the Solar System, it should have strong similarities to terrestrial microbes, which is a testable hypothesis. We expect that they have the same nucleic acids (DNA and RNA) and similar mechanisms of transcription and translation as in terrestrial prokaryotes. The ability to survive interstellar transfer was the major selection factor among prokaryotes on the cosmic scale. Thus, prokaryotic life forms were successful in colonizing the cosmos only if they were resilient to radiation, cold, drying, toxic substances, and highly adaptable to a broad range of planetary environments. In particular, photosynthesis or chemosynthesis is needed to be independent from organic resources. The similarity between terrestrial and extraterrestrial prokaryotes may appear sufficient to draw a unified evolutionary tree of life, though it may be complicated by later transfers between the planets, such as between Earth and Mars (Gordon and McNichol, 2012).

Third, we can anticipate repositories within our own solar system of samples of the life that might have first "contaminated" its planets. For example, the more distant, perhaps geologically inactive dwarf planets, beyond Pluto, such as Eris, Makemake, and Haumea (Holman and Rudenko, 2016), are close enough to be reached by current space technology. Their orbits have low enough eccentricity that they probably have never been close to the sun. Eris varies from 37.74 to 97.59 Au from the sun (Holman and Rudenko, 2016) and has a surface temperature (Alvarez-Candal et al., 2011; Howett et al., 2016) of 30°K. Thus, Eris is a deep freeze for any life that may have fallen on it, and probably also records, under pristine conditions, much of the history of our solar system. Rogue planets formed before our solar system could similarly record much of the history of our galaxy, including panspermia.

Fourth, attempts to reproduce the origin of life in laboratory conditions (Damer et al., 2012) may prove more difficult than generally expected because such experiments have to emulate many cumulative rare events that occurred during several billion years before organisms reached the complexity of the RNA world (Davies, 2003; Lineweaver and Davis, 2003; Sharov, 2009). Despite the success of copying an existing bacterial genome (Gibson et al., 2008), humans have so far failed to invent a single new functional enzyme from scratch (i.e., without copying it from nature) and have had limited success in imitating existing enzymes (Bjerre et al., 2008). Thus, it may prove hard to make a primitive living system that does not resemble anything that we observe on Earth.

Fifth, the environments in which life originated and evolved to the prokaryote stage may have been quite different from those envisaged on Earth. Thus, emulating conditions on the young Earth may not increase the chance of generating primordial living systems in the lab. Even a bigger mistake would be to use contemporary minerals in such experiments because most mineral species on Earth are directly or indirectly modified by life (Hazen, 2010). To define possible environments for the origin of life, we can extrapolate the evolution of minerals back in time to the initial cooling of the universe after the Big Bang. The major questions to ask are: (1) When did stars and planets form? (2) What was the elemental composition of stars and planets versus cosmic time? (3) How was the surface of those early planets stratified? (4) What atmospheres might those planets have had? It is reasonable to assume that life originated in the presence of water, as water is very abundant in space and is a byproduct of star formation. Young stars shoot jets of water into the interstellar space (Fazekas, 2011). Large quantities of water have been detected in space clouds (Glanz, 1998). Thus, it is reasonable to assume that water was present early in the young universe and could have supported the origin of life. Major chemical elements of living organisms (carbon, hydrogen, oxygen, and nitrogen) are among the most abundant in the Universe. Phosphorus is less abundant and may have been the limiting factor for the origin of life, although recent studies suggest that it can be effectively extracted from apatite or come from volcanic activity or space (Schwartz, 2006). Life may have started on planets around the first, low "metal" stars, where "metals" mean, to astronomers, higher atomic number elements that formed from hydrogen and helium during stellar nucleosynthesis. Such stars may have formed as early as 200 million years after the Big Bang (Zheng et al., 2012). High metallicity is a negative factor for life origin because Earth-like planets may be destroyed by giant planets (Lineweaver and Grether, 2002). The stars of such planets probably lasted only 4 billion years, so if life didn't have a "false start" in such systems (Johnson and Li, 2012), it may have been propagated into interstellar space during the star's supernova event (Gordon and Hoover, 2007). Therefore, the time scale we are proposing for the origin and complexifying of life requires panspermia mechanisms (McNichol and Gordon, 2012) for life to persist.

And sixth, the original Drake equation for guesstimating the number of civilizations in our galaxy (Wikipedia contributors, 2017c) may be wrong (Haqq-Misra and Kopparpu, 2017), as we conclude that intelligent life like us has just begun appearing in our universe. The Drake equation is a steady state model, and we may be at the beginning of a pulse of civilization. Emergence of civilizations is a non-ergodic process, and some parameters of the equation are therefore time-dependent. Because the cosmic transport of life is most likely limited to prokaryotes, young planets have not had enough time to develop intelligent life. Another time-dependent process is the probability of interstellar transfer of prokaryotes, which we expect to have become more frequent as the total pool of prokaryotes in the galaxy increased with time. There are many modifications of the Drake equation, but if civilizations have just begun to appear, any version is of limited use. The answer to the Fermi paradox (Wikipedia contributors, 2017d) may be that we are among the first, if not the only so far, civilization to emerge in our galaxy. The "Rare Earth" hypothesis (Ward and Brownlee, 2003) need not be invoked. The linking of civilization to the lifetime of a particular star, such as our Sun (Livio and Kopelman, 1990; Webb, 2002), is also not necessary.

9 GENETIC COMPLEXITY LAGS BEHIND THE FUNCTIONAL COMPLEXITY OF MIND

The idea that genetic complexity can be used as a generic scale of functional complexity of all organisms is intellectually attractive and works well at the lower end of complexity. However, the genome fails to capture correctly the complexity of higher-level organisms. Based on Fig. 1, the difference in genetic complexity is small between fish and mammals, and there is no difference between mouse and human. What makes humans superior to mice is not the genome but mind.

Eukaryotes progressively used epigenetic memory (i.e., stable chromatin modifications) to encode or modify their functions. In contrast to the genome, the epigenetic memory is rewritable, and therefore, can easily support phenotypic plasticity and development of habits. Eventually, epigenetic mechanisms led to the emergence of mind which is a tool for classifying and modeling of objects (Sharov, 2012). Mind operates at the level of holistically perceived objects, whereas primitive organisms (e.g., prokaryotes) regulate their functions via simple molecular level signal-response circuits. The power of mind has increased dramatically since the emergence of higher-level learning, which enabled organisms to distinguish and model new classes of objects. Thus, the functional complexity of organisms became encoded partially in the heritable genome and partially in the hardly transmissible mind. Despite the short life span of the individual mind, its advantages are tremendous because it allows an organism to develop complex behaviors such as running, flying, and communicating with peers and helps to adjust these behaviors to changing environments. As the complexity of mind increased, the role of the genome has shifted from direct coding and controlling of functions to creating favorable conditions for the development of mind, which can take care of these functions later in life. In other words, the role of the genome became to provide the informational infrastructure (e.g., nervous system) and initial training for the growing mind. Thus, the functional complexity of higher animals (e.g., birds and mammals) started growing much faster than in their ancestors, and this growth can no longer be captured by the growth of genetic complexity (Fig. 1). This is an example of accelerating (or hyperexponential) growth of complexity caused by the emergence of novel methods of information processing.

The increase of the functional complexity of mind is more difficult to measure than the increase of genetic complexity because we still do not know how mental memory is encoded. The size of the brain can be used as a first approximation of the complexity of mind. However, the size of brain is also correlated to body size (McHenry, 1975), which makes comparisons difficult. Plots of brain mass versus body mass, nevertheless, lead to clear classifications and trends (Jerison, 1973). The functionality of the brain may depend more strongly on its structure (e.g., on surface area, neuron density, or connectedness) than on the volume (Roth and Dicke, 2005). More reliable trends in brain size can be detected within a narrow taxonomic group. For example, the regression of log-transformed brain size to the time of origin in Hominids (data from Wikipedia contributors, 2017a) indicates the doubling time of 3.2 million years. If we use encephalization quotient, which adjusts brain volume to body size (Roth and Dicke, 2005), the doubling time from chimps to humans equals 3.0 million years. Thus, the rate of brain increase in evolution exceeds the rate of genetic evolution by a factor of ca. 100, which shows the advantage of switching from the genome code to the "mind code." Another factor that possibly accelerated the evolution of the brain was an epigenetic differentiation code that increases the number of kinds of cells in the brain beyond "inherited components" (Proposition 202 in Gordon, 1999).

The origin of humans marks another major transition in the evolution of functional complexity because humans invented methods to transfer information effectively between minds. Initially, information transfer was based on copying the behavior of other members of a social group, but later it was augmented by the development of speech and finally by the written language. Thus, the content of minds became shared between individuals and preserved for future generations. This transition fueled further acceleration for the growth of functional complexity. To get an idea of the doubling time of human social information, we consider the number of characters in the Chinese language, which increased from ca. 2500 at 3200 years ago (the Oracle bone script) to ca. 47,000 at the present time (New World Encyclopedia contributors, 2017). Thus, the rate of language doubling time appears to be 825 years, which exceeds the rate of brain increase in evolution by a factor of >3000. The fast growth of human culture may parallel that of language and its memes (Dennett, 2017).

10 EXTRAPOLATING THE GROWTH OF COMPLEXITY INTO THE FUTURE

Predicting the future was historically based on spiritual revelations or dreams (e.g., Joseph, Isaiah, John the Baptist), or analysis of ancient texts (Nostradamus). Nowadays, we can use science and statistics to extrapolate existing trends into the future, at least over the persistence time of our models (Pilkey and Pilkey-Jarvis, 2007). Moore's law (Moore, 1965) is an example of a relationship that can be extrapolated into the future. Similar exponential trends are known for other technologies (e.g., speed of DNA sequencing, hard disc capacity, and bandwidth of networks). High rates of exponential growth of these technologies with doubling time of ca. one year has led to the idea of a "technological singularity," which refers to the time when technology-based intelligence will emerge and possibly replace humans (Kurzweil, 2005).

However, growth rates of specific technologies should not be confused with the increase of functional complexity of the human civilization as a whole. Humans developed economy mechanisms (e.g., loans, bonds, stock market) used to redistribute resources to key industries that are perceived as bottlenecks to functional growth. Recent such industries include computer technology, Internet, wireless communication, and biotechnology. But we cannot expect that these industries will remain major limiting factors forever. For example, computation is no longer the key factor in most human applications; instead, it became an inexpensive commodity, which can be easily expanded on demand using cloud computing. Thus, the further progress in computer technology will not be as revolutionary as it was in the previous 3 decades. Similarly, the speed of DNA sequencing will soon reach its limits and cease to be the most critical factor. The rate of increase of functional complexity of human civilization can be better measured by indicators that are not linked to a single technology. For example, the doubling time of the number of scientific publications from 1900 to 1960 was only 15 years (de Solla Price, 1986). Interestingly, extrapolating the exponential increase of scientific publications backwards gives us an estimated origin of science at 1710, which is the time of Isaac Newton. The increase in the number of patents has the doubling time of ca. 25 years (Wikipedia contributors, 2017e). Thus, the functional complexity of human civilizations doubles approximately every generation (i.e., 15–25 years), which is ca. 20-fold slower than for most "critical" technologies. There may be analogies here between continuing differentiation (Gordon, 1999) and the ability to "perform operations that increase complexity" of our technology (Barroso and Luz, 2015).

Prediction of future events such as "technological singularity" (Kurzweil, 2005) is flawed if based on trends within a single technology (in this case, computer speed). There is no doubt that computers can outperform humans at specific tasks ranging from simple number crunching to automatic vehicle navigation and games (e.g., chess). It is also obvious that many human jobs will shift to fully automated devices, continuing a trend that started in the age of steam (Carnegie, 1905). But this does not mean that technology is going to replace humans because novel jobs are created to operate, program, and use each new technology. Technology cannot compete with humans because humans control the resources and would not allow technical agents to take over. Until technical devices become fully self-reproducing, which is not in the cards so far, they pose no danger for humanity. The only self-reproducing artificial agents are computer viruses, and they indeed can do substantial harm within information-processing networks and the equipment they control (Wikipedia contributors, 2017b). But computer viruses cannot displace humans, and most of their effects can be controlled.

Another interesting trend is that human intelligence can be augmented by technology, which includes molecular bioengineering and development of brain–computer interfaces (BCIs) (Wikipedia contributors, 2017b). In particular, genetic manipulation and the use of growth factors and hormones during brain development may artificially increase the number of neurons and enhance their connectivity (Chelen, 2012). In this way, an "organic singularity" may be achieved before the predicted "technological singularity." Mixed organic-technical systems based on BCIs also have promising prospects. Animals with implanted electrodes in their brain can learn to manipulate a robotic arm or computer screen (Carmena et al., 2003). Noninvasive BCI (EEG-based) allows humans to manipulate a computer (Farwell and Donchin, 1988). Functional MRI technology allows instantaneous reconstruction of videos watched by human subjects (Nishimoto et al., 2011), a form of mind reading. The immediate application for BCI is to compensate for various disabilities (e.g., loss of vision, hearing, movement), but in the future it may serve to augment the intelligence of physiologically normal people. In particular, it is attractive to use BCI to create additional vision fields that represent a computer screen or improve deteriorating memory with the help of external silicon memory chips. Further advance in the quality of BCI can be expected if the interface is established early in life, so that brain functions can better adjust to the external or implanted device. The evolution of augmented intelligence may eventually lead to direct memory sharing between people, chip-directed learning, and even partial immortality (Koene, 2012; Tipler, 1994) (e.g., by using pre-programmed external devices). Augmented intelligence is a controversial technology because it may easily violate existing ethical norms (Wikipedia contributors, 2017b). In particular, it should not do any harm to people, or violate people's right for privacy and personal identity. Also, developers of this technology have to share partial responsibility for the actions of the customers who use it. Thus, augmented intelligence would require strict regulations from society.

In summary, the functional complexity of human civilization grows exponentially with a doubling time ca. 20 years, but we do not see any signs of an approaching "technological singularity" when humans would be replaced by intelligent machines. Instead, we expect a stronger integration of human minds with technology that would result in augmented intelligence. Creation of new technologies, i.e., emergences, is the norm in the evolution of life (Reid, 2007), and we should not be afraid of it. Our technologies represent the functional envelope for human society, just as intracellular molecular machines make the functional envelope for the DNA molecule. Can we anticipate further reduction of the doubling time in the growth of functional complexity? There is no doubt that exponential "acceleration of returns" will continue because of the positive feedback from increasing functional complexity of human civilization. However, the doubling time, which is inverse of the exponential parameter, is likely to remain stable as it was stable for billions of years in the evolution of the genome. We have already passed through a period of hyperexponential acceleration caused by advances in science, computer technology, Internet, and molecular biology. New technologies will keep emerging, but they are unlikely to change the established doubling time for human civilization. There are too many factors that counteract the growth of technology, such as negative population growth in most developed countries, increasing unemployment, unsolved environmental problems, and threats of war and social unrest. Thus, a simple exponential growth seems to be the most realistic forecast.

11 A BIOSEMIOTIC PERSPECTIVE

One way to look at the phenomenon of complexity increase is through the eyes of the growing discipline of biosemiotics, in which organisms are considered to be *agents*, capable of self-reproduction and adaptation (Hoffmeyer, 2008; Sharov, 2010, 2016, 2017). Some kind of memory is necessary for self-communication within an agent to preserve its functions. For example, when we learn how to use a hammer and nails, we establish a message within our brain to our future self, which will help us to replicate our actions in the future. In the same way, organisms supply their descendants with genetic "memory" that helps subsequent generations of organisms to replicate their functions. The elements of memory can be called "signs" because they play a role similar to words in a human language. There are several levels of complexity in the interpretation of signs from direct triggering of an action in response to a signal to categorization of streaming visual images into object types which then become interconnected through mental semantic networks. The pathways of meaning transmission are also different: the meaning of words is established via learning and culture, whereas the meaning of genetic signs is preserved by heredity.

The natural sciences nowadays tend to deliberately avoid any talk of goals and meanings. Numerous terms have been invented as substitutes, e.g., teleonomy (de Laguna, 1962) or cybernetics (Ashby, 1963; Gordon and Stone, 2016). Nevertheless, we deem it necessary to have some straight talk about goals and meanings. According to a recent review "[...]attributes of an agent include goal-directedness, self-governed activity, processing of semiosis and choice of action, with these features being vital for the functioning of the living system in question" (Tønnessen, 2015). Goals in an agent can be self-generated, but they can also be externally programmed by parental agents or higher-level agents (Sharov, 2010). Meanings are stable responses of agents to certain signs. However, the stability of meanings can be supported by various processes from social interactions to assorted heritable functions. Each function supports the meaning of signs that encode other functions. For example, the meaning of coding genome regions is supported by the function of RNA polymerase and DNA polymerase, and the meaning of regulatory genome regions is supported by the function of various transcription factors. Although manufacturing of these subagents is encoded in some portion of the DNA, the cell requires a minimal number of physically existing subagents to interpret the DNA. Thus, organisms are not just digital, they require material subagents to process these digital signs. Because cellular functions establish meanings for each other, they form a *semiotic closure* (Joslyn, 2000; Pattee, 1995).

Because of the semiotic nature of life, we have to reconsider the notion of biological evolution. The most fundamental process in evolution is accumulation of novel functions, which provide organisms with tools and methods for more successful survival, reproduction, programming of their offspring, capturing resources, and recruiting or reprogramming other agents for their benefit. This process is accompanied by changes of heritable signs including the genome and epigenetic factors. Heritable signs should not be viewed as a description or a blueprint of an organism because their meanings are not deterministic. Instead, they establish regulatory networks that continuously control cellular processes and differentiation waves that control embryo development and coordinate these processes with each other and with changing environmental factors (Gordon and Gordon, 2016; Gordon and Stone, 2016; Sharov, 2014). Among heritable signs, genetic information supports most fundamental biological functions that are comparable between all lineages of life. This feature allowed us to use genome complexity as a scale of progressive evolution and estimate the age of life.

12 CONCLUSION

The change of genome complexity of organisms represents a new molecular clock for predicting the age of life. Genome complexity increased nearly exponentially and doubled in size every 340 million years. This increase can be explained by positive feedback mechanisms that include gene cooperation and duplication. An extrapolation of this trend to earlier times suggests that life originated ca. 10 billion years ago. This cosmic time scale for the evolution of life has important consequences for our vision of the universe and the place of our biosphere and human civilization in it.

ACKNOWLEDGMENTS

The contribution of A.S. to this chapter was supported by the Intramural Research Program of the National Institute on Aging (NIA/NIH), project Z01 AG000656-13.

REFERENCES

Abbot, D.S., Switzer, E.R., 2011. The Steppenwolf: a proposal for a habitable planet in interstellar space. Astrophys. J. Lett. 735 (2), L27.

Abergel, C., Legendre, M., Claverie, J.-M., 2015. The rapidly expanding universe of giant viruses: mimivirus, pandoravirus, pithovirus and mollivirus. FEMS Microbiol. Rev. 39 (6), 779–796.

Abrescia, N.G.A., Bamford, D.H., Grimes, J.M., Stuart, D.I., 2012. Structure unifies the viral universe. Annu. Rev. Biochem. 81, 795–822.

Adam, G., Delbrück, M., 1968. Reduction of dimensionality in biological diffusion processes. In: Rich, A., Davidson, N. (Eds.), Structural Chemistry and Molecular Biology, A Volume Dedicated to Linus Pauling by His Students, Colleagues, and Friends. W.H. Freeman, New York, pp. 198–215.

Adami, C., Ofria, C., Collier, T.C., 2000. Evolution of biological complexity. Proc. Natl. Acad. Sci. U. S. A. 97 (9), 4463–4468.

Alicea, B., Gordon, R., 2014. Toy models for macroevolutionary patterns and trends. BioSystems 122 (Special Issue: Patterns of Evolution), 25–37.

Alicea, B., Gordon, R., 2016. Quantifying mosaic development: towards an Evo-Devo Postmodern Synthesis of the evolution of development via differentiation trees of embryos [invited]. Biology (Basel) 33 (3). https://doi.org/10.3390/biology5030033. Special Issue: Beyond the Modern Evolutionary Synthesis- what have we missed?, Ed. John S. Torday.

Alvarez-Candal, A., Pinilla-Alonso, N., Licandro, J., Cook, J., Mason, E., Roush, T., Cruikshank, D., Gourgeot, F., Dotto, E., Perna, D., 2011. The spectrum of (136199) Eris between 350 and 2350 nm: results with X-Shooter. Astron. Astrophys. 532, A130.

Ashby, W.R., 1963. An Introduction to Cybernetics. John Wiley & Sons, Inc., New York.

Badescu, V., 2011. Constraints on the free-floating planets supporting aqueous life. Acta Astronaut. 69 (9-10), 788–808.

Bar-Even, A., Shenhav, B., Kafri, R., Lancet, D., 2004. The lipid world: from catalytic and informational head-groups to micelle replication and evolution without nucleic acids. In: Seckbach, J., ChelaFlores, J., Owen, T., Raulin, F. (Eds.), Life in the Universe: from the Miller Experiment to the Search for Life on Other Worlds. Springer, pp. 111–114.

Barroso, G.V., Luz, D.R., 2015. On the limits of complexity in living forms. J. Theor. Biol. 379, 89–90.

Beckage, B., Gross, L.J., Kauffman, S., 2011. The limits to prediction in ecological systems. Ecosphere 2 (11), 1–12.

Benner, S.A., Kim, H.J., Carrigan, M.A., 2012. Asphalt, water, and the prebiotic synthesis of ribose, ribonucle-osides, and RNA. Acc. Chem. Res. 45 (12), 2025–2034.

Benton, M.J., 2010. Evolutionary biology: new take on the Red Queen. Nature 463 (7279), 306–307.

Bernal, A., Ear, U., Kyrpides, N., 2001. Genomes OnLine Database (GOLD): a monitor of genome projects world-wide. Nucleic Acids Res. 29 (1), 126–127.

Bialy, S., Sternberg, A., Loeb, A., 2015. Water formation during the epoch of first metal enrichment. Astrophys. J. Lett. 804 (2), L29.

Bjerre, J., Rousseau, C., Marinescu, L., Bols, M., 2008. Artificial enzymes, "Chemzymes": current state and per-spectives. Appl. Microbiol. Biotechnol. 81 (1), 1–11.

Bond, H.E., Nelan, E.P., VandenBerg, D.A., Schaefer, G.H., Harmer, D., 2013. HD 140283: a star in the solar neighborhood that formed shortly after the big bang. Astrophys. J. Lett. 765 (1), L12.

Buonaccorsi, J.P., 2010. Measurement Error: Models, Methods, and Applications. Chapman & Hall/CRC Interdisciplinary Statistics, Taylor & Francis Group, Boca Raton, FL, USA.

Carmena, J.M., Lebedev, M.A., Crist, R.E., O'Doherty, J.E., Santucci, D.M., Dimitrov, D.F., Patil, P.G., Henriquez, C.S., Nicolelis, M.A.L., 2003. Learning to control a brain-machine interface for reaching and grasping by primates. PLoS Biol. 1 (2), 193–208.

Carnegie, A., 1905. James Watt. Doubleday, Page & Co., New York.

Carroll, L., 1875. Through the Looking-Glass, and What Alice Found There. Macmillan & Co., New York.

Cavalier-Smith, T., 2005. Economy, speed and size matter: evolutionary forces driving nuclear genome miniaturization and expansion. Ann. Bot. (Lond.) 95 (1), 147–175.

Cavalier-Smith, T., 2010. Origin of the cell nucleus, mitosis and sex: roles of intracellular coevolution. Biol. Direct 5, 7.

Cavalier-Smith, T., 2014. The Neomuran revolution and phagotrophic origin of eukaryotes and cilia in the light of intracellular coevolution and a revised tree of life. Cold Spring Harb. Perspect. Biol. 6 (9), a016006.

Chelen, J., 2012. Biology and computing: could the organic singularity occur prior to Kurzweil's technological singularity? http://scienceprogress.org/2012/06/could-the-organic-singularity-occur-prior-to-kurzweils-technological-singularity/.

Chong, D.T., Liu, X.S., Ma, H.J., Huang, G.Y., Han, Y.L., Cui, X.Y., Yan, J.J., Xu, F., 2015. Advances in fabricating double-emulsion droplets and their biomedical applications. Microfluid. Nanofluid. 19 (5), 1071–1090.

Colson, P., de Lamballerie, X., Fournous, G., Raoult, D., 2012. Reclassification of giant viruses composing a fourth domain of life in the new order *Megavirales*. Intervirology 55 (5), 321–332.

Copley, S.D., Smith, E., Morowitz, H.J., 2007. The origin of the RNA world: co-evolution of genes and metabolism. Bioorg. Chem. 35 (6), 430–443.

Cowen, R., 2009. History of Life. John Wiley & Sons.

Creevey, O.L., Thévenin, F., Berio, P., Heiter, U., von Braun, K., Mourard, D., Bigot, L., Boyajian, T.S., Kervella, P., Morel, P., Pichon, B., Chiavassa, A., Nardetto, N., Perraut, K., Meilland, A., Mc Alister, H.A., ten Brummelaar, T.A., Farrington, C., Sturmann, J., Sturmann, L., Turner, N., 2015. Benchmark stars for *Gaia*: fundamental properties of the Population II star HD 140283 from interferometric, spectroscopic, and photometric data. Astron. Astrophys. 575, A26.

Crick, F.H.C., Orgel, L.E., 1973. Directed panspermia. Icarus 19 (3), 341–346.

Damer, B., 2016. A field trip to the Archaean in search of Darwin's warm little pond. Life 6, 21.

Damer, B., Newman, P., Norkus, R., Gordon, R., Barbalet, T., 2012. Cyberbiogenesis and the EvoGrid: a twenty-first century grand challenge. In: Seckbach, J. (Ed.), Genesis—In the Beginning: Precursors of Life, Chemical Models and Early Biological Evolution. Springer, Dordrecht, pp. 267–288.

Darwin, C., 1866. Origin of Species, fourth ed. John Murray, London.

Dasgupta, S., Aich, A., Mukhopadhyay, S.K., 2005. Evolutionary arms race: a review on the Red Queen hypothesis. J. Environ. Sociobiol. 2 (1–2), 109–118.

Davies, P.C.W., 2003. Does life's rapid appearance imply a Martian origin? Astrobiology 3 (4), 673–679.

Dawkins, R., 1986. The Blind Watchmaker, Why the Evidence of Evolution Reveals a Universe Without Design. W.W. Norton & Co., New York.

de Farias, S.T., Rego, T.G., José, M.V., 2016. A proposal of the proteome before the last universal common ancestor (LUCA). Int. J. Astrobiol. 15 (1), 27–31.

de Laguna, G.A., 1962. The role of teleonomy in evolution. Philos. Sci. 29 (2), 117–131.

de Solla Price, D.J., 1986. Little Science, Big Science... and Beyond. Columbia University Press, New York.

Deacon, T.W., 2006. Reciprocal linkage between self-organizing processes is sufficient for self-reproduction and evolvability. Biol. Theory 1 (2), 136–149.

Deacon, T.W., 2011. Incomplete Nature: How Mind Emerged From Matter. W.W. Norton & Company, New York.

Deamer, D., 2011. First Life: Discovering the Connections Between Stars, Cells, and How Life Began. University of California Press.

Deamer, D.W., Georgiou, C.D., 2015. Hydrothermal conditions and the origin of cellular life. Astrobiology 15 (12), 1091–1095.

Debes, J., Sigurdsson, S., 2002. Rogue earth-moon systems: safe havens for life? Astrobiology 2 (4), 583.

Dennett, D.C., 2017. From Bacteria to Bach and Back: The Evolution of Minds. W.W. Norton.

DePamphilis, M.L., Bell, S.D., 2010. Genome Duplication. Garland Publishing.

Di Giulio, M., 2011. The Last Universal Common Ancestor (LUCA) and the ancestors of archaea and bacteria were progenotes. J. Mol. Evol. 72 (1), 119–126.

Doolittle, W.F., 2000. The nature of the universal ancestor and the evolution of the proteome. Curr. Opin. Struct. Biol. 10 (3), 355–358.

Doolittle, W.F., 2013. Is junk DNA bunk? A critique of ENCODE. Proc. Natl. Acad. Sci. U. S. A. 110 (14), 5294–5300.

Dutta, A., DuBois, D.L., Roberts, J.A., Shaw, W.J., 2014. Amino acid modified Ni catalyst exhibits reversible H_2 oxidation/production over a broad pH range at elevated temperatures. Proc. Natl. Acad. Sci. U. S. A. 111 (46), 16286–16291.

Eagle, H., Piez, K.A., Levy, M., 1961. The intracellular amino acid concentrations required for protein synthesis in cultured human cells. J. Biol. Chem. 236, 2039–2042.

Eddy, S.R., 2012. The C-value paradox, junk DNA and ENCODE. Curr. Biol. 22 (21), R898–R899.

Eddy, S.R., 2013. The ENCODE project: missteps overshadowing a success. Curr. Biol. 23 (7), R259–R261.

Edger, P.P., Heidel-Fischer, H.M., Bekaert, M., Rota, J., Glöckner, G., Platts, A.E., Heckel, D.G., Der, J.P., Wafula, E.K., Tang, M., Hofberger, J.A., Smithson, A., Hall, J.C., Blanchette, M., Bureau, T.E., Wright, S.I., dePamphilis, C.W., Eric Schranz, M., Barker, M.S., Conant, G.C., Wahlberg, N., Vogel, H., Pires, J.C., Wheat, C.W., 2015. The butterfly plant arms-race escalated by gene and genome duplications. Proc. Natl. Acad. Sci. U. S. A. 112 (27), 8362–8366.

Ehrlich, H.L., Newman, D.K., 2008. Geomicrobiology, fifth ed. Taylor & Francis.

Eigen, M., Schuster, P., 1979. The Hypercycle: A Principle of Natural Self-Organization. Springer-Verlag, Berlin.

Eldredge, N., Gould, S.J., 1972. Punctuated equilibria: an alternative to phyletic gradualism. In: Schopf, T.J.M. (Ed.), Models in Paleobiology. Freeman, Cooper and Company, San Francisco, pp. 82–115.

Farwell, L.A., Donchin, E., 1988. Talking off the top of your head: toward a mental prosthesis utilizing event-related brain potentials. Electroencephalogr. Clin. Neurophysiol. 70 (6), 510–523.

Fazekas, A., 2011. Star found shooting water "bullets". http://news.nationalgeographic.com/news/2011/06/110613-space-science-star-water-bullets-kristensen.

Fournier, G.P., 2015. Molecular clocks constrained by horizontal gene transfers predict an archaeal common ancestor ~3.9 Ga, coincident with the proposed late heavy bombardment, and consistent with the hypothesis of a thermophilic bottleneck for early life. In: Doran, P. (Ed.), AbSciCon2015: Astrobiology Science Conference 2015, Habitability, Habitable Worlds, and Life, June 15–18, 2015, Chicago, Illinois. http://www.hou.usra.edu/meetings/abscicon2015/pdf/7248.pdf.

Fournier, G.P., Alm, E.J., 2015. Ancestral reconstruction of a pre-LUCA aminoacyl-tRNA synthetase ancestor supports the late addition of Trp to the genetic code. J. Mol. Evol. 80 (3-4), 171–185.

Fournier, G.P., Andam, C.P., Gogarten, J.P., 2015. Ancient horizontal gene transfer and the last common ancestors. BMC Evol. Biol. 15, 70.

Frankel, J., 1989. Pattern Formation, Ciliate Studies and Models. Oxford University Press, New York.

Frohoff, T., Oriel, E., 2016. 17. Conversing with dolphins: the holy grail of interspecies communication? In: Gordon, R., Seckbach, J. (Eds.), Biocommunication: Sign-Mediated Interactions Between Cells and Organisms. World Scientific Publishing, London, pp. 575–597.

Germain, P.L., Ratti, E., Boem, F., 2014. Junk or functional DNA? ENCODE and the function controversy. Biol. Philos. 29 (6), 807–831.

Gibson, C.H., 2000. Turbulent mixing, viscosity, diffusion, and gravity in the formation of cosmological structures: the fluid mechanics of dark matter. J. Fluids Eng.-Trans. ASME 122 (4), 830–835.

Gibson, D.G., Benders, G.A., Andrews-Pfannkoch, C., Denisova, E.A., Baden-Tillson, H., Zaveri, J., Stockwell, T.B., Brownley, A., Thomas, D.W., Algire, M.A., Merryman, C., Young, L., Noskov, V.N., Glass, J.I., Venter, J.C., Hutchison III, C.A., Smith, H.O., 2008. Complete chemical synthesis, assembly, and cloning of a *Mycoplasma genitalium* genome. Science 319 (5867), 1215–1220.

Gilbert, W., 1986. Origin of life: the RNA world. Nature 319 (6055), 618.

Gladman, B., Chan, C., 2006. Production of the extended scattered disk by rogue planets. Astrophys. J. 643 (2), L135–L138.

Glansdorff, N., Xu, Y., Labedan, B., 2008. The Last Universal Common Ancestor: emergence, constitution and genetic legacy of an elusive forerunner. Biol. Direct 3, 29.

Glanz, J., 1998. Infrared astronomy—a water generator in the Orion nebula. Science 280 (5362), 378.

Gordon, N.K., Gordon, R., 2016. Embryogenesis Explained. World Scientific Publishing, Singapore.

Gordon, R., 1992. The fractal physics of biological evolution. In: Beysens, D., Boccara, N., Forgacs, G. (Eds.), Dynamical Phenomena at Interfaces, Surfaces and Membranes. NOVA Science Publishers, Commack, NY, pp. 99–111.

Gordon, R., 1999. The Hierarchical Genome and Differentiation Waves: Novel Unification of Development, Genetics and Evolution. World Scientific & Imperial College Press, Singapore & London.

Gordon, R., 2013. The differentiation tree as a source of novelty and evolvability, comment on: "Beyond Darwin: evolvability and the generation of novelty" by Marc Kirschner. BMC Biology 11, https://bmcbiol. biomedcentral.com/articles/10.1186/1741-7007-11-110/comments.

Gordon, R., Hanczyc, M.M., Denkov, N.D., Tiffany, M.A., Smoukov, S.K., 2017. Emergence of polygonal shapes in oil droplets and living cells: the potential role of tensegrity in the origin of life. In: Gordon, R., Sharov, A.A., (Eds.), Habitability of the Universe Before Earth. In: Rampelotto, P.H., Seckbach, J., Gordon, R. (Eds.), Astrobiology: Exploring Life on Earth and Beyond. Elsevier B.V., Amsterdam, pp. 427–490.

Gordon, R., Hoover, R.B., 2007. Could there have been a single origin of life in a Big Bang universe? Proc. SPIE 6694. https://spie.org/Publications/Proceedings/Paper/10.1117/12.737041.

Gordon, R., McNichol, J., 2012. Recurrent dreams of life in meteorites. In: Seckbach, J. (Ed.), Genesis—In the Beginning: Precursors of Life, Chemical Models and Early Biological Evolution. Springer, Dordrecht, pp. 549–590.

Gordon, R., Stone, R., 2016. Cybernetic embryo. In: Gordon, R., Seckbach, J. (Eds.), Biocommunication: Sign-Mediated Interactions Between Cells and Organisms. World Scientific Publishing, London, pp. 111–164.

Gray, M.W., 1994. Split RNAs and modified nucleosides in ribosome evolution. In: Hartman, H., Matsuno, K. (Eds.), The Origin and Evolution of the Cell. World Scientific Publishing, Singapore, pp. 333–358.

Gregory, T.R., 2001. Coincidence, coevolution, or causation? DNA content, cell size, and the C-value enigma. Biol. Rev. Camb. Philos. Soc. 76 (1), 65–101.

Griffith, J.S., 1967. Self-replication and scrapie. Nature 215 (5105), 1043–1044.

Grimes, G.W., Aufderheide, K.J., 1991. Cellular Aspects of Pattern Formation: The Problem of Assembly. Karger, Basel.

Grosch, E.G., Hazen, R.M., 2015. Microbes, mineral evolution, and the rise of microcontinents-origin and coevolution of life with early Earth. Astrobiology 15 (10), 922–939.

Gruzewska, K., Michno, A., Pawelczyk, T., Bielarczyk, H., 2014. Essentiality and toxicity of vanadium supplements in health and pathology. J. Physiol. Pharmacol. 65 (5), 603–611.

Haqq-Misra, J., Kopparapu, R., 2017. The Drake equation as a function of spectral type and time. In: Gordon, R., Sharov, A.A., (Eds.), Habitability of the Universe Before Earth. In: Rampelotto, P.H., Seckbach, J., Gordon, R. (Eds.), Astrobiology: Exploring Life on Earth and Beyond. Elsevier B.V., Amsterdam, pp. 307–319.

Hazen, R.M., 2010. Evolution of minerals. Sci. Am. 302 (3), 58–65.

Hedges, S.B., 2002. The origin and evolution of model organisms. Nat. Rev. Genet. 3 (11), 838–849.

Herschy, B., Whicher, A., Camprubi, E., Watson, C., Dartnell, L., Ward, J., Evans, J.R.G., Lane, N., 2014. An origin-of-life reactor to simulate alkaline hydrothermal vents. J. Mol. Evol. 79 (5-6), 213–227.

Hoeppner, M.P., Gardner, P.P., Poole, A.M., 2012. Comparative analysis of RNA families reveals distinct repertoires for each domain of life. PLoS Comput. Biol. 8 (11). e1002752.

Hoffmeyer, J., 2008. Biosemiotics: An Examination into the Signs of Life and the Life of Signs. University of Chicago Press, Chicago.

Hoffmeyer, J., 2011. Epilogue: biology is immature biosemiotics. In: Emmeche, C., Kull, K. (Eds.), Towards a Semiotic Biology. Life Is the Action of Signs. Imperial College Press, London, pp. 43–65.

Holland, L.Z., 2013. Evolution of new characters after whole genome duplications: insights from amphioxus. Semin. Cell Dev. Biol. 24 (2), 101–109.

Holman, M.J., Rudenko, M., 2016. Dwarf Planets. http://minorplanetcenter.net/dwarf_planets.

Howett, C.J.A., Spencer, J.R., Hurford, T., Verbiscer, A., Segura, M., 2016. Thermal properties of Rhea's poles: evidence for a meter-deep unconsolidated subsurface layer. Icarus 272, 140–148.

Hu, P., Peng, L., Zhen, S., Chen, L., Xiao, S., Huang, C., 2010. Homochiral expression of proteins: a discussion on the natural chirality related to the origin of life. Sci. China-Chem. 53 (4), 792–796.

Hull, P., 2015. Life in the aftermath of mass extinctions. Curr. Biol. 25 (19), R941–R952.

Hunding, A., Kepes, F., Lancet, D., Minsky, A., Norris, V., Raine, D., Sriram, K., Root-Bernstein, R., 2006. Compositional complementarity and prebiotic ecology in the origin of life. Bioessays 28 (4), 399–412.

Huskey, W.P., Epstein, I.R., 1989. Autocatalysis and apparent bistability in the formose reaction. J. Am. Chem. Soc. 111 (9), 3157–3163.

Inger, A., Solomon, A., Shenhav, B., Olender, T., Lancet, D., 2009. Mutations and lethality in simulated prebiotic networks. J. Mol. Evol. 69 (5), 568–578.

Isambert, H., Singh, P.P., 2016. OHNOLOGS: a repository of genes retained from whole genome duplications in the vertebrate genomes. http://ohnologs.curie.fr.

Jablonka, E., Szathmáry, E., 1995. The evolution of information storage and heredity. Trends Ecol. Evol. 10 (5), 206–211.

Jablonski, D., 2005. Mass extinctions and macroevolution. Paleobiology 31 (2), 192–210.

Jarosik, N., Bennett, C.L., Dunkley, J., Gold, B., Greason, M.R., Halpern, M., Hill, R.S., Hinshaw, G., Kogut, A., Komatsu, E., Larson, D., Limon, M., Meyer, S.S., Nolta, M.R., Odegard, N., Page, L., Smith, K.M., Spergel, D.N., Tucker, G.S., Weiland, J.L., Wollack, E., Wright, E.L., 2011. Seven-year Wilkinson Microwave Anisotropy Probe (WMAP) observations: sky maps, systematic errors, and basic results. Astrophys. J. Suppl. Ser. 192 (2), 1–15.

Jerison, H.J., 1973. Evolution of the Brain and Intelligence. Academic Press, New York.

Johnson, J.L., Li, H., 2012. The first planets: the critical metallicity for planet formation. Astrophys. J. 751 (2), 81.

Johnson, S.S., Hebsgaard, M.B., Christensen, T.R., Mastepanov, M., Nielsen, R., Munch, K., Brand, T., Gilbert, M.T., Zuber, M.T., Bunce, M., Ronn, R., Gilichinsky, D., Froese, D., Willerslev, E., 2007. Ancient bacteria show evidence of DNA repair. Proc. Natl. Acad. Sci. U. S. A. 104 (36), 14401–14405.

Johnston, W.K., Unrau, P.J., Lawrence, M.S., Glasner, M.E., Bartel, D.P., 2001. RNA-catalyzed RNA polymerization: accurate and general RNA-templated primer extension. Science 292 (5520), 1319–1325.

Jørgensen, S.E., 2007. Evolution and exergy. Ecol. Model. 203 (3-4), 490–494.

Joseph, R., 2009. Life on Earth came from other planets. J. Cosmol. 1, 1–56.

Joslyn, C., 2000. Levels of control and closure in complex semiotic systems. Ann. N. Y. Acad. Sci. 901, 67–74.

Katz, C., 2012. Bugs in the ice sheet. Sci. Am. 306 (5), 20.

Kauffman, S.A., 1986. Autocatalytic sets of proteins. J. Theor. Biol. 119, 1–24.

Kauffman, S.A., Farmer, J.D., Packard, N.H., 1986. Autocatalytic sets of proteins. Orig. Life Evol. Biosph. 16 (3-4), 446–447.

Kazan, C., 2010. Could the Universe be older than we think? New findings point that way. http://www.dailygalaxy.com/my_weblog/2010/06/could-the-universe-be-far-older-than-we-think-new-findings-point-that-way.html.

Keller, G., 2005. Impacts, volcanism and mass extinction: random coincidence or cause and effect? Aust. J. Earth Sci. 52 (4-5), 725–757.

Kellis, M., Wold, B., Snyder, M.P., Bernstein, B.E., Kundaje, A., Marinov, G.K., Ward, L.D., Birney, E., Crawford, G.E., Dekker, J., Dunham, I., Elnitski, L.L., Farnham, P.J., Feingold, E.A., Gerstein, M., Giddings, M.C., Gilbert, D.M., Gingeras, T.R., Green, E.D., Guigo, R., Hubbard, T., Kent, J., Lieb, J.D., Myers, R.M., Pazin, M.J., Ren, B., Stamatoyannopoulos, J.A., Weng, Z.P., White, K.P., Hardison, R.C., 2014. Defining functional DNA elements in the human genome. Proc. Natl. Acad. Sci. U. S. A. 111 (17), 6131–6138.

Klevenz, V., Sumoondur, A., Ostertag-Henning, C., Koschinsky, A., 2010. Concentrations and distributions of dissolved amino acids in fluids from Mid-Atlantic Ridge hydrothermal vents. Geochem. J. 44 (5), 387–397.

Koene, R.A., 2012. The society of neural prosthetics and whole brain emulation science. http://www.minduploading.org/.

Koonin, E.V., Galperin, M.Y., 2003. Sequence—Evolution—Function: Computational Approaches in Comparative Genomics. Springer-Science+Business Media B.V., Dordrecht.

Kritsky, M.S., Telegina, T.A., 2004. Role of nucleotide-like coenzymes in primitive evolution. In: Seckbach, J. (Ed.), Origins: Genesis, Evolution and Diversity of Life. Kluwer, Dordrecht, The Netherlands, pp. 215–231.

Kümmerle, R., Kyritsis, P., Gaillard, J., Moulis, J.M., 2000. Electron transfer properties of iron-sulfur proteins. J. Inorg. Biochem. 79 (1-4), 83–91.

Kurzweil, R., 2005. The Singularity Is Near: When Humans Transcend Biology. Penguin Books, New York.

Lambert, L.H., Cox, T., Mitchell, K., Rosselló-Mora, R.A., Del Cueto, C., Dodge, D.E., Orkand, P., Cano, R.J., 1998. *Staphylococcus succinus* sp. nov., isolated from Dominican amber [erratum: 49, 933]. Int. J. Syst. Bacteriol. 48 (2), 511–518.

Lane, N., Martin, W.F., Raven, J.A., Allen, J.F., 2013. Energy, genes and evolution: introduction to an evolutionary synthesis. Phil. Trans. R. Soc. B-Biol. Sci. 368 (1622), 20120253.

Lang, S.Q., Fruh-Green, G.L., Bernasconi, S.M., Butterfield, D.A., 2013. Sources of organic nitrogen at the serpentinite-hosted Lost City hydrothermal field. Geobiology 11 (2), 154–169.

Laurent, M., 1997. Autocatalytic processes in cooperative mechanisms of prion diseases. FEBS Lett. 407 (1), 1–6.

Lefebvre, L., 2013. Brains, innovations, tools and cultural transmission in birds, non-human primates, and fossil hominins. Frontiers in Human. Neuroscience 7, 245.

Line, M.A., 2002. The enigma of the origin of life and its timing. Microbiology 148, 21–27.

Line, M.A., 2007. Panspermia in the context of the timing of the origin of life and microbial phylogeny. Int. J. Astrobiol. 6 (3), 249–254.

Lineweaver, C.H., Davis, T.M., 2003. Does the rapid appearance of life on Earth suggest that life is common in the Universe? Orig. Life Evol. Biosph. 33 (3), 311–312.

Lineweaver, C.H., Grether, D., 2002. The observational case for Jupiter being a typical massive planet. Astrobiology 2 (3), 325–334.

Lipowski, A., 2005. Periodicity of mass extinctions without an extraterrestrial cause. Phys. Rev. E 71 (5). 052902.

Lisman, J.E., Fallon, J.R., 1999. Neuroscience—what maintains memories? Science 283 (5400), 339–340.

Livio, M., Kopelman, A., 1990. Life and the Sun's lifetime. Nature 343 (6253), 25.

Loeb, A., 2014. The habitable epoch of the early Universe. Int. J. Astrobiol. 13 (4), 337–339.

Luisi, P.L., 2015. Chemistry constraints on the origin of life. Israel J. Chem. 55 (8), 906–918.

Lundstrom, M., 2003. Applied physics. Moore's law forever? Science 299 (5604), 210–211.

Luo, L.-f., 2009. Law of genome evolution direction: coding information quantity grows. Front. Phys. China 4 (2), 241–251.

Mannige, R.V., Brooks, C.L., Shakhnovich, E.I., 2012. A universal trend among proteomes indicates an oily last common ancestor. PLoS Comput. Biol. 8 (12). e1002839.

Marcano, V., Benitez, P., Palacios-Prü, E., 2003. Acyclic hydrocarbon environments $\geq n$-C_{18} on the early terrestrial planets. Planet. Space Sci. 51 (3), 159–166.

Markov, A.V., 2000. The return of the Red Queen, or the law of increase of mean duration of genera in the course of evolution [Russian with English abstract]. Zh. Obshch. Biol. 61 (4), 357–370.

Markov, A.V., Anisimov, V.A., Korotayev, A.V., 2010. Relationship between genome size and organismal complexity in the lineage leading from prokaryotes to mammals. Paleontol. J. 44 (4), 363–373.

Martin, C.C., Gordon, R., 1995. Differentiation trees, a junk DNA molecular clock, and the evolution of neoteny in salamanders. J. Evol. Biol. 8, 339–354.

Marzban, C., Viswanathan, R., Yurtsever, U., 2014. Earth before life. Biol. Direct 9, 1.

Marzban, C., Viswanathan, R., Yurtsever, U., 2017. Earth before life [Reprint of: Marzban, C., Viswanathan, R., Yurtsever, U., 2014. Earth before life. *Biology Direct* **9**, 1.]. In: Gordon, R., Sharov, A.A. (Eds.), Habitability of the Universe Before Earth. In: Rampelotto, P.H., Seckbach, J., Gordon, R. (Eds.), Astrobiology: Exploring Life on Earth and Beyond. Elsevier B.V., Amsterdam, pp. 297–305.

Massingham, T., Davies, L.J., Liò, P., 2001. Analysing gene function after duplication. BioEssays 23 (10), 873–876.

Maury, C.P.J., 2009. Self-propagating β-sheet polypeptide structures as prebiotic informational molecular entities: the amyloid world. Orig. Life Evol. Biosph. 39 (2), 141–150.

Mayr, E., 2002. What Evolution Is. Basic Books, New York.

McDonald, G.D., 2014. Biochemical pathways as evidence for prebiotic synthesis. In: Kolb, V.M. (Ed.), Astrobiology: An Evolutionary Approach. CRC Press, Taylor & Francis Group, Boca Raton, Florida, USA, pp. 119–148.

McHenry, H.M., 1975. Fossil hominid body weight and brain size. Nature 254 (5502), 686–688.

McNichol, J., Gordon, R., 2012. Are we from outer space? A critical review of the panspermia hypothesis. In: Seckbach, J. (Ed.), Genesis—In the Beginning: Precursors of Life, Chemical Models and Early Biological Evolution. Springer, Dordrecht, pp. 591–620.

Mikhailovsky, G.E., Gordon, R., 2017. Symbiosis: why was the transition from microbial prokaryotes to eukaryotic organisms a cosmic gigayear event? In: Gordon, R., Sharov, A.A. (Eds.), Habitability of the Universe Before Earth. In: Rampelotto, P.H., Seckbach, J., Gordon, R. (Eds.), Astrobiology: Exploring Life on Earth and Beyond. Elsevier B.V., Amsterdam, pp. 355–405.

Moore, G.E., 1965. Cramming more components onto integrated circuits. Electronics 38 (8), 114–117.

Morowitz, H.J., Kostelnik, J.D., Yang, J., Cody, G.D., 2000. The origin of intermediary metabolism. Proc. Natl. Acad. Sci. U. S. A. 97 (14), 7704–7708.

Nadkarni, R.A.K., 2009. Determination of inorganic species in petroleum products and lubricants. In: Rand, S.J. (Ed.), Significance of Tests for Petroleum Products, eighth ed. ASTM International, West Conshohocken, PA, USA, pp. 283–298.

Nelson, K.E., Levy, M., Miller, S.L., 2000. Peptide nucleic acids rather than RNA may have been the first genetic molecule. Proc. Natl. Acad. Sci. U. S. A. 97 (8), 3868–3871.

New World Encyclopedia Contributors, 2017. Chinese character. http://www.newworldencyclopedia.org/entry/chinese_character.

Newman, M., 2000. Statistical models of mass extinction. In: BarYam, Y. (Ed.), Unifying Themes in Complex Systems. Perseus Publishing, Cambridge, MA, USA, pp. 373–384.

Nielsen, J.T., Guffanti, A., Sarkar, S., 2016. Marginal evidence for cosmic acceleration from Type Ia supernovae. Sci Rep 6, 35596.

Nishimoto, S., Vu, A.T., Naselaris, T., Benjamini, Y., Yu, B., Gallant, J.L., 2011. Reconstructing visual experiences from brain activity evoked by natural movies. Curr. Biol. 21, 1641–1646.

Niu, D.-K., Jiang, L., 2013. Can ENCODE tell us how much junk DNA we carry in our genome? Biochem. Biophys. Res. Commun. 430 (4), 1340–1343.

Nutman, A.P., Bennett, V.C., Friend, C.R., Van Kranendonk, M.J., Chivas, A.R., 2016. Rapid emergence of life shown by discovery of 3,700-million-year-old microbial structures. Nature 537 (7621), 535–538.

Ohno, S., 1970. Evolution by Gene Duplication. Springer-Verlag, New York.

Oparin, A.I., 1936. The Origin of Life, 1953 reprint ed. Dover Publications, Mineola, NY.

Orgel, L., 2000. Origin of life—a simpler nucleic acid. Science 290 (5495), 1306–1307.

Patrushev, L.I., Minkevich, I.G., 2008. The problem of the eukaryotic genome size. Biochem. Mosc. 73 (13), 1519–1552.

Pattee, H.H., 1995. Evolving self-reference: matter, symbols, and semantic closure. Commun. Cogn.-Artif. Intell. 12 (1-2), 9–28.

Patthy, L., 1999. Genome evolution and the evolution of exon-shuffling—a review. Gene 238 (1), 103–114.

Pennisi, E., 2012. Human genome is much more than just genes. http://www.sciencemag.org/news/2012/09/human-genome-much-more-just-genes.

Perez-Moral, N., Watt, S., Wilde, P., 2014. Comparative study of the stability of multiple emulsions containing a gelled or aqueous internal phase. Food Hydrocoll. 42 (Part 1), 215–222.

Pilkey, O.H., Pilkey-Jarvis, L., 2007. Useless Arithmetic: Why Environmental Scientists Can't Predict the Future. Columbia University Press, New York.

Powner, M.W., Gerland, B., Sutherland, J.D., 2009. Synthesis of activated pyrimidine ribonucleotides in prebiotically plausible conditions. Nature 459 (7244), 239–242.

Reid, R.G.B., 2007. Biological Emergences: Evolution by Natural Experiment. MIT Press, Cambridge.

Rollins, R.A., Haghighi, F., Edwards, J.R., Das, R., Zhang, M.Q., Ju, J.Y., Bestor, T.H., 2006. Large-scale structure of genomic methylation patterns. Genome Res. 16 (2), 157–163.

Root-Bernstein, M., Root-Bernstein, R., 2015. The ribosome as a missing link in the evolution of life. J. Theor. Biol. 367, 130–158.

Roth, G., Dicke, U., 2005. Evolution of the brain and intelligence. Trends Cogn. Sci. 9 (5), 250–257.

Samuel, E., 2001. Free spirits—is the cosmos teeming with rogue planets without stars to call their own? New Sci. 170 (2287), 12.

Schmalhausen, I.I., 1949. Factors of Evolution: The Theory of Stabilizing Selection, 1987 translation ed. University of Chicago Press, Chicago.

Schroeder, G.L., 1990. Genesis and the Big Bang: The Discovery of Harmony Between Modern Science and the Bible. Bantam Books, New York.

Schwartz, A.W., 2006. Phosphorus in prebiotic chemistry. Phil. Trans. R. Soc. B-Biol. Sci. 361 (1474), 1743–1749.

Segré, D., Ben-Eli, D., Deamer, D.W., Lancet, D., 2001. The lipid world. Orig. Life Evol. Biosph. 31 (1-2), 119–145.

Segré, D., Ben-Eli, D., Lancet, D., 2000. Compositional genomes: prebiotic information transfer in mutually catalytic noncovalent assemblies. Proc. Natl. Acad. Sci. U. S. A. 97 (8), 4112–4117.

Segré, D., Lancet, D., 1999. A statistical chemistry approach to the origin of life. Chemtracts-Biochem. Mol. Biol. 12, 382–397.

Sharma, V., Colson, P., Pontarotti, P., Raoulti, D., 2016. Mimivirus inaugurated in the 21st century the beginning of a reclassification of viruses. Curr. Opin. Microbiol. 31, 16–24.

Sharov, A.A., 1991. Self-reproducing systems: structure, niche relations and evolution. BioSystems 25 (4), 237–249.

Sharov, A.A., 2006. Genome increase as a clock for the origin and evolution of life. Biol. Direct 1, 17.

Sharov, A.A., 2009. Coenzyme autocatalytic network on the surface of oil microspheres as a model for the origin of life. Int. J. Mol. Sci. 10 (4), 1838–1852.

Sharov, A.A., 2010. Functional information: towards synthesis of biosemiotics and cybernetics. Entropy (Basel) 12 (5), 1050–1070.

Sharov, A.A., 2012. The origin of mind. In: Maran, T., Lindström, K., Magnus, R., Tønnensen, M. (Eds.), Biosemiotics Turning Wild: Essays in Honour of Kalevi Kull on the Occasion of His 60th Birthday. University of Tartu Press, Tartu, Estonia, pp. 63–69.

Sharov, A.A., 2014. Evolutionary constraints or opportunities? BioSystems 123, 9–18.

Sharov, A.A., 2016. Coenzyme world model of the origin of life. BioSystems 144, 8–17.

Sharov, A.A., 2017. Coenzyme world model of the origin of life [Reprint of: Sharov, A.A., 2016. Coenzyme world model of the origin of life. *BioSystems* 144, 8–17.]. In: Gordon, R., Sharov, A.A. (Eds.), Habitability of the Universe Before Earth. In: Rampelotto, P.H., Seckbach, J., Gordon, R. (Eds.), Astrobiology: Exploring Life on Earth and Beyond. Elsevier B.V, Amsterdam, pp. 407–426.

Sharov, A.A., Gordon, R., 2013. Life before earth. http://arxiv.org/abs/1304.3381.

Shenhav, B., Bar-Even, A., Kafri, R., Lancet, D., 2005. Polymer GARD: computer simulation of covalent bond formation in reproducing molecular assemblies. Orig. Life Evol. Biosph. 35 (2), 111–133.

Shenhav, B., Oz, A., Lancet, D., 2007. Coevolution of compositional protocells and their environment. Phil. Trans. R. Soc. Lond. B, Biol. Sci. 362 (1486), 1813–1819.

Simons, C., Pheasant, M., Makunin, I.V., Mattick, J.S., 2006. Transposon-free regions in mammalian genomes. Genome Res. 16 (2), 164–172.

Sinnott, E.W., 1962. Matter, Mind and Man: The Biology of Human Nature. Atheneum, New York.

Steele, A., Baross, J., 2006. Are prions relevant to astrobiology? Astrobiology 6 (1), 284.

Suzan-Monti, M., La Scola, B., Raoult, D., 2006. Genomic and evolutionary aspects of Mimivirus. Virus Res. 117 (1), 145–155.

Szathmáry, E., 1999. The first replicators. In: Keller, L. (Ed.), Levels of Selection in Evolution. Princeton University Press, pp. 31–52.

Theobald, D.L., 2010. A formal test of the theory of universal common ancestry. Nature 465 (7295), 219–222.

Tilman, D., Kareiva, P. (Eds.), 1997. Book Spatial Ecology: The Role of Space in Population Dynamics and Interspecific Interactions. Princeton University Press, Princeton, NJ, USA.

Tipler, F.J., 1994. The Physics of Immortality: Modern Cosmology, God and the Resurrection of the Dead. Doubleday, New York.

Tønnessen, M., 2015. The biosemiotic glossary project: agent, agency. Biosemiotics 8 (1), 125–143.

Trail, D., Tailby, N.D., Sochko, M., Ackerson, M.R., 2015. Possible biosphere-lithosphere interactions preserved in igneous zircon and implications for Hadean Earth. Astrobiology 15 (7), 575–586.

Tuller, T., Birin, H., Gophna, U., Kupiec, M., Ruppin, E., 2010. Reconstructing ancestral gene content by coevolution. Genome Res. 20 (1), 122–132.

Unrau, P.J., Bartel, D.P., 1998. RNA-catalysed nucleotide synthesis. Nature 395 (6699), 260–263.

Valentine, J.W., 2004. On the Origin of Phyla. University of Chicago Press, Chicago.

Valley, J.W., 2005. A cool early Earth? Sci. Am. 293 (4), 58–65.

Valley, J.W., Peck, W.H., King, E.M., Wilde, S.A., 2002. A cool early Earth. Geology 30 (4), 351–354.

Van Valen, L.M., 1973. A new evolutionary law. Evol. Theory 1, 1–30.

Vanhamäki, H., 2011. Emission of cyclotron radiation by interstellar planets. Planet. Space Sci. 59 (9), 862–869.

Voje, K.L., Nolen, Ø.H., Liow, L.H., Stenseth, N.C., 2015. The role of biotic forces in driving macroevolution: beyond the Red Queen. Proc. R. Soc. B-Biol. Sci. 282 (1808), 20150186.

Wächtershäuser, G., 1988. Before enzymes and templates: theory of surface metabolism. Microbiol. Rev. 52 (4), 452–484.

Wallis, M.K., Wickramasinghe, N.C., 2004. Interstellar transfer of planetary microbiota. Mon. Not. R. Astron. Soc. 348 (1), 52–61.

Ward, P., Brownlee, D., 2003. Rare Earth: Why Complex Life Is Uncommon in the Universe. Copernicus, Springer-Verlag, New York.

Watzky, M.A., Morris, A.M., Ross, E.D., Finke, R.G., 2008. Fitting yeast and mammalian prion aggregation kinetic data with the Finke-Watzky two-step model of nucleation and autocatalytic growth. Biochemistry 47 (40), 10790–10800.

Webb, S., 2002. If the Universe Is Teeming with Aliens... Where Is Everybody? Fifty Solutions to Fermi's Paradox and the Problem of Extraterrestrial Life. Springer, New York.

Weiss, M.C., Sousa, F.L., Mrnjavac, N., Neukirchen, S., Roettger, M., Nelson-Sathi, S., Martin, W.F., 2016. The physiology and habitat of the last universal common ancestor. Nat. Microbiol. 1, 16116.

West-Eberhard, M.J., 2002. Developmental Plasticity and Evolution. Oxford University Press, Oxford.

Wikipedia Contributors, 2017a. Brain size. http://en.wikipedia.org/wiki/Brain_size.

Wikipedia Contributors, 2017b. Brain–computer interface. http://en.wikipedia.org/wiki/Brain-computer_interface.

Wikipedia Contributors, 2017c. Drake equation. https://en.wikipedia.org/wiki/Drake_equation.

Wikipedia Contributors, 2017d. Fermi paradox. http://en.wikipedia.org/wiki/Fermi_paradox.

Wikipedia Contributors, 2017e. Patent. https://en.wikipedia.org/wiki/Patent.

Yarus, M., 2011. Getting past the RNA world: the initial Darwinian ancestor. Cold Spring Harb. Perspect. Biol. 3 (4), a003590.

Zheng, W., Postman, M., Zitrin, A., Moustakas, J., Shu, X., Jouvel, S., Host, O., Molino, A., Bradley, L., Coe, D., Moustakas, L.A., Carrasco, M., Ford, H., Benitez, N., Lauer, T.R., Seitz, S., Bouwens, R., Koekemoer, A., Medezinski, E., Bartelmann, M., Broadhurst, T., Donahue, M., Grillo, C., Infante, L., Jha, S.W., Kelson, D.D., Lahav, O., Lemze, D., Melchior, P., Meneghetti, M., Merten, J., Nonino, M., Ogaz, S., Rosati, P., Umetsu, K., van der Wel, A., 2012. A magnified young galaxy from about 500 million years after the Big Bang. Nature 489 (7416), 406–408.

EARTH BEFORE LIFE*

Caren Marzban*,#, Raju Viswanathan†, Ulvi Yurtsever‡
*University of Washington, Seattle, WA, United States
†Technome, Clayton, MO, United States
‡MathSense Analytics, Altadena, CA, United States

CHAPTER OUTLINE

1 BACKGROUND

Sharov (2006), and more recently, Sharov and Gordon (2013) reported an analysis of data on the evolution of genetic complexity during the history of life on Earth. These two works (hereafter denoted SG) use the functional genome size of major phylogenetic lineages, as a measure of genetic complexity, and show that it has an exponential relationship with the estimated dates of the transitions where these lineages first originated. As such, there exists a linear relationship between the logarithm of genome size (y) and the transition date (x). SG performed regression on a dataset on y vs. x, and

---

*Reprinted from Marzban, C., Viswanathan, R., Yurtsever, U., 2014. Earth before life. Biol. Direct 9, 1. This article was reviewed by Yuri Wolf, Peter Gogarten, and Christoph Adami. When originally published as: Marzban, C., Viswanathan, R., Yurtsever, U., 2014. Earth before life. Biology Direct 9, #1. Reprinted with permission under the terms of the Creative Commons Attribution License.
#CM performed the statistical analysis, with significant feedback and discussion provided by RV and UY.

proposed that the *x*-intercept of the fit (i.e., where genome size is zero) provides an estimate for the age of life.

The work was criticized on many levels, ranging from the manner in which the data was produced, to the way in which the data was analyzed. A fundamental problem is the paucity of data over the first 2 billion years or so of Earth's history, resulting in large uncertainties in functional genome size at specific times. For instance, for prokaryotes, the size of the functional genome is guessed from the smallest present-day prokaryote genome. Exactly when this genome size evolved is a matter of conjecture; although an approximate date can be estimated from molecular clock type evolution rates based on more recent organisms (see e.g., the reviews of Sharov, 2006), rates of increase of functional genome size could have been very different in the distant past. Fitting the data with an extrapolation based on a single, fixed rate of increase could lead to possibly incorrect conclusions. Likewise, the use of only coding regions of the genome as a measure of genome complexity has been pointed out as a potential problem, as noncoding regions could play a regulatory role and the associated complexity is unaccounted for when only coding regions are measured. Thus estimating genome complexity of extinct organisms based on an uncertain estimate of functional genome size of present-day organisms could be doubly flawed.

In addition to all of the above criticisms, there are additional concerns over the statistical analysis in SG. First, and foremost, is the way in which the regression fit is used to extrapolate far beyond the range of *x* values appearing in the data. It is well known that extrapolation can lead to misleading conclusions Perrin, 1904), and so, any conclusions regarding the age of life, based on extrapolation, should be considered with extreme caution. A second aspect of the SG regression fit is that it does not incorporate statistical uncertainty due to sampling variability, e.g., through confidence or prediction intervals. The inclusion of such intervals can lessen the misleading impacts of extrapolation, because they generally widen as one moves away from the mean of the data (Ryan, 1997). When an interval estimate is produced for the *x*-intercept, then the age of life can be estimated to within a range of possible values. Values outside of the interval may be rejected (with some confidence), but all of the values within the interval are possible, and in fact, equally "likely." As such, consideration of interval estimates is important because it can mitigate misleading conclusions.

Another limitation of the SG regression analysis is that it does not systematically account for uncertainty in the dates at which the transitions occurred (i.e., the *x*-values of the data), also known as *measurement errors*. As explained here, measurement errors generally reduce the slope of the regression fit, and consequently increase the value of the *x*-intercept. As such, failure to account for measurement errors leads to overestimates for the age of life. Regression models which systematically take measurement errors into account are called *measurement error models* (Fuller, 1987; Buonaccorsi, 2010).

In this chapter, a simple measurement error model is developed for the SG data, and rudimentary interval estimates are produced for the fit. In such a framework, the age of life is estimated to be within a range of possible values, with the range itself depending on a quantity proportional to the variance of the measurement errors. In other words, an estimate of the age of life is contingent on an estimate of the typical error in the lineage transition dates. An attempt is made to estimate the variance of the measurement errors, and it is shown that the proposed model is consistent with life having formed around 4.5-billion years ago. In short, we find that when the regression analysis involves interval estimates, and incorporates measurement errors, then the data used by SG provide no evidence to support the claim that life must have formed prior to the formation of the Earth.

2 METHOD
2.1 REGRESSION EFFECT

Consider a scatterplot of y vs. x, displaying some amount of association between the two variables (e.g., Fig. 1). It is well known that as the spread of the data increases, the slope of a least-squares fit approaches zero. This effect is known by a variety of names, including *the regression effect* (Bland and Altman, 1994). It is demonstrated in Fig. 1, where the black, red, and blue circles are fictitious data with increasing error in x; i.e., the black circles have less scatter than the red circles, etc. The straight lines are the ordinary least-squares fits to the respective datasets. It can be seen that increasing scatter leads to lower values of the slope.

The mathematics underlying the regression effect is straightforward. It is easy to show

$$\frac{\hat{y}(x) - \bar{y}}{s_y} = r \left(\frac{x - \bar{x}}{s_x} \right), \tag{1}$$

where $\hat{y}(x)$ is the predicted/fitted value, $\bar{x}$ and $\bar{y}$ are the sample mean of x and y, respectively, and s_x, s_y are their sample standard deviations. The quantity r is Pearson's correlation coefficient, and measures the amount of scatter on the scatterplot. As r approaches zero (from either side), the predicted value $\hat{y}(x)$ tends to the sample mean of y. Indeed, this "regression to the mean" is the reason why the least-squares fit is called regression (Galton, 1886). In summary, as the amount of scatter in the scatterplot of y vs. x increases, the least-squares fit converges to a horizontal line with slope zero, and y-intercept equal to $\bar{y}$.

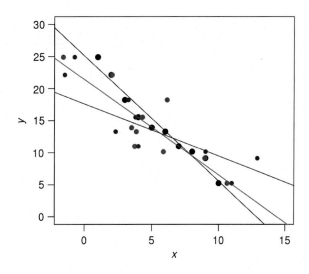

FIG. 1

Demonstration of regression effect. The *black, red, and blue circles* show datasets with increasing errors in x-values. As a result, their ordinary least-squares fits have progressively smaller slopes.

2.2 REGRESSION DILUTION

The aforementioned scatter may be due to errors in x, in y, or both. In the most common form of regression, the predictor x is assumed to be error-free, and only the response y is assumed to be subject to errors. Measurement error models (Fuller, 1987; Buonaccorsi, 2010) are designed to allow for both x and y to be subject to errors. Consequently, as expected from the previous example, measurement errors tend to "flatten" the least-squares line—a phenomenon called *regression dilution* (Fuller, 1987; Buonaccorsi, 2010). Moreover, if the measurement errors can be estimated, then one can undo the dilution.

A simple measurement error model is as follows: Let $(X_i, Y_i), i = 1, \ldots, a$, denote the true, error-free, values of two continuous random variables satisfying the relation

$$Y_i = \alpha^* + \beta^* X_i. \tag{2}$$

The corresponding observed values (x_i, y_i) can then be written as

$$x_i = X_i + \omega_i, \quad y_i = Y_i + \epsilon_i, \tag{3}$$

where ω_i and ϵ_i are the measurement error in X and the error in Y, respectively. For simplicity, it is assumed that both errors are normally distributed with zero mean, and variances given by σ_w^2 and σ_ϵ^2. That is, $\omega_i \sim N(0, \sigma_w^2)$ and $\epsilon_i \sim N(0, \sigma_\epsilon^2)$. In the *functional model* the X is assumed fixed (nonrandom), but in the *structural model* X is assumed to be a random variable (Fuller, 1987; Buonaccorsi, 2010). In other words, in the former, the a values of X_i are assumed to be fixed quantities, while in the latter they are considered to be a random sample taken from a population (or a distribution). The latter is adopted here because it is more appropriate for the problem at hand, and for simplicity we assume $X_i \sim N(\mu, \sigma_b^2)$. The subscripts "$b$" and "$w$" are motivated by "between-group" and "within-group" variances—language common to the analysis of variance formulation of regression (Montgomery, 2009).

If one mistakenly ignores measurement errors (in X), and instead performs regression on (x_i, y_i), i.e.,

$$y_i = \alpha + \beta x_i + \epsilon_i, \tag{4}$$

then it can be shown that the least-squares estimate of the regression slope and y-intercept are given by (Fuller, 1987; Buonaccorsi, 2010; Draper and Smith, 1998):

$$\beta = \beta^* / \lambda, \quad \alpha = \bar{y} - (\beta^*/\lambda)\bar{x} \tag{5}$$

with

$$\lambda = 1 + \frac{\sigma_w^2}{\sigma_b^2}. \tag{6}$$

Given that $\lambda > 1$, it follows that $\beta < \beta^*$, i.e., the slope is "diluted" relative to the slope that would have been obtained if measurement errors were zero. Therefore, as mentioned previously, measurement errors tend to flatten the least-squares fit, and thereby lead to an overestimate of the x-intercept. For the SG data, then, measurement errors lead to an overestimate for the age of life.

Eq. (5) implies that one can correct this overestimation by simply multiplying the observed regression coefficient β by λ (Fuller, 1987; Buonaccorsi, 2010; Draper and Smith, 1998). In other words, the quantity $(\beta \lambda)$ is an estimator of β^*. Similarly, the least-squares estimate of α^* is $(\bar{y} - (\beta \lambda)\bar{x})$. Note that in a measurement error model of the SG data, the estimate for the age of life is given by the corrected x-intercept $\bar{x} - \bar{y}/(\beta \lambda)$. In order to make any of these corrections, however, one must estimate λ.

Frost and Thompson (2000) discuss six methods for estimating λ and the corresponding variance. One of the simpler methods examined there identifies λ as the inverse of the intraclass correlation coefficient (also known as the reliability ratio). One advantage of that estimator is that its variance has a simple expression:

$$\frac{(\lambda^2 - 1)^2}{a}. \tag{7}$$

In the following not only an attempt is made to estimate λ itself, we also consider the "inverse problem" of finding the range of λ values which lead to x-intercepts consistent with 4.5 billion years as the age of life.

It is not necessary to find a specific λ value which leads to an x-intercept of 4.5 billion years. A regression fit whose confidence or prediction interval includes an x-intercept of 4.5 billion years is sufficient, in the sense that it does not contradict the null hypothesis that life began after the formation of the Earth. In order to construct such an interval, one must compute the variance for the corrected regression slope, a quantity which has been derived in (Frost and Thompson, 2000):

$$V[(\beta\,\lambda)] = \lambda^2 V[\beta] + \frac{1}{a}(\beta^2 + V[\beta])(\lambda^2 - 1)^2, \tag{8}$$

where Eq. (7) has been used.

There is an ambiguity in whether the appropriate interval for this problem is a confidence interval or a prediction interval (Ryan, 1997). The former is designed to cover the true conditional mean of y, given x, a certain percentage of time, e.g., 95%. The latter is designed to cover a single prediction of y, a certain percentage of time. By construction, the prediction interval is wider than the confidence interval. An argument in favor of using a prediction interval is that the x-intercept corresponds to a single prediction of y. One can also argue that the appropriate interval is a confidence interval, because the x-intercept is technically a population parameter. The choice between the two intervals is of secondary importance. What is more important than the choice of the two intervals is that *some* interval must be considered. Here, a prediction interval is used; using confidence intervals leads to qualitatively similar conclusions.

The construction of prediction intervals in measurement error models is itself a complex issue and is considered by Buonaccorsi (1995). One relatively simple 95% prediction interval is given by $\hat{y}(x) \pm 1.96\sigma_{pe}$, where σ_{pe}^2 is the variance of the prediction error, given by

$$\sigma_{pe}^2 = \sigma_\epsilon^2 + \frac{\sigma_\epsilon^2}{a} + (X - \overline{X})^2 V[\beta\lambda] + [(\beta\,\lambda)^2 + V[\beta\,\lambda]]\,\sigma_b^2, \tag{9}$$

where $V[(\beta\,\lambda)]$ is given by Eq. 7, and σ_ϵ^2 is estimated by the variance of the residuals. This is the expression derived in Buonaccorsi (1995) for the special case where the value of X at which the prediction is made is a known (nonrandom) quantity. The first three terms on the right-hand side of Eq. (9) are the variance of the prediction error in the error-free case (Draper and Smith, 1998); the last term is the result of measurement errors.

2.3 ESTIMATING MEASUREMENT ERRORS

One may wonder what is a typical value of λ for the data at hand. Given Eq. (6), σ_b and σ_w must be estimated. To that end, consider a situation where each X_i is measured n times. Denoting the resulting data as x_{ij}, $i = 1, \ldots, a, j = 1, \ldots, n$, it is known that unbiased estimates of σ_b^2 and σ_w^2 are

$$\left(\frac{s_b^2}{n} - \frac{s_w^2}{n}\right), \quad s_w^2, \tag{10}$$

respectively, with s_b^2 and s_w^2 defined as

$$s_b^2 = \frac{n}{a-1}\sum_i^a (\overline{x_{i.}} - \overline{x_{..}})^2, \quad s_w^2 = \frac{1}{a(n-1)}\sum_{i,j}^{a,n}(x_{ij} - \overline{x_{i.}})^2, \tag{11}$$

where an overline denotes averaging over the index with a dot (Montgomery, 2009). For large n, the quantity s_b^2/n converges to the sample variance of the X_i, i.e., $s_X^2 = \frac{1}{a-1}\sum_i^a (X_i - \overline{X})^2$, which in turn can be estimated with the sample variance of the x_i. For the data at hand $s_b^2/n \sim 1.86$ (billion years)^2. In the large-n limit, the term s_w^2/n converges to zero, because in that limit s_w^2 itself converges to the constant σ_w^2. Therefore, asymptotically, $\sigma_b \sim \sqrt{1.86} \sim 1.36$ billion years.

The within-group standard deviation s_w reflects the spread in values or uncertainty of the dates of appearance of the respective functional genomes (e.g., prokaryote, eukaryote, worms, fish, mammals). While the statistical analysis presented here assumes that the a measurements all have common variance (i.e., homoscedastic), in reality the uncertainty in the time of appearance of a functional genome increases from present to past. Thus the largest errors or uncertainties are found in the oldest functional genome considered. As an example of dating uncertainty, while the earliest mammals are believed to have arisen about 225 million years ago, based on early fossils (Rose, 2006), molecular clock studies based on genomes place mammalian origins around 100 million years ago (Dawkins, 2005). There is thus an uncertainty of the order of 100 million years or more in setting the time of the mammalian functional genome. The origin of eukaryotes has been identified to lie in the time interval between 2.3 billion and 1.8 billion years ago, thus with an uncertainty of 250 million years around the mean estimate of 2.05 billion years ago (Seilacher et al., 1998). Early fossil evidence for prokaryotes in lava beds has been dated to a time around 3.5 billion years ago (Furnes et al., 2004); however, it is unclear exactly when the functional genome size reached the present-day minimum value of around 5×10^5; the uncertainty in this time value could easily be of the order of 1 billion years.

Within-group standard deviations in the dates at which respective functional genome sizes were attained, therefore, have an order of magnitude spread in range of values, from 100 million to 1000 million years, with most standard deviation values being of the order of a few hundred million years. In a homoscedastic model of the type assumed in the present article, we will use as a rough (weighted) estimate a value of $s_w \sim 500$ million years.

Therefore, with $\sigma_b \sim 1.36$ billion and $\sigma_w \sim 0.5$ billion, we have $\lambda \sim 1.14$. For uncertainties around 100 million years, λ is around 1.00, and it is around 1.54 if uncertainty is around 1 billion years.

3 RESULTS AND DISCUSSION

The formula for the prediction interval shown above depends on the quantity λ. In the previous subsection we arrived at a rough estimate for that quantity. Now, we examine the range of λ values which lead to conclusions consistent with the hypothesis that life did not begin prior to the formation of the Earth.

Fig. 2 shows all of the results. The black line shows the ordinary least-squares fit to the data. It is the x-intercept of this line, i.e., about 9.5 billion years, which led SG to conclude that life must have begun prior to the formation of the Earth (i.e., about 4.5 billion years ago). The region between the black,

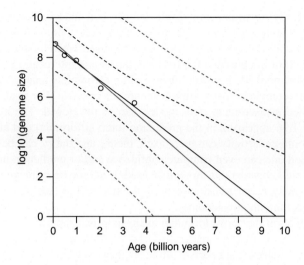

FIG. 2

Regression dilution. The *circles* denote the data from SG, and the *black, solid line* is the ordinary least-squares fit to that data. The x-intercept of this line (about 9.5 billion years), is the estimate for the age of life according to SG. The region between the *black, dashed curves* is the 95% prediction interval for the fit. The *red, solid line* is the least-squares fit according to a measurement error model with $\lambda = 1.14$; the *red, dashed curves* outline the 95% prediction interval/region. Note that this region includes 4.5 billion years (i.e., the Earth's age).

dashed curves is the 95% prediction interval for the ordinary least-squares fit. According to this prediction interval (without taking measurement errors into account) life may have originated as early as 7 billion years ago.

The analogous results based on the above measurement error model are shown in red. The value of λ for this fit is 1.14—the value estimated in the previous section. Note that the resulting prediction interval includes 4.5 billion. In other words, if λ is about 1.14, then the results of the analysis do not reject the hypothesis that life began after the formation of the Earth. Even a λ as small as 1.1 leads to results (not shown here) consistent with 4.5 billion years as the age of life.

Although, a proper interpretation of prediction intervals correctly draws all focus away from the "center" of the interval, it is possible to arrange for the corrected fit itself to have an x-intercept of 4.5. The result is not shown here, but the corresponding value of λ is about 2.7. Values of λ in the 1.1–2.7 range are not uncommon; Frost and Thompson (2000) even consider λ values as large as 5. More importantly, that range includes the λ values estimated in the previous section.

Based on Fig. 2, note that the genome size at Age $= 0$ may be as low as 10^5 (i.e., the lower limit of the prediction interval at Age $= 0$); one may be inclined to conclude that such a low value is unlikely. Also note that Age $= 4.5$ is only at the lowest limit of the possible x-intercepts; one may then be inclined to conclude that Age $= 8.5$ (the x-intercept of the red line itself) is the more likely estimate for the age of life. Although both premises are correct, the conclusions do not follow. The fallacy in both arguments is to attribute a likelihood to different regions within the prediction intervals. The correct interpretation of prediction (and confidence) intervals does not allow that type of interpretation

(Ryan, 1997). Specifically, regarding Fig. 2, all one can conclude is that about 95% of prediction intervals will cover genome size, at a given value of Age. More loosely stated, one can be "confident" that the true fit lies *somewhere* within the region between the red, dashed curves; nothing can be said about which of the possible fits (or their slopes and intercepts) are more, or less, likely.

The original conclusion of SG is a consequence of an incomplete analysis. Although the analysis presented here is more complete, many improvements are possible. For instance, nonlinear fits can made, and more refined measurement error models can be developed. The inference component of our analysis (i.e., the 1.96 appearing in the prediction interval formula) can also be improved upon. The assumption of homoscedasticity can be relaxed, the σ_w and the σ_b can be estimated without the large-n assumption, and one can even compute confidence and/or prediction intervals for the x-intercept itself. Lastly, datasets considerably larger than used by SG can be employed as genome size data is readily available for many major transitions along the tree of life (see, e.g., http://itol.embl.de/itol.cgiandhttp://www.ncbi.nlm.nih.gov/genome).

In short, many aspects of the above formulation are simplistic, approximate, or even controversial. As such, they offer avenues of further research. These limitations have not been of concern here because the main goal of the chapter has been to introduce measurement error models and to highlight the importance of producing interval estimates of the fit. The details of the measurement error model, the manner in which the interval estimates are generated, or whether the appropriate interval is a confidence or prediction interval, are all of secondary importance because they affect the conclusions only in degree, not in kind.

4 CONCLUSIONS

A naïve regression model relating genome size to the age of life suggests that life may have formed prior to the Earth's formation. Here we have shown that measurement errors lead to biased (i.e., over-) estimates for the age of life, and that the bias can be corrected/removed. Additionally, a more refined regression analysis is performed which (1) takes into account measurement errors, and (2) generates interval estimates of the fit. The analysis depends on a parameter, λ, which is related to the variance of the measurement errors. We find that a wide range of plausible λ values lead to intervals that allow for life to have been formed more recently than 4.5 billion years ago, In short, the data analyzed by SG are consistent with the hypothesis that life may have formed after the Earth's formation.

ACKNOWLEDGMENTS

We thank A. Sharov and R. Gordon for valuable discussions. We are grateful to Sining Wang for identifying sources of data for further studies. This work was done under no funding.

REFERENCES

Bland, J.M., Altman, D.G., 1994. Statistic notes: regression towards the mean. Br. Med. J. 308, 1499.
Buonaccorsi, J.P., 1995. Prediction in the presence of measurement error: general discussion and an example predicting defoliation. Biometrics 51 (4), 1562–1569.

Buonaccorsi, J.P., 2010. Measurement Error: Models, Methods, and Applications. Chapman & Hall/CRC Interdisciplinary Statistics, Boca Raton (FL), London, New York.

Dawkins, R., 2005. The Ancestor's Tale. Mariner Books, Boston, New York.

Draper, N.R., Smith, H., 1998. Applied Regression Analysis, third ed. John Wiley and Sons, Inc, New York.

Frost, C., Thompson, S.G., 2000. Correcting for regression dilution bias: comparison of methods for a single predictor variable. J. R. Stat. Soc. A. Stat. Soc. 163, 173–189.

Fuller, W.A., 1987. Measurement Error Models. John Wiley, New York.

Furnes, H., Banerjee, N.R., Muehlenbachs, K., Staudigel, H., de Wit, M., 2004. Early life recorded in archean pillow lavas. Science 304, 578581.

Galton, F., 1886. Regression towards mediocrity in hereditary stature. J. Anthropol. Inst. 15, 246–263.

Montgomery, D.C., 2009. Design and Analysis of Experiments, seventh ed. John Wiley & Sons, New York.

Perrin, E., 1904. On some dangers of extrapolation. Biometrika 3, 99–103.

Rose, K., 2006. The Beginning of the Age of Mammals. The Johns Hopkins University Press, Baltimore, MD.

Ryan, T.P., 1997. Modern Regression Methods. John Wiley & Sons, New York.

Seilacher, A., Bose, P.K., Pfluger, F., 1998. Triploblastic animals more than 1 billion years ago: trace fossil evidence from India. Science 282, 80–83.

Sharov, A.A., 2006. Genome increase as a clock for the origin and evolution of life. Biol. Direct 1, 17–26.

Sharov, A.A., Gordon, R., 2013. Life before earth. http://arxiv.org/abs/1304.3381.

THE DRAKE EQUATION AS A FUNCTION OF SPECTRAL TYPE AND TIME

Jacob Haqq-Misra*,†, Ravi K. Kopparapu*,†,‡,§
Blue Marble Space Institute of Science, Seattle, WA, United States
†*NASA Astrobiology Institute's Virtual Planetary Laboratory, Seattle, WA, United States*
‡*NASA Goddard Space Flight Center, Greenbelt, MD, United States*
§*University of Maryland, College Park, MD, United States*

CHAPTER OUTLINE

1 INTRODUCTION

The Drake equation is a probabilistic expression for the number of communicative civilizations in the galaxy (Drake, 1965). This equation typically takes the form

$$N = R_* \cdot f_p \cdot n_e \cdot f_l \cdot f_i \cdot f_c \cdot L, \tag{1}$$

where R_* is the rate of star formation, f_p is the fraction of stars with planets, n_e is the number of habitable planets per system, f_l is the fraction of habitable planets that develop life, f_i is the fraction of inhabited planets that develop intelligence, f_c is the fraction of planets with intelligent life that develop

Habitability of the Universe Before Earth, editors: Richard Gordon & Alexei Sharov, Volume 1 in the series:
Astrobiology: Exploring Life on Earth and Beyond, series editors: Pabulo Henrique Rampelotto,
Joseph Seckbach & Richard Gordon. ISSN 2468-6352. https://doi.org/10.1016/B978-0-12-811940-2.00013-7

technology capable of interstellar communication, and L is the average communicative lifetime of technological civilizations (c.f. Ćirković and Vukotić, 2008). The original formulation of this equation was developed by Frank Drake during a 1961 conference at the National Radio Astronomy Observatory in Green Bank, West Virginia, with the goal of identifying the key factors needed for a planet to develop life and technology that could communicate with Earth. Since this discussion at Green Bank, the Drake equation has seen wide use among scientists and educators, with some studies proposing methods for constraining the Drake equation with observable quantities (Frank and Sullivan, 2016) or statistical methods (Maccone, 2010; Glade et al., 2012) in order to guide future search for extraterrestrial intelligence (SETI) surveys. Others have extended the traditional representation of the Drake equation to include time-dependent quantities, in an effort to understand the distribution of civilizations across the history of the galaxy (Ćirković, 2004; Glade et al., 2012; Prantzos, 2013).

In this chapter, we discuss the functional dependence of the Drake equation parameters on the spectral type of the host star and the time since the Milky Way galaxy formed. We discuss observations, planetary formation models, and climate models that provide constraints on the values of R_*, f_p, and n_e, as well as looser bounds on f_l and f_i. We also examine common assumptions for estimating the parameter L and develop two approaches for calculating the maximum value, L_{max}, as a function of spectral type. We use these constraints to develop an expression for $N(s,t)$, the number of communicative civilizations in the galaxy as a function of spectral type s and time since galaxy formation t.

We also discuss and critique the possibility that N has distinct phases in time, such that certain events are impossible or unlikely until a particular galactic era. Ćirković and Vukotić (2008), following Annis (1999), describe this behavior as a "phase transition," suggesting that phenomena such as supernovae, gamma ray bursts, or other physical phenomena could have been more prevalent during the early phase of the galaxy compared to today. The idea of a phase transition indicates that astrophysical factors may have inhibited the emergence of communicative civilizations until relatively recently in galactic history. We argue that phase transitions for N are most likely to emerge from the dependence of L_{max} on stellar spectral type.

2 CONSTRAINTS FROM OBSERVATIONS

In this section, we discuss observational constraints on the parameters R_*, f_p, and n_e along with their dependence on spectral type and time. We consider F-, G-, K-, and M-dwarf stars in the majority of our analysis, so that $s \in \{F, G, K, M\}$, as these present the most likely candidates for supporting intelligent life. However, casting the Drake equation as a function of spectral type also allows for consideration of O-, B-, and A-dwarf stars (which will explicitly show why such stars are deemed unlikely for advanced life) and even planets around evolved stars.

2.1 RATE OF STAR FORMATION

Common analyses of the Drake equation cite a value of 7–10 stars/year as an estimate of R_*. Because R_* is difficult to measure directly, one way of calculating this rate is to write $R_* = n_*/t_0$, where n_* is the current number of stars in the galaxy and $t_0 \approx 13$ Gyr is the current age of the galaxy. However, this uniform approach assumes that the rate of star formation has remained constant over the history of the galaxy (Ćirković, 2004), proportional to the stellar population today. As an improvement to previous estimates that offer a single value of R_*, we calculate values of R_* that depend on spectral type.

Table 1 Summary of Drake Equation Parameters Discussed in This Chapter

Star	R_* (star/year)	f_p	n_e	$f_l \cdot f_i \cdot f_c$	L_* (Gyr)	L_{max}^{EET} (Gyr)	L_{max}^{PET} (Gyr)
F-dwarf	1	1	0.2	1	4	0	2
G-dwarf	1	1	0.2	1	10	6	5
K-dwarf	2.2	1	0.2	1	30	26	15
M-dwarf	10.5	1	0.2	1	100	96	50

Analysis of *Spitzer* data (Robitaille and Whitney, 2010) finds that the stellar mass formation rate is 0.68 to 1.45 M_{sun}/year (where M_{sun} is the mass of the sun, a G-dwarf star). To convert this number into a value for $R_*(s)$, we separate the spectral types into mass ranges based on the initial mass function categories of Kroupa (2001). We assume an initial mass of 1 M_{sun} for F- and G-dwarfs, 0.5 M_{sun} for K-dwarfs, and 0.1 M_{sun} for M-dwarfs. We then divide the observed stellar mass formation rate by this initial mass function to give $R_*(s)$, with units of number of stars formed per year. F- and G-dwarfs have the lowest rate of $R_*(F) \sim R_*(G) \sim 1$ star/year. K-dwarfs have a rate of $R_*(K) \sim 1.3$–3 star/year (we use $R_*(K) = 2.2$ star/year in our analysis), while M-dwarfs have the highest rate of $R_*(M) \sim 7$–14 star/year (we use $R_*(M) = 10.5$ star/year in our analysis). We summarize these values for R_* in Table 1.

2.2 FRACTION OF STARS WITH PLANETS

Ground- and space-based observations all suggest that terrestrial planets are commonplace around all stars. Microlensing observations indicate that typical stellar systems contain one or more bounded planets (Cassan et al., 2012). Furthermore, analysis of *Kepler* data indicates that there are ∼2 planets per cool star ($T_{eff} < 4000$ K) with periods <150 days (Morton and Swift, 2014). This orbital period encompasses the habitable zone (HZ) of cool stars. Therefore, we can safely assume that $f_p = 1$ for all spectral types F, G, K, and M (c.f. Frank and Sullivan, 2016).

Lineweaver et al. (2004) developed a model for the evolution of the galaxy and found that rocky planets could not have formed until about $t = 4$ Gyr, due to the lower metallicity of stars at the time. If planet formation were lower during the early history of the galaxy, then this would suggest a phase transition for the emergence of planets in the galaxy. However, these results of Lineweaver et al. (2004) are inconsistent with observations of planets orbiting stars nearly as old as the galaxy itself. The Kepler-444 system is one of the oldest known to host exoplanets, with five planets smaller than Earth orbiting this 11.2 Gyr old K-dwarf. At least for the long-lived K- and M-dwarf stars, $f_p = 1$ for $t > 1$ Gyr, which is close to the age of the galaxy. Even if F- and G-dwarf stars develop planets slightly later than this due to metallicity limitations, we remain skeptical that differences in metallicity over the history of the galaxy are responsible for a phase transition in the Drake equation. We therefore maintain the expression $f_p(s, t) = 1$.

2.3 NUMBER OF HABITABLE PLANETS PER SYSTEM

A common approach to estimating n_e is to focus on the related quantity "eta-Earth," η_{earth}, which is defined as the fraction of stars with at least one terrestrial mass/size planet within the HZ. The value of η_{earth} can be estimated from *Kepler* data (Dressing and Charbonneau, 2013, 2015; Kopparapu, 2013; Gaidos, 2013; Petigura et al., 2013; Morton and Swift, 2014; Foreman-Mackey et al., 2014), with current estimates varying between 22% (Petigura et al., 2013) and 2% (Foreman-Mackey et al., 2014) for G- and K-dwarfs, and 20% for M-dwarf stars (Dressing and Charbonneau, 2015). In reality, n_e and η_{earth} are describing the same quantity: the average number of habitable planets (which we will assume means terrestrial planets located within the HZ) per system is the same as the fraction of stars with at least one terrestrial planet in the HZ. This means that n_e is in the range of 0.02–0.22 for G- and K-dwarfs, and 0.20 for M-dwarfs. Following Frank and Sullivan (2016), we will assume a constant value $n_e = 0.2$ for all spectral types.

Lineweaver et al. (2004) also argue that habitability in the early era of the galaxy would have been complicated by the intense radiation from supernova activity near the galactic core, effectively sterilizing any planets that could have formed until about 4 Gyr. However, other studies (Morrison and Gowanlock, 2015) suggest that the galactic center may be ideal locations to search for intelligent life, even with higher supernova rates. Furthermore, intense supernova activity near the galactic core does not preclude habitability at farther distances from the galactic center. Other simulations of galactic habitability suggest that the edge of a galaxy's stellar disk provides an optimal location for habitable planets during the early phase of the galaxy, while this region migrates inward at later times (Forgan et al., 2017). We therefore conclude that sterilization by supernovae at the early era of the galaxy is unlikely to result in a phase transition for N, so we maintain the expression $n_e(s, t) = 0.2$.

We note that gamma ray bursts have also been suggested as antithetical to habitability, particularly the suggestion that the early phase of the universe may have shown a higher rate of gamma ray bursts than observed today (Piran and Jimenez, 2014). However, other studies have argued that the radiation environment of the early universe would not have been any more intense than today, particularly for galaxies like the Milky Way, suggesting that Earth-like habitability cannot be precluded even for the early phase of the galaxy (Li and Zhang, 2015; Gowanlock, 2016).

3 CONSTRAINTS FROM THEORY

In this section, we discuss theoretical constraints on the parameters f_l, f_i, and f_c along with their dependence on spectral type and time. Other analyses (e.g., Frank and Sullivan, 2016) assume that f_l and f_i are wholly unconstrained; however, we present insights from theoretical models of planetary habitability that can help to place bounds on these two terms. We also speculate on theoretical limits for f_c, although f_c is much more formidable to constrain.

3.1 FRACTION OF HABITABLE PLANETS THAT DEVELOP LIFE

The value of f_l is difficult to estimate observationally, although remote characterization of exoplanet environments with spectral signatures through missions such as the James Webb Space Telescope (JWST) could provide one way to eventually constrain f_l. Spectral evidence of life on an exoplanet could be determined by observations of chemical disequilibrium, such as the simultaneous detection of O_2 and CH_4, which is maintained by biology on Earth (Des Marais et al., 2002). Some calculations with climate and photochemistry models suggest abiotic sources for O_2 and O_3 (Segura et al., 2007; Domagal-Goldman et al., 2014; Wordsworth and Pierrehumbert, 2014), although such a "false positive" spectral signature might be discernable from inhabited planets by the detection of species such as CO and O_4 (Schwieterman et al., 2016). Observations by *JWST* and future direct imaging missions will begin to provide an estimate for f_l by examining the statistics of terrestrial exoplanets that show possible spectral biosignatures.

Origin of life research, both laboratory experiments and computational modeling, provides another avenue for further constraining f_l. This may not directly tell us about processes occurring on other planets, but origin of life research seeks to elucidate the range of possibilities that could give rise to biological phenomenon. Perhaps life can form in many possible ways other than how it occurred on Earth, and origin of life researches in fields such as experimental evolution and synthetic biology (e.g., Ray, 1993; Benner and Sismour, 2005; Luisi, 2016; Kacar, 2016) are potential ways of examining different pathways toward making a habitable world actually inhabited. One approach by Scharf and Cronin (2016) suggests an analytic framework, inspired by the form of the Drake equation for quantifying the probability of biogenesis by considering the availability of chemical and environmental building blocks for a given set of parameters. This particular method does not presently provide a reliable estimate of f_l for planets in general, although it represents an important step toward reaching this goal.

Computational models of planetary habitability can also provide insight into f_l. Several authors have suggested that planets orbiting late M-dwarf stars may have experienced runaway greenhouse effects during the star's extended premain sequence phase, rendering such planets dry and uninhabitable (Ramirez and Kaltenegger, 2014; Luger and Barnes, 2015; Tian and Ida, 2015). This implies that some planets in the HZ of M-dwarf stars may be unable to develop life unless they begin with large water inventories or sufficient water is later delivered by impacts. Other calculations suggest that extreme X-ray and extreme ultraviolet radiation from active M-dwarf stars could induce hydrogen escape and further contribute to water loss on orbiting exoplanets (Airapetian et al., 2017). We cannot rule out the possibility that some of these planets will regenerate a habitable atmosphere during the star's main sequence lifetime, but this mechanism serves as an upper limit on the fraction of planets in the HZ of M-dwarf stars that could actually develop life. Although we cannot assign a number to f_l based on these results, we can at least hypothesize that $f_l(M) < f_l(\{F,G,K\})$. Even if M-dwarfs can host habitable planets, they may be less likely to develop life than other spectral types due to their limited propensity to retain water.

3.2 FRACTION OF LIFE-BEARING PLANETS THAT DEVELOP INTELLIGENCE

Assigning a particular value to f_i can be somewhat controversial, and this value is generally considered to be unconstrained. However, insights from climate models can also assist with linking physical factors to this term.

Recent climate calculations suggest that some planets in the habitable zone may be prone to "limit cycles" with punctuated periods of warmth lasting about 10 Myr, followed by extended periods of global glaciation lasting about 100 Myr (Kadoya and Tajika, 2014, 2015; Menou, 2015; Batalha et al., 2016; Haqq-Misra et al., 2016; Abbot, 2016). Shorter duration changes in Earth's ice coverage on timescales ranging from thousand to hundreds of thousands of years may have accelerated the pace of evolution by opening up ecological niche space for new species to occupy. Such ice age cycles are driven by long-term variations in Earth's orbit (known as "Milankovitch cycles") or other Earth system processes (Haqq-Misra, 2014). By contrast, limit cycling occurs due to changes in the rate of weathering in the carbonate-silicate cycle, on the scale of tens to hundreds of millions of years. With such prolonged 100 Myr periods of glaciation, the evolution of complex (animal-like) life could be difficult, which could therefore preclude the development of any forms of intelligence. The presence of limit cycles is a function of spectral type, due to the wavelength dependence of ice-albedo feedback, and also depends on the volcanic outgassing rate of the particular planet.

Planets of F-dwarf stars are more prone to limit cycling, but both F- and G stars show an expansion of the limit cycle region of the HZ when volcanic outgassing rates are lower than those on Earth today. For example, a volcanic outgassing rate a tenth that of Earth today would reduce the HZ of F-dwarfs by about 75% and G-dwarfs by about 50% (Haqq-Misra et al., 2016). By contrast, planets of K-type and M-type stars do not experience limit cycles at all due to the reduced effects of ice-albedo feedback. We cannot predict the presence of limit cycles based on spectral type alone, but we can write the relationship $f_i(F) < f_i(G) < f_i(\{K,M\})$. Additional modeling studies of volcanic outgassing and seafloor weathering rates expected for terrestrial exoplanets can place further constraints upon f_i, although observationally confirming such predictions will remain a daunting task.

Limit cycles are only one possible mechanism that could restrict the development of complex, and therefore intelligent, life on a planet located in the HZ. However, limit cycles provide a bound on f_i that depends upon the net outgassing rate of the planet. This formulation links the previously unconstrained value of f_i to properties of the planet itself. Other theoretical approaches may also provide insight on how the physical environment of a planet can constrain our expectations of f_i.

3.3 FRACTION OF INTELLIGENCE-BEARING PLANETS THAT BECOME COMMUNICATIVE

Few theoretical studies have attempted to estimate values for f_c, as this parameter remains extremely difficult to constrain. Anthropological research provides a potential way for approaching a value for f_c. The human species is incredibly diverse, and many existing indigenous people groups live in small rural communities where they depend on only a fraction of technology that the modern world requires. Although developed nations today are able to engage in spacefaring activities, it remains unclear as to how common such desires would be among intelligent beings. Until the search for extraterrestrial life succeeds, studying our fellow human beings may be the best way toward understanding the factors that drive us toward communicative technology (e.g., Finney and Jones, 1986; DeVito, 2011; Denning, 2011). It remains possible that f_c is dependent upon spectral type, but any such attribution remains pure speculation this point. We therefore leave f_c as an unconstrained parameter.

4 RETHINKING THE LONGEVITY PARAMETER

The average lifetime of a communicative civilization is perhaps the most controversial value of the Drake equation. Drake's own estimate of his equation is $N = L = 10,000$ civilizations, which suggests that the other factors multiply approximately to one (Drake, 2011). Historically, Carl Sagan and others (Sagan, 1973) have argued that L is the determining factor in the prevalence of communicative civilizations; however, much of this early speculation of the value of L occurred in a geopolitical environment locked in a Cold War and facing an imminent nuclear catastrophe. Pessimism over humanity's ability to survive its own technology may have subsided in recent years, although humanity still holds the capacity to destroy itself several times over with nuclear weapons. However, we also note that L represents the length of time that a civilization remains communicative, not necessarily its timescale for existence. A long-lived civilization could potentially become undetectable through the development of new technology, or perhaps even the abandonment of spacefaring technology altogether (Haqq-Misra et al., 2013).

Grinspoon (2004) provides an alternative and more optimistic interpretation of L, suggesting that long-lived civilizations will necessarily overcome any developmental challenges and become "immortals" in the sense that they are unconstrained by existential threats. Under this interpretation, Grinspoon (2004) suggests that $L = f_{IC} \cdot T$, where f_{IC} is the fraction of intelligent civilizations that become immortal and T is the length of time that this process has been occurring. The value of T is likely to be a fraction that may approach the age of the universe, although it is unclear how to empirically or theoretically resolve the value of f_{IC}.

One feature that we consider a significant omission from the Drake equation is the expected main sequence lifetime of the host star. G-dwarf stars have a typical main sequence lifetime of 10 Gyr, while F-dwarf stars evolve faster with a typical main sequence lifetime of about 4 billion years. K-dwarf stars are longer lived, with a main sequence lifetime of about 20 Gyr, while M-dwarf stars can live up to 50–100 Gyr or longer. The habitable lifetime of a planet depends on the evolutionary trajectory of its host star, which can be calculated with computational models. We therefore suggest a maximum value, L_{max}, which depends upon the evolutionary history of the star itself, as the expected lifetime of a communicative civilization cannot be any longer than the habitable lifetime of its planet.

Once a star reaches the end of its main sequence lifetime and transitions into its giant phase, a space-faring civilization could potentially survive by migrating outward to farther regions of the stellar system (Finney and Jones, 1986). We set aside this possibility in our analysis below, limiting our consideration to main sequence F-, G-, and K-dwarfs that are the target of many exoplanet and SETI surveys. However, we acknowledge that any civilization able to survive into the postmain sequence phase of its host star will have a larger value of L_{max} than we calculate, possibly approaching the age of the universe as with the case of Grinspoon's (2004) "immortal" civilizations.

4.1 EQUAL EVOLUTIONARY TIME

The first representation of L_{max} we consider assumes that the maximum communicative lifetime is simply the difference between the host star's main sequence lifetime and the time required for the prerequisite evolutionary steps. We also note that we should expect $L_{max} = 0$ if evolutionary timescales are shorter than the host star's main sequence lifetime. Under these assumptions, we can therefore write

$$L_{max}(s, t) = \begin{cases} L_*(s) - t_{evo}, & \text{for } t > t_{evo} \\ 0 & \text{for } t < t_{evo} \end{cases}, \tag{2}$$

where L_* is the main sequence lifetime of the host star and t_{evo} is the average time for the evolution of intelligent and communicative life. This simplifies a formerly unconstrained parameter into one that depends upon stellar evolutionary histories. We refer to the assumptions of Eq. (2) as the *equal evolutionary time* (EET) hypothesis. EET implies that t_{evo} is independent of spectral type and thus approximately constant for all environments.

EET defines L_{max} in terms of the known parameter L_*, which we can estimate from observations and models. We are left with the unknown parameter t_{evo}, which is often estimated as 4 billion years (e.g., Lineweaver et al., 2004) with Earth as our only example of communicative intelligent life. The assumption of EET with $t_{evo} = 4$ Gyr, in particular, implies that certain spectral types will have longer-lived communicative civilizations than others. F-dwarfs have relatively short stellar lifetimes of about 4 Gyr, which by Eq. (4) gives us $L_{max}(F) = 0$. G-dwarfs are longer lived with a main sequence lifetime of about 10 Gyr, which gives a positive value of $L_{max}(G) = 6$ Gyr. Similarly, K-dwarfs with a 30 Gyr lifetime have $L_{max}(K) = 26$ Gyr, while M-dwarfs with a 100 Gyr lifetime have $L_{max}(M) = 96$. The logic of EET takes the timescale for the evolution of communicative life on Earth to be a static factor that occurs on average after t_{evo} years, regardless of spectral type. These assumptions imply that M-dwarfs are likely to host the most long-lived civilizations because L_{max} increases for later spectral types.

Various forms of EET are regularly invoked in astrobiology as a way of coping with our lack of information on evolutionary timescales in other stellar environments. However, when we include a dependence on spectral type in our version of EET, as described by Eq. (2), the overwhelming preference toward later spectral types raises questions of the suitability of a constant evolutionary timescale t_{evo}. We next consider an alternative formulation of L_{max} that may be better suited to represent dependences on spectral type and time.

4.2 PROPORTIONAL EVOLUTIONARY TIME

Instead of assuming a constant evolutionary timescale, we can improve upon our expression for L_{max} by making t_{evo} dependent upon spectral type. Rather than presuming that the 4 Gyr evolutionary timescale of Earth is typical, we note that humans appeared on Earth roughly halfway through the Sun's main sequence lifetime. As an alternative EET, we suggest that the relative timing of intelligent communicative life is proportional to the lifetime of the star. If we assume that communicative civilizations typically tend to arise approximately halfway through the lifetime of their host star, then $t_{evo} = L_*/2$. As a replacement to Eq. (2), we now write the expression

$$L_{max}(s, t) = \begin{cases} \dfrac{L_*(s)}{2}, & \text{for } t > t_{evo}, \\ 0 & \text{for } t < t_{evo} \end{cases} \tag{3}$$

which depends only on spectral type. We refer to the assumptions of Eq. (3) as the *proportional evolutionary time* (PET) hypothesis. PET implies that t_{evo} depends on the main sequence lifetime of the host star and suggests that the emergence of communicative civilizations may occur at different temporal eras of the universe for each spectral type. PET suggests that F- and G-dwarfs, which are shorter lived than M-dwarfs, are more likely to host communicative civilizations today, at the present era of the universe. Eq. (3) gives $L_{max}(F) = 2$ Gyr and $L_{max}(G) = 5$ Gyr, which suggests that communicative civilizations could have already arisen around these spectral types in the history of the galaxy. However, Eq. (3) also gives $L_{max}(K) = 15$ Gyr and $L_{max}(M) = 50$ Gyr, which further suggests that communicative civilization should not yet be prominent around these later spectral types. PET predicts that M-dwarf planets are more likely to be habitable in the future than today. This could be because of physical factors, such as the high stellar activity of M-dwarfs during the early main sequence phase or the enhanced contribution of infrared radiation in the stellar spectrum—any of which could alter the temporal trajectory of evolution in such an environment. Likewise, PET suggests that the present era is probably the best time for G- dwarf stars to provide habitable conditions because the current age of the galaxy is approximately half that of the main sequence lifetime for these spectral types.

The assumptions of PET do not require that all civilizations are necessarily long-lived, and our choice of focusing on the maximum lifetime L_{max} acknowledges that other factors, not necessarily linked to spectral type, could cause a civilization to achieve a value of L much lower than L_{max}.

However, such short-lived civilizations would be poor targets for SETI (Grinspoon, 2004; Haqq-Misra and Baum, 2009), so any logic that can provide upper bounds with stellar dependence on the emergence of communicative civilizations can help to constrain SETI efforts today.

5 DISCUSSION

Our formulation now provides numerical constraints for R_*, f_p, n_e, and L as well as comparative bounds for f_l and f_i that may be further enhanced through additional theoretical research. Only f_c remains elusive, with few ties to observables or theory. We also provide expressions for L_{max} that allow us to predict possible habitable trajectories over the history of the galaxy.

With the assumptions of EET, along with constraints from observations, we can summarize our expression for the maximum value of the Drake equation, N_{max}, as:

$$
N_{max}^{EET}(s, t) = \begin{cases} \frac{1}{5} R_*(s)[L_*(s) - t_{evo}] f_l f_i f_c, & \text{for } t > t_{evo} \\ 0 & \text{for } t < t_{evo} \end{cases} \tag{4}
$$

Likewise, the assumptions of PET allow us to write the maximum value of the Drake equation as:

$$
N_{max}^{PET}(s, t) = \begin{cases} \frac{1}{10} R_*(s) L_*(s) f_l f_i f_c, & \text{for } t > t_{evo} \\ 0 & \text{for } t < t_{evo} \end{cases}. \tag{5}
$$

It also remains possible that one of the terms f_l, f_i, or f_c includes another abrupt phase transition at a later time in the history of the galaxy. Nevertheless, our expressions in Eqs. (4), (5) do represent phase transitions in the emergence of communicative civilization due to the functional dependence of L_{max} on spectral type.

We proceed by assuming that the product f_l, f_i, and f_c is unity, which will allow us to sketch historical trajectories for N_{max} under the EET and PET hypotheses. Although we argue in Section 3.1 that f_l should be lower for M-dwarfs than other spectral types due to water loss and other factors that could preclude life altogether, we also acknowledge in Section 3.2 that f_i could be higher for M-dwarfs due to the reduced propensity toward limit cycling. Lacking any further observational constrains, we consider these two effects to balance each other and proceed with the assumption that $f_l \cdot f_i \cdot f_c = 1$. Our analysis is thus an optimistic case, where we make best-case-scenario assumptions regarding the emergence and longevity of communicative civilizations. Additional factors not considered in our analysis would only reduce our estimates of N by lowering the value of f_l, f_i, f_c, or L. The values we use for the Drake equation parameters in each of the hypotheses are summarized in Table 1.

The temporal evolution described by Eqs. (4), (5) are plotted in Fig. 1, which shows EET assumptions in the left panel and PET assumptions in the right panel. EET shows a single phase transition representing the origin of communicative civilizations at all spectral types (except for short-lived F-dwarfs). By contrast, PET shows phase transitions with timing that depends upon spectral type. EET suggests that all spectral types are equally likely to develop communicative civilizations, with M-dwarfs favored due to their larger number. However, PET suggests that the later era of the galaxy is better suited for the emergence of communicative civilizations around late type stars.

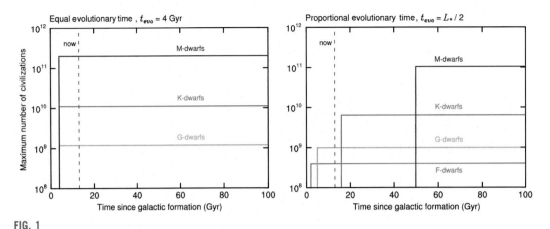

FIG. 1

Maximum number of communicative civilizations in our galaxy, N_{max}, as a function of time since galactic formation. For the EET assumptions (left panel), communicative civilizations should be most numerous around M-dwarfs today, whereas the PET assumptions (right panel) show that communicative civilizations should be most numerous around G-dwarfs today.

The present era of the universe, under EET, should host a higher proportion of communicative civilizations around M-dwarfs compared to K- or G-dwarfs, whereas PET predicts that the present era is near the peak for communicative civilizations around G-dwarfs. Although some analyses of the galactic habitable zone make assumptions similar to EET, such that the emergence of life requires about 4 Gyr to arise on any star type (e.g., Lineweaver et al., 2004), other analyses predict that the emergence of communicative civilizations depends upon spectral type (Loeb et al., 2016). We argue that the agreement between the cosmological model of Loeb et al. (2016) and our simplified Drake equation analysis with the PET hypothesis suggests that PET should be preferred over EET. More generally, even if PET is itself too limited, we recommend that other alternatives to EET should continue to be explored.

6 CONCLUSION

The Milky Way galaxy has conceivably provided opportunity for other communicative civilizations to arise prior to the formation of Earth. We provide an analysis of the Drake equation that uses contemporary observations and theoretical models, arguing that none of these astrophysical parameters show obvious phase transition behavior. However, we also consider two possibilities for the timing of the emergence of communicative civilization, which do introduce phase transition behavior that depends upon spectral type of the host star.

The EET hypothesis permits the emergence of life around G-, K-, and M-dwarf stars about 4 Gyr after the formation of the galaxy. Other civilizations could conceivably have developed between this period and the formation of Earth (about 9 Gyr after the galaxy formed). This 5 Gyr period in the history of the galaxy provides a wide window for the development of biology, intelligence, and communicative civilization on other worlds. Unless other factors contribute to a more recent phase transition for the emergence of intelligence, EET suggests a history where previous civilizations in the galaxy may have risen and fallen (or perhaps even still exist today). EET suggests that G-, K-, and M-dwarfs would make good targets for SETI today, but M-dwarfs should be preferred because they have a greater value of N_{max}.

By contrast, the PET hypothesis permits the emergence of life around all spectral types, with a timing that occurs halfway through the main sequence lifetime of the host star. PET predicts that F-dwarfs could develop communicative civilizations as early as 2 Gyr after the galaxy formed, while G-dwarfs civilizations would have emerged at 5 Gyr. This again provides a window of about 7 Gyr between the first F-dwarf civilizations and the formation of Earth. PET predicts an even greater historical window than EET for the past emergence of communicative civilization. PET also predicts that the emergence of communicative civilization on K- and M-dwarfs remains in the future, so G- and possibly F-dwarfs are the best targets for SETI today.

Although both EET and PET make simplifying assumptions, we argue that PET is preferable for guiding SETI observations. The conspicuous absence of signals detected by SETI thus far suggests that communicative civilizations are rare enough that a much more prolonged search is required—or else, perhaps our detection methods are flawed. EET would indeed suggest that the silence of SETI means that intelligence is rare and we are alone—after all, if life can arise with equal timing regardless of spectral type (and if technological intelligence is common), then we should expect to soon see signs of communicative civilization on M-dwarf systems with all-sky surveys such as the Breakthrough Listen initiative. Instead, PET predicts that the peak of civilizations for K- and M-dwarfs remains in the future, so the lack of signals from these type of systems is to be expected. PET suggests that G-dwarf systems, perhaps some older than our solar system, are the best targets to search for signs of intelligent communicative life today.

REFERENCES

Abbot, D.S., 2016. Analytical investigation of the decrease in the size of the habitable zone due to a limited CO_2 outgassing rate. Astrophys. J. 827, 117.

Airapetian, V.S., Glocer, A., Khazanov, G.V., Loyd, R.O.P., France, K., Sojka, J., Danchi, W.C., Liemohn, M.W., 2017. How hospitable are space weather affected habitable zones? The role of ion escape. Astrophys. J. Let. 836, L3.

Annis, J., 1999. An astrophysical explanation for the great silence. J. Br. Interplanet. Soc. 52, 19–22.

Batalha, N.E., Kopparapu, R.K., Haqq-Misra, J., Kasting, J.F., 2016. Climate cycling on early Mars caused by the carbonate–silicate cycle. Earth Planet. Sci. Lett. 455, 7–13.

Benner, S.A., Sismour, A.M., 2005. Synthetic biology. Nat. Rev. Genet. 6, 533–543.

Cassan, A., Kubas, D., Beaulieu, J.P., Dominik, M., Horne, K., Greenhill, J., et al., 2012. One or more bound planets per Milky Way star from microlensing observations. Nature 481, 167–169.

Ćirković, M.M., 2004. The temporal aspect of the Drake equation and SETI. Astrobiology 4, 225–231.

Ćirković, M.M., Vukotić, B., 2008. Astrobiological phase transition: towards resolution of Fermi's paradox. Orig. Life Evol. Biosph. 38, 535–547.

Denning, K., 2011. Ten thousand revolutions: conjectures about civilizations. Acta Astronaut. 68, 381–388.

Des Marais, D.J., Harwit, M.O., Jucks, K.W., Kasting, J.F., Lin, D.N., Lunine, J.I., Schneider, J., Seager, S., Traub, W.A., Woolf, N.J., 2002. Remote sensing of planetary properties and biosignatures on extrasolar terrestrial planets. Astrobiology 2, 153–181.

DeVito, C.L., 2011. On the universality of human mathematics. In: Vakoch, D.A. (Ed.), Communication with Extraterrestrial Intelligence. SUNY Press, Albany, NY, pp. 439–448.

Domagal-Goldman, S.D., Segura, A., Claire, M.W., Robinson, T.D., Meadows, V.S., 2014. Abiotic ozone and oxygen in atmospheres similar to prebiotic Earth. Astrophys. J. 792, 90.

Drake, F., 1965. The radio search for intelligent extraterrestrial life. In: Mamikunian, G., Briggs, M.H. (Eds.), Current Aspects of Exobiology. Pergamon, New York, NY, pp. 323–345.

Drake, F., 2011. The search for extra-terrestrial intelligence. Philos. Trans. R. Soc. Lond. A Math. Phys. Eng. Sci. 369, 633–643.

Dressing, C.D., Charbonneau, D., 2013. The occurrence rate of small planets around small stars. Astrophys. J. 767, 95.

Dressing, C.D., Charbonneau, D., 2015. The occurrence of potentially habitable planets orbiting M dwarfs estimated from the full Kepler dataset and an empirical measurement of the detection sensitivity. Astrophys. J. 807, 45.

Finney, B.R., Jones, E.M., 1986. Interstellar Migration and the Human Experience. University of California Press, Berkeley, California.

Foreman-Mackey, D., Hogg, D.W., Morton, T.D., 2014. Exoplanet population inference and the abundance of Earth analogs from noisy, incomplete catalogs. Astrophys. J. 795, 64.

Forgan, D., Dayal, P., Cockell, C., Libeskind, N., 2017. Evaluating galactic habitability using high-resolution cosmological simulations of galaxy formation. Int. J. Astrobiol. 16, 60–73.

Frank, A., Sullivan III, W.T., 2016. A new empirical constraint on the prevalence of technological species in the universe. Astrobiology 16, 359–362.

Gaidos, E., 2013. Candidate planets in the habitable zones of Kepler stars. Astrophys. J. 770, 90.

Glade, N., Ballet, P., Bastien, O., 2012. A stochastic process approach of the drake equation parameters. Int. J. Astrobiol. 11, 103–108.

Gowanlock, M.G., 2016. Astrobiological effects of gamma-ray bursts in the Milky Way galaxy. Astrophys. J. 832, 38.

Grinspoon, D., 2004. Lonely Planets: The Natural Philosophy of Alien Life. Harper Collins, New York, NY.

Haqq-Misra, J., 2014. Damping of glacial-interglacial cycles from anthropogenic forcing. J. Adv. Model. Earth Sys. 6, 950–955.

Haqq-Misra, J.D., Baum, S.D., 2009. The sustainability solution to the Fermi paradox. J. Br. Interplanet. Soc. 62, 47–51.

Haqq-Misra, J., Busch, M.W., Som, S.M., Baum, S.D., 2013. The benefits and harm of transmitting into space. Space Policy 29, 40.

Haqq-Misra, J., Kopparapu, R.K., Batalha, N.E., Harman, C.E., Kasting, J.F., 2016. Limit cycles can reduce the width of the habitable zone. Astrophys. J. 827, 120.

Kacar, B., 2016. Rolling the dice twice: evolving reconstructed ancient proteins in extant organisms. In: Pence, C.H., Ramsey, G. (Eds.), Chance in Evolution. Chicago Press, Chicago, IL, pp. 264–276.

Kadoya, S., Tajika, E., 2014. Conditions for oceans on Earth-like planets orbiting within the habitable zone: importance of volcanic CO_2 degassing. Astrophys. J. 790, 107.

Kadoya, S., Tajika, E., 2015. Evolutionary climate tracks of Earth-like planets. Astrophys. J. Let. 815, L7.

Kopparapu, R.K., 2013. A revised estimate of the occurrence rate of terrestrial planets in the habitable zones around Kepler M-dwarfs. Astrophys. J. Let. 767, L8.

Kroupa, P., 2001. On the variation of the initial mass function. Mon. Not. R. Astron. Soc. 322, 231–246.

Li, Y., Zhang, B., 2015. Can life survive Gamma-Ray Bursts in the high-redshift universe? Astrophys. J. 810, 41.

Lineweaver, C.H., Fenner, Y., Gibson, B.K., 2004. The galactic habitable zone and the age distribution of complex life in the Milky Way. Science 303, 59–62.

Loeb, A., Batista, R.A., Sloan, D., 2016. Relative likelihood for life as a function of cosmic time. J. Cosmol. Astropart. Phys. 2016, 40.

Luger, R., Barnes, R., 2015. Extreme water loss and abiotic O_2 buildup on planets throughout the habitable zones of M dwarfs. Astrobiology 15, 119–143.

Luisi, P.L., 2016. The Emergence of Life: From Chemical Origins to Synthetic Biology. Cambridge University Press, Cambridge, UK.

Maccone, C., 2010. The statistical Drake equation. Acta Astronaut. 67, 1366–1383.

Menou, K., 2015. Climate stability of habitable Earth-like planets. Earth Planet. Sci. Lett. 429, 20–24.

Morrison, I.S., Gowanlock, M.G., 2015. Extending galactic habitable zone modeling to include the emergence of intelligent life. Astrobiology 15, 683–696.

Morton, T.D., Swift, J., 2014. The radius distribution of planets around cool stars. Astrophys. J. 791, 10.

Petigura, E.A., Howard, A.W., Marcy, G.W., 2013. Prevalence of Earth-size planets orbiting Sun-like stars. Proc. Natl. Acad. Sci. 110, 19273–19278.

Piran, T., Jimenez, R., 2014. Possible role of gamma ray bursts on life extinction in the universe. Phys. Rev. Lett. 113, 231102.

Prantzos, N., 2013. A joint analysis of the Drake equation and the Fermi paradox. Int. J. Astrobiol. 12, 246–253.

Ramirez, R.M., Kaltenegger, L., 2014. The habitable zones of pre-main-sequence stars. Astrophys. J. Let. 797, L25.

Ray, T.S., 1993. An evolutionary approach to synthetic biology: zen and the art of creating life. Artif. Life 1, 179–209.

Robitaille, T.P., Whitney, B.A., 2010. The present-day star formation rate of the Milky Way determined from Spitzer-detected young stellar objects. Astrophys. J. Let. 710, L11.

Sagan, C. (Ed.), 1973. Communication with Extraterrestrial Intelligence (CETI). MIT Press, Cambridge, MA.

Scharf, C., Cronin, L., 2016. Quantifying the origins of life on a planetary scale. Proc. Natl. Acad. Sci. 113, 8127–8132.

Schwieterman, E.W., Meadows, V.S., Domagal-Goldman, S.D., Deming, D., Arney, G.N., Luger, R., Harman, C.E., Misra, A., Barnes, R., 2016. Identifying planetary biosignature impostors: spectral features of CO and O_4 resulting from abiotic O_2/O_3 production. Astrophys. J. Let. 819, L13.

Segura, A., Meadows, V.S., Kasting, J.F., Crisp, D., Cohen, M., 2007. Abiotic formation of O_2 and O_3 in high-CO_2 terrestrial atmospheres. Astron. Astrophys. 472, 665–679.

Tian, F., Ida, S., 2015. Water contents of Earth-mass planets around M dwarfs. Nat. Geosci. 8, 177–180.

Wordsworth, R., Pierrehumbert, R., 2014. Abiotic oxygen-dominated atmospheres on terrestrial habitable zone planets. Astrophys. J. Let. 785, L20.

ARE WE THE FIRST: WAS THERE LIFE BEFORE OUR SOLAR SYSTEM?

Pauli E. Laine*, Sohan Jheeta[†]

*University of Jyväskylä, Jyväskylä, Finland
[†]Network of Researchers on Horizontal Gene Transfer and the Last Universal Common Ancestor, Leeds, United Kingdom

CHAPTER OUTLINE

Abbreviations

amu	atomic mass unit
eV	electron volt
IDP	interstellar dust particle
ISM	Interstellar Medium
S	Svedberg unit
UV light	Ultra-Violet light

Habitability of the Universe Before Earth, editors: Richard Gordon & Alexei Sharov, Volume 1 in the series:
Astrobiology: Exploring Life on Earth and Beyond, series editors: Pabulo Henrique Rampelotto,
Joseph Seckbach & Richard Gordon. ISSN 2468-6352. https://doi.org/10.1016/B978-0-12-811940-2.00014-9

1 INTRODUCTION

The universality of life in the Cosmos is not a known phenomenon as yet. This is simply because we have no proof as to whether there ever has been, or still is, life elsewhere other than on Earth. However, evidence for its existence could soon come to light because of advances being made both in space technologies such as the new types of telescopes which are on the horizon (e.g., 39 m European Extremely Large Telescope, E-ELT, Davis, 2012) and the emergence and development of new scientific fields such as astrobiology. These instruments and sciences may also help to elucidate whether life existed elsewhere in the Universe and was subsequently delivered onto the Earth after the accretion of our Solar System began 4.6 billion (10^9) years ago; that is, was life made de novo elsewhere in the Universe? And, how feasible is this? In order to discover whether there was a possibility of life existing before the formation of our Solar System, we need to examine the criteria and mechanisms for life's initial emergence on Earth, as this process which commenced about 4.0 billion years ago took less than 300 million years—an exceptionally small period of time when compared to the age of the Universe (Lazcano and Miller, 1994). This demonstrates a potential for life to have emerged multiple times in numerous different solar systems within the early Universe.

Since we do not know of any other life form(s) to compare and contrast with, the most we can do is to make physical, biological, and chemical models, as well as computer simulations and evolutionary observations, ultimately arriving at the "best fit" picture of life in the Universe. What this chapter will not do is to speculate on that which is not carbon/water-based life, meaning that silicon or other constructs will not be brought into focus here. Readers can learn more about other weird and wonderful chemistry from a paper by Bains (2003).

In this chapter, we review the main ideas, hypotheses, and theories in chronological order and attempt to draw viable conclusions from them.

2 THE BIG BANG AND THE ELEMENTS

It is well-known that the Universe began with the Big Bang, out of which the Universe came into being 13.8 billion (10^9) years ago (Wikipedia, 2017a). The age of the Universe may well be re-evaluated in the future, as the numbers assigned to various constants are "tweaked" and the techniques used to measure the changes in the Universe's infrastructures (e.g., dark matter and energy) continue to develop.

Without going into the evolutionary history of the Cosmos, we will fast forward to the point when the necessary elements (and molecules) became available for the "kick" starting of life. About 400 million years after the Big Bang, elements started to be made, initially within massive stars (Wikipedia, 2017b), or so-called element-making factories. The general name for this process is stellar nucleosynthesis (Busso et al., 1999). Note that under this title there are also many subprocesses. By the time the subprocess "C—N—O" (carbon, nitrogen, and oxygen) cycle was taking place (Hoyle, 1954), elements up to and including iron had been synthesized (Green and Jones, 2003). This is an important milestone in that, at this point, all the necessary elements required for life had been formed.

These elements were then distributed further within the Cosmos when stars which were at least 10 times the mass of the Sun exploded (supernovae). Such stars have a short life span as their mass determines their longevity; the more massive the star, the shorter its lifespan. The fact that these massive stars exploded within relatively short timeframes helped speed up the process of the distribution of elements. The technical details on the life of stars are not within the scope of this chapter; readers may wish to consult The Open University book entitled: "The Sun and Stars—Part 2" edited by Green and Jones (2003).

3 INTERSTELLAR MEDIUM—HOLES IN THE SKY

When the renowned astronomer William Herschel (1738–1822) first saw dark patches in space (e.g., the Barnard 68 nebula), he referred to them as "holes in the sky" (Greenberg, 2002). These holes are formed when massive stars explode and spew out newly made elements and molecules along with Interstellar Dust Particles (IDPs) (Henning and Salama, 1998) into the surrounding space at an exceptionally high velocity causing a shockwave front. When eventually the ejecta comes to a halt, the resulting clouds of elements, molecules, and IDPs are collectively termed Interstellar Medium (ISM). Scientists have discovered ISM almost at the edge of the Universe at 13 billion light years from Earth (Rho et al., 2008), indicating that the elements, molecules and IDPs were available less than 1 billion years after the Big Bang took place (noting that light years equate to "time-years").

These large scale ISM structures contain one IDP for every 10^6 H_2 per cm^3, resulting in them being dark (Snow and McCall, 2006) and referred to as dark molecular clouds or nebulae (e.g., the Horsehead Nebula, Fig. 1). Such places are extremely cold (10–50 K) with pressures equivalent to 10^{-13} torr (Strazzulla et al., 2001; Mason et al., 2014). They are composed broadly of 80% molecular hydrogen (H_2) and 19% helium (He), with other elements and molecules making up less than 1% by weight (Herbst, 2001). In addition, IDPs, e.g., Cassiopeia A supernova ejecta, are composed of proto-silicates, silicon dioxide, iron oxide, pyroxene, carbon, and aluminum oxide (Rho et al., 2008). They have largely two functions: (a) allowing solid phase astrochemistry to take place on the surface as below and (b) as material for the formation of solar systems, achieved by the process of coalescing, first into small "chunks" and then into larger and larger boulders, eventually resulting in the formation of M dwarf and G class stars as well as their orbiting planets and satellites. Even by 1–2 billion years after the Big Bang, there would have existed ample quantities of these solar systems on which life could have had a chance to emerge, remembering that it took less than 300 million years for life to emerge on Earth (Lazcano and Miller, 1994).

4 MAKING ORGANIC MOLECULES—CRADLE FOR LIFE?

Since life is a result of the chemical evolution of the necessary simpler molecules being organized into more and more complex ones and then subsequently into three-dimensional life forms, life is a pinnacle of ultimate complex chemistry. Still, a question we need to ask is where and how were the necessary molecules of life made? We know that biogenic molecules could be made by astrochemistry; at atmospheric boundaries; between the bilayers of clay surfaces and mineral surfaces and via atmospheric lightning.

4.1 ASTROCHEMISTRY

To date ~190 molecules have been detected in the ISM (University of Koeln, February, 2017), ranging from the simplest diatoms such as H_2 and CO to polyatomic ones, as in hydrogen cyanide (HCN), formyl radical (HCO), carbonyl sulfide (OCS), isocyanic acid (HNCO), then onto even more

FIG. 1

Example of a dark molecular cloud—the Horsehead Nebula.

Courtesy: http://www.eso.org/gallery/v/ESOPIA/Nebulae/phot-02a-02.tif.html.

complex ones like cyanoacetylene (C_3HN, 5 atoms), methenamine (CH_3N, 6 atoms), acetaldehyde (CH_3CHO, 7 atoms), and even molecules with 13 atoms. Exotic molecules such as fullerenes and isotopes like D, ^{13}C, ^{17}O, and ^{18}O were also discovered. The reason that there are only 190 discoveries of molecules in the ISM to date is that, as the molecules become polyatomic, heterogenous, aromatic etc., with various types of functional groups (e.g., carbonyl, aldose, alcohol groups etc.), the spectrum arising from them becomes indecipherable. This is one of the reasons as to why amino acids, for example, are difficult to detect in space.

It is now understood that the molecules in the ISM can be made both in gas (75%) and solid (25%) phase, even when the pressure is as low as 10^{-13} torr. The precursor molecules—CO, CO_2, ammonia (NH_3), methane (CH_4)—within the ISM are processed by both energetic particle radiation (e.g., e^-, H, H^+, D^+, He, He^+) via radiolysis and via photolysis with UV light as well as thermal processing—for example, Jheeta et al., (2013) showed that during the irradiation of pure CH_3OH ices at 30 K using 1 keV (at 10 μA) electrons, new molecular products were detected and monitored using Fourier Transform InfraRed spectroscopy. The organics produced in radiolysis (Jheeta et al., 2013) and photolysis (Gerakines et al., 1996) experiments are compared with organics found in comet's comas (Ferris, 2006) and are listed in Table 1.

Table 1 The Results of the Jheeta et al. (2013) and Gerakines et al. (1996) Experiments Show That Organic Molecules Can Be Made by Particle and Electromagnetic Radiation in Simulated Space Conditions

	Jheeta et al. (2013) Radiology via 1 keVe⁻	Gerakines et al. (1996) Photolysis via UV Light	Ferris (2006) Organics in the Comas of Comets
Methylformate	✓	✓	✓
Methane	✓	✓	✓
Hydroxymethyl	✓	✓	
Formic acid	✓		✓
Formaldehyde	✓	✓	
Formyl radical	✓	✓	
Carbon monoxide	✓	✓	
Carbon dioxide	✓	✓	
Ethanol		✓	
Methanol			✓
Formamide			✓
Ethylene			✓
Methylacetylene			✓
Acetonitrile			✓
Acetylene			✓
Ethane			✓
Hydrogen cyanide			✓

Ferris (2006) reports the organic molecules contained in the coma of comets originating in the ISM.

The presence of these molecules in the ISM (as well as in other galaxies, e.g., Andromeda Galaxy, Hudson, 2006; Onaka et al., 2008) can be confirmed by taking radio-astronomical infra-red spectra using a radio astronomical dish, although the potential signal/noise ratio arising from distant galaxies reduces to indecipherable levels, thereby impeding identification of relevant molecules. Also, in 2004, the Stardust spacecraft mission brought back IDPs from the coma of the comet 82P/Wild 2; on analysis of these IDPs, the nonchiral amino acid, glycine, was detected (Kwok, 2016; Sandford et al., 2016); glycine being the first amino acid in a series of 20 used in the biology on Earth. The outcome of such experiments and investigation confirms the presence of organic molecules just about everywhere in the galaxies and that these molecules were delivered onto the Earth via impactors such as carbonaceous meteorites, comets, and asteroids (Cronin and Pizarello, 1986; Cronin, 1989; Sephton, 2002).

4.2 ATMOSPHERIC BOUNDARIES

There exists a region of ionosphere in the upper atmosphere of a planet or in the atmosphere of its satellite(s). This region contains a population of charged organic molecular ions generated due to the action of the energetic plasma from the planet's magnetosphere (e.g., as from Saturn), as well as the action of solar and high energy cosmic radiation. Such organic ions play an important role in the rich chemistry of the atmosphere of the planet—for example on Titan, one of Saturn's largest Moons, the ionosphere begins at an altitude of ~ 800 km (c.f., Earth's ionosphere at 60 km) and it contains ions of organic molecules with a mass/charge ratios of up to 100 amu (Cravens et al., 2006; Coates et al., 2007). By way of illustrating Titan's atmospheric photochemistry, the following are good examples of the formation of methylene amidogen (H_2CN) and HCN, respectively, from reactions with $N(^4S)$ atom and methyl group (CH_3). (Hébrard et al., 2012):

$$N(^4S) + CH_3 \rightarrow H_2CN + H \tag{1}$$

$$H_2CN + H \rightarrow HNC + H_2 \tag{2}$$

These molecules were confirmed by the unmanned Huygens spacecraft as it traversed Titan's atmosphere. When the probe landed on the satellite's surface on January 14, 2005, it also confirmed the presence of an "orange haze" of atmospheric hydrocarbons such as CH_3OH, ethanol (C_2H_5OH), benzene (C_6H_6) ring and nitrogen containing compounds such as cyanopolyyne (HC_3N) (Teanby et al., 2007), indicating that Titan's surface is rich in potentially life forming organic chemistry, increasing the potential for the emergence of life on Titan (Shawn et al., 2016). Experimental laboratory work by Khare and Sagan as long ago as 1981 predicted the presence of a complex dark brown solid of a class called tholins, which they suggested is present on the surface of Titan and gives Titan its characteristic "orange haze" atmosphere. Analysis of tholins revealed its composition to be a wide range of aliphatics, aromatic nitriles, alkanes, alkenes, aromatic hydrocarbons, pyrrole, and pyridine. Later experiments yielded similar results (Khare et al., 2002).

4.3 CLAY AND MINERAL SURFACES

Important biological molecules, such as chains of peptides and polynucleotides, can be synthesized on clay surfaces from their respective monomers, namely amino acids and nucleotides (Jheeta and Joshi, 2014): such polymers are important in kick starting the origins of life as posited by the RNA world hypothesis (i.e., genetics first, Crick, 1968). Clays such as montmorillonite (which are made from volcanic ash) are excellent ion-exchangers; they can also swell, forming bilayers in the presence of water; thus, they can catalyze polymerization reactions between their negatively charged bilayers. In the presence of positively charge sodium ions (1.5–2.0 M), for example, phosphorimidazolides of the nucleosides can be made to react, resulting in the formation of RNA oligomers (Jheeta and Joshi, 2014). More importantly, montmorillonite bilayers expand to accommodate the growing oligomer chains, making a plausible polymerization "factory." In addition, Hansma (2014) proposes that polymerization of prebiotic molecules can occur in the spaces between Muscovite mica sheets. Most hypotheses of the origins of life suggest that an RNA with chain lengths in the range of 30–50 nucleotides is needed to initiate the catalysis that makes a genetic system viable (Briones et al., 2009). Ferris et al. (1996) showed the formation of up to 50 nucleotide long chains on the surface of clays.

Other than clays, mineral surfaces in general can also aid in the polymerization of monomers of polynucleotides and of peptides by acting as catalysts. Iqubal et al. (2017) showed that metal ferrite with the general formula MFe_2O_4, (where M = Ni, Co, Cu, and Mn) can bring about thermodynamically unfavorable condensation reactions between two nucleotides or two amino acids. He showed that the overall trend in the yield of the oligomeric products obtained was $NiFe_2O_4 > CoFe_2O_4 > CuFe_2O_4 > ZnFe_2O_4 > MnFe_2O_4$ (Fig. 2).

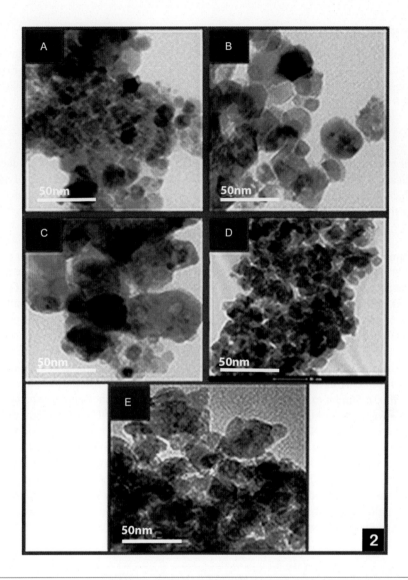

FIG. 2

Transmission electron microscope (TEM) images of nano-sized particles of: (A) $NiFe_2O_4$, (B) $CoFe_2O_4$, (C) $CuFe_2O_4$, (D) $ZnFe_2O_4$, and (E) $MnFe_2O_4$ minerals showing the surface area on which reactions can occur. From Iqubal et al. (2017).

4.4 ATMOSPHERIC LIGHTNING

Electric discharge in the form of lightning during the heavy bombardment epoch of the Earth would have been incredibly intense, resulting in a high amount of energy being discharged (Thomas et al., 2007; Johnson et al., 2008). This intense lightning can break apart stubborn bonds like those of N_2 (requiring 9.8 eV) via photolysis as in Eq. (3).

$$N_2 + hv \rightarrow N^\bullet + N^\bullet \quad \text{at} \quad \lambda = 126.5 \, nm \quad \text{or} \quad 9.80 \, eV \tag{3}$$

$$O_2 + hv \rightarrow O^\bullet + O^\bullet \quad \text{at} \quad \lambda = 240 \, nm \quad \text{or} \quad 5.2 \, eV \tag{4}$$

Atmospheric photolytic reactions in Eqs. (3) and (4) are termed homolytic and the products $N^\bullet$ and $O^\bullet$ are free radicals of nitrogen and oxygen, respectively. Eq. (5) is an example of heterolytic photolysis and its products are ions. Both the free radicals and ions are highly reactive and so are very important in determining the chemistry of an atmosphere.

$$H_2O + hv \rightarrow H^+ + OH^- \quad \text{at} \quad \lambda = 239.8 \, nm \quad \text{or} \quad 5.7 \, eV \tag{5}$$

The initial atmosphere of the early Earth was thought to be a reducing one, consisting of H_2, H_2O, CH_4, and NH_3 as promulgated by Harold Urey of the University of Chicago, although some scientists now think that the atmosphere could have been composed of either H_2, N_2, CO_2 or N_2, H_2O, CO_2 or even CH_4, N_2, H_2O. Whichever hypothesized atmosphere it was, one thing is certain in that all of them contain a source of the elements H, C, O, and N; these elements being vital in the formation of HCN (see below), as well as the necessary amino acids and nitrogenous bases (pyrimidines and purines). The initial electric discharge experiments were first carried out by Stanley Miller, Harold Urey's PhD student. The amino acids formed in Miller's experiments are shown in Table 2 and are compared with those found in proteins in living entities on Earth as well as in the Murchison meteorite (Gilmour and Wright, 2003). Similar experiments by Oró and Kimball (1961) showed that when HCN is heated at 70°C over several days, it combines with ammonium hydroxide (NH_4OH) to form the nitrogenous base, adenine (A). Thus, HCN is a versatile compound in that it can form both pyrimidines and purines (Roy et al., 2007). Saladino et al. (2007) synthesized cytosine (C), guanine (G), uracil (U), and thymine (T), noting that the latter is a methylated version of uracil. The C, G, A, and U bases are used in RNA, and when U is replaced with T (and ribose with deoxyribose sugar), this becomes part of DNA's make up. Both of these nucleic acids are essential for life on Earth, as they are needed in the conveying of genetic information in the first instance, noting that RNA has additional functions as described in the genetic first hypothesis below.

5 ORIGIN OF LIFE *PER SE*: CURRENT HYPOTHESES

Although there are more than a dozen hypotheses claiming to have the answer to the question of the origin of life; however, of these the four front runners are panspermia, metabolism (e.g., alkaline hydrothermal vents), genetic (RNA world), and vesicles first.

5.1 PANSPERMIA HYPOTHESIS

The central tenet of this hypothesis is that life was made elsewhere in the Universe (Burchell, 2010; Wickramasinghe and Smith, 2014) and then delivered on to the Earth "ready-made;" however, life's actual creation in the ISM (Adams and Spergel, 2005) is not vital to the hypothesis, which concentrates

Table 2 Amino Acids Could Have Been Made Both on the Earth As Well As in Space and Then Delivered onto the Earth via Impactors (Gilmour and Wright, 2003).

Amino Acid	Synthesized in the Miller-Urey Experiments	Found in Proteins on Earth	Found in the Murchison Meteorite
Glycine	✓	✓	✓
Alanine	✓	✓	✓
α-Amino-*N*-butyric acid	✓		✓
α-Aminoisobutyric acid	✓		✓
Valine	✓	✓	✓
Norvaline	✓		✓
Isovaline	✓		✓
Proline	✓	✓	✓
Pipecolic acid	✓		✓
Aspartic acid	✓	✓	✓
Glutamic acid	✓	✓	✓
β-Alanine	✓		✓
β-Amino-*N*-butyric acid	✓		✓
β-Aminoisobutyric acid	✓		✓
Θ-Aminobutyric acid	✓		✓
Sarcosine	✓		✓
N-ethylglycine	✓		✓
N-methylalanine	✓		✓

These amino acids could then have been used in the chemical evolution of life.

more on the mechanisms of delivery rather than the origin of life. Presuming that some sections of space did harbor life then, as our Solar System traversed through space within the Milky Way Galaxy, this life might have been "rained" onto the sterile Earth, thereby seeding it. The traversing of our Solar System through the Galaxy is just one mode of delivery of life onto the Earth, impactors being the other. The earliest fossil evidence of microbial life dates to 3.5 billion years ago (Schopf, 1993; Dodd et al., 2017), so this presumes that life could have been made from organic molecules in the chemical factories within the interstellar media prior to that. Moreover, some scientists posit that life is still being made in space and delivered onto the Earth on a daily basis, introducing new genetic variations which bring about bio-diversity on the Earth (Wickramasinghe, 2003; Wainwright et al., 2003).

Is there any persuasive evidence offered by panspermia supporters that life emerged, albeit elsewhere in the Universe, prior to the formation of our Solar System? Explanations offered appear to be mostly conjecture—for example, organic carbon in the ISM consists of biological breakdown products (Rauf and Wickramsinghe, 2010) or micro-organisms which can survive for millions of years

(Vreeland et al., 2000) while in transit from one solar system to another. However, there is no consideration given to the fact that DNA would be thermodynamically unstable for such long durations (Graur and Pupko, 2001). So, in short, panspermia does not offer us any clues to whether there was life elsewhere in the Universe before our Solar System except promulgating conjectured delivery mechanisms.

5.2 METABOLISM FIRST HYPOTHESIS

The main contenders for this include: the methanethiol, CH_3SH, world (Huber and Wächtershäuser, 1997), the zinc world (Mulkidjanian and Galperin, 2009), thermosynthesis (Muller, 2005), and deep sea vents. Of these, the latter is generally referred to as the alkaline hydrothermal vent hypothesis (Jheeta, 2013a; Russell et al., 2014) and is more credible than the others since continuing research delivers well-thought out experiments (Herschy et al., 2014) and the resulting evidence in favor of it.

Biogenic reactions occur in the clay bubbles of the white smoker vents found on the sea bed (e.g., Fig. 3). The clays (e.g., montmorillonite) contain the mineral Fe—S (Chatzitheodoridis et al., 2014) which forms a network of miniscule, millimeter-sized bubbles in which the synthesis reactions take

FIG. 3

White smoker chimneys at Champagne vent site, NW Eifuku volcano, Mariana Arc region, Western Pacific Ocean. April, 2004 similar to vents where chemical evolution may have occurred which led to the emergence of the three domains of life, namely Archaea, Bacteria, and Eukarya.

Courtesy: NOAA Office of Ocean Exploration; Dr. Bob Embley, NOAA PMEL, Chief Scientist.

place. These bubbles can be deemed temporary "holding chambers" during the process of chemical evolution (Kelley et al., 2005), and due to the fact that they eventually erode, any resultant cellular life form would be released into the water. Proponents of this hypothesis have been able to simulate the conditions within these vents, i.e., the white smokers known as Lost City (Kelley et al., 2001), found about 1 km from the main black smokers of the mid-Atlantic ridge and have synthesized the organic molecules in a reactor that could possibly have gone on to make life, e.g., CH_3OH, formic acid ($HCOOH$), and formaldehyde (H_2CO) from the H_2 and CO_2 dissolved in water at a given pH (Herschy et al., 2014). $HCOOH$, H_2CO, and formamide ($HCONH_2$) can be made as follows:

$$FeS + H_2S \rightarrow FeS_2 + 2H^+ + 2e^- \qquad (6)$$

$$FeS + H_2S + CO_2 \rightarrow FeS_2 + HCOOH \qquad (7)$$

$$2H + HCOOH \rightarrow H_2C(OH)_2 \rightarrow H_2CO + H_2O \qquad (8)$$

Eq. (6) and (7) are from Martin and Russell (2007) and Cody (2004), respectively. Formamide is relatively easy to make via an acid/base ($HCOOH/NH_3$) reaction (Eq. 9).

$$HCOOH + NH_3 \rightarrow HCONH_2 + H_2O \qquad (9)$$

Such biogenic compounds, initially having been delivered in huge quantities by carbonaceous meteorites impacting the surface of early Earth (including the oceans), eventually formed a layer anywhere up to 1.6 m thick. Given the vast quantities of these deposits, some of them would have been preserved in the sediment on the sea bed (Sephton, 2003). These compounds could have been instrumental during the process of chemical evolution—for example, five carbon sugar, ribose ($C_5H_{10}O_5$), can be made from H_2CO via the formose reaction sequence (Wikipedia, 2017c). Ribose forms the backbone (along with phosphate ions, PO_4^{3-}, and nitrogenous bases) in the polyribonucleotide chain.

Hydrothermal vents are also featured on Saturn's moon, Enceladus, (Fig. 4), which could therefore potentially be harboring reservoirs of life due to the presence of active cryovolcanoes ejecting ice crystals, water vapor, CO_2, NH_3, and organic volatiles such as acetylene (C_2H_2), HCN, H_2CO, and CH_4

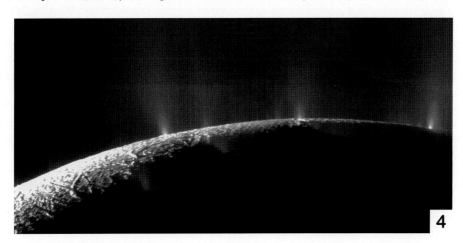

FIG. 4

Flyby image of Saturn's moon, Enceladus, taken by Cassini Spacecraft, showing the emission of plumes of gas.

Courtesy: https://www.nasa.gov/feature/jpl/an-ice-worldwith-an-ocean.

(Waite et al., 2006). The latter molecule may be due to the activity of extremophiles such as obligate anaerobes, methanogens (McKay et al., 2008). Io, another moon of Saturn, has active cryovolcanoes as evidenced by plumes of sulfur dioxide (SO_2) and spewing out of molten sulfur—could this be yet another potential candidate for life elsewhere in the Solar System? (Schulze-Makuch, 2010). Jupiter's moons Ganymede, Callisto, and especially Europa are believed to have active surfaces and ample supplies of liquid water and they may even harbor life, in the form of chemolithoautotrophs (Gaidos et al., 1999), in the water beneath their icy surfaces (Jheeta, 2013b).

5.3 GENETICS FIRST HYPOTHESIS

This hypothesis maintains that an information system in the form of RNA came first, followed by proteins and then DNA (Crick, 1968). Before elaborating on this point, it is stressed here that RNA-catalyzed reactions do not occur in central metabolic reactions, pathways such as glycolysis and cycles (e.g., Kreb's cycle) of cellular biology; these are carried out by specific proteinous catalysts known as enzymes (Keller et al., 2014). Having stated that, RNA is unique in that it exhibits three very important properties. Firstly, it can form four levels of structural integrities, beginning with primary structures observed when tRNA decomposes into small double-stranded RNAs. The next level being secondary, as exemplified by 5S, 120 nucleotides RNA motifs of ribosomes, tRNA cloverleaf, and circular RNAs. Tertiary or third level structures which are broadly composed of RNAs and protein complexes termed ribonucleoproteins (RNP), as denoted by RNase P and RNase MRP (**M**itochondrial **R**NA **P**rocessing) complexes. The final level being quaternary, and again in conjunction with RNAs, proteins, various co-factors, and metal ions (e.g., Zn), RNA forms huge ncRNA-nanomachines as in spliceosomes, the Varkud Satellite (VS) ribozyme, and ribosomes. The essence of structural integrity defines the functions and range of activities that an RNA can carry out, considering there are no more than 8 nitrogenous (with 5 most common ones being A, C, G, T and U) bases being used compared with proteins using 20 different amino acids (Roth and Breaker, 2009)—for instance acting as a riboswitch, thereby enabling gene silencing and preventing the expression of a gene (Dambach and Winkler, 2009). Secondly, ncRNA can undertake catalytic activity as in ribozymes (Gilbert, 1986; Cech et al., 1981; Altman, 2000). For example, ribozymatic activity is important during the translation process, whereby a codon in the mRNA synchronizes with the relevant anticodon in the tRNA carrying activated amino acid; this amino acid is added by the formation of a peptide bond [—C(O)—N(H)—] to the growing peptide chain. Finally, the two types of RNAs, namely coding and noncoding (nc), can carry chemical information. The coding mRNA is used during protein synthesis and the latter has multifaceted actions within all cellular life forms on Earth—for example, ncRNAs are involved in protein synthesis as typified by tRNAs; they are involved in, as mentioned above, gene regulation; as RNA riboswitches (Roth and Breaker, 2009); and as "sensors" (Lee et al., 2010) for detecting invading (via mechanism of transformation) "parasitic" nucleic acids (Whangbo and Hunter, 2008). The range of activities covered by ncRNA is far too wide to detail here as new ncRNAs are discovered and characterized almost monthly and so cannot be fully covered in print. The genetic role of RNAs does not end here, as there are even RNA viruses such as the flu viruses. In addition, as RNA is unique in relation to its catalytic and code carrying activities, it is possible that it went on to form the first theoretical living entity or Last Universal Common Ancestor (LUCA), (Lazcano and Forterre, 1999; Jheeta, 2013a), which took place approximately 4.0 billion years ago, as evidenced by fractionation studies of the carbon and sulfur isotopes, ^{13}C and ^{34}S, respectively (Bell et al., 2015; Jheeta, 2016). So accordingly, it seems that life may have emerged shortly after the formation of Earth, somewhere between 4.3–4.0 billion years ago

Papineau, D. World's oldest fossils unearthed - https://www.ucl.ac.uk/news/newsarticles/0217/010317-Worlds-oldest-fossils-unearthed/#sthash.b22JQPTl.dpuf (accessed on a May 2017).

It is now generally accepted that two LUCAs gave rise to the origin of the two domains of life, namely Archaea and Bacteria, as it is believed to be "alive" and possessing a DNA-based repository genetic code; a DNA replication system; and the ability to carry out aminoacylation, transacylation, and peptide synthesis, i.e., it possessed the all-important functional ribosome (Farias et al., 2016) for protein synthesis, a very important "workhorse" of all living cellular entities on Earth (Leipe and Aravind, 2000). However, the RNA world hypothesis does not explain the origin of vesicles (i.e., membranes) within which RNA activity would have needed to occur and this hypothesis will be explored next.

5.4 VESICLES FIRST HYPOTHESIS

In recent times, this hypothesis has been gaining ground on the premise that some sort of "bag" was necessary in order to bring most of the necessary organic molecules into the vicinity of one another and cause them to react with each other (Hanczyc et al., 2003; Szostak, 2012). The initial vesicles were in no way as sophisticated as modern bio-membranes are today. These initial vesicles could either have been made on the sea shores of early Earth, during hydration/dehydration cycles following the ebb and flow of tides leading to concentration of organic matter while forming "bubbles" or that long, up to 10C atoms, chains of *n*-monocarboxylic acids were delivered onto the Earth via impactors, in particular carbonaceous chondrite, e.g., Murchison meteorite (Deamer, 1985, 2017; McCollom and Ritter, 1999; Apel et al., 2002). In the latter scenario, these *n*-monocarboxylic acid chains with their hydrophilic (water loving) and hydrophobic (water hating) moieties spontaneously transformed into double layered simple compartments, termed bilayer vesicles (Sephton, 2003; Chen and Walde, 2010). Bilayer vesicles (Fig. 5) would then go on to acquire, during the RNA world period, the necessary biogenic molecules that would have then led to further chemical evolution (Ichihashi et al., 2013; Jheeta, 2015).

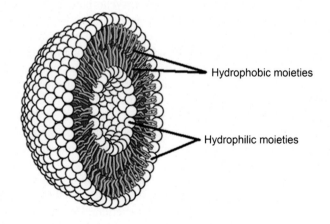

Hydrophobic moieties

Hydrophilic moieties

FIG. 5

Diagrammatic likeness of a bilayered vesicle. These represent the simplest "bag" required to contain the necessary organic molecules prior to cell development.

Courtesy: https://pt.khanacademy.org/science/biology/membranes-and-transport/the-plasma-membrane/a/structure-of-the-plasma-membrane.

It would be quite a feat to expand this approach to the emergence of a sustainable protocell as purported Szostak. In this regard, researchers have been trying to synthesize chemical systems where nonenzymatic RNA polymerization can occur in model protocells (Szostak, 2012).

6 THE VIRUS CONNECTION

Having considered these hypotheses as well as introducing the construct, LUCA—we can now introduce the role played by viruses in the origins of life. If LUCA was almost a cellular entity at the dawn of emergence of life, as it had almost all the working components of cellular biology on Earth, then what role did viruses play? And, where did they come from? The answer to these questions may be as follows: Forterre (2006, 2010) suggested that viruses probably predated LUCA based on biochemical/genetic analysis and further added that these first viruses were composed solely of RNAs, with DNA and DNA replication machineries emerging later (Leipe and Aravind, 2000). He reached this conclusion because some archaea and bacteria share homologous capsid proteins and/or ATPase proteins for packaging, suggesting that they have all evolved from a common virus that existed at the time of LUCA. The scientific community is still divided on the origin of RNA viruses, but do agree that they had a parallel evolution alongside LUCA. However, the reader may wish to read the following papers for opposing view points (Forterre, 2005, 2006; Claverie, 2006).

7 EXTREMOPHILES—THE RESILIENCE OF LIFE

Harsh conditions on early Earth, or indeed elsewhere in the Universe, called for any emergent life forms to be extremely hardy. We now have evidence that such entities still inhabit the Earth today, demonstrating high levels of resilience and adaptability. Life has been found in the most inhospitable and extreme places imaginable on Earth. Micro-organisms (psychrophiles) have been found to exist hundreds of meters below the polar ice cap in freezing temperatures as well as in permafrost and tar pits (ScienceDaily, May, 2007). Endoliths live within rock pores and between aggregate grains; communities of various micro-organisms (cryptoendoliths) have been found thriving in fissures, aquifers, and faults filled with groundwater in the deep subsurface. Some species of micro-organisms have been found to survive in hot springs (Fig. 6) where the temperature is in excess of 121°C and are called hyperthermophiles (Baker, 2010), deep in the ocean (in the vicinity of black smokers as well as in the oceanic mud floor and trenches) with pressures in excess of 38 MPa; and, there are some micro-organisms which live in two or more extreme environments simultaneously—for instance, *Sulfolobus acidocaldarius* (acidophile), an archaea, which flourishes at pH3 and at 80°C (Rothschild and Mancinelli, 2001).

On land, arid deserts harbor a variety of flora and fauna including well-established communities of bacteria and archaea. There are species of micro-organisms which can thrive in situations with high alkalinity, salinity, or sugar concentration. Moreover, the bacterium, *Deinococcus*, can tolerate nuclear reactor radiation (Huyghe, 1998), even though the survival rate in these conditions would be expected to be virtually nil. In addition, various microbial spores have been isolated from the outermost reaches of the Earth's atmosphere (Wainwright et al., 2003). Lastly, exobiological experiments (BIOPAN and EXPOSE), (Horneck et al., 2010; Meeßen et al., 2015) executed in the International Space Station,

FIG. 6

Example of an extreme environment where extremophiles can survive quite comfortably. These micro-organisms have special heat-resistant proteins which maintain the cell's internal environment.

Reproduced with permission from: Sohan Jheeta (location: Wai O Tapu Rotorua, New Zealand, 2016).

demonstrated that spores of bacteria (*Bacillus subtilis* and *Chroococcidiopsis*) and of lichen (*Xanthoria elegans*) could survive being exposed to a space environment which included extreme low temperatures and pressures (acidophile), solar radiation, and cosmic rays (Nicholson et al., 2000; Mancinelli, 2014).

Extremophiles which are chemolithoautotrophs (meaning that these microbes obtain their energy from inorganic compounds, e.g., H_2, NH_3, S^0, S^{2-}, Fe^{2+} etc) make their "living" at the edges of extreme environments. In addition, some extremophiles can live for long durations—for example, the oldest micro-organism to have been isolated is a halotolerant bacterium from a 250 million-year-old primary salt crystal (Vreeland et al., 2000). Chemolithoautotrophic extremophiles utilize inorganic molecules, for example, H_2 and CO_2, for energy and as a source of carbon and thus are thought to be the vanguard of the origins of life (Lane, 2005), and probably emerged from equally resilient LUCAs. Eq. (10) (Thauer, 1998) shows the utilization of inorganic molecules, hydrogen gas (H_2) in this respect, as a source of energy.

$$4H_2 + 2CO_2 \rightarrow CH_4 + 2H_2O \quad \Delta G'_o = -131\,\text{kJ}\,\text{mol}^{-1} \tag{10}$$

8 BALANCE OF PROBABILITY: LIFE BEFORE OUR SOLAR SYSTEM

Now that we have shown the making of the necessary molecules; the possible chemical evolutions (c.f. hypotheses), and the resilience of microbes on Earth, we can ask: what is the likelihood that there was any life in the Cosmos prior to 4.6 billion years ago? There are at least three environments where life could emerge from chemical evolution and then proceed to evolve. These are the solar systems around the G class of stars to which our Sun belongs; and M-dwarf stars, 75% of main sequence stars (c.f. ~8%

for G class), are those having a low mass of between 0.08 and 0.8 $M_{\odot}$, noting that the Sun makes up 99.86% of the Solar System's mass denoted as $M_{\odot}$. Finally, the exomoons, of both the G class and M-dwarf stars, are probably more important than both of first two categories put together because there are more moons orbiting the planets in solar systems than there are planets orbiting stars, for instance, excluding Mercury, Venus, and Earth (with its one Moon), the other planets in our Solar System, including the dwarf planet Pluto, all have numerous moons. In effect, each planet within any solar system is a smaller "planetary solar system" version of the parent system. So typically, any solar system could potentially have many possible places where life could take a foothold. The reasoning being this is that such exomoons (to take an example from our Solar System, Enceladus) will be heated primarily by tidal gravitational forces of the host planet (Saturn), neighboring moons (Mimas and Tethys), and planets (Jupiter, Uranus and Neptune), as well as the star itself (the Sun), and due to the effect of all these bodies, the moon will wax and wane, in the process providing much needed warmth which would help chemolithoautotrophic life to emerge. Additional warmth would also have been provided by radiogenic elements as these would be part and parcel of inventory of elements of any planetary system formation. Therefore, if life were to be found on Enceladus, this would indicate that it is not too difficult to make and so this is the reasoning behind the plausibility of its existence on Earth prior to 4 billion years ago as well as increasing the probability of life arising in the Universe enormously and dramatically (further reading: Kasting and Kirschvink, 2012; Jheeta, 2013b).

How many exoplanets and exomoons are there in the Universe? As of March 23, 2017, as detailed by NASA's Kepler space-based observatory, the number of exoplanets confirmed in the Milky Way Galaxy is 3604 and those having planetary systems is 2699 (European exoplanet catalog, March, 2017). There may be countless more habitable places to be found, but at present these can't be detected as observational instrumentations surpassing the current capabilities of the Atacama Large Millimeter Array (ALMA) radio telescopes (Mason et al., 2014) are needed; that is, telescopes with the technology to observe planets which are at least 6 billion light years away, preferably spaced-based, where there would be no interference from Earth's atmosphere, resulting in improved image resolution. If we can ascertain biomarkers of life (Jheeta, 2013b) at such distances, this could be definitive proof that life existed before our Solar System was accreted but how likely is this to happen... unfortunately not anytime soon!

9 FINAL SAY—BEST FIT SOLUTION?

In this section, we will address two major questions, (a) was there any life in the Universe prior to the formation of our Solar System and the subsequent emergence of life on Earth? Then (b) is there life elsewhere in the Universe *per se*?

Question (a): to attempt to answer the question of the origin of life prior to 5 billion years ago is a real challenge and the fact that the origin of life is multifaceted means that we just don't know how big the puzzle is and what has been outlined in this chapter is merely the tip of the iceberg. Further, we have no examples of life for comparison other than that which emerged on Earth. We begin by noting the time frame of events. It was observed that by the time our Solar System was accreted some 4.6 billion years ago, the Universe was already over 9.0 billion years old. It took between 1–2 billion years after the Big Bang for the initial elements (up to iron) and biogenic molecules necessary for life to be made (Hudson, 2006; Onaka et al., 2008). This means that there were at least 7 billion years prior to the formation of our Solar System during which life could have begun to emerge and, as it may have taken no more than 300 million years for life (LUCA) to emerge on Earth after this, noting that Lazcano and Miller's (1994) paper speculates this time frame based on the history of the Earth's heavy bombardment period. Some scientists, including the authors of this chapter, think that it may not even have taken 10,000 years, intimating that the opportunities for life to emerge are multiplied exponentially, therefore, it is not beyond the realms of possibility that life could have arisen countless times elsewhere in the Universe, taking into account the presence of exomoons.

On the balance of probability, it is likely that the environments and components were in place to enable life to emerge in the Universe before it did on the Earth, and so we can conclude that there could have been life elsewhere on other solar systems.

Question (b): it is posited that life emerged on Earth independently at least two times to represent each of the two domains of life, Archaea and Bacteria (Dibrova et al., 2015). Taken together with what we know about LUCA, viruses and extremophiles, it is highly probable that life exists elsewhere in the Universe. Developments in new sciences and technologies (e.g., bio-informatics, synthetic biology, systems chemistry and computer modeling, etc.) as well as exploratory missions throughout our Solar System have greatly increased the chances that, at some point, life will be found (using biomarkers, Jheeta, 2013b) in the Universe. If we do find life within our Solar System—e.g., on Saturn's moon, Enceladus—then that could be taken as potential evidence of life elsewhere in the Universe because it would have developed in Enceladus' hydrothermal vent systems, which are located below thick slabs of ice and indicates that such life could not have come about via contamination from space missions launched by humans. Further afield, the problem will be confirming the presence and extent of life because we will only be able to provide tentative proof; to fully prove life's existence, we would need to bring it back to Earth to fully examine it. Only after that has been done would we be able to state for certain that there is life elsewhere in the Universe, unless that is, ET happens to phone and say: "Hey guys, I'm over here!"

REFERENCES

Adams, F.C., Spergel, D.N., 2005. Lithopanspermia in star-forming clusters. Astrobiology 5, 497–514.

Altman, S., 2000. Structural biology. Nature 7, 827–828.

Apel, C.L., Deamer, D.W., Mautner, M.N., 2002. Self-assembled vesicles of monocarboxylic acids and alcohols: conditions for stability and for the encapsulation of biopolymers. Biochim. Biophys. Acta 1559, 1–9.

Bains, W., 2003. Many chemistries could be used to build living systems. Astrobiology 4, 137–167.

Baker, D., 2010. The Sulfur-Lovers: Archaea. Crabtree Publishing Company, Hove. p. 26.

Bell, E.A., Boehnke, P., Harrison, M., Mao, W.L., 2015. Potentially biogenic carbon preserved in a 4.1 billion-year-old zircon. PNAS 112, 14518–14521.

Briones, C., Stich, M., Manrubia, S.C., 2009. The dawn of the RNA world: toward functional complexity through ligation of random RNA oligomer. RNA 15, 743–749.

Burchell, M.J., 2010. Why do some people reject panspermia? Journal of Cosmology 5, 828–832.

Busso, M., Gallino, R., Wasserberg, G.J., 1999. Nucleosynthesis in asymptotic giant branch stars: relevance for galactic enrichment and solar system formation. Annu. Rev. Astron. Astrophys. 37, 239–309.

Cech, T.R., Zaug, A.J., Grabowski, P.J., 1981. In vitro splicing of the ribosomal RNA precursor of Tetrahymena: involvement of a guanosine nucleotides in the excision of the intervening sequence. Cell 27, 487–496.

Chatzitheodoridis, E., Haigh, S., Lyon, I., 2014. A conspicuous clay ovoid in Nakhla: evidence for subsurface hydrothermal alteration on Mars with implications for astrobiology. Astrobiology 14, 651–693.

Chen, I.A., Walde, P., 2010. From self-assembled vesicles to protocells. Cold Spring Harb. Perspect. Biol. 2, a002170.

Claverie, J.M., 2006. Viruses take center stage in cellular evolution. Genome Biol. 7, 110.

Coates, A.J., Crary, F.J., Lewis, G.R., Young, D.T., Waite Jr., J.H., Sittler Jr., E.C., 2007. Discovery of heavy negative ions in Titan's ionosphere. Geophys. Res. Lett. 34, L22103.

Cody, G.D., 2004. Transition metal sulfides and the origins of metabolism. Annu. Rev. Earth Planet. Sci. 32, 569–599.

Cravens, T.E., Robertson, P., Waite Jr., J.H., Yelle, R.V., Kasprzak, W.T., Keller, C.N., Ledvina, S.A., Niemann, H.B., Luhmann, J.G., McNutt, R.L., Ip, W.-H., De La Haye, V., Mueller-Wodarg, I., Wahlund, J.-E., Anicich, V.G., Vuitton, V., 2006. Composition of Titan's ionosphere. Geophys. Res. Lett. 33, L07105.

Crick, F.H., 1968. The origin of the genetic code. J. Mol. Biol. 38, 367–379.

Cronin, J.R., 1989. Origin of organic compounds in carbonaceous chondrites. Adv. Space Res. 9, 59–65.

Cronin, J.R., Pizzarello, S., 1986. Amino acids of the Murchison meteorite. III. Seven carbon acyclic primary alpha-amino alkanoic acids. Cosmochim. Acta 50, 2419–2427.

Dambach, M.D., Winkler, W.C., 2009. Expanding roles for metabolite-sensing regulatory RNAs. Curr. Opin. Microbiol. 12, 161.

Davis, R., 2012. Telescopes of the future. Astron. Geophys. 53, 4.15–4.18.

Deamer, D.W., 1985. Boundary structures are formed by organic components of the Murchison carbonaceous chondrite. Nature 317, 792–794.

Deamer, D., 2017. The role of lipid membranes in life's origin. Life 7, 5. 1–17.

Dibrova, D.V., Galperin, M.Y., Koonin, E.V., Mulkidjanian, A.Y., 2015. Ancient systems of sodium/potassium homeostasis as predecessors of membrane bioenergetics. Biochemistry (Moscow) 80, 495–516.

Dodd, M.S., Papineau, D., Grenne, T., Slack, J.F., Rittner, M., Pirajno, F., O'Neil, J., Little, C.T.S., 2017. Evidence for early life in Earth's oldest hydrothermal vent precipitates. Nature 543, 60–64.

European exoplanet catalog, 2017. http://exoplanet.eu/catalog/.

Farias, S.T., Rêgo, T.G., José, M.V., 2016. tRNA core hypothesis for the transition from the RNA world to the ribonucleoprotein world. Life 23, E15.

Ferris, J.P., 2006. Montmorillonite-catalzed formation of RNA oligomers: the possible role of catalysis in the origin of life. Philos. Trans. R. Soc. B 361, 1777–1786.

Ferris, J.P., Hill, A.R., Liu, R., 1996. Synthesis of long prebiotic oligomers on mineral surfaces. Nature 381, 59–61.

Forterre, P., 2005. The two ages of the RNA world, and the transition to the DNA world: a story of viruses and cells. Biochimie 87, 793–803.

Forterre, P., 2006. The origin of viruses and their possible roles in major evolutionary transitions. Virus Res. 117, 5–16.

Forterre, P., 2010. Giant viruses: conflicts in revisiting the virus concept. Intervirology 53, 362–378.

Gaidos, E.J., Nealson, K.H., Kirschvink, J.L., 1999. Life in ice-covered oceans. Science 284, 1631–1633.

Gerakines, P.A., Schutte, W.A., Ehrenfreund, P., 1996. Ultraviolet processing of interstellar ice analogs.1. pure ices. Astron. Astrophys. 312, 289–305.

Gilbert, W., 1986. Origin of life: the RNA world. Nature 319, 618.

Gilmour, I., Wright, I., 2003. The window of opportunity. In: Wright, J. (Ed.), Origins of Earth and Life. MRM Graphics, Winslow, pp. 100–102.

Graur, D., Pupko, T., 2001. The Permian bacterium that isn't. Mol. Biol. Evol. 18, 1143–1146.

Green, S., Jones, M., 2003. The main sequence life of stars. In: Green, S., Jones, M. (Eds.), The Sun and Stars. Bath Press, Bath, pp. 175–196.

Greenberg, J.M., 2002. Cosmic dust and our origins. Surf. Sci. 500, 793–822.

Hanczyc, M.M., Fujikawa, S.M., Szostak, J.W., 2003. Experimental models of primitive cellular compartments: encapsulation, growth, and division. Science 302, 618–622.

Hansma, H.G., 2014. The power of crowding for the origins of life. Orig. Life Evol. Biosph. 44, 307–311.

Hébrard, E., Dobrijevic, M., Loison, J.C., Bergeat, A., Hickson, K.M., 2012. Neutral production of hydrogen isocyanide (HNC) and hydrogen cyanide (HCN) in Titan's upper atmosphere. Astron. Astrophys. 541, A21.

Henning, T., Salama, F., 1998. Carbon in the universe. Science 282, 2204–2210.

Herbst, E., 2001. The chemistry of interstellar space. Chem. Soc. Rev. 30, 168–176.

Herschy, B., Whicher, A., Camprubi, E., Watson, C., Dartnell, L., Ward, J., Evans, J.R., Lane, N., 2014. An origin-of-life reactor to simulate alkaline hydrothermal vents. J. Mol. Evol. 79, 213–227.

Horneck, G., Klaus, M.D., Mancinelli, R.L., 2010. Space microbiology. Microbiol. Mol. Biol. Rev. 74, 121–156.

Hoyle, F., 1954. On nuclear reactions occurring in very hot stars. 1. The synthesis of elements from carbon to nickel. J. Astrophys. J. Suppl. 1, 121–145.

Huber, C., Wächtershäuser, G., 1997. Activated acetic acid by carbon fixation on (Fe, Ni) S under primordial conditions. Science 276, 670–672.

Hudson, R.L., 2006. Astrochemistry examples in the classroom. J. Chem. Educ. 83, 1611–1616.

Huyghe, P., 1998. Conan the bacterium. Science 38, 16–19.

Ichihashi, N., Usui, K., Kazuta, Y., Sunami, T., Matsuura, T., Yomo, T., 2013. Darwinian evolution in a translation-coupled RNA replication system within a cell-like compartment. Nat. Commun. 4. Article number 2494.

Iqubal, M.A., Sharma, R., Jheeta, S., Kamaluddin, 2017. Thermal condensation of glycine and alanine on metal ferrite surface: primitive peptide bond formation scenario. Life 7, 15.

Jheeta, S., 2013a. Horizontal gene transfer and its part in the reorganisation of genetics during the LUCA epoch. Life 3, 518–523.

Jheeta, S., 2013b. Final frontiers: the hunt for life elsewhere in the universe. Astrophys. Space Sci. 348, 1–10.

Jheeta, S., 2015. The routes of emergence of life from LUCA during the RNA and viral world: a conspectus. Life 5, 1445–1453.

Jheeta, S., 2016. The landscape of the emergence of life. Life 6, 20.

Jheeta, S., Joshi, P.C., 2014. Prebiotic RNA synthesis by montmorillonite catalysis. Life 4, 318–330.

Jheeta, S., Domaracka, A., Ptasinska, S., Sivaram, B., Mason, N., 2013. The irradiation of pure CH_3OH and 1:1 mixture of $NH_3:CH_3OH$ ices at 30 K using low energy electrons. Chem. Phys. Lett. 556, 359–364.

Johnson, A.P., Cleaves, H.J., Dworkin, J.P., Glavin, D.P., Lazcano, A., Bada, J.L., 2008. The miller volcanic spark discharge experiment. Science 322, 404.

Kasting, J., Kirschvink, J., 2012. Evolution of a habitable planet. In: Impey, C., Lunine, J., Funes, J. (Eds.), Frontiers of Astrobiology. Cambridge University Press, Cambridge, pp. 115–131.

Keller, M.A., Turchyn, A.V., Ralser, M., 2014. Non-enzymatic glycolysis and pentose phosphate pathway-like reactions in a plausible Archean ocean. Mol. Syst. Biol. 10, 725.

Kelley, D.S., Karson, J.A., Blackman, D.K., Früh-Green, G.L., Butterfield, D.A., Lilley, M.D., Olson, E.J., Schrenk, M.O., Roe, K.K., Lebon, G.T., Rivizzigno, P., 2001. An off-axis hydrothermal vent field near the Mid-Atlantic Ridge at 308 N. Nature 412, 145–149.

Kelley, D.S., Karson, J.A., Früh-Green, Yoerger, D.R., Shank, T.M., Butterfield, D.A., Hayes, J.M., Schrenk, M.O., Olson, E.J., Proskurowski, G., Jakuba, M., Bradley, A., Larson, B., Ludwig, K., Glickson, D., Buckman, K., Bradley, S.A., Brazelton, W.J., Roe, K.K., Elend, M.J., Delacour, A., Bernasconi, S.M., Lilley, M.D., Baross, J.A., Summons, R.E., Sylva, S.P., 2005. A serpentinite-hosted ecosystem: the Lost City hydrothermal field. Science 307, 1428–1434.

Khare, B.N., Sagan, C., 1981. Organic solids produced by electrical discharge in reducing atmospheres: tholin molecular analysis. Icarus 48, 290–297.

Khare, B.N., Bakes, E.L.O., Imanaka, H., McKay, C.P., Cruikshank, D.P., Arakawa, E.T., 2002. Analysis of the time-dependent chemical evolution of titan haze tholin. Icarus 160, 172–182.

Kwok, S., 2016. Complex organics in space from solar system to distant galaxies. Astron. Astrophys. Rev. 24, 1–27.

Lane, N., 2005. The hydrogen hypothesis: Power, Sex, Suicide. Oxford Press Publication, Oxford. pp. 51–64.

Lazcano, A., Forterre, P.T., 1999. The molecular search for the last common ancestor. J. Mol. Evol. 49, 411–412.

Lazcano, A., Miller, S.L., 1994. How long did it take for life to begin and evolve to cyanobacteria? J. Mol. Evol. 39, 546–554.

Lee, E.R., Baker, J.L., Weinberg, Z., Sudarsan, N., Breaker, R.R., 2010. An allosteric self-splicing ribozyme triggered by a bacterial second messenger. Science 329, 845–848.

Leipe, D.D., Aravind, L., 2000. The bacterial replicative helicase DnaB evolved from a RecA duplication. Genome Res. 10, 5–16.

Mancinelli, R.L., 2014. The affect of the space environment on the survival of halorubrum chaoviator and synechococcus (nägeli): data from the space experiment OSMO on EXPOSE-R. Int. J. Astrobiol. 14, 123–128.

Martin, W., Russell, M.J., 2007. On the origin of biochemistry at an alkaline hydrothermal vent. Phil. Trans. R. Soc. B 362, 1887–1925. https://doi.org/10.1098/rstb.2006.1881.

Mason, J.M., Nair, B., Jheeta, S., Szymańska, E., 2014. Electron induced chemistry: a new frontier in astrochemistry. Faraday Discuss. 168, 235–247.

McCollom, T.M., Ritter, G., 1999. Lipid synthesis under hydrothermal conditions by Fischer-Tropsch-type reactions. Orig. Life. Evol. Biosph. 29, 153–166.

McKay, C.P., Porco, C.C., Altheide, T., Davis, W.L., Kral, T.A., 2008. The possible origin and persistence of life on Enceladus and detection of biomarkers in the plume. Astrobiology 8, 909–919.

Meeßen, J., Wuthenow, P., Schille, P., Rabbow, E., de Vera, J.-P.P., Ott, S., 2015. Resistance of the lichen *Buellia frigida* to simulated space conditions during the preflight tests for BIOMEX—Viability assay and morphological stability. Astrobiology 15, 601–615.

Mulkidjanian, A.Y., Galperin, M.Y., 2009. On the origin of life in the zinc world. 2. Validation of the hypothesis on the photosynthesizing zinc sulfide edifices as cradles of life on Earth. Biol. Direct 4, 1–37.

Muller, A.W.J., 2005. Thermosynthesis as energy source for the RNA world: a model for the bioenergetics of the origin of life. Biosystems 82, 93–102.

Nicholson, W.L., Munakata, N., Horneck, G., Melosh, H.J., Setlow, P., 2000. Resistance of Bacillus endospores to extreme terrestrial and extraterrestrial environments. Microbiol. Mol. Biol. Rev. 64, 548–572.

Onaka, T., Matsumoto, H., Sakon, I., Kaneda, H., 2008. Organic compounds in galaxies. In: Proceedings IAU Symposium, 251, pp. 229–236.

Oró, J., Kimball, A.P., 1961. Synthesis of purines under possible primitive earth conditions, I. Adenine from hydrogen cyanide. Arch. Biochem. Biophys. 94, 217–227.

Rauf, K., Wickramsinghe, C., 2010. Evidence for biodegradation products in the interstellar medium. Int. J. Astrobiol. 9, 29–34.

Rho, J., Kozasa, T., Reach, W.T., Smith, J.D., Rudnick, L., DeLaney, T., Ennis, J.A., Gomez, H., Tappe, A., 2008. Freshly formed dust in the Cassiopeia A supernova remnant as revealed by the Spitzer space telescope. Astrophys. J. 673, 271–282.

Roth, A., Breaker, R.R., 2009. The structural and functional diversity of metabolite-binding riboswitches. Annu. Rev. Plant Physiol. Plant Mol. Biol. 78, 305–334.

Rothschild, L.J., Mancinelli, R.L., 2001. Life in extreme environments. Nature 409, 1092–1101.

Roy, D., Najafian, K., von Rague, P., 2007. Chemical evolution: the mechanism of the formation of adenine under prebiotic conditions. PNAS 104, 17272–17277.

Russell, M.J., Barge, L.M., Bhartia, R., Bocanegra, D., Bracher, P.J., Branscomb, E., Kidd, R., McGlynn, S., Meier, D.H., Nitschke, W., Shibuya, T., Vance, S., White, L., Kanik, I., 2014. The drive to life on wet and icy worlds. Astrobiology 14, 308–343.

Saladino, R., Crestini, C., Ciciriello, F., Costanzo, G., Di Mauro, E., 2007. Formamide chemistry and the origin of informational polymers. Chem. Biodivers. 4, 694–720.

Sandford, S.A., Engrand, C., Rotundi, A., 2016. Organic matter in cosmic dust. Elements 12, 185–189.

Schopf, J.W., 1993. Microfossils of the early archean apex chert: new evidence of the antiquity of life. Science 260, 640–646.

Schulze-Makuch, D., 2010. Io: is life possible between fire and ice? J. Cosmol. 5, 912–919.

ScienceDaily, 2007. Bacteria Found in Tar Pits. http://www.sciencedaily.com/releases/2007/05/070510151916. htm, May.

Sephton, M.A., 2002. Organic compounds in carbonaceous meteorites. Nat. Prod. Rep. 19, 292–311.

Sephton, M.A., 2003. Formation of boundary layers. In: Gilmour, I., Sephton, M.A. (Eds.), An Introduction to Astrobiology. The Cambridge University Press, Cambridge, pp. 31–32.

Shawn, G.A., Domagal-Goldman, D., Meadows, V.S., Wolf, E.T., Schwieterman, E., Charnay, B., Claire, M., Hébrard, E., Trainer, M.G., 2016. The pale orange dot: The spectrum and habitability of hazy archean Earth. Earth Planet. Astrophys. https://doi.org/10.1089/ast.2015.1422.

Snow, T.P., McCall, B.J., 2006. Diffuse atomic and molecular clouds. Annu. Rev. Astron. Astrophys. 44, 367–414.

Strazzulla, G., Baratta, G.A., Palumbo, M.E., 2001. Vibrational spectroscopy of ion-irradiated ices. Spectrochim. Acta A Mol. Biomol. Spectrosc. 57, 825–842.

Szostak, J.W., 2012. The eightfold path to non-enzymatic RNA replication. J. Syst. Chem. 3, 1–6.

Teanby, N.A., Irwin, R.J., de Kok, R., Vinatier, S., Bezard, B., Nixon, C.A., Flasar, F.M., Culcutt, S.B., Bowles, N.E., Fletcher, L., Howett, C., Taylor, F.W., 2007. Vertical profiles of HCN, HC3N, and C2H2 in Titan's atmosphere derived from Cassini/CIRS data. Icarus 186, 364–384.

Thauer, R.K., 1998. Biochemistry of methanogenesis: a tribute to Marjory Stephenson. Microbiology 144, 2377–2406.

Thomas, R.J., Krehbiel, P.R., Rison, W., Edens, H.E., Aulich, G.D., Winn, W.P., McNutt, S.R., Tytgat, G., Clark, E., 2007. Electrical activity during the 2006 Mount St. Augustine volcanic eruptions. Science 315, 1097.

University of Koeln, 2017. cdms/molecules. http://www.astro.uni-koeln.de/cdms/molecules.

Vreeland, R.H., Rosenzweig, W.D., Powers, D.W., 2000. Isolation of a 250 million-year-old halotolerant bacterium from a primary salt crystal. Nature 407, 897–900.

Wainwright, M., Wickramasinghe, N.C., Narlikar, J.V., Rajaratnam, P., 2003. Microorganisms cultured from stratospheric air samples obtained at 41 km. Fems Microbiol. Lett. 218, 161–165.

Waite, J. Jr, Combi, M.R., Ip, W.-H., Cravens, T.E., McNutt Jr., R.L., Kasprzak, W., Yelle, R., Luhmann, J., Niemann, H., Gell, D., Magee, B., Fletcher, G., Lunine, J., Tseng, W.-L., 2006. Cassini ion and neutral mass spectrometer: enceladus plume composition and structure. Science 311, 1419–1422.

Whangbo, J.S., Hunter, C.P., 2008. Environmental RNA interference. Trends Genet. 643, 1–9.

Wickramasinghe, C., 2003. SARS—a clue to its origins? Lancet 361, 1832.

Wickramasinghe, C., Smith, W.E., 2014. Convergence to panspermia. Hypothesis 12, 1–4.

Wikipedia, 2017a. Age of the universe. https://en.wikipedia.org/wiki/Age_of_the_universe.

Wikipedia, 2017b. Universe/media/File: CMBTimeline300. https://en.wikipedia.org/wiki/Universe#/media/File: CMB_Timeline300_no_WMAP.jpg.

Wikipedia, 2017c. Formose reaction. https://en.wikipedia.org/wiki/Formose_reaction.

LIFE BEFORE ITS ORIGIN ON EARTH: IMPLICATIONS OF A LATE EMERGENCE OF TERRESTRIAL LIFE

Julian Chela-Flores[*,†]
The Abdus Salam ICTP, Trieste, Italy
†*IDEA, Caracas, Venezuela*

CHAPTER OUTLINE

1 INTRODUCTION

The established view that life on Earth is represented by a single-rooted phylogenetic tree with three domains (Woese et al., 1990) is confronted by alternative scenarios that view life in the cosmos, not as a tree of life, but better described by the metaphor of a universe with a "cosmic forest of life" (cf., Section 3). We attempt to follow-up the possibility that life on Earth is a relatively recent phenomenon in geologic terms. There have been some suggestions that this may be the case, for instance,

Habitability of the Universe Before Earth, editors: Richard Gordon & Alexei Sharov, Volume 1 in the series:
Astrobiology: Exploring Life on Earth and Beyond, series editors: Pabulo Henrique Rampelotto,
Joseph Seckbach & Richard Gordon. ISSN 2468-6352. https://doi.org/10.1016/B978-0-12-811940-2.00015-0

Lineweaver (2001) has supported this view with his estimate on the age distribution of terrestrial planets in the Milky Way. His calculations show that Earth-like planets began forming more than 9.2 billion years ago and that their median age is $t_{med} = (6.4 \pm 0.9) \times 10^9$ years. This estimate is much greater than the age of the Solar System itself.

Living phenomena that emerged prior to terrestrial life has been suggested implicitly, though it has only been clearly stated in fiction. For example, Carl Sagan, a distinguished astronomer, astrobiologist, and confirmed enthusiast for science communication, militated in favor of life elsewhere being more ancient than life on Earth: This opinion is evident in his novel Cosmos. After a first contact had been made with another intelligent species, the more evolved beings assert (Sagan, 1985):

> In our case, we emerged a long time ago on many different worlds in the Milky Way. The first of us developed interstellar space-flight.

In the same context, the English writer, Sir Arthur C. Clarke, remarkable for his writings on science fiction and his vision on the progress of science, advocated that life on Earth was a late cosmic phenomenon. This is especially clear in the short story The Sentinel (Clarke, 1951): "When our world was half its present age, something from the stars swept through the Solar System... and went again upon its way."

In terms of a traditional scientific approach that is more appropriate for the present work, theoretical hypotheses have only gone as far as arguing in favor of terrestrial life being a rare case in a galactic scale (Carter, 1983). Additional theoretical discussions support the rare life hypothesis, but accept that the "final scientific assessment on life in the universe will only come from biologists and observers" (Livio, 2008).

The question of how ancient life in the cosmos really emerged can be tested, in principle, with present instrumentation and with future feasible technologies that are now being developed. These open questions force us to reconsider life before its origin on Earth. Some implications of the existence of extraterrestrial life as an astrobiology phenomenon have been covered in the cases of theology (Haught, 1998; Lovin, 2015) and philosophy (Lupisella, 2015).

1.1 TIME AVAILABLE BEFORE THE EMERGENCE OF LIFE ON EARTH

The age of the Solar System is about half the age of our galaxy. But the Sun itself is about half way through its lifetime. Consistent with past estimates, Earth was formed after 80% of Earth-like planets already existed (Behroozi & Peeples, 2015). Hence, the time available for life to have evolved before Earth is considerable.

The afterglow radiation of the cosmic microwave background (CMB) has steadily decreased in temperature as the universe expands. Today, the expansion has induced a CMB temperature close to the absolute zero. This is equivalent to a red shift of $z \sim 100$. Yet going back in time, we can estimate the time after the Big Bang as being 15 Myr when the temperature was close to room temperature, a condition that is favorable to life (Loeb, 2014). The implication is that in such circumstances any planet, or satellite around an early star, could have sustained oceans of liquid water.

This phenomenon would be independent of the particular orbit, and it may even happen on planets that have escaped from their original stellar orbit. The only question that remains open is whether the early star system would have generated sufficient basic chemical elements (produced from earlier supernovae) for supporting biosynthesis. In the absence of sufficient data, such an early onset of life in the cosmos remains an open possibility.

Thermal gradients are needed for life. These can be supplied by geological variations on the surface of rocky planets. Examples for sources of free energy are geothermal energy powered by the planet's gravitational energy at the time of its formation and radioactive energy from unstable elements produced by the earliest supernova. These internal heat sources (in addition to possible heating by a nearby star) may have kept planets warm even without the CMB, extending the habitable epoch from such an early age to later times.

The CMB temperature at a time well before the moment of cosmic decoupling may have allowed ice to form on objects that delivered water to a planet's surface. This phenomenon may have helped to maintain the cold trap of water in the planet's stratosphere. Planets could have kept a blanket of molecular hydrogen that maintained their warmth (Stevenson, 1999; Pierrehumbert & Gaidos, 2011), allowing life to persist on internally warmed planets at late cosmic times. If life persisted at $z \sim 100$, it could have been transported to newly formed objects through panspermia (McNichol & Gordon, 2012). Under the assumption that interstellar panspermia is plausible, the redshift of $z \sim 100$ can be regarded as the earliest cosmic epoch after which life was possible in our Universe. Searching for atmospheric biomarkers in planets, or satellites around low-metallicity stars in the Milky Way galaxy, or its dwarf galaxy satellites can test the feasibility of life in the early universe. Such stars represent the closest analogs to the first generation of stars at early cosmic times.

1.2 RATIONALIZING OUR ORIGINS IN TERMS OF THERMODYNAMICS

A recent original approach to the origin of life is due to Jeremy England, which applies to any terrestrial-like planet. He emphasizes rationalization of the origin of life in terms of thermodynamics (England, 2013), rather than on chemical evolution, as we have persevered exhaustively in the past (Ponnamperuma & Chela-Flores, 1995; Chela-Flores et al., 1995, 2001). We assume only the occurrence of biological phenomena that are independent of any contingent facts here on Earth. We further assume that such conditions may hold in other places, such as in exoplanets, where life might develop, since the expected number of Earth-like planets is sufficiently large (cf., Section 1.1). We will refer to certain biological phenomena that make no reference to any contingent facts here on Earth. In addition, such special conditions could be expected to hold in other places where life might develop. As a consequence of the two previous assumptions, universal biology may be a robust hypothesis (Mariscal, 2015).

To a certain extent, such an insight into a universal biology may be testable, especially concerning the case of evolutionary convergence (Chela-Flores, 2003). This is in principle possible to test in the foreseeable future, due to the approval of funding by the European Space Agency (ESA) for a mission to the Jovian icy satellites in the 2020s (Grasset et al., 2013). Another factor in favor of testing convergence is the formidable progress in instrument miniaturization that has already an appreciable heritage in previous successful exploration of the Solar System (Tulej et al., 2015). These two factors facilitate the search for feasible biomarkers that we have repeatedly suggested in the past (Chela-Flores & Kumar, 2008; Chela-Flores et al., 2015).

---

2 HOW WOULD WE SEE OURSELVES IF EARLY ORIGINS ARE IDENTIFIED?

Biocentrism is the view adopted by some philosophers and scientists that life has evolved exclusively on Earth. This has been a concept deeply rooted into our cultural history. Since the ancient times of Greek civilization, it has persevered to relatively recent times (Henderson, 1913; Chela-Flores, 2005).

It has been reconsidered, firstly by the astronomy of Copernicus and Galileo that displaced our planet from the center of the cosmos. With Darwin, we reevaluated biocentrism once more.

2.1 APPROACHING THE END OF BIOCENTRISM IF LIFE ON EARTH IS A LATECOMER

The displacement of humans took place from a privileged position that otherwise would have been radically removed from the evolution of life on Earth. Now, Darwinism is widely accepted, even in a theological context. We should now turn to the restricted topic of cultural implications that detection of an extraterrestrial civilization, either contemporary, but especially preceding us, might have on the ever-receding position of biocentrism.

Two different cases are relevant for the theme of life emerging in the universe. One aspect concerns firstly, whether the origin of the extraterrestrial civilization is contemporary or less developed than ours, or secondly, whether there are further implications if life on Earth is a latecomer in the evolution of life in the cosmos.

2.2 ANTHROPOCENTRISM

This doctrine maintains that man is the center of everything, the ultimate end of nature. But there are topics that argue in favor of a wider point of view. One is the surprising evolution of brain size in humans; for the knowledge we have today is very narrow when viewed in geologic time. Recently, in these terms dolphins on Earth had brains that were comparable with our ancestors (Marino, 2000), but more surprisingly, dolphin EQ actually exceeded the EQ of our ancestors for times greater than 2 Mya.

This is relevant to the evolution of intelligence, including human intelligence, many aspects of which may not be exceptional. Lori's findings have important implications for the probability of the emergence of human level of intelligence in an independent lineage. In the case of an early evolution of life elsewhere, the evolution of intelligence comparable to our experience on Earth is not excluded and anthropocentrism recedes as a valid doctrine.

The second point to keep in mind is by the Australian ethicist Peter Singer (Singer, 1993), who has argued that the implications of the fundamental principle of equality for humans should not be seen from the limited point of view of anthropocentrism and equality should be extended to animals as well. But, once again if life on Earth is a latecomer, the Singer extrapolations should be extended to all life, independent of both their evolutionary stage and their cosmic location.

3 PHILOSOPHICAL COMMENTS ON AN EARLY "FOREST OF LIFE"

The metaphor of a "cosmic forest of life" (Martini & Chela-Flores, 1999) is a convenient phrase that summarizes the possibility that life exists on the large number of exoplanets discovered by the Kepler mission. From early philosophical doctrines, beginning with the Greek philosopher Protagoras (born around 500 BC, in Abdera), there has been the conviction that "man is the measure of all things" (Russell, 1991). Sadly, this has still not been removed from our thinking when we search for life in cosmos.

3.1 PROCESS PHILOSOPHY

Process philosophy holds that what exists in nature is not just originated and sustained by processes, but in fact processes specify what exists. Similarly, process theology is an approach to natural theology based on the metaphysics of Alfred North Whitehead, who rejects Divine Action in terms of causality, especially regarding events that are mostly determined by their past. This philosophical system is considered to be particularly helpful in the task of constructing an evolutionary theology that may throw some further insights into Darwinism (Haught, 1998) and consequently into astrobiology.

Another approach along these lines points out that theologians already have the concept of Divine action continuous creation, which can be used to explore the implications of modern science for religious belief (Coyne, 2005). In this view, God is working with the universe (and consequently, life in the cosmos). The universe has a certain vitality of its own like a child does. It has the ability to respond to words of endearment and encouragement. You may discipline a child, but you try to preserve and enrich the individual character of the child and its own passion for life. A parent must allow the child to grow into adulthood, to come to make its own choices, to go on its own way in life.

In such a way, we can imagine that Divine action may deal with the universe and its life contents. There are no compulsory theological arguments pointing towards a limit, in this sense, to single out a planet in the periphery of a single galaxy such as ours. Along the above lines, it is possible that evolution may also provide a way in which the tradition of natural theology may undergo a renewal. Within the scope of process philosophy, instead of discussing design without a designer, such as in terms of Darwinism (Ayala, 1998), both a renewed philosophy and a revived natural theology may take place if we interpret correctly the origin, evolution, and distribution of life (astrobiology).

3.2 STELLAR EVOLUTION

Stellar evolution predicts that the lifetime of the Sun will continue for a few billion years, while the science of anthropology suggests that humans are a fairly recent addition to an Earth biota, whose origins may be traced back to microorganisms that emerged on the early Earth from 3 to 4 billion years ago. We expect that philosophers and theologians will continue their independent search for a deeper understanding.

Science will be confronted with ever-increasing mysteries, as our scientific instruments become better and more accurate to allow confrontation of theory with experiment. The questions of how the universe and life in it were formed will reveal new insights. Some questions, however, escape the scope of astrobiology. We expect that future developments in philosophy and theology will gradually give us better insights into the question that was raised by the philosopher Gottfried Wilhelm Leibniz (1714):

> Why is there something rather than nothing?. ...

Further, assuming that things must exist, it must be possible to give a reason why they should exist as they do and not otherwise. Would the presence of a cosmic forest of life guide us towards a partial answer to Leibnitz queries?

3.3 CULTURAL COMMENTS ON AN EARLY FOREST OF LIFE

With the eventual discovery of other life in the cosmos, either contemporary or before the Sun was formed, we would have a deeper understanding of human beings. Such an event would lead to a careful reappraisal of the above biblical quotation. In order to find out what is the position of our tree of life in the universe, we have to appeal to astrobiology. Evolutionary convergence is an aspect of the evolutionary phenomenon that constrains its randomness. In his book, Chance and Necessity Monod put too much emphasis on the role played by pure randomness in Darwinism (Monod, 1972).

Inserting the word "blindness" in this context is, according to Polkinghorne (1996), a tendentious adjective. He has exposed the limitation of this point of view: It could be interpreted that a world of chance and necessity is necessarily one that lacks purpose. Alternatively, although it is possible to defend such a metaphysical statement, "it is a claim that should not be presented as if it were a scientific conclusion."

The possibility that evolutionary convergence has actually taken place elsewhere in the universe is a question that can also be tested observationally by means of radio astronomy (cf., Section 1.2 in relation with testing the restricted aspect of universal biology in the Solar System). Since the middle of the last century, signals have been searched from evolutionary processes that have actually led to intelligence in the sense of beings that have actually developed communication technologies. The question still remains open, but this astronomical search would be a direct way to get profound insights into the question raised in the title of this chapter. The search for other intelligences assumes that the tendencies of terrestrial biological evolution (towards more complex neural networks such as our brains) with the consequent intelligences, if eventually successful, could serve as a first stage in our search for the place of our tree of life in the universe.

4 TERRESTRIAL LIFE AS A LATECOMER IN COSMIC EVOLUTION

To conclude, in this section we discuss what would be the cultural changes that would be forced upon us if terrestrial life were a latecomer in cosmic evolution. From the point of view of the present work, we focus, firstly, on what were the impressions of the presence of life on Earth in the middle of the 20th century, well-summarized by the English philosopher Bertrand Russell (1985). Secondly, we will review the cultural perspectives in the foreseeable future, now that new insights are open due to the existence of planetary systems around other stars. This additional information will be possible by probing exoplanets in the next few years, after the completion of better, much improved astronomical instrumentation that will go beyond the Kepler mission.

The Earth is one of the smaller planets of a not particularly important star, a very minor portion of the Milky Way, which is one of the very large number of galaxies. This is still a fair description of our position in the cosmos. Russell continues with a stronger statement that there might be no life at all except on this planet or now in agreement with the "rare Earth" hypothesis that the origin of life and the evolution of biological complexity on Earth (and, subsequently, human intelligence) required an improbable combination of astrophysical and geological events and circumstances (Ward & Brownlee, 2000).

Quite impressively for a writer in the middle of the 20th century, Russell continues to consider the possibility of extra-solar planets "in such stages of development this one," but from the point of view of this work, we disagree with his anticipation of the existence being confined to few parts of the universe.

For several years, the American agency, NASA, sponsored the Kepler mission with a satellite in a solar orbit similar to the terrestrial one. With the Kepler results, there is now a growing conviction among astrobiologists for the existence of life in the two thousand, or so, planets known to be orbiting around other stars. Several of these are Earth-like. In the coming years, among the components of this restricted group, we are convinced that they will provide examples, firstly of a second genesis with a second tree of life, and secondly, evidence would be forthcoming for a full forest of life. The scientific basis for this conviction is the imminent arrival of various instruments that are now under construction, including new huge telescopes. But life as a latecomer in evolution raises new questions beyond Russell's impressive foresight that was limited by his early approach. We discuss in turn the two cases of the solar system and other solar systems.

There is some preliminary conviction among astrobiologists that at least in the solar system, there may have been two trees of life in the cosmic forest of life and the terrestrial tree may have been preceded by one on Mars: Firstly, we survived here on Earth. But on Mars, there is ample evidence that bacterial life may have started off contemporarily with the emergence of our own bacteria some 4 billion years ago. Yet, our dynamic planet with earthquakes and atmosphere worked in favor of preserving bacterial life. These aspects of the terrestrial geophysics allowed Darwinian evolution to raise the bacterial blueprint to the human one. But Mars lacked (and still lacks) the life-friendly dynamic features of our planet. It gave no opportunity for evolution to raise the bacterial blueprint, inevitably leading to the extinction of Martian life that most likely emerged only as an episode early in the solar system.

Yet, we still cannot exclude extant microbial life under the polar ice or in underground sites. Others have discussed the possibility that the possible episodic biological event may have preceded the terrestrial one (Davies, 1998). If it did happen, that event would encourage further theological discussions beyond the literal interpretation of Genesis. In the Leibnitz quotation (cf., Section 3), it would be possible to give a reason why things should exist as they do, and not otherwise. This question of existence can be appreciated in a different light if we were latecomers in the universe.

Granted the hypothesis of life being a latecomer in cosmic evolution, philosophy would have a fresh point of view to face the Leibnitz queries. Indeed, astrobiology would provide an explanation why life would exist on Earth, especially if the above-mentioned general thermodynamic rationalization is valid. In this case, life would be the inevitable outcome on Earth-like planets in a process that would be traced back to planetary formation in earlier cycles of life. Philosophy would be presented with the new challenge for the interpretation of why the living process should exist as it does, and not otherwise.

5 CONCLUSION

New cultural perspectives will arise in the post-Kepler era, when we would be facing the eventual recognition of a multitude of terrestrial-like worlds in our galaxy and beyond. The likelihood of the emergence of habitable planets, before the emergence of the Solar planetary system, is now a possibility that deserves consideration, given the forthcoming short-term development of improved astronomical instrumentation that will clarify what are the specific biomarkers that would be recognizable. We feel that the above sketch of the cultural novelty in Section 4 deserves more thorough attention, beyond the preliminary comments of the present work.

REFERENCES

Ayala, F.J., 1998. Darwin's devolution: design without designer. In: Russell, R.J., Stoeger, W.R., Ayala, F.J. (Eds.), Evolutionary and Molecular Biology: Scientific Perspectives on Divine Action. Vatican Observatory and the Center for Theology and the Natural Sciences (CTNS), Vatican City State/Berkeley, CA, pp. 101–116.

Behroozi, P.S., Peeples, M., 2015. On The History and Future of Cosmic Planet Formation. MNRAS 454, 1811–1817.

Carter, B., 1983. The Anthropic Principle and its Implications for Biological Evolution. Phil. Trans. R. Soc. Lond. A 310, 347.

Chela-Flores, J., 2003. Testing evolutionary convergence on Europa. Int. J. Astrobiol. 2 (4), 307–312.

Chela-Flores, J., 2005. Fitness of the universe for a second genesis: is it compatible with science and Christianity? Sci. Christ. Belief 17 (2), 187–197.

Chela-Flores, J., Kumar, N., 2008. Returning to Europa: can traces of surficial life be detected? Int. J. Astrobiol. 7, 263–269.

Chela-Flores, J., Chadha, M., Negron-Mendoza, A., Oshima, T. (Eds.), 1995. In: Chemical Evolution: Self-Organization of the Macromolecules of Life (A Cyril Ponnamperuma Festschrift), Vol. 139. A. Deepak Publishing, Hampton, VA.

Chela-Flores, J., Owen, T., Raulin, F. (Eds.), 2001. The First Steps of Life in the Universe. Kluwer Academic Publishers, Dordrecht, The Netherlands.

Chela-Flores, J., Cicuttin, A., Crespo, M.L., Tuniz, C., 2015. Biogeochemical fingerprints of life: earlier analogies with polar ecosystems suggest feasible instrumentation for probing the Galilean moons. Int. J. Astrobiol. 14, 427–434. http://users.ictp.it/~chelaf/IJA2015.pdf.

Clarke, A. C., 1951. The Sentinel. Reprinted in: 2001 A Space Odyssey. Arrow Books, 1988, pp. 239–250.

Coyne, S.J.G., 2005. God's chance creation. Tablet 6. http://www.ictp.it/~chelaf/Coyne.pdf.

Davies, P.C.W., 1998. Did Earthlife come from Mars? In: Chela-Flores, J., Raulin, F. (Eds.), Exobiology: Matter, Energy, and Information in the Origin and Evolution of life in the Universe. Kluwer Academic Publishers, The Netherlands, pp. 241–244.

England, J.L., 2013. Statistical physics of self-replication. J. Chem. Phys. 139, 121923-1–121923-8.

Grasset, O., Dougherty, M.K., Coustenis, A., Bunce, E.J., Erd, C., Titov, D., Blanc, M., Coates, A., Drossart, P., Fletcher, L.N., Hussmann, H., Jaumann, R., Krupp, N., Lebreton, J.-P., Prieto-Ballesteros, O., Tortora, P., Tosi, F., Van Hoolst, T., 2013. JUpiter ICy moons Explorer (JUICE): an ESA mission to orbit Ganymede and to characterize the Jupiter system. Planet. Space Sci. 78, 1–21.

Haught, J.F., 1998. Darwin's gift to theology. In: Russell, R.J., Stoeger, W.R., Ayala, F.J. (Eds.), Evolutionary and Molecular Biology: Scientific Perspectives on Divine Action. Vatican Observatory and the Center for Theology and the Natural Sciences (CTNS), Vatican City State/Berkeley, CA, pp. 393–418.

Henderson, L.J., 1913. The Fitness of the Environment: An Inquiry into the Biological Significance of the Properties of Matter. MacMillan, New York. Reprinted in 1958 by Beacon Press, Boston, 1970.

Leibniz, G., 1714. The principles of nature and grace, based on reason. In: Loemker, L. (Ed.), Philosophical Papers and Letters. D. Reidel, Dordrecht, pp. 636–642. 1969.

Lineweaver, C.H., 2001. An estimate of the age distribution of terrestrial planets in the Universe: quantifying metallicity as a selection effect. Icarus 151, 307–313.

Livio, M., 2008. The interconnections between cosmology and life. In: Barrow, J.D., Conway Morris, S., Freeland, S.J., Harper Jr., C.L. (Eds.), Fitness of the Cosmos for Life Biochemistry and Fine-Tuning. Cambridge University Press, Cambridge, pp. 114–131.

Loeb, A., 2014. The habitable epoch of the early Universe. Int. J. Astrobiology 13 (4), 337–339.

Lovin, R.W., 2015. Astrobiology and theology. In: Dick, S.J. (Ed.), The Impact of Discovering Life Beyond Earth. Cambridge University Press, Cambridge, pp. 222–232.

Lupisella, M., 2015. Life, intelligence and the pursuit of value in cosmic evolution. In: Dick, S.J. (Ed.), The Impact of Discovering Life Beyond Earth. Cambridge University Press, Cambridge, pp. 159–174.

Marino, L., 2000. Turning the empirical corner on Fi. In: Bioastronomy 99 A New Era in Bioastronomy, Astronomical Society of the Pacific Conference Series, pp. 431–435.

Mariscal, C., 2015. Universal biology: assessing universality. In: Dick, S.J. (Ed.), The Impact of Discovering Life Beyond Earth. Cambridge University Press, Cambridge, pp. 113–126.

Martini, C.M., Chela-Flores, J., 1999. Dialogo. In: Sindoni, E., Sinigaglia, C. (Eds.), Carlo Maria Martini Orizzonti e limiti della scienza (Decima Cattedra di non credenti). "Scienze e Idee," Directed by G. Giorello, Raffaello Cortina Editore, Milano, pp. 65–68.

McNichol, J., Gordon, R., 2012. Are we from outer space? A critical review of the Panspermia hypothesis. In: Seckbach, J. (Ed.), Genesis—In the Beginning: Precursors of Life, Chemical Models and Early Biological Evolution. Springer, Dordrecht, pp. 591–620.

Monod, J., 1972. Chance and Necessity. Collins, London. p. 187.

Pierrehumbert, R., Gaidos, E., 2011. Hydrogen greenhouse planets beyond the habitable zone. Astrophys. J. Lett. 734, 1–5. https://doi.org/10.1088/2041-8205/734/1/L13.

Polkinghorne, J., 1996. Scientists as Theologians. SPCK, London.

Ponnamperuma, C., Chela-Flores, J. (Eds.), 1995. Chemical Evolution: The Structure and Model of the First Cell. Kluwer Academic Publishers, Dordrecht, The Netherlands.

Russell, B., 1985. The existence and nature of God. In: Rampel, R., Brink, A., Moran, M. (Eds.), Collected Papers of Bertrand Russell, vol. 12. Allen and Unwin, London. Reproduced in: Greenspan, L., Andersson, S. (Eds.) (1999). Russell on Religion. Selections from the writings of Bertrand Russell, Routledge, London and New York, pp. 92–106.

Russell, B., 1991. History of Western Philosophy and its Connection with Political and Social Circumstances from the Earliest Times to the Present Day. Routledge, London. pp. 91–97.

Sagan, C., 1985. Contact. Pocket Books, a division of Simon & Schuster Inc., New York. (Chapter 20), p. 366.

Singer, P., 1993. Practical Ethics, second ed. CUP, Cambridge.

Stevenson, D.J., 1999. Life sustaining planets in interstellar space? Nature 400, 32–33.

Tulej, M., Neubeck, A., Ivarsson, M., Riedo, A., Neuland, M.B., Meyer, S., Wurz, P., 2015. Chemical composition of micrometer-sized filaments in an Aragonite host by a miniature laser ablation/ionization mass spectrometer. Astrobiology 15 (8), 1–14.

Ward, P., Brownlee, D., 2000. Rare Earth: Why Complex Life is Uncommon in the Universe. Springer, New York.

Woese, C.R., Kandler, O., Wheelis, M., 1990. Towards a natural system of organisms: proposal for the domains Archaea, Bacteria, and Eucarya. Proc. Natl. Acad. Sci. U.S.A. 87 (12), 4576–4579.

FURTHER READING

John Paul II., 1992. Discorso di Giovanni Paolo alla Pontificia Academia delle Scienze. L'Osservatotre Romano, 1 Novembre, p. 8.

SYSTEM PROPERTIES OF LIFE

SYMBIOSIS: WHY WAS THE TRANSITION FROM MICROBIAL PROKARYOTES TO EUKARYOTIC ORGANISMS A COSMIC GIGAYEAR EVENT?

George Mikhailovsky*, Richard Gordon[†,‡]

**Global Mind Share, Norfolk, VA, United States*
[†]*Gulf Specimen Marine Laboratories, Panacea, FL, United States*
[‡]*Wayne State University, Detroit, MI, United States*

CHAPTER OUTLINE

1 INTRODUCTION

The transition from prokaryotes to eukaryotes was an event of hierarchogenesis or megaevolution (Arthur, 2008; Jagers op Akkerhuis, 2010; Jagers op Akkerhuis, 2017; Mikhailovsky and Levich, 2015) on a cosmic time scale. We will use Ga to designate "years ago from the present" and Gy for durations of time, both measured in billions (10^9) of years. Calculating from the earliest evidence of prokaryotes (3.8–3.5 Ga) to the appearance of eukaryotes (about 2 Ga), eukaryogenesis took or lasted 1.5–1.8 Gy (Knoll, 2015), perhaps longer. Why did it take so long?

Habitability of the Universe Before Earth, editors: Richard Gordon & Alexei Sharov, Volume 1 in the series:
Astrobiology: Exploring Life on Earth and Beyond, series editors: Pabulo Henrique Rampelotto,
Joseph Seckbach & Richard Gordon. ISSN 2468-6352. https://doi.org/10.1016/B978-0-12-811940-2.00016-2

2 EUKARYOGENESIS AS SYMBIOSIS

The embryologist Paul Weiss wrote, explaining the role of symbiosis in (mega)evolution in the last point of his "canon":

> "12. Although I have emphasized for didactic reasons the relatively conservative features of systems, the unidirectional change of systems must not be overlooked. We find it expressed, for instance, in the mutability of systemic patterns in evolution, ontogeny, maturation, learning, etc., as well as in the capacity to combine systems into what then appear as super-systems with the emerging properties of novelty and creativity".

(Weiss, 1973)

Weiss highlights the inextricable connection between the emergence of real novelties and symbiotically combining systems into a kind of super-system; in other words, the appearance of a new hierarchical level (hierarchogenesis).

The term "symbiosis" was introduced almost 150 years ago by Heinrich Anton de Bary (de Bary, 1878; Gontier, 2015). But as rightly pointed out by Natalie Gontier (Gontier, 2015), "de Bary was inspired by the zoologist Van Beneden (van Beneden, 1873, 1875), who a couple of years earlier had distinguished between 'commensalism,' 'mutualism,' and 'parasitism' to characterize the 'social lives' of animals." Then, after a quarter of a century, Petr Kropotkin (Kropotkin, 1902) came to the conclusion that mutual aid in addition to competition plays an essential role in evolution. Symbiosis is a form of mutual aid. Kropotkin's ideas were further developed in the context of symbiosis by Konstantin Mereschkowsky, Boris Kozo-Polyansky, and Andrey S. Famitsin (Khakhina, 1992; Lewin, 1994; Provorov, 2016).

However, orthodox Darwinism and the modern evolutionary synthesis (Huxley et al., 2010; Huxley, 1942) left these ideas at the far periphery of the scientific mainstream for almost all of the 20th century. Only during the last few decades, it became more and more obvious that symbiosis, as a kind of win-win strategy, is one of the important factors of evolution (Margulis and Fester, 1991; Sapp, 1994; Watson and Pollack, 1999), particularly reticulate evolution (Gontier, 2015; Wikipedia, 2015b). Numerous symbiotic interrelations occurred throughout evolution after the emergence of multicellular eukaryotes (Douglas, 2010; Sapp, 2009).

The theory of symbiogenesis, according to which eukaryotes appeared as a result of a symbiosis of different prokaryotes, was formulated at the beginning of last century by Mereschkowsky (Mereschkowsky, 1910). His ideas were developed by Kozo-Polyansky (Kozo-Polyansky, 1924). A few years later, Ivan Wallin (Wallin, 1927) presented his endosymbiotic theory that was very close to the concept by Mereschkowsky. Wallin was the first researcher who provided experimental foundations of the idea that eukaryotic organelles originated from prokaryotic bacteria.

For more than 40 years, Mereschkowsky's, Kozo-Polyansky's, and Wallin's ideas were not popular or even well-known until Lynn Margulis published her famous book *Origin of Eukaryotic Cells* that gave endosymbiotic theory new life (Margulis, 1970). However, the path of Margulis' conception to recognition was not strewn with roses. Her formative paper, "On the origin of mitosing cells" (Sagan, 1967), written under her married surname, appeared in the *Journal of Theoretical Biology* (JTB) only after being rejected by fifteen journals (Schaechter, 2012). (This is no record, as the number of rejections before acceptance was 30 for (Maruyama, 1963) and (Gordon, 1999).) At that time, the editor of JTB was James Danielli (Rosen, 1985; Stein, 1986), who erred on the side of letting new ideas see the light of day (personal communication to RG, about 1970). In the 1970s, endosymbiotic theory thereby gained wide recognition. For recent reviews and nuances, see (Cavalier-Smith, 2006; Curtis and Archibald, 2010; Emelyanov, 2001; Ettema, 2016; Garg and Martin, 2016; He et al., 2016; Lang and Burger, 2012; Okie et al., 2016; Pittis and Gabaldón, 2016; Shiflett and Johnson, 2010; Szklarczyk and Huynen, 2010; van der Giezen, 2011).

3 ORDER OF EVENTS RESULTING IN EUKARYOTES

The current debate centers not on whether eukaryotes are symbionts, but on the order of events by which this came about and when (Lang and Burger, 2012). For example, the order in which the three domains, Archaea, Bacteria, and Eukaryotes, appeared is far from settled:

1. Archaea first (Caetano-Anollés et al., 2014).
2. Bacteria first (Cavalier-Smith, 2014).
3. Archaea and Bacteria separately, then Eukaryotes (Fuerst and Sagulenko, 2012; Spang et al., 2013).
4. All three originated separately, none "first" (Kandler, 1994).
5. Eukaryota first (Glansdorff et al., 2008; Mariscal and Doolittle, 2015).
6. "Eukaryotes are just a peculiar kind of archaea" (Zimmer, 2009). In a way (discussed below), this is the current consensus (Dacks et al., 2016).
7. Some researchers (Baum and Baum, 2014) suggest that the eukaryotic cell emerged in an "inside-out" manner.

Models for the origin of Eukaryotes include:

1. Archaea $\Rightarrow$ Eukaryotes (Spang et al., 2015), followed by "a gradual transfer of bacterial genes and membranes" (Dey et al., 2016). Both Bacteria and Eukaryotes have unipolar membrane lipids, whereas Archaea have bipolar membrane lipids (reviewed in (Gordon et al., 2017a)).
2. Bacteria $\Rightarrow$ Eukaryotes (Cavalier-Smith, 2014): Eukaryotes and Archaeobacteria are sister clades evolved from Posibacteria.
3. FECA: the first eukaryotic common ancestor $=>$ LECA: the last eukaryotic common ancestor (Butterfield, 2015; Field and Dacks, 2009; Koumandou et al., 2013).
4. Slow-drip: there is no sudden moment of symbiosis leading to eukaryogenesis. The mechanism is rather one of slow transfer of genes, especially from bacteria to Archaea, probably followed by endosymbiosis incorporating bacterial cells into an Archaea cell (Butterfield, 2015; Lester et al., 2006; Martijn and Ettema, 2013; Moreira and López-García, 1998).

Matej Vesteg and Juraj Krajčovič suggest: "The origin of each domain was likely characterized by bottlenecks followed by rapid radiation…" (Vesteg and Krajčovič, 2011); without suggesting why there were bottlenecks nor how long it might have taken to overcome them. Taxonomy, molecular dating, and biomarkers give quite different estimates of when the first eukaryotes appeared (Fig. 1). The estimates range from 0.95 Ga to 3.5 Ga (Fig. 1), including 1.5 Ga (Fig. 2) and 1.8 to 2.8 Ga (Fig. 3), an overall range of 2.8 Ga to 0.95 Ga. This is hardly a consensus. Thus, it took prokaryotes of the order of 2 Gy to have a successful symbiosis that resulted in Eukaryotes, perhaps as long as 3.25 Gy, or as short as 0.2 Gy for the LUCA to LECA transition, considering both extremes (Fig. 1). For a round figure, we may guesstimate eukaryogenesis at roughly 2 Ga and LUCA at 4 Ga (Figs. 1–3), "shortly" after formation of the Earth and the Moon (Fig. 1). The impact of this choice is that we confine our remarks to this period of Earth's history.

4 WHAT ON EARTH HAPPENED WHEN PROKARYOTES WERE ITS ONLY HABITANTS?

This question has a number of aspects:

1. What is the earliest reliable evidence for prokaryotes?
2. What is the earliest reliable evidence for eukaryotes?
3. How long was and when did the transition take place?
4. What other events were occurring before and during this transition?
5. Could any of these other events have been causal or necessary prerequisites of the transition?

To answer question #4 first, we have rescaled Figures 1–23 from many publications so that they are all on the same abscissa for time before present, starting with the formation of the Earth at about 4.5 Ga. The accretion of most of the material forming the Earth took 8 to 12 My, followed 40 to 100 My later by collision with a Mars-sized planet named Theia and formation of the Moon (Gordon and Gordon, 2016; Yu and Jacobsen, 2011) (Figs. 1 and 4). The Earth's surface consisted "of a global magma ocean that cover[ed] most of the Earth's surface" with a depth of 500 to 3000 km of molten rock at $4000°K$ (Maas and Hansen, 2015) that solidified within 10 My years (Valley, 2005). There is evidence for an anoxic, sulfur-poor, iron-rich liquid water ocean by 4.4 Ga (Figs. 4 and 5) (Nutman, 2006; Wilde et al., 2001) to 4.3 Ga (Harrison, 2009; Mojzsis et al., 2001), even though frequent impacts of large asteroids and meteors persisted until 4.1 Ga and then tapered off (Fig. 6). The Earth was spinning with a 6-h period or less compared to the present 24 h (Fig. 7). The Sun was at less than 75% of its present luminosity (Fig. 8) (Feulner, 2012).

The faint sun paradox is the notion that Earth's cooling should have continued to the point where the global ocean froze (Figs. 9 and 10), but it seems not to have done so. The main explanation of this paradox is the greenhouse effect (Lowe and Tice, 2004; Schaefer and Fegley, 2014; Shaw, 2008), perhaps as a result of methane produced by the Late Heavy Bombardment (LHB) event at $\sim$3.9 Ga (Figs. 6 and 11) and/or from methanogenic prokaryotes (Battistuzzi et al., 2004). However, according the latest data (Boehnke and Harrison, 2016), the LHB could be illusory. Another hypothesis explains the faint young Sun paradox by greater radiogenic heat emitted from the decay of the isotopes K_{40}, U_{235}, U_{238}, and Th_{232}, which was at that time about 4 times (Arevalo et al., 2009) to 6 times (Valley, 2005) more than now (Fig. 12). But the problem persists yet, as the faint sun paradox does not have any commonly accepted solution.

As for continents, by which we mean portions of Earth's crust above the water ocean, some microcontinents possibly existed as "the first sialic crust, which provided building blocks for the earliest microcontinents" at 4.02 to 3.9 Ga (Grosch and Hazen, 2015), to 3.4 Ga (Griffin et al., 2011; Lang and Burger, 2012), but most appeared only 0.8 Gy later at 3 Ga (Dhuime et al., 2015) (Figs. 4 and 13). The earliest microcontinents may have been a byproduct of early life (Grosch and Hazen, 2015). The Earth's crust at the beginning of the Archaean at 4.0 Ga was thus mostly covered with oceans and an atmosphere above (Fig. 5). If the early solar wind (Fig. 14) depleted the Earth (Tarduno et al., 2014) of much of its original water (Izidoro et al., 2013), one possible implication is that "the Earth may have had a greater water inventory at and prior to the Paleoarchaean, allowing for preservation of the modern oceans" (Tarduno et al., 2014). In that case, the early ocean would have been much deeper, possibly delaying the appearance of any above water land.

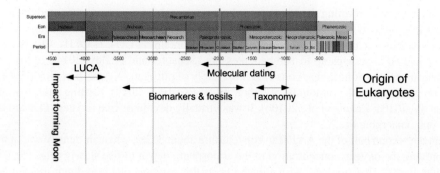

FIG. 1

What on Earth happened. In Figs. 1–23, we have taken drawings from many sources and rescaled and aligned them to fit on a common time axis. They may be printed and cut out to combine them in one long strip. Vertical lines crossing all figures are at Gy intervals. Some figures had to be left/right flipped to conform to the direction of the time axis. In these cases, we have added some legible labels. While some figures give overlapping information, their estimates of the timing of events often differ and so warrant comparisons. *Banner*: Names of geological periods. The leftmost date may be taken as the origin of the solar system (4.52 Ga (Zahnle et al., 2007), 4.566 Ga (Zhang, 2002)) or of the Earth (4.54 Ga (Wikipedia, 2015a), 4.550 Ga (Vázquez et al., 2010), 4.56 Ga (Moorbath, 2009), 4.568 Ga (Bourdon et al., 2008), 4.6 Ga (Cavalier-Smith, 2014)). From Wikipedia (2016c) with permission under a Creative Commons Attribution-ShareAlike License. This banner is repeated for each figure in this set. Cr. = Cryogenian, Ed. = Ediacaran, C = Cenozoic. *Bottom*: Three approaches to the timing of the origin of eukaryotes. *Taxonomy*: "...bracket the origin of eukaryotes between 0.95 and 1.45 Ga ago" (Cavalier-Smith, 2014). *Bangiomorpha* at 1.2 Ga (Butterfield, 2000) falls in this range. *Molecular dating*: as reviewed in (Butterfield, 2015). See Fig. 3 for an earlier estimate. *Biomarkers & fossils*: as reviewed in (Butterfield, 2015). The earliest evidence is for methane-producing Euryarchaeota at 3.46 Ga (Ueno et al., 2006). The right end of this interval is "The oldest known body fossils that can be positively assigned to the Eukaryota (Prasad et al., 2005; Ray, 2006)" at 1.631 + 0.001 Ga, after which there is "a clear and continuous record of eukaryotic life" (Butterfield, 2015). "With an early Archaean FECA [First Eukaryotic Common Ancestor] and an early–middle Palaeoproterozoic LECA [Last Eukaryotic Common Ancestor], the assembly of a modern-style eukaryotic cell appears to have taken a billion, maybe a billion and a half years" (Butterfield, 2015). One hypothesis that LUCA (Last Universal Common Ancestor) was a protoeukaryote with Archaea and Bacteria being later reduced forms (Glansdorff et al., 2008) would put the origin of eukaryotes in the Hadean (Fig. 2). A recent critical review would not extend this arrow any earlier than from 1.7 Ga (Dacks et al., 2016). The 2 Ga line is widened and colored, as it approximates the earliest molecular dating origin of eukaryotes. The date for LUCA ranges from 4.0 Ga to 3.7 Ga (reviewed in (Sharov and Gordon, 2017)), though from Fig. 2 this may be as early as 4.3 Ga, so we indicate a range of 4.3–3.7 Ga here. This corresponds to the oldest "putative fossilized microorganisms that are at least 3770 million and possibly 4280 million years old" (Dodd et al., 2017). Formation of the Moon occurred via impact of a Mars size object named Theia about 30–100 million years after formation of the Earth (Gordon and Gordon, 2016; Yu and Jacobsen, 2011). The orbital parameters of the Moon and the obliquity of the Earth settled down to current values by 50 million years after impact, with the initial period of rotation of the Earth after impact being 2.5 h (Ćuk et al., 2016). Earlier estimates put the impact at 4.5 Ga (Maas and Hansen, 2015). Biomarkers have now been found back to 3.95 Ga (Som, 2017; Tashiro et al., 2017).

Prior to the impact forming the Moon (Fig. 1), the atmosphere may have consisted of gases outgassed from the mixture of chondritic meteorites from which the Earth formed, suggesting that "the early atmosphere seems to have been composed of mixtures containing CO_2, H_2O, H_2, CO, N_2, NH_3, CH_4, SO_2, HCl, and HF," perhaps up to 100 bars pressure (Schaefer and Fegley, 2014). In the first half of the Archaean eon, the atmosphere consisted most likely of nitrogen (N_2), carbon dioxide (CO_2), water vapor (H_2O), and smaller amounts of other gases (Kasting, 2014; Kasting and Howard, 2006; Zahnle et al., 2010a; Zahnle et al., 2010b). It was virtually devoid of free oxygen (O_2) and thus could support only anaerobic life.

But in the second half of the Archaean eon (starting about 3 Ga), when air density was at most twice that at present, the oxygen concentration in the atmosphere began to rise slowly (Fig. 15). This put an end to "the Earth's first iron age, with reduced Fe as the principal electron donor for photosynthesis, oxidized Fe the most abundant terminal electron acceptor for respiration, and Fe a key cofactor in proteins" (Knoll et al., 2016) (Fig. 15). After about 0.5 Ga, this led to the Great Oxygenation Event (GOE) or Oxygen Catastrophe that started Proterozoic eon (2.5 Ga). It drastically changed the evolution of our planet presumably due to photosynthesis provided by cyanobacteria (Warren, 2016) (Fig. 16). It has to be noted, however, that there are alternative explanations of the GOE:

"The appearance of O_2 in the atmosphere ca. 2.3 Ga could be explained easily if cyanobacteria evolved at that time. However, this explanation has been rendered very unlikely by the discovery of biomarkers in 2.7- to 2.8-Ga sedimentary rocks that are characteristic of cyanobacteria (Brocks et al., 1999). An alternative explanation involves a change in the redox state of volcanic gases as the trigger for the change in the oxidation state of the atmosphere. This was proposed by Kasting et al. (Kasting et al., 1993)".

(Holland, 2002)

"The causes of the abrupt transition remain hotly debated. Published hypotheses include a change in the redox state of mantle gases (Kump and Barley, 2007; Kump et al., 2001), changes in marine nutrient supply (Campbell and Allen, 2008) and/or reduced sinks for O_2 (Bjerrum and Canfield, 2002), a switch between two feedback-stabilized steady states (Goldblatt et al., 2006), a decrease in atmospheric methane levels (Konhauser et al., 2009; Zahnle et al., 2006) and the evolution of oxygenic photosynthesis itself (Kopp et al., 2005)".

(Sessions et al., 2009)

"The gradual removal of hydrogen under reduced standoff conditions [Fig. 14] may also have been important for the transformation of Earth's atmosphere from one that was mildly reducing to one that was oxidizing".

(Tarduno et al., 2014)

Cf. (Zahnle and Catling, 2014). About 0.2 Gy after the GOE, limestones and dolostones appeared to be extraordinary enriched with the ^{13}C isotope of carbon. This was named the Lomagundi Event. It lasted from 130 to 250 My (Martin et al., 2013) and was a result of the burial of vast quantities of organic carbon that accompanied the production of correspondingly large amounts of oxygen (Bachan and Kump, 2015), as indicated by C and S isotope data (Fig. 17).

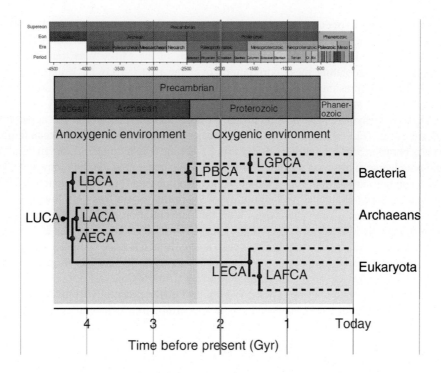

FIG. 2

"Schematic phylogenetic tree showing the geological time in which different extinct organisms lived. Dashed lines represent further bifurcations. Divergence times are compiled from multiple sources and are summarized in (Hedges and Kumar, 2009)." From Perez-Jimenez et al. (2011) with permission of Nature Publishing Group. Note that LUCA (Last Universal Common Ancestor) is placed here at 4.3 Ga and LECA (Last Eukaryotic Common Ancestor) at 1.5 Ga, a duration of 2.8 Gy. LBCA = Last Bacterial Common Ancestor, LACA = Last Archaeal Common Ancestor, LPBCA = last common ancestor of the cyanobacterial and deinococcus and thermus groups, LGPCA = last common ancestor of γ-proteobacteria, AECA = archaeal-eukaryotic common ancestor, LAFCA = last common ancestor of animals and fungi.

Free oxygen presumably acted as a poison for obligate anaerobic prokaryotes, and many prokaryotic species may have become extinct or retreated to anoxic niches. This might prove to be the most massive extinction event in Earth's history. In addition, the free oxygen reacted with atmospheric methane (CH_4), greatly reducing its concentration (Fig. 11). Because methane is one of the most important greenhouse gases, its reduction triggered the Huronian glaciation (from 2.4 Ga to 2.1 Ga), the longest one in the Earth's history (Frei et al., 2009). During these 0.3 Gy, the Earth looked like a snowball because its surface became entirely (or almost entirely) frozen. Ryuho Kataoka et al. (Kataoka et al., 2013; Kataoka et al., 2014) point out an additional possible cause of this and other "snowball" events in the upper Proterozoic: two major potentially climate changing starburst events happened in our galaxy at 2.4 to 2.0 Ga and 0.8 to 0.6 Ga (Marcos and Marcos, 2004; Rocha-Pinto et al., 2000) (Fig. 18). Astrophysical events such as supernovae may cause fluctuations in mutation rates on Earth and thus speed up molecular clocks (Melott, 2017), perhaps explaining the discrepancies in Fig. 1. Mass extinctions of eukaryotes may (Filipović et al., 2013; Leitch and Vasisht, 1997; Vukotić, 2017) or may not (Bailer-Jones, 2009; de Mello et al., 2009; Feng and Bailer-Jones, 2013) correlate with passage of our solar system through the spiral arms of the Milky Way, but for now we have no direct way of extrapolating these correlations over the past 0.5 Gy back into the Precambrian and Hadean. The only presumptive (but not proven) mass extinction during this period, of anaerobic prokaryotes, occurred at the GOE, providing little data for the question.

The history of the hydrosphere, and foremost the ocean, in the Archaean eon was apparently not full of significant events (Fig. 5). For a while, it was generally accepted that surface ocean temperature at that time was between 55°C and 80°C, salinity was 1.5–2 times higher than its modern value, and these values did not change significantly from 4 Ga to 2.5 Ga (Gaucher et al., 2008; Gouy and Chaussidon, 2008). Based on this supposition, supported by oxygen isotope data for early diagenetic cherts, L. Paul Knauth (Knauth, 2005) concluded that solubility of O_2 in that ocean was very low, and nobody but anaerobic organisms could exist there even after the concentration of O_2 in the atmosphere began to increase drastically. However, a few years later, new studies (Blake et al., 2010; Kasting and Howard, 2006; Som et al., 2012) combining oxygen and hydrogen isotope compositions of cherts made a case for Archaean ocean temperatures being no greater than 40°C (Figs. 9 and 10). This result supports a cooler, temperate Archaean ocean around 26°C to 35°C.

Generally, the Proterozoic ocean (after 2.5 Ga) didn't differ essentially from the Archaean ocean. Donald Canfield (Canfield, 1998) suggested a model according to which ventilation of the deep ocean lagged behind the GOE by more than a billion years, resulting in a vast, deep reservoir of hydrogen sulfide. The ocean water masses in the beginning of Proterozoic eon remained anaerobic with expansion of H_2S-rich regions occurring around 2.4 Ga (Fig. 5), but with one essential exception, where the ocean's surface contacted the atmosphere, in the very upper level of open ocean, shallow waters, estuaries, littoral baths, etc. They were the most suitable for life. This very small but the most populated part of the ocean had, as we saw above, a temperature only about 30°C and lower salinity due to input from rains and rivers. That allowed oxygen to dissolve. As a result, oxygenating of these ocean areas began just after the GOE, i.e., at 2.5 Ga.

Continental crust and the first real continents appeared around 3.0 Ga (Dhuime et al., 2015) and continental thickness grew for the next 2 Gy reaching 40 km, which is almost a quarter more than its contemporary average thickness (~32 km) (Fig. 13). During the period from 3.0 Ga to 2.5 Ga, the continental crust contained ~1500 species of minerals. Then after the GOE, the number of such species drastically increased up to ~4000, close to ~4500 mineral types that the crust has now (Hazen, 2012) (Fig. 19). This jump in mineral diversity happened due to a sharp increase in atmospheric oxygen, which followed several major glaciation events and ultimately gave rise to skeletal biomineralization, transforming Earth's surface mineralogy. Biochemical processes may thus be responsible, directly or indirectly, for most of Earth's 4500 known mineral species (Hazen et al., 2008).

The first eukaryotes were probably aerobic or at least aerotolerant (Margulis et al., 2006), especially taking into account that endosymbiotic theory suggests a close relationship between mitochondria and aerobic Rickettsiales or other α-proteobacteria (Curtis and Archibald, 2010; Emelyanov, 2001). If so, the eukaryotic step of hierarchogenesis took about 500 My from 2.5 Ga to 2.0 Ga, i.e., the period that lies between two peaks in large impacts and respective mantle plumes (Fig. 20).

We cannot, however, rule out the other possibilities. Anaerobic eukaryotes could have appeared at the end of the Archaean eon and then become extinct due to the GOE, although there is no evidence of this. Nevertheless, some researchers (Mentel and Martin, 2008) underscore the evolutionary

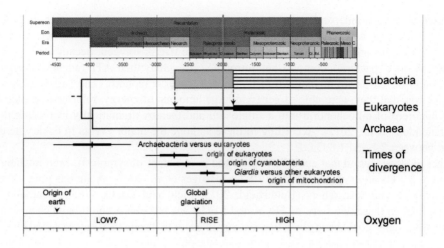

FIG. 3

"Summary diagram showing relationship between timing of evolutionary events and that of Earth and atmospheric histories. Time estimates are shown with ±1 standard error (thick line) and 95% confidence interval (narrow line). The phylogenetic tree illustrates the radiation of extant eubacterial lineages (blue), and dashed lines with arrows indicate the origin of eukaryotes.... The increasing thickness of the eukaryote lineage represents eubacterial genes added to the eukaryote genome through two major episodes of horizontal gene transfer. The rise in oxygen represents a change from <1% to >15% present atmospheric level, although the time of the transition period and levels have been disputed." From Hedges et al. (2001) with permission of BioMed Central Ltd.

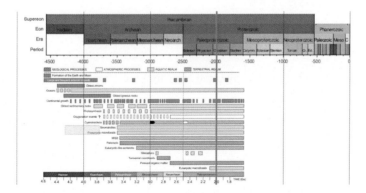

FIG. 4

"Suggested chronology of geological, atmospheric, and biological events during the Hadean, Archean, and Paleoproterozoic eons." From Beraldi-Campesi (2013) with permission of Springer and BioMed Central Ltd. under a Creative Commons Attribution License. Ocean data is from Nutman (2006). Here we have added a pale green rectangle indicating the latest discovery of the "oldest" fossils, which may have been stromatolite prokaryotes dated between 4.28 Ga and 3.77 Ga (Dodd et al., 2017).

significance of oxygen-independent ATP-generating pathways in mitochondria that could have been used by hypothetical Archaean anaerobic eukaryotes. In addition, Martin Van Kranendonk et al. highlighted that "eukaryote development is linked to the rise of oxygen, because these large (10^4–10^5 x the volume of bacteria), complex heterotrophic cells satisfy their ATP (adenosine triphosphate) needs through the oxidative breakdown of reduced organic compounds" and "a key advance of the eukaryotes was that they were able to generate large amounts of energy, relative to prokaryotes, by packing hundreds of mitochondria into a single cell and thereby dramatically increasing their size" (Van Kranendonk et al., 2012). According to these authors, the major events in eukaryote evolution occurred between 2.1 and 1.8 Ga.

Hedges et al. suggested that eukaryotes emerged in two steps of symbiosis: premitochondrial that happened at 2.7 Ga and mitochondrial at 1.8 Ga (Hedges et al., 2001). A similar point of view was expressed by several researchers (Butterfield, 2015; Field and Dacks, 2009; Koumandou et al., 2013) that marked these two steps as FECA and LECA where "FECA could have been a very much simpler system" and "could have given rise to many lineages that were then outcompeted by LECA and its descendants" (Field and Dacks, 2009). LECA, on the other hand, "possessed mitochondria, substantial internal differentiation and a well-defined nucleus" (Koumandou et al., 2013).

5 WHY DID IT TAKE SO LONG FOR EUKARYOTES TO APPEAR ON EARTH?

One class of answers to why it took so long for eukaryotes to appear on Earth could be that prokaryotes were "waiting for something" that was an absolutely necessary component or even a trigger of this hierarchogenetic step. Such possible events that eukaryogenesis had to wait for could be divided into two main categories: geophysiochemical and biological.

5.1 GEOPHYSIOCHEMICAL WAITING

The first category could, generally speaking, include waiting for:

1. Fresh water
2. Salinity changes
3. Protection from ultraviolet light, cosmic rays, and terrestrial radiation from the ground
4. Right climate
5. Nitrogen
6. Land
7. Oxygen

Let us consider each of these options:

1. *Fresh water.* Fresh water cannot have been the cause of a delay once the Earth had cooled below the boiling point of water. Boiling temperature was around 80°C in the Archaean because the atmospheric pressure at sea level (Marty et al., 2013) was at that time about half of the current pressure (Emspak, 2016; Som et al., 2016). So, because temperature in Archaean/early Proterozoic eons was not warmer than 50–70°C (Knauth, 2005) and possibly 20–30°C lower (Blake et al., 2010), water evaporated from the ocean shed rain above the ocean and continents as soon as the

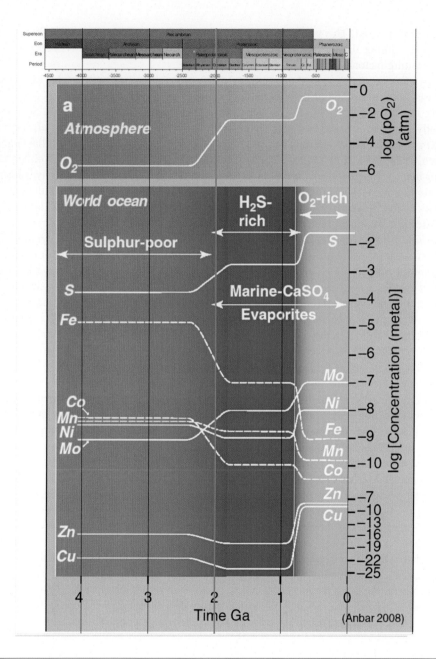

FIG. 5

"Oxygen, sulfur and metal levels in the world oceans across time. The Archaean ocean was sulfur-poor, an expansion in H_2S-rich oceanic regions occurred around 2.4 billion years, which then gave away to fully oxygenated oceans around 800 Ma ago. That is, marine-derived sulfate evaporites formed both from mother brines that were sourced in H_2S-rich oceans from 2.4 to 0.8 Ga and from O_2-enriched since 0.8 Ga." From Warren (2016) with kind permission of John K. Warren, as adapted from Anbar (2008) and Zerkle et al. (2005).

continents appeared. This was perhaps as early as 4.5 Ga to 4.3 Ga (Harrison, 2009). As a result, fresh water during all the possible period of eukaryogenesis (4 to 2 Ga) was in excess and couldn't be a "waiting factor."

2. *Salinity changes.* As we saw above, salinity of the ocean remained pretty stable during that period and was 1.5–2 times higher than its modern value (Gaucher et al., 2008; Gouy and Chaussidon, 2008), i.e., about 5.25%–7.0%. This is 5–7 times less than salinity of the Dead Sea, which is not actually dead and is populated with filamentous eukaryotic fungi (Buchalo et al., 1998; Mbata, 2008) and halophilic Archaea (Bodaker et al., 2010; Gordon et al., 2017a; Mullakhanbhai and Larsen, 1975; Oren et al., 1996). Thus, salinity cannot be the "waiting factor" either, especially taking into account that the hypotheses that Archaean life developed either near hydrothermal vents that are rich in H_2, CO_2, transition metals, and S (Weiss et al., 2016) where biota are adapted to salinity up to 8% even in the present (Fontaine et al., 2007), or in the very upper layer of the ocean, estuaries, littoral zones, etc., desalted by inflow from rain and rivers.

3. *Protection from ultraviolet light, cosmic rays, and terrestrial radiation from ground.* The ozone layer in the atmosphere presumably emerged as a result of the GOE. Waiting for an ozone layer in the atmosphere that protected the Earth's surface from harsh short-wave ultraviolet radiation, as well as cosmic rays, (oxygenation catastrophe at nearly 2.5 Ga) would make sense only for terrestrial biota. For planktonic creatures in the body of the ocean, this "waiting factor" cannot play an essential role because only 1 meter of water absorbs 0.99999% of ultraviolet light (Wikipedia, 2016b). On the other hand, while cosmic rays and terrestrial radiation from the ocean bottom penetrate the surface a bit further with an absorption length of the order of a meter (Dyer, 1953; Rollosson, 1952; White, 1950), that leaves the vast volume of the ocean effectively unirradiated. Hence, shallow waters and benthos (including hydrothermal vents ecosystems) could have been heavily irradiated. In favor of this assumption is that the earliest living organisms, probably including LUCA, had powerful mechanisms of DNA repair (DiRuggierro and Robb, 2004). It looks like life would not have emerged or developed without such mechanisms. Furthermore, these mechanisms in cells of Archaea, Bacteria, and then eukaryotes have been so effective that they allowed them to survive not only more intense terrestrial radiation (Fig. 12) and far more intensive harsh UV and cosmic rays (due to the absence of an ozone shield in anoxic atmosphere), but also numerous radiation surges due to geomagnetic reversals, supernovae, gamma-ray bursts, or solar proton events (Thomas et al., 2015) and starbursts (Fig. 18). (However, reversals of the Earth's magnetic field may have been less frequent early on (Vázquez et al., 2010).) For example, a rod-shaped bacterium *Desulforudis audaxviator* was recently found that lives in a gold mine in South Africa 2.8 kilometers underground in a habitat without light, oxygen, and carbon (Boddy, 2016). Instead, this bizarre creature gets energy from radioactive uranium in the depths of the mine. This allows us to assume that life elsewhere in the universe might also feed off radiation, especially radiation raining down from space. Coming back to our point, waiting for an ozone layer can possibly be considered as essential but not a crucial factor of eukaryogenesis.

4. *Right climate.* There is no doubt that climate could be so wrong that not only eukaryotes but even prokaryotes cannot survive it. For example, the Earth just after formation of the Moon was probably incompatible with life, with no safe spaces, if the surface indeed consisted "of a global magma ocean" (Maas and Hansen, 2015). After it cooled off, the Earth may have been habitable from that point forward (Harrison, 2009). On a finer point, it is hard to determine parameters and their

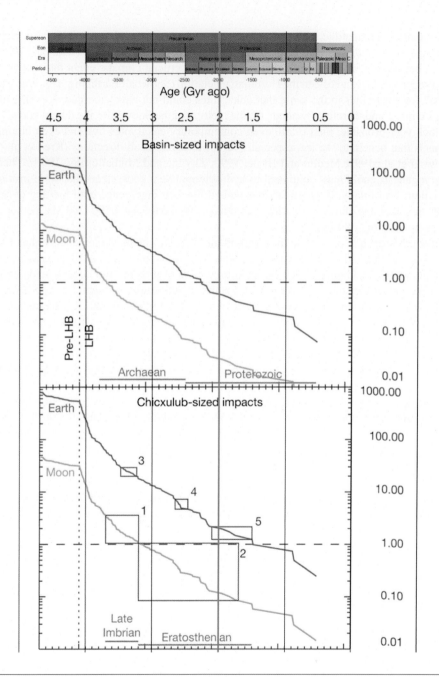

FIG. 6

Cumulative number of E-belt impacts. *Top*: "The number of E-belt impacts making basin-sized craters (diameters over 300 km) on Earth and the Moon." *Bottom*: "The number of E-belt impacts making Chicxulub-sized

(Continued)

values that are compatible with prokaryotes but incompatible with eukaryotes, especially in the Archaean or early Proterozoic eons. If an environment is suitable for *Archaea* and *Bacteria,* it should be suitable for their symbionts too. The most crucial climatic event during the eukaryogenesis may have been the Huronian glaciation (Snowball Earth, Fig. 18). Was it a "waiting factor" for eukaryotes in the sense that eukaryotes could not have emerged or even existed in such a climate and therefore "waited" for the melting of the Snowball? We do not know, but it does not look very probable. First of all, even contemporary eukaryotes live and even bloom in brine channels that penetrate sea ice, especially its lower layer, in all directions (Krell et al., 2008; Ligowski et al., 1992; Medlin and Hasle, 1990; Zhytina and Mikhailovsky, 1990). Then, eukaryogenesis could have continued in hydrothermal vent ecosystems that were insulated from glaciation. And finally, if the eukaryotes waited for their emergence until melting of Snowball Earth, then this step of megaevolution would have taken only about 100–200 My, perhaps too short compared with the other steps.

5. *Nitrogen.* Nitrogen is one of the most important biogenic chemical elements, and any kind of earthly life, prokaryotic or eukaryotic, is impossible without it because nitrogen is incorporated into both amino acids and nucleotides, the building bricks of proteins and nucleic acids, respectively. However, eukaryotes did not have to wait for the appearance of nitrogen because it was the main component of the secondary (3.8–2.5 Ga) atmosphere (Marty et al., 2013; Silverman et al., 2017; Som et al., 2016). In the third atmosphere (after GOE) and until now, it has remained the main component of the atmosphere (Fig. 21). The problem was not so much the presence, but the fixation of the atmospheric N_2 to other chemical forms. Archaean anaerobic life couldn't do this directly and used nitrogen oxides produced by lightning, which even today accounts for 5%–33% of nitrogen fixation (Raymond et al., 2004). Results of Navarro-González et al. (Navarro-González et al., 2001) indicate decreasing of NO production from the origin of life, that started consuming it, to about 2.2 Ga, i.e., just before GOE. They connect this reduction with reduced abiotic nitrogen fixation by lightning whose intensity decreased during the Archaean eon. This may have caused an ecological crisis and triggered development of biological nitrogen fixation. So, not nitrogen itself but its biological fixation potentially could be the "waiting factor;" but only if eukaryogenesis required nitrogen fixation. However, this is not quite obvious due to the hypothesis that nitrogenase (the enzyme responsible for biological nitrogen fixation) had already evolved in LUCA (Fani et al., 2000; Normand et al., 1992). Thereby, nitrogen is not a good candidate for a role as a "waiting factor."

6. *Land.* "It is quite possible that early eukaryotic evolution ... took place in terrestrial, not marine settings" (Wellman and Strother, 2015). Cf. (Lenton and Daines, 2016). However, given that land

---

FIG. 6—CONT'D

craters (diameters over 160 km) on Earth and the Moon." From Bottke et al. (2012) with permission of Nature Publishing Group. Note that "Cumulative" means here going backwards in time starting at the present. The steep slope between 4.1 Ga and 4.0 Ga indicates the beginning of the LHB, which these authors conclude lasted for 0.4 Gy until 3.7 Ga. The E-belt is "an extended and now largely extinct portion of the asteroid belt between 1.7 and 2.1 astronomical units from [the orbit of the] Earth." This model contradicts the notion that the LHB was a statistical fluctuation (Boehnke and Harrison, 2016). From the graphs, we estimate about 600 impacts over this period, the order of 1 every million years, not likely enough to keep the Earth sterile (Abramov and Mojzsis, 2009).

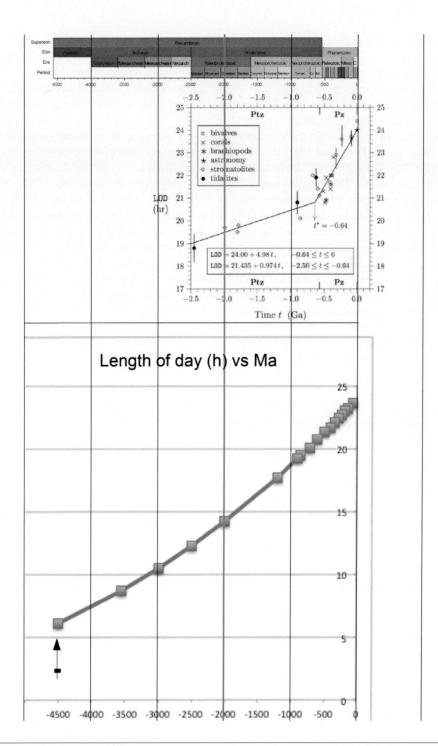

FIG. 7

Top: "Diagram of the empirical LOD [length of day] data for the Phanerozoic (Pz) and the Proterozoic (Ptz) (after (Varga et al., 1998))." From Denis et al. (2011) with permission of John Wiley and Sons. *Bottom*: Length of

(Continued)

above the ocean may have been present from Hadean times as microcontinents (Figs. 5 and 13) (Griffin et al., 2011), the existence of land above sea water is not likely to be the limiting factor. An argument has been made that a clade of prokaryotes, named Terrabacteria, diverged on land 3.54–2.83 Ga (Battistuzzi and Hedges, 2009), which would not exclude the remaining "Hydrobacteria" and Archaea from land-based bodies of fresh water or salterns that require the presence of land above the ocean.

7. *Oxygen.* The last but in no case the least option in the list is oxygen. It is generally presumed that the GOE (Figs. 4, 5, 11, 15, and 16), i.e., switching from second to third atmosphere, with the rapid (100 My (Schopf, 2012)) increase in the concentration of oxygen led to mass extinction of anaerobic organisms. This would seem to have been a big event in the history of life on Earth, probably occurring during eukaryogenesis. The GOE was the biggest of perhaps 7 stepwise increases in O_2 (Campbell and Allen, 2008). The timing of the GOE was estimated in a computer simulation (Hart, 1978) and formed the basis for an argument that this timing is a function of the lifetime of a star (Livio and Kopelman, 1990). But was oxygen the "waiting factor?" This depends on whether the first eukaryotes were aerobic or anaerobic. In the last decade, it was found that mitochondria can exercise anaerobic synthesis of ATP (Mentel and Martin, 2008). Moreover, a few years later, it was concluded that anaerobic energy metabolism occurred in the last eukaryote common ancestor (LECA) (Müller et al., 2012). If LECA was really anaerobic, then eukaryogenesis could have been completed before the GOE. In this case, oxygen evidently cannot be the "waiting factor." On the other hand, Martin van Kranendonk et al. (Van Kranendonk et al., 2012) bring very convincing arguments for an essential role of oxygenation in eukaryogenesis. Unfortunately, there is not yet any fossil evidence of eukaryotes before about 2 Ga, and some researchers (e.g., (Koumandou et al., 2013)) argue that the LECA was an aerobe and anaerobic eukaryotes have arisen from secondary losses of mitochondrial metabolic capacity. Thus, alas, we have no definite answer on a role of oxygen as the "waiting factor."

5.2 BIOLOGICAL WAITING

The second category could possibly include waiting for:

1. Introns
2. Sex reproduction
3. Rare event(s)

Consider now each of these three options:

1. *Introns.* There are no known prokaryotes whose genome contains spliceosomal introns (Koonin, 2006). On the other hand, there are no existing eukaryotes without introns in some of their genes. So, it would be natural to relate one to another. Nonetheless, immediately after

---

FIG. 7—CONT'D

day, plotted using data from Arbab (2008). The black bar is the day length estimate of 2.5 h just after impact of Earth with Theia (Ćuk et al., 2016). The vertical arrow is another estimate of day length after impact of 2–5 h (Maas and Hansen, 2015). Stromatolites give direct evidence of the daily rising and setting of the sun in the Precambrian (Zhang, 1986).

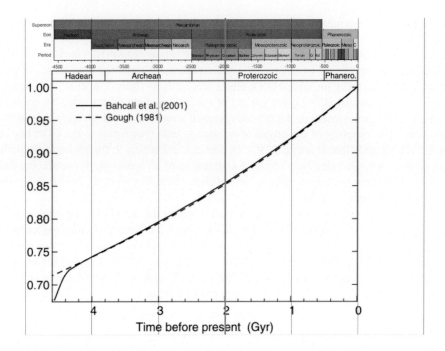

FIG. 8

Solar luminosity relative to today. "Evolution of solar luminosity over the four geologic eons for the standard solar model described in Bahcall et al. (2001) (solid line) and according to the approximation formula." From Feulner (2012) with permission of the American Geophysical Union.

the discovery of introns by Phillip Sharp (Berget et al., 1977) and Richard Roberts (Chow et al., 1977) in 1977, Ford Doolittle (Doolittle, 1978) and James E. Darnell (Darnell, 1978) put forward the Introns-Early idea, a concept that was later elaborated (Jeffares et al., 1998; Poole et al., 1998). According to this concept, introns as non-coding regions interspersed between RNA genes in an early RNA world appeared before the emergence of life itself in prebiological systems. John Mattick (Mattick, 1994) and John Logsdon, Jr. (Logsdon, 1998) formulated the opposite Introns-Late concept countering that introns emerged only in eukaryotes, and moreover, have been inserted into protein-coding genes continuously throughout the evolution of eukaryotes. Presently, both hypotheses coexist, although the Introns-Late one is more confirmed. Besides, there have been attempts either to incorporate aspects of the Intron-Late hypothesis into the Introns-Early hypothesis (Fedorov et al., 2001) or to reconcile both concepts (Koonin, 2006; Stoltzfus et al., 1994). Howbeit, the fact that only existing eukaryotes and no prokaryotes have introns in their genes makes us consider emergence of introns as an essential part of eukaryogenesis. But did eaukaryotes "wait" for introns or vice versa introns "waited" for emergence of eukaryotes in a sense that introns were byproducts of this stage of hierarchogenesis in biological megaevolution? Unfortunately, we cannot yet answer this question.

2. *Sex*. "Waiting" for emergence of sex reproduction does not look probable or even possible because, although it exists in a vast majority of contemporary eukaryotes and does not exist in prokaryotes, we have no evidence of sex-based reproduction before 1.2 Ga (Butterfield, 2000), i.e., 0.8 Gy after completion of eukaryogenesis. At the same time, the appearance of actual eukaryotic sex drastically improved DNA repair mechanisms (Bernstein et al., 2012) that allowed eukaryotes (after another 0.75 Gy) to get out of the radiation shield provided by water to the protection of the ozone shield of the new oxygenated atmosphere and occupy the land in addition to the ocean. The coupling of the origin of sex during eukaryogenesis to the GOE has been proposed (Gross and Bhattacharya, 2010). In contrast, according to the hydrogen hypothesis, the endosymbiosis was preceded by a physical association of an anaerobic H_2-producing fermenter with an H_2-dependent prokaryote (Martin and Müller, 1998; Martin et al., 2016). This would allow eukaryogenesis before the GOE.

3. *Rare events*. Eukaryogenesis may ". . . have been hard, in the sense of being against the odds in the available time . . ." (Carter, 2008). The idea has been put forth that eukaryogenesis is the result of between one (Koonin, 2007) and many (Brown and Doolittle, 1997) rare or rarely successful events. One of these, perhaps the first, was the symbiotic origin of the mitochondrion, which is now seen as universal in functional or reduced form in all tested extant eukaryotes (Burki, 2016; Karnkowska et al., 2016; Martin et al., 2016).

But we must still ask, why so long? Except for those hypotheses that key eukaryogenesis to a particular event, such as the GOE, none answer this question.

6 SEMANTIC APPROACHES TO EUKARYOGENESIS

We would like to offer a new answer to the question: "Why did it take so long for eukaryotes to appear on Earth?" This answer relates to internal system factors that could play their own role. To estimate the essentiality of this role, let us consider the following model situation.

In a well-known thought experiment (Borel, 1913), monkeys eventually type the text of Shakespeare's works (e.g., Hamlet) by randomly hitting the keys of typewriters. But then, it may be calculated that such a process would take far more time than the age of our universe (Wikipedia, 2017b). However, if we change experimental conditions and place behind a single monkey a reciter who knows the Hamlet text by heart and erases or prevents all the incorrect monkey's hits, the time would be incredibly reduced. The text of Hamlet contains 132,680 alphabetical letters and 199,749 characters overall. If we estimate the monkey's typing speed as 4 hits per second, we will need about 40 hits (including spaces and punctuation marks) or 10 seconds (on average) per correct character. This gives for a whole text of Hamlet:

$$(199749 \text{ characters} \times 10 \text{ s})/3600 \text{ s per h}/24 \text{ h per day} \cong 23 \text{ days or } 3.3 \text{ weeks}$$

Let's extend this thought experiment by replacing the typewriter with a computer keyboard and the Shakespeare reciter with an AI (artificial intelligence) program that estimates meaning of new typing by comparing it with already existing texts. This program eliminates the next character if the set of characters after the last space cannot belong to any word; the next word if its combination with the previous ones in a phrase doesn't make any sense, i.e., doesn't have at least approximate analogs among existing texts; and the next phrase if it completely drops out of the context. Of course, the monkey could

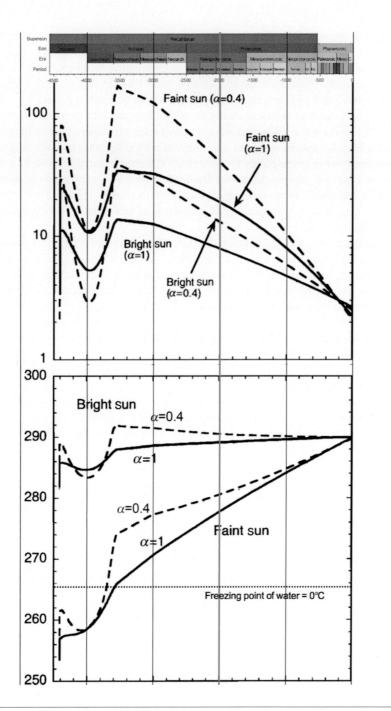

FIG. 9

Atmospheric carbon dioxide (top) and global surface temperature (bottom) predicted by modeling, with Faint Sun and Bright Sun scenarios, for two different values of seafloor weathering rates (α). CO_2 is given in PAL units

(Continued)

be also replaced with a generator of random characters. Such a program would be able to create essentially new texts (novelties) that are meaningful and at the same time completely unpredictable.

If the AI program were more sophisticated and, after accumulation of a rich enough vocabulary, would operate with randomly chosen words and phrases from this vocabulary rather than making up words from completely random characters, new phrases would be created faster.

This model allows us to look at the early evolution of life as a semiotic process (Barbieri, 2007; Emmeche and Kull, 2011) where the role of the AI program in selecting meaningful words and phrases is played by natural selection constrained by the environment, including other organisms. The buildup of a few words that make sense corresponds to the transition from presumably random polymers to informational RNA or DNA, i.e., the origin of life. Once a few "words" pass the test of natural selection, gene duplication and duplications of larger groups, up to ploidy, followed by drift apart in sequence and function, can take over the process of generation of new genes, a process that can generate novelty exponentially.

Partial DNA duplications followed by mutation thus offer a mechanism for our more abstract model of characters, words, and phrases that are reused, sometime with modification. The whole process is one of novelty based on previous novelties and is a branching process, which can accelerate as more novelties are generated. The reciter or AI program is replaced by natural selection: only novelties that work in the milieu of the novel organism are retained.

The problem is that these novel genes, in prokaryotes, occur in different individual cells that may be widely separated over the Earth. They may be brought together into single cells by horizontal gene transfers. The delay to eukaryogenesis, then, may be the time it takes for LUCA to generate a large array of novel genes, in both its Bacteria and Archaea descendants, and by horizontal gene transfers for enough of them to come together into the two cells that fused to form LECA. We are composing a computer simulation of this population-based model of the transition from a single cell called LUCA through many generations and species of prokaryotes to form, by symbiosis, the single cell named LECA (Gordon et al., 2017b), sketched below. The test of the model will be whether there is a set of reasonable parameters that explain the 2 Gy time for this event to occur.

This then could be the internal factor of the eukaryotic step of hierarchogenesis after the basic genetic vocabulary of eukaryotes had been created.

7 EVOLUTION OF PROKARYOTES PRIOR TO EUKARYOGENESIS

We can make the above abstract consideration concrete by asking if there was an accelerating evolution of prokaryotes prior to the event(s) of eukaryogenesis. If so, then the delay in the timing of eukaryogenesis could be attributed to the (postulated) need of sufficiently complex prokaryotes to pull off eukaryogenesis. In other words, the earlier prokaryotes simply weren't complex enough to generate this level of hierarchogenesis.

FIG. 9—CONT'D

(Present Atmospheric Level = 300 ppm) and temperature °K, respectively. "Despite its name, the Hadean climate would have been freezing unless tempered by other greenhouse gases." From Sleep and Zahnle (2001) with permission of the American Geophysical Union. "Bright Sun" lumps three possible scenarios, including other greenhouse gases being present, such as methane.

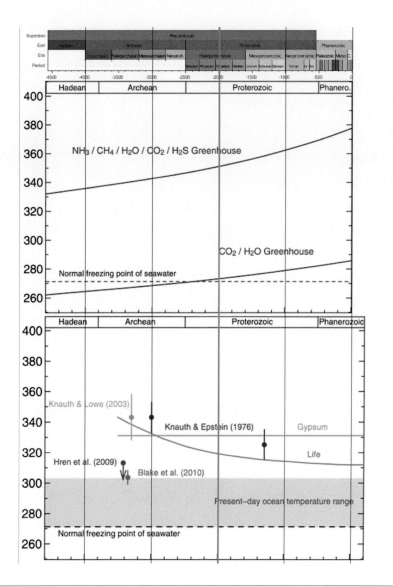

FIG. 10

Top: "Average surface temperature [°K] evolution of Earth as a function of time given the changes in solar luminosity and assuming present-day concentrations of carbon dioxide and water vapor (blue line) according to Sagan and Mullen (Sagan and Mullen, 1972). The calculations... assume a total pressure of 1 bar, an atmospheric composition constant with time and a fixed albedo of 0.35. Going into the past, the surface temperature drops below the normal freezing point of water ~2 Ga in this model. The solution of the faint young Sun problem suggested by Sagan and Mullen (Sagan and Mullen, 1972) in terms of greenhouse gas warming dominated by ammonia (NH_3) is shown as well (red line). In this scenario, volume mixing ratios 10^{-5} of NH_3, CH_4 and H_2S have been added to the CO_2-H_2O greenhouse. In terms of warming, ammonia is the dominant greenhouse gas in

(Continued)

We have reviewed the many morphological and behavioral features of prokaryotes that "anticipate" eukaryote properties (Gordon and Gordon, 2016), but made no attempt to order them by any complexity measure, nor when they might have first arisen, nor claimed that all carried over to eukaryotes. Here we just list these numerous features:

1. Differentiation of up to four kinds of cells
2. Mating
3. Chiral shape
4. Square colonies
5. Square cells
6. Gliding motility
7. Colonial motility
8. Quorum sensing
9. Alternate cell shapes
10. Colonial branching
11. Swarm swimming behavior
12. Cell polarity
13. Fruiting bodies
14. Multinucleoid
15. Magnetite inclusions
16. Colony size control
17. Huge cells with membrane electric potentials
18. Membrane surface waves during cell division
19. Cytoskeleton
20. Selected cell death in colonies and cell aging
21. Internal pressure
22. Membrane-bound vacuoles
23. Extracellular slime
24. Consortia with complementary capabilities
25. Mirror symmetric divisions

FIG. 10—CONT'D

this case." From Feulner (2012) with permission of the American Geophysical Union. *Bottom:* "Constraints on ocean temperatures [°K] during the Archaean. The existence of diverse life since about 3.5 Gar and the typical ranges of temperature tolerance of living organisms suggest the upper limit indicated by the green line (Walker, 1982). Evaporate minerals are present since about 3.5 Gar, and the fact that many were initially deposited as gypsum sets an upper limit at 58°C (cyan line) (Holland, 1978). The comparatively high (but controversial...) temperatures derived from oxygen isotope ratios in cherts are shown in blue (Knauth and Epstein, 1976; Knauth and Lowe, 2003). More recent estimates based on a combination of oxygen and hydrogen isotope ratios (Hren et al., 2009) and the oxygen isotope composition of phosphates (Blake et al., 2010) are shown in red and magenta, respectively. The range of present-day ocean temperatures is indicated in gray (Locarnini et al., 2010), and the freezing point of seawater at normal pressure and for present-day salinity is indicated by the dashed line." From Feulner (2012) with permission of the American Geophysical Union. Gar = Ga in our notation.

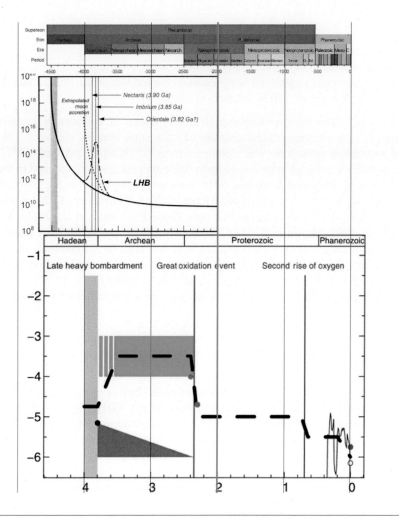

FIG. 11

Top: "Various interpretations of the mass flux (accretion rate) on the Moon. Ages of the most important impact basins are indicated. The solid line is the present-day background flux extrapolated back to the origin of the solar system. LHB indicates the spike in the accretion rate representing the Late Heavy Bombardment. The dotted line indicates an accretion curve that includes the masses of the basin-forming projectiles; this curve is unlikely to be correct, because it implies accretion of the Moon at 4.1 Ga. The gray band marks the age of the Moon obtained from measurement of radiogenic isotopes" (Koeberl, 2006). The vertical scale is mass in grams per year. The LHB may be a statistical fluctuation rather than a real phenomenon (Boehnke and Harrison, 2016). From Koeberl (2003) with permission of Springer. *Bottom*: "Estimates for the methane partial pressure in the atmosphere in various epochs in Earth's history. The period of frequent impacts during the Late Heavy Bombardment is shown in gray, with the estimate for methane produced in impacts by Kasting (2005) as a black circle. The green area indicates the range based on estimates of biological methane fluxes during the Archaean (Kasting, 2005). The brown triangle shows the contribution from abiogenic sources based on the present-day

(Continued)

To this, we can add:

26. Two or more chromosomes (Bavishi et al., 2010)
27. Multiple quorum sensing (Dunn and Stabb, 2007)
28. Fusion of prokaryotic cells to produce heterodiploids (Grandjean et al., 1996; Lake, 2009; Naor and Gophna, 2013; Rice et al., 2013; Swithers et al., 2011).

It has been noted that: "The proto-mitochondrial endosymbiont is confidently identified as an α-proteobacterium. In contrast, the archaeal ancestor of eukaryotes remains elusive …. We suggest that the archaeal ancestor of eukaryotes was a complex form…" (Koonin and Yutin, 2014). Shortly after, an extant candidate was found in an Arctic hydrothermal event site and assigned to the Archaea phylum Lokiarchaeota (Embley and Williams, 2015; Koonin, 2015; Spang et al., 2015; Wikipedia, 2017c). It has many characteristics of eukaryotes (Koonin, 2015; Mariotti et al., 2016; Spang et al., 2015; Surkont and Pereira-Leal, 2016) and is hydrogen-dependent (Sousa et al., 2016), as predicted by (Martin and Müller, 1998; Martin et al., 2016). So there had to be an evolutionary sequence to that archaeal symbiosis ancestor of eukaryotes, consistent with the "mito-late" hypothesis that the symbiosis event occurred after the evolution of "significant complexity" of the archaeal component (Pittis and Gabaldón, 2016). That may be what eukaryogenesis was "waiting for."

We choose here to give a shorthand notation for the two symbionts that combined to form the first eukaryote:

A-euk: the archaeal symbiosis ancestor of eukaryotes, a Lokiarchaeota
B-euk: the bacterial symbiosis ancestor of eukaryotes, an α-proteobacterium

For multicellular cyanobacteria, we can reconstruct the sequence of phenotypes from fossils (Schirrmeister et al., 2013) (Fig. 16). Perhaps, molecular fossils (Brocks et al., 1999), molecular divergence time estimates (Battistuzzi and Hedges, 2009; Hedges et al., 2001) (Figs. 3 and 22), or genome complexity measures (Sharov and Gordon, 2017) will allow us to reconstruct the detailed sequence of improvements in Archaea that led to the ability of one of its lineages to be the archaeal partner A-euk that formed eukaryotes. A similar research program may be needed for B-euk.

The organization of the genome of A-euk may be analogous to the differentiation tree of later eukaryotic organisms that went on the next hierarchogenetic step of continuing differentiation, an event

---

FIG. 11—CONT'D

estimate of (Emmanuel and Ague, 2007), including a possible increase up to a factor of 10 in earlier times due to faster creation of seafloor (Kharecha et al., 2005). The decrease with time is not based on any detailed model but only intended to give a rough indication of this possibility. Estimates for atmospheric methane content from a model of the Great Oxidation Event are indicated in magenta (Goldblatt et al., 2006). Phanerozoic CH_4 concentrations estimated in (Beerling et al., 2009) are represented by the thin black line. Finally, the preindustrial and present-day methane partial pressures are shown as open and filled red circles, respectively (Forster and Ramaswamy, 2007). The thick black dashed line is a highly idealized sketch of Earth's methane history based on these estimates. Methane fluxes have been converted to atmospheric mixing ratios using the relation shown in (Kasting, 2005), and a total pressure $p = 1$ bar is assumed for the conversion from volume mixing ratios to partial pressure values." Vertical units are Log_{10} of the partial pressure of methane measured in bars. From Feulner (2012) with permission of the American Geophysical Union.

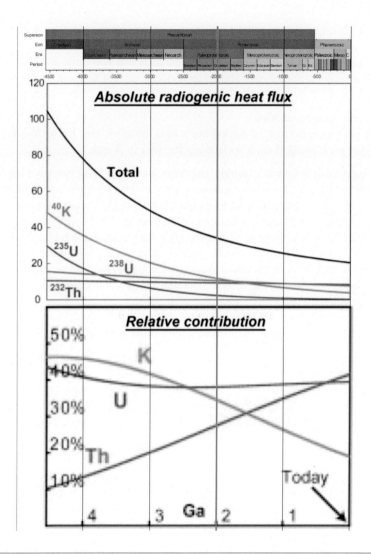

FIG. 12

"Earth's radiogenic heat production [in TW] from the decay of long-lived radionuclides through time. Prior to 2.5 Ga, K acted as the dominant radiogenic heat source within the planet. The exponential increase in radiogenic heat in the geologic past likely resulted in a higher convective Urey number in the ancient mantle." From Arevalo et al. (2009) with permission of Elsevier.

that also may have taken 1 Gy or so (Fig. 23). The growth of differentiation trees over evolutionary time may be due to duplication of portions of the genome that represent whole tissues (Gordon and Gordon, 2016; Gordon, 1999). The duplicated sections and their corresponding tissues would then diverge, resulting in tissues that would differ in function. Of course, lower level duplications, gene cascades down to the single gene level, horizontal DNA transfers, and polyploidy (Markov and Kaznacheev, 2016) (highest level whole genome duplications) could also occur.

Thus, what we are seeking in A-euk is a genome organization that was built up by a branching sequence of DNA duplications of assorted sizes and understanding, as best as possible, of that sequence. In particular, we seek:

1. A best estimate of the genome of A-euk.
2. The portion of the phylogenetic tree, including horizontal transfers, that led to A-euk.
3. The novelty that was introduced at each node along the way to A-euk.

The implication was made that A-euk is complex and therefore B-euk may not have been complex (Koonin and Yutin, 2014). But if this is not correct, a similar research program would be needed for B-euk. In any case, since most horizontal transfers have been from bacteria to Archaea, with many fewer from Archaea to Archaea or Archaea to bacteria (Akanni et al., 2015), the bacteria figure prominently in the history leading to eukaryogenesis, not merely as donators of the precursor to mitochondria. Some of this kind of work has been done for LUCA (Mat et al., 2008; Mushegian, 2008), so reconstructing the paths from LUCA to the much later A-euk and B-euk might be in our grasp. In that reconstruction, we may learn the sequence of novelties that led to eukaryotes and perhaps come to understand why it took so long.

Such a reconstruction of genome history will require going beyond the single gene level (Ghoshdastider et al., 2015) to the higher level organization of the genome:

> "In some perhaps more than metaphorical way, DNA may be both the book and the tree of life. With a bit more work, we may come to understand how... to read it hierarchically, chapter and verse, not just letter by letter, syllable by syllable, and word by word, i.e., until now we have been getting no further than nucleotides, codon and genes, respectively, plus a few "downstream controls". Our two meters of DNA... is a hierarchical, fractal structure, but we will have to learn to look at it at many levels to begin to see this".
>
> **(Gordon, 1999)**

For example, the *E. coli* genome is compartmentalized into at least three domains of its single chromosome (Fritsche et al., 2012) (Fig. 3.30 in (Gordon and Gordon, 2016)). It is here, we propose, that the solution may lie to our question of what took so long for prokaryotes to give rise to eukaryotes. At this writing, 150 Archaea genomes (Wikipedia, 2016d) and thousands of Bacteria genomes (Wikipedia, 2017a) have been completely sequenced, so there may be enough data to proceed with analyzing the higher order structures of their genomes and estimating the order in which novelties were introduced over geological time, and which ones came from DNA duplication or horizontal transfer. We might then be able to determine whether or not the GOE or the Huron snowball Earth (Figs. 3 and 18) slowed or set back (by extinction) the progress towards the opportunity to create the first eukaryote.

In any case, we suggest that early prokaryotes, while they may have had opportunities for extracellular and intracellular symbiosis, had too small a repertoire of components to lead to a successful symbiont, inaugurating the eukaryotes. We know roughly when that threshold was achieved. But it was a long slog during which prokaryotes evolved and reached that threshold. So symbiosis itself was not the trigger of eukaryogenesis. It was the long term; it was the rich evolution of prokaryotes that eventually set up the conditions. Whether particular geophysical, geochemical, atmospheric, and astrophysical events played an essential role or a hindrance might become clear as we dissect this long

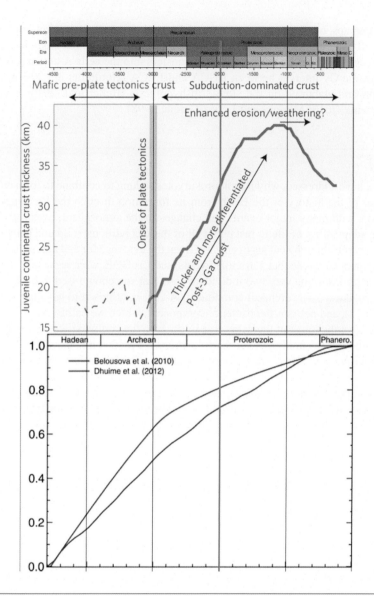

FIG. 13

Top: Thickness of continental crust. From Dhuime et al. (2015) with permission of Nature Publishing Group. The onset of plate tectonics is documented in (Shirey and Richardson, 2011), though "some sort of plate tectonics activity may have commenced as early as 4.5–4.4 Ga" (Wellman and Strother, 2015). "A review of continental growth models leaves open the possibilities that Earth during the Hadean Eon (~4.5–4.0 Ga) was characterized by massive early crust or essentially none at all. … . Hadean detrital zircons… are interpreted to reflect an early terrestrial hydrosphere, early felsic crust in which granitoids were produced and later weathered under high water activity conditions, and even the possible existence of plate boundary interactions—in strong contrast to the traditional view of an uninhabitable, hellish world" (Harrison, 2009). *Bottom*: Cumulative

(Continued)

stretch of evolution. Perhaps, our formulation of a refined model of the transition from the single cell LUCA to the single cell LECA, which includes accumulating novelty generation along many lineages with the novelties brought together by horizontal gene transfers into the cells A-euk and B-euk, may finally explain the cosmic Gy interval between LUCA and LECA (Gordon et al., 2017b) (Fig. 24). Thus, we concur with and elaborate on the slow-drip hypothesis (Butterfield, 2015; Lester et al., 2006; Martijn and Ettema, 2013; Moreira and López-García, 1998).

8 CONCLUSION

The main question we have addressed, why it took prokaryotes so long to combine to form eukaryotes, has taken us on a tour of the history of the Earth from its formation through the Precambrian. That period was a wild ride, with many major events and changes in the atmosphere, ocean, and forming continents. Given the adaptability of life to just about all of these conditions, mild and harsh, it is hard to conclude that these events retarded eukaryogenesis. Given the evidence for and against an anaerobic LECA, we likewise cannot conclude that particular events, such as GOE, were necessary for eukaryogenesis. We have thus turned to the very two individual cells that somehow fused to become one, the initial event of eukaryogenesis. The archaeal component, A-euk, is surmised to have been "complex" (Koonin and Yutin, 2014), and perhaps the bacterial component, which we call B-euk, was also complex. We therefore suggest that most of the delay was intrinsic to the complexification from LUCA to A-euk and B-euk and the rare event of their successful fusion. We predict that the steps along the way will be found by examining the higher order genome structure of many Archaea and Bacteria. These steps may be described in semiotic terms, especially if their higher order genome structures have many levels. It is then the multiple innovations and horizontal gene transfers culminating in A-euk and B-euk that determined the delay in eukaryogenesis. This notion can be tested by examining Archaea and Bacteria genomes at many levels, finding their phylogenetic relationships and trying to estimate when each innovation occurred.

While the mechanisms may have been different, we can anticipate that two earlier events, the origin of life and the transition from the first protocell to LUCA, may have also required cosmological time scales. The LUCA to LECA transition thus hints that these earlier events may have occurred before the formation of the Earth (Sharov and Gordon, 2017).

ACKNOWLEDGMENTS

RG would like to thank the US Corp of Army Engineers for providing a peaceful campsite atop the 800 km wide Sabine microcontinent in Pike County, Arkansas (Griffin et al., 2011), where this chapter was completed and life on Earth may have begun. We would like to thank Alexei Sharov for his editorial suggestions.

FIG. 13—CONT'D

volume of continental crust relative to today. "Examples for recent results on the growth of the volume of continental crust over time derived from isotopic data (Belousova et al., 2010; Dhuime et al., 2012)." From Feulner (2012) with permission of the American Geophysical Union.

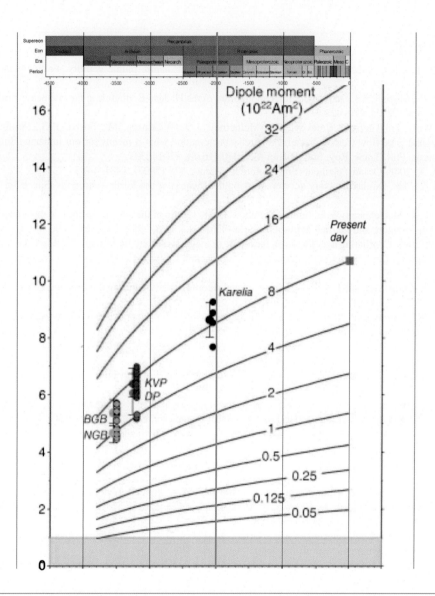

FIG. 14

Magnetopause standoff distance of the solar wind from the sunlit side of the Earth. The magnetopause is the point towards the Sun where the solar wind pressure is balanced by the Earth's magnetic field. The vertical scale is in units of the Earth's present radius. The fit of measured dipole moments versus age suggests that the solar wind standoff distance was never less than half that at present. From Tarduno et al. (2014) with permission of Elsevier.

REFERENCES

Abbott, D.H., Isley, A.E., 2002. Extraterrestrial influences on mantle plume activity. Earth Planet. Sci. Lett. 205 (1–2), 53–62.

Abramov, O., Mojzsis, S.J., 2009. Microbial habitability of the Hadean Earth during the late heavy bombardment. Nature 459 (7245), 419–422.

Akanni, W.A., Siu-Ting, K., Creevey, C.J., McInerney, J.O., Wilkinson, M., Foster, P.G., Pisani, D., 2015. Horizontal gene flow from Eubacteria to Archaebacteria and what it means for our understanding of eukaryogenesis. Phil. Trans. Roy. Soc. B-Biol. Sci. 370 (1678), #20140337.

Anbar, A.D., 2008. Oceans: elements and evolution. Science 322 (5907), 1481–1483.

Arbab, A.I., 2008. On the planetary acceleration and the rotation of the Earth. Astrophys. Space Sci. 314 (1–3), 35–39.

Arevalo Jr., R., McDonough, W.F., Luong, M., 2009. The K/U ratio of the silicate Earth: Insights into mantle composition, structure and thermal evolution. Earth Planet. Sci. Lett. 278 (3–4), 361–369.

Arthur, W., 2008. Conflicting hypotheses on the nature of mega-evolution. In: Minelli, A., Fusco, G. (Eds.), Evolving Pathways: Key Themes in Evolutionary Developmental Biology. Cambridge University Press, New York, USA, pp. 50–61.

Bachan, A., Kump, L.R., 2015. The rise of oxygen and siderite oxidation during the Lomagundi Event. Proc. Natl. Acad. Sci. U. S. A. 112 (21), 6562–6567.

Bahcall, J.N., Pinsonneault, M.H., Basu, S., 2001. Solar models: Current epoch and time dependences, neutrinos, and helioseismological properties. Astrophys. J. 555 (2), 990–1012.

Bailer-Jones, C.A.L., 2009. The evidence for and against astronomical impacts on climate change and mass extinctions: a review. Int. J. Astrobiol. 8 (3), 213–239.

Barbieri, M., 2007. Is the cell a semiotic system? In: Barbieri, M. (Ed.), Introduction to Biosemiotics: The New Biological Synthesis. Springer, Dordrecht, pp. 179–208.

Barley, M.E., Bekker, A., Krapez, B., 2005. Late Archean to Early Paleoproterozoic global tectonics, environmental change and the rise of atmospheric oxygen. Earth Planet. Sci. Lett. 238 (1–2), 156–171.

Battistuzzi, F.U., Hedges, S.B., 2009. A major clade of prokaryotes with ancient adaptations to life on land. Mol. Biol. Evol. 26 (2), 335–343.

Battistuzzi, F.U., Feijao, A., Hedges, S.B., 2004. A genomic timescale of prokaryote evolution: insights into the origin of methanogenesis, phototrophy, and the colonization of land. BMC Evol. Biol. 4, #44.

Baum, D.A., Baum, B., 2014. An inside-out origin for the eukaryotic cell. BMC Biol. 12, #76.

Bavishi, A., Abhishek, A., Lin, L., Choudhary, M., 2010. Complex prokaryotic genome structure: rapid evolution of chromosome II. Genome 53 (9), 675–687.

Beerling, D., Berner, R.A., Mackenzie, F.T., Harfoot, M.B., Pyle, J.A., 2009. Methane and the CH_4-related greenhouse effect over the past 400 million years. Am. J. Sci. 309 (2), 97–113.

Bekker, A., Holland, H.D., 2012. Oxygen overshoot and recovery during the early Paleoproterozoic. Earth Planet. Sci. Lett. 317, 295–304.

Belousova, E.A., Kostitsyn, Y.A., Griffin, W.L., Begg, G.C., O'Reilly, S.Y., Pearson, N.J., 2010. The growth of the continental crust: Constraints from zircon Hf-isotope data. Lithos 119 (3–4), 457–466.

Beraldi-Campesi, H., 2013. Early life on land and the first terrestrial ecosystems. Ecol. Process. 2 (1), 1–17.

Berget, S.M., Moore, C., Sharp, P.A., 1977. Spliced segments at the 5′ terminus of adenovirus 2 late mRNA. Proc. Natl. Acad. Sci. U. S. A. 74 (8), 3171–3175.

Berner, R.A., 2006. Geological nitrogen cycle and atmospheric N_2 over Phanerozoic time. Geology 34 (5), 413–415.

Bernstein, H., Bernstein, C., Michod, R.E., 2012. DNA repair as the primary adaptive function of sex in bacteria and eukaryotes. Int. J. Med. Biol. Front. 18 (2/3), 111–146.

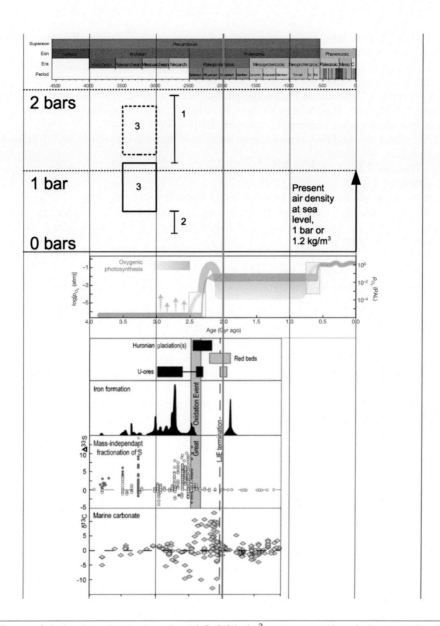

FIG. 15

Top: (1) Range of air density estimate at sea level 1.3–2.3 kg/m^3, as measured by raindrop cratering of volcanic ash (Som et al., 2012). (2) Range 0.23–0.5 bar using the size distribution of gas bubbles in basaltic lava flows that solidified at sea level (Som et al., 2016). (3) N_2 range of 0.5–1.1 bar (solid rectangle) supplemented with a maximum of 0.7 bar CO_2 (dashed rectangle) (Marty et al., 2013). *Middle*: Atmospheric oxygen. The darker gray

(Continued)

Bjerrum, C.J., Canfield, D.E., 2002. Ocean productivity before about 1.9 Gyr ago limited by phosphorus adsorption onto iron oxides. Nature 417 (6885), 159–162.

Blake, R.E., Chang, S.J., Lepland, A., 2010. Phosphate oxygen isotopic evidence for a temperate and biologically active Archaean ocean. Nature 464 (7291), 1029–1032.

Bodaker, I., Sharon, I., Suzuki, M.T., Feingersch, R., Shmoish, M., Andreishcheva, E., Sogin, M.L., Rosenberg, M., Maguire, M.E., Belkin, S., Oren, A., Béjà, O., 2010. Comparative community genomics in the Dead Sea: an increasingly extreme environment. ISME J. 4 (3), 399–407.

Boddy, J., 2016. Alien life could feed on cosmic rays. http://www.sciencemag.org/news/2016/10/alien-life-could-feed-cosmic-rays.

Boehnke, P., Harrison, T.M., 2016. Illusory late heavy bombardments. Proc. Natl. Acad. Sci. U. S. A. 113 (39), 10802–10806.

Borel, É., 1913. Mécanique Statistique et Irréversibilité. J. Phys. 5e Sér. 3, 189–196.

Bottke, W.F., Vokrouhlický, D., Minton, D., Nesvorný, D., Morbidelli, A., Brasser, R., Simonson, B., Levison, H.F., 2012. An Archaean heavy bombardment from a destabilized extension of the asteroid belt. Nature 485 (7396), 78–81.

Bourdon, B., Touboul, M., Caro, G., Kleine, T., 2008. Early differentiation of the Earth and the Moon. Phil. Trans. Roy. Soc. A Math. Phys. Eng. Sci. 366 (1883), 4105–4128.

Brocks, J.J., Logan, G.A., Buick, R., Summons, R.E., 1999. Archean molecular fossils and the early rise of eukaryotes. Science 285 (5430), 1033–1036.

Brown, J.R., Doolittle, W.F., 1997. *Archaea* and the prokaryote-to-eukaryote transition. Microbiol. Mol. Biol. Rev. 61 (4), 456–502.

Buchalo, A.S., Nevo, E., Wasser, S.P., Oren, A., Molitoris, H.P., 1998. Fungal life in the extremely hypersaline water of the Dead Sea: first records. Proc. Roy. Soc. B-Biol. Sci. 265 (1404), 1461–1465.

Burki, F., 2016. Mitochondrial evolution: Going, going, gone. Curr. Biol. 26 (10), R410–R412.

Butterfield, N.J., 2000. *Bangiomorpha pubescens* n. gen., n. sp.: implications for the evolution of sex, multicellularity, and the Mesoproterozoic/Neoproterozoic radiation of eukaryotes. Paleobiology 26 (3), 386–404.

Butterfield, N.J., 2015. Early evolution of the Eukaryota. Palaeontology 58 (1), 5–17.

Caetano-Anollés, G., Nasir, A., Zhou, K.Y., Caetano-Anollés, D., Mittenthal, J.E., Sun, F.J., Kim, K.M., 2014. Archaea: The first domain of diversified life. Arch. Int. Microbiol. J. #590214.

Campbell, I.H., Allen, C.M., 2008. Formation of supercontinents linked to increases in atmospheric oxygen. Nat. Geosci. 1 (8), 554–558.

Canfield, D.E., 1998. A new model for Proterozoic ocean chemistry. Nature 396 (6710), 450–453.

FIG. 15—CONT'D

represents the current model. Vertical arrows denote possible "whiffs" of O_2. From Lyons et al. (2014) with permission of Nature Publishing Group. The latest estimate is that the GOE started between 2.460 Ga and 2.426 Ga. "Furthermore, the rise of atmospheric oxygen was not monotonic, but was instead characterized by oscillations, which together with climatic instabilities may have continued over the next ~200 My until ≤" 2.250 Ga to 2.240 Ga (Gumsley et al., 2017). *Bottom:* "Schematic depiction of major events during the Archaean to Proterozoic transition. The dark and light boxes in the upper panels represent reducing and oxidizing atmospheres, respectively;... S = sulfur (after (Bekker and Holland, 2012; Holland, 1994)).... The distribution of Fe deposits is after (Isley and Abbott, 1999). Plot of $\Delta^{33}S$ versus age after (Farquhar and Wing, 2003) and (Reinhard et al., 2013); white circles are bulk rock data and grey circles are secondary ion mass spectrometry generated data. Carbon isotopic evolution of marine carbonates is after (Shields and Veizer, 2002)." From Martin et al. (2013) with permission of Elsevier.

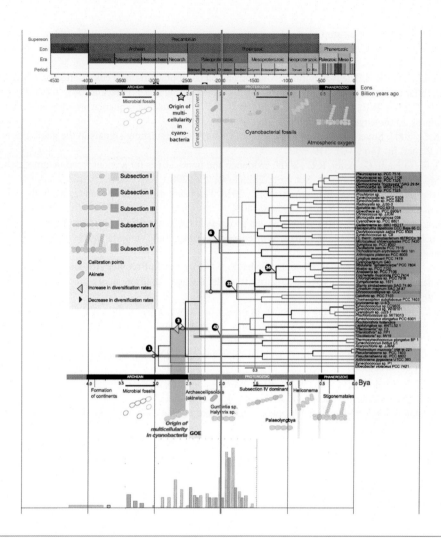

FIG. 16

Top: "Timeline of the evolution of Earth. Continents formed ~3.8 Bya [Ga], after temperatures had cooled and meteorite bombardment had stopped. In ~3.46 billion year old rocks (Australia) first evidence for life is found. Around 2.45–2.32 Bya oxygen accumulated in the atmosphere during the Great Oxidation Event produced by cyanobacterial oxygenic photosynthesis. Oldest cyanobacterial fossils can be found in the Gunflint Chert (1.88–2.0 Bya). First eukaryotes evolve around 1.6–1.8 Bya." Bya = Ga. From Marshall and Schirrmeister (2013) with kind permission of Bettina Schirrmeister. The "Microbial fossils" would be bacteria, as "No one, to the best of my knowledge, has yet reported an archaeal microfossil" (Knoll, 2015). *Middle*: "Time calibrated phylogeny of cyanobacteria displaying divergence time estimates.... Morphological features of taxa are marked by colored boxes and listed in the inset.... Branches with posterior probabilities >0.9 in all analyses are presented as thick lines. Gray circles mark points used for calibration of the tree.... A significant increase in diversification rate (yellow triangle) [9.66-fold (average of all analyses)] can be detected at node 3 and a minor

(Continued)

Carter, B., 2008. Five- or six-step scenario for evolution? Int. J. Astrobiol. 7 (2), 177–182.

Cavalier-Smith, T., 2006. Origin of mitochondria by intracellular enslavement of a photosynthetic purple bacterium. Proc. Roy. Soc. B-Biol. Sci. 273 (1596), 1943–1952.

Cavalier-Smith, T., 2014. The Neomuran revolution and phagotrophic origin of eukaryotes and cilia in the light of intracellular coevolution and a revised tree of life. Cold Spring Harb. Perspect. Biol. 6 (9), #a016006.

Chow, L.T., Gelinas, R.E., Broker, T.R., Roberts, R.J., 1977. An amazing sequence arrangement at the 5' ends of adenovirus 2 messenger RNA. Cell 12 (1), 1–8.

Condie, K.C., 2004. Supercontinents and superplume events: distinguishing signals in the geologic record. Phys. Earth Planet. In. 146 (1–2), 319–332.

Condie, K.C., O'Neill, C., Aster, R.C., 2009. Evidence and implications for a widespread magmatic shutdown for 250 My on Earth. Earth Planet. Sci. Lett. 282 (1–4), 294–298.

Ćuk, M., Hamilton, D.P., Lock, S.J., Stewart, S.T., 2016. Tidal evolution of the Moon from a high-obliquity, high-angular-momentum Earth. Nature 539, 402–406.

Curtis, B.A., Archibald, J.M., 2010. Problems and progress in understanding the origins of mitochondria and plastids. In: Seckbach, J., Grube, M. (Eds.), Symbioses and Stress: Joint Ventures in Biology. Springer, Dordrecht, pp. 39–62.

Dacks, J.B., Field, M.C., Buick, R., Eme, L., Gribaldo, S., Roger, A.J., Brochier-Armanet, C., Devos, D.P., 2016. The changing view of eukaryogenesis—fossils, cells, lineages and how they all come together. J. Cell Sci. 129 (20), 3695–3703.

Damer, B., 2016. A field trip to the Archaean in search of Darwin's warm little pond. Life 6, #21.

Darnell, J.E., 1978. Implications of RNA-RNA splicing in evolution of eukaryotic cells. Science 202 (4374), 1257–1260.

de Bary, H.A., 1878. Über symbiose, Tageblatt 51 Versamml. Deutscher Naturforscher und Aerzte, Cassel, pp. 121–126.

de Mello, C.F.P., Lepine, J.R., Dias, W.D., 2009. Astrobiologically interesting stars near the Sun: Galactic orbits and mass extinctions. In: Meech, K.J., Keane, J.V., Mumma, M.J., Siefert, J.L., Werthimer, D.J. (Eds.), Bioastronomy 2007: Molecules, Microbes, and Extraterrestrial Life. Astronomical Society of the Pacific, San Francisco, CA, USA, pp. pp. 349–352.

Denis, C., Rybicki, K.R., Schreider, A.A., Tomecka-Suchoń, S., Varga, P., 2011. Length of the day and evolution of the Earth's core in the geological past. Astronomische Nachrichten 332 (1), 24–35.

Dey, G., Thattai, M., Baum, B., 2016. On the archaeal origins of eukaryotes and the challenges of inferring phenotype from genotype. Trends Cell Biol. 26 (7), 476–485.

FIG. 16—CONT'D

decrease (red triangle) at 33/34. The earlier shift close to node 3 coincides with the origin of multicellularity. Schematic drawings of cyanobacterial fossils are provided under the timeline, with the ones used for calibration of the tree marked in red. Our results indicate that multicellularity (green shade) originated before or at the beginning of the GOE." From Schirrmeister et al. (2013) with kind permission of Proceedings of the National Academy of Sciences USA. *Bottom*: Cyanobacteria are a major component of stromatolites. "Distribution of reported number of occurrences of total stromatolites and of microdigitate stromatolites during the Precambrian." From Condie (2004) with permission of Elsevier. From right to left, the distribution has been extended here to 3.5 Ga with data from Schopf (2006), to 3.7 Ga with the discovery of shallow water stromatolites of that age (Nutman et al., 2016), and to between 4.28 Ga to 3.77 Ga with the latest discovery (Dodd et al., 2017). Vertical scale is counts of sites. Stromatolites continue to exist to the present (Damer, 2016) (Figs. 4 and 7).

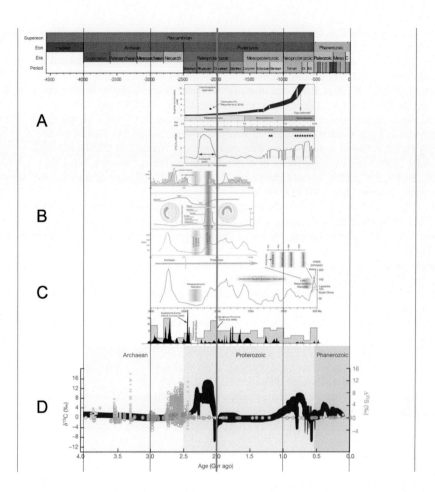

FIG. 17

(A) The Lomagundi event. From Warren (2016) with kind permission of John K. Warren. (B) "The transition from Protopangaea to Palaeopangaea illustrating the features of prolonged quiescence in late Archaean and early Palaeoproterozoic times followed by resurgence of tectonic activity after ~2.2 Ga accompanying reconfiguration of the continental crust on the surface of the Earth and coinciding with the Longamundi–Jatulian isotopic event (Melezhik et al., 2007). Geological signatures included here comprise a histogram of arc greenstones, the global frequency of granitoids, the duration of sedimentary unconformities, and variations in sea level and isotopic carbon (Barley et al., 2005; Condie et al., 2009). The figures of continental distribution (earth brown) on projections of the whole globe show the configuration of the crust with respect to the geomagnetic field (and presumed rotation) axis before and after 2.2 Ga." From Piper (2015) with permission of Taylor & Francis.
(C) "Variation in V_{RMS} [continental velocity] for the continental crust during Precambrian times (after Piper, 2013). Individual values of velocity are qualitative and based on calculation of mean poles within 50 million year time windows; between 0.8 and 0.6 Ga, poles are calculated at 25 million year intervals to emphasize the slowing down followed by rapid increase in the rate of APW [Apparent Polar Wander] in the Ediacaran Period after ~0.6 Ga; the
(Continued)

Dhuime, B., Hawkesworth, C.J., Cawood, P.A., Storey, C.D., 2012. A change in the geodynamics of continental growth 3 billion years ago. Science 335 (6074), 1334–1336.

Dhuime, B., Wuestefeld, A., Hawkesworth, C.J., 2015. Emergence of modern continental crust about 3 billion years ago. Nat. Geosci. 8 (7), 552–555.

DiRuggierro, J., Robb, F.T., 2004. Early evolution of DNA repair mechanisms. In: de Poupkna, L.R. (Ed.), Genetic Code and the Origin of Life. Landes Bioscience, Georgetown, TX, USA, pp. 169–182.

Dodd, M.S., Papineau, D., Grenne, T., Slack, J.F., Rittner, M., Pirajno, F., O'Neil, J., Little, C.T.S., 2017. Evidence for early life in Earth's oldest hydrothermal vent precipitates. Nature 543 (7643), 60–64.

Doolittle, W.F., 1978. Genes in pieces: were they ever together? Nature 272 (5654), 581–582.

Douglas, A.E., 2010. The Symbiotic Habit. Princeton University Press, Princeton, NJ, USA.

Dunn, A.K., Stabb, E.V., 2007. Beyond quorum sensing: the complexities of prokaryotic parliamentary procedures. Anal. Bioanal. Chem. 387 (2), 391–398.

Dyer, A.J., 1953. The absorption of the hard component of cosmic rays in water. Aust. J. Phys. 6 (1), 60–66.

Embley, T.M., Williams, T.A., 2015. Steps on the road to eukaryotes. Nature 521 (7551), 169–170.

Emelyanov, V.V., 2001. Rickettsiaceae, *Rickettsia*-like endosymbionts, and the origin of mitochondria. Biosci. Rep. 21 (1), 1–17.

Emmanuel, S., Ague, J.J., 2007. Implications of present-day abiogenic methane fluxes for the early Archean atmosphere. Geophys. Res. Lett. 34, L15810.

Emmeche, C., Kull, K., 2011. Towards a Semiotic Biology: Life Is the Action of Signs. Imperial College Press, London, UK.

Emspak, J., 2016. Early Earth's Atmosphere Was Surprisingly Thin. https://www.scientificamerican.com/article/early-earth-s-atmosphere-was-surprisingly-thin/.

Ettema, T.J.G., 2016. Mitochondria in the second act. Nature 531 (7592), 39–40.

Eyles, N., 2008. Glacio-epochs and the supercontinent cycle after ∼3.0 Ga: Tectonic boundary conditions for glaciation. Palaeogeogr. Palaeoclimatol. Palaeoecol. 258 (1–2), 89–129.

Fani, R., Gallo, R., Lio, P., 2000. Molecular evolution of nitrogen fixation: The evolutionary history of the *nifD*, *nifK*, *nifE*, and *nifN* genes. J. Mol. Evol. 51 (1), 1–11.

Farquhar, J., Wing, B.A., 2003. Multiple sulfur isotopes and the evolution of the atmosphere. Earth Planet. Sci. Lett. 213 (1–2), 1–13.

Fedorov, A., Cao, X.H., Saxonov, S., de Souza, S.J., Roy, S.W., Gilbert, W., 2001. Intron distribution difference for 276 ancient and 131 modern genes suggests the existence of ancient introns. Proc. Natl. Acad. Sci. U. S. A. 98 (23), 13177–13182.

Feng, F., Bailer-Jones, C.A.L., 2013. Assessing the influence of the solar orbit on terrestrial biodiversity. Astrophys. J. 768 (2), #152.

Feulner, G., 2012. The faint young sun problem (corrected on 2 June 2016). Rev. Geophys. 50, RG2006.

Field, M.C., Dacks, J.B., 2009. First and last ancestors: reconstructing evolution of the endomembrane system with ESCRTs, vesicle coat proteins, and nuclear pore complexes. Curr. Opin. Cell Biol. 21 (1), 4–13.

Filipović, M.D., Horner, J., Crawford, E.J., Tothill, N.F.H., White, G.L., 2013. Mass extinction and the structure of the Milky Way. Serbian Astron. J. 187, 43–52.

FIG. 17—CONT'D

latter effect is illustrated for three individual APWPs [Apparent Polar Wander Paths] in this figure during the interval 0.6–0.5 Ma." From Piper (2015) with permission of Taylor & Francis. (D) "Summary of carbon (black) and sulfur (red and grey) isotope data through Earth's history." From Lyons et al. (2014) with permission of Nature Publishing Group.

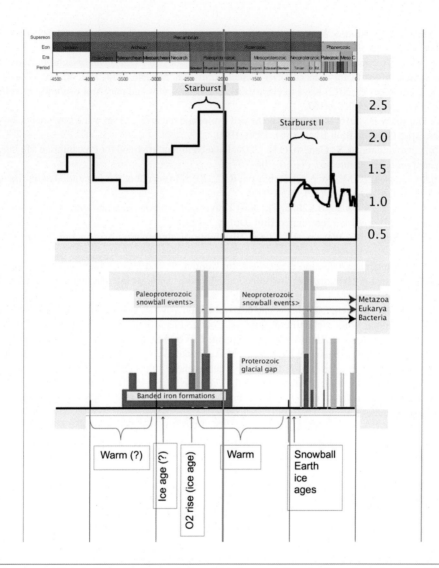

FIG. 18

Top: Starbursts and Snowball events. The scale top right shows relative star formation rate in the Milky Way Galaxy. From Kataoka et al. (2013) with permission of Elsevier. Left/right reversed and relabeled. See also (Kataoka et al., 2014). *Bottom*: Ice ages. The "O₂ rise (ice age)" is the Huronian. Adapted from Kasting and Howard (2006). See also (Eyles, 2008; Kataoka et al., 2013).

Fontaine, F.J., Wilcock, W.S.D., Butterfield, D.A., 2007. Physical controls on the salinity of mid-ocean ridge hydrothermal vent fluids. Earth Planet. Sci. Lett. 257 (1–2), 132–145.

Forster, P., Ramaswamy, V., 2007. Changes in atmospheric constituents and in radiative forcing. In: Solomon, S., Qin, D., Manning, M., Marquis, M., Averyt, K., Tignor, M.M.B., Miller, H.L., Chen, Z.L. (Eds.), Climate Change 2007: The Physical Science Basis. Cambridge University Press, New York, pp. 129–234.

Frei, R., Gaucher, C., Poulton, S.W., Canfield, D.E., 2009. Fluctuations in Precambrian atmospheric oxygenation recorded by chromium isotopes. Nature 461 (7261), 250–253.

Fritsche, M., Li, S., Heermann, D.W., Wiggins, P.A., 2012. A model for *Escherichia coli* chromosome packaging supports transcription factor-induced DNA domain formation. Nucleic Acids Res. 40 (3), 972–980.

Fuerst, J.A., Sagulenko, E., 2012. Keys to eukaryality: planctomycetes and ancestral evolution of cellular complexity. Front. Microbiol. 3, #167.

Garg, S.G., Martin, W.F., 2016. Mitochondria, the cell cycle, and the origin of sex via a syncytial eukaryote common ancestor. Genome Biol. Evol. 8 (6), 1950–1970.

Gaucher, E.A., Govindarajan, S., Ganesh, O.K., 2008. Palaeotemperature trend for Precambrian life inferred from resurrected proteins. Nature 451 (7179), 704–707.

Ghoshdastider, U., Jiang, S.M., Popp, D., Robinson, R.C., 2015. In search of the primordial actin filament. Proc. Natl. Acad. Sci. U. S. A. 112 (30), 9150–9151.

Glansdorff, N., Xu, Y., Labedan, B., 2008. The Last Universal Common Ancestor: emergence, constitution and genetic legacy of an elusive forerunner. Biol. Direct 3, #29.

Goldblatt, C., Lenton, T.M., Watson, A.J., 2006. Bistability of atmospheric oxygen and the Great Oxidation. Nature 443 (7112), 683–686.

Gontier, N., 2015. Reticulate evolution everywhere. In: Gontier, N. (Ed.), Reticulate Evolution: Symbiogenesis, Lateral Gene Transfer, Hybridization and Infectious Heredity. Springer International Publishing Switzerland, Cham, pp. 1–40.

Gordon, R., 1999. The Hierarchical Genome and Differentiation Waves: Novel Unification of Development, Genetics and Evolution. World Scientific & Imperial College Press, Singapore & London.

Gordon, N.K., Gordon, R., 2016. Embryogenesis Explained. World Scientific Publishing, Singapore.

Gordon, R., Hanczyc, M.M., Denkov, N.D., Tiffany, M.A., Smoukov, S.K., 2017a. Chapter 18: Emergence of polygonal shapes in oil droplets and living cells: The potential role of tensegrity in the origin of life. In: Gordon, R., Sharov, A.A. (Eds.), Habitability of the Universe Before Earth [in series: Astrobiology: Exploring Life on Earth and Beyond, eds. Pabulo Henrique Rampelott, Joseph Seckbach & Richard Gordon]. Amsterdam, Elsevier B.V, pp. 427–490.

Gordon, R., Mikhailovsky, G.E., Gordon, N.K., 2017b. LUCA to LECA: A model for the gigayear delay from the first prokaryote to eukaryogenesis. Genes (Spec. Iss: Horizont. Gene Transf.) in preparation.

Gouy, M., Chaussidon, M., 2008. Evolutionary biology: Ancient bacteria liked it hot. Nature 451 (7179), 635–636.

Grandjean, V., Hauck, Y., LeDérout, J., Hirschbein, L., 1996. Noncomplementing diploids from *Bacillus subtilis* protoplast fusion: Relationship between maintenance of chromosomal inactivation and segregation capacity. Genetics 144 (3), 871–881.

Griffin, W.L., Begg, G.C., Dunn, D., O'Reilly, S.Y., Natapov, L.M., Karlstrom, K., 2011. Archean lithospheric mantle beneath Arkansas: continental growth by microcontinent accretion. Bull. Geol. Soc. Am. 123, 1763–1775.

Grosch, E.G., Hazen, R.M., 2015. Microbes, mineral evolution, and the rise of microcontinents-origin and coevolution of life with early Earth. Astrobiology 15 (10), 922–939.

Gross, J., Bhattacharya, D., 2010. Uniting sex and eukaryote origins in an emerging oxygenic world. Biol. Direct 5, #53.

Gumsley, A.P., Chamberlain, K.R., Bleeker, W., Söderlund, U., de Kock, M.O., Larsson, E.R., Bekker, A., 2017. Timing and tempo of the great oxidation event. Proc. Natl. Acad. Sci. 114(8), 1811–1816.

Harrison, T.M., 2009. The Hadean crust: evidence from > 4 Ga zircons. Annu. Rev. Earth Planet. Sci. 37, 479–505.

Hart, M.H., 1978. The evolution of the atmosphere of the Earth. Icarus 33, 23–39.

Hazen, R.M., 2012. Mineral Evolution I. The Initial Idea for Mineral Evolution (December, 2006 to November, 2008). https://hazen.carnegiescience.edu/research/mineral-evolution.

Hazen, R.M., Ferry, J.M., 2010. Mineral evolution: mineralogy in the fourth dimension. Elements 6 (1), 9–12.

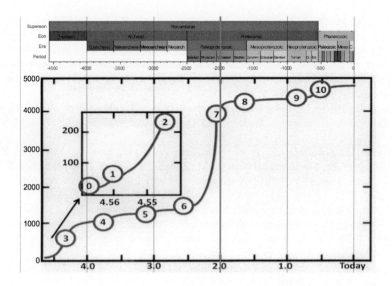

FIG. 19

Vertical scale is "Mineral diversity (cumulative # species)." "Earth's mineralogical diversity has increased through 10 stages." From Hazen (2012) with kind permission of Robert M. Hazen. The labels for stages during the Precambrian are from Table 1 in (Hazen and Ferry, 2010). The number of present day minerals on Earth has recently been raised to 5200 (Morrison et al., 2017).

Hazen, R.M., Papineau, D., Leeker, W.B., Downs, R.T., Ferry, J.M., McCoy, T.J., Sverjensky, D.A., Yang, H.X., 2008. Mineral evolution. Am. Mineral. 93 (11–12), 1693–1720.

He, D., Fu, C.J., Baldauf, S.L., 2016. Multiple origins of eukaryotic *cox15* suggest horizontal gene transfer from bacteria to Jakobid mitochondrial DNA. Mol. Biol. Evol. 33 (1), 122–133.

Hedges, S.B., Kumar, S. (Eds.), 2009. The Timetree of Life. Oxford University Press, Oxford.

Hedges, S.B., Chen, H., Kumar, S., Wang, D.Y., Thompson, A.S., Watanabe, H., 2001. A genomic timescale for the origin of eukaryotes. BMC Evol. Biol. 1 (1), #4.

Hirshfeld, J., 2014. File:Timeline showing the Boring Billion.png. https://commons.wikimedia.org/wiki/File: Timeline_showing_the_Boring_Billion.png.

Holland, H.D., 1978. The Chemistry of the Atmosphere and Oceans. Wiley, New York.

Holland, H.D., 1994. Early proterozoic atmospheric change. In: Bengtson, S. (Ed.), Early Life on Earth. Columbia University Press, New York, pp. 237–244.

Holland, H.D., 2002. Volcanic gases, black smokers, and the Great Oxidation Event. Geochim. Cosmochim. Acta 66 (21), 3811–3826.

Hren, M.T., Tice, M.M., Chamberlain, C.P., 2009. Oxygen and hydrogen isotope evidence for a temperate climate 3.42 billion years ago. Nature 462 (7270), 205–208.

Huxley, J.S., 1942. Evolution, The Modern Synthesis. George Allen & Unwin, London.

Huxley, J., Pigliucci, M., Müller, G.B., 2010. Evolution: The Modern Synthesis: The Definitive Edition. MIT Press, Cambridge.

Isley, A.E., Abbott, D.H., 1999. Plume-related mafic volcanism and the deposition of banded iron formation. J. Geophys. Res.-Solid Earth 104 (B7), 15461–15477.

Izidoro, A., de Souza Torres, K., Winter, O.C., Haghighipour, N., 2013. A compound model for the origin of Earth's water. Astrophys. J. 767 (1), #54.

Jagers op Akkerhuis, G.A.J.M., 2010. The Operator Hierarchy: A Chain of Closures Linking Matter, Life and Artificial Intelligence [Ph.D. Thesis]. Radboud University Nijmegen, Nijmegen, The Netherlands.

Jagers op Akkerhuis, G.A.J.M., 2017. Why on theoretical grounds it is likely that "life" exists throughout the universe. In: Gordon, R., Sharov, A.A. (Eds.), Habitability of the Universe Before Earth [in series: Astrobiology: Exploring Life on Earth and Beyond, eds. Pabulo Henrique Rampelott, Joseph Seckbach & Richard Gordon]. Elsevier B.V, Amsterdam, pp. 491–505.

Jeffares, D.C., Poole, A.M., Penny, D., 1998. Relics from the RNA world. J. Mol. Evol. 46 (1), 18–36.

Kandler, O., 1994. Cell wall biochemistry in Archaea and its phylogenetic implications. J. Biol. Phys. 20 (1–4), 165–169.

Karnkowska, A., Vacek, V., Zubáčová, Z., Treitli, S.C., Petrželková, R., Eme, L., Novák, L., Zárský, V., Barlow, L.D., Herman, E.K., Soukal, P., Hroudová, M., Doležal, P., Stairs, C.W., Roger, A.J., Eliáš, M., Dacks, J.B., Vlček, C., Hampl, V., 2016. A eukaryote without a mitochondrial organelle. Curr. Biol. 26 (10), 1274–1284.

Kasting, J.F., 2005. Methane and climate during the Precambrian era. Precambrian Res. 137 (3–4), 119–129.

Kasting, J.F., 2014. Atmospheric composition of Hadean-early Archean Earth: The importance of CO. In: Shaw, G.H. (Ed.), Earth's Early Atmosphere and Surface Environment. The Geological Society of America, Boulder, Colorado, USA, pp. 19–28.

Kasting, J.F., Howard, M.T., 2006. Atmospheric composition and climate on the early Earth (with Discussion). Phil. Trans. Roy. Soc. B-Biol. Sci. 361 (1474), 1733–1742.

Kasting, J.F., Eggler, D.H., Raeburn, S.P., 1993. Mantle redox evolution and the oxidation state of the Archean atmosphere. J. Geol. 101 (2), 245–257.

Kataoka, R., Ebisuzaki, T., Miyahara, H., Maruyama, S., 2013. Snowball Earth events driven by starbursts of the Milky Way Galaxy. New Astron. 21, 50–62.

Kataoka, R., Ebisuzaki, T., Miyahara, H., Nimura, T., Tomida, T., Sato, T., Maruyama, S., 2014. The Nebula Winter: The united view of the snowball Earth, mass extinctions, and explosive evolution in the late Neoproterozoic and Cambrian periods. Gondw. Res. 25 (3), 1153–1163.

Khakhina, L.N., 1992. Concepts of Symbiogenesis: A Historical and Critical Study of the Research of Russian Botanists. Yale University Press, New Haven.

Kharecha, P., Kasting, J., Siefert, J., 2005. A coupled atmosphere-ecosystem model of the early Archean Earth. Geobiology 3 (2), 53–76.

Knauth, L.P., 2005. Temperature and salinity history of the Precambrian ocean: implications for the course of microbial evolution. Palaeogeogr Palaeoclimatol Palaeoecol 219, 53–69.

Knauth, L.P., Epstein, S., 1976. Hydrogen and oxygen isotope ratios in nodular and bedded cherts. Geochim. Cosmochim. Acta 40 (9), 1095–1108.

Knauth, L.P., Lowe, D.R., 2003. High Archean climatic temperature inferred from oxygen isotope geochemistry of cherts in the 3.5 Ga Swaziland Supergroup, South Africa. Geol. Soc. Am. Bull. 115 (5), 566–580.

Knoll, A.H., 2015. Paleobiological perspectives on early microbial evolution. Cold Spring Harb. Perspect. Biol. 7 (7), #a018093.

Knoll, A.H., Bergmann, K.D., Strauss, J.V., 2016. Life: the first two billion years. Phil. Trans. Roy. Soc. B-Biol. Sci. 371 (1707), #20150493.

Koeberl, C., 2003. The Late Heavy Bombardment in the inner solar system: Is there any connection to Kuiper belt objects? Earth Moon Planets 92 (1–4), 79–87.

Koeberl, C., 2006. Impact processes on the early Earth. Elements 2 (4), 211–216.

Konhauser, K.O., Pecoits, E., Lalonde, S.V., Papineau, D., Nisbet, E.G., Barley, M.E., Arndt, N.T., Zahnle, K., Kamber, B.S., 2009. Oceanic nickel depletion and a methanogen famine before the Great Oxidation Event. Nature 458 (7239), 750–U785.

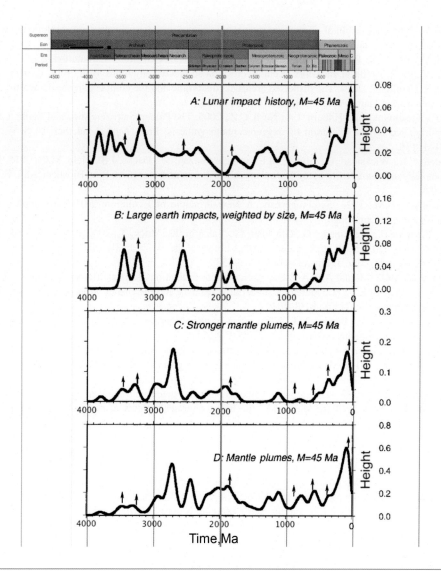

FIG. 20

"Height of time series versus age in Ma. All data is smoothed by adjusting the minimum age error (M) to 45 Ma. (A) Lunar impact history as derived from dating of impact spherules with age errors of ≤150 Ma. Arrows are at the same point in time as the tops of peaks in large impacts. Note that, within the mean error of the ages in the lunar time series (76 Ma), these peaks line up. (B) Terrestrial impact history derived from dating of known impact craters and inferred impact craters with diameters greater than 11 km. The peak heights in the time series are weighted by the size of the inferred impact crater for each impact event. Note the arrows denoting peaks that line up in the terrestrial and lunar impact time series. These arrows are throughout the time series, supporting the 97% confidence level of the cross-correlation between the two time series. (C) Stronger mantle plumes derived from dating of high-MgO extrusives and intrusives through time. Note the arrows showing seven

(Continued)

Koonin, E.V., 2006. The origin of introns and their role in eukaryogenesis: a compromise solution to the introns-early versus introns-late debate? Biol. Direct 1, #22.

Koonin, E.V., 2007. The Biological Big Bang model for the major transitions in evolution. Biol. Direct 2, #21.

Koonin, E.V., 2015. Archaeal ancestors of eukaryotes: not so elusive any more. BMC Biol. 13, #84.

Koonin, E.V., Yutin, N., 2014. The dispersed archaeal eukaryome and the complex archaeal ancestor of eukaryotes. Cold Spring Harb. Perspect. Biol. 6 (4), #a016188.

Kopp, R.E., Kirschvink, J.L., Hilburn, I.A., Nash, C.Z., 2005. The Paleoproterozoic snowball Earth: a climate disaster triggered by the evolution of oxygenic photosynthesis. Proc. Natl. Acad. Sci. U. S. A. 102 (32), 11131–11136.

Koumandou, V.L., Wickstead, B., Ginger, M.L., van der Giezen, M., Dacks, J.B., Field, M.C., 2013. Molecular paleontology and complexity in the last eukaryotic common ancestor. Crit. Rev. Biochem. Mol. Biol. 48 (4), 373–396.

Kozo-Polyansky, B.M., 1924. Symbiogenesis: A New Principle of Evolution, 2010, translation ed. Harvard University Press, Cambridge.

Krell, A., Beszteri, B., Dieckmann, G., Glockner, G., Valentin, K., Mock, T., 2008. A new class of ice-binding proteins discovered in a salt-stress-induced cDNA library of the psychrophilic diatom *Fragilariopsis cylindrus* (Bacillariophyceae). Eur. J. Phycol. 43 (4), 423–433.

Kropotkin, P.A., 1902. Mutual Aid, A Factor of Evolution. Extending Horizons Books, Boston.

Kump, L.R., Barley, M.E., 2007. Increased subaerial volcanism and the rise of atmospheric oxygen 2.5 billion years ago. Nature 448 (7157), 1033–1036.

Kump, L.R., Kasting, J.F., Barley, M.E., 2001. Rise of atmospheric oxygen and the "upside-down" Archean mantle. Geochem. Geophys. Geosyst. 2, #2000GC000114.

Lake, J.A., 2009. Evidence for an early prokaryotic endosymbiosis. Nature 460, 967–971.

Lang, B.F., Burger, G., 2012. Mitochondrial and eukaryotic origins: A critical review. In: Marechal Drouard, L. (Ed.), Mitochondrial Genome Evolution. Elsevier, Oxford, pp. 1–20.

Leitch, E.M., Vasisht, G., 1997. Mass extinctions and the sun's encounters with spiral arms. New Astron. 3 (1), 51–56.

Lenton, T.M., Daines, S.J., 2016. Matworld—the biogeochemical effects of early life on land. New Phytol. https://doi.org/10.1111/nph.14338.

Lester, L., Meade, A., Pagel, M., 2006. The slow road to the eukaryotic genome. Bioessays 28 (1), 57–64.

Lewin, R.A., 1994. Book review of: Concepts of Symbiogenesis: A Historical and Critical Study of the Research of Russian Botanists, by L.N. Khakhina. Ann. Sci. 51 (5), 567–569.

Ligowski, R., Godlewski, M., Łukowski, A., 1992. Sea ice diatoms and ice edge planktonic diatoms at the northern limit of the Weddell Sea pack ice. Proc. NIPR Symp. Polar Biol. 5, 9–20.

Livio, M., Kopelman, A., 1990. Life and the Sun's lifetime. Nature 343 (6253), 25.

FIG. 20—CONT'D

peaks that line up throughout geological time in series B, C and D, supporting the 96 to 99% confidence level of the cross-correlation between the two plume time series and the terrestrial impact time series. (D) All mantle plumes as derived from dating of the four mantle plume proxies: massive dike swarms, high-Mg extrusives, flood basalts, and ultramafic and mafic layered intrusions. Arrows are at the same point in time as the tops of peaks in large impacts that line up with peaks in the lunar impact record. Note that seven out of eight peaks line up within the mean error of the ages in the terrestrial impact time series (46 Ma). Note also that some of the Meso-Proterozoic and Early Archaean peaks in the lunar time series appear to line up with peaks in the two plume time series, suggesting that there are major terrestrial impact events that remain to be discovered." From Abbott and Isley (2002) with permission of Elsevier. Left/right reversed and relabeled. Ma = My in our notation.

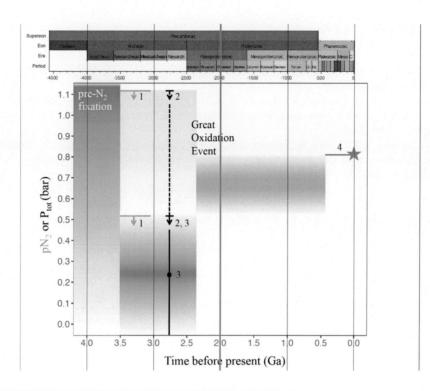

FIG. 21

"Current estimates for atmospheric N_2 levels throughout Earth's history. Darkly shaded regions indicate a higher degree of certainty to which these measurements have been constrained, and measurements beyond 2.4 Ga are presented as upper limit estimates, denoted by solid lines. Constraints are heavily data-limited, but overall pN_2 appears to have fluctuated in a U-shaped trend. Graph compiled by data from #1 (Marty et al., 2013), #2 (Som et al., 2012), #3 (Som et al., 2016), #4 (Berner, 2006)." From Silverman (2017) with kind permission of Shaelyn Nicole Silverman.

Locarnini, R.A., Mishonov, A.V., Antonov, J.I., Boyer, T.P., Garcia, H.E., Baranova, O.K., Zweng, M.M., Johnson, D.R., 2010. World Ocean Atlas 2009, vol. 1, Temperature. NOAA Atlas NESDIS, vol. 68. National Oceanic and Atmospheric Administration, Silver Spring, MD, USA.

Logsdon Jr., J.M., 1998. The recent origins of spliceosomal introns revisited. Curr. Opin. Genet. Dev. 8 (6), 637–648.

Lowe, D.R., Tice, M.M., 2004. Geologic evidence for Archean atmospheric and climatic evolution: Fluctuating levels of CO_2, CH_4, and O_2, with an overriding tectonic control. Geology 32 (6), 493–496.

Lyons, T.W., Reinhard, C.T., Planavsky, N.J., 2014. The rise of oxygen in Earth's early ocean and atmosphere. Nature 506 (7488), 307–315.

Maas, C., Hansen, U., 2015. Effects of Earth's rotation on the early differentiation of a terrestrial magma ocean. J. Geophys. Res.-Solid Earth 120 (11), 7508–7525.

Marcos, R.D., Marcos, C.D., 2004. On the correlation between the recent star formation rate in the Solar Neighbourhood and the glaciation period record on Earth. New Astron. 10 (1), 53–66.

Margulis, L., 1970. Origin of Eukaryotic Cells: Evidence and Research Implications for a Theory of the Origin and Evolution of Microbial, Plant, and Animal Cells on the Precambrian Earth. Yale University Press, New Haven.

Margulis, L., Fester, R., 1991. Symbiosis as a Source of Evolutionary Innovation: Speciation and Morphogenesis. MIT Press, Cambridge.

Margulis, L., Chapman, M., Guerrero, R., Hall, J., 2006. The last eukaryotic common ancestor (LECA): Acquisition of cytoskeletal motility from aerotolerant spirochetes in the Proterozoic Eon. Proc. Natl. Acad. Sci. U. S. A. 103 (35), 13080–13085.

Mariotti, M., Lobanov, A.V., Manta, B., Santesmasses, D., Bofill, A., Guigó, R., Gabaldón, T., Gladyshev, V.N., 2016. *Lokiarchaeota* marks the transition between the archaeal and eukaryotic selenocysteine encoding systems. Mol. Biol. Evol. 33 (9), 2441–2453.

Mariscal, C., Doolittle, W.F., 2015. Eukaryotes first: how could that be? Phil. Trans. Roy. Soc. B-Biol. Sci. 370 (1678), #20140322.

Markov, A.V., Kaznacheev, I.S., 2016. Evolutionary consequences of polyploidy in prokaryotes and the origin of mitosis and meiosis. Biol. Direct 11, #28.

Marshall, D., Schirrmeister, B., 2013. Episode 16: Multicellularity in cyanobacteria. http://www.palaeocast.com/episode-16-multicellularity-in-cyanobacteria/.

Martijn, J., Ettema, T.J.G., 2013. From archaeon to eukaryote: the evolutionary dark ages of the eukaryotic cell. Biochem. Soc. Trans. 41, 451–457.

Martin, W., Müller, M., 1998. The hydrogen hypothesis for the first eukaryote. Nature 392 (6671), 37–41.

Martin, A.P., Condon, D.J., Prave, A.R., Lepland, A., 2013. A review of temporal constraints for the Palaeoproterozoic large, positive carbonate carbon isotope excursion (the Lomagundi-Jatuli Event). Earth-Sci. Rev. 127, 242–261.

Martin, W.F., Neukirchen, S., Zimorski, V., Gould, S.B., Sousa, F.L., 2016. Energy for two: New archaeal lineages and the origin of mitochondria. Bioessays 38 (9), 850–856.

Marty, B., Zimmermann, L., Pujol, M., Burgess, R., Philippot, P., 2013. Nitrogen isotopic composition and density of the Archean atmosphere. Science 342 (6154), 101–104.

Maruyama, M., 1963. The second cybernetics: deviation-amplifying mutual causal processes. Amer. Sci. 51 (2), 164–179.

Mat, W.K., Xue, H., Wong, J.T., 2008. The genomics of LUCA. Front. Biosci. 13, 5605–5613.

Mattick, J.S., 1994. Introns: Evolution and function. Curr. Opin. Genet. Develop. 4 (6), 823–831.

Mbata, T.I., 2008. Isolation of fungi in hyper saline Dead Sea water. Sudanese J. Public Health 3 (4), 170–172.

Medlin, L.K., Hasle, G.R., 1990. Some *Nitzschia* and related diatom species from fast ice samples in the Arctic and Antarctic. Polar Biol. 10 (6), 451–479.

Melezhik, V.A., Huhma, H., Condon, D.J., Fallick, A.E., Whitehouse, M.J., 2007. Temporal constraints on the Paleoproterozoic Lomagundi-Jatuli carbon isotopic event. Geology 35 (7), 655–658.

Melott, A.L., 2017. A possible role for stochastic astrophysical ionizing radiation events in the systematic disparity between molecular and fossil dates. Astrobiology 17 (1), 87–90.

Mentel, M., Martin, W., 2008. Energy metabolism among eukaryotic anaerobes in light of Proterozoic ocean chemistry. Phil. Trans. Roy. Soc. B-Biol. Sci. 363 (1504), 2717–2729.

Mereschkowsky, C., 1910. Theorie der zwei Plasmaarten als Grundlage der Symbiogenesis, einer neuen Lehre von der Entstehung der Organismen. Biol. Centralbl. 30, 278–288.

Mikhailovsky, G., Levich, A., 2015. Entropy, information and complexity or which aims the arrow of time? Entropy 17 (7), 4863–4890.

Mojzsis, S.J., Harrison, T.M., Pidgeon, R.T., 2001. Oxygen-isotope evidence from ancient zircons for liquid water at the Earth's surface 4,300 Myr ago. Nature 409 (6817), 178–181.

Moorbath, S., 2009. The discovery of the Earth's oldest rocks. Notes Records Roy. Soc. 63 (4), 381–392.

Moreira, D., López-García, P., 1998. Symbiosis between methanogenic archaea and δ-proteobacteria as the origin of eukaryotes: The syntrophic hypothesis. J. Mol. Evol. 47 (5), 517–530.

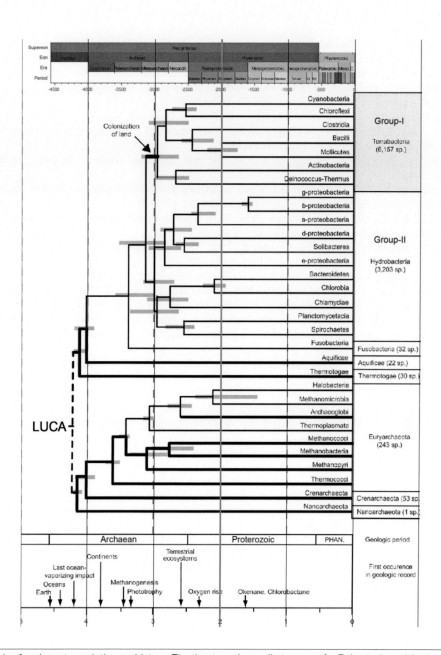

FIG. 22

"Timescale of prokaryote evolutionary history. The timetree shows divergences for Eubacteria and Archaebacteria (ML, protein data set) with particular attention to major groups: Hydrobacteria and Terrabacteria (Eubacteria) and Euryarchaeota and Crenarchaeota (Archaebacteria). First occurrences of major events in the geologic record

(Continued)

Morrison, S.M., Liu, C., Eleish, A., Prabhu, A., Li, C.R., Ralph, J., Downs, R.T., Golden, J.J., Fox, P., Hummer, D.R., Meyer, M.B., Hazen, R.M., 2017. Network analysis of mineralogical systems. American Mineralogist 102 (8), 1588–1596.

Mullakhanbhai, M.F., Larsen, H., 1975. *Halobacterium volcanii* spec. nov., a Dead Sea halobacterium with a moderate salt requirement. Arch. Microbiol. 104 (3), 107–114.

Müller, M., Mentel, M., van Hellemond, J.J., Henze, K., Woehle, C., Gould, S.B., Yu, R.-Y., van der Giezen, M., Tielens, A.G.M., Martin, W.F., 2012. Biochemistry and evolution of anaerobic energy metabolism in eukaryotes. Microbiol. Mol. Biol. Rev. 76 (2), 444–495.

Mushegian, A., 2008. Gene content of LUCA, the last universal common ancestor. Front. Biosci. 13, 4657–4666.

Naor, A., Gophna, U., 2013. Cell fusion and hybrids in Archaea: Prospects for genome shuffling and accelerated strain development for biotechnology. Bioengineered 4 (3), 126–129.

Navarro-González, R., McKay, C.P., Mvondo, D.N., 2001. A possible nitrogen crisis for Archaean life due to reduced nitrogen fixation by lightning. Nature 412 (6842), 61–64.

Normand, P., Gouy, M., Cournoyer, B., Simonet, R., 1992. Nucleotide sequence of *nifD* from *Frankia alni* strain ArI3: phylogenetic inferences. Mol. Biol. Evol. 9 (3), 495–506.

Nutman, A.P., 2006. Antiquity of the oceans and continents. Elements 2 (4), 223–227.

Nutman, A.P., Bennett, V.C., Friend, C.R., Van Kranendonk, M.J., Chivas, A.R., 2016. Rapid emergence of life shown by discovery of 3,700-million-year-old microbial structures. Nature 537 (7621), 535–538.

Okie, J.G., Smith, V.H., Martin-Cereceda, M., 2016. Major evolutionary transitions of life, metabolic scaling and the number and size of mitochondria and chloroplasts. Proc. Roy. Soc. B-Biol. Sci. 283 (1831), #20160611.

Oren, A., Duker, S., Ritter, S., 1996. The polar lipid composition of Walsby's square bacterium. FEMS Microbiol. Lett. 138 (2–3), 135–140.

Perez-Jimenez, R., Inglés-Prieto, A., Zhao, Z.M., Sanchez-Romero, I., Alegre-Cebollada, J., Kosuri, P., Garcia-Manyes, S., Kappock, T.J., Tanokura, M., Holmgren, A., Sanchez-Ruiz, J.M., Gaucher, E.A., Fernandez, J.M., 2011. Single-molecule paleoenzymology probes the chemistry of resurrected enzymes. Nat. Struct. Mol. Biol. 18 (5), 592–596.

Piper, J.D.A., 2013. Continental velocity through Precambrian times: The link to magmatism, crustal accretion and episodes of global cooling. Geosci. Front. 4 (1), 7–36.

Piper, J.D.A., 2015. The Precambrian supercontinent Palaeopangaea: two billion years of quasi-integrity and an appraisal of geological evidence. Int. Geol. Rev. 57 (11–12), 1389–1417.

Pittis, A.A., Gabaldón, T., 2016. Late acquisition of mitochondria by a host with chimaeric prokaryotic ancestry. Nature 531 (7592), 101–104.

Poole, A.M., Jeffares, D.C., Penny, D., 1998. The path from the RNA world. J. Mol. Evol. 46 (1), 1–17.

Prasad, B., Uniyal, S.N., Asher, R., 2005. Organic-walled microfossils from the Proterozoic Vindhyan supergroup of Son Valley, Madhya Pradesh, India. Palaeobotanist (Lucknow) 54 (1–3), 13–60.

Provorov, N.A., 2016. K.S. Merezhkovsky and the origin of the eukaryotic cell: 111 years of symbiogenesis theory. Agricult. Biol. 51 (5), 746–758.

Ray, J.S., 2006. Age of the Vindhyan Supergroup: A review of recent findings. J. Earth Syst. Sci. 115 (1), 149–160.

FIG. 22—CONT'D

are represented by arrows on the timescale.... Each horizontal line represents a class; exceptions are the phyla Bacteroidetes (which includes two classes), Cyanobacteria, and Nanoarchaeota. Thicker lines are lineages that include hyperthermophilic species." From Battistuzzi and Hedges (2009) with permission of Oxford University Press.

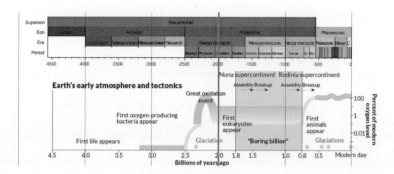

FIG. 23

A second billion year period in evolution between the first appearance of eukaryotes and the first animals (Wikipedia, 2016a). From Hirshfeld (2014) with permission under a Creative Commons Attribution-Share Alike 4.0 International license. This "Boring billion" may be the period during which the next hierarchogenic event occurred, namely continuing differentiation (Gordon, 1999), which allowed the radiation of multicellular plants and animals with multiple cell types.

Raymond, J., Siefert, J.L., Staples, C.R., Blankenship, R.E., 2004. The natural history of nitrogen fixation. Mol. Biol. Evol. 21 (3), 541–554.

Reinhard, C.T., Planavsky, N.J., Lyons, T.W., 2013. Long-term sedimentary recycling of rare sulphur isotope anomalies. Nature 497 (7447), 100–103.

Rice, D.W., Alverson, A.J., Richardson, A.O., Young, G.J., Sanchez-Puerta, M.V., Munzinger, J., Barry, K., Boore, J.L., Zhang, Y., dePamphilis, C.W., Knox, E.B., Palmer, J.D., 2013. Horizontal transfer of entire genomes via mitochondrial fusion in the angiosperm *Amborella*. Science 342 (6165), 1468–1473.

Rocha-Pinto, H.J., Scalo, J., Maciel, W.J., Flynn, C., 2000. Chemical enrichment and star formation in the Milky Way disk II. Star formation history. Astron. Astrophys. 358 (3), 869–885.

Rollosson, G.W., 1952. A study of penetrating cosmic-ray showers in water. Phys. Rev. 87 (1), 71–74.

Rosen, R., 1985. James F. Danielli: 1911–1984. J. Social Biol. Struct. 8 (1), 1–11.

Sagan, L., 1967. On the origin of mitosing cells. J. Theor. Biol. 14 (3), 255–274.

Sagan, C., Mullen, G., 1972. Earth and Mars: Evolution of atmospheres and surface temperatures. Science 177 (4043), 52–56.

Sapp, J., 1994. Evolution by Association: A History of Symbiosis. Oxford University Press, USA, New York.

Sapp, J., 2009. The New Foundations of Evolution: On the Tree of Life. Oxford University Press, New York.

Schaechter, M., 2012. Retrospective: Lynn Margulis (1938–2011). Science 335 (6066), 302.

Schaefer, L., Fegley Jr., B., 2014. Atmospheric composition of Hadean-early Archean Earth: The importance of CO: Comment. In: Shaw, G.H. (Ed.), Earth's Early Atmosphere and Surface Environment. The Geological Society of America, Boulder, Colorado, USA, pp. 29–31.

Schirrmeister, B.E., de Vos, J.M., Antonelli, A., Bagheri, H.C., 2013. Evolution of multicellularity coincided with increased diversification of cyanobacteria and the Great Oxidation Event. Proc. Natl. Acad. Sci. U. S. A. 110 (5), 1791–1796.

Schopf, J.W., 2006. Fossil evidence of Archaean life. Phil. Trans. Roy. Soc. B-Biol. Sci. 361 (1470), 869–885.

Schopf, J.W., 2012. The fossil record of cyanobacteria. In: Whitton, B.A. (Ed.), Ecology of Cyanobacteria II. Springer, London, pp. 15–36.

Sessions, A.L., Doughty, D.M., Welander, P.V., Summons, R.E., Newman, D.K., 2009. The continuing puzzle of the great oxidation event. Curr. Biol. 19 (14), R567–R574.

Sharov, A.A., Gordon, R., 2017. Life before Earth. In: Gordon, R., Sharov, A.A. (Eds.), Habitability of the Universe Before Earth [in series: Astrobiology: Exploring Life on Earth and Beyond, eds. Pabulo Henrique Rampelotto, Joseph Seckbach & Richard Gordon]. Elsevier B.V., Amsterdam, pp. 265–296.

Shaw, G.H., 2008. Earth's atmosphere—Hadean to early Proterozoic. Chem. Erde-Geochem. 68 (3), 235–264.

Shields, G., Veizer, J., 2002. Precambrian marine carbonate isotope database: Version 1.1. Geochem. Geophys. Geosyst. 3 (6), #1031.

Shiflett, A.M., Johnson, P.J., 2010. Mitochondrion-related organelles in eukaryotic protists. Annu. Rev. Microbiol. 64, 409–429.

Shirey, S.B., Richardson, S.H., 2011. Start of the Wilson Cycle at 3 Ga Shown by Diamonds from Subcontinental Mantle. Science 333 (6041), 434–436.

Silverman, S., 2017. Morphological and Isotopic Changes of *Anabaena cylindrica* PCC 7122 in Response to N_2 Partial Pressure [Honors Thesis]. Department of Molecular, Cellular and Developmental Biology, Honors Council of the College of Arts and Sciences. University of Colorado, Boulder, Colorado, USA.

Silverman, S.N., Kopf, S., Gordon, R., Bebout, B., Som, S., 2017. Measuring ancient N_2 pressure using fossilized cyanobacteria in: Desch, S.J. (Ed.), AbSciCon2017, Mesa, Arizona, April 24–28, 2017, http://www.hou.usra.edu/meetings/abscicon2017/pdf/3242.pdf.

Sleep, N.H., Zahnle, K., 2001. Carbon dioxide cycling and implications for climate on ancient Earth. J. Geophys. Res.-Planets 106 (E1), 1373–1399.

Som, S.M., Catling, D.C., Harnmeijer, J.P., Polivka, P.M., Buick, R., 2012. Air density 2.7 billion years ago limited to less than twice modern levels by fossil raindrop imprints. Nature 484 (7394), 359–362.

Som, S.M., Buick, R., Hagadorn, J.W., Blake, T.S., Perreault, J.M., Harnmeijer, J.P., Catling, D.C., 2016. Earth's air pressure 2.7 billion years ago constrained to less than half of modern levels. Nat. Geosci. 9 (6), 448–451.

Som S., Life May Be Even Older Than We Thought! Careful analyses of the composition of ancient rocks betrays the presence of life. http://sciworthy.com/life-may-be-even-older-than-we-thought/.

Sousa, F.L., Neukirchen, S., Allen, J.F., Lane, N., Martin, W.F., 2016. Lokiarchaeon is hydrogen dependent. Nature Microbiol. 1 (5), #16034.

Spang, A., Martijn, J., Saw, J.H., Lind, A.E., Guy, L., Ettema, T.J.G., 2013. Close encounters of the third domain: The emerging genomic view of archaeal diversity and evolution. Archaea-An International Microbiological Journal 2013, #202358.

Spang, A., Saw, J.H., Jorgensen, S.L., Zaremba-Niedzwiedzka, K., Martijn, J., Lind, A.E., van Eijk, R., Schleper, C., Guy, L., Ettema, T.J.G., 2015. Complex archaea that bridge the gap between prokaryotes and eukaryotes. Nature 521 (7551), 173–179.

Stein, W.D., 1986. James Frederic Danielli, 1911–1984, Elected F.R.S. 1957. Biogr. Memoirs Fellows Roy Soc. 32, 115–135.

Stoltzfus, A., Spencer, D.F., Zuker, M., Logsdon, J.M., Doolittle, W.F., 1994. Testing the exon theory of genes: the evidence from protein structure. Science 265 (5169), 202–207.

Surkont, J., Pereira-Leal, J.B., 2016. Are there Rab GTPases in Archaea? Mol. Biol. Evol. 33 (7), 1833–1842.

Swithers, K.S., Fournier, G.P., Green, A.G., Gogarten, J.P., Lapierre, P., 2011. Reassessment of the lineage fusion hypothesis for the origin of double membrane bacteria. PLoS One 6(8), #e23774.

Szklarczyk, R., Huynen, M.A., 2010. Mosaic origin of the mitochondrial proteome. Proteomics 10 (22), 4012–4024.

Tarduno, J.A., Blackman, E.G., Mamajek, E.E., 2014. Detecting the oldest geodynamo and attendant shielding from the solar wind: Implications for habitability. Phys. Earth Planet. In. 233, 68–87.

Tashiro, T., Ishida, A., Hori, M., Igisu, M., Koike, M., Méjean, P., Takahata, N., Sano, Y., Komiya, T., 2017. Early trace of life from 3.95 Ga sedimentary rocks in Labrador, Canada. Nature 549 (7673), 516–518.

Thomas, B.C., Neale, P.J., Snyder II, B.R., 2015. Solar irradiance changes and photobiological effects at Earth's surface following astrophysical ionizing radiation events. Astrobiology 15 (3), 207–220.

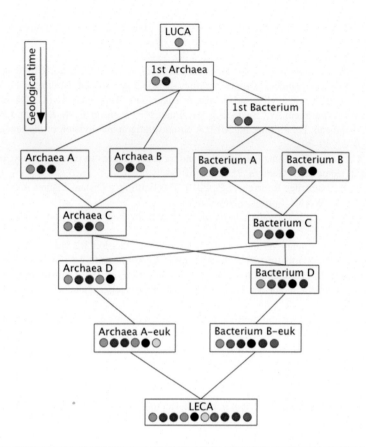

FIG. 24

This is a schematic depiction of the evolution from LUCA (Last Universal Common Ancestor) to LECA (Last Eukaryotic Common Ancestor). Each colored dot represents a novelty, such as a protein with a new function. LUCA is presumed to be exactly one cell whose descendants first give rise to the first Archaea and (perhaps via Archaea) to the first Bacterium. In both groups, many species evolve with gene duplication followed by gene specialization giving rise to further novelties, indicated by increasing numbers of dots of new colors (A and B). Occasionally, a horizontal gene transfer between two species of Archaea, or between two species of Bacteria, gives rise to new species (B to A resulting in C). Horizontal gene transfers between Archaea and Bacteria are also possible (Archaea C to Bacterium C resulting in Bacterium D, or Bacterium C to Archaea C resulting in Archaea D). Cell fusion events between Archaea and Bacteria occur occasionally, one of which eventually has enough novelties to give rise to LECA. The two cells A-euk and B-euk are those whose fusion is successful in producing LECA. In the computer simulation (under development), large numbers of species are simulated, with empirically derived parameters for rates of novelty production, horizontal gene transfer, and extinction of species. Frequency of polyploidy and cell fusions (most below the threshold for number of novelties needed for LECA, and thus presumed to leave no progeny) may have to be guesstimated. The opportunities for cell fusion depend of course on spatial proximity, which requires introduction of a spatial mixing model on the scale of the whole Earth. We cannot presume that all species have global ranges. Species not leading to LECA are not

(Continued)

Ueno, Y., Yamada, K., Yoshida, N., Maruyama, S., Isozaki, Y., 2006. Evidence from fluid inclusions for microbial methanogenesis in the early Archaean era. Nature 440 (7083), 516–519.

Valley, J.W., 2005. A cool early Earth? Sci. Am. 293 (4), 58–65.

van Beneden, P.J., 1873. Un mot sur la vie sociale des animaux inférieurs. Bull. Acad. R Belgique 2 (36), 779–796.

van Beneden, P.J., 1875. Les comensaux et les parasites dans le règne animal. Biblio Sci Int, Paris.

van der Giezen, M., 2011. Mitochondria and the rise of eukaryotes. Bioscience 61 (8), 594–601.

Van Kranendonk, M.J., Altermann, W., Beard, B.L., Hoffman, P.F., Johnson, C.M., Kasting, J.F., Melezhik, V.A., Nutman, A.P., Papineau, D., Pirajno, F., 2012. A chronostratigraphic division of the precambrian. Geologic Time Scale 1 & 2, 299–392.

Varga, P., Denis, C., Varga, T., 1998. Tidal friction and its consequences in palaeogeodesy, in the gravity field variations and in tectonics. J. Geodyn. 25 (1–2), 61–84.

Vázquez, M., Pallé, E., Montañés Rodríguez, P., 2010. The Earth as a Distant Planet: A Rosetta Stone for the Search of Earth-Like Worlds. Springer, New York

Vesteg, M., Krajčovič, J., 2011. The falsifiability of the models for the origin of eukaryotes. Curr. Genet. 57 (6), 367–390.

Vukotić, B., 2017. N-body simulations and galactic habitability, in: Gordon, R., Sharov, A.A. (Eds.), Habitability of the Universe Before Earth [in series: Astrobiology: Exploring Life on Earth and Beyond, eds. Pabulo Henrique Rampelott, Joseph Seckbach & Richard Gordon]. Elsevier B.V., Amsterdam, pp. 173–197.

Walker, J.C.G., 1982. Climatic factors on the archean Earth. Palaeogeogr. Palaeoclimatol. Palaeoecol. 40 (1–3), 1–11.

Wallin, I.E., 1927. Symbionticism and the Origin of Species. Williams & Wilkins Co., Baltimore.

Warren, J.K., 2016. Evaporites: A Geological Compendium, second ed. Springer, Cham.

Watson, R.A., Pollack, J.B., 1999. How symbiosis can guide evolution. In: Floreano, D., Nicoud, J.D., Mondada, F. (Eds.), Advances in Artificial Life Proceedings, pp. 29–38.

Weiss, P.A., 1973. The Science of Life: The Living System—A System for Living. Futura Publishing Company, Mt. Kisco, New York, USA.

Weiss, M.C., Sousa, F.L., Mrnjavac, N., Neukirchen, S., Roettger, M., Nelson-Sathi, S., Martin, W.F., 2016. The physiology and habitat of the last universal common ancestor. Nature Microbiol. 1, #16116.

Wellman, C.H., Strother, P.K., 2015. The terrestrial biota prior to the origin of land plants (embryophytes): a review of the evidence. Palaeontology 58 (4), 601–627.

White, G.R., 1950. The penetration and diffusion of Co^{60} gamma-rays in water using spherical geometry. Phys. Rev. 80 (2), 154–156.

Wikipedia, 2015a. Age of the Earth. https://en.wikipedia.org/wiki/Age_of_the_Earth.

Wikipedia, 2015b. Reticulate evolution. https://en.wikipedia.org/wiki/Reticulate_evolution.

Wikipedia, 2016a. Boring Billion. https://en.wikipedia.org/wiki/Boring_Billion.

Wikipedia, 2016b. Electromagnetic absorption by water. https://en.wikipedia.org/wiki/Electromagnetic_absorption_by_water.

FIG. 24—CONT'D

drawn here, but would be represented in the simulation. The number of events depicted is of course far fewer than we will simulate, by orders of magnitude. This sketch is only intended to give a sample of the simulated events. The number of novelties will be hundreds in the simulation, matching estimates of the number of protein families in LUCA and LECA. The total number of species we can simulate may prove to be limited by computer time and memory. The question to be addressed is whether or not a reasonable set of rate parameters put into the simulation would produce a 1 or 2 billion year delay in producing one LUCA cell from one LECA cell via such a web of events.

Wikipedia, 2016c. History of Earth. https://en.wikipedia.org/wiki/History_of_Earth.

Wikipedia, 2016d. List of sequenced archaeal genomes. https://en.wikipedia.org/wiki/List_of_sequenced_archaeal_genomes.

Wikipedia, 2017a. Bacteria. https://en.wikipedia.org/wiki/Bacteria.

Wikipedia, 2017b. Infinite monkey theorem. https://en.wikipedia.org/wiki/Infinite_monkey_theorem.

Wikipedia, 2017c. Lokiarchaeota. https://en.wikipedia.org/wiki/Lokiarchaeota.

Wilde, S.A., Valley, J.W., Peck, W.H., Graham, C.M., 2001. Evidence from detrital zircons for the existence of continental crust and oceans on the Earth 4.4 Gyr ago. Nature 409 (6817), 175–178.

Yu, G., Jacobsen, S.B., 2011. Fast accretion of the Earth with a late Moon-forming giant impact. Proc. Natl. Acad. Sci. U. S. A. 108 (43), 17604–17609.

Zahnle, K., Catling, D., 2014. Waiting for O_2, in: Shaw, G.H. (Ed.), Earth's Early Atmosphere and Surface Environment, pp. 37–48.

Zahnle, K., Claire, M., Catling, D., 2006. The loss of mass-independent fractionation in sulfur due to a Palaeoproterozoic collapse of atmospheric methane. Geobiology 4 (4), 271–283.

Zahnle, K., Arndt, N., Cockell, C.S., Halliday, A., Nisbet, E., Selsis, F., Sleep, N.H., 2007. Emergence of a habitable planet. Space Sci. Rev. 129 (1–3), 35–78.

Zahnle, K., Claire, M., Wing, B., 2010a. Biogenic sulfur gases, MIF-S, and the rise of free oxygen. Geochim. Cosmochim. Acta 74 (12), A1195.

Zahnle, K., Schaefer, L., Fegley, B., 2010b. Earth's earliest atmospheres. Cold Spring Harbor Perspectives in Biology 2 (10), #a004895.

Zerkle, A.L., House, C.H., Brantley, S.L., 2005. Biogeochemical signatures through time as inferred from whole microbial genomes. Am. J. Sci. 305 (6–8), 467–502.

Zhang, Z.Y., 1986. Solar cyclicity in the Precambrian microfossil record. Palaeontology 29, 101–111.

Zhang, Y.X., 2002. The age and accretion of the Earth. Earth-Sci. Rev. (1–4), 235–263.

Zhytina, L.S., Mikhailovsky, G.E., 1990. Ледовая и план-ктонная флора Белого моря как объект мониторинга [Ice and planktonic flora in the White Sea as an object of monitoring] [Russian], Биологический монито-ринг прибрежных вод Белого моря [Biological Monitoring in Coastal Waters of the White Sea]. P. P. Shirshov Institute of Oceanology, Academy of Sciences of the USSR, pp. 41–49.

Zimmer, C., 2009. On the origin of eukaryotes. Science 325 (5941), 666–668.

COENZYME WORLD MODEL OF THE ORIGIN OF LIFE

Alexei A. Sharov[1]

National Institute on Aging (NIA/NIH), Baltimore, MD, United States

CHAPTER OUTLINE

1 INTRODUCTION

Reconstruction of past evolutionary events is an exciting challenge; it requires integration of huge amounts of facts and theoretical approaches in order to filter out the most likely scenario of the origin of specific organs. It is even more difficult to approach the ultimate challenge—understanding the origin of life. Despite the large number of publications on the origin of life (sf. reviews Ikehara, 2016; Sponer et al., 2016; Weber, 2007), the problem is rarely addressed from the systems theory perspective. Instead of defining what life is, most publications discuss the possible origins of specific cellular components and functions. These papers certainly elucidate the later steps in the evolution

---

[1]Reprinted from Sharov A.A. 2016. Coenzyme world model of the origin of life. Biosystems 144: 8–17.

Habitability of the Universe Before Earth, editors: Richard Gordon & Alexei Sharov, Volume 1 in the series:
Astrobiology: Exploring Life on Earth and Beyond, series editors: Pabulo Henrique Rampelotto,
Joseph Seckbach & Richard Gordon. ISSN 2468-6352. https://doi.org/10.1016/B978-0-12-811940-2.00017-4
© 2018 Elsevier Inc. All rights reserved.

of cells, but they miss the origin of life itself. In particular, the RNA-world scenario (Orgel, 2004) has nothing to do with the origin of life because RNA is a product of a long earlier evolution rather than a "starting molecule" (Bernhardt, 2012). There are reasons to expect that primordial hereditary systems were organized in a radically different way than in contemporary cells and required neither nucleic acids nor proteins.

Life is often confused with self-organization or autocatalysis such as in crystal growth or dissipative structures (e.g., hurricanes, tornados, and flames) (Egel, 2012; Kauffman, 1986). Self-organization has one important feature that is common with life—the capacity to persist despite perturbations via positive feedback driven by dissipating energy. But neither self-organization nor auto-catalysis is sufficient for life. The main difference is that self-organized systems die out without affecting future self-organizing events. In contrast, living systems are *constructed by parental systems* and carry unique conditions that allow them to produce the next generation of systems (Deacon et al., 2014). This feature is *heredity* that distinguishes life from nonlife. In other words, heredity is a *recursive construction* (Bickhard, 2005): it is a capacity to make specific conditions (including resources, tools, scaffolds, codes) that are sufficient for both self-maintenance and making offspring systems that, in turn, carry the *same* capacity ("same" means that this capacity is stable in a sequence of generations[1]). Life creates unique self-supporting artificial structures and conditions that did not exist before. Besides stability, heredity has to support evolvability, which is a capacity to change and keep the changed state recursively across generations. Thus, the origin of life means the emergence of heredity and evolvability rather than the origin of nucleic acids or proteins (Jablonka and Szathmáry, 1995; Tessera, 2011).

The second problem with the origin of life stems from the notion of *chemical evolution*, which is supposed to provide an increasing diversity of organic molecules in primordial soup or inorganic compartments and supply first living systems with every resource they need (Benner et al., 2012; Ikehara, 2016; Koonin and Martin, 2005). However, nobody have described mechanisms of chemical evolution and proved that it could indeed provide these resources (e.g., sugars, aminoacids, nucleic bases, and lipids). Without heredity, past self-organizing events have no effect on the future events, and thus, chemical evolution is not "evolution" in biological sense. And if heredity indeed existed in certain chemical systems, then these systems were already alive because heredity is the essential feature of life. This is not an attempt to draw a demarcation line between life and nonlife; certainly, there is a gray transition zone, where heredity is weak and evolutionary potential is limited. But this transition zone already belongs to life as far as we are focused on heredity and evolution. Thus, instead of chemical evolution and soup, we need to focus on early phases of biological evolution.

In this paper, I first briefly discuss problems with existing models of the origin of life, and then describe an expanded version the *coenzyme world* model proposed earlier (Sharov, 2009). In particular, I added the discussion on the nature of primordial CLMs and their potential interaction with each other and with microelements that facilitate catalytic capacities. Also, I elaborated scenarios for

---

[1]Stability can be represented mathematically by an attractor in a model of branching dynamical systems, i.e., a quasispecies (Eigen and Schuster, 1979). However, any mathematical model is only an approximation, and thus, it never fully captures all details that remain conserved across generations. Hence, the notion of 'sameness' is not defined mathematically; instead it expands in content as we learn more about living organisms. For example, in the past biologists did not believe in epigenetic heredity and the notion of 'sameness' was restricted to genotypes or haplotypes.

transformation of surface metabolism into internal metabolism within cells. The model is discussed in the context of recent experiments and observations related to the origin of life.

2 PROBLEMS WITH EXISTING MODELS OF THE ORIGIN OF LIFE

The RNA-world hypothesis remains widely accepted in the field of the origin of life (Orgel, 2004; Robertson and Joyce, 2012). This model assumes that first living systems had self-replicating nucleic acids (Gilbert, 1986) or other kinds of similar heteropolymers[2] (Nelson et al., 2000; Orgel, 2000). Alternatively, the RNA world is viewed as an intermediate step of evolution that preceded contemporary prokaryotic cells with DNA-based heredity and protein synthesis (Bernhardt, 2012). Bernhardt mentioned the following main objections to the RNA world as the initial stage of life: RNA is too complex to have arisen prebiotically, it is highly unstable, and is generally a weak catalyst. Another problem is the lack of resources (i.e., nucleotides) for replication of RNA. Although nucleotides can be synthesized abiogenically (Benner et al., 2012; Powner et al., 2009), they are unlikely to get concentrated in quantities sufficient for RNA replication. Even if several nucleotides appear in a close proximity to each other due to a rare coincidence and produce a complimentary RNA strain, there would be no resources left for the next round of replication. Nucleotides can be synthesized from bases and sugars by RNA-mediated catalysis (Unrau and Bartel, 1998), but these precursors are unlikely to be supplied in sufficient quantities to support the reaction. For example, in the Murchison meteorite, nucleobases and sugars are present at concentration of ~1 ppm (Callahan et al., 2011; Cooper et al., 2001). Recent discovery of alcohol and sugar on the comet Lovejoy (Biver et al., 2015) is intriguing, but it does not prove that primordial organisms used hydrocarbons of abiotic origin as resources. It is very unlikely that life originated on small comets, and if a comet landed on a planet, organic chemicals would have degraded almost immediately or become diluted. Neither chemical evolution nor primordial soup can provide resources of self-replications (Lane et al., 2010). An alternative scenario is that replicating polymers such as RNA emerged much later in evolution after primordial living systems have developed heritable mechanisms for making monomers (Sharov, 2009; Wächtershäuser, 1988).

The most appealing point of the RNA-world hypothesis is the assumption that primordial replication systems were identical to the present-day nucleic acids, and thus, there is no need to reconstruct alternative mechanisms of heredity. However, self-reproduction[3] can be achieved in substantially more simple systems without nucleic acids. For example, synthetic peptide-based monomers selectively self-assemble into fibers, which then reproduce via fiber elongation-and-breakage mechanism (Colomb-Delsuc et al., 2015). Heredity is a systems-level autocatalysis, where parental systems construct the same kind of descendants. However, autocatalytic synthesis (in contrast to decay) is a rare property among organic molecules. Thus, it was suggested that self-reproduction could arise more easily in multi-component mixtures of molecules with random cross-catalysis (Kauffman, 1986). In particular, Kauffman proposed that a mixture of peptides makes an *autocatalytic set* if the synthesis of each component is catalyzed by some other member(s) of the same set. This hypothesis has no experimental

---

[2]For example, threose nucleic acid (TNA) or peptide nucleic acid (PNA).

[3]I use the term "replication" only for template-guided synthesis (as in nucleic acids); in other cases I use the term "self-reproduction" or "reproduction."

support and appears as weak as the RNA world from the theoretical point of view. It is hard to imagine a mechanism that preserves long peptide sequences over time without encoding; whereas short or random peptides (that do not need encoding) have no catalytic activity. Another problem is that autocatalytic sets easily dissipate if they are not enclosed in a membrane; but peptides are not likely to make a non-permeable membrane.

Although peptide cross-catalysis is certainly unrealistic as a model for the origin of life, the theoretical idea that autocatalytic sets can support heritable self-reproduction is valid. However, not every autocatalytic set, as defined by Kauffman, can support self-reproduction. Self-reproduction is possible only in autocatalytic sets with specific stoichiometry constraints, where a sequence of internal reactions can increase the number of all molecular species within the set (Sharov, 1991). It appears that such systems always have a unique minimum core with at least one "reproduction" reaction that generates an additional molecule within the cycle. Self-reproducing systems can propagate in space, which is similar to the growth and expansion of populations of living organisms (Gray and Scott, 1994; Tilman and Kareiva, 1997). Autocatalytic sets have two alternative steady states: "on" and "off"; thus, they represent the most simple hereditary system or memory unit (Jablonka and Szathmáry, 1995). A biochemical example of a self-reproducing autocatalytic set is a Calvin cycle (also known as a reverse citric acid cycle), which has a "reproduction" reaction: a ribulose 1 5 bisphosphate molecule divides into two molecules of 3-phosphoglycerate after acquiring carbon dioxide as a resource. It was speculated that the Calvin cycle without protein enzymes could have been the first self-reproducing system at the origin of life (Hartman, 1998; Morowitz et al., 2000). The advantage of this hypothesis is that it does not require soup or RNA world (Hartman, 1998). However, it seems unlikely that CO_2 fixation was the initial source of carbon in primordial systems because the Calvin cycle requires too much energy (e.g., ATP), which was not available in simple molecular assemblies. The network modeling approach suggested that the Calvin cycle had a low probability of a random realization and the length and cost of the cycle are close to the extreme values (Zubarev et al., 2015). This indicates that the Calvin cycle is likely to be a product of adaptive evolution via selection. Because natural selection requires heredity, we have to admit that heredity existed before the emergence of the Calvin cycle. Another example of a self-reproducing molecule is prion (Griffith, 1967; Laurent, 1997), and indeed prions have been invoked in various ways in discussions on the origin of life (Maury, 2009, 2015; Steele and Baross, 2006). However, prions cannot support the synthesis of the primary (i.e., unfolded) polypeptides, and therefore they had no role in the origin of life.

Another group of theories is based on the assumption that abundant organic molecules were generated by direct reduction of CO_2. In particular, Wächtershäuser (Wächtershäuser, 1988) assumed that the energy from oxidation of FeS to FeS_2 at the sea floor was used for organic synthesis. He further suggested that life started from autocatalytic coenzymes similar to nicotinamide NADP+ or thiamine pyrophosphate (TPP), and that negatively charged organic molecules were adsorbed and concentrated on positively charged pyrite mineral surfaces. This hypothesis is close to my idea that life originated from coenzyme-like molecules, although it assumes a different source of carbon for primordial systems. Another model of the origin of life is based on the idea that formaldehyde was synthesized using reduction of CO_2 via photo-catalysis mediated by ZnS (Mulkidjanian and Galperin, 2007). This model further assumed that other organic molecules and polymers were synthesized by various inorganic catalysts including Zn, preparing a soup where self-replicating systems (RNA-world) can emerge spontaneously. The common problem of these theories is that carbon fixation from CO_2 is too difficult energetically and could not produce sufficient resources for the RNA world.

The next class of theories known as "lipid-world" is based on the idea that lipid-like molecules played the key role in the origin of life (Segre et al., 2001, 1998, 2001; Tessera, 2011). These models are certainly more realistic than the RNA-world or protein-world models because (1) they consider a spatially heterogeneous system of lipid compositional assemblies in water, (2) lipids do not require coding, (3) lipid surfaces and their individual molecular components have catalytic activity, and (4) the dynamics of compositional assemblies depends on affinity and binding, which are common features in lipids. In particular, the GARD (Graded Autocatalysis Replication Domain) model of the origin of life assumes that compositional assemblies grow and eventually break down into two (or more) daughter assemblies (Segre et al., 1998). As a result, these assemblies can reproduce and form discrete quasispecies. Disproportional split of components between daughter assemblies can result in a heritable variation that may occasionally give rise to new quasispecies. The lipid-world model, however, has its own problems. First, the diversity and abundance of abiotic lipid-like molecules is low and hardly can support the propagation and evolution of molecular assemblies. Lipids *per se* do not exist in a non-living world,[4] and their synthesis requires glycerol, which is a rare and unstable molecule. For example, concentration of glycerol in Murchison meteorite is only 15 ppm (Cooper et al., 2001). Finally, the catalytic activity of lipids alone is low. Thus, they may require specialized catalysts to establish new covalent bonds.

Very few publications attempt to predict the functions of primordial living systems. Deacon developed a model of autocatalytic set called "autocell," which besides cross-catalysis included self-assembly of a multi-component unit that resembles a viral capsid (Deacon, 2006; Deacon, 2011). This structure may have additional functions such as long-term survival in unfavorable conditions. Deacon further argued that the specific feature of life is "teleodynamics," which is an intrinsic tendency to create higher-order constraints that are beneficial for the self-organization of the whole system. However, Deacon admitted that the emergence and self-propagation of autocells appears highly unlikely because it requires conditions that did not exist without life, such as "(1) presence of significant quantities and concentrations of large polymers; (2) some considerable degree of structural similarity among these molecules; and (3) sufficient variety and degeneracy of their stereochemical properties to support spontaneous autocatalysis and self-assembly" (Deacon, 2006: p. 145).

It appears that the problem of the origin of life cannot be adequately addressed by focusing on one specific component of cells as a sole leading factor. It is meaningless to separate metabolism from heredity because metabolism can persist and evolve only if it is heritable, and heredity requires metabolism as a source of organic molecules and energy (Jablonka and Szathmáry, 1995). Models of prebiotic networks suggest the importance of combinations of various kinds of molecules (Caetano-Anolles and Seufferheld, 2013; Nghe et al., 2015). Thus, we need to envision primordial systems as multi-component functional units capable of capturing resources, dispersion, and heritable self-reproduction. Functions emerged and changed simultaneously with chemical and structural changes. Organic molecules in Stanley Miller's experiments or in meteorites are not relevant for the discussion of the origin of life because these molecules are not functional alone. It is even more important to find a context/niche in which organic molecules may become functional. Heredity is the central feature of life, but it can be based on other molecules than nucleic acids, and this nongenetic information was likely distributed among various molecular components of primordial systems. In summary, the next turn in the study of the origin of life should involve a system's approach, where heredity and other functions emerge and coevolve in multi-component systems.

---

[4]Rare molecules of lipids in meteorites do not count because their concentrations are very low; also they may be artifacts of heating during meteorite entry into the atmosphere.

3 COMPONENTS, FUNCTIONS, AND EVOLUTION OF FIRST LIVING SYSTEMS
3.1 LIFE ON THE SURFACE

Surfaces play an extraordinary role in living cells (Hoffmeyer, 1998). Most functional surfaces are represented by bilayer phospholipid membranes, which separate cells from the environment, prevent the dissipation of cell resources, provide a sensorial interface with the outside world, support the structural scaffold for the cell, and carry tools for cell propulsion (e.g., flagella or pseudopodia). Eukaryotic cells have additional membrane functions associated with endoplasmic reticulum, nuclear envelope, and various organelles, such as mitochondria, plastids, golgi, vacuoles, and secretion vesicles. There are two main reasons why surfaces are so abundant and enriched in various functions: (1) dimensionality effect: chemical reactions go faster on the surface because molecules encounter each other more readily in two dimensions than in three dimensions (Adam and Delbrück, 1968), and (2) continuous surfaces can separate compartments.

Life could not have originated in a homogeneous three-dimensional space; thus, we need to consider what kinds of surfaces could have been important for the origin of life. Among surfaces available on Earth-like planets, two kinds can be considered as most likely places for the origin of life: water–mineral and water–oil surfaces. Here, by "oil" I mean various hydrocarbons, mostly alkenes, which are the most abundant organic molecules in the universe (Deamer, 2011) and are likely to exist on early terrestrial planets (Marcano et al., 2003). These two kinds of surfaces are selected here for discussion because they both include water. Life could not originate without water because most chemical reactions in living cells require water. We do not discuss air-water surface because it is too unstable and subject to wind and radiation. In this paper, we focus mostly on water-oil surfaces because (1) oil self-aggregates in water forming fine droplets or microspheres (Fig. 1A), which resemble living cells and can divide or merge, (2) oil-attached molecules are not fixed in space, but can move around the surface and have a chance to interact with other oil-attached molecules, (3) oil can be used as a source of carbon for primordial life, and (4) oil droplets could be eventually transformed into a (phospho)lipid membrane (see Section 4.2.).

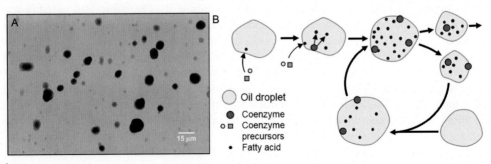

FIG. 1

Oil (hydrocarbon) droplets in water as a potential substrate for coenzyme-like molecules. (A) Emulsion of oil/petroleum in water, from http://petrowiki.org/Oil_emulsions. (B) Scenario of coenzyme self-reproduction on oil droplets: a coenzyme molecule makes the surface hydrophilic via oxidation of hydrocarbons; this change facilitates synthesis of coenzymes from precursors on the surface. Hydrophilic oil droplets easily divide and may coalesce with new oil droplets (i.e., capture new oil resource).

Both mineral and oil surfaces can be colonized from the water side by catalytically active self-reproducing simple molecules which could be precursors of life. I call them coenzyme-like molecules (CLMs) because they may resemble coenzymes, and the model of the origin of life is therefore named "coenzyme world" (Sharov, 2009). Catalysis is necessary for self-reproduction of CLMs as well as for modification of other molecules on the surface. Because many coenzymes (e.g., ATP, NADH, and CoA) are similar to nucleotides, CLMs can be viewed as predecessors of nucleotides. Some ancient organic molecules such as carotinoids and vitamin K are oleophilic and may be relics of the primordial metabolism on the surface of oil droplets.

Let us consider a hypothetical scenario of how CLMs can colonize the surface of oil droplets in water. Assume that simple precursors of CLMs exist in water, but cannot anchor to the hydrophobic oil surface. However, some droplets may include a few fatty acids with hydrophilic ends that allow the precursors to attach and make a functional CLM (Fig. 1B). This CLM can now catalyze the oxidation of outer ends of hydrocarbons in the oil droplet, making more fatty acids where additional CLMs can be assembled from precursors. Fatty acids produced by CLMs make the surface hydrophilic, and this increases the chance of a droplet to split into smaller ones. Later, small droplets can coalesce with other oil droplets, i.e., capture new oil resources (Fig. 1B). Some components of CLMs could come from the oil phase rather than from the water. These include microelements as well as organic molecules. Furthermore, some organic molecules can be obtained via degradation of oil itself. The hypothesis that oil droplets served not just as a habitat but also as a source of carbon for self-reproduction of CLMs seems very attractive because oil can provide abundant organic resources. In particular, some organic molecules could have been derived from fatty acids by pathways similar to the beta-oxidation pathway. The idea that hydrocarbons were used by primordial systems as nutrients was proposed by Oparin (1953 [1936]). Oparin assumed that hydrocarbons together with other organic molecules formed colloidal systems with spontaneous metabolism, which resembled living cells. More recently, Deemer wrote: "It seems likely that primitive cells incorporated lipid-like molecules from the environment as a nutrient, rather than undertaking the much more complex process of synthesizing complex lipids by an enzyme-catalyzed process" (Deamer, 1999).

The emergence of self-reproducing CLMs could be facilitated by interaction of organic molecules with nonorganic ions or crystals. Organic molecules produced by primordial systems were initially very simple, and simple molecules are not likely to act as strong catalysts. This limitation can be removed, however, if these small molecules acquired or modified their catalytic competence via interaction with nonorganic ions or crystals. For example, aminoacids readily interact with heavy metals such as cobalt, nickel, or zinc, and some of these complexes are catalysts (Dutta et al., 2014); vanadium has effects in carbohydrate metabolism (Gruzewska et al., 2014), whereas iron and sulfur facilitate electron transfer (Kummerle et al., 2000). Oxidation of hydrocarbons can be facilitated by iron oxide. These inorganic compounds are natural contaminants in petroleum, and thus, it is reasonable to assume that they may be present in abiogenic oil droplets.

Oil droplets with CLMs represent a two-level hierarchical system that includes self-reproduction at both levels: (1) level of catalytically active surface molecules and (2) level of oil droplets populated by communities of surface-bound molecules. Moreover, CLMs support self-reproduction at both levels because they change their local environment (i.e., surface properties of oil droplets) and this change enables them to capture resources, self-reproduce, and propagate to other oil droplets (Fig. 1B). As a result, CLMs play the role of hereditary signs (i.e., coding molecules) because they encode the surface properties of oil droplets (phenotype) and transfer the code to the next generation of colonized droplets in the form of additional copies of the same molecule type (Sharov, 2009). Note, that CLMs are not resources and do not have to exist in large quantities in the environment to support the primordial metabolism. A single molecule is sufficient to initiate a wave of self-reproduction that can then spread over the surface of an individual oil droplet and over the large population of oil droplets.

Besides oil droplets, CLMs can also colonize mineral surfaces (Wächtershäuser, 1988); however, minerals cannot be used as a source of carbon. Thus, Wächtershäuser assumed that CO_2 was the source of carbon for the synthesis of organic molecules. However, carbon fixation from CO_2 was hardly possible at the early stages of the origin of life (see Section 2). A more realistic habitat for primordial molecules could be minerals covered with a film of oil. Then, self-reproducing CLMs can spread along such surfaces and possibly travel on oil droplets to colonize other patches of oleophylic minerals. Thus, mineral surfaces and oil droplets could have been two alternative niches for the same population of self-reproducing CLMs.

3.2 EVOLUTIONARY POTENTIAL OF THE COENZYME WORLD

The next important condition for primordial life is a capacity to evolve adaptively towards more complex functional molecules and pathways. Self-reproduction of a single kind of molecules (e.g., autocatalysis or crystal growth) is not sufficient for evolution because there is no alternative way to reproduce. Without heritable variation, there is no Darwinian selection and no adaptive evolution. In rare cases, crystallization may follow multiple alternative pathways that provide a limited potential for change (Cairns-Smith, 1982). But the number of variations is always small and not sufficient to support the long-term evolution.

The only conceivable way for primordial self-reproducing molecules to increase their evolutionary potential was to establish *cooperation* with other self-reproducing molecules. Such cooperation can be facilitated by group selection, where groups of molecules either propagate or dissipate together. For example, if the interaction between two kinds of coding molecules (i.e., self-reproducing CLMs) facilitated the rates of self-reproduction of the jointly colonized oil droplets, then the proportion of droplets that harbor both kinds of coding molecules will increase with time. Thus, oil droplets likely harbored a community of self-reproducing molecules with cooperative and possibly conflicting relations. Mineral surfaces are probably less effective in supporting group selection as compared to oil droplets because their existence does not depend on the presence of CLMs. However, minerals can lose an oil film as a consequence of CLM activity and then become unsuitable for colonization; this would be functionally equivalent to the dissipation of an oil droplet.

The term "cooperation" in relation to molecules means that interaction (or simply coexistence) of two or more kinds of self-reproducing molecules is beneficial for each kind of molecules, where benefits are measured by the increase in the rate of reproduction. Cooperation is more than mutual catalysis (Conrad, 1982; Kauffman, 1986) because mutual catalysis is static: it is either present or not, whereas the notion of cooperation is meaningful only in the context of continuous evolutionary change. Continuous evolution appears very unlikely to happen in a system with small coding molecules. In particular, we cannot expect that CLMs easily change their chemical structure and keep the competence for self-reproduction. Thus, instead of looking for changes in the chemical structure of CLMs, we should consider changes in relations between molecules or changes in their local environment. By relations, I mean weak or transient interactions that do not cause irreversible change in molecule structure, as for example, in catalysis. Relations can be mediated by modification of local environments (e.g., surface properties of oil droplets). Potential relations between even simple molecules are very diverse: two molecules may turn relative to each other, approach each other from different sides, or interact via intermediate molecules that serve as signals or resources. There are many more potential cooperative relations as compared to the number of potential heritable modifications of a single molecule. These relations may become heritable signs if they reinforce each other or appear self-reinforced. Thus, heredity is supported not just by coding molecules, but also by coding relations between molecules, and in fact, by the *whole network of relations*. Note that the autopoiesis theory always emphasized the self-renewal of relations in addition to the self-renewal of components (Maturana and Varela, 1980).

Among all kinds of relations between molecules, the relations mediated by local environment (e.g., by surface properties of oil droplet or abundance of resources) seem most important in evolution. Each kind of self-reproducing CLMs performs some function, such as capturing resources, storing energy, or catalyzing a reaction; and these effects may appear beneficial not only for this particular kind of molecules, but also for other kinds of self-reproducing CLMs on the same oil droplet. If these beneficial effects are reciprocal and increase the rate of self-reproduction of the entire molecular community, then we can say that molecules have a cooperative relation. The main advantage of molecular cooperation compared to modification of single molecules is in the increase of potential variability of outcomes. Modification of oil droplets can be viewed as "niche construction" and was necessary for boosting the evolutionary potential of primordial systems. In summary, the evolvability of primordial molecular communities was supported mainly by subtle changes such as modification of local environment and cooperative relations between species of molecules. Cooperative molecular communities in perishable local environments (e.g., oil droplets) can be viewed as precursors of organisms/cells.

Evolvable primordial systems cross the threshold, where physical and chemical processes become integrated into potentially endless and branching chains of events that constitute life. Thus, we need a term that represents this potentially immortal and evolvable component of life, and this term is *sign*. Historically, the doctrine of signs (or semiotics) was applied almost exclusively to human communication. Biologists, however, noticed that meaningful signs exist in the behavior and communication of animals (Sebeok, 1972; Uexküll, 1982). Discovery of DNA and genetic code showed, however, that molecular processes in living cells follow programs encoded in a genome, which bring the experience of ancestral generations to new-formed organisms and help them to develop, survive, and reproduce in changing environments. Thus, the genome carries a set of inter-related signs that encode meaningful information. A few decades ago, a new discipline of *biosemiotics* emerged whose aim is integration of theoretical biology with semiotics (Sebeok, 2001; Hoffmeyer, 2008). The main challenge of biosemiotics is to depart from the anthropocentric interpretation of signs and replace it with an evolutionary approach, where signs and their meanings evolved from simple ones (as in bacteria) to more complex (as in animals and humans) (Sharov et al., 2015). Primitive sign process, or *protosemiosis*, do not require mind; instead signs are associated with specific actions of molecular agents (e.g., ribosome or RNA polymerase) or cells (Sharov and Vehkavaara, 2015). Protosemiosis may exist in very simple systems, such as molecular communities on the surface of oil droplets, where CLMs play the role of signs that encode surface properties of oil and future catalytic actions and disseminate this capacity to the progeny. Thus, the notion of protosemiosis is important for understanding the origin of life.

3.3 DIVERSIFICATION OF MOLECULAR COMMUNITIES

The major criterion of progress in biological evolution at its largest scale is the increase of functional complexity (Sharov, 2006). Thus, the increase in the number of components and relations in molecular communities is the main path towards full-fledged organisms. The core of the primordial self-reproducing system is a minimal subset of irreplaceable coding components and coding relations. If any of these components or relations is lost entirely, then it cannot be recovered based on other components and relations. Establishment of new coding molecules depends on their capacity to self-reproduce as well as on their compatibility with already existing molecules. If new molecules strongly compete with other components for resources, then the self-reproduction capacity of the whole system may appear compromised.

The first path for the emergence of new coding molecules is de-novo acquisition, similar to the emergence of the first self-reproducing CLMs. Such acquisition can be facilitated by the coalescence of droplets carrying different coding molecules. Consider that some population of oil droplets has coding molecules of type a, and another partially separated population has coding molecules of type b. If droplets from different populations coalesce, they make a new kind of droplets with two types of coding molecules: a and b (Fig. 2A). If the new droplet splits up due to mechanical agitation, then the progeny droplets are likely to carry both kinds of coding molecules because multiple copies of each are present in the parental system. Thus, combinations of coding molecules can be inherited and make their own

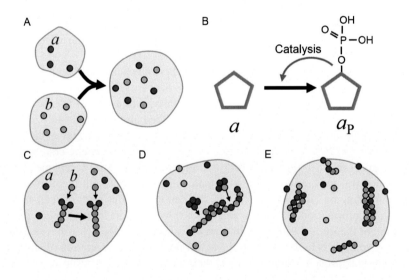

FIG. 2

Emergence of new coding molecules. (A) Coalescence of oil droplets carrying different self-reproducing coenzymes. (B) Phosphorylation of molecule a, which is stably produced within the chemical community on the surface of oil droplets, results in the emergence of a new coding molecule a_P if this reaction is autocatalytic. (C) Polymerization of molecules b catalyzed by self-reproducing molecules a. (D) Simple repetitive polymers facilitate polymerization of identical polymers aligned on the surface of oil droplets. (E) Template-based universal synthesis (i.e., replication) of coding polymers with any aperiodic sequence of monomers.

evolutionary lineages (Sharov, 2009). The combination is beneficial, if droplets with both coding molecules reproduce faster than droplets with just one kind of coding molecules.

Initially, coding molecules were not connected and were transferred to offspring systems in different combinations at random. The probability of transferring the full set of coding molecules to descendants by pure chance may be problematic, especially if droplets carry too many kinds of coding molecules and some of them are present in a small number of copies. This combinatorial problem can be partially meliorated by the "stochastic corrector" mechanism, which is a preferential propagation of systems with a full set of coding molecules (Szathmáry, 1999). Systems with an incomplete set of coding molecules are more likely to fail in surviving and reproduction because some of their functions appear missing. This kind of stochastic correction is a primordial version of the purifying selection; and like the purifying selection, it reduces the overall reproduction rate of the population.

An important consequence of the emergence of new kinds of coding molecules is the diversification of the downstream molecules. For example, if coding molecule A generates some products from available resources, then catalytic activities of the additional coding molecule B can be applied to the products of molecule A. Similarly, catalytic activities of A can be applied to the products of B. Thus, the increase of the number of coding molecules results in a much faster combinatorial increase in the diversity of downstream molecules. This diversity could in turn increase the chances of acquiring new coding molecules through paths discussed below.

The second path towards the emergence of new kinds of coding molecules is the modification of existing coding and downstream molecules (Fig. 2B). It is important that new coding molecules facilitate their self-reproduction. For example, if molecules of type a are already produced in the systems and the phosphorylated modification of a (i.e., a_P) facilitates the phosphorylation of a, then a single phosphorylated molecule a_P initiates a chain reaction that converts a into a_P. This reaction may initially be harmful due to a decrease in the number of molecules a and weakening of functions supported by a. However, the molecular community may eventually develop protecting mechanisms such as shielding some molecules a from phosphorylation or inhibiting the catalytic activity of a_P if it becomes too abundant. As a result, the system acquires a new kind of coding molecules a_P, which can self-reproduce and support a new chemical function (i.e., phosphorylation). This molecule is heritable because the descendant systems are likely to get copies of a_P after the division of oil droplets.

Group selection at the level of oil droplets is likely to facilitate cooperation of newly acquired coding molecules with the previously existing molecular community. However, group selection is effective only if the exchange of members between groups is rare compared to the rate of self-reproduction of groups. This condition is required for any kind of symbiogenesis (Guerrero et al., 2013). Frequent exchange of members between groups is beneficial for the emergence of molecular "parasites" that self-reproduce at the expense of other members in molecular community inhabiting oil droplets. Thus, primordial systems should have a selection pressure to develop mechanisms that reduce the chances of molecule exchange with other systems. For example, certain molecules produced within the system could inhibit the coalescence with droplets that already carry some CLM molecules on the surface.

Finally, the third path towards the emergence of new coding molecules is the establishment of novel relations, such as polymerization (Sharov, 2009). Polymers may initially appear as downstream products of non-polymeric coding molecules. For example, a coding molecule a could catalyze the polymerization of molecules b, and in this way, encode the formation of long polymers $b-b-b-b-\ldots$ (Fig. 2C), which could cover the surface of droplets and modify their physical properties. Some polymers may then become coding molecules if they could facilitate the synthesis of the same kind of polymers (Fig. 2D).

4 EVOLUTION FROM OIL DROPLETS TO LUCA
4.1 TEMPLATE-BASED REPLICATION

The main factor that restricted the rate of primordial evolution was the absence of universal methods for producing new hereditary molecules. Transformation of coding molecules via change of functional groups or polymerization increased the evolvability of primordial systems to some extent (Sharov, 2009). But the ultimate solution of this problem came only with the invention of template-based (or digital) replication. Replication can be viewed as a special case of autocatalysis applied to heteropolymers, where each monomer is added to the newly constructed polymer sequentially based on a uniform simple rule applied to each monomer in another linearly aligned polymer that serves as a template (Szathmáry, 1999). Digital replication is universal in the sense that the rule of replication works for polymers of various lengths and for any sequence of monomers from the allowed set. Using the terminology of Kauffman (2014), we can say that the invention of digital replication provided "enabling constraints" that expanded the boundaries of heritable "adjacent possible," and in this way, increased the evolutionary potential of life.

However, the notion of universal coding should be interpreted with caution because true universal properties exist only in mathematics. In the real world, there are always some limitations even within a universal coding system. For example, replication of the lagging DNA strand requires different molecular agents than replication of the leading strand. Replication and elongation of telomeres requires additional mechanisms that are not equivalent to simple template-based copying. Molecular machinery which is sufficient for replicating short DNA fragments (200–1000 bp) may not work for long sequences (e.g., >1 Mb).

The starting point for the origin of template-based replication was the presence of heteropolymeric coding molecules with repetitive or partially random sequence. Such polymers may initially stick to each other to perform some other functions (e.g., to increase mechanical stiffness of the surface or facilitate other reactions). Then, the shorter strand in a paired sequence can become elongated by adding monomers that weakly matched the longer strand (Fig. 2D). Natural selection may have supported the increase in fidelity of this process, which helped to produce better copies of existing polymers. Template-based replication probably started with copying short repeats (e.g., telomere-like), and then progressed into copying longer repeats and entirely aperiodic sequences (Sharov, 2009) (Fig. 2E). First replicating polymers were probably similar to nucleic acids; however, the sugar-phosphate backbone of RNA does not support anchoring to the surface of oil droplet. In contrast, peptide nucleic acids (PNAs) with a pseudopeptide backbone are able to absorb at the lipid-water surface (Weronski et al., 2007). Thus, PNA-like molecules could have been evolutionary predecessors of RNA (Nelson et al., 2000). Invention of digital replication was the turning point in the origin of life that substantially increased the evolutionary potential of primordial living systems (Jablonka and Szathmáry, 1995; Sharov, 2009).

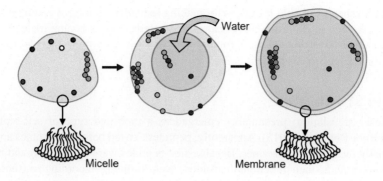

FIG. 3

The origin of bilayer membranes and transition from surface metabolism to intra-cellular metabolism.

4.2 BILAYER MEMBRANE

The surfaces of oil droplets provide plentiful local niches for the emerging life. However, subsequent evolution was constrained by the limited supply of oil, and by the two dimensions of a droplet surface. The lack of hydrocarbons slowed the rates of growth in primordial organisms, which in turn, affected negatively the rates of metabolism, which was surface-dependent. Although surface metabolism is helpful for the emergence of life, two-dimensional surfaces do not provide enough capacity for storing resources. Also, the mobility of polymers in a two-dimensional space is restricted because long molecules cannot easily pass each other.

To overcome these limitations, primordial systems presumably transformed oil droplets into an outer membrane via engulfing water (Fig. 3). Agitation of oil emulsion in water easily makes "nested" droplets, but they are unstable. Cell-like organization is unlikely to yield any functional advantage if the membrane breaks easily. Thus, the membrane has to be strong enough to sustain mechanical disturbances, and osmoregulation is needed to prevent bursting. Membranes can be stabilized by replacing fatty acids with lipids and phospholipids. Lipids require glycerol, which has to be synthesized within the system (it is not available in the environment). Thus, the emergence of metabolic pathways for glycerol synthesis was necessary for making stable membranes. Glycerol could have been used also to make sugars that are suitable for storing energy, regulating osmosis, and making nucleic acids. It is not clear if template-based replication of coding polymers appeared before or after the formation of stable cell membrane. But in any case, these two evolutionary events produced cells with ribozyme-based catalysts that match the RNA-world model (Gilbert, 1986).

4.3 CHROMOSOMES

As the number of coding molecules increased in evolution, the stochastic correction mechanism became less effective in keeping all coding components together. Stochastic corrector has a significant fitness load because daughter cells with incomplete set of coding molecules are not fully viable. This load increased with each additional type of coding molecules, which limited the increase in overall complexity of the system. In addition, the stochastic correction mechanism requires abundant copies of each type of coding molecules, so that at least one copy is transferred to each daughter cell. As a result, cells had to spend valuable resources for making redundant copies of coding molecules. These problems were resolved in evolution by concatenating heterogeneous coding molecules (e.g., RNAs) into one or very few large units (e.g., chromosomes) and by developing mechanisms of controlled transfer of chromosomes to descendent cells.

4.4 PROTEIN SYNTHESIS

Protein synthesis (translation) presumably emerged as a modified type of self-replication (Root-Bernstein and Root-Bernstein, 2015). Apparently, peptides existed before ribosomes and were synthesized individually by specific ribozymes. Because first peptides were likely short and included only a few kinds of aminoacids, their functional capacities were limited. According to Root-Bernstein and Root-Bernstein, the evolutionary fate of peptides changed when they were used to assist the self-replication of RNA. As a result, natural selection supported not only self-replication of the RNA itself, but also the production of helper peptides. In particular, peptides became longer and included a more specific sequence of aminoacids. The diversity of animoacids increased in evolution, and especially important was the emergence of aminoacids that are currently found in catalytically active portions of proteins: lysine, cysteine, serine, treonine, aspartate, histidine, and thyrosine (Holliday et al., 2009). The first four of them are rather simple and probably appeared earlier in evolution that the latter three. The evolution of peptide-assisted RNA replication also included the emergence of efficient catalysis of peptide bond formation and tRNA for transporting aminoacids. This hypothesis is supported by the fact that the ribosomal RNA of bacteria *Escherichia coli* carries remnants of the entire set of tRNAs for all aminoacids together with fragments of coding sequences for ribosomal proteins, polymerases, ligases, synthetases, and phosphatases (Root-Bernstein and Root-Bernstein, 2015). Thus, ribosomal RNA apparently originated from a primordial genome that encoded a self-organizing and self-replicating molecular assembly.

The correspondence between the triplets of nucleotides in the RNA molecule and corresponding aminoacids in the encoded protein is determined by the genetic code. Although genetic code is nearly universal in all living organisms, it is not likely to appear initially in its current form.

It is possible that the initial code was based on nucleotide doublets rather than triplets and encoded a smaller set of aminoacid types (Patel, 2005; Travers, 2006). It is also likely that genetic code evolved towards lower rates of translation errors (Novozhilov et al., 2007).

Reconstructions of the Last Universal Common Ancestor (LUCA) of all known living organisms (Doolittle, 2000; Theobald, 2010) indicate that LUCA was far more complex than RNA-world systems described above. The hypothetical LUCA had a DNA-based genome and a fully developed machinery for programmed synthesis of proteins. The integrity of the DNA was maintained by a group of diverse enzymes, including DNA helicase, topoisomerase, ligase, and DNA repair proteins. The RNA-polymerase complex was used to synthesize mRNA copies of each gene. LUCA-type cells synthesized fatty acids, and thus were no longer dependent on the supply of abiotic oil. The outer membrane of LUCA included ion pumps, receptors, and other proteins. Although the genome of LUCA is a product of phylogenetic reconstruction, it can be viewed as a bacteria- or archea-like organism with the number of genes in the range from 500 to 1000 (Koonin, 2003, 2009). The difference in complexity between early cells with first encoded proteins and LUCA-type cells is hard to overestimate, and it is comparable in scope to the difference between prokaryotes and multicellular eukaryotes. Emergence of each new function, and/or new gene, opened previously unavailable possibilities for subsequent evolution

(Sharov, 2014). This positive feedback supported the exponential increase of the overall complexity in most successful lineages of organisms (Sharov, 2006). These later evolutionary events, however, are not relevant for the problem of the origin of life (and for this paper) because the backbone of hereditary mechanisms and chemical construction networks has been finalized with the advent of protein synthesis.

5 DISCUSSION

Proposed scenario of the origin of life from catalytically active self-reproducing molecules (coenzymes) on the surface of oil droplets in water has multiple advantages as compared to alternative models. First, surface metabolism increases the chances of molecular interaction as compared to the open 3D space; an additional advantage of oil surfaces is that functional molecules can be covalently attached to oil (hydrocarbon) molecules and still float around. Second, surfaces provide ready access to the outside resources including both chemicals and energy. Third, there is no need for either prebiotic soup or energetically expensive and biochemically complex pathway of carbon fixation from CO_2; instead, oil was used as a source of carbon. Fourth, coenzymes played the role of hereditary signs that encoded surface properties of oil droplets. And fifth, the emergence of advanced features of life, such as template-based replication, bilayer outer membrane, and carbohydrate metabolism, is not required at the origin of life; instead, these features appeared much later in evolution. According to this model, novel chemical structures, functions, and heredity emerged simultaneously by reinforcing each other with each cycle of reproduction. Thus, it does not fit into traditional categories such as "replication first" or "metabolism first." The model also assumes that evolution of primordial systems was driven by natural selection from the very start of life, and long before the emergence of template-based replication of polymers.

For comparison, I will consider an alternative model of the origin of life in inorganic compartments within hydrothermal vents (Koonin, 2009; Koonin and Martin, 2005). Authors correctly reasoned that biochemical processes could have hardly evolved in free solution; but instead of considering surface metabolism, they selected another option—inorganic compartments. Indeed, compartments prevent free diffusion and can keep molecules together. In the presence of catalysts (e.g., FeS), simple organic molecules (e.g., formaldehyde) can become concentrated within compartments. However, there is no evidence that more complex organic molecules can be generated in sufficient amounts in such systems. Even the formose reaction is hardly possible due to the low concentration of formaldehyde. The model further assumes the spontaneous emergence of self-replicating nucleic acids and even protein synthesis. As we argued in Sections 1–2, self-replicating polymers were unlikely to emerge in evolution before the development of heritable pathways for producing monomers. Also, it is questionable if cooperation between different kinds of replicons could emerge in such systems. Isolated compartments do not allow the spread of replicons, whereas connected compartments cannot support group selection. In contrast to inorganic compartments with mostly static connections (pores), oil droplets provide a highly dynamic habitat for primordial life: they are fully isolated, and yet in certain conditions they may either coalesce or split into smaller droplets. This dynamic patchy environment supports both, propagation in space and group selection.

The origin of life is a very long multi-step process, which cannot be replicated in laboratory conditions. Thus, experimental studies should be focused on individual steps in this process and our expectations for the results of such experiments should be set lower. It would never be possible to mix inorganic chemicals and generate functional cells with a membrane, self-supporting metabolism, and nucleic acids as hereditary molecules. Experiments in artificially enriched medium with abundant complex organic molecules such as aminoacids, peptides, sugars, or phospholipids will not take us any closer to the understanding of the origin of life because these kinds of synthetic molecules have never been available in sufficient quantities in nature. More meaningful results can be expected from experiments that use nutrient-poor media, such as hydrocarbons in water supplemented with rare additional molecules that may become assembled into catalytically active units.

Analysis of the rates of increase in genomic complexity indicates that the doubling time of the non-redundant and functional fraction of the genome is about 340 ± 100 (c.i.)[5] million years (Sharov, 2006). Similar estimates (360 million years) were obtained independently by Markov et al. (2010). Based on this rate, the evolution from the origin of life to the level of bacteria-like organisms would require from 4 to 8 billion years. Considering that bacteria-like organisms flourished on Earth as early as 3.48 By ago (Noffke et al., 2013), and the age of earth is only 4.5 By, it is likely that life originated on a planet orbiting another star and then was transferred to earth with comets, asteroids, or rogue planets. The cosmic journey of life may have been shorter if the Solar System originated from the remnants of an exploded parental star that had life on its planets. In this case, LUCA could represent the genome of organisms (possibly from multiple lineages) that arrived to the primordial earth. This hypothesis is further supported by the facts that there are no traces of precellular life on earth,[6] no RNA-world organisms without protein synthesis, no alternative nucleotide composition, no alternative aminoacid composition, no alternative genetic code (minor modifications likely appeared after arrival to earth), and no free-living organisms with a genome <0.5 Mbp.

An alternative view is that primordial evolution was qualitatively different from subsequent evolution, and thus, extrapolations of the rates of complexity increase are not valid. In particular, it was suggested that the primordial evolution was unusually fast due to the lack of competition (Koonin and Galperin, 2003). However, primordial competition may have been very intensive because of the scarcity of organic resources; and there is no consensus on the inverse relationship between competition and rates of complexity increase. For example, competition was shown to increase the diversity and facilitate the emergence of new taxa in animals (Stanley, 1973). Carl Woese proposed another hypothesis that primordial organisms evolved via non-Darwinian process that was much faster than regular evolution because of unusually intensive horizontal gene transfer (HGT) (Woese, 2002). However, historical rates of HGT are measured indirectly based on mathematical models of the evolutionary process and are highly sensitive to model parameters (Koonin and Galperin, 2003: p. 241-242). There is no direct evidence that HGT rates >3.5 By ago were much higher than subsequent HGT rates in bacterial lineages. Moreover, HGT is substantially more frequent in bacteria than in eukaryotes, and yet the rates of complexity increase are lower in bacteria than in eukaryotes (Sharov, 2006). Thus, there seem to be no positive association between HGT frequency and the rates of increase in genome complexity. In summary, these hypotheses failed to explain the origin of life on earth.

---

[5]Confidence interval was determined by sensitivity analysis.
[6]Viruses likely arrived to earth together with their bacterial hosts. Thus, we cannot consider them as evidence of precellular life.

The most important novel component of the proposed model is that the evolvability of primordial systems was largely based on cooperation of molecules, which is the establishment of mutually beneficial relations between functional molecules colonizing the same local environment such as droplet of oil. Some relations became heritable if they reinforced each other, and their cooperative nature was supported by group selection. Thus, life originated from simple (not polymeric) but already functional molecules, and its gradual evolution towards higher complexity was driven by cooperation and natural selection.

ACKNOWLEDGMENTS

This project was supported entirely by the Intramural Research Program of the National Institute on Aging (NIA/NIH), project Z01 AG000656-13. Funding organization had no involvement in this study.

REFERENCES

Adam, G., Delbrück, M., 1968. Reduction of dimensionality in biological diffusion processes. In: Rich, A., Davidson, N. (Eds.), Structural Chemistry and Molecular Biology. W.H. Freeman, New York, pp. 198–215.

Benner, S.A., Kim, H.J., Carrigan, M.A., 2012. Asphalt, water, and the prebiotic synthesis of ribose, ribonucleosides, and RNA. Acc. Chem. Res. 45, 2025–2034.

Bernhardt, H.S., 2012. The RNA world hypothesis: the worst theory of the early evolution of life (except for all the others). Biol. Direct 7, 23.

Bickhard, M.H., 2005. Functional scaffolding and self-scaffolding. New Ideas Psychol. 23, 166–173.

Biver, N., Bockelee-Morvan, D., Moreno, R., Crovisier, J., Colom, P., Lis, D.C., Sandqvist, A., Boissier, J., Despois, D., Milam, S.N., 2015. Ethyl alcohol and sugar in comet C/2014 Q2 (Lovejoy). Sci. Adv. 1. e1500863.

Caetano-Anolles, G., Seufferheld, M.J., 2013. The coevolutionary roots of biochemistry and cellular organization challenge the RNA world paradigm. J. Mol. Microbiol. Biotechnol. 23, 152–177.

Cairns-Smith, A.G., 1982. Genetic Takeover and the Mineral Origins of Life. Cambridge University Press, Cambridge.

Callahan, M.P., Smith, K.E., Cleaves 2nd, H.J., Ruzicka, J., Stern, J.C., Glavin, D.P., House, C.H., Dworkin, J.P., 2011. Carbonaceous meteorites contain a wide range of extraterrestrial nucleobases. Proc. Natl. Acad. Sci. U. S. A. 108, 13995–13998.

Colomb-Delsuc, M., Mattia, E., Sadownik, J.W., Otto, S., 2015. Exponential self-replication enabled through a fibre elongation/breakage mechanism. Nat. Commun. 6, 7427.

Conrad, M., 1982. Bootstrapping model of the origin of life. Biosystems 15, 209–219.

Cooper, G., Kimmich, N., Belisle, W., Sarinana, J., Brabham, K., Garrel, L., 2001. Carbonaceous meteorites as a source of sugar-related organic compounds for the early Earth. Nature 414, 879–883.

Deacon, T.W., 2006. Reciprocal linkage between self-organizing processes is sufficient for self-reproduction and evolvability. Biol. Theory 1, 136–149.

Deacon, T.W., 2011. Incomplete Nature: How Mind Emerged From Matter. W. W. Norton and Company, New York.

Deacon, T.W., Srivastava, A., Bacigalupi, J.A., 2014. The transition from constraint to regulation at the origin of life. Front. Biosci. (Landmark Ed) 19, 945–957.

Deamer, D.W., 1999. How did it all begin? The self-assembly of organic molecules and the origin of cellular life. In: Scotchmoor, J., Springer, D.A. (Eds.), Evolution: Investigating the Evidence. Paleontological Society, Pittsburgh, PA.

Deamer, D., 2011. First Life: Discovering the Connections Between Stars, Cells, and How Life Began. University of California Press, Berkley, CA.

Doolittle, W.F., 2000. Uprooting the tree of life. Sci. Am. 282, 90–95.

Dutta, A., DuBois, D.L., Roberts, J.A., Shaw, W.J., 2014. Amino acid modified Ni catalyst exhibits reversible H_2 oxidation/production over a broad pH range at elevated temperatures. Proc. Natl. Acad. Sci. U. S. A. 111, 16286–16291.

Egel, R., 2012. Primal eukaryogenesis: on the communal nature of precellular States, ancestral to modern life. Life (Basel) 2, 170–212.

Eigen, M., Schuster, P., 1979. The Hypercycle, a Principle of Natural Self-Organization. Springer-Verlag, Berlin, New York.

Gilbert, W., 1986. The RNA world. Nature 319, 618.

Gray, P., Scott, S.K., 1994. Chemical Oscillations and Instabilities: Non-linear Chemical Kinetics. Oxford University Press, New York.

Griffith, J.S., 1967. Self-replication and scrapie. Nature 215, 1043–1044.

Gruzewska, K., Michno, A., Pawelczyk, T., Bielarczyk, H., 2014. Essentiality and toxicity of vanadium supplements in health and pathology. J. Physiol. Pharmacol. 65, 603–611.

Guerrero, R., Margulis, L., Berlanga, M., 2013. Symbiogenesis: the holobiont as a unit of evolution. Int. Microbiol. 16, 133–143.

Hartman, H., 1998. Photosynthesis and the origin of life. Orig. Life Evol. Biosph. 28, 515–521.

Hoffmeyer, J., 1998. Surfaces inside surfaces. On the origin of agency and life. Cyber. Human Knowing 5, 33–42.

Hoffmeyer, J., 2008. Biosemiotics: An Examination Into the Signs of Life and the Life of Signs. University of Scranton Press, Scranton, PA.

Holliday, G.L., Mitchell, J.B., Thornton, J.M., 2009. Understanding the functional roles of amino acid residues in enzyme catalysis. J. Mol. Biol. 390, 560–577.

Ikehara, K., 2016. Evolutionary steps in the emergence of life deduced from the bottom-up approach and GADV hypothesis (top-down approach). Life (Basel) 6. https://doi.org/10.3390/life6010006.

Jablonka, E., Szathmáry, E., 1995. The evolution of information storage and heredity. Trends Ecol. Evol. 10, 206–211.

Kauffman, S.A., 1986. Autocatalytic sets of proteins. J. Theor. Biol. 119, 1–24.

Kauffman, S.A., 2014. Prolegomenon to patterns in evolution. Biosystems 123, 3–8.

Koonin, E.V., 2003. Comparative genomics, minimal gene-sets and the last universal common ancestor. Nat. Rev. Microbiol. 1, 127–136.

Koonin, E.V., 2009. On the origin of cells and viruses: primordial virus world scenario. Ann. N. Y. Acad. Sci. 1178, 47–64.

Koonin, E.V., Galperin, M.Y., 2003. Sequence-Evolution-Function: Computational Approaches in Comparative Genomics. Kluwer Academic, Boston.

Koonin, E.V., Martin, W., 2005. On the origin of genomes and cells within inorganic compartments. Trends Genet. 21, 647–654.

Kummerle, R., Kyritsis, P., Gaillard, J., Moulis, J.M., 2000. Electron transfer properties of iron-sulfur proteins. J. Inorg. Biochem. 79, 83–91.

Lane, N., Allen, J.F., Martin, W., 2010. How did LUCA make a living? Chemiosmosis in the origin of life. Bioessays 32, 271–280.

Laurent, M., 1997. Autocatalytic processes in cooperative mechanisms of prion diseases. FEBS Lett. 407, 1–6.

Marcano, V., Benitez, P., Palacios-Pru, E., 2003. Acyclic hydrocarbon environments $>=$n-C18 on the early terrestrial planets. Planet. Space Sci. 51, 159–166.

Markov, A.V., Anisimov, V.A., Korotaev, A.V., 2010. Relationship between the genome size and organismal complexity in the lineage leading from prokaryotes to mammals [in Russian]. Paleontol. J. 4, 1–12.

Maturana, H., Varela, F., 1980. Autopoiesis and Cognition: The Realization of the Living. D. Reidel Publishing Co., Dordecht.

Maury, C.P., 2009. Self-propagating beta-sheet polypeptide structures as prebiotic informational molecular entities: the amyloid world. Orig. Life Evol. Biosph. 39, 141–150.

Maury, C.P., 2015. Origin of life. Primordial genetics: information transfer in a pre-RNA world based on self-replicating beta-sheet amyloid conformers. J. Theor. Biol. 382, 292–297.

Morowitz, H.J., Kostelnik, J.D., Yang, J., Cody, G.D., 2000. The origin of intermediary metabolism. Proc. Natl. Acad. Sci. U. S. A. 97, 7704–7708.

Mulkidjanian, A.Y., Galperin, M.Y., 2007. Physico-chemical and evolutionary constraints for the formation and selection of first biopolymers: towards the consensus paradigm of the abiogenic origin of life. Chem. Biodivers. 4, 2003–2015.

Nelson, K.E., Levy, M., Miller, S.L., 2000. Peptide nucleic acids rather than RNA may have been the first genetic molecule. Proc. Natl. Acad. Sci. U. S. A. 97, 3868–3871.

Nghe, P., Hordijk, W., Kauffman, S.A., Walker, S.I., Schmidt, F.J., Kemble, H., Yeates, J.A., Lehman, N., 2015. Prebiotic network evolution: six key parameters. Mol. Biosyst. 11, 3206–3217.

Noffke, N., Christian, D., Wacey, D., Hazen, R.M., 2013. Microbially induced sedimentary structures recording an ancient ecosystem in the ca. 3.48 billion-year-old Dresser Formation, Pilbara, Western Australia. Astrobiology 13, 1103–1124.

Novozhilov, A.S., Wolf, Y.I., Koonin, E.V., 2007. Evolution of the genetic code: partial optimization of a random code for robustness to translation error in a rugged fitness landscape. Biol. Direct 2, 24.

Oparin, A.I., 1953 [1936]. The Origin of Life. Dover Publications, Mineola, NY.

Orgel, L., 2000. Origin of life. A simpler nucleic acid. Science 290, 1306–1307.

Orgel, L.E., 2004. Prebiotic chemistry and the origin of the RNA world. Crit. Rev. Biochem. Mol. Biol. 39, 99–123.

Patel, A., 2005. The triplet genetic code had a doublet predecessor. J. Theor. Biol. 233, 527–532.

Powner, M.W., Gerland, B., Sutherland, J.D., 2009. Synthesis of activated pyrimidine ribonucleotides in prebiotically plausible conditions. Nature 459, 239–242.

Robertson, M.P., Joyce, G.F., 2012. The origins of the RNA world. Cold Spring Harb. Perspect. Biol. 4, a003608.

Root-Bernstein, M., Root-Bernstein, R., 2015. The ribosome as a missing link in the evolution of life. J. Theor. Biol. 367, 130–158.

Sebeok, T.A., 1972. Perspectives in Zoosemiotics. Mouton de Gruyter, The Hague.

Sebeok, T.A., 2001. Biosemiotics: its roots, proliferation and prospects. Semiotica 134, 61–78.

Segre, D., Lancet, D., Kedem, O., Pilpel, Y., 1998. Graded autocatalysis replication domain (GARD): kinetic analysis of self-replication in mutually catalytic sets. Orig. Life Evol. Biosph. 28, 501–514.

Segre, D., Ben-Eli, D., Deamer, D.W., Lancet, D., 2001. The lipid world. Orig. Life Evol. Biosph. 31, 119–145.

Sharov, A.A., 1991. Self-reproducing systems: structure, niche relations and evolution. Biosystems 25, 237–249.

Sharov, A.A., 2006. Genome increase as a clock for the origin and evolution of life. Biol. Direct 1, 17.

Sharov, A.A., 2009. Coenzyme autocatalytic network on the surface of oil microspheres as a model for the origin of life. Int. J. Mol. Sci. 10, 1838–1852.

Sharov, A.A., 2014. Evolutionary constraints or opportunities? Biosystems 123, 9–18.

Sharov, A.A., Vehkavaara, T., 2015. Protosemiosis: agency with reduced representation capacity. Biosemiotics 8, 103–123.

Sharov, A., Maran, T., Tønnessen, M., 2015. Towards synthesis of biology and semiotics. Biosemiotics 8, 1–7.

Sponer, J.E., Sponer, J., Novakova, O., Brabec, V., Sedo, O., Zdrahal, Z., Costanzo, G., Pino, S., Saladino, R., Di Mauro, E., 2016. Emergence of the first catalytic oligonucleotides in a formamide-based origin scenario. Chemistry. https://doi.org/10.1002/chem.201503906.

Stanley, M.S., 1973. Effects of competition on rates of evolution, with special reference to bivalve mollusks and mammals. Syst. Zool. 22, 486–506.

Steele, A., Baross, J., 2006. Are prions relevant to astrobiology? Astrobiology 6, 284.

Szathmáry, E., 1999. The first replicators. In: Keller, L. (Ed.), Levels of Selection in Evolution. Princeton University Press, Princeton, pp. 31–52.

Tessera, M., 2011. Origin of evolution versus origin of life: a shift of paradigm. Int. J. Mol. Sci. 12, 3445–3458.

Theobald, D.L., 2010. A formal test of the theory of universal common ancestry. Nature 465, 219–222.

Tilman, D., Kareiva, P., 1997. Spatial Ecology: The Role of Space in Population Dynamics and Interspecific Interactions. Princeton University Press, Princeton, NJ.

Travers, A., 2006. The evolution of the genetic code revisited. Orig. Life Evol. Biosph. 36, 549–555.

Uexküll, J.v., 1982. The theory of meaning. Semiotica 42, 25–82.

Unrau, P.J., Bartel, D.P., 1998. RNA-catalysed nucleotide synthesis. Nature 395, 260–263.

Wächtershäuser, G., 1988. Before enzymes and templates: theory of surface metabolism. Microbiol. Rev. 52, 452–484.

Weber, B.H., 2007. Emergence of life. Zygon 42, 837–856.

Weronski, P., Jiang, Y., Rasmussen, S., 2007. Molecular dynamics study of small PNA molecules in lipid-water system. Biophys. J. 92, 3081–3091.

Woese, C.R., 2002. On the evolution of cells. Proc. Natl. Acad. Sci. U. S. A. 99, 8742–8747.

Zubarev, D.Y., Rappoport, D., Aspuru-Guzik, A., 2015. Uncertainty of prebiotic scenarios: the case of the non-enzymatic reverse tricarboxylic acid cycle. Sci. Rep. 5, 8009.

EMERGENCE OF POLYGONAL SHAPES IN OIL DROPLETS AND LIVING CELLS: THE POTENTIAL ROLE OF TENSEGRITY IN THE ORIGIN OF LIFE

Richard Gordon*,[†], **Martin M. Hanczyc**[‡], **Nikolai D. Denkov**[§], **Mary A. Tiffany**[¶], **Stoyan K. Smoukov**[‖]

*Gulf Specimen Marine Laboratories, Panacea, FL, United States
[†]Wayne State University, Detroit, MI, United States
[‡]Università degli Studi di Trento, Povo, Italy
[§]Sofia University, Sofia, Bulgaria
[¶]Bainbridge Island, WA, United States
[‖]Queen Mary University of London, London, United Kingdom
University of Cambridge, Cambridge, United Kingdom
University of Sofia, Sofia, Bulgaria

CHAPTER OUTLINE

An organism with such a square shape is without precedent and it is not surprising, therefore, that the original report of the square bacterium was received with some skepticism
(Parkes & Walsby, 1981).

Habitability of the Universe Before Earth, editors: Richard Gordon & Alexei Sharov, Volume 1 in the series:
Astrobiology: Exploring Life on Earth and Beyond, series editors: Pabulo Henrique Rampelotto,
Joseph Seckbach & Richard Gordon. ISSN 2468-6352. https://doi.org/10.1016/B978-0-12-811940-2.00018-6

1 INTRODUCTION

A full model of the origin of life must be multifaceted to explain all features of life (e.g., self-reproduction, heredity, metabolism, sensing and response, capturing resources, and evolutionary mechanisms). Diversity in shape is a central property that influences a number of these features and is observed in all organisms, including bacteria and Archaea. Small changes in shape in prokaryotes can be associated with profound advantages in motility and nutrient uptake, avoiding predation and biofilm structure, which are top on the list of evolutionary pressures (Smith et al., 2017; Young, 2006, 2007, 2010). In this chapter, instead of relying on cellular cytoskeleton mechanisms or information-rich genetics to form or program such shapes, we hypothesize that shapes in protocells may have emerged simply as tensegrity structures in liquid droplets driven by phase-change mechanisms. Recent experiments show that regular geometric shapes with liquid cores may emerge simply in oil droplets from phase transitions upon cooling (Cholakova et al., 2016; Denkov et al., 2015; Haas et al., 2017). Our proposal is consistent with the hypothesis of a lipid world (Bar-Even et al., 2004; Segré et al., 2001) and an oil-droplet-based origin of life (Hanczyc, 2014; Sharov, 2016; Sharov & Gordon, 2017). We build on:

1. Shaped droplets: a new discovery, that cooled oil droplets with surfactants acquire flat, polygonal shapes (Denkov et al., 2015) (Fig. 1);
2. Oil-based protocells: models for the origin of life from oil droplets (Hanczyc, 2014; Sharov, 2016; Sharov & Gordon, 2017), preceding membrane-bound vesicles (Fiore & Strazewski, 2016), with genomic evidence that life started from an "oily" state (Mannige et al., 2012);

And we show how these shaped droplets and their potential role as protocells may have given rise to present-day organisms and their structures, including:

3. Polygonal prokaryotes: the fact that many halophilic Archaea (thought on discovery to have been bacteria) have flat, polygonal shapes (Walsby, 1980) (Fig. 1);
4. Molecular dating indicating the ancient origin of cytoskeleton, and recent evidence of its central role in cell functioning (Gordon & Gordon, 2016a);
5. Polygonal diatoms: we suggest that the valve silicalemma of polygonal diatoms has a flat, polygonal shape, resulting in precipitation of the silica within as a polygon (Fig. 1).

This adventure is new, and thus our sketch will leave many gaps in our understanding, some of which, however, are approachable experimentally. What we are interjecting into the dialogue on the origin of life is that the shape of protocells may have been adaptive, and that structure has evolutionary consequences. We have called cooled oil droplets with flat, polygonal shapes "shaped droplets." We do not discuss traditional topics related to the origin of life, such as the emergence of proto-heredity and proto-metabolism, though we suggest that the structure of shaped droplets may have provided new evolutionary opportunities for improving various cell functions, including heredity and metabolism.

2 SHAPED DROPLETS

It may seem strange that a three-dimensional spherical liquid drop of oil can take on a nearly two-dimensional, polygonal shape when cooled slowly, forming what we have called "shaped droplets" (Fig. 1). We are just beginning to understand the mechanism by which this happens (Azadi &

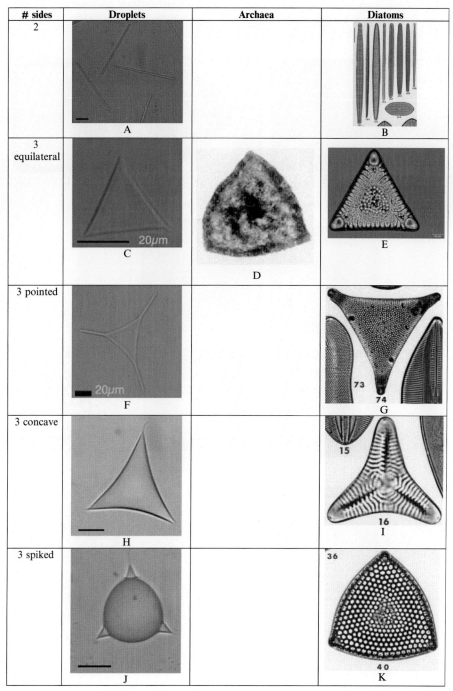

# sides	Droplets	Archaea	Diatoms
2	A		B
3 equilateral	C	D	E
3 pointed	F		G
3 concave	H		I
3 spiked	J		K

FIG. 1—Cont'd

See figure legend on next page.

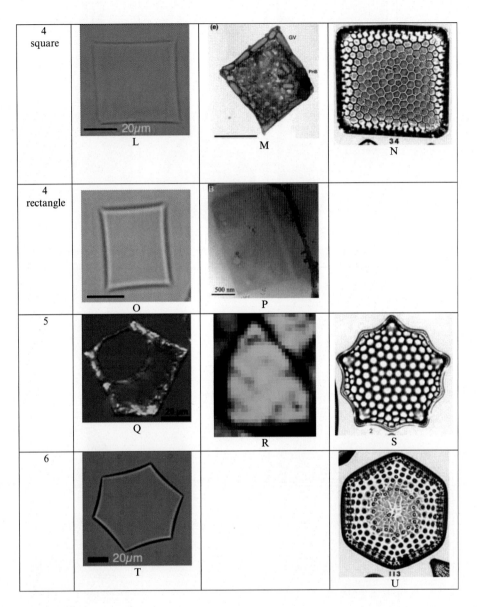

FIG. 1—Cont'd

See figure legend on opposite page.

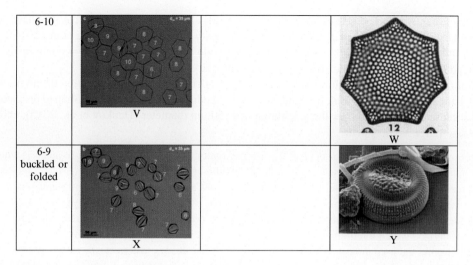

FIG. 1—Cont'd

Shaped droplets vs. Archaea and diatoms. *Left column:* samples of abiotic drops, matched with comparable prokaryotic Archaea (*middle column*) and eukaryotic diatom valves (*right column*). Archaea: D: *Haloarcula japonica* (Takao, 2006). M, P: *Haloquadratum walsbyi.* Diatoms: B: 153 *Tabularia gaillonii*, 154 *Nitzschia macilenta*, 155 *Alveus marinus*, 156 *Synedra fulgens*, 157 *Synedra baculus*, 158 *Synedra formosa*, 159 *Synedra superba*, and 160 *Hyalosynedra* (?) sp. indet. E: *Triceratium morlandii* var. *morlandii*. G: *Biddulphia spinosa*. I: *Schuettia annulata*. K: *Triceratium broeckii*. N: *Triceratium favus* fo. *quadrata*. S: *Triceratium campechianum*. U: *Stictodiscus parallelus* fo. *hexagona*. W: *Triceratium septangulatum*. Y: *Cyclotella litoralis*.

(A, C, F, H, J, L, Q, T, V, X) From Denkov, N., Tcholakova, S., Lesov, I., Cholakova, D., Smoukov, S.K., 2015. Self-shaping of oil droplets via the formation of intermediate rotator phases upon cooling. Nature 528(7582), 392–395, with permission of Nature Publishing Group. (B, G, I, K, N, S, U, W) From Stidolph, S.R., Sterrenburg, F.A.S., Smith, K.E.L., Kraberg, A., 2012. Stuart R. Stidolph Diatom Atlas: U.S. Geological Survey Open-File Report 2012–1163. http://pubs.usgs.gov/of/2012/1163/, in public domain. (E) From Mikhaltsov, A., 2014. File:Triceratium morlandii var. morlandii.jpg. https://commons.wikimedia.org/wiki/File:Triceratium_morlandii_var._morlandii.jpg with permission under the Creative Commons Attribution-Share Alike 4.0 International license. (M) From Burns, D.G., Camakaris, H.M., Janssen, P.H., Dyall-Smith, M.L., 2004. Cultivation of Walsby's square haloarchaeon. FEMS Microbiology Letters 238(2), 469–473, with permission of Oxford University Press. (P) From Comolli, L.R., Duarte, R., Baum, D., Luef, B., Downing, K.H., Larson, D.M., Csencsits, R., Banfield, J.F., 2012. A portable cryo-plunger for on-site intact cryogenic microscopy sample preparation in natural environments. Microscopy Research and Technique 75(6), 829–836, with permission of John Wiley and Sons. (Y) From Gordon, R., Tiffany, M.A., 2011. Possible buckling phenomena in diatom morphogenesis, in: Seckbach, J., Kociolek, J.P. (Eds.), The Diatom World. Springer, Dordrecht, The Netherlands, pp. 245–272, with permission of Springer. (R) is an enlargement of one cell in Fig. 4E.

Grason, 2016; Cholakova et al., 2016; Denkov et al., 2015, 2016; Grason, 2016; Guttman et al., 2016a,b; Haas et al., 2017), and much experimental, theoretical, and computer simulation (molecular dynamics) work lies ahead. Shape is a global property of a cell ((Boulbitch, 2000), as discussed in (Gordon & Gordon, 2016a)), presumed, but rarely demonstrated, to be the result of a balance of forces (Pomp et al., 2016; Sain et al., 2015; Sims et al., 1992). The same notion applies to the spherical liquid drop of oil and thus presumably to its shaped droplet state. A general framework in which to consider such problems is that of tensegrity structures, which have been applied at many hierarchical levels, from molecules to the human body (Levin, 2006a,b; Scarr, 2014), on up to human edifices. We have made a preliminary suggestion that shaped droplets are tensegrity structures (Cholakova et al., 2016), promising a full manuscript, which is this chapter. We provide an Appendix overviewing tensegrity structures and giving a toy model for flat, polygonal tensegrity structures.

Contrary to our former intuition of spherical shapes determined by minimization of interfacial energy, shaped droplets can be thin, flat, and have straight edges and sharp corners. Shaped droplets have been successfully obtained over the size range of 1–50 μm diameter (Denkov et al., 2015), about the same size range as most cells:

> Recently, we reported (Denkov et al., 2015) a novel bottom-up mechanism for morphogenesis within droplets of linear alkanes, which leads to the formation of micrometer sized particles with different shapes. This phenomenon is driven by the formation of a "skin" of intermediate rotator phases in cooled alkane droplets—a process which was triggered by freezing adsorption layers of long-chain surfactants. These rotator phases (called also "plastic crystals" or "highly ordered smectics") are characterized with a long-range translational order of the molecules, yet with some rotational freedom around the long molecular axis in the case of linear alkanes (Sirota et al., 1993). We showed (Denkov et al., 2015) how the formation of anisotropic rotator phases, within the confines of the micro-scale fluid droplets, results in a surprisingly wide array of regular geometric shapes, including regular polyhedra; hexagonal, tetragonal and trigonal platelets, with and without fibers protruding from their corners. Furthermore, we found that frozen solid particles with the respective shapes can be produced by selecting appropriate cooling rates. In this way, we succeeded in making a nontrivial link between a novel process for making complex shapes in materials science and the fundamental discovery that phase transitions confined inside a drop could be a driving force for morphogenesis (Denkov et al., 2015). The process opens new opportunities for both scalable particle synthesis and studying the structure/shape formation processes in amazingly simple chemical systems
>
> **(Cholakova et al., 2016).**

Similar "faceted liquid droplets" were soon after independently discovered by Guttman et al. (2016a,b), though rather than formation of a tensegrity structure to counteract a measurable surface tension in order to create the shapes, they ascribed the process to an ultra-low surface tension. Subsequent reports have shown that the surface tension in these systems is not ultra-low (Denkov et al., 2016), confirming the plastic crystal tensegrity structure argument.

Using the general tensegrity background knowledge (see Appendix), we can now consider the balance of forces in a shaped droplet. At the smallest scale, the individual atoms are stiff. Surfactant at the surface of an oil droplet in water consists of linear molecules that align perpendicularly to the surface and from the liquid oil template the interfacial freezing of a 2D plastic crystal solid, at temperatures slightly below, at, or even above the bulk freezing point of the oil. The formation of such a stiff plastic crystal layer may counteract the surface tension of the whole droplet and form shapes far from the area-minimizing sphere. Let us examine how a droplet of oil capable of forming plastic crystal phases with amphiphilic molecules (amphiphiles or surfactants) at the surface can be a tensegrity structure. The overall surface tension is the tension component, while the thin layers of plastic crystal forming edges at some parts of the droplet interface provide the stiff elements, acting as struts (Fig. 2A). These struts, however, are not fixed, but can grow as their formation can be favorable under conditions of supercooling.

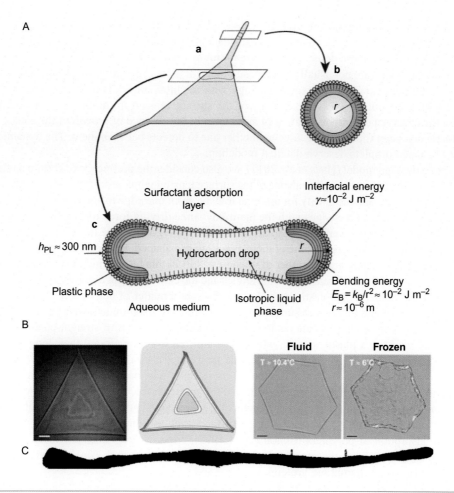

FIG. 2

(A) Schematic of alkane packing in the plastic crystal phase resulting in a cylindrical morphology. We don't yet know the reason the molecules pack in this fashion—with cylindrical packing—no curvature in one direction, a certain curvature in the other. In our case, the curvature is much lower than that of the nanotube on the right, so the diameter is microns instead of nanometers. The plastic phase of the alkane molecules results in a stiff element along each straight edge that balances the surface tension of the surfactant-coated oil droplet (the compressive element), making this a tensegrity structure (see Appendix). Inside the alkane (hydrocarbon), molecules are in a liquid phase. (B) Light micrographs showing the thicker border, and the thin film in the middle, of shaped droplets. In the triangle, a sizeable thin patch has formed inside, whereas in the hexagon it has stretched all the way. Sometimes in these shapes, because the middle film is so thin, they puncture before they can stretch all the way, creating holes. The second triangle is a sketch of the first. Scale bars = 20 μm.
(C) A binary mask of an electron micrograph cross section of a square Archaea made by hand using ImageJ from Fig. 6B in (Stoeckenius, 1981). The distance between the vertical tic marks is 1 μm. Mean thickness is $T = 0.204$ μm and length of the cross section is $L = 7.36$ μm.

(A) From Denkov, N., Tcholakova, S., Lesov, I., Cholakova, D., Smoukov, S.K., 2015. Self-shaping of oil droplets via the formation of intermediate rotator phases upon cooling. Nature 528(7582), 392–395, with permission of Nature Publishing Group.

In addition to the energetics of strut formation, there are discontinuities at the corners, with their own energies that are not fully clear. The persistence of common angles (60, 120 degrees) in these structures suggests that these defect energies are far higher than any local deformation energies during shape changes, so they are not usually influenced/changed. In the Appendix, we present a simple physical toy model that shows the tensegrity analogy with modular sidewalls built from magnetic discs, where the magnetic attraction is the analogy of the surface tension in the drops. In addition, it can illustrate the observed disproportionation of sides even in the absence of growth of the plastic crystal, as magnetic discs can flip from one edge to another due to defects between them. The dynamics at the corners of shaped droplets deserve detailed modeling.

In a more detailed model (Haas et al., 2017), we also consider the preference of certain angles in the observed structure and connect the balance of the surface tension with the stiffness of the rods of plastic crystal forming edges and the ability for material transfer from one edge to another. We also associated the overall growth rate of the edges with the thermodynamic driving force for formation of the plastic crystal. The model is able to predict the exact sequence of polygonal shape transformations, as well as the stable puncture that occurs in some of them, resulting in a shaped, topologically toroidal oil droplet (triangle in Fig. 2B). There are a number of unknowns still in this fascinating phenomenon, including the detailed origin of the preference for some angles and the exact packing of the molecules resulting in the observed curvature of the plastic crystal edges that are a matter of active investigation.

Usually, the stiffness of a frozen surfactant layer is orders of magnitude lower than the surface tension. However, when oil droplets are cooled slowly, a number of layers of straight-chain alkanes near the interface can form a plastic state in which they are free to rotate around their long axes but not translate ("rotator phase"). Though more flexible than other crystals, the stiffness of these regions is enough to counteract the surface tension of the droplet and impart a nonspherical shape to it (Figs. 2 and 3). Shaped droplets provide anisotropic flat surfaces which could have provided metabolic, mobility, and other advantages to oil-based protocells, as we'll see in the next section. Such shapes may have been stabilized further by reactions at the surface, such as mineralization by inorganic precipitation and adsorption of polymers (Simon et al., 1995), either hydrophobic polymers on the inside or hydrophilic polymers on the outside. Polymers with stretches of both could weave in and out of the surface like extant membrane proteins.

---

3 OIL-BASED PROTOCELLS

Our interest in the shape of oil droplets stems from increasing support for the hypothesis that life may have started from oil or lipid droplets in water (Sharov, 2016; Segré et al., 2001; Segré & Lancet, 2000). An oil droplet that absorbs a tar-like complex mixture of abiotic compounds has no membrane, yet is capable of movement in various ways (Hanczyc & Ikegami, 2010; Horibe et al., 2011) via internal oil flow (Hanczyc, 2011). Such oil droplets can hydrodynamically split themselves, imitating cell division without specific polymers being involved (Caschera et al., 2013; Tcholakova et al., 2017). See also (Banno et al., 2015; Turk-MacLeod et al., 2015).

We have demonstrated the interplay between stability and dynamics in a recursive spherical oil droplet division and fusion cycle (Caschera et al., 2013). In this system, starting far from equilibrium due to the solvation of a cationic surfactant in the organic phase and anionic surfactant in the aqueous phase, we observe a temporal window on the way to equilibrium due to surfactant diffusion at the

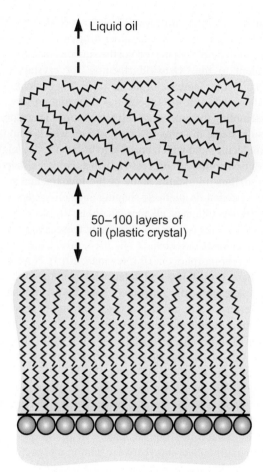

FIG. 3

Model of the plastic phase at the edges of shaped droplets. Tens of layers of molecules must order in these plastic crystal layers to overcome the interfacial tension keeping droplets otherwise circular. The molecules close the surface interdigitate with the surfactant tails, making a well-packed monolayer. That monolayer anchors the formation of the rotator (plastic crystal) phase, which progressively gets more and more disordered and eventually transitions to the liquid which is contained inside all these deformed droplets.

interface where the interfacial tension is minimized. During this time period, the droplet is easily affected by force, whether externally applied or internally generated. Often the droplet will spontaneously divide into daughter droplets due to internal fluid dynamic instability, which then round up as the system approaches a thermodynamic equilibrium.

Self-shaping droplets are nonlinear systems which convert the thermal fluctuations of the environment into phase-transition energy which, in its own turn, is used for droplet shape transformations and for drop division into smaller "daughter" droplets (Cholakova et al., 2016, 2017; Denkov et al., 2015; Tcholakova et al., 2017). The "feeding" of the daughter drops via coalescence with other drops or via capturing dissolved organic molecules transferred via the aqueous medium from smaller to larger droplets (in a process equivalent to the "Ostwald ripening" of crystals (Taylor, 1998; Tcholakova et al., 2011)) could lead to prebiotic increase of the drop size via basic physicochemical mechanisms. The freezing of the shaped droplets at low temperatures fixes and preserves their shape until the temperature is raised again, resembling the process of cell hibernation. Therefore, shaped droplets are probably the simplest chemical system, in terms of composition, which could appear naturally in water pools containing organic molecules and which are able to utilize the day/night (and other) natural thermal cycles for realization of basic protocell processes, such as complex shape-transformations, division and growth, and hibernation, thus mimicking some of the fundamental processes of living cells.

It is important to emphasize that droplet self-shaping and self-splitting have been observed with a wide variety of organic molecules (alkanes, alkanes, alcohols, triglycerides, esters, etc.) and with complex mixtures of such molecules. This is direct evidence that these are general phenomena, not bound to very specific classes of molecules (Cholakova et al., 2016, 2017), and shaped droplets could thus be called a "generic" mechanism for morphogenesis (Newman, 2014). Note that only basic physicochemical processes, such as phase transitions and interfacial phenomena, are involved in the above processes, without the complex molecular machinery used by evolutionary developed living organisms. Furthermore, all these physicochemical processes are highly reproducible and could serve as a basis for development of more subtle and complex processes, such as droplet attachment on specific substrates and accumulation of molecules on specific domains of the drop surface (e.g., on the edges, corners, broad surfaces, or tips of the shaped drops), thus allowing the beginning of a subsequent evolutionary development of more complex molecular machines.

Another remarkable feature of shaped droplets is that they adapt to the variations of the environmental energy and acquire shape and other related properties which correspond to equilibrium at the given temperature. Therefore, they do not need a constant influx of energy from the environment to maintain their "homeostasis," which is a typical feature of living organisms. From this viewpoint, shaped droplets appear as an intermediate state between natural inorganic matter, which has an entirely passive response to changes in the environment, and living organisms which use complex molecular mechanisms, requiring a regular influx and conversion of energy, to handle their homeostasis and to adapt to the variations in the external conditions.

One of the strongest arguments in favor of the oily origin of life is the relative abundance of oil-like organic molecules. There are potentially three main sources of organic material in places where life could have originated: arrival from space, energy-coupled chemistry in the atmosphere, and ocean floor vent chemistry. Investigations of the interstellar medium have produced evidence of short-chain alcohols, aldehydes, ketones, acids, aliphatic hydrocarbons, amines, amides, esters, ethers, cyanide derivatives of paraffin, as well as aromatic rings (e.g., benzene) and even large and complex carbon structures (e.g., cyanopolyynes and polyaromatic hydrocarbons (Henning & Salama, 1998; Kwok, 2007)). In the lab, the UV irradiation of interstellar ice analogs with water, methanol, CO, ammonia, and other components produced not only various organic molecules (Bernstein et al., 1995; Gerakines et al., 2001), but also structures resembling oil droplets and vesicles (Dworkin et al., 2001). Certain carbonaceous chondrite meteorites found on the Earth contain an organic crust with oily substances including aliphatic and aromatic hydrocarbons that may form oil droplets in water (Krishnamurthy et al., 1992; Yuen et al., 1984). Considering the ocean floor as a source of organic chemistry, two identified chemical processes may be at play: the coupled processes of serpentinization and Fischer-Tropsch Type synthesis. These chemical processes can produce short- and medium-chain alkanes (Holm & Charlou, 2001; Macleod et al., 1994; Schulte et al., 2006; Sherwood Lollar et al., 2002; Simoneit, 1995; Sleep et al., 2004), which could form oil droplets. Finally, the atmosphere of early Earth could have also been a source for organic chemistry and molecules. The spirit of this inquiry was captured in the famous Urey-Miller experiment first published in 1953 (Miller, 1953; Miller & Urey, 1959). Here, a simulated prebiotic Earth atmosphere was constructed in the lab and subjected to repeated electrical sparks to simulate lightening. Using such approaches and varying the initial conditions, studies have produced several different types of molecules such as amino acids and monocarboxylic acids (Miller, 1953; Rode, 1999), along with much more complex molecules including tar and oil phases (Allen & Ponnamperuma, 1967; Yuen et al., 1981). It is not clear if complex and relatively unstable organic molecules (e.g., amino acids and sugars) can reach biologically relevant concentrations without life (Sharov & Gordon, 2017). However, hydrocarbons, and especially alkanes, could become abundant on planets and form oil droplets in water. The addition of polar moieties in the synthesis of such organics could have produced a variety of simple surfactants (Hanczyc & Monnard, 2017).

There are some recent arguments emerging that support the idea of an oily origin of life. The reconstructed putative last common ancestral proteome appears to have high degree of hydrophobicity, with evolution following an "oil escape" towards more water-friendly proteins (Mannige, 2013; Mannige et al., 2012). This does not necessary mean that ancestral proteins were embedded in an oil droplet. It could simply indicate that proper protein folding largely depends on the hydrophobic effects in an aqueous medium. Nevertheless, the functioning of proteins has been demonstrated in nonaqueous solvents (Klibanov, 2001; Zaks & Klibanov, 1985). In some cases, the yield and longevity of an enzymatic reaction is improved when embedded in largely organic media (Stolarow et al., 2015). Further investigation of enzymes in organic solvents revealed that some enzymes acquire new properties including greater stability, altered selectivity, and molecular memory (Klibanov, 2003). Therefore, it is not only possible for some of the basic biochemical functionalities to be hosted in oil droplets, but perhaps advantageous.

Many researchers have contemplated a "lipid world" scenario for the origin of life where the importance of the container for the first protocells is considered (Anella & Danelon, 2014; Bar-Even et al., 2004; Bukhryakov et al., 2015; Cavalier-Smith, 2001; Kauffman, 2013; Lombard et al., 2012; Paleos, 2015; Paleos et al., 2004; Sharov & Gordon, 2017; Szathmáry, 2006; Szathmáry et al., 2005; Wieczorek, 2012). This is based on the spontaneous self-assembly of both primitive and evolved surfactant molecules into higher order structures. Such supramolecular structures include oil droplet emulsions, bilayer lamellar vesicles, micelles, and others. These supramolecular assemblies not only consist of single amphiphiles, but also mixtures that are often notably more stable under varying environmental conditions (Hanczyc & Monnard, 2017). Such mixtures of amphiphilic molecules in supramolecular structures could carry heritable compositional information as an early and primitive form of protocellular information (Gross et al., 2014; Hunding et al., 2006; Markovitch & Lancet, 2014; Naveh et al., 2004; Segré et al., 2001; Segré & Lancet, 2000; Shenhav et al., 2004, 2005; Wu & Higgs, 2008). Additionally, heritable information in such scenarios could have been represented by catalytically active self-reproducing molecules, inside or on the surfaces of oil droplets in water (Sharov, 2016).

One scenario to consider is a transition from an oil droplet-based protocell to a water-centric vesicle that reflects the current biological cell membrane architecture. Perhaps, the oil droplet protocell came first as it was easier to form and more robust. This oil droplet contained some essential biochemical machinery. Persistent shapes, whether repeatedly formed or stabilized, would have given such proto-cells advantages in hydrodynamic mobility, surface-area-based feeding/metabolism, resistance to being merged/diluted with other droplets, and driven a kind of protocell evolution (Young, 2006, 2007, 2010). This oil-first model was then overtaken by a cell architecture based on a thin lipid membrane with an aqueous interior, which became the obviously dominant form of known life. Such a scenario faces difficult challenges which still must be explored. Here, we simply note the similar lenticular shape of both shaped droplets and polygonal Archaea, including the development of holes or closely apposed membranes where the two surfaces meet (Figs. 2b and 4d). Stabilization of double droplets or "water-in-oil-in-water (W/O/W) emulsions" (Leong et al., 2017) affords a model for the transition from droplets to membrane-bound vesicles (Chong et al., 2015).

4 POLYGONAL PROKARYOTES

Polygonal shapes of cooled oil droplets described above appear surprisingly similar to the polygonal prokaryotes. With the first discovery of a square "bacterium" in a natural pond of evaporating marine water in the Sinai (Walsby, 1980), now recognized as an Archaea (Bolhuis, 2005), a worldwide search for halophilic polygonal prokaryotes began. The following geometrically regular shapes have been reported across taxa (Gupta et al., 2015):

Flat Archaea:

1. Triangles (Andrade et al., 2015; Baxter et al., 2005; Burns et al., 2004; Castillo et al., 2006; Chaban et al., 2006; Emerson et al., 1994; Grant & Larsen, 1989; Hamamoto et al., 1988; Horikoshi et al., 1993; Javor et al., 1982; Lasbury, 2013; Malfatti et al., 2009; Mehrshad et al., 2015, 2016; Miyashita et al., 2015; Mullakhanbhai & Larsen, 1975; Nishiyama et al., 1992; Oren, 1999; Oren et al., 1990, 1999; Otozai et al., 1991; Ozawa et al., 2000, 2005; Sabet et al., 2009; Takashina et al., 1990; Wakai et al., 1997; Walsby, 1980; Yang et al., 2007; Yatsunami et al., 2014)
2. Squares (Andrade et al., 2015; Bardavid & Oren, 2008a; Baxter et al., 2005; Bolhuis et al., 2006; Burns et al., 2004, 2007; Castillo et al., 2006, 2007; Chaban et al., 2006; Dyall-Smith et al., 2011; Fredrickson et al., 1989; Ghai et al., 2011; Grant & Larsen, 1989; Hamamoto et al., 1988; Javor et al., 1982; Kamekura, 1998; Lasbury, 2013; Lobasso et al., 2008; Malfatti et al., 2009; Mullakhanbhai & Larsen, 1975; Oh et al., 2010; Oren, 1994, 1999, 2005; Oren et al., 1990, 1996, 1999; Romanenko, 1981; Santos et al., 2012; Torrella, 1986; Tully et al., 2015; Walsby, 1980, 1994; Yang et al., 2007)
3. Rectangles (Dyall-Smith et al., 2011; Javor et al., 1982; Mullakhanbhai & Larsen, 1975; Oren, 1999, 2005; Oren et al., 1996)
4. Rhombi (Hamamoto et al., 1988; Oren, 1999; Takashina et al., 1990)
5. Trapezoids (Oren, 1999, 2005; Walsby, 1980)
6. Pentagons (Malfatti et al., 2009)
7. Circles, ovals, plates, disks (Duggin et al., 2015; Mehrshad et al., 2016; Mullakhanbhai & Larsen, 1975; Stetter, 1982)

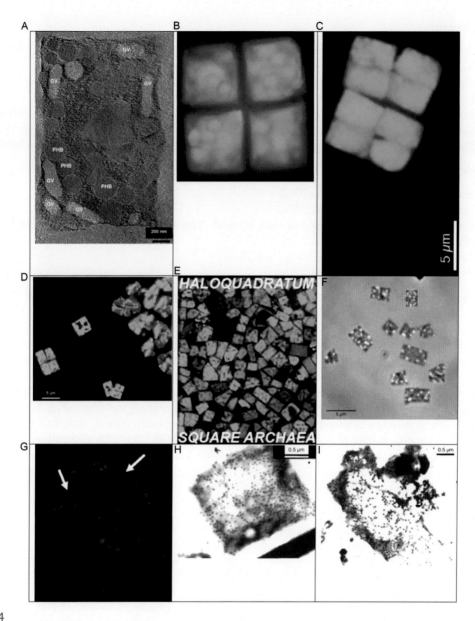

FIG. 4

Square Archaea. (A) Electron tomograph by H. Engelhardt of a "Spanish isolate of the square halophilic archaeon *Haloquadratum walsbyi* strain HBSQ001." GV=gas vesicle, others "most likely poly-3-hydroxy-butyric acid (PHB) polymers." (B) Four cells of *Haloquadratum walsbyi* in the "postage stamp" array. (C) 8-cell colony. Note the sharp corners and the parallel third divisions, suggesting cell-cell communication. Green fluorescence image with acridine orange staining per (Burns & Dyall-Smith, 2006; Burns et al., 2004). (D, E) *H. walsbyi*:

(Continued)

Flat Bacteria:

8. Triangles (Awramik & Barghoorn, 1977; Fritz et al., 2004; Lafitskaya & Vasilieva, 1976; Vasilyeva & Semenov, 1984, 1985)
9. Squares (Oren, 1999; Whang & Hattori, 1990)
10. Rectangles (Kuhn, 1981; Whang & Hattori, 1990)
11. Stars (Chernykh et al., 1988; Fritz et al., 2004; Hirsch, 1974; Hirsch et al., 1977; Hirsch & Schlesner, 1981; Reimer & Schlesner, 1989; Rusconi et al., 2013; Semenov & Vasilyeva, 1987; Staley, 1968; Vasilyeva, 1970, 1985; Vasilyeva et al., 1974; Vasilyeva & Semenova, 1986)

Three-dimensional bacteria:

12. Hexagonal columns (Wu et al., 2012)
13. Heptagonal columns (Wu et al., 2012)
14. Corrugated columns (star-shaped in cross section) (Lasbury, 2013; Wanger et al., 2008; Wu et al., 2012)
15. Tetrahedra (Fritz et al., 2004)
16. Pyramids (Baxter et al., 2005)

Some halophiles are reported as pleomorphic, yet do not show polygonal geometries (Liao et al., 2016; Mori et al., 2016; Shimane et al., 2015; Wang et al., 2016b; Xu et al., 2016; Yin et al., 2015), while others are rod-shaped (Mou et al., 2012) or sometimes so (Duggin et al., 2015). In Fig. 5, we have catalogued a number of three-dimensional polygonally faceted and corrugated shapes, because their structures may provide hints of mechanisms also applicable to flat cells. The straightness of edges can vary with fixation technique and resolution of microscopy (cf. Rusconi et al., 2013). We include flat, circular Archaea, as equivalent to polygons with an infinite number of edges, per Archimedes (Aktümen & Kaçar, 2007). Most of the bacteria have rounded corners and edges (Fig. 5).

There are now a few different polygonal Archaea species distinguished, including the original (Walsby, 1980) now named *Haloquadratum walsbyi* (Burns et al., 2007; Dyall-Smith et al., 2011), *Haloarcula quadrata* (Burns et al., 2004; Oren et al., 1999; Yang et al., 2007), *Haloarcula argentinensis* (Yang et al., 2007), *Haloarcula japonica* (Takashina et al., 1990; Yang et al., 2007), *Haloarcula*

FIG. 4—Cont'd

Note that in the same culture, triangular, square, rectangular, trapezoidal, quadrilateral, and pentagonal cells may be seen. Note that two cells in each image appear to have holes. (F) The white spots in the image are gas vesicles. (G) Arrows indicate two "holes" in the fluorescence image of an *H. walsbyi* cell stained with Nile Blue for its intracellular polyhydroxybutyrate granules. (H) "Square Archaea with phage particles, head diameter ca 40 nm." (I) "Square Archaea lysed by phages, head diameter ca 50 nm." Transmission electron micrographs.

(A) From Bolhuis, H., Palm, P., Wende, A., Falb, M., Rampp, M., Rodriguez-Valera, F., Pfeiffer, F., Oesterhelt, D., 2006. The genome of the square archaeon Haloquadratum walsbyi: life at the limits of water activity. BMC Genomics 7, #169, with permission from Biomed Central under a Creative Commons Attribution License (CC BY); (B) From Wikipedia, 2008. File:Haloquadratum walsbyi00.jpg. https://commons.wikimedia.org/wiki/File:Haloquadratum_walsbyi00.jpg, in public domain; (C, D, E) from Mike L. Dyall-Smith with his kind permission; (G) From Zenke, R., von Gronau, S., Bolhuis, H., Gruska, M., Pfeiffer, F., Oesterhelt, D., 2015. Fluorescence microscopy visualization of halomucin, a secreted 927 kDa protein surrounding Haloquadratum walsbyi cells. Front. Microbiol. 6, #249, with permission under a Creative Commons Attribution License (CC BY); (H, I) From Guixa-Boixereu, N., Calderón-Paz, J.I., Heldal, M., Bratbak, G., Pedrós-Alió, C., 1996. Viral lysis and bacterivory as prokaryotic loss factors along a salinity gradient. Aquat. Microb. Ecol. 11(3), 215–227, with permission under a Creative Commons by Attribution Licence (CC-BY).

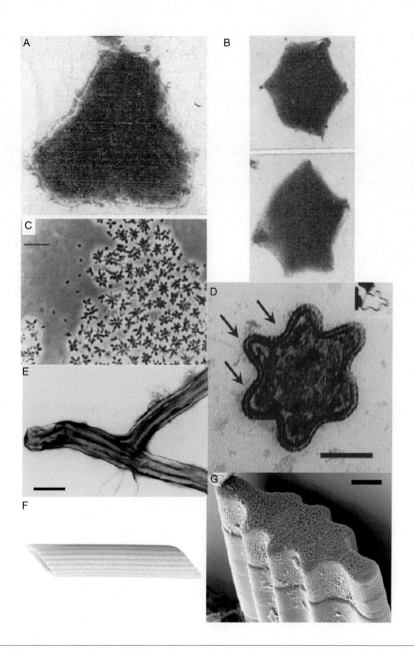

FIG. 5

Polygonal bacteria seem not to have straight edges, nor sharp corners. (A) Triangular bacterium, 0.7–0.9 μm in width. (B) *Stella humosa*, star-shaped bacterium. See also *Prosthecomicrobium*: "Each has six appendages extending in one plane" (Staley, 1968). (C) Rosette budding bacteria with "a stable protein envelope." Scale bar = 10 μm. (D) This is a star-shaped cross section of a bacterium (E) whose corrugated cell wall resembles penne rigate pasta (F). This example has 6 "points." Cells were reported with 4, 5, 6, 7, and 9 "points." "The Stars reproduce by forming branching structures. . . . How the Stars manage to hold their shape is a puzzling

(Continued)

marismortui (Oren et al., 1990), *Haloferax* sp. (Emerson et al., 1994), *Halovivax asiaticus* (Castillo et al., 2006), *Halosiccatus urmianus* (Mehrshad et al., 2016), *Halogeometricum borinquense* (Malfatti et al., 2009); most containing gas vesicles (Walsby, 1994), some not (Javor et al., 1982). The star-shaped *Stella vacuolata* and *Stella humosa* are gram-negative bacteria (Vasilyeva, 1985).

Square Archaea represent over 50%–60% of the cells found in crystallizer ponds of salterns worldwide (Antón et al., 1999; Bardavid et al., 2008; Bettarel et al., 2011; Bolhuis, 2005; Mutlu & Guven, 2015), though sampling may have been done at times of blooms (Ram-Mohan et al., 2016). They are present, but not predominant, in the Dead Sea (Bodaker et al., 2010) and the Great Salt Lake (Baxter et al., 2005). The square archaeon has a thickness of as little as 0.1 μm at the edge "or even less in the central regions" (Parkes & Walsby, 1981), "possibly even as thin as 0.07 μm" (Oren, 1999). Thus, it is lenticular in shape (Fig. 2C), like shaped droplets (Fig. 2A,B), although somewhat irregular in thickness away from the square edges, perhaps due to cell inclusions and/or fixation artifacts (Stoeckenius, 1981). In fact, the thickness seems to go to near zero in some places, "where the opposing cell membranes are in contact," without affecting the shape of the cell (Zenke et al., 2015), again as observed in shaped droplets (Figs. 2 and 4).

Square Archaea have been reported as 0.25 μm thick for all sizes, with squares of 2 μm and 4 or 5 μm or rectangles of 2 μm by 4 μm, and in electron microscopy cross section, with a lenticular shape (Stoeckenius, 1981). Rectangular cells have been imaged as small as 1.5 μm by 2.1 μm (Fig. 1P), and squares have been reported (Oren, 1999; Romanenko, 1981) as small as 1.4 μm. (A square/rectangular bacterium is reported even smaller as 0.3–2 μm on a side (Whang & Hattori, 1990).) Pure cultures yield square cells of different sizes. The variability in size with congruent shapes suggests that there is no rigid component of a specific length that accounts for the straight edges.

Except for a brief mention of "triangles/pyramids" and the presence of spheres in a collected population containing squares (Baxter et al., 2005), we found no reports that any of the flat Archaea have alternative three-dimensional shapes. While prokaryotes in general can come in many shapes (Gordon & Gordon, 2016a; Huang et al., 2008; Margolin, 2009; Yang et al., 2016; Young, 2007), the flat Archaea have only been observed flat. Star-shaped bacteria change their morphology in some culture conditions (Semenov et al., 1989; Semenov & Vasilyeva, 1985, 1987; Vasilyeva & Semenova, 1986), though the authors did not report that flat cells became three-dimensional.

FIG. 5—Cont'd

question. . ... No internal skeleton was observed within the Star. In some plasmolyzed cells. . ., the inner membrane detached from the outer [S-]layer, however, the star-shape was maintained." Scale bar for cross section = 0.1 μm. Scale bar for side view = 0.2 μm. (G) The columnar, corrugated-centric diatom *Terpsinoë̈ musica*, collected from Whitefield Creek (by the Salton Sea), California. It shows a flat valve face at the top. Scale bar = 20 μm. Cf. (Gordon & Tiffany, 2011).

(A) From Lafitskaya, T.N., Vasilieva, L.V., 1976. A new triangular bacterium. Mikrobiologiya 45(5), 812–816, with kind permission of Lina V. Vasilyeva; (B) From Vasilyeva, L.V., Lafitskaya, T.N., Aleksandrushkina, N.L., Krasilnikova, E.N., 1974. Physiological-biochemical peculiarities of the prosthecobacteria Stella humosa and Prosthecomicrobium sp. Izv. Akad. Nauk SSSR Seri. Biol. 5, 699–714, also with permission of Lina V. Vasilyeva; (C) From König, E., Schlesner, H., Hirsch, P., 1984. Cell wall studies on budding bacteria of the Planctomyces/Pasteuria group and on a Prosthecomicrobium sp. Arch. Microbiol. 138(3), 200–205, with permission of Springer; (F) From Wanger, G., Onstott, T.C., Southam, G., 2008. Stars of the terrestrial deep subsurface: a novel 'star-shaped' bacterial morphotype from a South African platinum mine (Corrigendum: (6), 421). Geobiology 6(3), 325–330, with permission of John Wiley and Sons.

5 MECHANISMS CONTROLLING THE SHAPES OF PROKARYOTE CELLS

Spherical oil droplets become shaped droplets when the temperature is lowered. Polygonal Archaea become rounded when the salinity is lowered. If we look at the plastic phase of shaped droplets and the polygonal shaping of Archaea as phase transitions or precipitations (Kim & Bae, 2003), then they may be examples of a generic process such as is seen in Hofmeister series ranking anions and cations according to their influence on macroscopic properties (Schwierz et al., 2016).

Archaea have a cell wall external to the cell membrane called the S-layer, and some internal cytoskeletal components. There are four general factors that may cause polygonal shapes in Archaea:

1. The S-layer creates the shape of the chamber within which the cell lives.
2. The cell membrane creates the shape.
3. The cytoskeleton creates the shape.
4. Two or three of the above, in concert, create the shape.

Most of these options have been proposed but not rigorously tested, so we are left with evaluating observations that may bear on the question. However, strangely, a nonstructural, physical "mechanism" has been suggested, namely turgor pressure. So we'll start with that.

In square Archaea living in saturated salt environments, it has been argued that there is no turgor pressure, and that this somehow "allows" the cells to take on nonspherical shapes (Kessel & Cohen, 1982; Walsby, 1980, 2005). The experiment of osmotically shocking (Berthaud et al., 2016) square cells to watch how they change shape has not yet been done, nor have they been subjected to atomic force microscopy and other physical probes. However, cells lysed by phage (viruses) retain sharp corners (Guixa-Boixereu et al., 1996) (Fig. 4I).

Next, let's consider the cytoskeleton. A form of actin, crenactin, and an associated protein form a helical microfilament (Braun et al., 2015) that correlates with the rod shape of the archaeon *Pyrobaculum calidifontis* (Ettema et al., 2011). Electron cryotomography shows no internal cytoskeletal structures adjacent to the cell membrane, including at corners in square *Haloquadratum walsbyi*, which have radii of curvature as low as 50 nm (measured from Fig. 1E in (Burns et al., 2007)). Furthermore:

> The extreme flatness of the cells may imply the complete absence of cell turgor but the forces that cause cell edges to be as straight as they are observed are currently enigmatic since no genes could be identified in the genome that might express structural proteins involved in maintaining the cell structure

(Zenke et al., 2015).

Next, let's consider the cell membrane as a possible source of polygonal structure. In general, the membrane lipids of bacteria are unipolar, while those of Archaea are bipolar (Fig. 6). The lipids of square Archaea, in particular, have been elucidated (Lobasso et al., 2008; Oren, 1993; Oren et al., 1996). The membrane lipids of polygonal Archaea could, by themselves, provide geometric structure in a couple of ways. For example, it is possible that specific lipids accumulate at or form membrane regions of high curvature (Boekema et al., 2013; Mukhopadhyay et al., 2008). The bipolar lipids of some Archaea (Figs. 6 and 7) by themselves tend to form flat (Chong, 2010), stiff (Jacquemet et al., 2009; Jain et al., 2014; Kates, 1992) structures, but when mixed with unipolar lipids, the latter can partition more to one side, creating a curved surface (Lelkes et al., 1983). This phase separation of lipid species (Liu et al., 2006) has been used to hypothesize how bacteria and

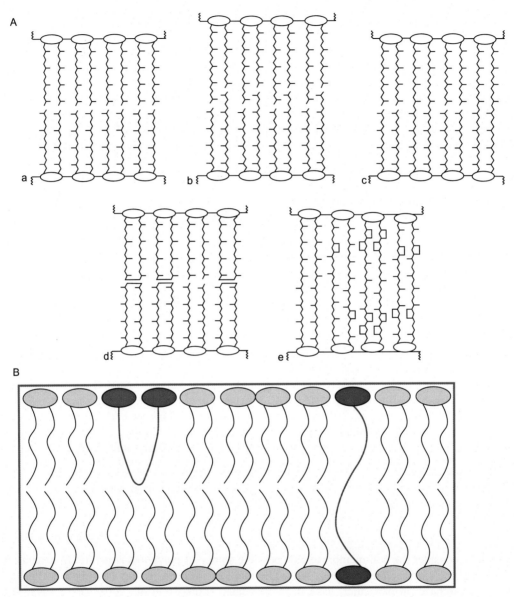

FIG. 6

(A) Various configurations of polar lipids found in Archaea. Those which covalently subtend the whole membrane are called bipolar and represent about 90% of Archaea membranes (Forbes et al., 2006). (B) An additional configuration for bipolar lipids, also called bolaamphiphiles, is the U-shape, where both polar groups are on the same side of the membrane.

(A) From Gambacorta, A., Gliozzi, A., Derosa, M., 1995. Archaeal lipids and their biotechnological applications. World J. Microbiol. Biotechnol. 11(1), 115–131, with permission; (B) From Forbes, C.C., DiVittorio, K.M., Smith, B.D., 2006. Bolaamphiphiles promote phospholipid translocation across vesicle membranes. J. Am. Chem. Soc. 128(28), 9211–9218, with permission of the American Chemical Society.

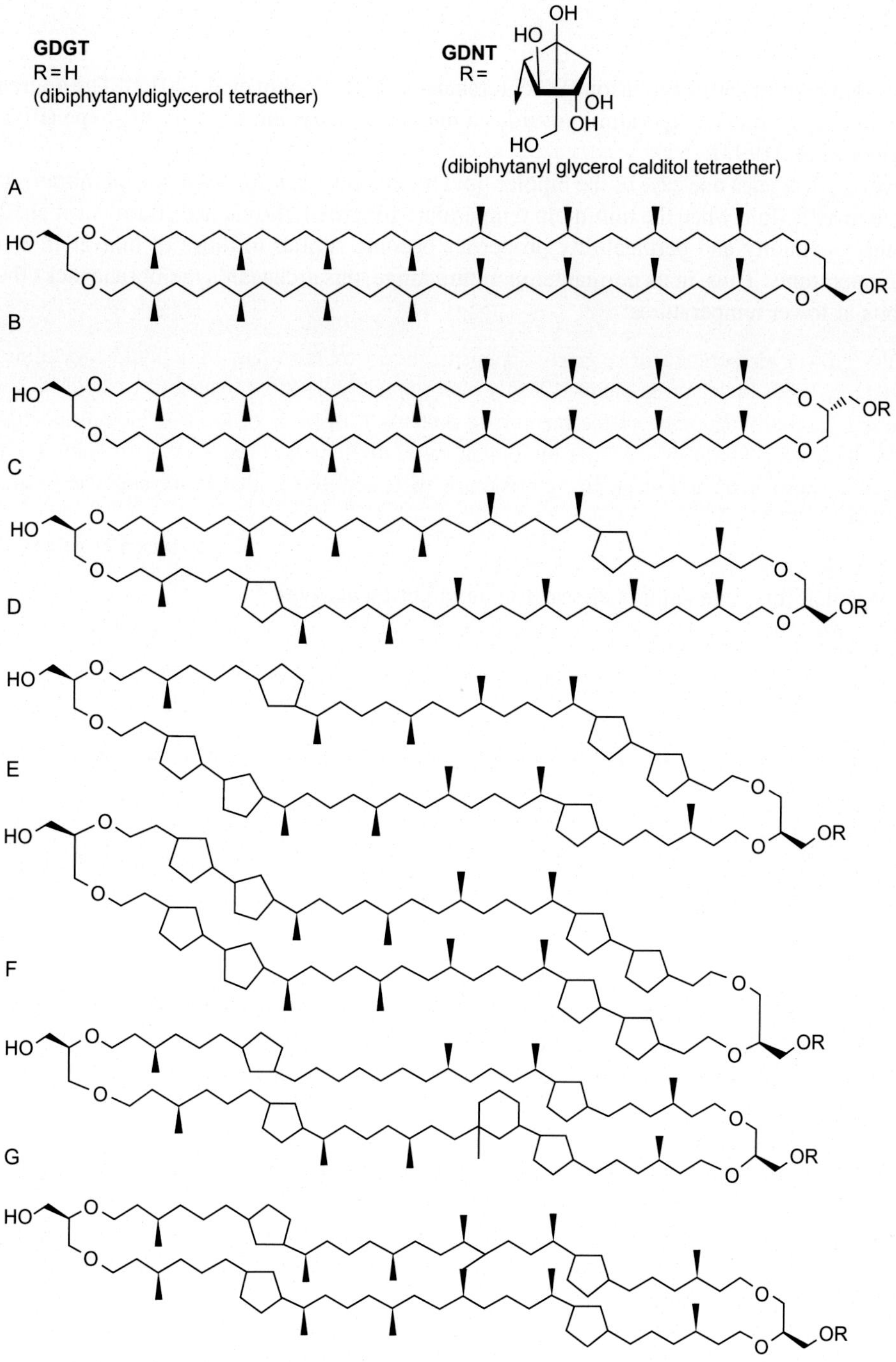

GDGT
R = H
(dibiphytanyldiglycerol tetraether)

GDNT
R =
(dibiphytanyl glycerol calditol tetraether)

FIG. 7

Examples of tetraether bipolar lipids in Archaea membranes.

From Jacquemet, A., Barbeau, J., Lemiègre, L., Benvegnu, T., 2009. Archaeal tetraether bipolar lipids: structures, functions and applications. Biochimie 91(6), 711–717, with permission of Elsevier Paris.

Archaea split apart in early evolution (Wächtershauser, 2003; Yokoi et al., 2012). The degree of fluctuations in shape versus temperature provides a means of analyzing stability of shape (Bivas, 2010; Döbereiner et al., 1997).

However, in at least one case of the bipolar lipid membrane of an Archaea, the membrane makes the transition to a 2D fluid when the minimum temperature for growth is reached (Bartucci et al., 2005). At that point, its fluidity and permeability properties become similar to those of bilayer membranes at lower temperatures. Thus, in its normal temperature range, this archaean's membrane lacks the rigidity it exhibits at lower temperatures:

> At the growth temperature of a given organism, the membranes are in a liquid-crystalline state (Melchior, 1982; Melchior & Steim, 1976) which implies substantial dynamics of lipid movement, ensuring optimal functioning of the membrane proteins.... Because of the low melting point of the archaeoland caldarchaeol-based polar lipid membranes of Archaea, these membranes are in a liquid-crystalline phase over a wide temperature range of 0 and 100°C that is physiologically relevant (De Rosa et al., 1986)
>
> **(Driessen & Albers, 2007).**

A caveat is that the same authors appear to contradict themselves:

> ...monolayer type of organization gives the membrane a high degree of rigidity (Bartucci et al., 2005; Elferink et al., 1994; Fan et al., 1995; Gliozzi et al., 1983; Mirghani et al., 1990; Thompson et al., 1992)
>
> **(Driessen & Albers, 2007).**

In any case, the liquidity of the 2D lipid array in polygonal Archaea needs further investigation, especially since it is possible that they may be able to adjust their membrane lipid composition to changing conditions:

> Moreover, bacteria can also stabilize their membranes at high temperatures through elongation of membrane lipids (Marr & Ingraham, 1962).... Surprisingly, little is known about membrane adaptation mechanisms of Archaea to temperature
>
> **(Andrade et al., 2015).**

If membrane proteins form condensations, i.e., two-dimensional phase transitions, this is the phenomenon of "capping" of membrane proteins that are otherwise mobile in the 2D liquid cell membrane which might provide structure. Capping appears to require attachment to the underlying cytoskeleton (Kindzelskii et al., 1994). Thus, it would seem an unlikely candidate for producing the long range, regular order of polygonal Archaea.

Prokaryotes are often surrounded by 2D crystalline glycoprotein cell walls called S-layers. Square Archaea have a glycoprotein S-layer (König, 1994) that appears to be a hexagonal array with a spacing of 20 nm (Parkes & Walsby, 1981; Stoeckenius, 1981; Sumper, 1993). Electron microscopy shows one S-layer evenly around a corner of *Haloquadratum walsbyi* and two S-layers around a new *Haloquadratum* species (Burns et al., 2007).

The S-layers, if most prokaryotes, are presumed, without proof, to cause and/or maintain the shape of the cell within (Akca et al., 2002; Sleytr et al., 1986). This notion has been contradicted in most cases:

> ...observations that the loss of S-layers is not accompanied by a morphological change of the progeny cells make it most unlikely that in organisms possessing a rigid cell envelope component, the paracrystalline arrays have an important function in determining cell shape
>
> **(Sleytr & Messner, 1983).**

An exception is the *Halobacterium salinarum* cell, rod-shaped at high salt concentrations (Sleytr & Messner, 1983), which becomes spherical on proteolysis or alteration of the S-layer (Mescher & Strominger, 1976) or at lower salt (Engelhardt, 2007b).

For polygonal Archaea, the following is the closest we have to a published model, which invokes the cell membrane and the S-layer:

> If the cells are not exposed to osmotic imbalance, and the lipids or other cellular factors... do not force strong prebending of the membrane, the S-layer-membrane assembly would assume a shape approaching the tension minimum, i.e., a flat, sheet-like arrangement in the ideal case. A model system shown in (Engelhardt, 2007a) illustrates the expected effect. The reconstituted S-layer of Delftia acidovorans (p4 symmetry, lattice constant 10.5 nm), interacts with the membrane via bound lipid molecules (Engelhardt et al., 1991) and flattens the symmetrical lipid layer of the vesicles. Interestingly, disk-like species such as H. volcanii and Methanoplanus limicola, or Archaea forming flat cellular boxes ("Square Bacterium") do possess S-layers with p6 symmetry and particularly short lattice constants (14.7–16.8 nm...). Although other, possibly unknown mechanisms might contribute to the architecture of Archaeal cells, it is conceivable that the biophysical properties of the S-layer-membrane assembly are sufficient to flatten cells (Engelhardt, 2007a)
>
> **(Engelhardt, 2007b).**

> The consensus idea at the moment is that the shape is most likely determined by the rigidity of the S-layer in combination with the absence of turgor pressure.... Kessel and Cohen (Kessel & Cohen, 1982) noted from their electron microscopic images that the edges are curved or rounded rather than straight, as would be expected for a perfect square or rectangular box with a very narrow width (cf. Fig. 2A). They envisioned the organism as a flattened cylinder, which, to my opinion, is the most likely explanation for the square shape
>
> **(Bolhuis, 2005).**

The idea then is that the source of flattening of the cell is indeed the S-layer, because it is intrinsically flat and attaches to the cell membrane (Engelhardt, 2007b; Engelhardt & Peters, 1998). One way of investigating this question is to see if there is a fixed or arbitrary angular relationship between the 2D crystal axes of the S-layer and the edges of the square cell. "However, it is generally thought that in halobacteria, the cell wall determines the cell shape, and it is difficult to reconcile the hexagonal lattice with the rectangular shapes of the cells" (Stoeckenius, 1981). A less definitive observation in this regard is that flagella of square cells of *Haloarcula quadrata* (Chaban et al., 2006; Grant & Larsen, 1989; Oren et al., 1999) and the "extracellular fibrils" of *Haloquadratum* sp. (Santos et al., 2012) occur at arbitrary positions along the cell's perimeter, not preferentially at an edge or corner (Alam et al., 1984). (*Haloquadratum walsbyi* lacks flagella (Bolhuis, 2005)). Damage to "the cell envelope" by the bile salt taurocholate (Bardavid & Oren, 2008b) might, at low doses, prove useful to separating the roles of the S-layer and the cell membrane. Note that "no direct genetic indication was found that can explain how this peculiar organism retains its square shape" (Bolhuis et al., 2006). S-layer self-assembly gives various cylindrical, tubular, and ribbon shapes (Pum & Sleytr, 2014), but nothing resembling a flat polygon. While the tubulin-related CetZ1-GFP "localizations are envelope associated" in *Haloferax volcanii*, this species is disc- or rod-shaped (Duggin et al., 2015), not polygonal.

In regard to membrane/S-layer interaction, we arrive at the conclusion that it is not sufficient to generate polygonal shapes:

The biophysical consequences of synergistic functions between S-layers and the cell membrane in Archaea, and the outer membrane or the peptidogylcan in bacteria are still largely unexplored. . . . Experimental data on natural systems are rare. . . . The lipid molecules are immobilized indirectly by non-specific association to the S-layer protein whereby the membranes become less fluid, less flexible, more stable and heat-resistant, and presumably more resistant to hydrostatic pressure. . . . Natural association of S-layers with the cytoplasmic membrane occur in Archaea only. . . . that up to 5% of the lipids may be immobilized by interactions with the S-layer anchor, disregarding other proteins. The important point is that the anchors, unlike common membrane proteins, do not freely float in the lipid phase. . . . Taken together, it becomes evident that the inherent properties of S-layers are probably insufficient to constitute a distinct shape-determining function by themselves, beyond that of a passive shape-modifying effect. There is obviously a need for additional structural or functional ingredients. . . . The conclusion from these considerations is that S-layers may be shape-maintaining and shape-modifying but they are not shape-determining in a strong (process-related) sense.

(Engelhardt, 2007a).

There are alternatives to the S-layer that have been considered. It has been suggested that: "A cross-linked matrix of poly-gamma-glutamate may contribute to the cell wall rigidity and maintenance of the unique square morphology" (Lobasso et al., 2008), though there is no proof this matrix exists in *Haloquadratum walsbyi* (Albers & Meyer, 2011; Bolhuis et al., 2006; Wu et al., 2012). Halomucin has been invoked in the "water enriched capsule," with speculation on its rigidity (Bolhuis et al., 2006). Later work showed halomucin only "loosely attached to cells" (Zenke et al., 2015).

Note that "the sensitivity of the cells to mechanical shearing" and "flexibility of the larger cell structures... rarely found in an unfolded state" (Bolhuis et al., 2004), also described as "extremely fragile" (Dyall-Smith et al., 2011), suggests rigidity, because each folded section remains flat (Fig. 8), with some sort of crease at the folding line, perhaps like stiff paper. Flat, disc-shaped Archaea are also fragile (Stetter, 1982). In contrast to square Archaea, bacterial cell walls can be highly elastic and capable of resisting turgor pressures up to 30 atmospheres (Hemmingsen & Hemmingsen, 1980).

Some bacteria are long ribbons that are star-shaped in cross section only (Fig. 5), resembling the presumed buckling patterns of some "corrugated" diatoms (Gordon & Tiffany, 2011). This led us to consider whether most of the patterns of single Archaea and bacteria listed above might be due to variations on a single mechanism, known as wrinkle/ridge (Jin et al., 2015) or buckling/folding (Schmalholz & Podladchikov, 1999) transitions (Fig. 9). Buckling has been invoked for shaping of one Archaea (Pum et al., 1991).

A cell is topologically a sphere. Here we would like to propose that polygonal Archaea may best be modeled as deflated spheres of a material, probably the S-layer, whose equilibrium shape, were it not constrained by a spherical geometry, would be planar. This property is evident in images of folded cells, which are creased rather than undulated (Fig. 8). A deflated soccer ball provides a qualitative model, from which we can see, at least for a material with the constitutive properties of a soccer ball, that various polygonal shapes with reasonably straight edges and sharp corners do develop (Fig. 10). This effect has recently been investigated (Quilliet et al., 2008) and polygons develop on the surfaces of spheres as their volume decreases below critical values (Knoche & Kierfeld, 2011, 2014b,c) (Fig. 11). Unfortunately, none of the simulations or experiments has yet been taken to the extreme found in shaped droplets and flat Archaea, so while cross-sectional sketches of the collapsed shapes are available (Fig. 11), we do not know if the full three-dimensional shapes will reflect those of polygonal Archaea or those of collapsing shaped droplets (Fig. 12).

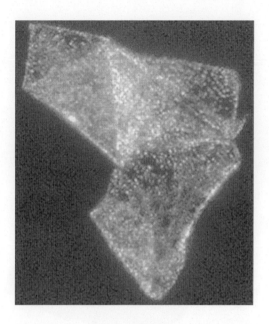

FIG. 8

"Darkfield microscopic image of a large folded sheet of 'Haloquadratum walsbyi,' revealing its sharp edges and straight corners. Size is ~40 × 40 μm. White spots are gas vesicles."

From Bolhuis, H., 2005. Walsby's square archaeon—it's hip to be square, but even more hip to be culturable. In: Gunde-Cimerman, N.,

Oren, A., Plemenitaš, A. (Eds.), Adaptation to Life at High Salt Concentrations in Archaea, Bacteria, and Eukarya. Springer, Dordrecht,

The Netherlands, pp. 185–199, with permission of Springer.

To check where on the volume reduction scale square Archaea fall, for instance, let's consider one cell for which we have both width (L) and thickness (T) measurements (Fig. 2C), approximating it as a flat, square box with rounded edges. The cell volume is:

$$V = T(L-T)^2 + 2\pi TL \approx TL^2,$$

and the surface area is:

$$A = 2(L-T)^2 + 2\pi TL \approx 2L^2$$

assuming a half cylindrical shape of the edges. Here, we are approximating the cell as a flat, rather than lenticular shape, so that $T = 2r$, where r is the radius of a cross section of an edge (Fig. 2A). A sphere of equivalent area has a radius R calculated from:

$$A = 4\pi R^2$$

Its volume is:

$$V_0 = (4/3)\pi R^3 = L^3 / \left(3\sqrt{\pi/2}\right)$$

Thus, the abscissa in Fig. 11B is:

$$V/V_0 = \left(3\sqrt{\pi/2}\right) T/L$$

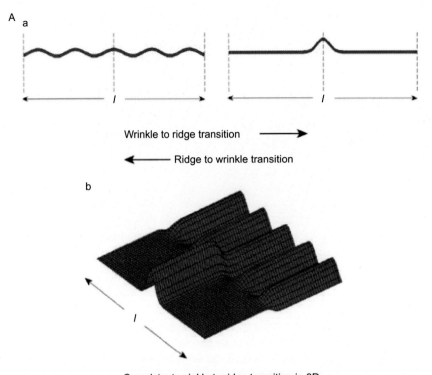

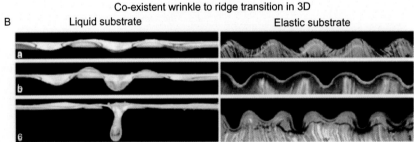

FIG. 9

(A) Whether a surface makes rolling wrinkles or produces sharp folds with nearly straight edges depends on prestress and nonlinear mechanical properties of the substrate under the sheet of buckling material, in the case of prokaryotes, the cell membrane/S-layer, and the cytoplasm mechanical properties. The transition is unstable, i.e., snapping occurs between the two states of a surface (Takei et al., 2014). (B) Folding (left) or buckling (right) depends on the viscosity or elasticity of the substrate, or more generally, on its viscoelasticity.

(A) From Jin, L., Takei, A., Hutchinson, J.W., 2015. Mechanics of wrinkle/ridge transitions in thin film/substrate systems. J. Mech. Phys. Solids 81, 22–40, with permission of Elsevier; (B) From Brau, F., Damman, P., Diamant, H., Witten, T.A., 2013. Wrinkle to fold transition: influence of the substrate response. Soft Matter 9(34), 8177–8186, with permission of the Royal Society of Chemistry.

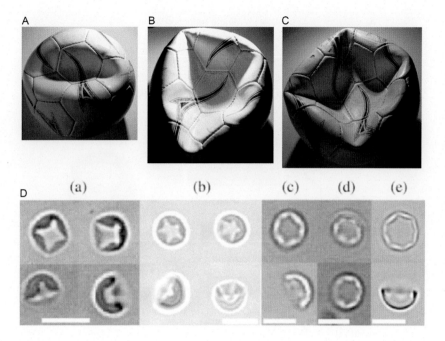

FIG. 10

Symmetry breaking in deflated balls. (A–C) Buckled soccer ball illustrating 2, 3, and 4 edges, with thanks to Diane Sucharyna. (D) "...colloidal spheres filled with oil, in a mixture of water and ethanol... showing 4–8 wrinkles (a–e). Each subfigure shows different transmission optical microscopy views of the same object. Scale bar 5 μm, except subfigure (c): 2 μm.... The oil consists of low molecular polydimethylsiloxane (PDMS) oligomers. By adding tetraethoxysilane (TEOS), a solid shell forms at the surface of the droplets, consisting mainly of PDMS with average oligomer length 4, cross-linked with hydrolyzed TEOS units." Computer simulations are reported there and in (Quilliet, 2012; Vliegenthart & Gompper, 2011), some reaching to a 14-sided polygon: "...the thinner the shell, the larger the number of wrinkles [polygonal edges] are to be expected" (Marmottant et al., 2011). At the lower limit, triangles are observed in stomatocytes (abnormally shaped red blood cell) (Lim et al., 2002), and buckling with 2-, 3-, or 4-sided polygons, the latter often irregular quadrilaterals, occur in some polymer particles (Okubo et al., 2001). For collapsed basketballs showing polygons see: 2-sided (Jicepix, 2015); 3-sided (Logan, 2015); 4-sided (Mitic, 2015; ShopAdvisor, 2015); 5-sided (Garnett, 2015).

From Quilliet, C., Zoldesi, C., Riera, C., van Blaaderen, A., Imhof, A., 2008. Anisotropic colloids through non-trivial buckling [Erratum: 32(4), 419–420]. Eur. Phys. J. E 27(1), 13–20, with permission of Springer.

For the square cell in Fig. 2C, we estimate $V/V_0 = 0.10$ (assuming the cross section was parallel to an edge), which coincides with category 4_c in Fig. 11B. Thus, its shape is consistent with the idea that it is due to spherical buckling caused by volume shrinkage with its area retained.

The transition between a spherical shape and a polygonal shape of an Archaea as salt concentration increases (Javor et al., 1982) may indeed involve exit of water from the cell and loss of cell volume, i.e., crenation. The transitions between shapes may be due to the various energy levels of different configurations, some of which are metastable (Knoche & Kierfeld, 2011, 2014a,b,c). The energy differences between states and the heights of their metastabilities may be guides to why certain polygons

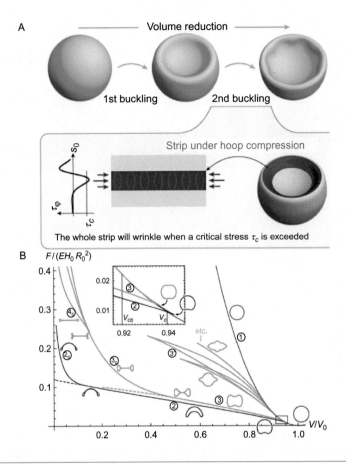

FIG. 11

(A) Buckling of a sphere whose surface area is retained but whose volume is decreased occurs in two stages. The second stage leads to wrinkling under hoop compression that results in a regular polygon around the rim. The authors estimated polygons of 6, 7, or 8 sides for various values of the parameters. (B) "Bifurcation diagram for given volume of a capsule [collapsed or crenated sphere with a thin, solid shell] with $\tilde{E}_B$ [dimensionless bending modulus]=0.001." Only midline cross sections are shown. The abscissa is the volume relative to a sphere, and the ordinate is a dimensionless ratio of the stored elastic energy F to the Young modulus E. Note that configurations 3_c and 4_c resemble the cross sections of shaped droplets and polygonal Archaea.

(A) Graphical abstract from Knoche, S., Kierfeld, J., 2014. The secondary buckling transition: wrinkling of buckled spherical shells. Eur. Phys. J. E 37(7), #62, with permission of Springer; (B) From Knoche, S., Kierfeld, J., 2011. Buckling of spherical capsules. Phys. Rev. E 84(4), #046608, with permission of the American Physical Society.

are more frequent in occurrence than others, as we have observed for shaped droplets (Haas et al., 2017).

Our crenated sphere buckling model for polygonal Archaea has a few peculiar characteristics:

1. The polygons formed can have any number of sides, independent of the 2D crystallinity of the S-layer.

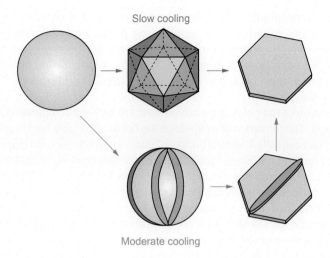

FIG. 12

Two modes of shape change of a sphere to a shaped droplet at constant volume: polyhedron as an intermediate phase, and folding as an intermediate phase. Cf. (Denkov et al., 2015; Dubois et al., 2001; Guttman et al., 2016a).

2. Nevertheless, when a polygon edge is parallel to a crystallographic direction, as in triangles or hexagons, the crystallinity of the S-layer may contribute to the stability of the polygon.

3. The S-layer is always somewhere in a state of stress, since at least in parts it is bent out of its equilibrium planar configuration.

4. Edges are in effect creases in the structure of the S-layer and may include changes in the basic 2D lattice that accommodate the high local stress via dislocations and other rearrangements of the S-layer proteins.

5. As the cell membrane is attached to the S-layer, it may develop regions of liquid crystal or plastic behavior, especially along the creases.

6. At corners of the polygons, we can anticipate even greater dislocations; in both S-layer and membrane lipids.

7. Thus, the total energy of the system includes the sum of the crease energy, the corner energy, and the deviations from planarity (curvature).

Thus, our 1D model for shaping in the Appendix, based on analogy with the linear arrays of magnets with periodic boundary conditions, may be of relevance in the crenated sphere buckling model, approximating molecular phenomena around the rim of the polygon. This is particularly so, since the sharp creasing of folded square Archaea (Fig. 8) suggests a nonlinear response of the S-layer (and perhaps the cell membrane) versus degree of bending. This nonlinearity may have to be added to the crenated sphere buckling model. In particular, we can anticipate its importance at edges and corners. It is probably a reversible nonlinearity, as suggested by squares changing to triangles in a time-lapse study of *Haloarcula japonicas* (Hamamoto et al., 1988). While the cells they observed roughly divided into two daughter cells of equal area, the angle of the plane of division ("cell plate") seems to have no fixed relationship to the sharp corners ("apices") at the perimeter, unlike the polar growth of rod-shaped prokaryotes (Kysela et al., 2013). In each case where there is a change in the number of corners, the cell shows a reduction from 4 to 3 corners.

In our model of shaped droplets, the plastic phase is visualized as occurring at the edges (Fig. 2A). The configurations at corners and over the broad, nearly planar top and bottom surfaces, are not considered. There is, however, no reduction in volume. Rather, shape change occurs with an increase in surface area. While this contrasts with the soccer ball model, where surface area is constant, both may be accommodated by adding a third axis to the phase diagram of Fig. 11, namely A/A_0. Shaped droplets would all be on the plane $V = V_0$.

The basic mechanism of flat polygon formation seems to be hoop compression around the rim of the indentation of a sphere. The formation of a cup shape has been called "first buckling" and the formation of polygonal structure has been called "second buckling" due to hoop compression (Fig. 11). Second buckling occurs over a limited range of mechanical parameters. Hoop compression is found in a number of contexts in biology, including the microfilament ring in eukaryotic cell division, the FtsZ ring in prokaryote cell division, and in the rim of the blastula during animal embryogenesis. It may well be worth looking carefully for polygon formation in these cases. Polygonal arrangements occur in microfilament rings in epithelial cells and at the perimeter of the silicalemma in polygonal diatoms, though in both cases it would seem there is a preexisting polygonal constraint. We can now recognize the formation of problastopores (Gordon & Gordon, 2016a; Gordon, 1999; Gordon et al., 1994) as a case of hoop compression, which ordinarily results in a normal embryo, but may be what sometimes generates a two-, three-, or four-headed embryo (Laale, 1984) (contrary to a priori considerations (Murray, 2012)). Armadillos form 4 to 12 monozygous embryos, which might come from a similar mechanism, followed by separation of the individuals (reviewed in (Gordon, 1999)). So what we may have here in hoop compression is what has been called a "generic" mechanism in development (Newman & Comper, 1990). While this is important and perhaps eye opening to know, it also presents a cautionary tale: things that look alike may not be related by evolutionary descent. This similarity results from so called "convergent evolution" (Powell & Mariscal, 2015).

6 POSSIBLE FUNCTIONS OF A POLYGONAL SHAPE OF CELLS

Small differences in shape have been associated with profound advantages in motility, drag reduction, surface-proportional nutrient uptake, and predator avoidance (here resistance to being merged/diluted with other droplets)—key drivers for evolutionary selection (Young, 2006, 2007, 2010). These differences could play a significant role in selecting and enhancing metabolic difference selection in protolife entities capable only of some of the functions of life, e.g., metabolism, replication, and heredity. Curiously, in addition to defying spherical form upon cooling, shaped oil droplets have recently been shown to exhibit spontaneous decrease in size, capturing the necessary energy to do so from thermal fluctuations (Tcholakova et al., 2017). The increase in interfacial energy shows that very simple chemical systems may be able to harvest energy from the environment in a fashion similar to life. Breakup of droplets has been seen previously due to various one-time chemical or thermal stimuli via different mechanisms. Yet self-shaping droplets can harness small thermal fluctuations for repeated droplet deformation and corresponding breakup, a likely common and repeated stimulus to have occurred on early Earth and elsewhere.

Additional potentialities for straight edges in shaped primordial oil droplets and polygonal Archaea may be that they can hypothetically accommodate long, straight polymers along their straight edges that may be predecessors of the later development of a cytoskeleton. This may simply be a wall effect (Aliabadi et al., 2015; Robledo & Rowlinson, 1986; van Roij et al., 2000). For example, gas vesicles are often found preferentially along the edges in square Archaea (Figs. 1: 4-sided polygons, 4A, 13).

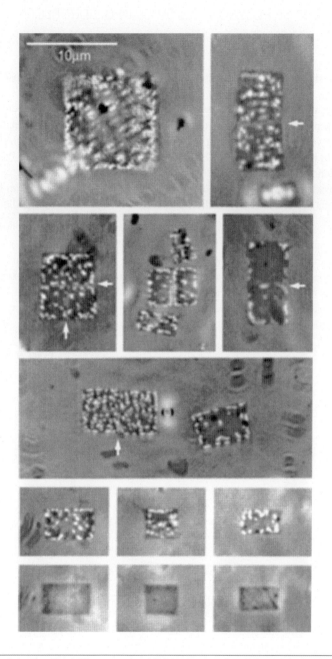

FIG. 13

"The square archeon from the Sinai. Phase contrast light micrographs of the original square haloarcheon discovered in a brine sample from a sabkha near Nabq, Sinai. Division lines (arrows) are visible in some of the cells but not in the largest, top left. Gas vesicles show as bright refractile granules in all cells except those in the last line, which have been exposed to pressure (the same cells as in the line above). Scale bar represents 10 μm." Note the frequent alignment of gas vesicles with the straight edges of the cell and the "division line." This figure is rearranged from figures in (Walsby, 1980) and copied in (Walsby, 2005) and here. See also Fig. 1C in (Parkes & Walsby, 1981).

From Walsby, A.E., 2005. Archaea with square cells. Trends Microbiol. 13(5), 193–195, with permission of Elsevier and Walsby, A.E., 1980. A square bacterium. Nature 283, 69–71, with permission of Nature Publishing Group.

Note that the cell shape does not change when the gas vesicles are collapsed by centrifugation (Stoeckenius, 1981).

At this juncture, the best we can do is speculate on what polymeric molecules might have been aligned in shaped protocells and review experiments on extant polymeric molecules confined in cells or small droplets or vesicles. A model for protocells has been advanced in which the inner surface of a vesicle is the site for polymer adsorption (Fig. 14). Shaped droplets would give the interface a specific structure onto which they could adsorb.

Oil vesicles enriched with polymer molecules aligned along the surface can be compared with a somewhat analogous system in extant cells: liposomes with cytoskeleton bundles (Fig. 15):

> We reconstitute a minimal model system of liposomes containing actin-fascin bundles to elucidate how a lipid membrane and a simplified actin cytoskeleton influence each other's organization through mechanical interactions. When the liposomes are sufficiently deformable, the bundles can deform the membrane into finger-like protrusions, reminiscent of cellular filopodia. The protrusions can reach lengths of up to 60 mm. In contrast, liposomes having a membrane with a large bending rigidity predominantly remain spherical and force the bundles to form cortical rings. . . . Half of the protruded stiff liposomes have bundles with sharp kinks arranged in a planar ring-like structure in the main body. . .
>
> **(Tsai & Koenderink, 2015).**

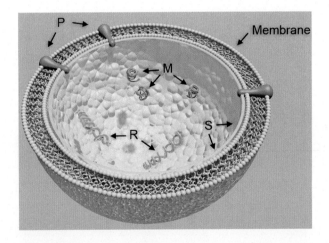

FIG. 14

The SPRM vesicle model for a protocell with adsorbed polymers. "An S-polymer arises by chance that stabilizes protocell membranes allowing them to survive to return their contents to the anhydrous phase. Protocells then evolve the P-polymer, which gives them access to nutrients through transmembrane pores. Access to nutrients supports the emergence of metabolism catalyzed by M-polymers. Metabolism will generate products that support replication (R-polymers)" (Damer & Deamer, 2015). Note that the hypothesized first component, the S-polymer, is in the place of future cortical cytoplasm, suggesting that a cytoskeletal component was the first step after the cell membrane in the origin of life. This model is guiding the search for the simplest molecules that can have the four functions. Note that the hypothesized S-protein is inside the membrane, whereas the Archaea S-layer is outside, so they may not be related.

Reproduced with kind permission of Bruce F. Damer and David W. Deamer.

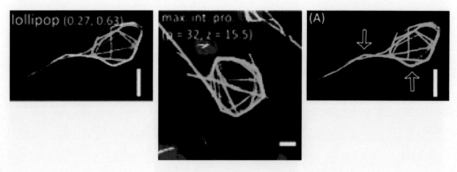

FIG. 15

Microfilaments confined in liposomes. The microfilaments often exhibit kinks, and the liposome bilayer membrane distorts with protrusions, so they are interacting. Sometimes nearly polygonal configurations result. "To address the competition between the deformability of the membrane and the enclosed actin bundles, we tune the [microfilament] bundle stiffness (through the fascin-to-actin molar ratio) and the membrane rigidity (through protein decoration). Scale bars 5 μm."

From Tsai, F.-C., Koenderink, G.H., 2015. Shape control of lipid bilayer membranes by confined actin bundles. Soft Matter 11(45), 8834–8847, with permission of the Royal Society of Chemistry.

Thus, confined linear polymers produce planar polygons, and we can imagine a positive feedback between shaped droplet protocells and their kinked polymeric contents. Computer simulations (Koudehi et al., 2016) could be extended to include such interactions. Once cytoskeletal molecules became associated with motor molecules, their behavior in a liquid-crystalline environment may have become important, by analogy to that of motile bacteria confined in small volumes of liquid crystals (Mushenheim et al., 2015). Propagating kinks (Fig. 16), if they also occur in confined microfilaments, would provide a further mechanism for polygonal interactions between cell shape and cytoskeletal shape.

7 POLYGONAL DIATOMS

We have demonstrated the uncanny rotational symmetry of some centric diatom valves that have n radially arranged sectors by rotating their scanning electron microscopy (SEM) images by $2\pi/n$ radians and subtracting the rotated image from the original (Sterrenburg et al., 2007). The valves are made of amorphous silica and thus, unlike snowflakes, they in themselves have no crystalline structure, despite their symmetry. Some diatoms, all in the Class Mediophyceae (Medlin & Kaczmarska, 2004), have polygonal shapes similar to shaped droplets (Fig. 1).

The precipitation of a new silica shell (the face of which is called a "valve") occurs inside flat membrane bags called silicalemmas (Crawford, 1981; Gordon & Drum, 1994). While the microtubule organizing center (MTOC) attached outside the valve silicalemma has been considered as a patterning device for the deposition of silica inside the silicalemma (Parkinson et al., 1999), we had no explanation for the high degree of symmetry of some diatoms. The exploratory behavior of microtubules (Gerhart & Kirschner, 1997, 2007; Kirschner & Gerhart, 1998; Schulze & Kirschner, 1988) seemed to be a

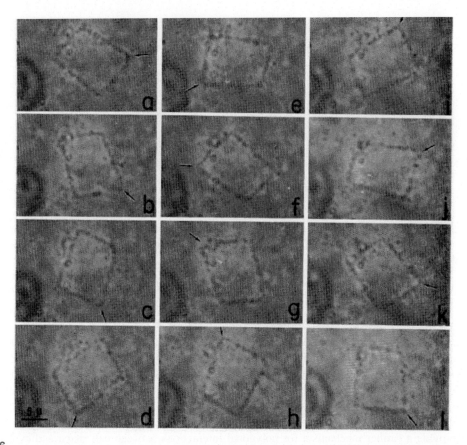

FIG. 16

"The propagation of angles of a polygon as waves. Particles attaching to the fibril do not change their positions. (a–l) Intervals of successive pictures: 1/2 s. Arrows show a corner which propagates as a wave. (Observed with an inverted brightfield microscope.)" Here there are 4 waves. Each is a corner that propagates through the fibril in a clockwise fashion. In other words, it is the sharp bend that propagates along the closed loop fibril, i.e., the fibril itself is not rotating.

From Kuroda, K., 1964. Behavior of naked cytoplasmic drops isolated from plant cells. In: Allen, R.D., Kamiya, N. (Eds.), Primitive Motile Systems in Cell Biology. Academic Press, New York, pp. 31–41, with permission from Elsevier.

possibility, but no mechanism was apparent that would lead to position them in a precise, equal angle arrangement. By analogy with the shaped droplets, we can now come up with a working model:

1. The microtubules in the MTOC attached to the silicalemma and the precipitating silica inside the silicalemma provide the rigid components of a tensegrity apparatus similar to the cell state splitter (Appendix, Fig. 24B) in developing embryos (Gordon & Gordon, 2016a,b; Gordon, 1999; Gordon & Brodland, 1987).

2. The microfilament ring around the outside of the perimeter of the silicalemma (Gordon et al., 2009; Pickett-Heaps & Kowalski, 1981; Pickett-Heaps et al., 1979a,b) and perhaps the silicalemma membrane and/or a radial network of microfilaments (Medlin, 2016) provide compressive elements of this tensegrity apparatus.
3. Under some conditions, the microfilament ring becomes polygonal, as in Figs. 15 and 16.

This then becomes another case of hoop compression (Fig. 11).

This model could be explored by computer simulation as a dynamic tensegrity structure (Caluwaerts et al., 2014; Shen & Wolynes, 2005), via simulations of buckling of the nascent diatom valve (Diaz Moreno et al., 2015; Gutiérrez et al., 2017), and via molecular dynamics (Annenkov & Gordon, 2017). Direct observation of the stages of formation of the valve inside live diatoms is also feasible, using fluorescently labeled microtubules and microfilaments. The nascent valve is formed in under 15 min, as demonstrated by fluorescent silica labeling (Hazelaar et al., 2005). This time is consistent with the dynamic instability of microtubules (Yenjerla et al., 2010). One must look through an older valve to see the newly-forming valve inside the silicalemma, which is inside the cell. The older valve could be cloaked (made "invisible" Koprowski, 2013) by using a nontoxic medium matching its refractive index ($n = 1.43$ (Fuhrmann et al., 2004)), for example, with methyl cellulose ($n = 1.4970$ (Scientific Polymer, 2013)) diluted to an appropriate concentration.

8 CONCLUSION

Our hypothesis is then that tensegrity is a universal mechanism for generating a variety of polygonal cell shapes. The discovery of such structures in self-shaping oil droplets and the potential advantages they can confer to protocell mobility, metabolism, and selection (Sharov, 2017) lends further support for droplets as primordial protocells in the lipid world scenario of the origin of life. Later in evolution, similar tensegrity generation mechanisms seem to have been utilized by Archaea, which individually can undergo a series of shape variations that closely follow the series of shapes in cooled oil droplets. Their structures agree on many points:

1. Sharp corners.
2. Straight edges.
3. No linear molecular structures supporting those edges.
4. Lenticular form, i.e., thinner in the middle.
5. Formation of holes.
6. Comparable size ranges.
7. Ability to change the number of edges.
8. Ability to reversibly adopt a spherical shape with the change of one environmental parameter.

A difference between them is that shaped oil droplets transform under constant volume, whereas polygonal Archaea tend to transform shape under constant area. Conditions that affect shape transformation in droplets and Archaea are temperature and salinity, respectively. The salinity parameter has yet to be explored for shaped droplets. We are at the beginning of understanding the behavior and dynamics of shaped droplets over time and whether these primitive droplet systems might display hystoresis or have mechanisms that confer robustness. At this point, we can say that such primitive-shaped droplets could have existed, but may be too primitive to preserve structure under environmental change or perturbation. Oil protocells apparently had neither genes nor proteins, but some molecules could have supported compositional and structural heredity by adsorption at edges and/or via the wall effect.

It is meaningless to talk about possibility of a direct descent of prokaryotes from the oil-based primordial self-reproducing systems, because of the enormous difference in their functional complexity. The evolutionary path towards the complexity of prokaryotes included numerous intermediate stages and possibly took several billion years (Sharov & Gordon, 2017). It is unlikely that all these intermediate steps required flat polygonal cells. However, we suggest that polygonal shapes recurrently appeared in evolution, and the physical implementation of these shapes was based on the same generic tensegrity principle with variations in the details.

The role of polygonal shapes in the origin of life is still obscure. Although there is a possibility that primordial polymers could have been constructed more effectively along the edges, we don't know which polymers formed the first adhesions to the edges of shaped protocells, possibly stabilizing their forms and providing a basis for the cytoskeleton and/or nucleic acids. Further, it is not clear how oil droplets had evolved into membrane-bound vesicles. We suggest double water-oil-water droplets as an intermediate step. Is it possible that the emergence of protein synthesis was facilitated by the presence of cell edges? There is a huge literature on the paths towards the origin of life. What we are then suggesting is that the existence of polygonal shapes in oil droplets rescales the possibilities of those scenarios that may permit us to hone in on a working hypothesis for the steps in the origin of life.

As for more recent ancestral organisms, it may be possible to evaluate the possibility whether, for example, LUCA (the Last Universal Common Ancestor) was a flat polygonal cell. Testing for such a possibility may require the analysis of genes that correlate with such shapes. This is a testable prediction, given that 355 protein families in LUCA have been deduced (Weiss et al., 2016) and we now have the complete genome sequences of some polygonal Archaea (Liu et al., 2011; Malfatti et al., 2009; Wikipedia, 2016c). The only halophilic Archaea used in ascertaining LUCA was *Haloferax volcanii* (Weiss et al., 2016), which occasionally has an indistinct triangular shape (Emerson et al., 1994), generally looking like a "potato chip" (Oren, 1999). A greater sampling of the genomes of truly polygonal Archaea might show a closer similarity with LUCA than currently reported (Weiss et al., 2016).

Our finding of polygon-shaped oil droplets opens new logical paths for the analysis of the origin of life. We surmise that some of these paths will eventually bring us closer to the understanding of the fundamental question: how did life come into being in the Universe? As oil droplets can be shaped in any aqueous environment subject to occasional or periodic slow cooling, we can anticipate shaped droplet protocells on many Earth-like exoplanets, including those that formed before the Earth. We extrapolated that the time interval from protocell to LUCA took so long that this process started before the Earth was formed (Sharov & Gordon, 2017). Thus the role of shaped droplets in the origin of life and its evolution may have begun prior to Earth's appearance.

ACKNOWLEDGMENTS

We would like to thank Stephen M. Levin for helpful comments on the overview of tensegrity, Diana Cholakova for providing the images of the shaped drops from the experiments, Jack Rudloe for discussions in the context of embryogenesis and his hosting of RG at Gulf Specimen Marine Laboratory & Aquarium while this was being written, and Alexei Sharov for his fine editorial job on our manuscript.

APPENDIX OVERVIEW OF TENSEGRITY STRUCTURES

Tensegrity structures were originally conceived as manmade systems of isolated components of two kinds: stiff parts (bars) that could bear compression with little distortion, and taut or prestressed parts (cables) attached to them that either hold the stiff elements from moving or additionally compress them to some extent (usually not to the point of buckling). Such simplest tensegrity structure is shown in Fig. 17. Tensegrity structures may be compounded into and actually originated in fine art (Snelson & Heartney, 2013) (Fig. 18). We can also find them in nature (Fig. 19).

In the mathematical theory of tensegrity, a third kind of element (or "member") is sometimes allowed that acts in the opposite manner to the cable. This element, called a strut, is straight when compressed and floppy when stretched. In simple mathematical terms, then, the three kinds of elements have the following behaviors:

Bar: $t = k(l - l_0)$
Cable: $t = k(l - l_0)$ for $l \geq l_0$, $t = 0$ otherwise
Strut: $t = -k(l - l_0)$ for $l \leq l_0$, $t = 0$ otherwise

where t is the internal energy of an element and l its length. The "initial" length of an element is l_0. Note that when $l = l_0$, $t = 0$, meaning that the element then has zero prestress. The spring constant k and the zero prestress lengths l_0 could be different for each kind of element, or even for each individual element. In one mathematical idealization, k approaches infinity (Fig. 20). Of course, these simple Hookean springs could be generalized to more complicated constitutive relationships (Rimoli, 2016).

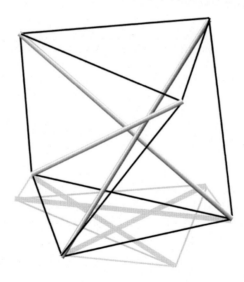

FIG. 17

This is the simplest nontouching rod tensegrity structure consisting of three stiff rods (bars) connected by 9 flexible strings (cables) that are stretched so that they are taut. It has been used as a stool (Passi, 2013).

From Dale, B.F., 2008. An SVG of a physically possible tensegrity structure in 3D, with a shadow. https://en.wikipedia.org/wiki/File:3-tensegrity.svg, with permission under a Creative Commons Attribution-Share Alike 3.0 Unported license.

FIG. 18

Examples of tensegrity structures as art. (A) Early X-Piece, 1948, wood and nylon, $29 \times 4.5 \times 4.5$ cm. Note that the three rigid parts are not simple rods, which became the archetype later. Also, two of the rigid parts are allowed to touch. The bottom platform could, alternatively, be regarded as a surface to which the elements are "pinned" (Connelly & Guest, 2015), rather than an element on its own. (B) Northwood I, 1969, painted steel and stainless steel, $3.65 \times 3.65 \times 3.65$ m. Collection: Northwood Institute, Dallas, TX. (C) Rainbow Arch, 2001, aluminum and stainless steel, $213.4 \times 386.1 \times 81.3$ cm. Cf. (Snelson, 1990; Snelson & Heartney, 2013).

(C) Reproduced with permission from the late artist Kenneth Snelson.

Snapping between configurations is possible (Gordon, 1999), as between the boat and chair shapes of cyclohexane (Wikipedia, 2016a) (Fig. 21). Sometimes small deviations in the length of elements can result in a large change in the shape of the whole configuration (Connelly & Gortler, 2015) (Fig. 22).

Tensegrity systems have been classified as follows:

> A tensegrity [system] that has no contacts between its rigid bodies [bars] is a class 1 tensegrity system, and a tensegrity system with as many as k rigid bodies in contact is a class k tensegrity system
>
> **(Skelton & de Oliveira, 2009).**

In our illustrations, the following are Class 1 tensegrity systems: Figs. 1, 18B and C, 19, 23. Class 2: Fig. 18A. Class 5: Fig. 21. Class 30: Fig. 22. Some systems with $k > 1$ are called "mechanisms" (Connelly & Gortler, 2015).

FIG. 19

A three-dimensional spider web made by a tangle-web spider (Wikipedia, 2016d) in Panacea, Florida. The elastic strands of web in this 3D tensegrity structure are decorated with fog dew drops. The webbing is the elastic component (cables) and the plant is the stiff component (bars).

When we deal with cells or organisms, we are closer to continuum mechanics than these tensegrity models suggest. Contrary to the open spaces in between human and spider-made tensegrity structure elements, in cells we are confronted with what has been called "the crowded cytoplasm" (Gnutt & Ebbinghaus, 2016). Similarly, the nucleus is crowded (Nakano et al., 2014), with 2 m of double-stranded DNA (Greulich, 2005) compacted into a human cell nucleus of 8 μm diameter (Greeley et al., 1978). Crowding itself has effects on the mechanical properties of cytoskeletal tensegrity structures (Zhou et al., 2009).

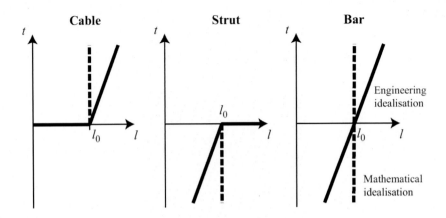

FIG. 20

Constitutive relationships for cables, struts, and bars. The slope of the tilted lines is the Hookean spring constant k. In the mathematical idealization shown by the dashed lines $k = \infty$.

From Connelly, R., Guest, S.D., 2015. Frameworks, Tensegrities and Symmetry: Understanding Stable Structures. http://www.math. cornell.edu/~web7510/framework.pdf, with kind permission of Robert Connelly and Simon Guest.

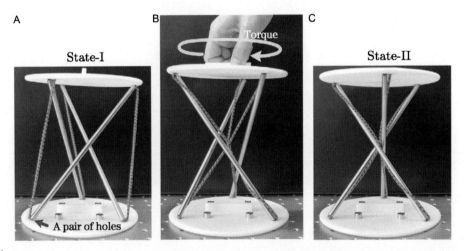

FIG. 21

This tensegrity structure suddenly snaps to a new configuration as a torque is applied on top. State-I is a "nonstandard" tensegrity structure, in that the top and bottom rigid components are not linear, as also in Fig. 20 Left. State-II is nonstandard in that the rigid linear elements are allowed to touch, and the elastic elements also touch the rigid linear elements along their lengths and are no longer single straight lines themselves. They may be regarded as kinked. Thus, these authors have generalized the tensegrity concept, allowing these "exceptions." "(A) Self-equilibrated and stable state with no interference of elements. (B) Loaded state under applied torque with only side strings intersecting each other. (C) Self-equilibrated and stable state with all bars and side strings intersecting with one another."

From Zhang, L.Y., Zhang, C., Feng, X.Q., Gao, H.J., 2016. Snapping instability in prismatic tensegrities under torsion. Appl. Math. Mech. (English Ed.) 37(3), 275–288, with permission of Shanghai University and Springer-Verlag Berlin Heidelberg.

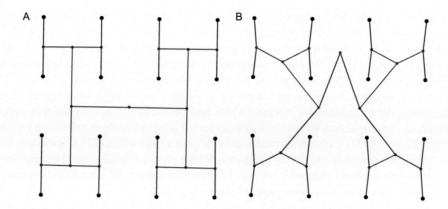

FIG. 22

A mathematically rigid structure can sometimes become nonrigid with a small change in parameters. This structure consists only of bars of fixed length. (A) "The large black vertices are pinned to the plane, and the whole framework is universally rigid...." (B) "...the same framework... but with the lengths of the bars increased by <0.5%." This may be thought of as analogous to buckling: "In practical terms, buckling can be defined as a sudden and dramatic increase in deformations for a relatively small increase in the loads" (Gutiérrez et al., 2017).

From Connelly, R., Gortler, S.J., 2015. Iterative universal rigidity. Discret. Comput. Geom. 53(4), 847–877, with permission of Springer.

FIG. 23

A NASA tensegrity robot named SUPERball. "Each rigid rod is a self-contained robotic system consisting of two smaller intelligent nodes" (Bruce et al., 2014), which can change the lengths of one or more attached cables, which are under tension, causing the structure to roll or climb hills.

From Vytas SunSpiral with his kind permission.

We can look at the Wurfel, for instance, a toddler's toy that we have used as a toy model for changes in gene expression in a cell nucleus (Gordon & Gordon, 2016a; Gordon, 1999), as a space filling tensegrity structure. A Wurfel consists of a set of wooden cubes connected via an elastic band through them that forms a closed loop. The elastic band enters and exits each cube at right angles, pulling them together face to face (Fig. 24). Each pair of connected blocks can be regarded as a strut. There is a discrete set of equivalent energy ground states of a Wurfel (Tromp & Gordon, 2006). However, if we replace the cubes by spheres, we still have a structure of stiff elements held together under tension ("if one imagines hard spherical billiard balls, the centers of any two touching balls form a natural strut" (Connelly & Guest, 2015)), but with a continuum set of ground states (Fig. 25). Cube and sphere-based Wurfels with many ground states have analogies in folded proteins. While most proteins have single, nondegenerate ground states (Khatib et al., 2011), some have many or even a continuum of ground states. The latter are called disordered proteins (Uversky, 2013).

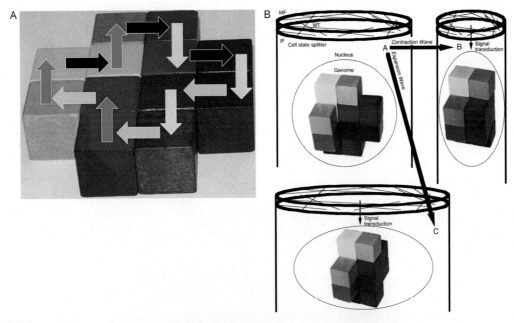

FIG. 24

(A) The Wurfel was invented by Peter Bell of Pappa Geppetto's Toys Victoria Ltd., Victoria, Canada (Flemons, 2016). Arrows show the path of the taut elastic band inside the blocks. The number of rectilinear configurations of a 2n-Wurfel ($n=6$ here) increases rapidly (Tromp & Gordon, 2006). Configurations not fitting on a cubic grid are also possible. (B) The same Wurfel in three 3D rectilinear configurations, being used as a toy model for changes in gene expression in a cell nucleus during cell differentiation. Note that the cell state splitter is also a tensegrity structure with the microtubules (MT) being the stiff elements, while the microfilament ring (MF) is in tension. The intermediate filament ring (IF) acts as elastic component. In an epithelium, the top (apical) end of each cell, and thus its cell state splitter, would be polygonal.

(B) From Gordon, N.K., Gordon, R., 2016. Embryogenesis Explained. World Scientific Publishing, Singapore, with permission of World Scientific Publishing.

FIG. 25

"Baby Beads" are topologically connected just like a Wurfel, but the spheres roll over one another easily. This is then a tensegrity structure with a continuum of equivalent energy ("degenerate") ground states.

If we allow a set of hard spheres to have attractive forces (nonzero prestress) between all near neighbors, we effectively have a hard sphere model for condensed matter (Camp, 2003) or dispersions when the spheres do not always touch (Gonzalez et al., 2014). This too could be considered a tensegrity structure, albeit a changing one as it flows or its atoms or molecules undergo Brownian motion relative to one another. Thus, any drop of liquid, with its molecules regarded as the stiff elements, is a tensegrity structure. Indeed, tensegrity models for the rigidity of packings of balls have been studied (Connelly, 2008; Connelly et al., 2014), which represent a step towards molecular tensegrity modeling of liquids and solids.

In most liquids, we have to deal only with interactions between near neighbors, to get an accurate picture of the statistics of their structure, such as the radial distribution function (Cockayne, 2008; Gotoh, 2012). Nevertheless, there can be a long-range order imposed by the network of elements under tension, as in packings and crystallization. Alternatively, the structure of long elements, i.e., elements that are far from spherical, can also lead to long-range effects and long-range order. A remarkable example is the case of microtubules, which are so long and thin that one would expect them to buckle like wet spaghetti (Gordon & Gordon, 2016a). However, when supported along their length by attached intermediate filaments, it takes the order of 10^4 times more compressive force to buckle them (Brodland & Gordon, 1990).

While this calculation has been used to justify microtubules as the stiff elements in a tensegrity model for cytoplasm (Ingber et al., 1994), there is some circular reasoning in doing so, since what anchors those particular intermediate filaments at their other ends (cf. pinning in Fig. 18) has not been worked out. Also, the multiple attachments (nodes) along the microtubule make for a structure that differs from standard tensegrity modeling, in which the elements are allowed to rotate freely to any angle about the "joints" or nodes. This is because each long, polymeric structure in the cytoplasm has a stiffness, characterized by a persistence length (Fig. 5.30 in (Gordon & Gordon, 2016a)), so that the amount of bending at each node would be constrained, and any bending would add to the energy (prestress) of the whole structure.

A further step in generalizing tensegrity structures was taken with the invention of tensegrity robots, in which element lengths are manipulated to make a tensegrity structure change shape (Piazza, 2015) and, for example, move over a rough planetary landscape by shifting the center of gravity of the robot or pushing against terrain (Bruce et al., 2014; SunSpiral, 2015) (Fig. 23). If one combines such force generation by the elements with the change of neighbors in dispersions, we approach a tensegrity model for cytoskeleton dynamically changing via motor molecules, polymerization, and depolymerization, and changing connections (nodes) via bifunctional attachment proteins (Perera et al., 2016). Getting beyond the stick and string tensegrity model for cytoskeleton has just begun (Ingber et al., 2014), for example, by looking at the bistable configurations of the cell state splitter (Gordon & Gordon, 2016a).

However, a tensegrity structure that can make and break connections and grow and dissolve elements is prone to instability and collapse. Indeed, such collapses may be important steps in cell differentiation (Gordon & Gordon, 2016a). A computer simulation framework for investigating such cytoskeletal instability phenomena is under construction based on PushMePullMe (Senatore, 2017).

In zero gravity, the taut strings of a simple tensegrity structure (Fig. 17), for example, would hold the structure together, but need not be under any prestress. If prestressed, the configuration would look much the same, except that the stiff rods would be slightly compressed and the strings slightly stretched or slightly buckled. In the biological literature, the prestress is assumed to be nonzero and essential to the maintenance of the structure (Ingber et al., 2014; Shen & Wolynes, 2005). In the mathematical literature on tensegrity structures, the concept of prestress includes allowing its value to be zero (Connelly & Guest, 2015). Thus, there is a conceptual contradiction here. Of course, mathematically, any small deviation from the equilibrium structure may generate a small prestress, usually driving the structure back towards its equilibrium shape, unless that equilibrium state is metastable, degenerate (Fig. 25), or sensitive to small perturbations (Fig. 22). Nevertheless, the presumption that nonzero prestress is essential to structure maintenance in biology is mathematically incorrect. While nonzero prestress may be present in most biological tensegrity structures, that does not imply the structure would collapse at zero prestress. Thus, the assumption "that the forces required for such a strained assembly in the cell are generated by nonequilibrium polymerizations and movements of motor proteins. . ." may be wrong, when cytoskeletal structure does not require such forces for its stability. For example, while microtubules may undergo frequent elongation and shortening, a process called dynamic instability (Gordon & Gordon, 2016a), they can also be stabilized against such behavior (van der Vaart et al., 2009) and thus act more like simple tensegrity bars. Prestress may be important in building cytoskeletal structures, but sometimes it may not be necessary for maintenance of those structures. These distinctions are important here, because in modeling a tensegrity origin of life, we cannot assume that continuous nonequilibrium, energy requiring processes of nascent cytoskeletal molecules, generating and maintaining prestress, played any role in the abiotic precursors to life.

Most tensegrity modeling ignores the buckling of elements under stress. Buckling can be quite important in cytoskeleton, varying from smooth Eulerian buckling (Brodland & Gordon, 1990) to kinking. Kinking in effect splits an element into two elements with a new node at the kink. However, kinks come in two kinds: stationary and propagating. If a cytoskeletal microtubule or microfilament is bent into a ring (Gordon & Brodland, 1987), so long as the radius of curvature is comparable to its persistence length, we can anticipate that the ring will be circular. Epithelia commonly have microfilament rings, but the cells are generally close-packed in a plane and polygonal in shape. Whether or not individual microfilaments in the bundle forming the polygonal ring end at the corners or are kinked there has apparently not yet been investigated. As the cell state splitter also has an intermediate filament ring

(Gordon & Gordon, 2016a; Martin & Gordon, 1997), the same question arises for its components. An epithelial cell is an example of confinement of a cytoskeletal structure (Gürsoy et al., 2014; Koudehi et al., 2016; Pinot et al., 2009; Soares e Silva et al., 2011; Vetter et al., 2014). Bundles of microfilaments confined to liposomes exhibit kinks and polygonal shapes (Tsai & Koenderink, 2015). Computer simulations have not yet revealed polygonal shapes, perhaps because spherical boundary conditions were imposed (Koudehi et al., 2016).

Details have been worked out for kinking of carbon nanotubes (Iijima et al., 1996; Wang et al., 2016a; Zeng et al., 2004). Analogies have been made between kinking of nanotubes and cytoskeleton (Cohen & Mahadevan, 2003).

Propagating kinks in cytoskeletal rings were discovered by Robert Jarosch (1956, 1957) in cytoplasm squeezed from *Chara foetida* and Kiyoko Kuroda in cytoplasm dripped out of cut *Nitella* cells (Kuroda, 1964) (Fig. 16). Kuroda observed:

> ...triangles, quadrangles, pentagons, hexagons and other polygons.... [Each] consists of a pair of straight lines running in parallel close to each other, their both ends being joined together by tiny circular arcs with a radius of approximately 1 μ. Of various polygons observed, pentagons and hexagons are found most frequently. The distribution of angles, measured in about 300 specimens, shows the sharp peak between 110° and 120°.... Corners of the polygon propagate as waves along the fibril all in the same direction with the same speed. Since the angle of each corner is also kept constant, the polygon maintains its definite shape while corners propagate successively in one direction. On pulling with two microneedles, the polygon is split into finer fibrils

(Kuroda, 1968).

These dynamic rings were later shown to consist of microfilaments (Higashi-Fujime, 1980), which are presumably the "finer fibrils." Whether the kink propagates with sliding or kinking of the individual microfilaments has not been investigated. Kink bends can propagate along microtubules (Tuszynski et al., 2005, 2009). It is worth noting that for shaped droplets: "A very large majority of the interior angles of the polygons are seen in experiments to have measures close to 60° or 120°..." (Haas et al., 2016). Electron microscopy of triangular Archaea shows unexpected 90° corners, made up by a slight rounding of the edges (Nishiyama et al., 1992, 1995; Takao, 2006) (Fig. 1D). All of these cases suggest specific molecular configurations at corners that warrant investigation and modeling.

This does not exhaust the phenomena we should anticipate in the tensegrity behavior of cytoskeleton and its precursors in protocells. A long molecule such as DNA, when supercoiled, exhibits nonlinear phenomena similar to that of a twisted rubber band (Marko & Neukirch, 2012). Supercoiling of microtubules, which are chiral, may alter the binding of motor molecules such as dynein (Gordon & Gordon, 2016a; Gordon, 1999) and perhaps attached bifunctional molecules.

A TOY MODEL FOR THE POLYGONAL SHAPE OF SHAPED DROPLETS

Here we give a physical toy model, based on what happens to a stack of "magnetic buttons" (Horizon Group, 2017b) arranged into a loop (Fig. 26). We can take such magnetic interactions as illustrative stand-ins for the van der Waals forces attracting the linear alkane molecules in the plastic crystal phases making the edges of the polygonal droplets, though their scaling is very different. Suppose there are N buttons arranged in a loop. Because each one is stiff and flat, in order to make a loop, there will be a

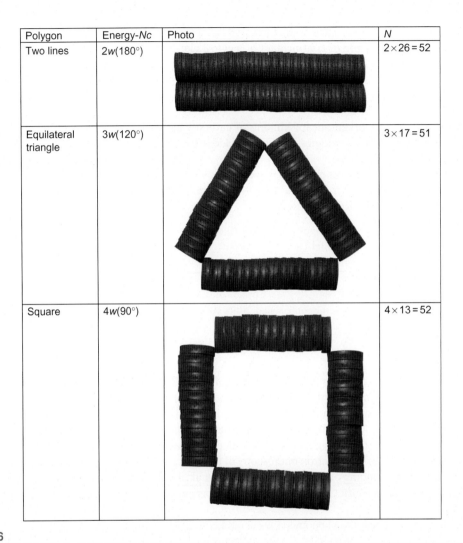

Polygon	Energy-Nc	Photo	N
Two lines	$2w(180°)$		$2 \times 26 = 52$
Equilateral triangle	$3w(120°)$		$3 \times 17 = 51$
Square	$4w(90°)$		$4 \times 13 = 52$

FIG. 26

Magnetic buttons of ¾″ (1.9 cm) diameter were arranged by hand into polygons. The angle $\theta = \theta_1 - \theta_2$ between magnets at the corners of the triangle, for instance, is 120°. Hexagons were difficult to make, as the groups of 8 magnets kept snapping together. While the two line configuration is unrealistic for a whole shaped droplet, as it encloses zero volume, it may present a model for the filaments that sometimes protrude from the corners of shaped droplets. To create the 12-sided dodecagon, we used thinner, weaker magnets of the same diameter that each has a white adhesive and plastic disc attached. This was about the limit for these weak magnets on this table surface, which provided static friction in all cases.

nonzero angle between certain consecutive buttons. It is clear that the sum of those angles around the loop, so long as a loop topology is retained and the loop resides in a plane, must be 360°. Let the energy of interaction between consecutive buttons be a monotonically decreasing function of the angle between them. Then the total energy of the system is:

Regular pentagon	5w(72°)	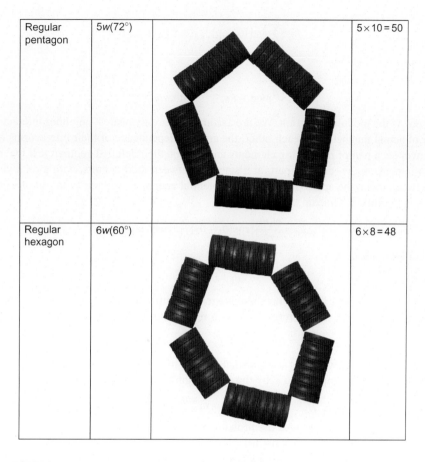	5 × 10 = 50
Regular hexagon	6w(60°)		6 × 8 = 48

| Regular dodecagon | 12w(30°) | 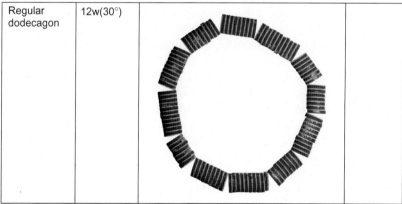 | |

FIG. 26—Cont'd

$$E = \sum_{i=0}^{N-1} w\left(\theta_{\mathrm{mod}(i+1,N)} - \theta_{\mathrm{mod}(i,N)}\right)$$

with the constraint that:

$$\sum_{i=0}^{N-1} \left(\theta_{\mathrm{mod}(i+1,N)} - \theta_{\mathrm{mod}(i,N)}\right) = 360°$$

where "mod" is the modulus function. We are assuming that only nearest neighbor interactions count. For a pair of actual magnets near each other, the angular dependence of their interactions requires integration over each pair of interacting elements and thus on their detailed geometry. If two magnets of magnetic moments m_1, m_2 are far apart, at distance r between their centers, with a relative angle $\theta = \theta_1 - \theta_2$ between them, the approximate mutual potential energy E_p of their point-dipole–point-dipole interaction is (Cullity & Graham, 2011):

$$E_p = \frac{m_1 m_2}{r^3}[\cos(\theta_1 - \theta_2) - 3\cos\theta_1 \cos\theta_2]$$

This falls to zero when:

$$\theta_1 - \theta_2 = 90°$$

but we find for the magnetic buttons that there is a residual attractive force when they are touching at an edge, at a right angle. We will therefore take as a representative energy of interaction:

$$w\left(\theta_{\mathrm{mod}(i+1,N)} - \theta_{\mathrm{mod}(i,N)}\right) = \cos\left(\theta_{\mathrm{mod}(i+1,N)} - \theta_{\mathrm{mod}(i,N)}\right) - 3\cos\theta_{\mathrm{mod}(i+1,N)}\cos\theta_{\mathrm{mod}(i,N)} + c$$

As a rough approximation we take c as a constant, though in principle it is a calculable function. We can now introduce absolute temperature T by the Boltzmann relationship:

$$P(E) = e^{-E/(kT)}$$

For any set of angles, this then gives the probability P of that configuration as a function of T.

Except for its geometry, this problem is not new. It is merely the one-dimensional lattice with nearest neighbor interactions, continuous state variable, and periodic boundary conditions. While the 1D Ising lattice does not exhibit a sharp phase transition (Brush, 1967; Wikipedia, 2016b) including in the continuous state case (Griffiths, 1969; van Beijeren & Sylvester, 1978) of our toy model, in our problem the global constraint on the sum of the angles makes a sharp phase transition possible, at low enough T. This in turn leads to the stability and metastability of polygonal configurations. Because of thermal fluctuations and/or under the influence of surface tension, transitions between the different polygon states can occur, even in the absence of an analog of plastic crystal edge growth in this toy model. These could be explored via Monte Carlo simulations (Gordon, 1980).

There are many other ways to alter a 1D Ising lattice to produce a phase transition. However, none of them involve the constraint of a finite length looped in 2D. Many of the modified Ising lattices invoke long-range interactions (Cassandro et al., 2014). In a way, a looped system has a long-range interaction: of each unit with itself around the loop. Multiple cycles around the ring may in effect give an infinite range to the interactions. The latter does lead to a phase transition for the 1D Ising lattice (van Beijeren & Sylvester, 1978). This may be the reason that rings exhibit sharp phase transitions. However, unlike any standard Ising systems, in our system the forces generated by neighbor-neighbor

FIG. 27

Once hexagons were made, tapping on the table rapidly produced various transitions to lower order polygons. Due to imperfect alignment, most of the original groups of 8 magnets can still be seen.

FIG. 28

Top row: a pentagon consisting of 5 groups of 10 magnetic buttons transitions to an isosceles triangle with consecutive hammerings of the surface on which they lie. The metastability was significantly stronger than that of the hexagons. Bottom row: a pentagon transitions to a quadrilateral.

interactions are allowed to alter the positions of the units in space, driving adjacent units towards being parallel to one another.

While setting up the magnetic buttons in various configurations (Fig. 26), it became clear that hexagons were metastable with small thresholds for transitions to lower order polygons. It was hard to achieve these configurations, and tapping on the table caused various transitions (Fig. 27). Pentagons were much more stable than hexagons, but could undergo transitions by hitting the surface nearby with a rubber hammer. Due to friction with the surface, intermediate states could be captured by camera (Fig. 28). Transitions between polygonal dimplings of spherical and cylindrical thin metal shells have been observed by high speed movies (Thompson & Sieber, 2016).

A set of weaker magnets was acquired (Horizon Group, 2017a). In addition to the configurations shown so far, we were able to get up to a 12-sided polygon (Fig. 26), not quite approximating a circular arrangement with many small consecutive angle differences. That would require either weaker magnets or a rougher surface.

The magnetic buttons toy model has a major limitation in that the energies change monotonically with the number of vertices, whereas this function has minima for shaped droplets, for which 60° and

120° appeared "to be the only possible internal angles," at least when they are not "shape-shifting" (Haas et al., 2017). The toy model does not capture features such as the flexibility of the linear molecules, their multiple layers, and the fact that many molecules (not just two, as with magnets) would be arrayed in specific configurations at each corner, nor three-dimensional effects such as the lenticular shape of the droplets, nor their constant volume.

REFERENCES

Akca, E., Claus, H., Schultz, N., Karbach, G., Schlott, B., Debaerdemaeker, T., Declercq, J.-P., König, H., 2002. Genes and derived amino acid sequences of S-layer proteins from mesophilic, thermophilic, and extremely thermophilic methanococci. Extremophiles 6 (5), 351–358.

Aktümen, M., Kaçar, A., 2007. Maplets for the area of the unit circle. J. Online Math. Appl. 7(May), #1549, http://www.maa.org/external_archive/joma/Volume1547/Aktumen/index.html.

Alam, M., Claviez, M., Oesterhelt, D., Kessel, M., 1984. Flagella and motility behaviour of square bacteria. EMBO J. 3 (12), 2899–2903.

Albers, S.-V., Meyer, B.H., 2011. The archaeal cell envelope. Nat. Rev. Microbiol. 9 (6), 414–426.

Aliabadi, R., Moradi, M., Varga, S., 2015. Orientational ordering of confined hard rods: the effect of shape anisotropy on surface ordering and capillary nematization. Phys. Rev. E 92 (3). #032503.

Allen, W.V., Ponnamperuma, C., 1967. A possible prebiotic synthesis of monocarboxylic acids. Curr. Mod. Biol. 1 (1), 24–28.

Andrade, K., Logemann, J., Heidelberg, K.B., Emerson, J.B., Comolli, L.R., Hug, L.A., Probst, A.J., Keillar, A., Thomas, B.C., Miller, C.S., Allen, E.E., Moreau, J.W., Brocks, J.J., Banfield, J.F., 2015. Metagenomic and lipid analyses reveal a diel cycle in a hypersaline microbial ecosystem. ISME J. 9 (12), 2697–2711.

Anella, F., Danelon, C., 2014. Reconciling ligase ribozyme activity with fatty acid vesicle stability. Life (Basel) 4 (4, Sp. Iss. SI), 929–943.

Annenkov, V.V., Gordon, R., 2017. Silsim—a program for simulating aggregation of siliceous nanoparticles involved in diatom morphogenesis. In preparation.

Antón, J., Llobet-Brossa, E., Rodriguez-Valera, F., Amann, R., 1999. Fluorescence *in situ* hybridization analysis of the prokaryotic community inhabiting crystallizer ponds. Environ. Microbiol. 1 (6), 517–523.

Awramik, S.M., Barghoorn, E.S., 1977. The Gunflint microbiota. Precambrian Res. 5 (2), 121–142.

Azadi, A., Grason, G.M., 2016. Neutral versus charged defect patterns in curved crystals. Phys. Rev. E 94 (1). #013003.

Banno, T., Kuroha, R., Miura, S., Toyota, T., 2015. Multiple-division of self-propelled oil droplets through acetal formation. Soft Matter 11 (8), 1459–1463.

Bardavid, R.E., Oren, A., 2008a. Dihydroxyacetone metabolism in *Salinibacter ruber* and in *Haloquadratum walsbyi*. Extremophiles 12 (1), 125–131.

Bardavid, R.E., Oren, A., 2008b. Sensitivity of *Haloquadratum* and *Salinibacter* to antibiotics and other inhibitors: implications for the assessment of the contribution of Archaea and Bacteria to heterotrophic activities in hypersaline environments. FEMS Microbiol. Ecol. 63 (3), 309–315.

Bardavid, R.E., Khristo, P., Oren, A., 2008. Interrelationships between *Dunaliella* and halophilic prokaryotes in saltern crystallizer ponds. Extremophiles 12 (1), 5–14.

Bar-Even, A., Shenhav, B., Kafri, R., Lancet, D., 2004. The lipid world: from catalytic and informational headgroups to micelle replication and evolution without nucleic acids. In: Seckbach, J., ChelaFlores, J., Owen, T., Raulin, F. (Eds.), Life in the Universe: From the Miller Experiment to the Search for Life on Other Worlds. Dordrecht, Netherlands, pp. 111–114.

Bartucci, R., Gambacorta, A., Gliozzi, A., Marsh, D., Sportelli, L., 2005. Bipolar tetraether lipids: chain flexibility and membrane polarity gradients from spin-label electron spin resonance. Biochemistry 44 (45), 15017–15023.

Baxter, B.K., Litchfield, C.D., Sowers, K., Griffith, J.D., Dassarma, P.A., Dassarma, S., 2005. Microbial diversity of Great Salt Lake. In: Gunde-Cimerman, N., Oren, A., Plemenitaš, A. (Eds.), Adaptation to Life at High Salt Concentrations in Archaea, Bacteria, and Eukarya. Springer, Dordrecht, The Netherlands, pp. 19–25.

Bernstein, M.P., Sandford, S.A., Allamandola, L.J., Chang, S., Scharberg, M.A., 1995. Organic compounds produced by photolysis of realistic interstellar and cometary ice analogs containing methanol. Astrophys. J. 454 (1), 327–344.

Berthaud, A., Quemeneur, F., Deforet, M., Bassereau, P., Brochard-Wyart, F., Mangenot, S., 2016. Spreading of porous vesicles subjected to osmotic shocks: the role of aquaporins. Soft Matter 12 (5), 1601–1609.

Bettarel, Y., Bouvier, T., Bouvier, C., Carré, C., Desnues, A., Domaizon, I., Jacquet, S., Robin, A., Sime-Ngando, T., 2011. Ecological traits of planktonic viruses and prokaryotes along a full-salinity gradient. FEMS Microbiol. Ecol. 76 (2), 360–372.

Bivas, I., 2010. Shape fluctuations of nearly spherical lipid vesicles and emulsion droplets. Phys. Rev. E 81 (6). #061911.

Bodaker, I., Sharon, I., Suzuki, M.T., Feingersch, R., Shmoish, M., Andreishcheva, E., Sogin, M.L., Rosenberg, M., Maguire, M.E., Belkin, S., Oren, A., Béjà, O., 2010. Comparative community genomics in the Dead Sea: an increasingly extreme environment. Isme J. 4 (3), 399–407.

Boekema, E.J., Scheffers, D.-J., van Bezouwen, L.S., Bolhuis, H., Folea, I.M., 2013. Focus on membrane differentiation and membrane domains in the prokaryotic cell. J. Mol. Microbiol. Biotechnol. 23 (4-5), 345–356.

Bolhuis, H., 2005. Walsby's square archaeon—it's hip to be square, but even more hip to be culturable. In: Gunde-Cimerman, N., Oren, A., Plemenitaš, A. (Eds.), Adaptation to life at high salt concentrations in Archaea, Bacteria, and Eukarya. Springer, Dordrecht, The Netherlands, pp. 185–199.

Bolhuis, H., te Poele, E.M., Rodriguez-Valera, F., 2004. Isolation and cultivation of Walsby's square archaeon. Environ. Microbiol. 6 (12), 1287–1291.

Bolhuis, H., Palm, P., Wende, A., Falb, M., Rampp, M., Rodriguez-Valera, F., Pfeiffer, F., Oesterhelt, D., 2006. The genome of the square archaeon *Haloquadratum walsbyi*: life at the limits of water activity. BMC Genomics 7. #169.

Boulbitch, A., 2000. Deformation of the envelope of a spherical Gram-negative bacterium during the atomic force measurements. J. Electron Microsc. (Tokyo) 49 (3), 459–462.

Braun, T., Orlova, A., Valegård, K., Lindås, A.C., Schröder, G.F., Egelman, E.H., 2015. Archaeal actin from a hyperthermophile forms a single-stranded filament. Proc. Natl. Acad. Sci. U. S. A. 112 (30), 9340–9345.

Brodland, G.W., Gordon, R., 1990. Intermediate filaments may prevent buckling of compressively-loaded microtubules. J. Biomech. Eng. 112 (3), 319–321.

Bruce, J., Sabelhaus, A., Chen, Y., Lu, D., Morse, K., Milam, S., Caluwaerts, K., Agogino, A., SunSpiral, V., 2014. SUPERball: exploring tensegrities for planetary probes. In: Dupuis, E. (Ed.), 12th International Symposium on Artificial Intelligence, Robotics and Automation in Space (i-SAIRAS). i-SAIRAS. http://robotics.estec.esa.int/i-SAIRAS/isairas2014/Data/Session%2205c/ISAIRAS_FinalPaper_0107.pdf.

Brush, S.G., 1967. History of the Lenz-Ising model. Rev. Mod. Phys. 39 (4), 883–893.

Bukhryakov, K.V., Almahdali, S., Rodionov, V.O., 2015. Amplification of chirality through self-replication of micellar aggregates in water. Langmuir 31 (10), 2931–2935.

Burns, D., Dyall-Smith, M., 2006. Cultivation of haloarchaea. Methods Microbiol. 35, 535–552.

Burns, D.G., Camakaris, H.M., Janssen, P.H., Dyall-Smith, M.L., 2004. Cultivation of Walsby's square haloarchaeon. FEMS Microbiol. Lett. 238 (2), 469–473.

Burns, D.G., Janssen, P.H., Itoh, T., Kamekura, M., Li, Z., Jensen, G., Rodriguez-Valera, F., Bolhuis, H., Dyall-Smith, M.L., 2007. Haloquadratum walsbyi gen. nov., sp nov., the square haloarchaeon of Walsby, isolated from saltern crystallizers in Australia and Spain. Int. J. Syst. Evol. Microbiol. 57, 387–392.

Caluwaerts, K., Despraz, J., Iscen, A., Sabelhaus, A.P., Bruce, J., Schrauwen, B., SunSpiral, V., 2014. Design and control of compliant tensegrity robots through simulation and hardware validation. J. R. Soc. Interface 11 (98). #20140520.

Camp, P.J., 2003. Phase diagrams of hard spheres with algebraic attractive interactions. Phys. Rev. E 67 (1).

Caschera, F., Rasmussen, S., Hanczyc, M.M., 2013. An oil droplet division-fusion cycle. ChemPlusChem 78 (1), 52–54.

Cassandro, M., Merola, I., Picco, P., Rozikov, U., 2014. One-dimensional Ising models with long range interactions: cluster expansion, phase-separating point. Commun. Math. Phys. 327 (3), 951–991.

Castillo, A.M., Gutierrez, M.C., Kamekura, M., Ma, Y., Cowan, D.A., Jones, B.E., Grant, W.D., Ventosa, A., 2006. *Halovivax asiaticus* gen. nov., sp nov., a novel extremely halophilic archaeon isolated from Inner Mongolia, China. Int. J. Syst. Evol. Microbiol. 56, 765–770.

Castillo, A.M., Gutierrez, M.C., Kamekura, M., Xue, Y., Ma, Y., Cowan, D.A., Jones, B.E., Grant, W.D., Ventosa, A., 2007. *Halovivax ruber* sp nov., an extremely halophilic archaeon isolated from Lake Xilinhot, Inner Mongolia, China. Int. J. Syst. Evol. Microbiol. 57, 1024–1027.

Cavalier-Smith, T., 2001. Obcells as proto-organisms: membrane heredity, lithophosphorylation, and the origins of the genetic code, the first cells, and photosynthesis. J. Mol. Evol. 53 (4-5), 555–595.

Chaban, B., Ng, S.Y.M., Jarrell, K.F., 2006. Archaeal habitats—from the extreme to the ordinary. Can. J. Microbiol. 52 (2), 73–116.

Chernykh, N.A., Vasilyeva, L.V., Semenov, A.M., Lysenko, A.M., 1988. DNA homology in prostecobacteria of *Stella* genus. Izvestiya Akademii Nauk SSSR Seriya Biologicheskaya (5), 776–779 (Russian).

Cholakova, D., Denkov, N., Tcholakova, S., Lesov, I., Smoukov, S.K., 2016. Control of drop shape transformations in cooled emulsions. Adv. Colloid Interface Sci. 235, 90–107.

Cholakova, D., Valkova, Z., Tcholakova, S., Denkov, N.D., Smoukov, S.K., 2017. "Self-shaping" of multicomponent drops. Langmuir. 33 (23), 5696–5706.

Chong, P.L.G., 2010. Archaebacterial bipolar tetraether lipids: physico-chemical and membrane properties. Chem. Phys. Lipids 163 (3), 253–265.

Chong, D.T., Liu, X.S., Ma, H.J., Huang, G.Y., Han, Y.L., Cui, X.Y., Yan, J.J., Xu, F., 2015. Advances in fabricating double-emulsion droplets and their biomedical applications. Microfluid. Nanofluid. 19 (5), 1071–1090.

Cockayne, D., 2008. The radial distribution function of amorphous materials. In: Ratinac, K.R. (Ed.), 50 Great Moments: Celebrating the Golden Jubilee of the University of Sydney's Electron Microscope Unit. Sydney University Press, Sydney, Australia, pp. 197–200.

Cohen, A.E., Mahadevan, L., 2003. Kinks, rings, and rackets in filamentous structures. Proc. Natl. Acad. Sci. U. S. A. 100 (21), 12141–12146.

Connelly, R., 2008. Rigidity of packings. Eur. J. Comb. 29 (8), 1862–1871.

Connelly, R., Gortler, S.J., 2015. Iterative universal rigidity. Discret. Comput. Geom. 53 (4), 847–877.

Connelly, R., Guest, S.D., 2015. Frameworks, Tensegrities and Symmetry: Understanding Stable Structures. http://www.math.cornell.edu/~web7510/framework.pdf.

Connelly, R., Shen, J.D., Smith, A.D., 2014. Ball packings with periodic constraints. Discret. Comput. Geom. 52 (4), 754–779.

Crawford, R.M., 1981. Valve formation in diatoms and the fate of the silicalemma and plasmalemma. Protoplasma 106 (1-2), 157–166.

Cullity, B.D., Graham, C.D., 2011. Introduction to Magnetic Materials, second ed. Wiley, Hoboken, New Jersey, USA.

Damer, B., Deamer, D., 2015. Coupled phases and combinatorial selection in fluctuating hydrothermal pools: a scenario to guide experimental approaches to the origin of cellular life. Life 5 (1), 872–887.

De Rosa, M., Gambacorta, A., Gliozzi, A., 1986. Structure, biosynthesis, and physicochemical properties of archaebacterial lipids. Microbiol. Rev. 50 (1), 70–80.

Denkov, N., Tcholakova, S., Lesov, I., Cholakova, D., Smoukov, S.K., 2015. Self-shaping of oil droplets via the formation of intermediate rotator phases upon cooling. Nature 528 (7582), 392–395.

Denkov, N., Cholakova, D., Tcholakova, S., Smoukov, S.K., 2016. On the mechanism of drop self-shaping in cooled emulsions. Langmuir 32 (31), 7985–7991.

Diaz Moreno, M., Ma, K., Schoenung, J., Dávila, L.P., 2015. An integrated approach for probing the structure and mechanical properties of diatoms: toward engineered nanotemplates. Acta Biomater. 25, 313–324.

Döbereiner, H.G., Evans, E., Kraus, M., Seifert, U., Wortis, M., 1997. Mapping vesicle shapes into the phase diagram: a comparison of experiment and theory. Phys. Rev. E 55 (4), 4458–4474.

Driessen, A.J.M., Albers, S.V., 2007. Membrane adaptations of (hyper)thermophiles to high temperatures. In: Gerday, C., Glansdorff, N. (Eds.), Physiology and Biochemistry of Extremophiles. American Society for Microbiology, Washington, DC, pp. 104–116.

Dubois, M., Demé, B., Gulik-Krzywicki, T., Dedieu, J.C., Vautrin, C., Désert, S., Perez, E., Zemb, T., 2001. Self-assembly of regular hollow icosahedra in salt-free catanionic solutions. Nature 411 (6838), 672–675.

Duggin, I.G., Aylett, C.H.S., Walsh, J.C., Michie, K.A., Wang, Q., Turnbull, L., Dawson, E.M., Harry, E.J., Whitchurch, C.B., Amos, L.A., Lowe, J., 2015. CetZ tubulin-like proteins control archaeal cell shape. Nature 519 (7543), 362–365.

Dworkin, J.P., Deamer, D.W., Sandford, S.A., Allamandola, L.J., 2001. Self-assembling amphiphilic molecules: synthesis in simulated interstellar/precometary ices. Proc. Natl. Acad. Sci. U. S. A. 98 (3), 815–819.

Dyall-Smith, M.L., Pfeiffer, F., Klee, K., Palm, P., Gross, K., Schuster, S.C., Rampp, M., Oesterhelt, D., 2011. *Haloquadratum walsbyi*: limited diversity in a global pond. PLoS One 6 (6). e20968, #e20968.

Elferink, M.G.L., de Wit, J.G., Driessen, A.J.M., Konings, W.N., 1994. Stability and proton permeability of liposomes composed of archaeal tetraether lipids. BBA-Biomembranes 1193 (2), 247–254.

Emerson, D., Chauhan, S., Oriel, P., Breznak, J.A., 1994. *Haloferax* SP D1227, a halophilic archaeon capable of growth on aromatic-compounds. Arch. Microbiol. 161 (6), 445–452.

Engelhardt, H., 2007a. Are S-layers exoskeletons? the basic function of protein surface layers revisited. J. Struct. Biol. 160 (2), 115–124.

Engelhardt, H., 2007b. Mechanism of osmoprotection by archaeal S-layers: a theoretical study. J. Struct. Biol. 160 (2), 190–199.

Engelhardt, H., Peters, J., 1998. Structural research on surface layers: a focus on stability, surface layer homology domains, and surface layer–cell wall interactions. J. Struct. Biol. 124 (2), 276–302.

Engelhardt, H., Gerblrieger, S., Santarius, U., Baumeister, W., 1991. The three-dimensional structure of the regular surface protein of Comamonas acidovorans derived from native outer membranes and reconstituted two-dimensional crystals. Mol. Microbiol. 5 (7), 1695–1702.

Ettema, T.J.G., Lindås, A.-C., Bernander, R., 2011. An actin-based cytoskeleton in archaea. Mol. Microbiol. 80 (4), 1052–1061.

Fan, Q., Relini, A., Cassinadri, D., Gambacorta, A., Gliozzi, A., 1995. Stability against temperature and external agents of vesicles composed of archaeal bolaform lipids and egg PC. BBA-Biomembranes 1240 (1), 83–88.

Fiore, M., Strazewski, P., 2016. Prebiotic lipidic amphiphiles and condensing agents on the early Earth. Life (Basel, Switzerland) 6 (2). #17.

Flemons, T., 2016. Undefining tensegrity, personal communication.

Forbes, C.C., DiVittorio, K.M., Smith, B.D., 2006. Bolaamphiphiles promote phospholipid translocation across vesicle membranes. J. Am. Chem. Soc. 128 (28), 9211–9218.

Fredrickson, H.L., Rijpstra, W.I.C., Tas, A.C., van der Greef, J., LaVos, G.F., de Leeuw, J.W., 1989. Chemical characterization of benthic microbial assemblages. In: Cohen, Y., Rosenberg, E. (Eds.), Microbial Mats: Physiological Ecology of Benthic Microbial Communities. American Society for Microbiology, Washington, DC, pp. 455–468.

Fritz, I., Strompl, C., Abraham, W.R., 2004. Phylogenetic relationships of the genera *Stella Labrys and Angulomicrobium within the 'Alphaproteobacteria' and description of Angulomicrobium amanitiforme sp*. Int. J. Syst. Evol. Microbiol. 54, 651–657.

Fuhrmann, T., Landwehr, S., El Rharbi-Kucki, M., Sumper, M., 2004. Diatoms as living photonic crystals. Appl. Phys. B Lasers Opt. 78 (3-4), 257–260.

Garnett, A., 2015. Basket Bowl. http://www.alexgarnett.com/product/basket-bowl.

Gerakines, P.A., Moore, M.H., Hudson, R.L., 2001. Energetic processing of laboratory ice analogs: UV photolysis versus ion bombardment. J. Geophys. Res. Planets 106 (E12), 33381–33385.

Gerhart, J.C., Kirschner, M.W., 1997. Cells, Embryos and Evolution: Toward a Cellular and Developmental Understanding of Phenotypic Variation and Evolutionary Adaptability. Blackwell Science, Malden, MA.

Gerhart, J., Kirschner, M., 2007. The theory of facilitated variation. Proc. Natl. Acad. Sci. U. S. A. 104 (Suppl. 1), 8582–8589.

Ghai, R., Pašić, L., Beatriz Fernández, A., Martin-Cuadrado, A.-B., Mizuno, C.M., McMahon, K.D., Papke, R.T., Stepanauskas, R., Rodriguez-Brito, B., Rohwer, F., Sánchez-Porro, C., Ventosa, A., Rodríguez-Valera, F., 2011. New abundant microbial groups in aquatic hypersaline environments. Sci. Rep. 1. #135.

Gliozzi, A., Rolandi, R., De Rosa, M., Gambacorta, A., 1983. Monolayer black membranes from bipolar lipids of archaebacteria and their temperature-induced structural changes. J. Membr. Biol. 75 (1), 45–56.

Gnutt, D., Ebbinghaus, S., 2016. The macromolecular crowding effect—from *in vitro* into the cell. Biol. Chem. 397 (1), 37–44.

Gonzalez, S., Thornton, A.R., Luding, S., 2014. Free cooling phase-diagram of hard-spheres with short- and long-range interactions. Eur. Phys. J.-Special Topics 223 (11), 2205–2225.

Gordon, R., 1980. Monte Carlo methods for cooperative Ising models. In: Karreman, G. (Ed.), Cooperative Phenomena in Biology. Pergamon Press, New York, pp. 189–241.

Gordon, R., 1999. The Hierarchical Genome and Differentiation Waves: Novel Unification of Development, Genetics and Evolution. World Scientific & Imperial College Press, Singapore & London.

Gordon, R., Brodland, G.W., 1987. The cytoskeletal mechanics of brain morphogenesis. Cell state splitters cause primary neural induction. Cell Biophys. 11 (1), 177–238.

Gordon, R., Drum, R.W., 1994. The chemical basis of diatom morphogenesis. Int. Rev. Cytol. 150 (243–372), 421–422.

Gordon, N.K., Gordon, R., 2016a. Embryogenesis Explained. World Scientific Publishing, Singapore.

Gordon, N.K., Gordon, R., 2016b. The organelle of differentiation in embryos: the cell state splitter [invited review]. Theor. Biol. Med. Model. 13, 11 (Special issue: Biophysical Models of Cell Behavior, Guest Editor: Jack A. Tuszynski).

Gordon, R., Tiffany, M.A., 2011. Possible buckling phenomena in diatom morphogenesis. In: Seckbach, J., Kociolek, J.P. (Eds.), The Diatom World. Springer, Dordrecht, The Netherlands, pp. 245–272.

Gordon, R., Björklund, N.K., Nieuwkoop, P.D., 1994. Dialogue on embryonic induction and differentiation waves. Int. Rev. Cytol. 150, 373–420.

Gordon, R., Losic, D., Tiffany, M.A., Nagy, S.S., Sterrenburg, F.A.S., 2009. The Glass Menagerie: diatoms for novel applications in nanotechnology. Trends Biotechnol. 27 (2), 116–127.

Gotoh, K., 2012. Particulate Morphology: Mathematics Applied to Particle Assemblies. Elsevier Science, Amsterdam, Netherlands.

Grant, W.D., Larsen, H., 1989. The genus *Haloarcula*. In: Staley, J.T., Bryant, M.P., Pfennig, N., Holt, J.G. (Eds.), Bergey's Manual of Determinative Bacteriology, nineth ed. Williams & Wilkins, Baltimore, pp. 2216–2233.

Grason, G.M., 2016. Perspective: geometrically frustrated assemblies. J. Chem. Phys. 145 (11). #110901.

Greeley, D., Crapo, J.D., Vollmer, R.T., 1978. Estimation of the mean caliper diameter of cell nuclei. 1. serial section reconstruction method and endothelial nuclei from human lung. J. Microsc.-Oxford 114, 31–39.

Greulich, K.O., 2005. Single-molecule studies on DNA and RNA. ChemPhysChem 6 (12), 2458–2471.

Griffiths, R.B., 1969. Rigorous results for Ising ferromagnets of arbitrary spin. J. Math. Phys. 10 (9), 1559–1565.

Gross, R., Fouxon, I., Lancet, D., Markovitch, O., 2014. Quasispecies in population of compositional assemblies. BMC Evol. Biol. 14. #265.

Guixa-Boixereu, N., Calderón-Paz, J.I., Heldal, M., Bratbak, G., Pedrós-Alió, C., 1996. Viral lysis and bacterivory as prokaryotic loss factors along a salinity gradient. Aquat. Microb. Ecol. 11 (3), 215–227.

Gupta, R.S., Naushad, S., Baker, S., 2015. Phylogenomic analyses and molecular signatures for the class *Halobacteria* and its two major clades: a proposal for division of the class *Halobacteria* into an emended order Halobacteriales and two new orders, *Haloferacales* ord. nov and *Natrialbales* ord. nov., containing the novel families *Haloferacaceae* fam. nov and *Natrialbaceae* fam. nov. Int. J. Syst. Evol. Microbiol. 65, 1050–1069.

Gürsoy, G., Xu, Y., Kenter, A.L., Liang, J., 2014. Spatial confinement is a major determinant of the folding landscape of human chromosomes. Nucleic Acids Res. 42 (13), 8223–8230.

Gutiérrez, A., Gordon, R., Dávila, L.P., 2017. Deformation modes and structural response of diatom shells. J. Mater. Sci. Eng. Adv. Technol. 15 (2), 105–134.

Guttman, S., Ocko, B.M., Deutsch, M., Sloutskin, E., 2016a. From faceted vesicles to liquid icoshedra: where topology and crystallography meet. Curr. Opin. Colloid Interface Sci. 22, 35–40.

Guttman, S., Sapir, Z., Schultz, M., Butenko, A.V., Ocko, B.M., Deutsch, M., Sloutskin, E., 2016b. How faceted liquid droplets grow tails. Proc. Natl. Acad. Sci. 113 (3), 493–496.

Haas, P.A., Goldstein, R.E., Smoukov, S.K., Cholakova, D., Denkov, N., 2016. A Theory of Shape-Shifting Droplets. https://arxiv.org/abs/1609.00584.

Haas, P.A., Goldstein, R.E., Smoukov, S.K., Cholakova, D., Denkov, N., 2017. Theory of shape-shifting droplets. Phys. Rev. Lett. 118. #088001.

Hamamoto, T., Takashina, T., Grant, W.D., Horikoshi, K., 1988. Asymmetric cell division of a triangular halophilic archaebacterium. FEMS Microbiol. Lett. 56 (2), 221–224.

Hanczyc, M.M., 2011. Metabolism and motility in prebiotic structures. Philos. Trans. R. Soc. B 366 (1580), 2885–2893.

Hanczyc, M.M., 2014. Droplets: unconventional protocell model with life-like dynamics and room to grow. Life-Basel 4 (4, Sp. Iss. SI), 1038–1049.

Hanczyc, M.M., Ikegami, T., 2010. Chemical basis for minimal cognition. Artif. Life 16 (3), 233–243.

Hanczyc, M.M., Monnard, P.-A., 2017. Primordial membranes: more than simple container boundaries. Curr. Opin. Chem. Biol. 40 (October), 78–86.

Hazelaar, S., van der Strate, H.J., Gieskes, W.W.C., Vrieling, E.G., 2005. Monitoring rapid valve formation in the pennate diatom *Navicula salinarum* (Bacillariophyceae). J. Phycol. 41 (2), 354–358.

Hemmingsen, B.B., Hemmingsen, E.A., 1980. Rupture of the cell envelope by induced intracellular gas-phase expansion in gas vacuolate bacteria. J. Bacteriol. 143 (2), 841–846.

Henning, T., Salama, F., 1998. Carbon in the Universe. Science 282 (5397), 2204–2210.

Higashi-Fujime, S., 1980. Active movement in vitro of bundle of microfilaments isolated from *Nitella* cell. J. Cell Biol. 87 (3), 569–578.

Hirsch, P., 1974. Budding bacteria. Annu. Rev. Microbiol. 28, 391–444.

Hirsch, P., Schlesner, H., 1981. The genus *Stella*. In: Starr, M.P., Stolp, H., Trüper, H.G., Balows, A., Schlegel, H.G. (Eds.), The Prokaryotes: A Handbook on Habitats, Isolation, and Identification of Bacteria. Springer, Heidelberg, pp. 461–465.

Hirsch, P., Müller, M., Schlesner, N., 1977. New aquatic budding and prosthecate bacterin and their laxonomic position. In: Skinner, F.A., Shewan, J.M. (Eds.), Aquatic Microbiology. Academic Press, London, pp. 107–133.

Holm, N.G., Charlou, J.L., 2001. Initial indications of abiotic formation of hydrocarbons in the Rainbow ultramafic hydrothermal system, Mid-Atlantic Ridge. Earth Planet. Sci. Lett. 191 (1-2), 1–8.

Horibe, N., Hanczyc, M.M., Ikegami, T., 2011. Mode switching and collective behavior in chemical oil droplets. Entropy 13 (3), 709–719.

Horikoshi, K., Aono, R., Nakamura, S., 1993. The triangular halophilic archaebacterium *Haloarcula japonica* strain TR-1. Experientia 49 (6-7), 497–502.

Horizon Group, 2017a. Adhesive Magnetic Buttons. http://craftprojectideas.com/products/magnetic-buttons-with-foam-adhesive/.

Horizon Group, 2017b. Magnetic Buttons. http://craftprojectideas.com/products/magnetic-buttons/.

Huang, K.C., Mukhopadhyay, R., Wen, B., Gitai, Z., Wingreen, N.S., 2008. Cell shape and cell-wall organization in Gram-negative bacteria. Proc. Natl. Acad. Sci. U. S. A. 105 (49), 19282–19287.

Hunding, A., Kepes, F., Lancet, D., Minsky, A., Norris, V., Raine, D., Sriram, K., Root-Bernstein, R., 2006. Compositional complementarity and prebiotic ecology in the origin of life. Bioessays 28 (4), 399–412.

Iijima, S., Brabec, C., Maiti, A., Bernholc, J., 1996. Structural flexibility of carbon nanotubes. J. Chem. Phys. 104 (5), 2089–2092.

Ingber, D.E., Dike, L., Liley, H., Hansen, L., Karp, S., Maniotis, A.J., McNamee, H., Mooney, D., Plopper, G., Sims, J., Wang, N., 1994. Cellular tensegrity: exploring how mechanical changes in the cytoskeleton regulate cell growth, migration, and tissue pattern during morphogenesis. Int. Rev. Cytol. 150, 173–224.

Ingber, D.E., Wang, N., Stamenović, D., 2014. Tensegrity, cellular biophysics, and the mechanics of living systems. Rep. Prog. Phys. 77 (4). #046603.

Jacquemet, A., Barbeau, J., Lemiègre, L., Benvegnu, T., 2009. Archaeal tetraether bipolar lipids: structures, functions and applications. Biochimie 91 (6), 711–717.

Jain, S., Caforio, A., Driessen, A.J.M., 2014. Biosynthesis of archaeal membrane ether lipids. Front. Microbiol. 5. #641.

Jarosch, R., 1956. Plasmaströmung und Chloroplastenrotation bei Characeen. Phyton (Argentina) 6, 87–107.

Jarosch, R., 1957. Zur Mechanik der Protoplasmafibrillenbewegung. Biochim. Biophys. Acta 25 (1), 204–205.

Javor, B., Requadt, C., Stoeckenius, W., 1982. Box-shaped halophilic bacteria. J. Bacteriol. 151 (3), 1532–1542.

Jicepix, 2015. vieux ballon crevé. https://eu.fotolia.com/id/33775052; https://stock.adobe.com/ca/search?k=33775052&filters%5Bcontent_type%3Aphoto%5D=1&filters%5Bcontent_type%3Aillustration%5D=1&filters%5Bcontent_type%3Azip_vector%5D=1&filters%5Bcontent_type%3Avideo%5D=1&filters%5Bcontent_type%3Atemplate%5D=1&filters%5Bcontent_type%3A3d%5D=1&load_type=homepage.

Jin, L., Takei, A., Hutchinson, J.W., 2015. Mechanics of wrinkle/ridge transitions in thin film/substrate systems. J. Mech. Phys. Solids 81, 22–40.

Kamekura, M., 1998. Diversity of extremely halophilic bacteria. Extremophiles 2 (3), 289–295.

Kates, M., 1992. Archaebacterial lipids: structure, biosynthesis and function. Biochem. Soc. Symp. 58, 51–72.

Kauffman, S., 2013. What is life, and can we create it? Bioscience 63 (8), 609–610.

Kessel, M., Cohen, Y., 1982. Ultrastructure of square bacteria from a brine pool in Southern Sinai. J. Bacteriol. 150 (2), 851–860.

Khatib, F., DiMaio, F., Cooper, S., Kazmierczyk, M., Gilski, M., Krzywda, S., Zabranska, H., Pichova, I., Thompson, J., Popovic, Z., Jaskolski, M., Baker, D., Foldit Contenders, G., Foldit Void Crushers, G., 2011. Crystal structure of a monomeric retroviral protease solved by protein folding game players. Nat. Struct. Mol. Biol. 18 (10), 1175–1177.

Kim, S.G., Bae, Y.C., 2003. Salt-induced protein precipitation in aqueous solution: single and binary protein systems. Macromol. Res. 11 (1), 53–61.

Kindzelskii, A.L., Xue, W., Todd, R.F., Boxer, L.A., Petty, H.R., 1994. Aberrant capping of membrane proteins on neutrophils from patients with leukocyte adhesion deficiency. Blood 83 (6), 1650–1655.

Kirschner, M., Gerhart, J., 1998. Evolvability. Proc. Natl. Acad. Sci. U. S. A. 95 (15), 8420–8427.

Klibanov, A.M., 2001. Improving enzymes by using them in organic solvents. Nature 409 (6817), 241–246.

Klibanov, A.M., 2003. Asymmetric enzymatic oxidoreductions in organic solvents. Curr. Opin. Biotechnol. 14 (4), 427–431.

Knoche, S., Kierfeld, J., 2011. Buckling of spherical capsules. Phys. Rev. E 84 (4). #046608.

Knoche, S., Kierfeld, J., 2014a. Osmotic buckling of spherical capsules. Soft Matter 10 (41), 8358–8369.

Knoche, S., Kierfeld, J., 2014b. The secondary buckling transition: wrinkling of buckled spherical shells. Eur. Phys. J. E 37 (7). #62.

Knoche, S., Kierfeld, J., 2014c. Secondary polygonal instability of buckled spherical shells. Epl 106 (2). #24004.

König, H., 1994. Analysis of archaeal cell envelopes. In: Goodfellow, M., O'Donnell, A.G. (Eds.), Chemical Methods in Prokaryotic Systematics. John Wiley & Sons, Chichester, pp. 85–119.

Koprowski, G.J., 2013. 'Invisible' airplanes: Chinese, US race for cloaking tech. http://www.foxnews.com/tech/2013/12/17/invisible-airplanes-chinese-us-scramble-for-cloaking-tech.html.

Koudehi, M.A., Tang, H., Vavylonis, D., 2016. Simulation of the effect of confinement in actin ring formation. Biophys. J. 110 (3), 126A.

Krishnamurthy, R.V., Epstein, S., Cronin, J.R., Pizzarello, S., Yuen, G.U., 1992. Isotopic and molecular analyses of hydrocarbons and monocarboxylic acids of the Murchison meteorite. Geochim. Cosmochim. Acta 56 (11), 4045–4058.

Kuhn, D.A., 1981. The genera *Simonsiella* and *Alysiella*. In: Starr, M.P., Stolp, H., Trüper, H.G., Balows, A., Schlegel, H.G. (Eds.), The Prokaryotes: A Handbook on Habitats, Isolation, and Identification of Bacteria. Springer-Verlag, Berlin, pp. 390–399.

Kuroda, K., 1964. Behavior of naked cytoplasmic drops isolated from plant cells. In: Allen, R.D., Kamiya, N. (Eds.), Primitive Motile Systems in Cell Biology. Academic Press, New York, pp. 31–41.

Kuroda, K., 1968. Protoplasmic streaming in a giant plant cell. Saibō Kagaku Shimpojiumu 19, 37–43 (Japanese).

Kwok, S., 2007. Physics and Chemistry of the Interstellar Medium. University Science Books, Sausalito, California, USA.

Kysela, D.T., Brown, P.J.B., Huang, K.C., Brun, Y.V., 2013. Biological consequences and advantages of asymmetric bacterial growth. Annu. Rev. Microbiol. 67, 417–435.

Laale, H.W., 1984. Polyembryony in teleostean fishes: double monstrosities and triplets. J. Fish Biol. 24, 711–719.

Lafitskaya, T.N., Vasilieva, L.V., 1976. A new triangular bacterium. Mikrobiologiya 45 (5), 812–816.

Lasbury, M., 2013. How Prokaryotes Shape Up. http://biologicalexceptions.blogspot.ca/2013/08/how-prokaryotes-shape-up.html.

Lelkes, P.I., Goldenberg, D., Gliozzi, A., Derosa, M., Gambacorta, A., Miller, I.R., 1983. Vesicles from mixtures of bipolar archaebacterial lipids with egg phosphatidylcholine. Biochim. Biophys. Acta 732 (3), 714–718.

Leong, T.S.H., Zhou, M.F., Kukan, N., Ashokkumar, M., Martin, G.J.O., 2017. Preparation of water-in-oil-in-water emulsions by low frequency ultrasound using skim milk and sunflower oil. Food Hydrocoll. 63, 685–695.

Levin, S., 2006a. Tensegrity: the new biomechanics. In: Hutson, M., Ellis, R. (Eds.), Textbook of Musculoskeletal Medicine. Oxford University Press, Oxford, pp. 69–80.

Levin, S.M., 2006b. Biotensegrity & Dynamic Anatomy [DVD]. Ezekiel Biomechanics Group, McLean, VA.

Liao, Y., Williams, T.J., Ye, J., Charlesworth, J., Burns, B.P., Poljak, A., Raftery, M.J., Cavicchioli, R., 2016. Morphological and proteomic analysis of biofilms from the Antarctic archaeon, *Halorubrum lacusprofund*. Sci. Rep. 6. #37454.

Lim, H.W.G., Wortis, M., Mukhopadhyay, R., 2002. Stomatocyte–discocyte–echinocyte sequence of the human red blood cell: Evidence for the bilayer– couple hypothesis from membrane mechanics. Proc. Natl. Acad. Sci. 99 (26), 16766–16769.

Liu, J., Kaksonen, M., Drubin, D.G., Oster, G., 2006. Endocytic vesicle scission by lipid phase boundary forces. Proc. Natl. Acad. Sci. U. S. A. 103 (27), 10277–10282.

Liu, H.L., Wu, Z.F., Li, M., Zhang, F., Zheng, H.J., Han, J., Liu, J.F., Zhou, J., Wang, S.Y., Xiang, H., 2011. Complete genome sequence of *Haloarcula hispanica*, a model Haloarchaeon for studying genetics, metabolism, and virus-host interaction. J. Bacteriol. 193 (21), 6086–6087.

Lobasso, S., Lopalco, P., Mascolo, G., Corcelli, A., 2008. Lipids of the ultra-thin square halophilic archaeon *Haloquadratum walsbyi*. Archaea 2 (3), 177–183.

Logan, G.-L., 2015. Deflated Ball. http://www.acclaimimages.com/_gallery/_pages/0017-0309-1921-5000.html.

Lombard, J., López-García, P., Moreira, D., 2012. The early evolution of lipid membranes and the three domains of life. Nat. Rev. Microbiol. 10 (7), 507–515.

Macleod, G., McKeown, C., Hall, A.J., Russell, M.J., 1994. Hydrothermal and oceanic pH conditions of possible relevance to the origin of life. Orig. Life Evol. Biosph. 24 (1), 19–41.

Malfatti, S., Tindall, B.J., Schneider, S., Fähnrich, R., Lapidus, A., LaButtii, K., Copeland, A., Del Rio, T.G., Nolan, M., Chen, F., Lucas, S., Tice, H., Cheng, J.-F., Bruce, D., Goodwin, L., Pitluck, S., Anderson, I., Pati, A., Ivanova, N., Mavromatis, K., Chen, A., Palaniappan, K., D'Haeseleer, P., Göker, M., Bristow, J., Eisen, J.A., Markowitz, V., Hugenholtz, P., Kyrpides, N.C., Klenk, H.-P., Chain, P., 2009. Complete genome sequence of *Halogeometricum borinquense* type strain (PR3(T)). Stand Genomic Sci. 1 (2), 150–158.

Mannige, R.V., 2013. Two modes of protein sequence evolution and their compositional dependencies. Phys. Rev. E 87 (6). #062714.

Mannige, R.V., Brooks, C.L., Shakhnovich, E.I., 2012. A universal trend among proteomes indicates an oily last common ancestor. PLoS Comp. Biol. 8 (12). e1002839, #e1002839.

Margolin, W., 2009. Sculpting the bacterial cell. Curr. Biol. 19 (17), R812–822.

Marko, J.F., Neukirch, S., 2012. Competition between curls and plectonemes near the buckling transition of stretched supercoiled DNA. Phys. Rev. E 85 (1). #011908.

Markovitch, O., Lancet, D., 2014. Multispecies population dynamics of prebiotic compositional assemblies. J. Theor. Biol. 357, 26–34.

Marmottant, P., Bouakaz, A., de Jong, N., Quilliet, C., 2011. Buckling resistance of solid shell bubbles under ultrasound. J. Acoust. Soc. Am. 129 (3), 1231–1239.

Marr, A.G., Ingraham, J.L., 1962. Effect of temperature on the composition of fatty acids in *Escherichia coli*. J. Bacteriol. 84 (6). 1260.

Martin, C.C., Gordon, R., 1997. Ultrastructural analysis of the cell state splitter in ectoderm cells differentiating to neural plate and epidermis during gastrulation in embryos of the axolotl *Ambystoma mexicanum*. Russ. J. Dev. Biol. 28 (2), 71–80.

Medlin, L.K., 2016. Evolution of the diatoms: major steps in their evolution and a review of the supporting molecular and morphological evidence. Phycologia 55 (1), 79–103.

Medlin, L.K., Kaczmarska, I., 2004. Evolution of the diatoms: V. Morphological and cytological support for the major clades and a taxonomic revision. Phys. Chem. Chem. Phys. 43 (3), 245–270.

Mehrshad, M., Amoozegar, M.A., Makhdoumi, A., Rasooli, M., Asadi, B., Schumann, P., Ventosa, A., 2015. *Halovarius luteus* gen. nov., sp nov., an extremely halophilic archaeon from a salt lake. Int. J. Syst. Evol. Microbiol. 65, 2420–2425.

Mehrshad, M., Amoozegar, M.A., Makhdoumi, A., Fazeli, S.A.S., Farahani, H., Asadi, B., Schumann, P., Ventosa, A., 2016. *Halosiccatus urmianus* gen. nov., sp nov., a haloarchaeon from a salt lake. Int. J. Syst. Evol. Microbiol. 66, 725–730.

Melchior, D.L., 1982. Lipid phase transitions and regulation of membrane fluidity in prokaryotes. Curr. Top. Membr. Trans. 17, 263–316.

Melchior, D.L., Steim, J.M., 1976. Thermotropic transitions in biomembranes. Annu. Rev. Biophys. Bioeng. 5, 205–238.

Mescher, M.F., Strominger, J.L., 1976. Structural (shape-maintaining) role of the cell surface glycoprotein of *Halobacterium salinarium*. Proc. Natl. Acad. Sci. U. S. A. 73 (8), 2687–2691.

Miller, S.L., 1953. A production of amino acids under possible primitive earth conditions. Science 117 (3046), 528–529.

Miller, S.L., Urey, H., 1959. Organic compound synthesis on the primitive earth. Science 130 (3370), 245–251.

Mirghani, Z., Bertoia, D., Gliozzi, A., De Rosa, M., Gambacorta, A., 1990. Monopolar-bipolar lipid interactions in model membrane systems. Chem. Phys. Lipids 55 (2), 85–96.

Mitic, S., 2015. Photo by Slobo Mitic/iStock. http://www.stthomas.edu/news/failure-youth-sports/.

Miyashita, Y., Ohmae, E., Nakasone, K., Katayanagi, K., 2015. Effects of salt on the structure, stability, and function of a halophilic dihydrofolate reductase from a hyperhalophilic archaeon, Haloarcula japonica strain TR-1. Extremophiles 19 (2), 479–493.

Mori, K., Nurcahyanto, D.A., Kawasaki, H., Lisdiyanti, P., Yopi Suzuki, K., 2016. *Halobium palmae* gen. nov., sp nov., an extremely halophilic archaeon isolated from a solar saltern. Int. J. Syst. Evol. Microbiol. 66, 3799–3804.

Mou, Y.Z., Qiu, X.X., Zhao, M.L., Cui, H.L., Oh, D., Dyall-Smith, M.L., 2012. *Halohasta litorea* gen. nov sp nov., and *Halohasta litchfieldiae* sp nov., isolated from the Daliang aquaculture farm, China and from Deep Lake, Antarctica, respectively. Extremophiles 16 (6), 895–901.

Mukhopadhyay, R., Huang, K.C., Wingreen, N.S., 2008. Lipid localization in bacterial cells through curvature-mediated microphase separation. Biophys. J. 95, 1034–1049.

Mullakhanbhai, M.F., Larsen, H., 1975. *Halobacterium volcanii* spec. nov., a Dead Sea halobacterium with a moderate salt requirement. Arch. Microbiol. 104 (3), 107–114.

Murray, J.D., 2012. Why are there no 3-headed monsters? mathematical modeling in biology. Not. Am. Math. Soc. 59 (6), 785–795.

Mushenheim, P.C., Trivedi, R.R., Roy, S.S., Arnold, M.S., Weibel, D.B., Abbott, N.L., 2015. Effects of confinement, surface-induced orientations and strain on dynamical behaviors of bacteria in thin liquid-crystalline films. Soft Matter 11 (34), 6821–6831.

Mutlu, M.B., Guven, K., 2015. Bacterial diversity in Çamalti Saltern, Turkey. Pol. J. Microbiol. 64 (1), 37–45.

Nakano, S.-i., Miyoshi, D., Sugimoto, N., 2014. Effects of molecular crowding on the structures, interactions, and functions of nucleic acids. Chem. Rev. 114 (5), 2733–2758.

Naveh, B., Sipper, M., Lancet, D., Shenhav, B., 2004. Lipidia: An Artificial Chemistry of Self-Replicating Assemblies of Lipid-Like Molecules.

Newman, S.A., 2014. Physico-genetics of morphogenesis: the hybrid nature of developmental mechanisms. In: Minelli, A., Pradeu, T. (Eds.), Towards a Theory of Development. Oxford University Press, Oxford, pp. 95–113.

Newman, S.A., Comper, W.D., 1990. 'Generic' physical mechanisms of morphogenesis and pattern formation. Development 110 (1), 1–18.

Nishiyama, Y., Takashina, T., Grant, W.D., Horikoshi, K., 1992. Ultrastructure of the cell wall of the triangular halophilic archaebacterium *Haloarcula japonica* strain TR-1. FEMS Microbiol. Lett. 99 (1), 43–48.

Nishiyama, Y., Nakamura, S., Aono, R., Horikoshi, K., 1995. Electron microscopy of halophilic Archaea. In: DasSarma, S., Fleischmann, E.M. (Eds.), Archaea: A Laboratory Manual. Halophiles. Cold Spring Harbor Laboratory, Cold Spring Harbor, NY, pp. 29–33.

Oh, D., Porter, K., Russ, B., Burns, D., Dyall-Smith, M., 2010. Diversity of *Haloquadratum* and other haloarchaea in three, geographically distant, Australian saltern crystallizer ponds. Extremophiles 14 (2), 161–169.

Okubo, M., Minami, H., Morikawa, K., 2001. Production of micron-sized, monodisperse, transformable rugby-ball-like-shaped polymer particles. Colloid Polym. Sci. 279 (9), 931–935.

Oren, A., 1993. Characterization of the halophilic archaeal community in saltern crystallizer ponds by means of polar lipid analysis. Int. J. Salt Lake Res. 3, 15–29.

Oren, A., 1994. The ecology of the extremely halophilic archaea. FEMS Microbiol. Rev. 13 (4), 415–439.

Oren, A., 1999. The enigma of square and triangular halophilic archaea. In: Seckbach, J. (Ed.), Enigmatic Microorganisms and Life in Extreme Environmental Habitats. Kluwer Academic Publishers, Dordrecht, pp. 337–355.

Oren, A., 2005. Microscopic examination of microbial communities along a salinity gradient in saltern evaporation ponds: a 'halophilic safari'. In: Gunde-Cimerman, N., Oren, A., Plemenitas, A. (Eds.), Adaptation to Life at High Salt Concentrations in Archaea, Bacteria, and Eukarya. Springer, Dordrecht, The Netherlands, pp. 41–57.

Oren, A., Ginzburg, M., Ginzburg, B.Z., Hochstein, L.I., Volcani, B.E., 1990. *Haloarcula marismortui* (Volcani) sp. nov., nom. rev., an extremely halophilic bacterium from the Dead Sea. Int. J. Syst. Bacteriol. 40 (2), 209–210.

Oren, A., Duker, S., Ritter, S., 1996. The polar lipid composition of Walsby's square bacterium. FEMS Microbiol. Lett. 138 (2-3), 135–140.

Oren, A., Ventosa, A., Gutiérrez, M.C., Kamekura, M., 1999. *Haloarcula quadrata* sp nov., a square, motile archaeon isolated from a brine pool in Sinai (Egypt). Int. J. Syst. Bacteriol. 49, 1149–1155.

Otozai, K., Takashina, T., Grant, W.D., 1991. A novel triangular archaebacterium, *Haloarcula japonica*. In: Horikoshi, K., Grant, W.D. (Eds.), Superbugs: Microorganisms in Extreme Environments. Springer-Verlag, Berlin, pp. 61–75.

Ozawa, K., Yatsunami, R., Nakamura, S., 2000. Cloning and sequencing of *ftsZ* homolog from extremely halophilic archaeon *Haloarcula japonica* strain TR-1. Nucleic Acids Symp. Ser 44, 155–156.

Ozawa, K., Harashina, T., Yatsunami, R., Nakamura, S., 2005. Gene cloning, expression and partial characterization of cell division protein FtsZ1 from extremely halophilic archaeon *Haloarcula japonica* strain TR-1. Extremophiles 9 (4), 281–288.

Paleos, C.M., 2015. A decisive step toward the origin of life. Trends Biochem. Sci. 40 (9), 487–488.

Paleos, C.M., Tsiourvas, D., Sideratou, Z., 2004. Hydrogen bonding interactions of liposomes simulating cell-cell recognition. a review. Orig. Life Evol. Biosph. 34 (1-2), 195–213.

Parkes, K., Walsby, A.E., 1981. Ultrastructure of a gas-vacuolate square bacterium. J. Gen. Microbiol. 126, 503–506.

Parkinson, J., Brechet, Y., Gordon, R., 1999. Centric diatom morphogenesis: a model based on a DLA algorithm investigating the potential role of microtubules. Biochim. Biophys. Acta, Mol. Cell Res. 1452 (1), 89–102.

Passi, S., 2013. Tensegrity Stool. https://www.behance.net/gallery/12453419/Tensegrity-Stool.

Perera, N., Qiao, J., Blostein, D., Flemons, T., Senatore, G., Gordon, R., 2016. Biotensegrity Simulation at the Full Body and Cytoskeleton Scale. *****.

Piazza, S., 2015. In-tense robots: Motorized sculptures may represent our best chance for exploring the surfaces of other worlds. Am. Sci. 103 (4), 264–267.

Pickett-Heaps, J., Kowalski, S.E., 1981. Valve morphogenesis and the microtubule center of the diatom *Hantzschia amphioxys*. Eur. J. Cell Biol. 25 (1), 150–170.

Pickett-Heaps, J.D., Tippit, D.H., Andreozzi, J.A., 1979a. Cell division in the pennate diatom *Pinnularia*. III—the valve and associated cytoplasmic organelles. Biochem. Cell Biol. 35 (2), 195–198.

Pickett-Heaps, J.D., Tippit, D.H., Andreozzi, J.A., 1979b. Cell division in the pennate diatom *Pinnularia*. IV—valve morphogenesis. Biochem. Cell Biol. 35 (2), 199–203.

Pinot, M., Chesnel, F., Kubiak, J.Z., Arnal, I., Nedelec, F.J., Gueroui, Z., 2009. Effects of confinement on the self-organization of microtubules and motors. Curr. Biol. 19 (11), 954–960.

Pomp, W., Schakenraad, K.K., van Hoorn, H., Balcioğlu, H.E., Danen, E.H.J., Giomi, L., Schmidt, T., 2016. Balance of isotropic and directed forces determines cell shape. Biophys. J. 110 (3, Suppl. 1). 305A.

Powell, R., Mariscal, C., 2015. Convergent evolution as natural experiment: the tape of life reconsidered. Interface Focus 5 (6). #20150040.

Pum, D., Sleytr, U.B., 2014. Reassembly of S-layer proteins. Nanotechnology 25 (31). #312001.

Pum, D., Messner, P., Sleytr, U.B., 1991. Role of the S layer in morphogenesis and cell division of the archaebacterium *Methanocorpusculum sinense*. J. Bacteriol. 173 (21), 6865–6873.

Quilliet, C., 2012. Numerical deflation of beach balls with various Poisson's ratios: from sphere to bowl's shape. Eur. Phys. J. E 35 (6). #48.

Quilliet, C., Zoldesi, C., Riera, C., van Blaaderen, A., Imhof, A., 2008. Anisotropic colloids through non-trivial buckling [Erratum: 32(4), 419-420]. Eur. Phys. J. E 27 (1), 13–20.

Ram-Mohan, N., Oren, A., Papke, R.T., 2016. Analysis of the bacteriorhodopsin-producing haloarchaea reveals a core community that is stable over time in the salt crystallizers of Eilat, Israel. Extremophiles 20 (5), 747–757.

Reimer, B., Schlesner, H., 1989. Isolation of 11 strains of star-shaped bacteria from aquatic habitats and investigation of their taxonomic position. Syst. Appl. Microbiol. 12 (2), 156–158.

Rimoli, J.J., 2016. On the impact tolerance of tensegrity-based planetary landers. In: 57th AIAA/ASCE/AHS/ASC Structures, Structural Dynamics, and Materials Conference. https://doi.org/10.2514/2516.2016-1511.

Robledo, A., Rowlinson, J.S., 1986. The distribution of hard rods on a line of finite length. Mol. Phys. 58 (4), 711–721.

Rode, B.M., 1999. Peptides and the origin of life. Peptides 20 (6), 773–786.

Romanenko, V.I., 1981. Square micro colonies in the surface saline water film of the Saxkoye Lake Ukrainian-SSR USSR. Mikrobiologiya 50 (3), 571–574 (Russian).

Rusconi, B., Lienard, J., Aeby, S., Croxatto, A., Bertelli, C., Greub, G., 2013. Crescent and star shapes of members of the Chlamydiales order: impact of fixative methods. Anton. Leeuw. Int. J. Gen. Mol. Microbiol. 104 (4), 521–532.

Sabet, S., Diallo, L., Hays, L., Jung, W., Dillon, J.G., 2009. Characterization of halophiles isolated from solar salterns in Baja California, Mexico. Extremophiles 13 (4), 643–656.

Sain, A., Inamdar, M.M., Julicher, F., 2015. Dynamic force balances and cell shape changes during cytokinesis. Phys. Rev. Lett. 114 (4). #048102.

Santos, F., Yarza, P., Parro, V., Meseguer, I., Rosselló-Móra, R., Antón, J., 2012. Culture-independent approaches for studying viruses from hypersaline environments. Appl. Environ. Microbiol. 78 (6), 1635–1643.

Scarr, G., 2014. Biotensegrity: The Architecture of Life. Handspring Publishing, Pencaitland, East Lothian.

Schmalholz, S.M., Podladchikov, Y., 1999. Buckling versus folding: importance of viscoelasticity. Geophys. Res. Lett. 26 (17), 2641–2644.

Schulte, M., Blake, D., Hoehler, T., McCollom, T., 2006. Serpentinization and its implications for life on the early Earth and Mars. Astrobiology 6 (2), 364–376.

Schulze, E., Kirschner, M.W., 1988. New features of microtubule behaviour observed in vivo. Nature 334 (6180), 356–359.

Schwierz, N., Horinek, D., Sivan, U., Netz, R.R., 2016. Reversed Hofmeister series-the rule rather than the exception. Curr. Opin. Colloid Interface Sci. 23, 10–18.

Scientific Polymer, 2013. Refractive Index of Polymers by Index. http://scientificpolymer.com/technical-library/refractive-index-of-polymers-by-index.

Segré, D., Lancet, D., 2000. Composing life. EMBO Rep. 1 (3), 217–222.

Segré, D., Ben-Eli, D., Deamer, D.W., Lancet, D., 2001. The lipid world. Orig. Life Evol. Biosph. 31 (1-2), 119–145.

Semenov, A.M., Vasilyeva, L.V., 1985. Morphophysiological characteristic of *Labrys monachus* growth—budding prosthecate bacterium with radial cell symmetry under periodical and continuous cultivation. Izv. Acad. Sci. SSSR Ser. Biol. [Russian], 288.

Semenov, A.M., Vasilyeva, L.V., 1987. Stella vacuolata growth upon periodical and continuous cultivation. Izv. Akad. Nauk SSSR Ser. Biol. 2, 307–311.

Semenov, A.M., Hanzlikova, A., Jandera, A., 1989. Quantitative estimation of poly-3-hydroxybutyric acid in some oligotrophic polyprosthecate bacteria. Folia Microbiol. 34 (3), 267–270.

Senatore, G., 2017. PUSHMEPULLME 3D. http://expeditionworkshed.org/workshed/push-me-pull-me-3d/.

Sharov, A.A., 2016. Coenzyme world model of the origin of life. BioSystems 144, 8–17.

Sharov, A.A., 2017. Coenzyme world model of the origin of life. In: Gordon, R., Sharov, A.A. (Eds.), Habitability of the Universe Before Earth. In: Rampelott, P.H., Seckbach, J., Gordon, R. (Series Eds.), Astrobiology: Exploring Life on Earth and Beyond. Elsevier B.V., Amsterdam, pp. 407–426 (Chapter 17).

Sharov, A.A., Gordon, R., 2017. Life before Earth. In: Gordon, R., Sharov, A.A. (Eds.), Habitability of the Universe Before Earth. In: Rampelott, P.H., Seckbach, J., Gordon, R. (Series Eds.), Astrobiology: Exploring Life on Earth and Beyond. Elsevier B.V., Amsterdam, pp. 265–296 (Chapter 11).

Shen, T.Y., Wolynes, P.G., 2005. Nonequilibrium statistical mechanical models for cytoskeletal assembly: Towards understanding tensegrity in cells. Phys. Rev. E 72 (4). #041927.

Shenhav, B., Kafri, R., Lancet, D., 2004. Graded artificial chemistry in restricted boundaries. In: Pollack, J., Bedau, M., Husbands, P., Ikegami, T., Watson, R.A. (Eds.), Artificial Life IX: 9th International Conference on the Simulation and Synthesis of Artificial Life (ALIFE9), Boston, MA, September 12-15, pp. 501–506.

Shenhav, B., Bar-Even, A., Kafri, R., Lancet, D., 2005. Polymer GARD: computer simulation of covalent bond formation in reproducing molecular assemblies. Orig. Life Evol. Biosph. 35 (2), 111–133.

Sherwood Lollar, B., Westgate, T.D., Ward, J.A., Slater, G.F., Lacrampe-Couloume, G., 2002. Abiogenic formation of alkanes in the Earth's crust as a minor source for global hydrocarbon reservoirs. Nature 416 (6880), 522–524.

Shimane, Y., Minegishi, H., Echigo, A., Kamekura, M., Itoh, T., Ohkuma, M., Tsubouchi, T., Usui, K., Maruyama, T., Usami, R., Hatada, Y., 2015. Halarchaeum grantii sp nov., a moderately acidophilic haloarchaeon isolated from a commercial salt sample. Int. J. Syst. Evol. Microbiol. 65, 3830–3835.

ShopAdvisor, 2015. Spalding Deflated TF-1000 Classic Basketball—Size 7. https://www.shopadvisor.com/p/ERUKRHAD2MJHMMLE0VDJVJN9WKWZ/deflated-tf-1000-classic-basketball-size-7-29-5-spalding.

Simon, J., Kühner, M., Ringsdorf, H., Sackmann, E., 1995. Polymer-induced shape changes and capping in giant Iiposomes. Chem. Phys. Lipids 76, 241–258.

Simoneit, B.R.T., 1995. Evidence for organic-synthesis in high temperature aqueousmedia—facts and prognosis. Orig. Life Evol. Biosph. 25 (1-3), 119–140.

Sims, J.R., Karp, S., Ingber, D.E., 1992. Altering the cellular mechanical force balance results in integrated changes in cell, cytoskeletal and nuclear shape. J. Cell Sci. 103 (Pt 4), 1215–1222.

Sirota, E.B., King, H.E., Singer, D.M., Shao, H.H., 1993. Rotator phases of the normal alkanes: an x-ray-scattering study. J. Chem. Phys. 98 (7), 5809–5824.

Skelton, R.E., de Oliveira, M.C., 2009. Tensegrity Systems. Springer, Dordrecht, Netherlands.

Sleep, N., Meibom, A., Fridriksson, T., Coleman, R., Bird, D., 2004. H2-rich fluids from serpentinization: geochemical and biotic implications. Proc. Natl. Acad. Sci. U. S. A. 101 (35), 12818–12823.

Sleytr, U.B., Messner, P., 1983. Crystalline surface layers on bacteria. Annu. Rev. Microbiol. 37, 311–339.

Sleytr, U.B., Messner, P., Sara, M., Pum, D., 1986. Crystalline envelope layers in archaebacteria. Syst. Appl. Microbiol. 7 (2-3), 310–313.

Smith, W.P.J., Davit, Y., Osborne, J.M., Kim, W., Foster, K.R., Pitt-Francis, J.M., 2017. Cell morphology drives spatial patterning in microbial communities. Proc. Natl. Acad. Sci. U. S. A. 114 (3), E280–E286.

Snelson, K., 1990. Letter from Kenneth Snelson to R. Motro. Int. J. Space Struct. 7, N2.

Snelson, K., Heartney, E., 2013. Kenneth Snelson; Art and Ideas. Marlborough Gallery, New York, NY.

Soares e Silva, M., Alvarado, J., Nguyen, J., Georgoulia, N., Mulder, B.M., Koenderink, G.H., 2011. Self-organized patterns of actin filaments in cell-sized confinement. Soft Matter 7 (22), 10631–10641.

Staley, J.T., 1968. Prosthecomicrobium and Ancalomicrobium: new prosthecate freshwater bacteria. J. Bacteriol. 95 (5), 1921–1942.

Sterrenburg, F.A.S., Gordon, R., Tiffany, M.A., Nagy, S.S., 2007. Diatoms: living in a constructal environment. In: Seckbach, J. (Ed.), Algae and Cyanobacteria in Extreme Environments. Cellular Origin, Life in Extreme Habitats and Astrobiology, vol. 11. Springer, Dordrecht, The Netherlands, pp. 141–172.

Stetter, K.O., 1982. Ultrathin mycelia-forming organisms from submarine volcanic areas having an optimum growth temperature of 105°C. Nature 300 (5889), 258–260.

Stoeckenius, W., 1981. Walsby's square bacterium: fine structure of an orthogonal procaryote. J. Bacteriol. 148 (1), 352–360.

Stolarow, J., Heinzelmann, M., Yeremchuk, W., Syldatk, C., Hausmann, R., 2015. Immobilization of trypsin in organic and aqueous media for enzymatic peptide synthesis and hydrolysis reactions. BMC Biotechnol. 15 (1). #77.

Sumper, M., 1993. S-layer glycoproteins from moderately and extremely halophilic archaeobacteria. In: Beveridge, T.J., Koval, S.F. (Eds.), Advances in Bacterial Paracrystalline Surface Layers. Plenum Press, New York, pp. 109–117.

SunSpiral, V., 2015. NASA ARC—Super Ball Bot—Structures for Planetary Landing and Exploration (@ 33:03 into the video). http://livestream.com/viewnow/NIAC2015/videos/75238510.

Szathmáry, E., 2006. The origin of replicators and reproducers. Philos. Trans. R. Soc., B 361 (1474), 1761–1776.

Szathmáry, E., Santos, M., Fernando, C., 2005. Evolutionary potential and requirements for minimal protocells. In: Walde, P. (Ed.), Prebiotic Chemistry: From Simple Amphiphiles to Protocell Models. Springer, Berlin, pp. 167–211.

Takao, 2006. Haloarcula Japonica TR-1 T (= NBRC 101032 T). http://www.nite.go.jp/nbrc/genome/project/annotation/ongoing/hj1.html.

Takashina, T., Hamamoto, T., Otozai, K., Grant, W.D., Horikoshi, K., 1990. Haloarcula japonica sp. nov., a new triangular halophilic archaebacterium. Syst. Appl. Microbiol. 13 (2), 177–181.

Takei, A., Jin, L., Hutchinson, J.W., Fujita, H., 2014. Ridge localizations and networks in thin films compressed by the incremental release of a large equi-biaxial pre-stretch in the substrate. Adv. Mater. 26 (24), 4061–4067.

Taylor, P., 1998. Ostwald ripening in emulsions. Adv. Colloid Interface Sci. 75 (2), 107–163.

Tcholakova, S., Mitrinova, Z., Golemanov, K., Denkov, N.D., Vethamuthu, M., Ananthapadmanabhan, K.P., 2011. Control of Ostwald ripening by using surfactants with high surface modulus. Langmuir 27 (24), 14807–14819.

Tcholakova, S., Valkova, Z., Cholakova, D., Vinarov, Z., Lesov, I., Denkov, N., Smoukov, S.K., 2017. Efficient self-emulsification via cooling-heating cycles. Nat. Commun. 8. #15012.

Thompson, J.M.T., Sieber, J., 2016. Shock-sensitivity in shell-like structures: with simulations of spherical shell buckling. Int. J. Bifurcation Chaos 26 (2). #1630003.

Thompson, D.H., Wong, K.F., Humphrybaker, R., Wheeler, J.J., Kim, J.M., Rananavare, S.B., 1992. Tetraether bolaform amphiphiles as models of archaebacterial membrane lipids: Raman spectroscopy, 31PNMR, x-ray scattering, and electron microscopy. J. Am. Chem. Soc. 114 (23), 9035–9042.

Torrella, F., 1986. Isolation and adaptive strategies of haloarculae to extreme hypersaline habitats. In: Abstracts of the Fourth International Symposium on Microbial Ecology. Slovene Society for Microbiology, Ljubljana, Slovenia. Slovene Society for Microbiology, Ljubljana, Slovenia. p. 59.

Tromp, J.T., Gordon, R., 2006. The Number of 3D Configurations of a Labeled Size 2*n "Wurfel" [A117613: The On-Line Encyclopedia of Integer Sequences]. http://oeis.org/A117613.

Tsai, F.-C., Koenderink, G.H., 2015. Shape control of lipid bilayer membranes by confined actin bundles. Soft Matter 11 (45), 8834–8847.

Tully, B.J., Emerson, J.B., Andrade, K., Brocks, J.J., Allen, E.E., Banfield, J.F., Heidelberg, K.B., 2015. De novo sequences of Haloquadratum walsbyi from Lake Tyrrell, Australia, reveal a variable genomic landscape. Archaea. #875784.

Turk-MacLeod, R., Nghe, P., Woronoff, G., Schnettler, D., Szathmáry, E., Griffiths, A.D., 2015. Compartmentalization of the formose reaction to test metabolism-first theories on the origin of life. In: Doran, P. (Ed.), Astrobiology Science Conference 2015, Habitability, Habitable Worlds, and Life, June 15–19, Chicago, Illinois. Lunar and Planetary Institute, Houston. p. #7162.

Tuszynski, J., Portet, S., Dixon, J., 2005. Nonlinear assembly kinetics and mechanical properties of biopolymers. Nonlinear Anal. Theory Methods Appl. 63 (5-7), 915–925.

Tuszynski, J.A., Portet, S., Dixon, J.M., Nishino, M., Yu-Lee, L.Y., 2009. Propagation of localized bending deformations in microtubules. J. Comput. Theor. Nanosci. 6 (3), 525–532.

Uversky, V.N., 2013. Unusual biophysics of intrinsically disordered proteins. Biochim. Biophys. Acta 1834 (5), 932–951.

van Beijeren, H., Sylvester, G.S., 1978. Phase transitions for continuous-spin Ising ferromagnets. J. Funct. Anal. 28 (2), 145–167.

van der Vaart, B., Akhmanova, A., Straube, A., 2009. Regulation of microtubule dynamic instability. Biochem. Soc. Trans. 37 (Pt 5), 1007–1013.

van Roij, R., Dijkstra, M., Evans, R., 2000. Orientational wetting and capillary nematization of hard-rod fluids. Europhys. Lett. 49 (3), 350–356.

Vasilyeva, L.V., 1970. A star-shaped soil microorganism. Izv. Akad. Nauk SSSR Ser. Biol. 2 (1), 308–310 (Russian).

Vasilyeva, L.V., 1985. Stella, a new genus of soil prosthecobacteria, with proposals for Stella humosa sp. nov. and Stella vacuolata sp. nov. Int. J. Syst. Bacteriol. 35 (4), 518–521.

Vasilyeva, L.V., Semenov, A.M., 1984. New budding prosthecate bacterium Labrys monahos with radial cell symmetry. Microbiology (English translation of Mikrobiologiya) 53, 68–75.

Vasilyeva, L.V., Semenov, A.M., 1985. Labrys monachus sp. nov. in: validation of the publication of new names and new combinations previously effectively published outside the IJSB, List no. 18. Int. J. Syst. Bacteriol. 35, 375–376.

Vasilyeva, L.V., Semenova, A.M., 1986. Prostecobacteria of Stella genus and description of a new Stella inoculata species. Izv. Akad. Nauk SSSR Seri. Biol. 4, 534–540.

Vasilyeva, L.V., Lafitskaya, T.N., Aleksandrushkina, N.L., Krasilnikova, E.N., 1974. Physiological-biochemical peculiarities of the prosthecobacteria Stella humosa and Prosthecomicrobium sp. Izv. Akad. Nauk SSSR Seri. Biol. 5, 699–714.

Vetter, R., Wittel, F.K., Herrmann, H.J., 2014. Morphogenesis of filaments growing in flexible confinements. Nat. Commun. 5. (#4437). #4437.

Vliegenthart, G.A., Gompper, G., 2011. Compression, crumpling and collapse of spherical shells and capsules. New J. Phys. 13 (24). #045020.

Wächtershauser, G., 2003. From pre-cells to Eukarya: a tale of two lipids. Mol. Microbiol. 47 (1), 13–22.

Wakai, H., Nakamura, S., Kawasaki, H., Takada, K., Mizutani, S., Aono, R., Horikoshi, K., 1997. Cloning and sequencing of the gene encoding the cell surface glycoprotein of Haloarcula japonica strain TR-1. Extremophiles 1 (1), 29–35.

Walsby, A.E., 1980. A square bacterium. Nature 283, 69–71.

Walsby, A.E., 1994. Gas vesicles. Microbiol. Rev. 58 (1), 94–144.

Walsby, A.E., 2005. Archaea with square cells. Trends Microbiol. 13 (5), 193–195.

Wang, C.G., Liu, Y.P., Al-Ghalith, J., Dumitrica, T., Wadee, M.K., Tan, H.F., 2016a. Buckling behavior of carbon nanotubes under bending: from ripple to kink. Carbon 102, 224–235.

Wang, Z., Xu, J.Q., Xu, W.M., Li, Y., Zhou, Y., Luu, Z.Z., Hou, J., Zhu, L., Cui, H.L., 2016b. Salinigranum salinum sp nov., isolated from a marine solar saltern. Int. J. Syst. Evol. Microbiol. 66, 3017–3021.

Wanger, G., Onstott, T.C., Southam, G., 2008. Stars of the terrestrial deep subsurface: a novel 'star-shaped' bacterial morphotype from a South African platinum mine (Corrigendum: (6), 421). Geobiology 6 (3), 325–330.

Weiss, M.C., Sousa, F.L., Mrnjavac, N., Neukirchen, S., Roettger, M., Nelson-Sathi, S., Martin, W.F., 2016. The physiology and habitat of the last universal common ancestor. Nat. Microbiol. 1. #16116.

Whang, K., Hattori, T., 1990. A square bacterium from forest soil. Bull. Jpn. Soc. Microb. Ecol. 5 (1), 9–11.

Wieczorek, R., 2012. On prebiotic ecology, supramolecular selection and autopoiesis. Orig. Life Evol. Biosph. 42 (5), 445–450.

Wikipedia, 2016a. Cyclohexane conformation. https://en.wikipedia.org/wiki/Cyclohexane_conformation-Boat_conformation.

Wikipedia, 2016b. Ising Model. https://en.wikipedia.org/wiki/Ising_model.

Wikipedia, 2016c. List of Sequenced Archaeal Genomes. https://en.wikipedia.org/wiki/List_of_sequenced_archaeal_genomes.

Wikipedia, 2016d. Theridiidae. https://en.wikipedia.org/wiki/Theridiidae.

Wu, M., Higgs, P.G., 2008. Compositional inheritance: comparison of self-assembly and catalysis. Orig. Life Evol. Biosph. 38 (5), 399–418.

Wu, M.L., van Teeseling, M.C.F., Willems, M.J.R., van Donselaar, E.G., Klingl, A., Rachel, R., Geerts, W.J.C., Jetten, M.S.M., Strous, M., van Niftrik, L., 2012. Ultrastructure of the denitrifying methanotroph "Candidatus Methylomirabilis oxyfera," a novel polygon-shaped bacterium. J. Bacteriol. 194 (2), 284–291.

Xu, W.M., Xu, J.Q., Zhou, Y., Li, Y., Lü, Z.Z., Hou, J., Zhu, L., Cui, H.L., 2016. Halomarina salina sp nov., isolated from a marine solar saltern. Anton. Leeuw. Int. J. Gen. Mol. Microbiol. 109 (8), 1121–1126.

Yang, Y., Cui, H.-L., Zhou, P.-J., Liu, S.-J., 2007. Haloarcula amylolytica sp nov., an extremely halophilic archaeon isolated from Albi salt lake in Xin-Jiang, China. Int. J. Syst. Evol. Microbiol. 57, 103–106.

Yang, D.C., Blair, K.M., Salama, N.R., 2016. Staying in shape: the impact of cell shape on bacterial survival in diverse environments. Microbiol. Mol. Biol. Rev. 80 (1), 187–203.

Yatsunami, R., Ando, A., Yang, Y., Takaichi, S., Kohno, M., Matsumura, Y., Ikeda, H., Fukui, T., Nakasone, K., Fujita, N., Sekine, M., Takashina, T., Nakamura, S., 2014. Identification of carotenoids from the extremely halophilic archaeon Haloarcula japonica. Front. Microbiol. 5. #100.

Yenjerla, M., Lopus, M., Wilson, L., Oroudjev, E., 2010. Analysis of dynamic instability of steady-state microtubules in vitro by video-enhanced differential interference contrast microscopy. Methods Cell Biol. 95, 189–206.

Yin, S., Wang, Z., Xu, J.Q., Xu, W.M., Yuan, P.P., Cui, H.L., 2015. Halorubrum rutilum sp nov isolated from a marine solar saltern. Arch. Microbiol. 197 (10), 1159–1164.

Yokoi, T., Isobe, K., Yoshimura, T., Hemmi, H., 2012. Archaeal phospholipid biosynthetic pathway reconstructed in Escherichia coli. Archaea. #438931.

Young, K.D., 2006. The selective value of bacterial shape. Microbiol. Mol. Biol. Rev. 70 (3), 660–703.

Young, K.D., 2007. Bacterial morphology: why have different shapes? Curr. Opin. Microbiol. 10 (6), 596–600.

Young, K.D., 2010. Bacterial shape: two-dimensional questions and possibilities. Annu. Rev. Microbiol. 64, 223–240.

Yuen, G.U., Lawless, J.G., Edelson, E.H., 1981. Quantification of monocarboxylic acids from a spark discharge synthesis. J. Mol. Evol. 17 (1), 43–47.

Yuen, G., Blair, N., Des Marais, D.J., Chang, S., 1984. Carbon isotope composition of low molecular weight hydrocarbons and monocarboxylic acids from Murchison meteorite. Nature 307 (5948), 252–254.

Zaks, A., Klibanov, A.M., 1985. Enzyme-catalyzed processes in organic solvents. Proc. Natl. Acad. Sci. U. S. A. 82 (10), 3192–3196.

Zeng, M.Y., Wu, C.X., Sun, G.Y., 2004. Kink angle of multiwall carbon nanotubes due to interlayer interaction as viewed by elastic theory. Phys. Rev. B 70 (13). #132409.

Zenke, R., von Gronau, S., Bolhuis, H., Gruska, M., Pfeiffer, F., Oesterhelt, D., 2015. Fluorescence microscopy visualization of halomucin, a secreted 927 kDa protein surrounding Haloquadratum walsbyi cells. Front. Microbiol. 30 (6), 249.

Zhou, E.H., Trepat, X., Park, C.Y., Lenormand, G., Oliver, M.N., Mijailovich, S.M., Hardin, C., Weitz, D.A., Butler, J.P., Fredberg, J.J., 2009. Universal behavior of the osmotically compressed cell and its analogy to the colloidal glass transition. Proc. Natl. Acad. Sci. U. S. A. 106 (26), 10632–10637.

FURTHER READING

Brau, F., Damman, P., Diamant, H., Witten, T.A., 2013. Wrinkle to fold transition: influence of the substrate response. Soft Matter 9 (34), 8177–8186.

Comolli, L.R., Duarte, R., Baum, D., Luef, B., Downing, K.H., Larson, D.M., Csencsits, R., Banfield, J.F., 2012. A portable cryo-plunger for on-site intact cryogenic microscopy sample preparation in natural environments. Microsc. Res. Tech. 75 (6), 829–836.

Dale, B.F., 2008. An SVG of a Physically Possible Tensegrity Structure in 3D, With a Shadow. https://en. wikipedia.org/wiki/File:3-tensegrity.svg.

Gambacorta, A., Gliozzi, A., Derosa, M., 1995. Archaeal lipids and their biotechnological applications. World J. Microbiol. Biotechnol. 11 (1), 115–131.

König, E., Schlesner, H., Hirsch, P., 1984. Cell wall studies on budding bacteria of the *Planctomyces*/*Pasteuria* group and on a *Prosthecomicrobium* sp. Arch. Microbiol. 138 (3), 200–205.

Mikhaltsov, A., 2014. File:*Triceratium morlandii* var. *morlandii*.jpg. https://commons.wikimedia.org/wiki/File: Triceratium_morlandii_var._morlandii.jpg.

Stidolph, S.R., Sterrenburg, F.A.S., Smith, K.E.L., Kraberg, A., 2012. Stuart R. Stidolph Diatom Atlas: U.S. Geological Survey Open-File Report 2012–1163. http://pubs.usgs.gov/of/2012/1163.

Wikipedia, 2008. File:Haloquadratum walsbyi00.jpg. https://commons.wikimedia.org/wiki/File:Haloquadratum_ walsbyi00.jpg.

Zhang, L.Y., Zhang, C., Feng, X.Q., Gao, H.J., 2016. Snapping instability in prismatic tensegrities under torsion. Appl. Math. Mech. (English Ed.) 37 (3), 275–288.

WHY ON THEORETICAL GROUNDS IT IS LIKELY THAT "LIFE" EXISTS THROUGHOUT THE UNIVERSE

Jagers op Akkerhuis

Wageningen Environmental Research (Alterra), Wageningen, The Netherlands

CHAPTER OUTLINE

1 INTRODUCTION

It is an open question of whether or not life existed in the universe before the Earth formed, and how life forms on other planets would look like. Such ideas remain speculative for two reasons. First, no observations of living beings elsewhere in the universe have been made. Second, due to the lack of

Habitability of the Universe Before Earth, editors: Richard Gordon & Alexei Sharov, Volume 1 in the series:
Astrobiology: Exploring Life on Earth and Beyond, series editors: Pabulo Henrique Rampelotto,
Joseph Seckbach & Richard Gordon. ISSN 2468-6352. https://doi.org/10.1016/B978-0-12-811940-2.00019-8

a consensus definition of "life," it is difficult to prove its existence anywhere. This study will refer to these two topics as the "observation gap" and the "definition gap" and deal with these topics in this introduction. Later in this study, a way will be suggested to make thoughts about life in the universe more specific.

To close the observation gap, one needs measurements. Such measurements can be obtained, for example, during a visit to another planet (or moon), or can take the form of signals from extra-terrestrial cultures. Measurements can also be obtained from meteorites, at least in principle. While a study of a Mars meteorite yielded five phenomena that in combination are viewed as evidence for primitive life on Mars, an extensive historical overview suggests that meteorite studies so far have offered hardly any convincing support for life in meteorites (Gordon and McNichol, 2012). While the presence of non-Earthen organisms in a meteorite can in principle prove extra-terrestrial life, it would "kick the can down the alley" with respect to the origin of life question, because the presence of organisms in the meteorite is not explained. Finally, a probabilistic way to circumvent the observation gap is the use of statistical approaches calculating the probability that life emerges on the many planets in the universe (McKay et al., 1996).

The closing of the definition gap asks for a commonly acceptable definition of life. Agreement on a definition of life is necessary, because without a definition one cannot formulate falsifiable hypotheses to base research on. Interestingly, the question of "What is life?" is an age old question and many studies have discussed the possibilities or impossibilities of defining the concept of life (e.g., Oliver and Perry, 2006; Bedau and Cleland, 2010). Some authors have suggested that it is either not possible or even damaging to science to define life (e.g., van der Steen, 1997; Cleland and Chyba, 2007; Hengeveld, 2010; Machery, 2012; Cleland, 2012). On the other hand, if one cannot agree on a specific definition, this may indicate that the concept is not well-understood scientifically, and that there is work to be done, as is indicated, e.g., by Emmeche (1997).

Researchers such as Popa (2003) and Trifonov (2011) reviewed more than a hundred historical and recent scientific definitions of life. A major reason why so many definitions exist seems to be that "life" can be viewed as an umbrella term that groups homonyms of different ontological kinds, including life as a *period* between birth and death, life as a *spark* that "illuminates" the body, life referring to the organismic *state of matter* (e.g., Jeuken, 1975), life as a *system* capable of evolution, life as an *out of equilibrium thermodynamic phenomenon*, life as a *metaphor* for the presence of organisms, etc. As the result of such differences in ontological kinds, there is little reason to expect that a single comprehensive definition is possible. However, for the identification of "life" on Mars or elsewhere in the universe, it remains relevant to have access to a definition of life that allows for the formulation of falsifiable hypotheses.

In this study, the options will be explored of how the observation gap and the definition gap can be closed and of how the results can be applied to answer questions about life existing throughout the universe.

2 CLOSING THE OBSERVATION GAP

2.1 MEASUREMENTS

When the aim is to perform measurements on another planet, there are objects in our own solar system, such as Mars or Saturnus' moon Europa, that are relatively close and that have had, or currently have, conditions that potentially support the emergence of bacteria. Important information about Mars was gained by vehicles such as the Spirit Rover, the Opportunity Rover, the Phoenix Mars lander, and the Mars science laboratory. No convincing proof for the presence of organisms has been found so far. Beyond our solar system, the universe is a vast place that cannot easily be explored because distances are measured in lightyears. For example, the nearest stars are clustered in the Alpha Centauri group, some 4.2 lightyears away (one lightyear is about 9.5×10^{12} km). By comparison, in the 39 years since it was launched in 1977, the Voyager 1 has travelled no more than 2.2×10^{10} km. Light uses about 0.002 lightyear (0.7 days) to travel the same distance. At the current speed, the Voyager 1 will need about 79,000 years to reach Alpha Centauri. Travelling times like these imply that, in spite of the beautiful episodes of Star Trek, it will be difficult for humans to visit planets outside our solar system for exploring the existence of organisms. A robotic computer-intelligence that could sleep for 79,000 years before it arrived at Alpha Centauri could improve on this situation. Researchers on Earth would, however, not live long enough to learn about the data obtained by the robot.

As an alternative to visiting another planet, researchers have since about 1900 tried to detect radio signals produced by extra-terrestrial intelligences. As far as the author of this article knows, the "search for extra-terrestrial intelligence" (SETI) has not yet been successful. A limitation to the method is that one can only observe a civilization when it has send its signals in a timeframe that—after a journey through space of many lightyears—fits in with our own timeframe. If a distant civilization did send its signals too early, no culture existed on the Earth to observe the signal. And if and when a civilization sends signals in the far future, one can only speculate whether there will be intelligences on Earth to observe such signals. And because the universe expands, any signal of a civilization "at the other end of the universe" will be too faint to detect or too slow to ever reach us.

Another method for observing life on other planets makes use of the general observation that living beings through their metabolism can alter the composition of a planet's atmosphere. For example, the oxygen level of the Earth atmosphere is 20%, which is far above thermodynamic equilibrium. Also the presence of methane can be an indication of the metabolic activity of early cells, such as the anoxic archaea that live, e.g., in undersea vents (Tung et al., 2005).

2.2 STATISTICS

Another way of dealing with questions about the existence of extra-terrestrial life is to calculate a probability of its existence. The idea is that life has emerged on Earth simply because this planet offered thermodynamically favorable conditions, and that for this reason, it is likely that life (in whatever form) will also have emerged on other planets offering thermodynamically favorable conditions. Recent observations of fossils in the oldest rocks on Earth indeed seem to suggest that bacteria already emerged 4.3 billion years ago (Dodd et al., 2017). Their early appearance seems to indicate that bacteria can form rapidly on a planet with chemistry like the early Earth. If bacteria indeed can form rapidly, and based on the fact that the universe is a vast place, it is statistically unlikely that the Earth would be the only spot for bacteria to emerge. Following reasoning like this, Drake suggested a statistical equation that stimulated scientific dialogue during the first meeting of the SETI project in 1961 (Search for Extra-Terrestrial Intelligence). The Drake equation calculates the chances for intelligent life in the Milky Way galaxy. Basically, the idea is that even when the probability of intelligent life to emerge on a "suitable" planet may be slim, the probability for it to emerge anywhere in the Milky Way depends on the product of that chance and the presumably large number of "suitable" planets (Seager, 2013).

3 CLOSING THE DEFINITION GAP

3.1 OPERATOR HIERARCHY

"Life" represents an umbrella term that refers to concepts and entities of different kinds. When studying life in the universe, however, one needs testable hypotheses based on well-defined concepts. When discussing hypotheses about life in the universe, the main interest is not to focus on abstractions, such as life as a period or as a "spark." Instead, the interest is in life as a metaphor for the presence of living beings. This focus demands an answer to the question of how the term "living being" can be defined in a necessary and sufficient way, such that all relevant cases are included, while all non-relevant cases are excluded. To define the term "living being," this study assumes that all living beings are organisms, while the organism concept is defined using the Operator Hierarchy (Jagers op Akkerhuis and van Straalen, 1999).

In simple wording, the Operator Hierarchy can be viewed as a (branching) complexity ladder, as is explained in Jagers op Akkerhuis (2001, 2008, 2010a, 2010b, 2012a, 2012b, 2014), and most recently and extensively, in Jagers op Akkerhuis et al. (2016). This ladder is the product of processes through which lower level objects contributed to the formation of next-higher level objects (the lower level objects frequently become part of the next-higher level objects, but this is not a requirement). Typical for every step on the ladder is that the next new system possesses an obligatory interaction between a new structurally and new functionally "closed" property. Such a situation is referred to as *dual closure*. If the cell is taken as an example, the structural closure takes the form of the membrane. A membrane is a layer of molecules that encloses a volume (hence "closure") and mediates its interactions with the world. The cell's functional closure takes the form of set-wise autocatalytic chemistry. In a set-wise autocatalytic process, every molecule mediates the formation of any next molecule, finally allowing for a closed process cycle. The obligatory interaction implies that the autocatalytic chemistry produces molecules for the membrane, while the membrane mediates the functioning of the autocatalytic set by controlling the passage of chemicals and protons, while keeping the elements of the autocatalytic set inside. Dual closure is a stringent criterion when it comes to the elimination of counterexamples. For example, a Belousov-Zhabotinsky reaction (autocatalysis) in a plastic bag (membrane) does not comply as dual closure, because the Belousov-Zhabotinsky reaction does neither produce the plastic bag, nor does the plastic bag mediate the passage of molecules in a way that contributes to the sustenance of the Belousov-Zhabotinsky reaction.

Every object that possesses dual closure is called an "operator," and the ranking of operators with increasingly complex kinds of dual closure is called the Operator Hierarchy. The Operator Hierarchy includes quarks, hadrons, atoms, molecules, cells, endosymbiont cells, multicellulars (of cells and of endosymbiont cells), and neural network operators called "memons." Memons emerge as the consequence of a functional closure between neurons, while the interface that creates structural closure is represented by sensors that allow for perceptions and muscles that allow for activity. The neural network and sensors necessarily interact to make sense of perceptions and act accordingly. Also, higher level—technical—operators can be predicted through extrapolation of the Operator Hierarchy (Jagers op Akkerhuis, 2001, 2010a).

Every step on the ladder implies a new dual closure. Dual closure thus specifies a minimal necessary change in the organizational state of matter that is typical for any next kind of operator. Dual closure steps may be the result of physically interacting operators of the immediately preceding level, such as cells that in interaction create a multicellular operator, but may also simply demand the "involvement" of the preceding level. For example, future technical memons cannot be constructed from biological cells or multicellular organisms and must be constructed using technical components. As long as technical memons possess the relevant next dual closure, this step complies with the Operator Hierarchy.

The Operator Hierarchy has aspects in common with several other important hierarchy theories. One example is the "constructable meta-hierarchy" of Alvarez de Lorenzana (1993). Another example is the "metasystem transition" approach advocated by Turchin (1977). As is discussed in Jagers op Akkerhuis (2016a) and Stoelhorst (2016), the Operator Hierarchy also has aspects in common with the Major Evolutionary Transitions theory (METT) of Szathmáry and Maynard Smith (1995). It is relevant that while METT is selectively based on functional criteria, notably cooperation, competition reduction, and reproduction as part of a larger unit (Szathmáry and Maynard Smith, 1995), the Operator Theory adds a focus on structural criteria, in the form of structural closure. The reason is that structural closure allows a clear distinction between an operator and its environment. For example, an organism has dual closure, while the ecosystem hasn't. Even when the ecosystem, or parts of it, may possess closures, these cannot—by definition—be of the kind dual closure (thus based on the preceding level dual closure), because as soon as dual closure is realized, the part with dual closure would classify as an operator. To emphasize their difference in status, systems consisting of interacting operators, e.g., ecosystems, are named "interaction systems" in the Operator Theory. And objects that consist of physically lumped operators, without being an operator, are called compound objects. Examples of compound objects are mussel banks, the slug of a slime mould, etc. Strictly speaking, also an operator with attached symbionts and parasites, such as a horse with bacteria and fungi on the skin and with digestive flora and fauna in the intestines, classifies as a compound object. At the same time, the horse and attached organisms represent separate evolutionary units that evolve at different speeds and along separate genetic lineages.

A practical aspect of the Operator Hierarchy in the context of definitions of life is that this hierarchy allows one to define the organism concept in a stringent way by postulating that only those operators classify as organisms that have a kind of dual closure that is equally or more complex than that of the cell (see, e.g., Jagers op Akkerhuis, 2010a, 2010c, 2012b, 2012c). According to this criterion, the following systems are organisms: the cell, the endosymbiont cell (a single cell with an obligatory endosymbiont), the multicellular operator (requiring that cells are connected through plasma connections, Jagers op Akkerhuis, 2010b), the endosymbiont multicellular operator (requiring plasma connections between endosymbiont cells), the neural network operator called the "memon," and in the future, higher level technical memons ("robots") that do not yet exist on Earth (Jagers op Akkerhuis, 2001, 2010a). Using this definition, it is easy to decide about whether or not a system is an organism. A long list of examples, difficult cases, and objects that—according to this viewpoint—are (not) organisms is offered in Jagers op Akkerhuis (2010a). For example, an unpacked viral DNA molecule qualifies as a molecular operator, which is too low a level to be an organism. And a viroid with capsid represents a compound object. A viroid is not an organism because it lacks a proper metabolism that produces a system-own boundary mediating the metabolism. And while the cells of a kidney in a culturing apparatus show plasma connections (representing functional closure), the cells lack the context of the multicellular organism to continue their functioning as a kidney.

If one uses the above method for defining the organism concept, the results can be used for suggesting definitions of a "living being" and "life." Here it is proposed that only organisms classify as beings, and when they show activity, they classify as living beings. Subsequently, the water-life analogy discussed by, e.g., Benner et al. (2004) and Cleland (2012), can be used to suggest organism-based definitions of life. Water can refer to: (1) water as a general term for all water molecules (H_2O), or (2) water as a name for a fluid consisting of interacting water molecules. By analogy, Jagers op Akkerhuis (2016a) suggests two definitions of "life," referring either to: (1) life as a general term for a property shared by all organisms, or (2) life as a name for a dynamic system consisting of interacting organisms. These two definitions of life will be referred to as organismic life, or O-life, and systemic life or S-life, respectively.

3.2 O-LIFE

What is the exact interpretation of the concept of O-life? Based on the above approach, all organisms are operators. And every operator has dual closure. The focus of O-life, however, is not on any dual closure, but only on the presence of dual closure in organisms. The definition of O-life thus makes use of three criteria: (1) it focuses on the hierarchy of the operators, (2) it classifies complex operators as organisms, and (3) it addresses selectively the presence of dual closure in organisms. It is stressed that O-life refers to an abstract *organizational property* of any and all organisms, namely their dual closure. This property can also be present during a frozen or dried state, which was discussed by Broca (1860–1861) and Becquerel (1950). As long as the necessary structure remains intact such that the organism can be re-activated, the non-active state can be referred to as suspended animation, anabiosis, viable lifelessness, or latent life. Successful re-activation of an organism proves that it is not irreversibly damaged by the freezing process, and that it still has its dual closure and represents O-life. In contrast to O-life, which refers to a *class property*, the organism concept refers to a *material object* that has the organismic state of matter. Representing a class property, O-life cannot be found on a planet. What can be found on a planet are organisms. The process of losing one's O-life properties is synonymous with dying. What remains is a corpse.

3.3 S-LIFE

The second life-concept, S-life, is inspired by interacting water molecules forming liquid water (Jagers op Akkerhuis, 2016a). In the case of S-life, two or more interacting organisms create a system termed an ecosystem. Consequently, one can say that if a planet hosts a dynamically active ecosystem, S-life is physically present on a planet. Just as frozen water classifies as ice, not water, any non-active part of an ecosystem does not comply with the definition of S-life. For a system to comply with S-life, the only criterion is the presence of interacting organisms. It is neither necessary that these organisms reproduce nor that reproduction and differences in survival lead to the pattern of Darwinian evolution. Reproduction and Darwinian evolution are viewed by the Operator Theory as special patterns of events that may be realized in a system of interacting organisms.

4 ANALYZING THE USE OF EPOCHS

Epochs are a well-known way for organizing the periods when "life" is present in the universe, possibly before Earth. A classic example of a timeline that includes seven epochs is offered by Chaisson (2001) (Table 1, column 2). Similar kinds of rankings have been proposed by, e.g., Young (1976) and Christian

Table 1 A Comparative Representation of: (1) Objects that Formed in the Universe, (2) Epochs According to Chaisson (2001), and (3) the Formation of Higher Level Operators According to the Operator Hierarchy

Objects in the Universe	Epochs (According to Chaisson)	Levels in the Operator Hierarchy (Indicating Epochs)
Primordial universe	Particulate evolution	Fundamental particles
Expanding universe, galaxies	Galactic evolution	Hadrons
Stars, black holes, white and brown dwarfs (some stars may explode and the debris may aggregate again forming second-generation stars, etc.)	Stellar evolution	Atoms
Planets	Planetary evolution	Molecules
	Chemical evolution	
	Biological evolution	Cells
		Endosymbiont cells
		Endosymbiont multicellular
	Cultural evolution	Neural network organisms (hardwired memon)
		Softwired memon

(2005). Rankings like these can be compared with the sequence of formation of larger objects in the universe (Table 1 column 1) and with the levels in the Operator Hierarchy (Table 1 column 3).

When comparing the ranking of objects/epochs/levels in Table 1, one can observe a trend that the level of detail increases towards the right. The least differentiated column is that of the major kinds of objects that have appeared in the universe. In the second column, the epochs by Chaisson (2001) offer more detail with respect to events that occur on planets. And, in the third column the Operator Hierarchy adds further detail by including organisms of increasing levels of organization, while also including current and future organisms with brains.

A second observation is that the kinds of things that are listed by the three rankings are different. In the first column, the things that are listed are physical celestial objects. The second column ranks epochs that are linked to entities of different kinds. For example, planetary evolution is linked to a planet (a physical object) and biological evolution is linked to organisms (which also are physical objects, but of a different kind than a planet). Culture is based on conscious or subconscious group-behavior and convictions. Accordingly, culture refers to mental/behavioral patterns. Cultural aspects can be reflected in the production of artifacts. The third column exclusively ranks operators of increasingly complex kinds, as indicated by a succession in dual closure.

5 WHY ON THEORETICAL GROUNDS IT IS LIKELY THAT "LIFE" EXISTS THROUGHOUT THE UNIVERSE

The question of why on theoretical grounds it is likely that "life" exists throughout the universe is related to a number of sub-question: (1) What does the concept of "life" refer to?, (2) Can Earth-based analyses of properties of organisms be generalized to extra-terrestrial situations?, (3) Can the concept of "life as we don't know" be specified?, and (4) What theoretical grounds support that it is likely that "life" exists throughout the universe?

5.1 WHAT DOES THE CONCEPT OF "LIFE" REFER TO?

To prove the existence of life in the universe, one needs a definition of life. Based on earlier work, it is advocated in the current study that a definition of life starts with a definition of the organism. The concept of the organism is a much debated subject, as is apparent from the different viewpoints held by, e.g., Hull (1980), Maynard Smith and Szathmáry (1995), Pepper and Herron (2008), and Godfrey-Smith (2013). The Operator Theory contributes to the field by suggesting that only the operators that have a dual closure that is equally or more complex than the dual closure of the cell classify as organisms. This way of defining the organism concept is used as a basis for defining two concepts of life: O-life and S-life. O-life is defined as a term that refers in a general way to the presence of dual closure in organisms. O-life thus can be viewed as a class property all organisms necessarily comply with. In addition, this study suggests that S-life can be defined as a system of interacting organisms. These two definitions refer to terms that are of different ontological kinds (a property of organisms, and a system, respectively). For this reason, one cannot merge them into a single definition of life. Moreover, representing a class property, O-life cannot be physically present on a planet. What can be physically present on a planet are organisms, and the system of interacting organisms referred to as S-life. It is relevant here to mention also the well-known homonym of "life" referring in a metaphorical way to the presence of organisms, e.g., "the deep sea is teeming with life." This homonym implies again that one should look for organisms. All these different lines of reasoning thus support that a search for life either starts with, or runs parallel with a search for organisms.

5.2 CAN DEFINITIONS OF O-LIFE OR S-LIFE BE GENERALIZED TO EXTRA-TERRESTRIAL SITUATIONS?

When speculating about extra-terrestrial life, our observations are limited to examples of living beings on Earth. To extrapolate to non-Earth life-forms, one needs a framework that defines life/organisms in a manner that is independent of conditions on Earth, at least in part.

In principle, any definition that is based on examples on Earth is restricted to Earthly life. Cleland (2012) remarks that: "Without additional examples of life, one cannot discriminate features that are universal to life, wherever it may be found, from features deriving from mere physical and chemical contingencies on the early Earth." The Operator Theory seems to offer a way to circumvent this problem along two lines: (1) The hierarchy makes use of topological criteria that allow for many physical realizations; (2) Generality of the logic of the hierarchy can be proven for low level operators, as these are formed everywhere in the universe.

The ranking of the Operator Hierarchy depends on dual closure, representing a topological criterion. The use of topology implies that the logic is more general than the specific structures of the operators on Earth from which the logic was derived. For example, when the Operator Theory speaks about a cell, the major criterion for the cell is its dual closure based on the obligatory interaction between the autocatalytic set and the membrane. Such a way of defining the cell allows much freedom with respect to the kind of molecules that constitute the autocatalytic set and the membrane. One can observe a similar generality in the concept of autopoiesis that Maturana and Varela (1980) suggested as the fundamental property of organisms. The Operator Theory contributes by indicating basic aspects of autopoiesis, because it makes use of dual closure, in this way specifying the processes that maintain organisms as well as the structural limit to the organism. Moreover, the use of cells, endosymbiont cells, multicellulars, and endosymbiont multicellulars specifies important classes of organization in a more detailed way than first-order autopoiesis (the cell) and second-order autopoiesis (multicellular) of Maturana and Varela (1980). As dual closure is based on topology, all cells that exist on Earth and (in principle) any cells on extra-terrestrial worlds are covered by the approach. Moreover, due to the dual closure criterion, a thing that is not a cell cannot be classified as a cell. This ontological stringency assists in the exclusion of counterexamples. For example, a droplet of viscous material in water that moves around due to chemical reactions has been named a protocell (Chen et al., 2004), which is a kind of cell. However, as long as a droplet/vesicle does not have dual closure, the Operator Theory would not classify it as a cell. A droplet would classify as a vesicle, when it has a membrane that does not interact in an obligatory way with the autocatalytic process. Or as a droplet, when it lacks a membrane.

Presuming the Operator Hierarchy represents a valid logic, the emergence of organisms on Earth is embedded in a larger, potentially universal scheme that results from the succession of dual closures. To demonstrate that the Operator Hierarchy does not just have validity on Earth, one needs non-Earthly confirmation of its logic. Such confirmation is available for low levels in the hierarchy and can be obtained through measurements of absorption spectra. When light shines through a plasma or gas of low level operators, each different kind of atom and molecule will cause a unique absorption spectrum (Hearnshaw, 1986). Measurements of the absorption spectra of light that arrives at the Earth after a long journey from a distant star have demonstrated that operators as complex as molecules exist everywhere in the universe. Such measurements offer evidence from outside the Earth for the general validity of the early steps in the Operator Hierarchy.

5.3 CAN THE CONCEPT OF "LIFE AS WE DON'T KNOW" BE SPECIFIED?

One of the recurring questions in astrobiology is whether or not it is possible to recognize "Life as we don't know?" Interestingly, as long as the concept of life is not well-defined, any discussion about "life as we don't know" is doomed to remain speculative. By lack of a definition offering falsifiable criteria, there are no limits to speculations about forms of "life" or "living beings" in the universe. And while Star Treck deserves praise for their imaginative episodes, the series also has offered a rich substrate for speculations about alien "life" forms as diverse as large revolving crystals, interstellar plasm clouds, robots, cyborgs, and the Borg.

The Operator Theory contributes to this field by suggesting falsifiable definitions that provide limits to what kinds of systems can be or cannot be organisms. And from this basis, further steps are made towards falsifiable definitions of O-life and S-life. Such definitions can contribute to discussions of the concept of "life as we don't know." As was explained above, the definitions of both O-life and S-life depend on the organism concept. And the kinds of entities classifying as organism can be identified with the help of the Operator Hierarchy. This reasoning limits speculations about "life as we don't know" to entities that are organisms. However, while the operator hierarchy offers focus, there still remains a lot of space for variation. On the one hand, an operator of the kind cell, such as a bacterium, must have a membrane and an autocatalytic set, which in interaction maintain each other. But, on the other hand, there is still room for bacteria of different sizes, colors, membrane compositions, autocatalytic chemistry and related thermodynamics, survival at different temperature, etc. Similar reasoning can be held for higher level operators.

5.4 WHAT THEORETICAL REASONING SUPPORTS THE LIKELIHOOD OF "LIFE'S" EXISTENCE THROUGHOUT THE UNIVERSE?

When thinking about theoretical grounds for life's existence throughout the universe, the approach in this study adds a range of novel insights. Firstly, it makes use of a well-defined concept of all the different kinds of organisms, from bacteria to technical memons, and related definitions of life. Secondly, the definition of organisms when based on dual closure is of a topological kind, and for this reason, not limited to the examples that are found on Earth. Thirdly, the lower part of the Operator Hierarchy can be shown to be universally valid. Fourthly, the entire universe has formed from a single, physically uniform state and currently presents itself to researchers in a more evolved uniform state, in which identical processes of matter formation can be observed to occur everywhere. The combination of such insights offers a theoretical foundation supporting that "life," the organismic state of matter instantiated by organisms, can in principle be expected to form anywhere in the universe where conditions allow. The chance that such conditions occur and organisms emerge addresses a different question that is not of an "in principle" but of a probabilistic kind.

6 DISCUSSION

The proposition of this study is that it is likely on theoretical grounds that "life" exists throughout the universe. To analyze the likelihood of the universal existence of life, the subject is approached from different angles, focusing on how to define life and on the universal validity of definitions. In this section, I will analyze how the current results relate to existing literature. The focus will be on the following aspects: (1) How does a focus on organisms, O-life, and S-life contribute to discussions about life in the universe?, (2) What contributions can be made to discussions about "life as we don't know?," (3) What contributions can be made to approaches based on epochs?

6.1 ORGANISMS, O-LIFE AND S-LIFE

Classically, several definitions of organisms have been suggested. Some approaches focus on lists of criteria, such as the "seven pillars of life" by Koshland (2002) which include program, improvization, compartmentalization, energy, regeneration, adaptability, and seclusion. Other approaches such as of Maturana and Varela focus on the autopoietic capacity of organisms, referring to the ability of organisms to be "self-making." Related approaches focus on autonomy based on closure (Ruiz-Mirazo et al., 2004). In many cases, there is a strong focus on the first organisms, e.g., archaeal/bacterial cells, as typical representatives of the organism concept (e.g., Ganti, 2003a,b). However, if one selects examples as a basis for defining the organism concept, it is problematic that such examples are selected while no proper definition is available yet. Accordingly, there is a risk of the exclusion of relevant examples and the inclusion of things that are not organisms.

If one uses the Operator Hierarchy as a basis, various definitional challenges can be dealt with at the same time. The Operator Hierarchy links the organism concept to operators and dual closure. This viewpoint allows for the integration of existing definitions. For example, when, e.g., Gánti (2003a, b) derived criteria for a definition of life/organism from cells (bacteria/archaea), the results can be linked to that specific level in the Operator Hierarchy. Another integrational aspect of the Operator Theory is that many classical criteria can be viewed as consequences/products of dual closure. For example, the functional closure of autocatalysis can be viewed as the origin of homeostasis, reproduction, and metabolism. And the structural closure of the membrane leads to properties such as seclusion, individuality, compartmentation, etc. In this way, many classical perspectives can be organized.

The Operator Theory also contributes to definitional aspects of life. Due to the new definitions of organisms and O-life, an organism that loses its dual closure ceases to be an organism and no longer has the properties relevant for O-life. When the organism is active in the sense that it shows dynamics, it is living. And when activity is arrested because the organism is frozen or desiccated, it is not living. But as long as its dual closure remains intact, an organism that is in a frozen or desiccated state still possesses the property of O-life. The reason is that O-life typically depends on the presence of dual closure, not on activity. Whether dual closure has stayed intact can be tested after thawing or remoistening. And the Operator Theory-based definition of the different kinds of organisms also offers a sounding board for explorations of, e.g., "limping life," of single-celled organisms as suggested by Luisi et al. (2006): "... a cell that produces proteins and does not reproduce itself; or one that does reproduce for a few generations and then dies out of dilution; or a cell that reproduces only parts of itself; and/or one characterized by very poor specificity and metabolic rate."

Another consequence of the use of O-life or S-life is that reproduction or evolution are no longer necessary criteria for life. Such criteria are, for example, suggested by NASA's definition of life: "a self-sustaining chemical system capable of Darwinian evolution" (Joyce, 1994). Of course, reproduction is a way of introducing more organisms in a system, while the pattern of Darwinian evolution will almost necessarily occur when lots of organisms reproduce and compete for limited resources. Yet, reproduction and evolution can be viewed as consequences of activities of organisms, instead of as requirements for life. For example, the definition of O-life focuses on the presence of dual closure in organisms, not on reproduction or evolution. And in the case of S-life, the focus is on any kind of interaction between organisms, not specifically reproduction or evolution. It can also be questioned whether a technical, robotic society would comply with NASA's criterion of "chemical system."

6.2 LIFE AS WE DON'T KNOW

People sometimes speculate that "life as we don't know" exists on other planets in forms that are quite different from Earthly examples (e.g., Stevenson et al., 2015). Interestingly, the current study advocates that life is coupled to organisms, while the organism concept can be defined with help of the Operator Hierarchy. Such insights offer guidance for narrowing down the concept of "life as we don't know." It is advocated that a focus on organisms supports scientific reasoning, because it can preclude people from entering a realm of discussions in which both the concepts of life and living suffer from definitional problems.

It is relevant in this context that Cleland (2012) has warned that a defective definition may introduce the risk of "misidentifying an example of truly "alien" life as a nonliving physical system. ...." The viewpoint of the current study is that the chance of a defective definition can be reduced with the help of the Operator Theory. After all, a defective definition of the organism is only possible when several steps in the Operator Hierarchy have been misinterpreted, or the criterion of dual closure proves not to be consistently applicable. The chances for this to occur are slim, because every level in the hierarchy offers an additional check of the construction rules.

The logic of the Operator Hierarchy is based on closure, a principle that was shown to be universal for low level operators up to the molecules. Based on the proven universal existence of low level operators, it seems a reasonable assumption that the emergence of all operators will follow the logic of the Operator Hierarchy anywhere in the universe. Accordingly, the concept of life as we don't know can be linked to organisms anywhere in the universe. Such organisms may either be biological, "artificial," or technical, which include man made bacteria and specific forms of technical A-life (Langton, 1989, 1995). Meanwhile, the Operator Theory supports the reasoning by Boden (2003) that also strong A-life will not be complete without a metabolism. Apart from dual closure, the current approach sets no limits to the properties of (extra-)terrestrial organisms.

6.3 WHAT CAN BE ADDED TO CURRENT EPOCH SYSTEMS?

An epoch/era is a period that has a beginning and, in many cases, an ending. With respect to the beginning of an epoch, it is relevant that, e.g., fluctuations in density may be the cause that different parts of the universe may develop at a different pace. For this reason, the start of an epoch can best be based on models, indicating when globally or locally the first entity of a kind is formed.

It is relevant to remark that epochs such as used by Chaisson (2001) rank era's based on entities/ concepts of different kinds. One kind of entity is large, easily observable things in the universe, such as dust clouds of galactic size, and the celestial objects forming in these clouds, such as black holes, stars, and planets. Another kind are the operators. Yet another kind is culture. The operator theory does not view culture as a physical object. Culture is viewed as resulting from interrelated memories that Blackmore (1999) calls memeplexes. As culture depends on memories, and memories are linked to memons, culture can be viewed as an emergent property of a system of interacting memons.

To preclude the mixing of kinds of criteria, epochs can be based on the Operator Hierarchy and its successive emergence of new kinds of operators, from fundamental particles (gluons, photons, quarks, etc.) to neural network organisms (column 3 in Table 1). Using the levels of the Operator Hierarchy, every next phase in the universe can be connected to the earliest appearance of any next level operator. And larger objects in the universe can be classified after the most complex kind of operator they harbor. For example, a planet that is not inhabited by organisms is classified as a compound object at the molecular level, and a planet that is inhabited by protozoa is classified as a compound object at the level of endosymbiont cells.

When focusing on operators, we do not mean to say that operators are the only relevant objects in the history of the universe. On the contrary, the operators are a basis for systems of interacting operators, called "interaction systems" by the Operator Theory. And any newly formed interaction system will form the basis for the formation of the next kind of operator, etc. What I like to emphasize, however, is that the use of a ranking that is based on kinds of operators can preclude that operators and interaction systems are ranked in an interchanging way, with a confounded logic as a result (see also Jagers op Akkerhuis, 2016b, Sections 16.1.3–16.1.8).

7 CONCLUSIONS

The reasoning in the current publication contributes to discussions about life and extra-terrestrial life in the following ways:

- A fundamental requirement of any definition of O-life and S-life is the definition of the organism. The organism concept is defined by the Operator Theory as: any operator that in the operator hierarchy has a position equal to or higher than the cell.
- By analogy with water, two definitions of life were suggested that focus on material systems and that are relevant for the life sciences: O-life and S-life. O-life focuses in a general way on the presence of dual closure in organisms. S-life focuses on systems of interacting organisms.
- As the logic of the Operator Hierarchy is based on dual closure throughout its entire range, measurements that prove the existence of low level operators in the entire universe suggest that also higher level operators are to be expected to represent universally valid classes. In turn, this implies that—theoretically speaking—organisms may exist throughout the universe.
- With the help of the Operator Hierarchy, it is possible to add new and detailed viewpoints to classical classifications based on epochs.

REFERENCES

Alvarez de Lorenzana, J.M., 1993. The constructive universe and the evolutionary systems framework. In: Salthe, S.N. (Ed.), Development and Evolution. Complexity and Change in Biology. MIT Press, Cambridge, pp. 291–308. Appendix.

Becquerel, P., 1950. La suspension de la vie des spores des bactéries et des moisissures desséchées dans le vide, vers le zéro absolu. Ses conséquences pour la dissémination et la conservation de la vie dans l'Univers. C. R. Acad. Sci. Paris 1274, 1392–1394.

Bedau, M.A., Cleland, C.E., 2010. The Nature of Life. Classical and Contemporary Perspectives from Philosophy and Science. Cambridge University Press, Cambridge, New York.

Benner, S.A., Ricardo, A., Carrigan, M.A., 2004. Is there a common chemical model for life in the universe? Curr. Opin. Chem. Biol. 8, 672–689.

Blackmore, S., 1999. The Meme Machine. Oxford University Press, Oxford.

Boden, M., 2003. Alien life: how would we know? Int. J.Astrobiol. 2, 121–129.

Broca, M.P., 1860–1861. Rapport sur la question soumise à la Société de Biologie au sujet de la reviviscence des animaux desséchés. Mém. Soc. Biol., 3me Série II 1860, 1–139.

Chaisson, E.J., 2001. Cosmic Evolution: The Rise of Complexity in Nature. Harvard University Press, Cambridge.

Chen, I.A., Roberts, R.W., Szostak, J.W., 2004. The emergence of competition between model protocells. Science 305, 1474–1476.

Christian, D., 2005. Maps of time: An introduction to big history. University of California press, Berkeley.

Cleland, C.E., 2012. Life without definitions. Synthese 185, 125–144.

Cleland, C.E., Chyba, C.F., 2007. Does "life" have a definition? In: Woodruff, T., Sullivan, I.I.I., Baross, J.A. (Eds.), Planets and Life: The Emerging Science of Astrobiology. Cambridge University Press, Cambridge.

Dodd, M.S., Papineau, D., Grenne, T., Slack, J.F., Rittner, M., Pirajno, F., O'Neil, J., Little, C.T.S., 2017. Evidence for early life in Earth's oldest hydrothermal vent precipitates. Nature 543, 60–64.

Emmeche, C., 1997. Autopoietic systems, replicators, and the search for a meaningful biological definition of life. Ultimate Reality Meaning 20, 244–264.

Gánti, T., 2003a. Chemoton Theory Vol. 1. Theoretical Foundation of Fluid Machineries. Kluwer Academic/Plenum Publishers, New York.

Gánti, T., 2003b. Chemoton Theory Vol. 2. Theory of Living Systems. Kluwer Academic/Plenum Publishers, New York.

Godfrey-Smith, P., 2013. Darwinian individuals. In: Bouchard, F., Huneman, P. (Eds.), From Groups to Individuals. Evolution and Emerging Individuality. MIT press, Cambridge, USA.

Gordon, R., McNichol, J., 2012. Recurrent dreams of life in meteorites. In: Seckbach, J. (Ed.), Genesis—In the Beginning: Precursors of Life, Chemical Models and Early Biological Evolution. Springer, Dordrecht, pp. 549–590.

Hearnshaw, J.B., 1986. The Analysis of Starlight. Cambridge University Press, Cambridge, Cambridge.

Hengeveld, R., 2010. Definitions of Life are not only unnecessary, but they can do harm to understanding. Found. Sci. 16, 323–325.

Hull, D.L., 1980. Individuality and selection. Annu. Rev. Ecol. Evol. Syst. 11, 311–332.

Jagers op Akkerhuis, G.A.J.M., 2001. Extrapolating a hierarchy of building block systems towards future neural network organisms. Acta Biotheor. 49, 171–189.

Jagers op Akkerhuis, G.A.J.M., 2008. Analyzing hierarchy in the organization of biological and physical systems. Biol. Rev. 83, 1–12.

Jagers op Akkerhuis, G.A.J.M., 2010a. Towards a hierarchical definition of life, the organism, and death. Found. Sci. 15, 245–262.

Jagers op Akkerhuis, G.A.J.M., 2010c. The Operator Hierarchy. A chain of closures linking matter, life and artificial intelligence. PhD thesis, Radboud University Nijmegen.

Jagers op Akkerhuis, G.A.J.M., 2010b. Explaining the origin of life is not enough for a definition of life. Found. Sci. 16, 327–329.

Jagers op Akkerhuis, G.A.J.M., 2012a. The role of logic and insight in the search for a definition of life. J. Biomol. Struct. Dyn. 29 (4), 619–620.

Jagers op Akkerhuis, G.A.J.M., 2012b. The Pursuit of Complexity. The Utility of Biodiversity from an Evolutionary Perspective. KNNV Publisher. 120 pp.

Jagers op Akkerhuis, G.A.J.M., 2012c. Contributions of the Operator Hierarchy to the field of biologically driven mathematics and computation. In: Simeonov, P.L., Smith, L.S., Ehresmann, Andrée C. (Eds.), Integral Biomathics: Tracing the Road to Reality. Springer Heidelberg, New York.

Jagers op Akkerhuis, G.A.J.M., 2014. General laws and centripetal science. Eur. Rev. 22, 113–144.

Jagers op Akkerhuis, G.A.J.M., 2016a. Learning from water: Two complementary definitions of the concept of life. In: Jagers op Akkerhuis GAJM (2016) ed. Springer, Switzerland, Evolution and transitions in complexity. The science of hierarchical organization in nature.

Jagers op Akkerhuis, G.A.J.M., 2016b. ed. Evolution and transitions in complexity. The science of hierarchical organization in nature Springer.

Jagers op Akkerhuis, G.A.J.M., van Straalen, N.M., 1999. Operators, the Lego–bricks of nature: evolutionary transitions from fermions to neural networks. World Futures, J. Gen. Evol. 53, 329–345.

Jagers op Akkerhuis, G.A.J.M., Spijkerboer, H.P., Koelewijn, H.-P., 2016. Introducing the operator theory. In: Jagers op Akkerhuis, G.A.J.M. (Ed.), Evolution and Transitions in Complexity. The Science of Hierarchical Organization in Nature. Springer, Switzerland.

Jeuken, M., 1975. The biological and philosophical definitions of life. Acta Biotheor. 24, 14–21.

Joyce, G., 1994. Foreword. In: Deamer, D.W., Fleischaker, G.R. (Eds.), Origins of Life: The Central Concepts. Jones and Bartlett Publishers, Boston.

Koshland Jr., D.E., 2002. The seven pillars of life. Science 295, 2215–2216.

Langton, C.G., 1989. Artificial Life. Addison-Wesley, Redwood City CA.

Langton, C.G., 1995. Artificial Life: An Overvie. MIT Press, Cambridge MA.

Luisi, P.L., Ferri, F., Stano, P., 2006. Approaches to semi-synthetic minimal cells: a review. Naturwissenschaften 93, 1–13.

Machery, E., 2012. Why I stopped worrying about the definition of life... and why you should as well. Synthese 185, 145–164.

Maturana, H.R., Varela, F.J., 1980. Autopoiesis and Cognition. The Realization of the Living. D. Reidel, Dordrecht (also in Boston Studies in the Philosophy of Science, 42).

Maynard Smith, J., Szathmáry, E., 1995. The Major Transitions in Evolution. W.H. Freeman Spektrum, Oxford.

McKay, D.S., Gibson Jr., E.K., Thomas-Keprta, K.L., Vali, H., Romanek, C.S., Clemett, S.J., Chillier, X.D.F., Maechling, C.R., Zare, R.N., 1996. Search for past life on Mars: Possible relic biogenic activity in Martian meteorite ALH84001. Science 273 (5277), 924–930.

Oliver, J.D., Perry, R.S., 2006. Definitely life but not definitely. Orig. Life Evol. Biosphere 36, 515–521.

Pepper, J.W., Herron, M.D., 2008. Does biology need an organism concept? Biol. Rev. 83, 621–627.

Popa, R., 2003. Between Necessity and Probability. Searching for the Definition and the Origin of Life. Advances in Astrobiology and Biogeophysics. Springer Verlag, Berlin, Heidelberg.

Ruiz-Mirazo, K., Pereto, J., Moreno, A., 2004. A universal definition of life: autonomy and open-ended evolution. Orig. Life Evol. Biosph. 34, 323–345.

Seager, S., 2013. The Drake Equation Revisited: An interview with Sara Seager. Astrobiology. Sept 3.

Stevenson, J., Lunine, J., Clancy, P., 2015. Membrane alternatives in worlds without oxygen: Creation of an azotosome. Sci. Adv. https://doi.org/10.1126/sciadv.1400067.

Stoelhorst, J.-W., 2016. Major transitions, operator theory, and human organization. In: Jagers op Akkerhuis, G.A. J.M. (Ed.), Evolution and Transitions in Complexity. The Science of Hierarchical Organization in Nature. Springer, Switzerland.

Szathmáry, E., Maynard Smith, J., 1995. The major evolutionary transitions. Nature 374, 227–232.

Trifonov, E.N., 2011. Vocabulary of definitions of life suggests a definition. J. Biomol. Struct. Dyn. 29, 259–266.

Turchin, V.F., 1977. The phenomenon of science. Columbia University Press, New York.

Tung, H.C., Bramall, N.E., Price, P.B., 2005. Microbial origin of excess methane in glacial ice and implications for life on Mars. Proc. Natl. Acad. Sci. 102 (51), 18292.

van der Steen, W.J., 1997. Limitations of general concepts: a comment on Emmeche's definition of "life". Ultim. Real. Mean. 20, 317–320.

Young, A.M., 1976. The reflexive universe. A Merloyd Lawrence Book, Delacorte, New York.

FURTHER READING

Weber, B., 2003. Life. Stanford Encyclopedia of Philosophy.

Zhuravlev, Y.N., Avetisov, V.A., 2006. The definition of life in the context of its origin. Biogeosci. Discuss. 3, 155–181.

Glossary

A

Abiogenesis The formation of living organism(s) from inorganic substances; the term is applied mostly to the origin of life.

Acidophiles Microorganisms capable of growth at low or acid pH values, typically below pH 6.

A-euk The possible archaeal symbiosis ancestor of eukaryotes, can be represented by *Lokiarchaeota*, a recently discovered phylum of Archaea.

AGB-phase star Asymptotic giant branch-phase star, according to the Hertzsprung-Russell diagram, is usually a low- or intermediate mass star in the terminal stages of its evolution, and which eventually transforms into a planetary nebula.

AGN Active galactic nucleus, a compact region at the center of a galaxy that has a much higher than normal luminosity not produced by stars, but a result of accretion of matter by a black hole at the center of its galaxy.

Albedo Percentage of incident light or radiation reflected by a surface.

Allometry Relationship between body size and other characteristics of the organism that deviates from isometry; the term is used to describe the variation in proportions of body parts in one species or in large taxa.

Allotrope One or more structural variant formed from the same element that can exist in the same physical state.

Alfvénic surface The surface of the region surrounding a spinning magnetized star within which ionized gas is pulled around by the magnetic field.

amu Atomic mass unit—an atomic mass unit is one twelfth of the mass of an atom of carbon-12.

Angular momentum Quantity related to rotations which remains constant in isolated systems. For a particle, it is equal to the cross product $(r \wedge v)m$ where r is the particle's position vector, v its velocity vector, and m its mass. For an extended body, the total angular momentum is the vector sum of the angular momenta for every particle of the body.

Anisotropic A material with different physical or mechanical properties or values when measured in different directions; not the same in every direction.

Anthropocentrism This is a doctrine that maintains that *Homo sapiens* are the center of everything, the ultimate end of nature. The term is often applied to notions that were originally associated with humans or human habitat.

Archaea (cf., archaebacteria) One of the three domains introduced in the taxonomic classification of all life on Earth by Carl Woese. Belongs to Prokaryotes; differs from Bacteria mostly by the composition of the cell membrane and enzymes involved in DNA replication.

Archaean A geologic era that spans from 4.5 to 2.5 Gyr BP. (The Hadean is the first of its suberas). We refer to this era together with the Proterozoic (2.5–0.57 Gyr BP) as the Precambrian Eon. (The eon in which multicellular eukaryotic life arose is called the Phanerozoic and it ranges from the end of the Precambrian till the present).

Astrobiological synthesis Synthesis of traditional scientific disciplines for the purpose of conducting multidisciplinary astrobiological research.

Habitability of the Universe Before Earth, editors: Richard Gordon & Alexei Sharov, Volume 1 in the series:
Astrobiology: Exploring Life on Earth and Beyond, series editors: Pabulo Henrique Rampelotto,
Joseph Seckbach & Richard Gordon. ISSN 2468-6352. https://doi.org/10.1016/B978-0-12-811940-2.09986-X
© 2018 Elsevier Inc. All rights reserved.

Astrobiology This term has replaced exobiology in the field of biological aspects of the subjects of the origin, evolution, and distribution of life in the universe. Astrobiology is currently in a period of rapid advancement due to the many space missions that are in their planning stages, or indeed already in operation.

Astronomical unit A distance frequently used by astronomers corresponding to the average Earth-Sun distance, nearly 150 million kilometers (149,597,870,700 m). Its abbreviation is a.u. or AU.

Astropause The boundary layer around a star or binary star where the stellar wind is slowed and compressed by the surrounding interstellar medium.

Astrosphere The roughy spherical circumstellar region defined by the astropause.

ATP Adenosine triphosphate, coenzyme used as an energy carrier in the cells of all known organisms.

au Astronomical unit. It corresponds to the average Earth-Sun distance, nearly 150 million kilometers (149,597,870,700 m).

Autocatalytic set Collection of entities (e.g., molecules), each of which can be created catalytically by other entities within the set, such that as a whole, the set is able to catalyze its own production. The notion was introduced by Kauffman in relation to autocatalytic sets of proteins.

Autopoietic A complex system that is capable of both creating and maintaining itself. The term is applied to self-reproducing living systems.

Azotomes A theoretical membrane predicted computationally to be stable in a nonpolar solvent, conditions thought to exist in ethane lakes on Titan, for example. This membrane is made of small organic nitrogen compounds, and under the physiochemical conditions on Titan, was modeled to serve analogous functions as the phospholipid cell membrane. Formation of the lipid bilayer membrane was a critical step in the formation of life on Earth.

B

B-euk The possible bacterial symbiosis ancestor of eukaryotes similar to α-*proteobacterium*, a class of bacteria in a phylum *Proteobacteria*.

BH Black hole.

Biocentrism An ethical point of view that extends inherent value to all living things

biomarker (or biosignature) A characteristic biochemical substance or mineral that is a reliable indicator of the biological origins of, or the presence of, biologic material within a given sample.

Biomineralization The process by which living organisms produce minerals, often to harden or stiffen existing tissues.

Biosemiotics Interdisciplinary approach that integrates theoretical biology with semiotics and supports biological grounding of signification and communication.

Black body Idealized physical body that absorbs all incident light, regardless of frequency. Its spectral thermal emission follows Planck's law, which is the characteristic emission of a body in thermodynamic equilibrium with its environment. The total emission (integration of Planck's law over frequency) follows the Stefan-Boltzmann law.

Black hole Gravitationally collapsed object with an escape velocity which is greater than the speed of light, so that no light or matter can escape.

Blazar Any of several classes of active galactic nuclei with jets pointed towards Earth.

BL Lac object A type of radio quiet galaxy with an active galactic nucleus.

bp Base pair, as in DNA or RNA.

BP Before present.

Bremsstrahlung radiation Electromagnetic radiation produced by the deceleration of a charged particle when deflected by another charged particle (usually an electron).

Byr Duration of a billion years.

Bya Billion years ago.

C

Calvin cycle Circular metabiolic pathway which is a light-independent component of photosynthesis that allows carbon fixation from CO_2 and produces precursors of sugar.

Carbon planet A hypothesized type of terrestrial planet comprised predominantly of carbon. These types of planets would form if protoplanetary disc were carbon-rich and oxygen-poor. Such a planet will probably be devoid of water, because oxygen delivered by comets or asteroids would react with the carbon on the surface.

Carbon star Asymptotic giant phase star that emits huge volumes of carbon, produced by its innate triple-alpha process, forming a massive and dusty-cum-sooty circumstellar envelope.

CEMP star Carbon-enhanced metal poor star, a subset of metal-poor and extremely metal poor stars that exhibits elevated carbon-to-iron ratio $[C/Fe] \geq +0.7$.

Carboxydotrophic A group of bacteria that obtain their energy and carbon requirements from the oxidation of carbon monoxide (CO).

Catalyst A material that speeds the kinetics of chemical reactions via structural features and is not consumed in the process. In a living organism, enzymes are proteins that act as catalysts.

Cellular automata A modeling platform realized as a grid of cells whose states evolve in discrete time steps according to a defined set of rules. The state of each cell depends on the states of the neighboring cells. Example: Conway's game of life.

Centromere Location of a chromosome used for attachment of microtubules during cell division (mitosis); centromeres usually have no genes and include DNA repeats.

Chandrasekhar limit The maximum possible mass for a white dwarf star. If a white dwarf gains enough mass to exceed this limit of 1.4 solar masses, it will explode as a supernova of type SN 1a.

Chaotrophic A solution that contains molecules (chaotropic agents) which disrupt the hydrogen bonding of water molecules. This in turn affects the stability of other molecules in solution such as proteins or nucleic acids by weakening the hydrophobic effect that maintains the stabilities of these molecules, potentially causing denaturation.

Chemosynthetic The biological conversion of inorganic carbon compounds and nutrients into organic matter using the energy from inorganic chemicals such as reduced hydrogen, sulfur, or iron.

Chert A hard, dark, opaque rock composed of silica (chalcedony) with an amorphous or microscopically fine-grained texture.

Chirality Properties of a molecule that are not superimposable on their "mirror image." Possessing handedness.

CHNOPS The group of elements deemed essential to life on Earth: carbon, hydrogen, nitrogen, oxygen, phosphorus, and sulfur.

Chromatin Protein scaffold for the DNA molecule; chromatin units (nucleosomes) include several proteins called histones. Histones are modified in order to change the structure of chromatin, which may open or close the access of signaling molecules to the DNA.

CLM Coenzyme-like molecule.

CLUES Constrained Local UniversE Simulations. The cosmological computer simulation of the local Universe.

CNO Carbon, nitrogen, oxygen.

CNO cycle CNO cycle is a set of fusion reactions that convert hydrogen into helium in the stars.

Coenzyme (cofactor) Catalytically active nonpolymeric molecule which often functions as a cofactor for proteins (enzymes). Examples of coenzymes are coenzyme A, ATP, NADH, flavin, and heme. Many coenzymes are known as vitamins.

Coenzyme world Hypothetical scenario of the origin of life that is focused on coenzyme-like molecules as primitive catalysts and hereditary agents that are assumed to be predecessors of nucleic acids as well as extant coenzymes.

Collapsar The collapse of a very massive star directly forming a black hole without an associated supernova explosion.

Coma The luminous cloud surrounding the nucleus in the head of a comet.

Complex system System composed of many components which interact with each other in a nonlinear way, showing the emergence of nontrivial collective behaviors, such as chaos and its spontaneous organization.

Convection The motion of gas or liquid caused by differences in pressure.

Convergent evolution This is an independent evolution of similar genetic or morphological features in different taxonomic groups.

Cooperative effects When two or more elements of a system work together in a way that allow for cooperation between those elements, often leading to multiplicative effects on a given system.

Coronal mass ejection Cloud of solar plasma blown away from the Sun during strong, long-duration solar flares and filament eruptions.

Cosmic microwave background radiation The remnants of the hot expanding Universe during Big Bang.

Cosmic rays (CRs) Particles originating from beyond Earth and moving at high to extremely high, even relativistic, velocities. Cosmic rays impacting the atmosphere of the Earth are called primary cosmic rays, while those produced in the atmosphere as by products of a higher energy primary cosmic ray are called secondary cosmic rays.

Cosmic ray spallation The formation of nucleons as the result of the fission of a heavier nucleus after the impact of a high energy cosmic ray particle.

Cosmobiology The study of life in the Universe on vast scales and over cosmic time. It is the extrapolation of astrobiology to the past, present, and future of life in the Universe.

Chromatin Complex of genomic DNA with scaffolding proteins and other components in the nucleus of cells. Chromatin is needed mostly to fold (condense) and unfold DNA, prevent DNA damage, control gene expression, and DNA replication. The primary protein components of chromatin are histones.

Cryovolcanoes Volcanoes located in extremely cold environments (such as those found on Saturn's moon, Enceladus) which erupt low temperature ($\sim$273 K) icy magmas, consisting of ice crystals, water, and other volatiles such as ammonia or methane.

Cryo-magmas Material released from a cryovolcano, which is a theoretical volcano that erupts water, ammonia, and/or methane volatiles. These cryomagmas or ice-volcanic substances are typically in the liquid or vapor form.

C-value paradox Lack of correspondence between the total amount of DNA in a single cell (C-value) and the level of organism complexity.

Cyanobacteria A phylum of bacteria that obtain their energy through photosynthesis. The name "cyanobacteria" comes from the color of the bacteria.

Cybernetics An interdisciplinary approach to modeling and discovery of large-scale mechanistic structures involved in regulation and control.

Cyclobutane pyrimidine dimers Molecular lesions formed from thymine or cytosine bases in DNA via photochemical reactions.

D

Data virtualization Accessing data, both offline and real-time, and from diverse sources to ultimately integrate it for unified analyses using virtual reality.

Darwinism A theory that explains the Darwinian kind of evolution of organisms, defined as: "descent with modification through variation and natural selection." Darwinian evolution differs in kind from the shorthand definition of the modern synthesis (or neo-Darwinism) which is focused on changes of gene frequencies (unknown to Darwin) in a population over generations via natural selection.

Degrees of freedom Object's attribute that equals the number of parameters that are allowed to vary independently with respect to each other.

Deliquescence The humidity-driven transition from solid to an aqueous phase. On Mars, this entails (perchlolorate) salts which can readily absorb water vapor that then subsequently deliquesces into an aqueous phase at low temperatures (surface Martian temperatures).

Diatom (1) A molecule consisting of 2 atoms (pronunciation: di-atom). (2) A eukaryotic single cell alga with a silica shell, extant on Earth for the past 200 million years (pronunciation: dia-tom).

Differentiation tree A graph that is a tree in the graph theory sense, where each edge represents a distinct tissue in an embryo, and each node represents the splitting of the tissue into two new tissues.

Differentiation wave A wave of contraction or expansion of the apical (outer) surface of cells in an epithelium (sheet of cells one cell thick) in an embryo that propagates from cell to cell and changes the kind of each cell that propagates it to a new cell type.

DNA repair A collection of processes by which a cell identifies and corrects damage to the DNA molecules that encode its genome.

Dolostones A sedimentary carbonate rock that contains a high percentage of the mineral dolomite, $CaMg(CO_3)_2$.

Domains of life Largest three evolutionary lineages of life (here referring to kinds of organisms) represented by Bacteria, Archaea, and Eukaryotes. Bacteria and Archaea are often considered together as Prokaryotes because they have similar organization.

Dual closure In the *Operator Theory,* dual closure refers to the pair of criteria that define different kinds of operators. A dual closure implies an interaction between a functional closure (e.g., the autocatalytic set in a cell) and a structural closure (e.g., the cell membrane).

DSFG Dusty star formation galaxy, a nonspecific term for a star-forming galaxy whose rest-frame optical or ultraviolet light is highly obscured due to substantial amount of inherent dust.

E

Eddington limit (or luminosity) The maximum luminosity a body may achieve when there is a balance between the outward force of radiation and the inward gravitational force. This balance imposes a limit on the rate of spherical accretion onto a star or other compact object. More complex accretion geometries allow for super-Eddington luminosities.

EeV Exa-electronvolt equal to 10^{18} eV, see electronvolt.

Electron volt (eV) Noninternational system of unit measure of energy that equals to its amount gained (or lost) by the charge of a single electron moving across an electric potential difference of 1 V, which equals 1.602×10^{-19} J.

EGCR Extra-Galactic cosmic ray. EGCRs are particles accelerated to high energies from sources within galaxies external to the Milky Way Galaxy.

Embryogenesis Process in which phenotypes emerge from a single cell due to cell division and differentiation.

Emergent Any phenomenon in which the interactions between entities, or between parts inside an entity, allow for new properties that the individual entities or parts did not have, or could not perform. For example, water can produce waves based on interacting water molecules, but a single water molecule cannot do that.

Endolithic Living inside rock or in the pores between mineral grains of a rock.

Endosymbiosis *Symbiosis* (see below) in which one of the symbiotic organisms lives inside the other.

Entropy In thermodynamics: a thermodynamic quantity representing the amount of energy in a system that is no longer available for doing mechanical work. In statistical mechanics: a measure of the number of configurations that correspond to a system in a certain state. Hence, it can be understood as a measure of disorder within the system.In information theory: a numerical measure of the uncertainty of an outcome.In summary: the entropy of the macrostate measures the degree of ignorance about which the microstate system is in, by counting the number of bits of additional information needed to specify it, with all of the microstates in the macrostate treated as equally probable (M. Gell-Mann).Configurational entropy: a measure of the number of configurational states available to a given thermodynamic or informational system.

Enzyme A protein that acts as a catalyst.

Epigenetic *Sensu* Conrad Waddington, traits of organisms that cannot be explained merely in terms of genetic encoding (e.g., gene activity/expression, protein interaction). However, there are also stable epigenetic traits which are passed down for a limited number of cell divisions and even generations that can also affect the development of phenotype.

Epigenetic heredity Heredity that is not based on the DNA sequence. Epigenetic heredity is stored in DNA changes such as methylation that do not alter the sequence, chromatin modifications, protein structure, and membrane structure, which are preserved after cell division. Epigenetic heredity can be passed to the next generation of multicellular organisms, but it is less stable than genetic heredity.

Ergodic Stochastic processes are called ergodic if they keep their parameters (including the mean and variance) stable indefinitely and every sequence or sizable sample is equally representative of the whole.

Escape velocity The minimum speed needed for an object to escape from the gravitational attraction of a massive body. It is proportional to the square root of the ratio *body mass/distance to the body center*.

Eukaryogenesis The first appearance of the eukaryotes, usually believed to be partly due to the process of symbiosis, probably during the Late Archaean 2.7 billion years before the present (Gyr BP), according to current views; but certainly during the Proterozoic, some 1.8 Gyr BP, eukaryotes were coexisting with prokaryotes.

Eukaryotes These are either single-celled or multicellular organisms in which the genetic material is enclosed inside a double membrane, which is called nuclear envelope. The evolutionary advanced domain of life, which is characterized by large cells with a nucleus, chromosomes, endoplasmic reticulum, well-developed cytoskeleton, and mitochondria. It is assumed that mitochondria and chloroplasts (organelles in plant cells) originated from symbiotic bacteria.

Evolution of life In the case of biology, it is a theory that assumes various types of animals and plants to have their origin in other preexisting types. In addition, evolutionary theory assumes that distinguishable differences between living organisms are due to modifications that occurred in previous generations. See: Darwinism.

Evolvability The ability of a population of organisms to not merely generate genetic diversity, but to generate adaptive genetic diversity, and thereby evolve through natural selection. Also, the degree to which a specific lineage can evolve and the scope of this change with respect to a single common ancestor.

Exergy Energy that is available to be used (e.g., by living cells). In the equilibrium, exergy equals zero.

Exoplanet Planet orbiting around a star other than our Sun.

Extremophile An organism that grows optimally under one or more chemical or physical extremes, such as high or low temperature or pH.

F

Faint young Sun paradox The apparent contradiction between observations of liquid water early in Earth's history and the astrophysical expectation that the Sun's output would be only 70% as intense during that epoch as it is during the modern epoch.

Far-IR Long wavelength infrared radiation.

Feedback An element in a cybernetic system that recycles (feeds back) information to already informed elements in the system.The return to the input of a part of the output of a machine, system, or process.

FGK-type star According to their spectral properties, the main sequence stars are classified into 7 basic types designated with letters O, B, A, F, G, K, and M. The O stars are the most massive, short-lived (only a few million years) bright stars, while most of the M type stars are a long-lived dimm dwarfs. The Sun is a G type star and has similar properties as F and K type stars so these stars are usually considered as the best candidates for hosting habitable planets.

Field galaxies Galaxies that are not part of clusters or galaxy groups.

Fractal Physical or mathematical object where a fragment of it has the same appearance as that of the whole object.

FR-I galaxy A classification for massive galaxies associated with X-ray emitting gas and seen often with collimated particle jets.

Function (biological) Repeatable activity that is initiated by living agents for the purpose of achieving (or keeping) life-related goals. Functions are encoded in the memory and/or genome to be preserved over time.

Fundamental forces Gravity plus electromagnetism and the strong and weak nuclear forces.

G

Ga A billion (10^9) years ago.

GADGET A computer code developed by Volker Springel for calculating gravitational interactions for a large number of gravitationally interacting particles.

Galactocentric radius Radial distance from a galactic center.

Galileo mission (1995–2003) The mission has changed the way we look at the Solar System. The spacecraft was the first to fly past an asteroid and the first to discover a moon of an asteroid. It provided the only direct observations of a comet colliding with a planet. Galileo was the first to measure Jupiter's atmosphere with a descent probe and the first to conduct long-term observations of the Jovian system from orbit. It found evidence of subsurface saltwater on Europa, Ganymede, and Callisto and revealed the intensity of volcanic activity on Io.

Gamma-ray burst A very intense short burst of gamma radiation resulting from the gravitational collapse of a huge mass star or from the mergers of neutron stars and black holes.

Gas giant Planet composed mainly of gas, especially hydrogen and helium, with mass ranging between 100 and 10,000 Earth's masses.

GCR Galactic cosmic ray. GCRs are particles accelerated to high energies from sources within the Milky Way Galaxy.

Genome complexity The length of functional and nonredundant DNA sequences in the genome.

Geothermal Characterized by high heat flow typically due to some geographical feature of tectonics or volcanism.

GeV Billion electron volts.

GHz Gigaherz, a billion oscillations per second.

GHZ Galactic habitable zone. A region within the galaxy with significant probability for hosting the habitable stellar systems.

GRB Gamma-ray burst.

Greenhouse effect Warming of the surface of a planet to a temperature above what it would be without its atmosphere.

Gy, Gyr A billion (10^9) years.

H

Habitable = inhabitable Suitable for life. Depending on context, habitable may refer to the potential for simple life or for complex life to exist, especially as we know it on Earth.

Habitable zone The orbital region around a star where a planet is capable of sustaining liquid water on its surface.

Hadean This term denotes the earliest subera of the *Archaean*.

Hadron Subatomic particles subjected to the strong nuclear force: e.g., protons, neutrons; and also the charge-carrying particles of the strong nuclear force itself: e.g., pions.

Heliopause The boundary layer between the Solar magnetosphere, known as the heliosphere, and the interstellar plasma, see magnetosphere.

Heterotrophic A lifestyle that requires the ingestion of preformed organic molecules, a heterotroph cannot fix inorganic matter for biomolecule construction.

Heteropolymer A polymer derived from two or more different (but often similar) types of monomer.

Hierarchogenesis A process of appearance of new levels of *systems hierarchy* when aggregation of many or at least several essentially different systems into a whole supersystem emerges a fundamental novelty. A fundamental kind of hierarchogenesis is represented by the transitions in the *operator hierarchy*.

Hierarchogenetic step An elementary act of *hierarchogenesis* that creates a new, highest level of *systems hierarchy* in our universe.

HMXB A stellar mass black hole or neutron star actively accreting matter from a high mass stellar companion as part of a binary star system.

Holistic The whole of a system or phenomenon or the practice of examining the operation of entire systems rather than the serial examination of individual parts. Holism and reductionism offer complementary, instead of opposing, perspectives when analyzing a system.

Homeothermy Thermoregulation by an organism in order to maintain a specific internal body temperature.

Hydrothermal vent A fissure in a planet's surface (commonly found near volcanically active places, areas where tectonic plates are moving apart, and hotspots at the ocean floor) from which geothermally heated water issues.

Hypersaline Liquid that contains a salt concentration that is higher than that of seawater (which is 3.5%).

Hyperthermophilic A type of organism that is capable of growth temperature optimums of 80°C or higher.

HZ Habitable zone around stars.

I

IDPs Interstellar dust particles are fine particles of matter found in interstellar and interplanetary spaces.

Inhabitable = habitable Suitable for sustaining the activities of organisms.

Inflationary age/epoch Period in the evolution of the early universe when the universe underwent an extremely rapid exponential expansion. This happen before the first 10^{-32} s after the Big Bang.

Information In a general sense, information can be viewed as data or message that can be given an interpretation. Information *sensu* Claude Shannon refers to the disambiguation of uncertainty. In the context of Shannon's information theory, a greater amount of information correspondingly reduces the amount of uncertainty in the same message.

Ionizing radiation Electromagnetic or particle radiation of sufficient energy to ionize atoms (remove electrons).

IR Infrared radiation, extends from 700 nm to 1 mm in the electromagnetic spectrum.

Immersion Alternative term for virtual reality.

Impact gardening A process where impact events (meteorite bombardment, micrometeorites, etc.) disturb the outermost crust of moons or other solar system objects that lack atmospheres. These objects often contain few erosional processes (besides volcanism), thus surface debris that accumulates these impact events is an important means of surficial material redistribution.

Instantiated How an instance of a phenomena is achieved in a given context.

Interstellar dust particle (IDP) These are tiny (<0.1 μm in size) carbonaceous particles which are present just about everywhere in space.

Interstellar medium Matter in the interstellar space. It exists in three phases (hot, warm, and cold) that are in pressure equilibrium. The cold phase consists of large-scale molecular clouds, e.g., the dark cloud called Horsehead Nebula. In the phases of higher temperature, the matter is in atomic or ionized state.

Intergalactic medium Matter in intergalactic space (between galaxies). Much less dense than the interstellar medium, the intergalactic medium nevertheless contains enormous amounts of warm and hot plasma.

Intron A nucleotide (DNA or RNA) sequence within a gene that does not encode proteins and is removed by RNA splicing during maturation of the final RNA product.

Inverse-Compton scattering Process by which lower energy photons are scattered to higher energies by relativistic electrons.

Ising model A simplified discrete model of ferromagnetism where atomic spins can have values -1 and 1. The spins are usually arranged in a lattice allowing each spin to interact with its neighbors.

ISM Interstellar medium, gas and dust between stars.

Isometry Relationship between body size and other characteristics of the organism that preserves proportionality.

K

Kardashev type Represents the ability of an advanced civilization to harness the energy. The lowest types are capable of utilizing energy sources of the home planet, while the highest types correspond to utilizing energy on a galactic scale.

Kilonova Supernova from the merger of two neutron stars or a neutron star and a black hole.

kpc Kiloparsec, 10^3 pc: 1 kpc $= 3.086 \times 10^{16}$ m.

L

ΛCDM Dark energy plus cold dark matter cosmology.

LECA Last eukaryotic common ancestor: the hypothesized single cell from which all eukaryotes are descended.

Lepton A subatomic particle that is not subject to the strong nuclear force and has a half-integer spin: e.g., electron, neutrino.

Life Life generally refers to living organisms (e.g., animals, plants, fungi, protists, bacteria), which are self-reproducing systems with self-sustained metabolism, heredity, and regulation. In the context of the origin of life, it refers to simpler autonomous systems that may have been predecessors of organisms, including viruses. In the context of Operator Theory, an organism-based concept of life (O-life) is opposed to a systems-based concept of life (S-life).

Limestones A hard sedimentary rock, composed mainly of calcium carbonate or dolomite.

Lipid world Scenario of the origin of life that is focused on lipid-like molecules as predecessors of cell membranes and carriers of heredity.

Liposome A sphere-shaped vesicle having at least one lipid bilayer. These can be artificially produced and used for drug delivery to cells. Because lipids in water spontaneously form these stable structures, liposomes are a model for the first membranes to appear in the origin of life.

LMXB A stellar mass black hole or neutron star actively accreting matter from a low mass stellar companion as part of a binary star system.

Local group Distinct gravitationally bound group of galaxies that includes Milky Way, Andromeda galaxy, and tens of smaller galaxies.

Local universe The near-field universe approximately within 15 Mpc that can be resolved by the extremely large telescopes of the present-day.

LUCA Last universal common ancestor is a hypothetical common ancestor of all living organisms on Earth. The genome of LUCA is a reconstruction based on stochastic mathematical models.

Lyman-alpha (α) Spectral line at 121.6 nm in the far ultraviolet region of electromagnetic spectrum that is characteristic of atomic hydrogen in its lowest excited state.

M

$M_\odot$ Mass of the Sun, equals 1.989×10^{30} kg.

M class star A spectral class of stars that includes red dwarfs (almost 90% of M class stars), red giants, and red supergiants. Surface temperatures are cool relative to other stars, around 2500–4000°K.

Ma Million years ago.

MeV Million electron volts.

Mpc Megaparsec, 10^6 pc: 1 Mpc $= 3.086 \times 10^{19}$ m.

My Million years.

Macroevolution Long-term evolutionary changes that take millions of years and include qualitative changes such as emergence of new species and higher-level taxons. Alternatively, any change in the topology of the differentiation tree.

Magnetopause The boundary layer between the magnetosphere and the surrounding plasma.

Magnetosphere The spherical or ellipsoidal magnetic field associated with a satellite, planetary, or stellar body.

Main sequence A distinctive and continuous strip-like structure emerging on the stellar brightness vs stellar color plot (the Hertzsprung-Russell diagram), with brighter stars having higher temperatures and vice versa. The stars spend most of their life on the Main Sequence during which they support themselves against gravitational collapse by fusing hydrogen in their cores.

Main sequence star The star that is located at the *Main Sequence*. The parameters of the star (temperature, luminosity, spectral distribution of radiation, etc.) do not change significantly while the star is on the Main Sequence, which makes some of the Main Sequence stars very good candidates for hosting habitable planets.

Measurement error model Regression models that account for measurement errors in the independent variables.

Megaevolution The evolution of our universe at the most coarse scale (of mega- and gigayears) implemented by *hierarchogenesis*.

Metal In astrophysics, any chemical element besides hydrogen or helium. (This is entirely distinct from the chemistry definition.)

Metallicity In astrophysics, the proportion of "metals" in a star or exoplanet.

Methanogenesis A metabolic process mediated by microorganisms which results in the formation of methane. These microorganisms are called methanogens and, to-date, have only been identified from the prokaryotic Domain Archaea.

Microcontinent An isolated fragment of continental crust forming part of a small crust plate.

Microevolution Short-term evolutionary changes that can be traced in natural or laboratory populations. These changes are mostly qualitative and can be characterized by the frequency of certain alleles. Alternatively, any change to a differentiation tree that preserves its topology.

Microquasar A stellar mass black hole accreting from a disk and ejecting bipolar jets, in contrast to the quasars which are accreting supermassive black holes in the nuclei of some galaxies.

MW Milky Way.

Mitosis Cell division in eukaryotes, which includes resorption of nuclear membrane, duplication of chromosomes, their condensation, pairing, and equal distribution between daughter cells.

Montmorillonite An aluminum-rich clay mineral of the smectite group, containing some sodium and magnesium.

Moore's law An empirical trend in the development of computer technology which states that the number of transistors per chip increased exponentially, doubling approximately every 12–18 months.

mRNA Messenger RNA is a single-strand RNA sequence, which is a copy of genomic DNA coding sequence for an individual gene. mRNA is synthesized directly on the single-strand DNA template, and this process is called transcription.

Muon Elementary particle similar to an electron, negatively charged, but with mass about 207 times the mass of an electron.

N

N(^{4}S) One of the electron states of a nitrogen atom. In this case, the electron is present in the 4S level.

Natural selection Darwin describes natural selection as follows: "This preservation of favorable individual differences and variations, and the destruction of those which are injurious, I have called Natural Selection, or the Survival of the Fittest" (Darwin, 1876). While selection is commonly viewed as a process, the differential nature of the term (favorable versus injurious variations) indicates that it refers to a pedigree in which in any one generation differences in heritable properties cause some organisms to survive until reproduction, while others die before that moment.

Neptune Unofficial name frequently used to denominate ice giants, a class of planet with mass ranging between 10 and 100 Earth's masses, which is mainly composed of gas but with proportions of hydrogen and helium much smaller than gas giants (only about 20% in mass) and a rocky core which is an important fraction of its mass (10%–20%). The planet Neptune in the Solar System is a representative case of ice giants, hence the name "neptune" for those planets.

Neutron star One of the possible end states of stellar evolution after a core collapse supernova. Neutron stars have the same density as an atomic nucleus and are made almost entirely of neutrons.

Nova (pl. novae) A thermonuclear runaway explosion on the surface of a white dwarf star. A nova does not destroy the star and may occur many times during the evolution of close interacting binaries with accreting white dwarfs.

NS Neutron star.

Nuclear lamina The outer layer of cell nucleus which supports the interaction of chromatin with nuclear membrane and pores.

O

OB-super-bubble Hot gas surrounding high mass O and B type star.

O-life (abbreviation for organismic life) The concept of O-life refers in a generic way to the presence of dual closure in organisms. This definition is based on the concepts of organism and dual closure as defined by the *Operator Theory*.

Oligomer A molecule consisting of a chain of relatively small and specifiable number of monomers, usually <5.

Operator (as used in the *operator theory*) Any physically united object that has a place in the *operator hierarchy* because it has uniform dual closure. Uniform means that (in principle) only objects of the immediate lower level are involved in the closure. Dual closure refers to the combination of structural and functional closure. An example of structural closure is the cell membrane. An example of functional closure is the autocatalytic set of chemicals in the cell. The interaction between the membrane and the autocatalytic chemistry now defines the cell.

Operator hierarchy The ranking of all the different kinds of operators according to the *operator theory*. With fundamental particles as the basis, the different kinds of operators that have formed in the universe are: hadrons, atoms, molecules, cells, endosymbiont cells, multicellulars, endosymbiont multicellulars, neural network organisms.

Operator theory A recently developed complexity theory that offers rules for how interactions between low level building block systems form higher level building block systems. The specific building block systems the theory deals with are called "operators" because they "operate" (in a general sense) in their environment as individually countable units.

Organic molecules Molecules which contain carbon, excluding carbon dioxide, monoxide, and carbonates.

Organism Classically, the word "organism" is used as a container concept for many things that consist of cells, and for this reason, has one or more of the following properties in common: metabolism, mobility, structural unity, reproduction, etc. A recent approach based on the *Operator Theory* narrows down the organism concept in the following way. A system is an organism if and only if the system is an *operator* (as defined by the Operator Theory) and if the kind of complexity of the system is at least equal to the cell.

Originability Refers to a planet that has conditions suitable for an origin of life.

P

PAH Polycyclic aromatic hydrocarbon.

Panspermia The theory that states life or the chemical precursors of life had an extraterrestrial origin and was transported to a suitable environment (such as Earth) on a meteor or some other means.

PAR Photosynthetically active radiation.

Parsec (pc) Length unit convenient for astronomical scales. One pc equals $3.086 + e16$m (3.26 light years) and is defined as the distance at which the mean Earth-Sun distance is seen at the angle of one arcsecond.

PCA Probabilistic cellular automata (see cellular automata). The states of the cells change according to probabilistic rules.

PeV Peta-electronvolt equal to 10^{15} eV, see electronvolt.

Phanerozoic The most recent eon of geologic time extending from the Paleozoic (570–230 Myr BP) to the Mesozoic (230–62 Myr BP) and the Cenozoic (since 62 Myr BP).

Polymerization Process of reacting monomer molecules together to form polymer chains or three-dimensional networks.

Polypeptide Chains of amino acid monomers linked by peptide (amide) bonds, forming part (or the whole) of a protein molecule.

Polyploidy Cells and organisms that have more than two pairs of all homologous chromosomes. Polyploidy often results from a failure to distribute chromosomes in cell division (mitosis or meiosis).

Population I/II/III stars Observed stars can be coarsely classified as either Population I (metal-rich, Sun-like stars) or Population II (metal-poor). Hypothesized Population III stars are massive stars that formed from metal-free hydrogen and helium gas left over from the Big Bang.

Power law Functional relationship between two quantities where one quantity varies as a power of another, i.e., $y = x^a$ where a is the exponent relating the two quantities x and y.

Prokaryote Small single-cell organisms without a membrane-bound nucleus, mitochondria, and other membrane-bound organelles. Prokaryotes include two domains: Bacteria and Archaea. In contrast, cells with nuclei and organelles are placed in the domain Eukaryota.

Proterozoic In geologic time, this is an era that ranges from the Archaean era (4.5–2.5 Gyr BP) to the Phanerozoic Eon (570 Myr BP to the present time).

Psychrophiles Organisms capable of growth at low temperatures and showing a cardinal growth temperature that is $< 15°C$.

Punctuated equilibrium A model of macroevolution proposed by Eldredge and Gould in 1972, which assumes that long periods of stasis are alternated with short periods of rapid evolutionary changes, usually associated with intensive speciation. This concept should not be confused with catastrophism of Cuvier or quantum evolution of Simpson. The concept first appears in the fourth edition of Darwin's *Origin of Species*.

Pyroxene Any of a large class of rock-forming silicate minerals, generally containing calcium, magnesium, and iron and typically occurring as crystals.

R

r-process elements A group of elements formed by rapid neutron capture built from iron or more massive nuclei. Multiple neutrons are captured before the nucleus has time to decay, so a high neutron flux is required.

Radial velocity In astronomy, the speed with which a body moves away from or approaches the Earth. Its measurement in stars gives a direct estimation of the lower bound of the mass of the body producing this movement in the star (as a planet or another star).

Recombination In astrophysics: the epoch $\sim$378,000 years after the Big Bang when free charged electrons and protons first became bound to form electrically neutral hydrogen atoms. In molecular biology: the formation of new combinations of genes.

Red dwarf Small and cool stars with masses between 0.1 and 0.5 solar masses and surface temperatures under 4000 K. They include M class stars and late K class stars.

Relativistic (Doppler) boosting The observed radiation from a relativistically moving source moving near the line of sight increases by a relativistic factor, γ, that depends on the velocity of the source.

Replication Making a copy of another object. When applied to heteropolymer molecules, it means template-based replication.

Reticulate evolution Evolution involving horizontal gene transfer, which turns the tree of life into a web of life.

Ribosome Organelle of all living cells (prokaryotes and eukaryotes) formed by two subunits of RNA molecules integrated with specific proteins, which had evolved for a programmed synthesis of proteins. Messenger RNA (mRNA) serves as a program that determines the order of amino acids in the synthesized protein.

Rickettsial bacteria Nonmotile, gram-negative, nonspore-forming, highly pleomorphic bacteria that can be present as cocci, rods, or thread-like. They were named after Howard Taylor Ricketts. Being obligate intracellular parasites, survival of these bacteria depends on entry, growth, and replication within the cytoplasm of eukaryotic host cells.

RNA world Hypothetical stage of primordial evolution representing cells with heredity based on self-replicating RNA molecules, which were not capable of programmed peptide synthesis via genetic code. The RNA world was initially viewed as the first step in the origin of life, but later it was often considered as an intermediate step. A world of the self-replicating ribonucleic acid (RNA) molecules hypothesized to have been the precursors to all current life on Earth.

S

SDSS Sloan Digital Sky Survey. The most comprehensive sky survey to date. It covered about one third of the sky in various optical bands.

SEP Solar or stellar energetic particle.

SFR Star formation rate. The rate at which clouds of galactic gas gravitationally collapse to form stars. Usually expressed in Solar masses per unit time and unit spatial volume.

SGHZ Super-Galactic habitable zone.

S-life Abbreviation for systemic life. The concept of S-life refers to a system of interacting organisms. This definition is based on the organism concept as defined by the *Operator Theory*.

SMBH Supermassive black hole.

SN Supernova, a generic stellar explosion.

SNe Supernovae, plural of supernova.

SN ia White dwarf explosion after exceeding the Chandrasekhar mass limit, no star remains.

SN Ib/c Massive star explosion after nuclear fusion, a neutron star or a black hole is formed.

SN II   Core collapse supernova.

Solar/stellar wind   Flow of gas ejected from the upper atmosphere of the sun or a star.

SPH   Smoothed particle hydrodynamics. A computational method usually used for simulating the dynamics of fluids. The fluid is represented as a collection of discrete elements called particles, which properties are smoothed by some kernel function. This way, each particle can influence the properties of all other particles that are within its smoothing radius (kernel width).

Starburst galaxy   A galaxy experiencing a relatively brief period of intense star formation, typically 1000 times the star formation rate of a normal star forming galaxy.

Stellar Nucleosynthesis   The process due to fusion reactions that occur in the core of the stars, changing the stars abundances of chemical elements.

Stellar remnant   A compact massive object formed during the terminal gravitational collapse of a star. Depending on the mass of the progenitor star, the remnant is a white dwarf, neutron star, or a black hole.

Stochastic   A probabilistic process leading to a randomly determined outcome.

Super-Earth   Planet with mass ranging between 1 and 10 Earth's masses. The proportion of mass corresponding to gas or rock is not clear, but probably they are of similar magnitude, maybe ranging the ratio rock mass/gas mass between 1/5 and 5. Probably, many of them are telluric planets.

Supermassive black hole   A black hole with a mass of millions to as much as a trillion Solar masses, found in the center or nucleus of many galaxies.

Supernova (pl. supernovae)   Terminal explosion of a star characterized with very intense electromagnetic and particle radiation. SN type Ia is the explosion of a white dwarf upon exceeding the Chandrasekhar limit. SN Ib and SN IIa,b, and c are the result of the core collapse of a high mass star.

Supernova remnant   Usually a shell-like or nebular structure with a central neutron star resulting from the supernovae explosion. A material ejected in the explosion forms an expanding collisionless shock wave sweeping over the interstellar space causing intense *synchrotron radiation*.

Symbiogenesis   An evolutionary theory based on the assumption that cohabitation and cooperation of unrelated organisms may result in the emergence of a larger organism type with original organisms as subunits. In particular, eukaryotic cells presumably originated from a symbiosis of prokaryotic cells.

Symbiosis   Interaction between different organisms or populations physically associating and typically to the advantage of both.

Synchrotron emission/radiation   Electromagnetic radiation caused by the circular or elliptical acceleration of charged particles perpendicular to the direction of the magnetic field.

Systems theory   An interdisciplinary endeavor tasked with the discovery of patterns and the elucidation of principles.

T

Taxonomic group (taxon)   Group of organisms linked by evolutionary relations (e.g., common origin) and overall similarity, which is seen by taxonomists to form a unit of various rank such as a species, family, or class.

Taxonomy, taxonomic classification   This is the study of the theory, practice, and rules of classification of living or extinct organisms into groups, according to a given set of relationships.

Teleonomy (vs teleology)   The quality of apparent purposefulness and goal-directedness of structures and functions in living organisms brought about by natural laws (like natural selection). The term derives from two Greek words, "telos" (end, purpose) and "nomos" (law) and means "end-directed" (literally "purpose-law"). Teleonomy is sometimes posited instead of teleology, where the latter is understood as a purposeful goal-directedness brought about through human or divine intention.

Telluric body   Planet or satellite composed primarily of rocky materials, that is, similar to the Earth, Mars, or Venus. The name comes from Tellus, latin word for the Earth (synonym of Terra).

Telomere Ends of chromosomes that contain specific satellite DNA repeats. The function of telomeres is to maintain the integrity of chromosome and prevent its shortening during cell division.

Template-based replication (=sequence copying) Synthesis of a complimentary aligned copy of a linear polymer composed of few different monomers via sequential addition of individual monomers. For example, replication of DNA is based on template-based synthesis and executed by DNA-polymerase.

TeV Tera-electronvolt equal to 10^{12} eV, see electronvolt.

Terminal velocity The speed at which an object falling through an atmosphere ceases to accelerate.

Terraforming Making a planet habitable for humans. Generalized to: any organism making a planet habitable for itself and its descendents.

Terrestrial Referring to planet Earth.

Terrestrial (organisms) In astrobiology, this term refers to organisms on the planet Earth. In biology, it refers to land-dwelling organisms as opposed to aquatic organisms. The meaning depends on the context.

Terrestrial radiation Long-wave electromagnetic radiation originating from Earth and its atmosphere. The Earth is heated up by the Solar radiation and radiation emitted by naturally radioactive materials on Earth including uranium, thorium, and radon. Assuming the Earth is in a thermodynamic equilibrium with its environment, the absorbed energy equals the energy that Earth radiates away as a black body. According to Stefan-Boltzmann law, the bulk of this energy is radiated as a long-wave radiation.

Thermoneutral conditions Conditions where an organism is in thermal comfort. In these conditions, the organism does not need to use energy to maintain its temperature, either cooling its body or heating it.

Tholin Tholins are classes of heteropolymer molecules formed by ultraviolet irradiation of simple organic compounds such as methane or ethane, often in combination with nitrogen. Usually of a reddish-brown color, they are thought to be found in the atmosphere of one of Saturn's moon: Titan.

Torr This is noninternational system of unit measure of pressure; whereas the universal SI unit is pascal (Pa). The torr measure is exclusively used in the United States; in Europe the preferred unit is: mbar. 1 torr = 133.322 Pa = 1.333 mbar. Torr is derived from Torricelli, an Italian physicist and mathematician who discovered the principle of the barometer in 1644.

Transdisciplinarity A strategy of inquiry that requires transcending several disciplines to meet a certain comprehension.

Transposon Repetitive DNA originated from parasitic viruses that often have a capacity to make additional copies in other genome locations.

tRNA Transport RNA which is used to bring amino acids to ribosomes and to control protein synthesis. tRNA carries anticodon site (3 nucleotides) that determines the specificity of the genetic code, i.e., the association between nucleotide triplets and certain types of amino acids.

U

UHECRs Ultra-high energy cosmic rays. Particles that have been accelerated to an energy >1 EeV $= 10^{18}$ eV.

ULX Ultra-luminous X-ray source.

Universal coding Association between parts of heteropolymers that is true for all kinds of sequences. Two examples of universal coding are in constructing of a reverse-complementary strand of nucleic acids (DNA or RNA), and synthesis of protein based on mRNA sequence.

Universe All of space and time (spacetime) and its contents, which includes galaxies, stars, planets, moons, minor planets, the contents of intergalactic space, and all matter and energy.

UV Ultraviolet radiation.

UVB Ultraviolet B-band, lower energy than UVA.

V

Vitrification The transformation of a substance into a noncrystalline, amorphous solid (i.e., a glass state).

W

WD White Dwarf star.
Weak interaction One of the four known fundamental interactions, responsible for radioactive decay.

X

XUV X-ray combined with ultraviolet radiation.

Z

z (red shift measure) As the universe expands, the wavelengths of emitted photons (λ_{emitted}) become longer ($\lambda_{\text{observed}}$). A measure of the redshift of photons, $z = (\lambda_{\text{observed}} - \lambda_{\text{emitted}})/\lambda_{\text{emitted}}$. Objects at the highest redshifts are the most distant objects.

Index

Note: Page numbers followed by "*f*" indicate figures, "*t*" indicate tables, "*b*" indicate boxes, and "*np*" indicate footnotes.

Habitability of the Universe Before Earth, editors: Richard Gordon & Alexei Sharov, Volume 1 in the series:
Astrobiology: Exploring Life on Earth and Beyond, series editors: Pabulo Henrique Rampelotto,
Joseph Seckbach & Richard Gordon. ISSN 2468-6352. https://doi.org/10.1016/B978-0-12-811940-2.09991-3
© 2018 Elsevier Inc. All rights reserved.